CONCISE ENCYCLOPEDIA OF
MEDICAL & DENTAL MATERIALS

ADVANCES IN MATERIALS SCIENCE AND ENGINEERING

This is a new series of Pergamon scientific reference works, each volume providing comprehensive, self-contained and up-to-date coverage of a selected area in the field of materials science and engineering. The series is being developed primarily from the highly acclaimed *Encyclopedia of Materials Science and Engineering*, published in 1986. Other titles in the series are listed below.

NOTICE TO READERS

Dear Reader
If your library is not already a standing order/continuation order customer to the series **Advances in Materials Science and Engineering**, may we recommend that you place a standing/continuation order to receive immediately upon publication all new volumes. Should you find that these volumes no longer serve your needs, your order can be cancelled at any time without notice.

ROBERT MAXWELL
Publisher at Pergamon Press

CONCISE ENCYCLOPEDIA OF
MEDICAL & DENTAL MATERIALS

Editor

DAVID WILLIAMS
University of Liverpool, UK

Executive Editor

ROBERT W CAHN
University of Cambridge, UK

Senior Advisory Editor

MICHAEL B BEVER
MIT, Cambridge, MA, USA

PERGAMON PRESS
Member of Maxwell Macmillan Pergamon Publishing Corporation
OXFORD • BEIJING • FRANKFURT
SYDNEY • TOKYO

THE MIT PRESS
CAMBRIDGE
MASSACHUSETTS

Distributed exclusively in North and South America by The MIT Press, Cambridge, Massachusetts, USA

Published and distributed exclusively throughout the rest of the world by Pergamon Press:

U.K.	Pergamon Press plc, Headington Hill Hall, Oxford OX3 0BW, England
PEOPLE'S REPUBLIC OF CHINA	Pergamon Press, Room 4037, Qianmen Hotel, Beijing, People's Republic of China
FEDERAL REPUBLIC OF GERMANY	Pergamon Press GmbH, Hammerweg 6, D-6242 Kronberg, Federal Republic of Germany
AUSTRALIA	Pergamon Press Australia Pty Ltd., P.O. Box 544, Potts Point, N.S.W. 2011, Australia
JAPAN	Pergamon Press, 5th Floor, Matsuoka Central Building, 1-7-1 Nishishinjuku, Shinjuku-ku, Tokyo 160, Japan

Copyright © 1990 Pergamon Press plc

First edition 1990

Library of Congress Cataloging in Publication Data
Concise encyclopedia of medical & dental materials / editor, David Williams ; executive editor, Robert W Cahn ; senior advisory editor, Michael B Bever. — 1st ed.
    p. cm. — (Advances in materials science and engineering)
    Includes index.
    1. Biomedical materials—Encyclopedias.   2. Dental materials—Encyclopedias.   I. Williams, D. F. (David Franklyn)   II. Cahn, R. W. (Robert W.), 1924–   .   III. Bever, Michael B. (Michael Berliner)   IV. Title: Concise encyclopedia of medical and dental materials.   V. Series.
    [DNLM: 1. Biocompatible Materials—encyclopedias.   2. Dental Materials—Encyclopedias.   QT 13 C744]
R857.M3C64   1990
610'.28–dc20a   90–7505
DNLM/DLC

ISBN 0–262–23149–2 (MIT Press)

British Library Cataloguing in Publication Data
Concise encyclopedia of medical & dental materials.
    1. Medical materials   2. Dental materials
    I. Williams, David
    610.28

ISBN 0–08–036194–3 (Pergamon Press)

Printed and bound in Great Britain by BPCC Wheatons Ltd, Exeter

CONTENTS

HONORARY EDITORIAL ADVISORY BOARD

FOREWORD

In the short time since its publication, the *Encyclopedia of Materials Science and Engineering* has been accepted throughout the world as the standard reference about all aspects of materials. This is a well-deserved tribute to the scholarship and dedication of the Editor-in-Chief, Professor Michael Bever, the Subject Editors and the numerous contributors.

During its preparation, it soon became clear that change in some areas is so rapid that publication would have to be a continuing activity if the Encyclopedia were to retain its position as an authoritative and up-to-date systematic compilation of our knowledge and understanding of materials in all their diversity and complexity. Thus, the need for some form of supplementary publication was recognized at the outset. The Publisher has met this challenge most handsomely: both a continuing series of Supplementary Volumes to the main work and a number of smaller encyclopedias, each covering a selected area of materials science and engineering, will be published in the next few years.

Professor Robert Cahn, the Executive Editor, was previously the editor of an important subject area of the main work and many other people associated with the Encyclopedia will contribute to its Supplementary Volumes and derived Concise Encyclopedias. Thus, continuity of style and respect for the high standards set by the *Encyclopedia of Materials Science and Engineering* are assured. They have been joined by some new editors and contributors with knowledge and experience of important subject areas of particular interest at the present time. Thus, the Advisory Board is confident that the new publications will significantly add to the understanding of emerging topics wherever they may appear in the vast tapestry of knowledge about materials.

The appearance of Supplementary Volumes and the new series *Advances in Materials Science and Engineering* is an event which will be welcomed by scientists and engineers throughout the world. We are sure that it will add still more luster to a most important enterprise.

Walter S Owen
Chairman
Honorary Editorial Advisory Board

EXECUTIVE EDITOR'S PREFACE

As the publication of the *Encyclopedia of Materials Science and Engineering* approached, Robert Maxwell resolved to build upon the immense volume of work which had gone into its creation by embarking on a follow-up project. This project had two components. The first was the creation of a series of Supplementary Volumes to the Encyclopedia itself. The second component of the new project was the creation of a series of Concise Encyclopedias on individual subject areas included in the Main Encyclopedia to be called *Advances in Materials Science and Engineering*.

These Concise Encyclopedias are intended, as their name implies, to be compact and relatively inexpensive volumes (typically 300–600 pages in length) based on the relevant articles in the Encyclopedia (revised where need be) together with some newly commissioned articles, including appropriate ones from the Supplementary Volumes. Some Concise Encyclopedias will offer combined treatments of two subject fields which were the responsibility of separate Subject Editors during the preparation of the parent Encyclopedia (e.g., dental and medical materials).

At the time of writing, ten Concise Encyclopedias have been contracted and others are being planned. These and their editors are listed below.

Concise Encyclopedia of Advanced Ceramic Materials	Prof. Richard J Brook
Concise Encyclopedia of Building & Construction Materials	Prof. Fred Moavenzadeh
Concise Encyclopedia of Composite Materials	Prof. Anthony Kelly CBE, FRS
Concise Encyclopedia of Electronic & Optoelectronic Materials	Dr Lionel C Kimerling
Concise Encyclopedia of Magnetic & Superconducting Materials	Dr Jan E Evetts
Concise Encyclopedia of Materials Economics, Policy & Management	Prof. Michael B Bever
Concise Encyclopedia of Medical & Dental Materials	Prof. David Williams
Concise Encyclopedia of Mineral Resources	Dr Donald D Carr & Prof. Norman Herz
Concise Encyclopedia of Polymer Processing & Applications	Mr P J Corish
Concise Encyclopedia of Wood & Wood-Based Materials	Prof. Arno P Schniewind

All new or substantially revised articles in the Concise Encyclopedias will be published in one or other of the Supplementary Volumes, which are designed to be used in conjunction with the Main Encyclopedia. The Concise Encyclopedias, however, are "free-standing" and are designed to be used without necessary reference to the parent Encyclopedia.

The Executive Editor is personally responsible for the selection of topics and authors of articles for the Supplementary Volumes. In this task, he has the benefit of the advice of the Senior Advisory Editor and of other members of the Honorary Editorial Advisory Board, who also exercise general supervision of the entire project. The Executive Editor is responsible for appointing the Editors of the various Concise Encyclopedias and for supervising the progress of these volumes.

Robert W Cahn
Executive Editor

ACKNOWLEDGEMENTS

I am very grateful to all the authors who responded to my invitation to write for this Concise Encyclopedia. I would like to acknowledge the very considerable help and encouragement of Professor Robert Cahn, the Executive Editor, and the experience and assistance of Dr Colin Drayton, Mr Michael Mabe and Mr Peter Frank of Pergamon Press.

David Williams
Editor

GUIDE TO USE OF THE ENCYCLOPEDIA

This Concise Encyclopedia is a comprehensive reference work covering all aspects of medical and dental materials. Information is presented in a series of alphabetically arranged articles which deal concisely with individual topics in a self-contained manner. This guide outlines the main features and organization of the Encyclopedia, and is intended to help the reader to locate the maximum amount of information on a given topic.

Accessibility of material is of vital importance in a reference work of this kind and article titles have therefore been selected, not only on the basis of article content, but also with the most probable needs of the reader in mind. An alphabetical list of all the articles contained in this Encyclopedia is to be found on p. xv.

Articles are linked by an extensive cross-referencing system. Cross-references to other articles in the Encyclopedia are of two types: in-text and end-of-text. Those in the body of the text are designed to refer the reader to articles that present in greater detail material on the specific topic under discussion. They generally take one of the following forms:

...which is fully described in the article *Biocompatibility of Dental Materials*.

...other aspects of biomaterials (see *Surface Structure and Properties*).

The cross-references listed at the end of an article serve to identify broad background reading and to direct the reader to articles that cover different aspects of the same topic.

The nature of an encyclopedia demands a higher degree of uniformity in terminology and notation than many other scientific works. The widespread use of the International System of Units has determined that such units be used in this Encyclopedia. It has been recognized, however, that in some fields Imperial units are more generally used. Where this is the case, Imperial units are given with their SI equivalent quantity and unit following in parentheses. Where possible the symbols defined in *Quantities, Units, and Symbols*, published by the Royal Society of London, have been used.

All articles in the Encyclopedia include a bibliography giving sources of further information. Each bibliography consists of general items for further reading and/or references which cover specific aspects of the text. Where appropriate, authors are cited in the text using a name/date system as follows:

...as was recently reported (Smith 1988).

Jones (1984) describes...

The contributor's name and the organization to which they are affiliated appear at the end of each article. All contributors can be found in the alphabetical List of Contributors, along with their full postal address and the titles of the articles of which they are authors or co-authors.

The Introduction provides an overview of the field of medical and dental materials and discusses in brief the issues covered by the articles in the body of the work.

The most important information source for locating a particular topic in the Encyclopedia is the multilevel Subject Index, which has been made as complete and fully self-consistent as possible.

ALPHABETICAL LIST OF ARTICLES

AN INTRODUCTION TO MEDICAL AND DENTAL MATERIALS
by David Williams

Many of the results of the endeavors of materials scientists are visible for us all to see and we can either watch with admiration or take for granted the late twentieth-century examples of high-performance materials in the aerospace, nuclear, off-shore, petrochemical, architectural and other industries. There is one materials-science-based industry, however, that, despite often utilizing state-of-the-art materials and advanced engineering design and manufacturing techniques, is hidden from the view of most people and has largely gone unnoticed. This is the industry of surgical implants and materials, the so-called spare parts used in surgery. It is now a large industry, but is unnoticed because we cannot see the products. Tens or hundreds of thousands of patients each year worldwide receive artificial hip or knee joints, pacemakers or heart valves, intraocular lenses or arteries, but we are unlikely to know if the person sitting next to us has any such device functioning inside his or her body.

This whole subject of implantable materials and devices is both intriguing and paradoxical. It may be seen, and indeed this Concise Encyclopedia will show all too well, that both sophisticated, highly specialized materials and very simple, low-technology materials are employed. In some situations the materials function efficiently and reliably, while in other cases success and reliability are very difficult to achieve. In some situations the precise nature of the material is the single most important arbiter of success or failure, while in others the material itself is almost an irrelevance, while surgical technique, engineering design or idiosyncratic behavior of the recipients are far more important.

It is because of these enormously varied factors, the highly significant role that materials science can play in these developments and the rapidly growing interest in the subject that this Concise Encyclopedia of Medical and Dental Materials has been compiled.

In order to understand the rationale for the selection of topics included in this volume, it is necessary to consider the scope of the subject from both the surgical/medical/dental point of view and from the materials science standpoint.

1. Clinical Issues

The most fundamental question that may be asked in relation to medical and dental materials concerns the reasons why they have to be used. The clinical reasons can be classified or described in several different ways, but it is usually convenient to consider them in the light of the function that the materials have to perform.

The first reason that materials are used is to physically replace tissues that have become damaged or destroyed through some pathological process. It is well known that the tissues and structures of the body, however well they are able to perform in most people for much of the time, do suffer from a variety of destructive processes, including infection and cancer, that cause pain, disfigurement or loss of function. Under these circumstances, it may be possible to remove the diseased tissue and replace it with some suitable synthetic material.

There are some very obvious examples. Osteoarthritis or rheumatoid arthritis both affect the structure and function of synovial joints, such as the hip, knee, shoulder, ankle and elbow. The pain in such joints subsequent to the destruction of the cartilage and underlying bone can be considerable and the effects on mobility quite devastating. It has been possible to replace these joints with prostheses for some time, and the relief of pain and restoration of movement is well known to thousands of patients.

The tissues of the eye can suffer from several diseases, leading to reduced vision and eventually blindness. Cataracts, for example, cause cloudiness of the lens. This may be very easily replaced with an intraocular lens, restoring vision. Similarly, destructive changes in the middle ear that lead to hearing impairment can be remedied by implants that replace the ossicular bones and provide a new sound-conduction pathway.

In the cardiovascular system, problems can arise with heart valves and arteries, both of which may be successfully treated with implants. The heart valves suffer from structural changes that prevent the valve from either fully opening or fully closing and the diseased valve can be replaced with a variety of substitutes. Arteries, particularly the coronary arteries and the vessels in the lower limbs, become occluded, or blocked, by deposits in atherosclerosis and it is possible in some (but by no means all) cases to replace segments with artificial arteries.

Within the mouth, the site of the most prevalent diseases in the West, both the tooth and the supporting tissues can be readily destroyed by bacterially controlled diseases. Dental caries, the demineralization and dissolution of teeth associated with the metabolic activity of bacteria in plaque, can cause extensive tooth loss. This, however, is becoming of secondary importance to periodontal disease, which involves destructive changes to the supporting mucosa (the gum tissue) and eventually the bone. This is the major cause of tooth extraction.

Both teeth in their entirety and segments of teeth can be replaced or restored by a variety of materials.

These are only a few, albeit the most common, examples of tissue damage or loss that may provide the reasons for using biomaterials for reconstructive purposes. It is an important principle that this type of reconstruction must always be considered as a secondary alternative to disease prevention or conservative/pharmacological methods of disease control. If reconstruction has to be performed, it is also a sound principle that natural tissues, that is tissue grafts or transplants, should be used wherever possible; bone grafts, skin grafts, corneal transplants and kidney transplants all provide good evidence as to the better performance achieved with natural reconstructive tissues. Synthetic materials are used for reconstruction when no natural tissue or transplant is available or appropriate.

The second reason for using implantable devices is rather similar, but involves the introduction of shape or function to a patient who has been born deficient or who has developed abnormally. There are a few life-saving uses of materials in this context, such as patches used to repair cardiac defects, and in some cases very significant deformities (e.g., of the spine) can be corrected. In many cases, however, it is not so much function as appearance (particularly in the face) that provides the driving force for such corrective surgery.

Third, implants are used in the treatment of some injuries. The oldest use of implantable materials can be traced back to the introduction of sutures or stitches for wound closure. Fracture plates, used to secure the displaced segments of bone after complex fractures, were in use in the nineteenth century. More recently, sports injuries (especially in skiing and American football) have led to ligament injuries requiring surgical intervention and repair, while the tragic consequences of both war and peacetime disasters bring with them new techniques and new materials for the treatment of the severely injured.

Fourth, moving towards more complex functions, there is an increasing interest in the use of electrically active devices for the transmission of signals into and out of the body. Cardiac pacemakers provide stimuli to the heart, implantable hearing aids provide the transduction of sound to electrical signals and stimulation of the inner ear, and so on. Neurological electrodes are able to detect activity within the nervous system, possibly allowing for continuous patient monitoring.

Finally, there is strong interest in the use of implantable devices for the controlled and targeted delivery of drugs. It has been recognized that the administration of drugs to patients by the traditional oral route results in widely varying levels of the drug in the patient between doses and also usually provides a systemic, rather than a focused, distribution. Thus, many attempts have been made to incorporate drug reservoirs into implantable devices for a sustained and preferably controlled release. Some of these technologies utilize new materials as vehicles for the drug delivery.

2. Materials

It is clear from the previous summary of the reasons for the use of materials that the functions required of medical devices are extremely varied. It is also self-evident that these functions must be performed within the environment of the human body—an environment that is surprisingly hostile and aggressive, yet one that is extremely sensitive. These two necessities, the need to perform some function and the need to perform this funciton *in vivo*, are the broad considerations in the selection of materials, the relevant properties being described under the headings of biofunctionality and biocompatibilty, respectively.

At the present time, the functions required are largely mechanical and physical, and are, in general, not too onerous. Although the stresses generated within the body are surprisingly high, they are not excessive in terms of the performance of high-strength materials and there are very few situations where a lack of appropriate mechanical properties represents a significant difficulty. There are, of course, a few exceptions, and the tribological properties of the materials used in joint prostheses and the mechanical performance of composites as dental restorative materials are good examples where improvements are still required. The same situation prevails with physical properties, such as optical transparency, acoustic conduction and electrical conduction/insulation. Generally, the properties available from a range of materials can satisfy these requirements.

The difficulties faced in the selection of medical and dental materials are mostly concentrated in the area of biocompatibility. This term is defined carefully in the Encyclopedia; for purposes of introduction, it can be said that biocompatibility refers to a collection of phenomena that are involved with the interaction between materials and the tissues of the body, phenomena that can result in adverse and undesirable effects on either the material, the tissues or both, but that ultimately lead to a failure of the material or device to perform the required function.

As far as the effects on the material are concerned, it is well known that the tissue environment is very aggressive, and metallic corrosion and polymer degradation are of great significance. Many of the failures of materials to perform adequately in the body result from such effects, particularly those concerned with combined environmental–mechanical behavior. Corrosion fatigue, stress-corrosion cracking, environmental crazing and other mechanisms can all occur.

Table 1
Examples of medical and dental materials and their applications

Material category	Material	Principal applications
Metals and alloys	316 stainless steel	bone and joint replacement
		spinal instrumentation
		fracture fixation
	titanium, Ti–Al–V, Ti–Al–Nb	bone and joint replacement
		fracture fixation
		dental implants
		pacemaker encapsulation
	cobalt–chromium alloys	bone and joint replacment
		dental implants
		dental restorations
		heart valves
	gold alloys	dental restorations
	silver products	antibacterial agents
	platinum and PGM	electrodes
	Hg–Ag–Sn amalgam	dental restorations
Ceramics and glasses	alumina	joint replacement
		dental implants
	zirconia	joint replacement
	calcium phosphates	bone repair and augmentation
		surface coatings on metals
	bioactive glasses	bone replacement
	porcelain	dental restorations
	carbons	percutaneous devices
		coatings for blood-contacting devices
Polymers	polyethylene	joint replacement
	polypropylene	sutures
	PTFE	soft-tissue augmentation
		vascular prostheses
	polyesters	vascular prostheses
		drug delivery systems
	polyurethanes	blood-contacting devices
	PVC	tubing
	PMMA	dental restorations
		intraocular lenses
		joint replacement
	silicones	soft-tissue replacement
		ophthalmology
	hydrogels	ophthalmology blood-contacting devices
Composites	dimethacrylate resin–quartz	dental restorations
	carbon fiber–thermosetting resins	bone repair
	carbon fiber–thermoplastics	bone repair
	carbon–carbon	bone and joint replacement

Perhaps of greater importance is the potential effect of the corrosion or degradation products on the tissues. Any soluble or particulate matter has the ability to irritate the tissue. The extent of this may range from the very mild to the severe and chronic, and it is quite possible for implant materials, or dental restorative materials, to cause the destruction of tissue through a variety of mechanisms.

The ability of the material and the device to perform their function for as long as is necessary within the body, being dependent on intrinsic physical properties and the chemical/biochemical stability in the body, is therefore a main requirement of biomaterials. It is usual for the intrinsic property requirements to dictate the broad type of material to be used (e.g., high-strength alloy, thermoplastic elastomer, rigid ceramic, tough compsite) and for biocompatibility considerations to dictate the exact material from within this group (e.g., a titanium alloy or platinum-group metal from among the metallic/ alloy systems).

It should not be surprising to find examples of all material types within the portfolio of currently used medical and dental materials. A broad, but by no means exclusive, list is given in Table 1. These materials perform with varying degrees of success; most are covered in some detail in this Encyclopedia.

3. Future Developments

When the reasons for using implants and/or dental materials are considered alongside the nature of the materials actually used, an illogicality can be detected. The main reason for using these devices was described as the need to replace, physically or functionally, tissues of the body that have been or are being destroyed. These tissues, although varying enormously in structural detail (e.g., compare nerve and bone tissue), have certain essential characteristics. They are, generally, living, cellular, water-containing, anisotropic, viscoelastic substances of very complex ultrastructure and architecture, often providing a multiplicity of functions. The materials used for their replacement, however, are usually nonviable, acellular, dehydrated, isotropic, elastic substances, possibly of complex microstructure, but usually of simple architecture, and almost invariably providing only one function. Generally, therefore, the replacement materials display structures and properties quite different from the tissues themselves.

It is for this reason that natural tissue grafts and transplants are preferred, but this is rarely possible. It is highly probable that the biomaterials of the future will bear much greater similarity to the tissues they are replacing. This, however, is an extremely difficult concept and it is unlikely that complex biochemistry and tissue architecture can be achieved in a replacment material. Some advances are already being made with a genetic engineering approach, but the difficulties of generating functional devices in this way are very great.

In the short and medium term, therefore, it will be a mixture of synthetic materials and tissues and tissue analogues that will lead the way in these developments, with the so-called hybrid organs or, more realistically in the shorter term, with biologically modified synthetic materials. It is now quite plausible to use synthetic "engineering" materials for the main functional performance of a device (e.g., to provide strength or rigidity) and to use natural materials, in the form of reconstituted tissue, cell populations or biologically active molecules, at or near the surface to control the reaction with the tissues and incorporation into the body.

4. Structure of the Encyclopedia

The diverse nature of the contributions in this Encyclopedia reflects the variety of materials and clinical applications. They can be broadly divided into five groups.

First, there are contributions concerned with individual materials, ranging from alumina to zirconia. Examples will be found of metals and alloys, ceramics and glasses, and a wide variety of polymer-based materials.

Second, there are contributions dealing with groups of materials used for specific applications, where the required properties and the problems are the same, but where a diverse group has evolved. Most of these examples relate to dental materials, such as dental investments and dental elastomers.

Third, there are several contributions that describe some of the important phenomena associated with material performance, including biocompatibility, biodegradation and corrosion.

Fourth, although the Encyclopedia is not providing a catalog of clinical applications, there are several entries that describe important and controversial areas. These include dental implants, prosthetic heart valves, invasive sensors and wound dressings.

Finally, there are contributions dealing with new techniques and technologies related to biomaterials. These especially include surface treatments and surface modifications, including beam ion implantation, heparinization and the attachment of other drugs to polymer surfaces.

With a subject that is so diverse and rapidly expanding in many different directions, the Encyclopedia cannot address every issue. It does, however, provide a thorough background to the broad science that underpins this very important use of materials.

A

Acoustic Measurements of Bone and Bone–Implant Systems

In addition to their biological functions, the principal purposes of both bones and teeth are mechanical; they support load, absorb the energy of impact, provide cutting edges and shearing forces, and give protection to other life-support organs. For brevity, this article deals mostly with the long bones (the load-supporting elements) of the body. Such bones as the femur (thigh bone) and fibula (shin bone) are quasihollow cylinders in nature. The dense or compact cortical bone in the diaphysis (the central section), as well as all the other mineralized tissues of the body, exhibit a definitive microstructure with associated anisotropic elastic properties. From the materials science standpoint, bone is a complex composite material based on several successive hierarchical levels. Most mineralized tissues are made up of collagen (1/3), which is the principal connective tissue protein, and an inorganic material (2/3) known as hydroxyapatite or OHAp ($Ca_5(PO_4)_3OH$, see *Calcium Phosphates and Apatites*). The only exception to this is enamel which is almost entirely OHAp with a small amount of a different protein. These two components are intimately interrelated structurally and eventually form, at the microstructural level, two types of lamellar units which pack together in regular arrangements. These are a concentric superposition of lamellae called osteons or Haversian systems found in humans and in some other adult mammals (see Fig. 1a), and dense parallel lamellar plates called plexiform bone found, for example, in young bovine and porcine bone (see Fig. 1b). Moreover, bone is a living tissue that undergoes continual remodelling. Locally, this process always results in a secondary Haversian structure, even in a primarily plexiform bone. Since these two types of bone do not generally differ in collagen–OHAp composition, their difference in microstructural arrangements results in different anisotropic elastic properties.

It is this structure–function relationship whose elucidation and understanding is particularly enhanced by the use of acoustic wave propagation techniques. The usual mechanical measurements performed in tension, compression, bending or torsion cannot provide the same precision in the small specimens necessarily available from such bones, especially in directions other than the longitudinal. Moreover, it is not generally possible to mechanically determine the full set of anisotropic properties on a limited number of specimens as is inherent with acoustic methods.

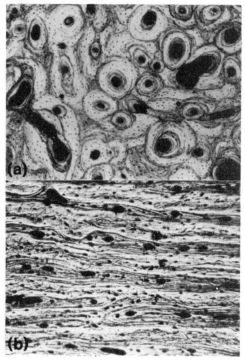

Figure 1
Cortical bone microstructure: (a) human femur and (b) bovine femur

1. Bulk Acoustic Wave Propagation Techniques

Such acoustic wave propagation techniques provide fundamental information about the elastic properties of biological tissues and organs. This is not to be confused with the ultrasonic imaging of internal organs as in cardiology or obstetrics, where the imaging process is similar to radar. Of interest here is the measurement of the properties of bone and bone–implant systems by the use of acoustic wave propagation at ultrasonic frequencies. Techniques include both bulk and surface wave measurements as well as scanning acoustic microscopy (SAM); parameters of interest include velocity, attenuation, scattering, dispersion and acoustic impedance. Such bulk and surface wave measurements provide information about the anisotropic elastic properties of both normal and pathological tissues as well as information about their viscoelastic properties, while the SAM studies provide maps of the local inhomogeneities on a scale much finer than that obtained with either bulk ultrasonic or standard mechanical measurements.

1.1 Technical Background

The ultrasonic techniques that have been used to study the anisotropic properties of bone are pulse-through (PT) transmission (including the pulsed-differential method), pulse-echo overlap (PEO), the continuous wave (CW) method and the right-angle reflector (RAR) method.

All of these techniques measure the transit time of propagation of the ultrasonic signal through the material, usually cut and shaped into small rectangular parallelepipeds (or possibly more complex shapes) along principal bone directions. The length of the specimen divided by the transit time provides the acoustic wave velocity.

Each of these methods can be used to provide both transverse (equivoluminal) and longitudinal (dilatational) waves of various frequencies through specimens of specific orientations in well-defined bone directions. In addition, the RAR technique permits measurements of surface wave velocities.

In the above measurements, the longitudinal wave velocity determination requires that the ultrasonic wavelengths used be much smaller than the specimen dimensions to ensure a pure dilational wave. If lower frequency transducers are used, such that the wavelength becomes smaller than the specimen dimensions, an extensional longitudinal wave is propagated. This case reduces to a bar wave, so that it is possible to obtain the Young's modulus of the specimen directly if the density is known.

Although the previously described mineralized tissues are actually viscoelastic, they can be treated as anisotropic elastic materials with little or no loss of understanding. In this case, the relationship between stress and strain is defined by the generalized Hooke's law: $\sigma_{ij} = C_{ijkl}\varepsilon_{kl}$.

For a material with orthogonal (orthotropic) symmetry, the fourth-rank tensor C_{ijkl} can be reduced to the following second-order matrix:

$$
\begin{bmatrix} \sigma_1 \\ \sigma_2 \\ \sigma_3 \\ \sigma_4 \\ \sigma_5 \\ \sigma_6 \end{bmatrix} = \begin{bmatrix} C_{11} & C_{12} & C_{13} & 0 & 0 & 0 \\ C_{12} & C_{22} & C_{23} & 0 & 0 & 0 \\ C_{13} & C_{23} & C_{33} & 0 & 0 & 0 \\ 0 & 0 & 0 & C_{44} & 0 & 0 \\ 0 & 0 & 0 & 0 & C_{55} & 0 \\ 0 & 0 & 0 & 0 & 0 & C_{66} \end{bmatrix} \begin{bmatrix} \varepsilon_1 \\ \varepsilon_2 \\ \varepsilon_3 \\ \varepsilon_4 \\ \varepsilon_5 \\ \varepsilon_6 \end{bmatrix}
$$

where 11 reduces to 1, 22 reduces to 2, 33 reduces to 3, 23 and 32 reduce to 4, 13 and 31 reduce to 5, and 12 and 21 reduce to 6. In the case of hexagonal (transverse isotropic) symmetry, the nine independent elastic stiffnesses C_{ij} reduce to five as

$$
C_{11} = C_{22}, \qquad C_{13} = C_{23}, \qquad C_{44} = C_{55},
$$
$$
C_{66} = (C_{11} - C_{12})/2
$$

In the isotropic case only two independent elastic constants are required.

From this elastic stiffnesses matrix, it is possible to calculate the different technical moduli (i.e., Young's modulus and the shear and bulk moduli) as well as the Poisson's ratios.

These elastic stiffnesses are directly related to the acoustic velocities of both longitudinal and transverse waves as well as to the density of the specimen (usually measured by an immersion technique). An extensive description of the relationship existing between elastic stiffnesses and acoustic wave velocities in bone is given by Yoon and Katz (1976a,b).

1.2 Ultrasonic Properties of Normal Bone and Bone-Related Materials

Ultrasonic methods have been used by researchers to study the anisotropic properties of various animal and human bones and teeth including, in the human case, both normal and pathological specimens. Various ceramics, both those which are inert and those to which bone will chemically attach, have also been studied using these techniques. Typical values of the technical moduli obtained for several calcified tissues and bone-related materials are given in Table 1.

Since it has been shown that ultrasonically measured and mechanically measured moduli agree over a wide range of Young's moduli, data from both types of experiments can generally be correlated. However, the scatter inherent in mechanical testing of the small specimens necessarily involved in bone measurements, in directions other than longitudinal, may obscure or mask the anisotropic results found ultrasonically. Therefore, two sets of mechanical measurements on bone are included for comparison in Table 1.

Just as in crystals where it is the atomic arrangement that determines the anisotropy in properties, so it is the microstructure of the bone that determines the symmetry of ultrasonic wave propagation (Neumann's principle: "Any kind of symmetry which is possessed by the crystallographic form of a material is possessed by the material in every respect of every physical quality" as applied to the textural symmetry found in the organization of the lamellar units of the bone).

This is readily observed by examining the elastic stiffness coefficients in Table 1. The fluorapatite (FAp, hexagonal crystal symmetry) and bone data are based on full sets of anisotropic velocity measurements. The OHAp (hexagonal crystal symmetry) data are calculated based on a model by Katz and Ukraincik (1971) from acoustic measurements in dense isotropic compacts; the same model is used to calculate the off-diagonal stiffness coefficients for both enamel and dentin from incomplete sets of anisotropic velocity measurements.

An even more dramatic verification of the structure–function relationship can be observed by comparing the elastic stiffness coefficients of the two

Table 1
Anisotropic elastic stiffness coefficients (GPa) of various calcified tissues and apatites

Material	C_{11}	C_{22}	C_{33}	C_{44}	C_{55}	C_{66}	C_{12}	C_{13}	C_{23}
FAp	151		185	42.4		50.9	48.8	62.2	
OHAp	137		172	39.6		47.3	42.5	54.9	
Dentin	37.0		39.0	5.70		10.2	16.6	8.70	
Enamel	115		125	22.8		36.6	42.4	30.0	
Human femur (dried)[a]	23.4		32.5	8.71		7.17	9.06	9.11	
Bovine femur[b,c]	17.0		29.6	3.60		3.40	10.2	9.80	
Human tibia[b,d]	11.6	14.4	22.2	4.91	3.56	2.41	7.95	6.10	6.92
Bovine femur[e]	14.1	18.4	25.0	7.00	6.30	5.28	6.34	4.84	6.94
Human femur[e]	20.0	21.7	30.0	6.56	5.85	4.74	10.9	11.5	11.5
Bovine femur (Haversian)[f]	21.2	21.0	29.0	6.30	6.30	5.40	11.7	12.7	11.1
Bovine femur (plexiform)[f]	22.4	25.0	35.0	8.20	7.10	6.10	14.0	15.8	13.6

a Yoon and Katz 1976a,b   b mechanical tests   c Reilly and Burstein 1975   d Knets 1978   e Van Buskirk and Ashman 1981   f Katz et al. 1983

structural forms of compact bone, plexiform and Haversian, provided in the last two rows of Table 1. These data are obtained through PT measurements made on specimens of fresh bovine femora (five to six year old animals) cut along three well-defined orthogonal axes for the bone (longitudinal–radial–tangential directions). The two different types of bone microstructure that are measured depend on the aspect from which the specimens are taken around the periphery; plexiform bone is found in the anteromedial aspect (AM) while Haversian bone is found in the posterolateral aspect (PL).

It is seen that the plexiform bone is significantly stiffer than the Haversian bone in all the principal directions of symmetry. Moreover, while the plexiform bone is clearly orthotropic in nature, the Haversian bone seems to be very close to being transversely isotropic. These differences demonstrate the effect of microstructural symmetry on elastic anisotropy.

As might be expected, since an important constituent of bone is the connective tissue protein collagen, bone exhibits a slight temperature dependency for the ultrasonic velocities and thus for the Young's modulus both along and perpendicular to the long-bone axis. For example, Young's modulus decreases approximately 0.071 GPa °C^{-1} in the long-bone axis direction.

While dry bone exhibits some linear dispersion of velocities (2.9% along the bone axis and 10.7% along the radial direction) over the frequency range 2–10 MHz, wet bone does not over the range 0.5–16 MHz. In addition, the acoustic velocities of dry bone are always greater in all directions than those in fresh bone.

Ultrasonic attenuation in bone has been measured by a number of investigators. It appears to increase almost linearly over the frequency range of 1–15 MHz; attenuation in the radial direction is greater than attenuation in the tangential direction which is greater than attenuation in the longitudinal direction. At 5 MHz in human Haversian bone the attenuation is 0.23 nepers mm^{-1}, while for bovine plexiform bone it is 3.4 times smaller in the longitudinal direction. This can be compared with 0.018 nepers mm^{-1} (at 1 MHz, 21.1 °C) found for poly(methyl methacrylate) (PMMA) or 0.14 nepers mm^{-1} (at 2.75 MHz) for a fibrous composite.

1.3 Pathological Bone

Since the ultrasonic velocities clearly depend on the morphology and organization of the bone, it has been suggested that such techniques would be useful in elucidating structure–function relationships in pathological bone as well. Both osteoporotic (decreased bone density) and osteopetrotic (calcified cartilaginous unformed Haversian systems) bone specimens were measured along with normal bone specimens. In this instance, the osteoporotic bone had densities approximately 10% below that of normal bone, while the osteopetrotic bone exhibited almost normal density. The increase in porosity in osteoporotic bone results in a decrease in the ultrasonic velocities measured in all directions. There is also a significant decrease observed in the osteopetrotic bone velocities, which is reflected in the corresponding decrease in respective elastic stiffness coefficients well below the linear relationship found over the measured range of bone densities. However, in this case, the decrease must be due to the disorganized bone structure. Figure 2 is a polar curve showing the angular dependence of Young's modulus for each of the three types of bone in the longitudinal–radial plane. The ratio of the major axis to the minor axis of each ellipse like curve is an indication of the degree of the anisotropy of the material; a polar curve with an almost circular shape reflects the near isotropy of this material. Thus,

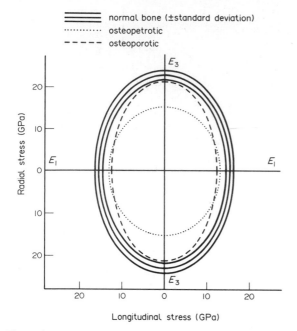

Figure 2
Polar curve showing the angular dependence of
Young's modulus of cortical bone in the longitudinal–
radial plane (for normal, osteoporotic and
osteopetrotic bone)

while osteoporotic bone shows the highest anisotropy (indeed slightly greater than that of normal
bone) osteopetrotic bone is nearly isotropic. This
lack of anisotropy is consistent with the light microscopic examination which shows no observable microstructural symmetry. Such data illustrate the
power and versatility of ultrasonic wave propagation
techniques, leaving the specimens undamaged for
additional or collateral measurements.

2. Scanning Acoustic Microscopy

A recent addendum to the armamentarium of ultrasonic wave propagation in materials is scanning
acoustic microscopy (SAM). A number of different
types of scanning acoustic microscope are available
operating either in the reflection or transmission

modes, using spherically focused or line-focused
transducers to measure various acoustic properties
over significant areas with a resolution limited only
by the frequency and quality of the transducers.

2.1 Technical Aspects

In reflection SAM, a single spherically focused
transducer, immersed in distilled water (or any other
appropriate fluid) transmits an ultrasonic pulse onto
the surface of the specimen coupled through the
same medium. This signal is reflected back to the
same transducer acting now as a receiver; a peak
detector is used to measure the amplitude of the
reflected signal. A computerized scanning system
permits measurement of this signal over a selected
surface of the specimen. Real-time data acquisition
is performed by a computer which is also used for
creating pseudo images and image analysis. This
method requires that one surface be flat and
polished to the degree of surface regularity imposed
by the resolution of the system.

The signal amplitude recorded at the peak detector is proportional to the reflection coefficient R_s,
which for a specimen immersed in water is given by
$R_s = (Z_s - Z_w)/(Z_s + Z_w)$, where Z_s is the acoustic
impedance of the specimen at the point being measured and Z_w is the acoustic impedance of water.
Acoustic impedance itself is a property of the material closely related to other physical properties
important in characterizing the material behavior of
the specimen. It is obtained as the product of the
local density ρ and the longitudinal velocity V_{long} in
the material (i.e., $Z = \rho . V_{long}$).

With the exception of soft tissues and polymers,
which have values of Z close to that of water, most
of the materials of interest here (bone and ceramic
or metallic biomaterials) have $Z_s \gg Z_w$ so that the
reflection coefficients are rather high (see Table 2).

Specimens to be analyzed can be prepared either
fresh or embedded. While quantitative measurements of true acoustic properties require fresh specimens, embedded specimens that have been or will
be used in histomorphological evaluation will
provide reliable, if not accurate, data.

A variation of this SAM technique involves transmission using a pair of spherically confocal transducers that are aligned colinearly in a water tank
perpendicularly to the surface of the specimen. This

Table 2
Values of Z_s (in Mrayl) and R for various materials in water

Material	H$_2$O	PMMA	Bone	Magnesium	CaCO$_3$	Aluminum	OHAp	Titanium	Steel
Z_s	1.50	3.17	8[a]	10.0	15.8	17.4	21.1	27.3	45.7
R	0.	0.358	0.684[a]	0.739	0.838	0.841	0.867	0.896	0.936

a typical maximum value

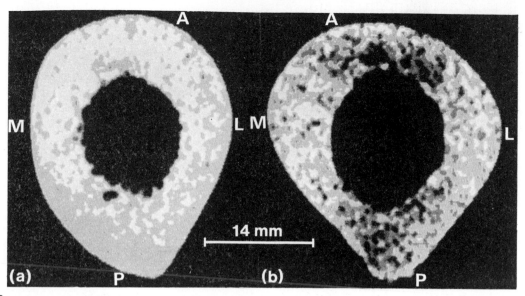

Figure 3
SAM images of normal human femur: (a) 22 year old male and (b) 77 year old male, with posterior (P), medial (M), lateral (L) and anterior (A) positions shown

requires that the specimen be thin with a pair of flat parallel surfaces. In this mode, rather than obtaining measurements related to acoustic impedance, variations in both wave attenuation and phase shift may be obtained simultaneously.

2.2 Applications

There are several different areas in which SAM provides information that cannot be obtained by other techniques or in which its use adds additional insight into the behavior of the material when correlated with collateral studies such as optical histomorphology, scanning electron microscopy (including backscattered electron imaging or contact x-ray microradiography) and bulk acoustic wave measurements.

(a) *Bone aging*. While mechanical measurements or bulk acoustic wave techniques can provide an average value of a technical modulus (e.g., Young's modulus) needed for the usual engineering purposes, the SAM image provides details of the local inhomogeneities on a scale much finer than is possible with either of these methods. Thus, in a study of changes in bone with age it is possible to obtain maps, related to acoustic properties, that show the dramatic changes occurring in the anterior and posterior aspects with age (see Fig. 3). When calibration of the signals is used, it is possible to show a direct correlation between the acoustic impedance measured with SAM and the appropriate elastic stiffness coefficient measured by bulk ultrasonic wave propagation. Table 3 presents a list of

Table 3
Comparison of the values of the longitudinal elastic properties in young and old human femurs

Age	Quadrant	ρ (10^3 kg m^{-3})	Z (Mrayl)	V (km s^{-1})	C_{33}[a] (GPa)	C_{33}[b] (GPa)
25	posterior	1.94	7.52±0.41	3.90	29.51	29.33
25	anterior	1.90	7.78±0.53	3.92	29.20	30.49
25	medial	1.97	7.56±0.49	3.89	29.75	29.41
25	lateral	1.96	7.60±0.41	3.86	29.20	29.33
74	posterior	1.79	5.83±1.96	3.60	23.32	21.00
74	anterior	1.78	6.40±1.68	3.74	25.00	23.94
74	medial	1.93	7.06±0.89	3.70	26.42	26.12
74	lateral	1.92	7.17±0.80	3.73	26.71	26.74

a measurements done on bulk specimens using PT technique   b calculations based on SAM measurments

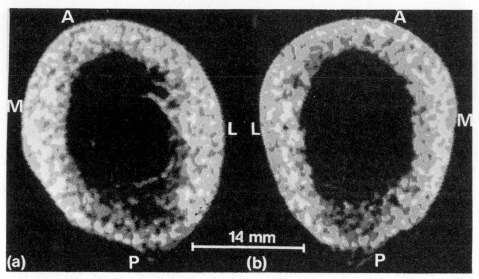

Figure 4
SAM images of (a) an implanted femur and (b) the contralateral bone with posterior (P), medial (M), lateral (L) and
anterior (A) positions shown

values of Z_{avg} obtained by averaging over individual
Z values within the pixel distribution corresponding
to the area covering the bulk specimen that is used
to measure both the density ρ and the dilational
velocity V_{dil}. These are used to calculate values of
C_{33} by $C_{33} = Z_{avg} V_{dil}$ which are compared with the C_{33}
values obtained directly from ultrasonic wave velo-
city measurements in the longitudinal direction in
the bulk specimens from each aspect ($C_{33} = \rho V_{dil}^2$).

(b) Implant–bone interactions. Figure 4a is a two-
dimensional map of a transverse cross section of an
adult female femur (72 years old at death) that had
had a successful Austin–Moore hip prosthesis
implanted in it during the two years prior to death.
Note the new bone formation around the implant as
well as the geometrical changes of the diaphysis.
Moreover, the acoustic impedance of the endosteal
region of the implanted femur has been reduced
when compared to the intact contralateral femur
(see Figure 4b).

A two-dimensional reflection SAM map of a
transverse cross section of a canine femur that has
had a bone plate fixed to the intact bone is shown in
Fig. 5a. Here, the bone directly under the plate
(right-hand side) has clearly reduced acoustic
properties when compared to the left-hand side in
which the bone properties have been maintained.
An x-ray microradiograph of the same specimen is
shown in Fig. 5b′ in order to show the alteration in
mineral content and increase in porosity which is, in
part, responsible for the local variations in bone

acoustic properties seen in Fig. 5a. These variations
are attributed to the stress shielding caused by the
greater stiffness of the implant relative to that of the
bone.

Finally, Fig. 6a is a pseudo-three-dimensional
mapping (where the apparent height or depth is a
direct representation of the amplitude of the reflected signal) of an Al_2O_3 cylindrical implant
($d = 6$ mm) with surrounding bone after 12 months
implantation in an ovine femur. The narrow dark
band, from left to front, clearly delineates a resorption area adjacent to the implant. The central circular portion in white represents a value of Z much
greater than that for bone; the absolute flatness of
this central region reflects the uniformity of the
isotropic material properties of the alumina.

This is in sharp contrast with Fig. 6b, the pseudo-
three-dimensional mapping of a SiC-coated carbon-
fiber-reinforced C–SiC composite plug ($d = 6$ mm)
also after 12 months in an ovine femur. Here, the
clearly seen banding of the variations in the signal
amplitudes across the surface of the implant corresponds to the optical microscopic observation of the
orientation of the fiber bundles in the composite on
and just below the surface.

While the previous illustrations are based on
reflection SAM techniques, corresponding implant
studies can also be done in transmission.

Besides the low frequency SAM studies described
previously, which are appropriate to evaluate the
variations in the local inhomogeneities caused by
macroscopic changes in mineralized tissues, high-
frequency instruments (up to 2 GHz) can be used to

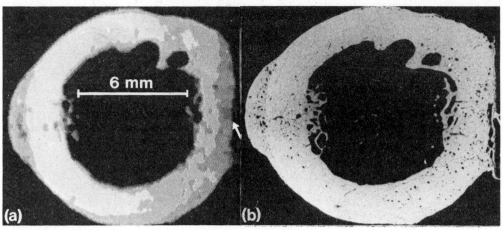

Figure 5
Plated canine femur: (a) SAM image and (b) microradiograph (arrow indicates location of the plate)

investigate microstructural modifications of the same tissues with a resolution in the range of 0.1 μm.

In addition, there are still several other SAM techniques. These include the method whereby the

signal is detected optically (scanning laser acoustic microscopy (SLAM)) rather than acoustically. There is also line focus transducer SAM, in which direct measurements of elastic moduli can be obtained with suitable materials. These methods have also been used with biological specimens, but applications to bone and bone biomaterials have not yet been fully realized.

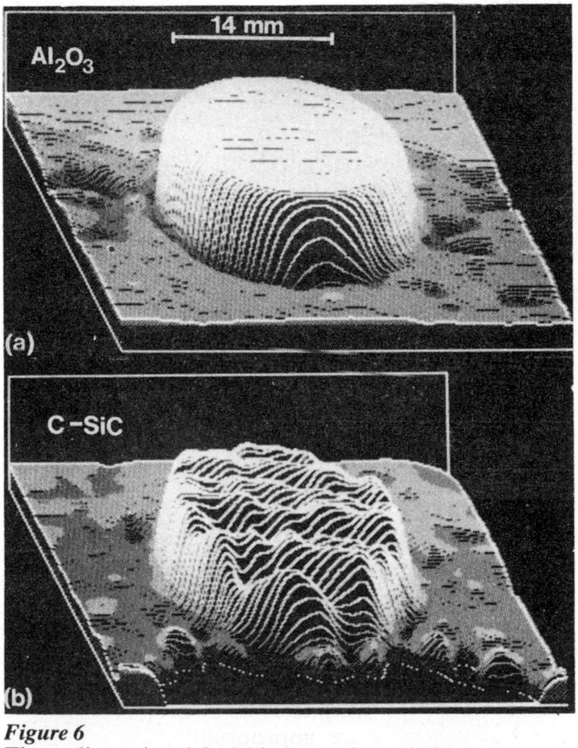

Figure 6
Three-dimensional SAM images of a cylindrical implant and the surrounding bone: (a) Al₂O₃ implant and (b) implant made of carbon fibers with SiC reinforcement

Bibliography

Ashman R B, Cowin S C, Van Buskirk W C, Rice J C 1984 A continuous wave technique for the measurement of the elastic properties of cortical bone. *J. Biomech.* 17: 349–61
Bonfield W, Tully A E 1982 Ultrasonic analysis of the Young's modulus of cortical bone. *J. Biomed. Eng.* 4: 23–7
Briggs G A D 1985 *An Introduction to Scanning Acoustic Microscopy*. Oxford University Press, Oxford
Edmonds P D (ed.) 1981 *Ultrasonics*, Methods of Experimental Physics, Vol. 19. Academic Press, New York
Fry F J, Barger J E 1978 Acoustic properties of the human skull. *J. Acoust. Soc. Am.* 63: 1576–90
Hearmon R F S 1961 *An Introduction to Applied Anisotropic Elasticity*. Oxford University Press, London
Katz J L 1980 The structure and biomechanics of bone. In: Vincent J F, Currey J D (eds.) 1980 *The Mechanical Properties of Biological Materials*. Cambridge University Press, Cambridge, pp. 137–68
Katz J L, Ukraincik K 1971 On the anisotropic elastic properties of hydroxyapatite. *J. Biomech.* 4: 221–7
Katz J L, Yoon H S, Lipson S, Maharidge R, Meunier A, Christel P 1983 The effects of remodeling on the elastic properties of bone. *Calcif. Tissue Int.* 36: 531–6
Kessler L W, Yuhas D E 1979 Acoustic microscopy. *Proc. IEEE* 67: 526–36

Knets I V 1978 Mechanics of biological tissues. A review. *Mekh. Polim.* 13: 434–40

Lakes R, Yoon H S, Katz J L 1986 Ultrasonic wave propagation and attenuation in wet bone. *J. Biomed. Eng.* 8: 143–8

Lang S B 1970 Ultrasonic method for measuring elastic coefficients of bone and results on fresh and dried bovine bones. *IEEE Trans. Biomed. Eng.* 17: 101–5

Lees S, Rollins Jr F R 1972 Anisotropy in hard dental tissues. *J. Biomech.* 5: 557–66

Lemons R A, Quate C F 1979 Acoustic microscopy. In: Mason W P, Thurston R N (eds.) 1979 *Physical Acoustics*, Vol. 14, Academic Press, London, pp. 1–92

Mason W P (ed.) 1964 *Physical Acoustics. Principles and Methods.* Academic Press, New York

Meunier A, Katz J L, Das P, Biro L, Yoon H S, Christel P 1986 Bone–implant interface evaluation using a scanning ultrasonic transmission imaging system. In: Williams D F (ed.) 1986 *Techniques of Biocompatibility Testing*, Vol. 1. CRC Press, Boca Raton, FL, pp. 181–92

Nikoonahad M 1984 Recent advances in high resolution acoustic microscopy. *Contemp. Phys.* 25: 129–58

Reilly S B, Burstein A H 1975 The elastic and ultimate properties of compact bone tissue. *J. Biomech.* 8: 293–405

Van Buskirk W C, Ashman R D 1981 The elastic moduli of bone. In: Cowin S C (ed.) 1981 *Mechanical Properties of Bone.* American Society for Mechanical Engineers, New York, pp. 131–43

Yoon H S, Katz J L 1976a Ultrasonic wave propagation in human cortical bone I. Theoretical considerations for hexagonal symmetry. *J. Biomech.* 9: 407–11

Yoon H S, Katz J L 1976b Ultrasonic wave propagation in human cortical bone II. Measurements of elastic properties and microhardness. *J. Biomech.* 9: 459–64

A. Meunier
[University of Paris, Paris, France]

J. L. Katz
[Rensselaer Polytechnic Institute, Troy, New York, USA]

Acrylic Dental Polymers

There are numerous dental applications for acrylic polymers, based on monofunctional monomers (principally methyl methacrylate, but also *n*-butyl methacrylate and isobutyl methacrylate) as well as on difunctional and trifunctional monomers. The resultant polymers are used for denture construction, repair and relining, temporary crown and bridge materials, the prevention of tooth decay and as a basis of materials for the restoration of teeth. These polymers are all formed by free-radical addition polymerization. Additionally, water soluble polymers and copolymers are important constituents of modern adhesive dental cement systems.

1. Monofunctional Monomers

Methyl methacrylate can be prepared from acetone:

$$(CH_3)_2CO \xrightarrow{HCN} (CH_3)_2COHCN \xrightarrow[H_2SO_4]{CH_3OH} CH_2{=}\overset{\displaystyle CH_3}{\underset{\displaystyle CO_2CH_3}{C}} \quad (1)$$

The monomers of higher alkyl methacrylates can be prepared from methyl methacrylate by reaction with the appropriate alcohol:

$$CH_2{=}\overset{\displaystyle CH_3}{\underset{\displaystyle CO_2CH_3}{C}} + ROH \longrightarrow CH_2{=}\overset{\displaystyle CH_3}{\underset{\displaystyle CO_2R}{C}} + CH_3OH \quad (2)$$

2. Free-Radical Addition Polymerization

The polymerization of methyl methacrylate involves three principal stages:

(a) the production of free radicals capable of reacting with the monomer,

(b) the propagation reaction, and

(c) termination reactions in which free radicals are destroyed.

Free radicals are produced by additives called initiators, examples of which are organic peroxides and azo compounds. The organic peroxides are of particular importance for dental applications, benzoyl peroxide often being the material of choice. A chloro-substituted benzoyl peroxide may be employed (Asmussen 1980b).

Benzoyl peroxide forms free radicals by the homolytic decomposition of the weak peroxy bond:

$$ (3) $$

Initiation of the polymerization reaction can be caused either by these radicals or by phenyl radicals produced from them:

$$ (4) $$

The decomposition of benzoyl peroxide can be achieved either thermally or in the presence of another chemical, usually a tertiary amine such as *N*,*N*-dimethyl-*p*-toluidine (Bolker 1974 p. 211), which acts as a promotor or activator:

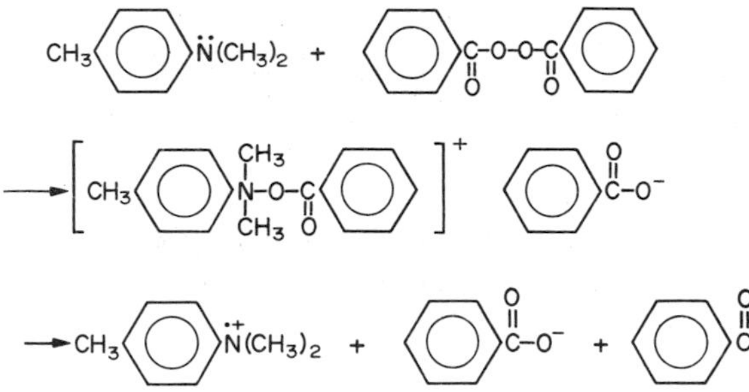

$$CH_3-\!\!\bigcirc\!\!-\ddot{N}(CH_3)_2 \;+\; \bigcirc\!\!-\overset{O}{\underset{}{C}}-O-O-\overset{O}{\underset{}{C}}-\!\!\bigcirc$$

$$\longrightarrow \left[CH_3-\!\!\bigcirc\!\!-\overset{CH_3}{\underset{CH_3}{N}}-O-\overset{O}{\underset{}{C}}-\!\!\bigcirc \right]^{+} \;\; \bigcirc\!\!-\overset{O}{\underset{}{C}}-O^{-} \qquad (5)$$

$$\longrightarrow CH_3-\!\!\bigcirc\!\!-\overset{\cdot+}{N}(CH_3)_2 \;+\; \bigcirc\!\!-\overset{O}{\underset{}{C}}-O^{-} \;+\; \bigcirc\!\!-\overset{O}{\underset{}{C}}-O\cdot$$

However, not all of the radicals produced will initiate polymerization. Induced decomposition may occur as follows (Margerison and East 1967):

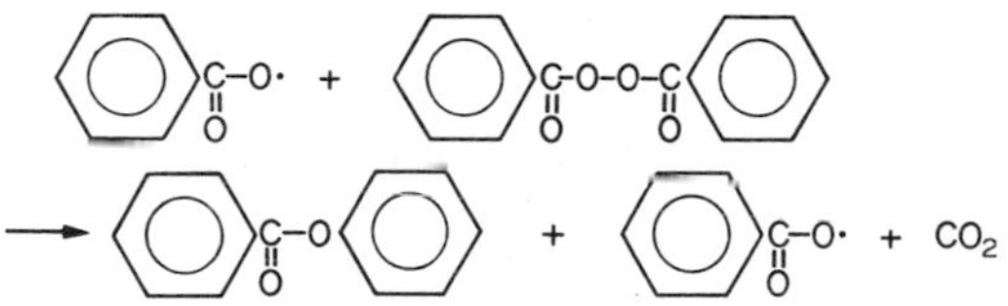

$$\bigcirc\!\!-\overset{O}{\underset{}{C}}-O\cdot \;+\; \bigcirc\!\!-\overset{O}{\underset{}{C}}-O-O-\overset{O}{\underset{}{C}}-\!\!\bigcirc$$

$$\longrightarrow \bigcirc\!\!-\overset{O}{\underset{}{C}}-O-\!\!\bigcirc \;+\; \bigcirc\!\!-\overset{O}{\underset{}{C}}-O\cdot \;+\; CO_2 \qquad (6)$$

Now, consider the reaction of a free radical (represented by R·) with methyl methacrylate. Addition occurs to yield a larger radical by two possible routes: either

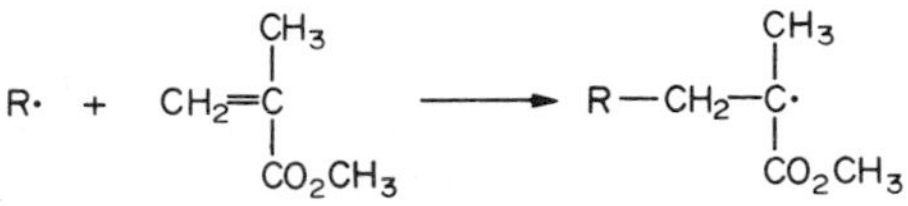

$$R\cdot \;+\; CH_2\!\!=\!\!\overset{CH_3}{\underset{CO_2CH_3}{C}} \longrightarrow R-CH_2-\overset{CH_3}{\underset{CO_2CH_3}{\overset{|}{C}}}\cdot \qquad (7)$$

or

$$R\cdot \;+\; \overset{CH_3}{\underset{CO_2CH_3}{C}}\!\!=\!\!CH_2 \longrightarrow R-\overset{CH_3}{\underset{CO_2CH_3}{\overset{|}{C}}}-CH_2\cdot \qquad (8)$$

The former reaction predominates, since the CO_2CH_3 group sterically hinders the approach of R· to a substituted carbon atom. Propagation continues as follows:

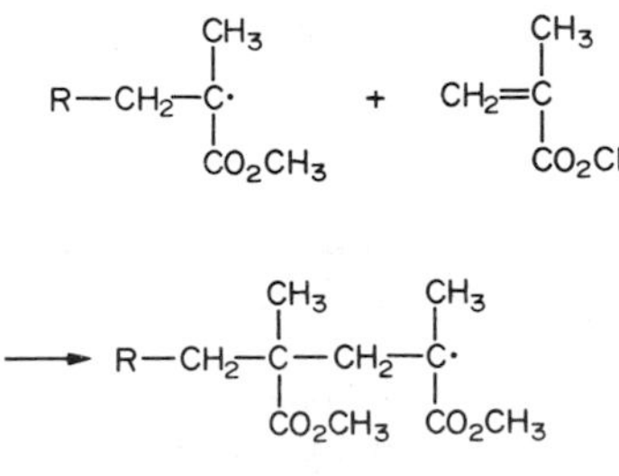

$$R-CH_2-\overset{CH_3}{\underset{CO_2CH_3}{\overset{|}{C}}}\cdot \;+\; CH_2\!\!=\!\!\overset{CH_3}{\underset{CO_2CH_3}{C}}$$

$$\longrightarrow R-CH_2-\overset{CH_3}{\underset{CO_2CH_3}{\overset{|}{C}}}-CH_2-\overset{CH_3}{\underset{CO_2CH_3}{\overset{|}{C}}}\cdot \qquad (9)$$

The reaction proceeds to produce a chain compound $R(M)_xM\cdot$, where M represents the monomer units.

The possibility exists for a "head-to-tail" arrangement of molecules, as shown above, or for a "head-to-head" arrangement, namely:

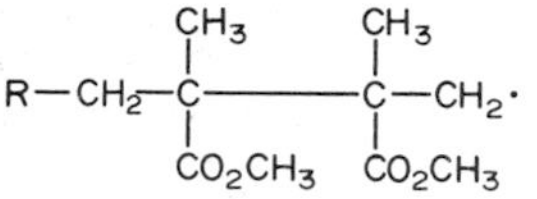

$$R-CH_2-\overset{CH_3}{\underset{CO_2CH_3}{\overset{|}{C}}}-\overset{CH_3}{\underset{CO_2CH_3}{\overset{|}{C}}}-CH_2\cdot \qquad (10)$$

The former reaction predominates, by analogy with the reaction of addition of R· to the monomer, as given above.

Termination of the reaction can occur by combination of free radicals to give a head-to-tail linkage:

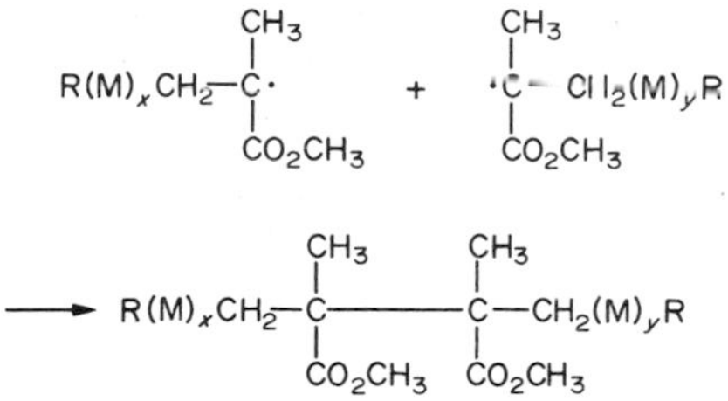

$$R(M)_xCH_2-\overset{CH_3}{\underset{CO_2CH_3}{\overset{|}{C}}}\cdot \;+\; \cdot\overset{CH_3}{\underset{CO_2CH_3}{\overset{|}{C}}}-CH_2(M)_yR$$

$$\longrightarrow R(M)_xCH_2-\overset{CH_3}{\underset{CO_2CH_3}{\overset{|}{C}}}-\overset{CH_3}{\underset{CO_2CH_3}{\overset{|}{C}}}-CH_2(M)_yR \qquad (11)$$

Alternatively, transfer of a hydrogen atom from one radical to another (disproportionation) can take place:

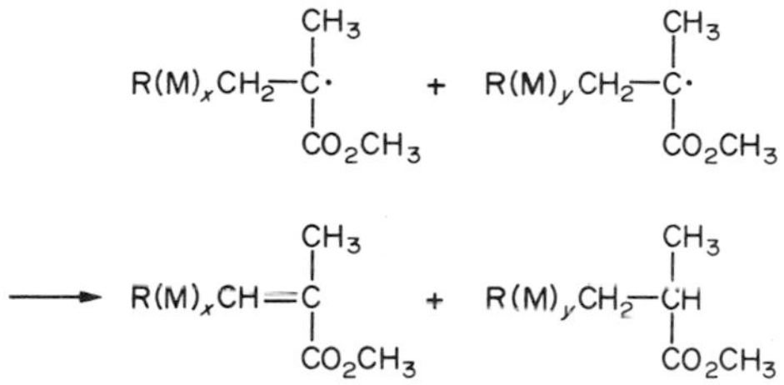

$$R(M)_xCH_2-\overset{CH_3}{\underset{CO_2CH_3}{\overset{|}{C}}}\cdot \;+\; R(M)_yCH_2-\overset{CH_3}{\underset{CO_2CH_3}{\overset{|}{C}}}\cdot$$

$$\longrightarrow R(M)_xCH\!\!=\!\!\overset{CH_3}{\underset{CO_2CH_3}{C}} \;+\; R(M)_yCH_2-\overset{CH_3}{\underset{CO_2CH_3}{\overset{|}{C}}}H \qquad (12)$$

The presence of added chemicals can influence polymerization. Inhibitors and retarders reduce the rate of reaction, while chain transfer agents affect the molecular weight without altering the reaction rate.

Hydroquinone and its monomethyl ether are useful inhibitors since they can be added to reactive monomers to prevent polymerization on storage. These compounds react with radicals, to form more stable radicals, followed by disproportionation (Allinger et al. 1976 p. 595):

$$(13)$$

The presence of oxygen reduces the rate of free-radical addition polymerization. Oxygen can react to give a peroxide radical:

$$R(M)_xM \cdot + O_2 \rightarrow R(M)_xM{-}O{-}O \cdot \qquad (14)$$

This radical has low reactivity, but can add to monomer units such as methyl methacrylate to give oxygen-containing polymers:

$$R(M)_xM{-}O{-}O \cdot + M \rightarrow R(M)_xM{-}O{-}O{-}M \cdot \quad (15)$$

However, in such reactions the degree of polymerization is low, typically 10–40 (Flory 1953 p. 168).

A transfer agent is a substance that can react with a polymer radical to convert it to a stable molecule, while producing a radical which can initiate a new polymer chain (TH is the transfer agent):

$$R(M)_xM \cdot + TH \rightarrow R(M)_xMH + T \cdot$$
$$T \cdot + M \rightarrow T{-}M \cdot \qquad (16)$$

In addition to chemicals added specifically for the purpose, transfer agents can sometimes include the initiator, the monomer itself or a solvent, if present.

3. Poly(Methyl Methacrylate)

Poly(methyl methacrylate) (PMMA) is a thermoplastic polymer that can be molded at 400 K, significantly higher temperatures causing depolymerization to occur. However, dental techniques of molding do not employ the thermoplastic nature of the material. Instead, a mixture of monomer, powdered polymer and other compounds (see Table 1) is made, and dissolution of polymer in the monomer results in the formation of a plastic dough. Distinction must be made between two types of materials: those which require application of heat to

Table 1
Components of dental PMMA systems

Component	Remarks
Powder	
PMMA	In the form of beads, produced from monomer by emulsion polymerization
Benzoyl peroxide	An initiator of polymerization—some may remain from initial preparation of PMMA beads
Pigment	Sometimes added during production of polymer, or may be milled into the polymer beads; the cadmium compounds which are used are believed to present no systemic toxicological problems (McCabe et al. 1978)
Fibers	Colored fibers (acrylic or rayon) may be added for aesthetic effect
Liquid	
Monomer	Methyl methacrylate
Ethylene glycol dimethacrylate	This or other similar difunctional monomer may be added to give a degree of cross-linking in the polymer (Hill 1981)
Hydroquinone or its monomethyl ether	Prevents polymerization on storage: concentration must be kept low to avoid undue interference with the rate of polymerization
Dibutyl phthalate	Such a plasticizer is sometimes added (it may also be added to polymer during ball milling stage)
Tertiary amine	Only used in autopolymerizing resins

decompose the benzoyl peroxide initiator (contained in the polymer powder) and cause the monomer to polymerize; and those which polymerize at room temperature, because of the inclusion of a tertiary amine in the monomer liquid—these are autopolymerizing resins.

The heat-cured materials are widely used as denture base materials. Autopolymerizing resins are occasionally used for this purpose, but also for the repair and relining of dentures. They were also used at one time as tooth filling materials (both with and without inorganic filler particles) and as temporary crown and bridge materials. The major components of these materials are listed in Table 1.

In attempting to obtain denture materials with better mechanical properties (e.g., resistance to impact), two approaches have been tried. The incorporation of carbon fibers (Schreiber 1971, Manley et al. 1979), and the incorporation of a rubbery phase. Rodford (1986) has compared various commercial products based on butadiene–methyl methacrylate, butadiene–styrene or butadiene–styrene–methyl methacrylate. She has also reported on the effects of the incorporation of low-molecular-weight butadiene–styrene rubbers in PMMA on impact and other properties.

Optimum physical and mechanical properties of a polymer are only obtained if the average molecular weight is greater than 10^5. Beech (1975) used gel permeation chromatography to measure the molecular weight and molecular-weight distribution for a dental PMMA powder as supplied and produced from polymer–monomer doughs under different conditions (i.e., heat curing at 343 K and 373 K, and room-temperature polymerization in the presence of a tertiary amine). Typical values of molecular weight were 5.4×10^5 for PMMA powder, and 6.0–7.3×10^5 for the cured materials. In all cases, the percentage of polymer with molecular weight less than 10^5 was in the range 8–10%. Beech concluded that the inferior mechanical properties observed in autopolymerizing resins could not be accounted for in terms of molecular-weight distributions.

One important difference between heat-cured and autopolymerized resins is the quantity of monomer remaining after polymerization has ceased. The residual monomer content of autopolymerized materials may be four times greater than that of correctly heat-cured resins (McCabe and Basker 1976). The plasticizing effect of the monomer is important in terms of mechanical properties. The amounts of monomer in heat-cured materials vary with the curing cycle. For example, acrylic specimens polymerized for 7 h at 348 K followed by 3 h at 373 K contained 0.20–0.40% monomer. Shorter curing cycles yielded polymers with 1.2–2.0% monomer. Dentures containing 1.7–3.2% monomer were found to cause mucosal damage to patients (Austin and Basker 1980).

Recent attention has been given to techniques involving the use of microwaves as the energy source for polymerization, to give rapid, efficient and effective curing (Clerck 1987, Al Doori et al. 1988).

Ruyter (1980) has shown the presence of formaldehyde in acrylic resin by high-performance liquid chromatography. It is postulated to be the product of oxidation of residual methyl methacrylate monomer:

$$\underset{\underset{CO_2CH_3}{|}}{\overset{\overset{CH_3}{|}}{H_2C{=}C}} \;+\; O_2 \longrightarrow H_2C{=}O \;+\; CH_3\overset{O}{\underset{O}{\overset{\|}{C}}C\overset{\|}{O}CH_3}$$

$$\text{(formaldehyde)} \qquad \text{(methyl pyruvate)}$$

$$(17)$$

The quantity of formaldehyde is greater in autopolymerized resins.

4. Higher Methacrylates

In some instances, soft polymers are required for lining dentures. One approach is to use either heat-cured or autopolymerizing methacrylate polymers containing plasticizers. These are supplied as two-component systems, consisting of a powder and a liquid. The powder is based on, for example, poly(ethyl methacrylate) or an acrylic copolymer. The liquid contains a polymerizable monomer, and also an aromatic ester (e.g., butylphthalyl butylglycolate). These reline materials should be distinguished from products which contain no monomer. The latter are resilient polymers supplied for tissue conditioning and the recording of functional impressions; such products contain poly(ethyl methacrylate) plus an alcohol and an ester, and setting occurs through gel formation.

Higher methacrylates are also used in rigid polymers as substitutes for autopolymerizing acrylic resins. These products are used for temporary crowns and bridges. They consist of poly(ethyl methacrylate) powder and either *n*- or iso-butyl methacrylate liquid. Polymerization occurs by the free-radical reactions discussed in Sect. 2. Advantages of such systems are lower polymerization exothermicity, the absence of the irritant methyl methacrylate monomer and the production of more flexible and tougher polymers (Braden et al. 1976). These higher methacrylate systems do not appear to have been as extensively studied in terms of, for example, structure, molecular weight and residual monomer as the comparable PMMA systems.

5. Systems Based on Difunctional Monomers

The systems discussed so far primarily contain monofunctional monomers, although some difunctional cross-linking agents may be present (Table 1).

Difunctional methacrylate systems, which may contain lesser quantities of monofunctional monomers, are also employed, both as the basis of sealants (used to protect pits and fissures of teeth from decay) and as constituents of polymer–ceramic composites (often used as tooth restorative materials).

An important example of a dimethacrylate monomer is Bowen's resin or bis-glycidyl methacrylate (bis-GMA), shown in Fig. 1. The high viscosity of bis-GMA necessitates the use of a lower-viscosity diluent monomer. Asmussen (1975) conducted nuclear magnetic resonance (NMR) analysis on 24 brands of composite material. Most contained bis-GMA (34–86% of the total monomer content). This was often diluted with triethylene glycol dimethacrylate (14–66%). Other monomers present in some products included 1,3-butanediol dimethacrylate, cyclohexyl methacrylate and ethylene glycol dimethacrylate. A similar monomer to the above, but without the OH groups, has also been used.

Urethane dimethacrylates provide an alternative to the use of bis-GMA (see Fig. 2). According to Bowen (1979) these monomers are structurally very similar to the aromatic dimethacrylates. They are viscous because of intermolecular hydrogen bonding.

Peroxide–amine systems are often used to initiate polymerization, and these have been extensively researched. Analysis was carried out on 16 restorative materials, and these were found to have between 0.3 wt% and 2.6 wt% of the monomer (Asmussen 1980b). These products used three different amine types: 12 contained *N*,*N*-diethanol-*p*-toluidene, two had *N*,*N*-diethanol-3,5-di-*t*-butylaniline, and two employed *N*,*N*-diethanol-3,4-dimethylaniline (Asmussen 1980a). Dulik (1979) studied 15 different amine compounds. The choice of amine may adversely affect the color stability of the composite (Bowen and Argentar 1971). The color stability of a set composite material is improved by incorporation of a compound capable of absorbing light just below the blue region of the spectrum. Such a compound is 2-hydroxy-4-methoxybenzophenone (see Fig. 3).

A polymerization inhibitor such as the methyl ether of hydroquinone not only enhances storage stability by preventing polymerization, but also delays the onset of reaction of the mixed material, giving the operator time to position the material while it is still capable of plastic deformation.

In the search for "command setting" of restorative materials, ultraviolet (uv) activation has been employed. Some products contained a benzoin compound as an alternative to the peroxide–amine initiation system; benzoin methyl ether was particularly favored for this application. On irradiation with uv light of wavelength 320–365 nm, free radicals are generated thus:

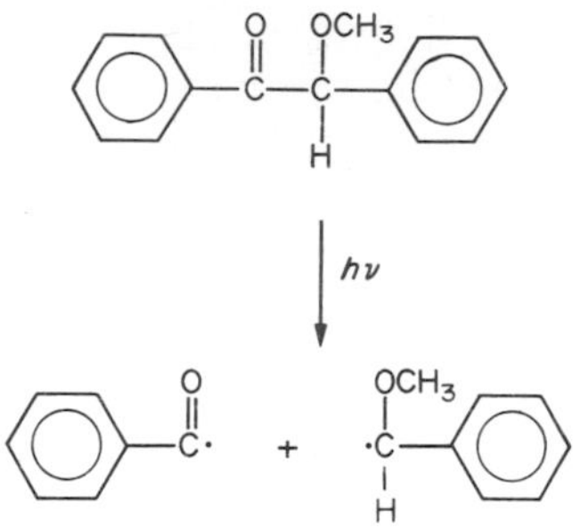

$$\text{(18)}$$

Detailed mechanistic studies of this reaction have been carried out by Pappas et al. (1976).

It is now recognized that visible-light curing systems are safer to use, and give more efficient polymerization—that is, greater depth of cure. Such products contain an α-diketone and an amine and are activated by the application of light of wavelength 480 nm, to react as follows:

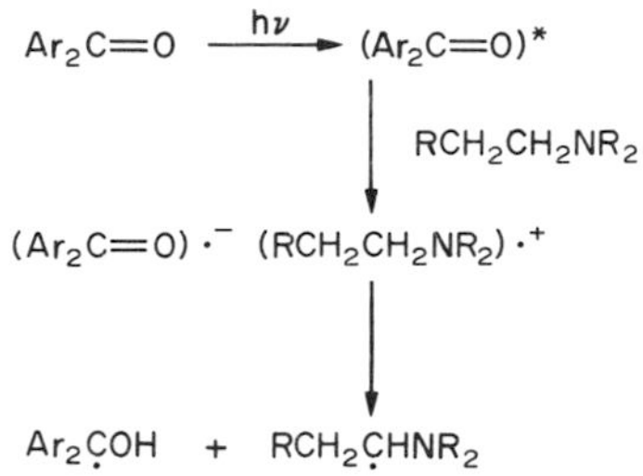

$$Ar_2C{=}O \xrightarrow{h\nu} (Ar_2C{=}O)^*$$

$$\downarrow RCH_2CH_2NR_2$$

$$(Ar_2C{=}O)\cdot^- \ (RCH_2CH_2NR_2)\cdot^+ \quad \text{(19)}$$

$$\downarrow$$

$$Ar_2\overset{\cdot}{C}OH \ + \ RCH_2\overset{\cdot}{C}HNR_2$$

The following factors are important in commenting on the efficiency of polymerization.

(a) A potential problem for pit and fissure sealants is the presence of an unpolymerized surface layer because of oxygen inhibition of free-radical polymerization—see Eqn (14). Ruyter (1981) studied this problem and found the depth of unpolymerized resin to vary from 7 μm to 84 μm. In comparing uv and chemically activated materials of comparable viscosity, he found that the former had less unpolymerized material. Resins of higher viscosity were less affected by oxygen inhibition. The choice of

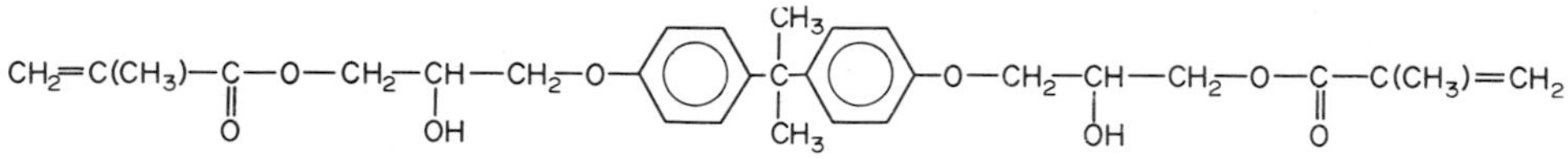

$$CH_2{=}C(CH_3)-\overset{O}{\underset{\|}{C}}-O-CH_2-\underset{OH}{\overset{}{CH}}-CH_2-O-\underset{CH_3}{\overset{CH_3}{\underset{|}{\overset{|}{C}}}}-O-CH_2-\underset{OH}{\overset{}{CH}}-CH_2-O-\overset{}{\underset{\|}{\underset{O}{C}}}-C(CH_3){=}CH_2$$

Figure 1
Structural representation of bis-GMA

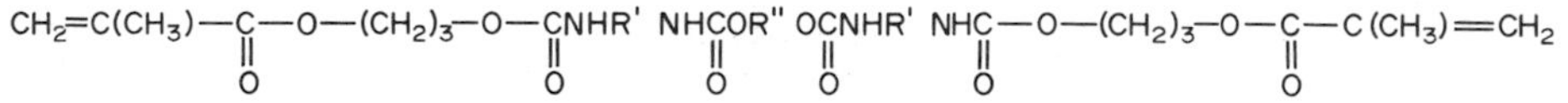

$$CH_2=C(CH_3)-\underset{\underset{O}{\|}}{C}-O-(CH_2)_3-O-\underset{\underset{O}{\|}}{C}NHR'\ \underset{\underset{O}{\|}}{NHCOR''}\ \underset{\underset{O}{\|}}{OCNHR'}\ \underset{\underset{O}{\|}}{NHC}-O-(CH_2)_3-O-\underset{\underset{O}{\|}}{C}-C(CH_3)=CH_2$$

Figure 2
Structural representation of urethane dimethacrylates

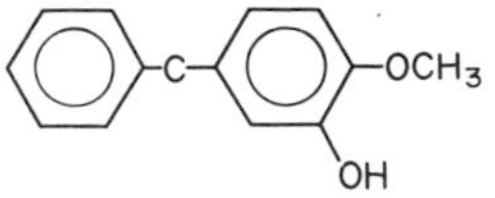

Figure 3
Structural representation of 2-hydroxy-4-methoxybenzophenone

tertiary amine for chemically activated materials was also found to be important.

(b) In the case of composite filling materials, oxygen inhibition of polymerization does not pose a great problem, since the surface of the setting composite is protected from air by the presence of a matrix.

(c) Ruyter and Svendsen (1978) have studied the structure of composites, and have reported that from about 25% up to nearly 50% of the methacrylate groups do not react. The higher values were obtained for materials with the greatest concentration of aromatic monomers. The proposed structure of the composites is that cross-linking is not efficient, there being many pendant unreacted methacrylate groups in the final product.

(d) With light-cured materials, it is important to ensure that an adequate depth of material is polymerized (Watts et al. 1984); the techniques necessary to ensure this have been discussed by Combe (1985).

A wide range of inorganic filler systems has been used in composite filling materials, including glass fibers and beads, lithium aluminosilicates, quartz, barium glass and strontium glass. These can comprise 78–86 wt% of the total material, are typically of particle size 1–$50\,\mu m$ and are treated with α-methacryloxypropyltrimethoxysilane coupling agent. Alternatively, composites have been introduced containing pyrogenic silica of particle size $0.05\,\mu m$, in quantities varying from under 30 wt% to over 60 wt%.

6. Water Soluble Polymers and Copolymers

Acrylic acid can be prepared by the two-stage oxidation of propylene:

$$CH_2=CHCH_3 + O_2 \rightarrow CH_2=CHCHO + H_2O \quad (20)$$

$$CH_2=CHCHO + \tfrac{1}{2}O_2 \rightarrow CH_2=CHCO_2H \quad (21)$$

An alternative process, starting with acetylene, is sometimes employed:

$$4C_2H_2 + Ni(CO)_4 + 4H_2O + 2HCl$$
$$\rightarrow 4CH_2=CHCO_2H + H_2 + NiCl_2 \quad (22)$$

Acrylic acid can be polymerized in aqueous system, using redox initiators, or in nonaqueous media such as benzene in the presence of benzoyl peroxide. Poly(acrylic acid) and copolymers of acrylic acid with either itaconic acid or maleic acid are widely used as components of adhesive dental polyalkenoate cements. One of these is based on the reaction product of zinc oxide with an aqueous polymer solution (Smith 1968). More recently, glass–ionomer cements, based on the reaction of an ion-leachable aluminosilicate glass with a water soluble copolymer, have been developed (Wilson and Kent 1972). Such products are widely recognized as being of value because of their ability to adhere to tooth enamel and dentine.

Bibliography

Al Doori D, Huggett R, Bates J F, Brooks S C 1988 A comparison of denture base acrylic resins polymerised by microwave irradiation and by conventional water bath systems. *Dent. Mater.* 4: 25–32

Allinger N L, Cava M P, DeJongh D C, Johnson C R, Lebel N A, Stevens C L 1976 *Organic Chemistry*, 2nd edn. Worth, New York

Asmussen E 1975 NMR-analysis of monomers in restorative resins. *Acta Odontol. Scand.* 33: 129–34

Asmussen E 1980a A qualitative and quantitative analysis of tertiary amines in restorative resins. *Acta Odontol. Scand.* 38: 95–9

Asmussen E 1980b Quantitative analysis of peroxide in restorative resins. *Acta Odontol. Scand.* 38: 269–72

Austin A T, Basker R M 1980 The level of residual monomer in acrylic denture base materials with particular reference to a modified method of analysis. *Br. Dent. J.* 149: 281–6

Beech D R 1975 Molecular weight distribution of denture base acrylic. *J. Dent.* 3: 19–24

Bolker H I 1974 *Natural and Synthetic Polymers: An Introduction.* Dekker, New York

Bowen R L 1979 Compatibility of various materials with oral tissues. I: The components in composite restorations. *J. Dent. Res.* 58: 1493–503

Bowen R L, Argentar H 1971 Amine accelerators for methacrylate resin systems. *J. Dent. Res.* 50: 923–8

Braden M, Clarke R L, Pearson G J, Keys W C 1976 A new temporary crown and bridge resin. *Br. Dent. J.* 141: 269–72

Clerck De J P 1987 Microwave polymerisation of acrylic resins used in dental prostheses. *J. Prosthet. Dent.* 57: 650–8

Combe E C 1985 Creating successful composite restorations. In Derrick D D (ed.) 1985 *The 1985 Dental Annual*. Wright, Bristol

Dulik D M 1979 Evaluation of commercial and newly synthesized amine accelerators for dental compositions. *J. Dent. Res.* 58: 1308–16

Flory P J 1953 *Principles of Polymer Chemistry*. Cornell University Press, Ithaca, New York

Hill R G 1981 The crosslinking agent ethylene glycol dimethacrylate content of the currently available acrylic denture base resins. *J. Dent. Res.* 60: 725–6

Manley T R, Bowman A J, Cook M 1979 Denture base reinforced with carbon fibre *Br. Dent. J.* 146: 25

Margerison D, East G C 1967 *An Introduction to Polymer Chemistry*. Pergamon, Oxford

McCabe J F, Basker R M 1976 Tissue sensitivity to acrylic resin. *Br. Dent. J.* 140: 347–50

McCabe J F, Wilson S J, Wilson H J 1978 Cadmium in denture base materials. *Br. Dent. J.* 144: 167–70

Pappas S P, Chattopadhyay A K, Carlblom L H 1976 Benzoin ether photoinitiated polymerization of acrylates. In: Labanna S S (ed.) 1976 *Ultraviolet Light Induced Reactions in Polymers*. American Chemical Society, Washington, DC, pp. 12–18

Rodford R 1986 The development of high impact strength denture-base materials. *J. Dent.* 14: 214–17

Ruyter I E 1980 Release of formaldehyde from denture base polymers. *Acta Odontol. Scand.* 38: 17–27

Ruyter I E 1981 Unpolymerized surface layers on sealants. *Acta Odontol. Scand.* 39: 27–32

Ruyter I E, Svendsen S A 1978 Remaining methacrylate groups in composite restorative materials. *Acta Odontol. Scand.* 36: 75–82

Schreiber C K 1971 Poly(methyl methacrylate) reinforced with carbon fibres *Br. Dent. J.* 130: 29–30

Smith D C 1968 A new dental cement. *Br. Dent. J.* 125: 381–4

Watts D C, Amer O, Combe E C 1984 Characteristics of visible light-activated composite resins. *Br. Dent. J.* 156: 209–15

Wilson A D, Kent B E 1972 A new translucent cement for dentistry. *Br. Dent. J.* 132: 133–5

E. C. Combe
[University of Manchester Dental Hospital, Manchester, UK]

Acrylics for Implantation

Acrylics or acrylic resins are trivial names for polymers of derivatives of acrylic acid and methacrylic acid, particularly their esters with a variety of alcohols. Poly(methyl methacrylate) (PMMA) was among the first plastics to be developed in the second and third decades of the twentieth century. It was successfully employed as a substitute for the heavy and vulnerable inorganic glasses in such applications as windscreens for motor vehicles and airplanes. At this time the material was first used by the dental profession for the fabrication of dentures and maxillofacial prostheses such as artificial eyes, noses and ears. The contemporary definition of an implant is "a mechanical device made from one or more biomaterials that is intentionally placed within the body, totally or partially buried beneath an epithelial surface." Although the above-mentioned prostheses and devices are not implants according to this definition, it is in the dental field that PMMA—or the acrylic resins—evolved as a biomaterial.

1. History of Clinical Applications

The potentials of acrylic resins as a material able to be buried in the body was evaluated by various investigators in the late 1930s and early 1940s. One of the first surgical applications of the material in the repair of cranial defects was made by Zander in 1941. In addition, the numerous casualties of World War II created a large demand for reconstructive and cosmetic surgery, which was reflected by the many reports on the use of acrylic resins for such purposes in those years.

In orthopedic surgery, PMMA was first used on a large scale in 1946 when Robert and Jean Judet introduced their Perspex or Plexiglass femoral-head prosthesis. In 1940 and 1943, German and French patents were granted for the use of tertiary amines as accelerators for the peroxide-initiated polymerization of methyl methacrylate. The autopolymerization or "cold-curing" acrylic resins that could thus be formulated, enhanced the possibilities in the already explored fields and created new applications. At the beginning of the 1950s several dental restorative resins were available, providing the dentist with an *in situ* curing material for tooth-colored fillings. Cranioplastic operations and other reconstructive procedures were simplified since it was no longer necessary to prefabricate the inserts in gypsum molds.

Ridley (1952) found that fragments of splintered airfighter canopies made of high-purity PMMA, found in the eyes of veteran pilots, were tolerated fairly well by the eye tissues. This discovery initiated a successful development of contact lenses and intraocular lenses. In the mid-1950s several published articles, for example Oppenheimer et al. (1953, 1955) discussed tumor induction by polymers after implantation in animals. The concern that was subsequently raised about the clinical use of polymers and thus of PMMA as implants was eased when it appeared that possible carcinogenicity was related to physical aspects of the material such as shape (foils) and dimensions (particulate material) rather than to its chemical composition.

One of the most important phases in the development of biomedical acrylic resins was the introduction of cold-curing PMMA as "bone cement" by

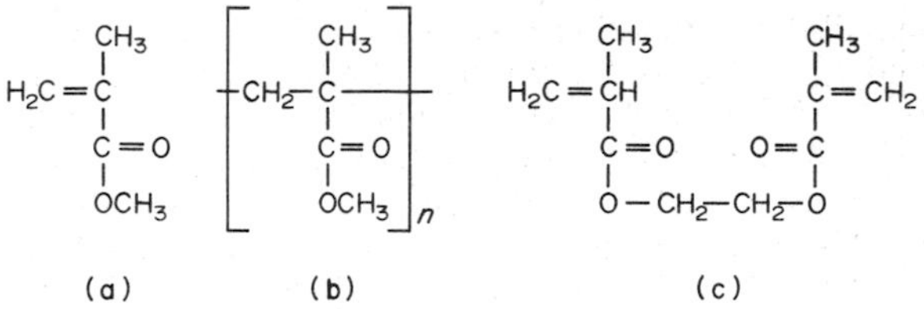

Figure 1
Structural repesentations of: (a) methyl methacrylate, (b) poly(methyl methacrylate) and (c) ethylene glycoldimethacrylate

Charnley (1960, 1961)—advised in his choice of plastic by D C Smith—for the stabilization of metallic femoral hip endoprostheses. Over the next twenty years a variety of investigators undertook refinements of implantation techniques, improvements of the material and assessments of possible toxicological aspects of PMMA and its components. In other surgical fields, the potential of acrylic polymers, either in prepolymerized form or as an *in situ* curing resin, was explored. Acrylic resins served such purposes as correction of bony deformities, internal fixation of traumatic and pathological fractures, spinal fixation and reinforcing inserts. Today, however, the main applications have stabilized in the areas of denture fabrication, filling materials, intraocular lenses, fixation of joint replacements and repair of cranial defects.

2. Chemistry of Acrylic Resins

Acrylic resins are polymers, the repeating units (monomers) of which are alkylesters from 1,2-propenoic acid (acrylic acid) or 2-methyl-1,2-propenoic acid (methacrylic acid). A common monomer from which acrylic polymers are synthesized is the methyl ester of methacrylic acid. Frequently, however, comonomers such as ethyl methacrylate, butyl methacrylate or methyl acrylate arc built into the polymer chains to modify the properties of the resulting plastic material. "Difunctional" esters such as ethylene glycoldimethacrylate (see Fig. 1) are sometimes used to "cross-link" the molecular chains to a three-dimensional network, which improves the material's insolubility and mechanical properties.

The polymerization of (meth)acrylate monomers is an addition reaction initiated through intermediate free radicals. When free radicals are generated in the presence of, for example, methyl methacrylate, the radical adds to the double bond under formation of another radical. The reaction commonly used to produce free radicals for the initiation of this type of polymerization is the decomposition of dibenzoylperoxide $(C_6H_5CO_2)_2$ into, among others, phenyl $C_6H_5\cdot$ radicals (see Fig. 2).

Each initiated chain radical is capable of adding on another monomer molecule, thus forming a growing chain. The propagation of chains continues until the reaction is terminated, for example, by two growing chains combining their unpaired electrons to form one single bond. The decomposition reaction proceeds spontaneously at temperatures above $60\,°C$ or at room temperature by a redox reaction with easily oxidizable compounds such as tertiary amines. Thus polymerization of alkyl methacrylates can be achieved by heating a monomer and a small amount of dissolved dibenzoylperoxide. When this heat curing is carried out in a mold, plastic products are formed *in situ* (e.g., dentures, intraocular lenses and contact lenses).

In cold-curing or autopolymerizing systems an accelerator is added to the reaction mixture. This is usually N,N-dimethyl-p-toluidine, $(CH_3)_2NC_6H_4CH_3$. Once initiated, these polymerizations proceed rapidly under the development of considerable heat ($58–67\,kJ\,mol^{-1}$) so that the temperature of the mixture may rise considerably; this in turn speeds up the reaction. This autoaccelerating effect may be used advantageously to "snap cure" small amounts of material such as dental fillings. It may, however, cause harmfully high temperatures and deleterious porosity in larger volumes of material (e.g., bone cements).

Moldable medical and dental acrylic resins are available as two-component systems: a powder that consists mainly of small PMMA spheres and a liquid containing the monomer(s). The peroxide initiator is added to the powder component and, in cold-curing formulations, the amine accelerator is dissolved in the monomer. The powder and liquid are mixed in a

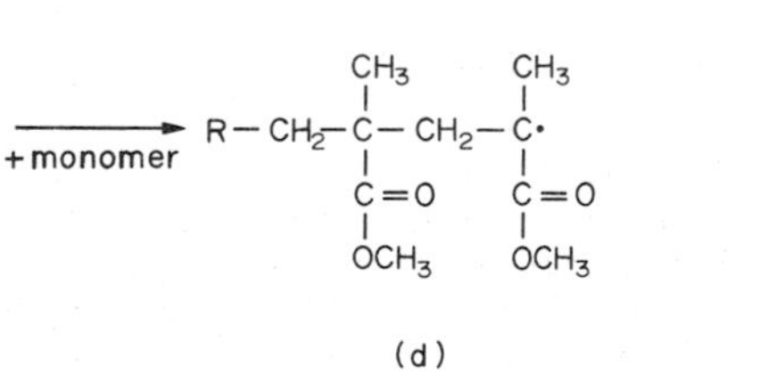

Figure 2
Reaction mechanism for the polymerization of methyl methacrylate showing: (a) the free radical, (b) the monomer, (c) the initiated chain radical and (d) the growing chain radical

Table 1
Composition of moldable room-temperature-curing acrylic resin

Component	Quantity	Component	Quantity
Powder		Liquid	
Poly(methyl methacrylate) spherical powder (10–40 µm)	>90%	Methylmethacrylate	>85%
		Comonomers	10–15%
Dibenzoylperoxide	1–2%	N,N-dimethyl-p-toluidine	1–2%
BaSO$_4$, ZrO$_2$ or other pigments	4–8%	Hydroquinone stabilizer	50–100 ppm

ratio of approximately two to one (by weight), respectively, to form a moldable dough which cures in about ten minutes or after heating overnight in a gypsum mold. The monomer polymerizes and binds together the pre-existing polymer particles. This offers several advantages over a polymerizing monomer alone: the initial mixture is easier to handle and heat evolution and volume shrinkage are reduced. For dental purposes, pigments and fillers can be added to the powder component. Surgical bone cements often contain some barium sulfate or zirconium oxide as radiopacifiers since the polymer itself is translucent for x rays. A typical formulation of cold-curing acrylic resin is given in Table 1. Heat-curing dental formulations are similar apart from the absence of toluidine accelerators.

When optically clear products such as intraocular lenses are desired, the liquid monomers have to be polymerized in metal molds under carefully controlled circumstances to avoid voids or discoloration. Some of the merits of these pure homogeneous PMMAs such as their good optical transparency and color stability have been recognized in areas of technology other than medicine and dentistry.

3. Biological Reactions

Biological reactions to the presence of an acrylic implant (usually PMMA) are provoked by low-molecular-weight constituents of the polymer (e.g., monomers, initiators, accelerators and stabilizers) and physical properties and aspects of the material (e.g., particle size of wear debris, temperature rise during polymerization *in situ*, and the rigidity of the material). Most of the reactions to low-molecular-weight compounds occur shortly after implantation and can be discussed in terms of toxicity, systemic effects and hypersensitivity. In the longer term, histological reactions may occur owing to low-molecular-weight compounds or to physical aspects of the implant.

3.1 Low-Molecular-Weight Constituents

(*a*) *Monomers*. Generally, when monomer is converted to polymer, the conversion will not be complete and thus the polymeric material will contain some residual monomer. In the case of biomedical acrylic resins the amount of residual monomer depends strongly on the conditions during polymerization and varies from less than 1% in heat-cured dental and ophthalmological prostheses to about 2–4% in *in situ* cured cements. The use of *in situ* curing cements involves a temporary contact of living tissue with the still-uncured material, so that some contamination of the biological system with monomeric material is inevitable.

In numerous experiments on rats, guinea pigs and dogs, the LD50 value for methyl methacrylate has been established as 1–2 g kg^{-1}, when administered intravenously or intraperitoneally. Oral LD50s were in the range of 10 g kg^{-1} body weight for methacrylic esters and 0.2–1.0 g kg^{-1} for the lower acrylic esters. Similar values apply for dermal LD50. Methyl methacrylate impairs the pulmonary functions at doses of 0.075 g kg^{-1} intravenously in dogs, the cause of death in the LD50 experiments being respiratory failure. The liver and kidneys appear to be attacked to a lesser degree, although some degenerative changes have been observed in animals exposed to vapors or high doses of the compound. Furthermore, it is beyond doubt that (meth)acrylates are cytotoxic. In cell culture experiments, methyl methacrylate has been reported to kill cells or at least impair culture growth at concentrations as low as 1–10 mg per 100 g serum (0.001–0.01 wt%).

In the early days of bone-cement applications for the anchoring of total hip prostheses, there were numerous reports of moderate to severe drops in blood pressure during operations. These were usually transient but were lethal in some cases. The effect occurred shortly after insertion of the cement in the reamed medullary canal of the femur. Some, notably Charnley, were of the opinion that the fall in blood pressure could not be related to absorption of the monomer in the bloodstream but that a fat or air embolism caused by forcing the cement into the femoral cavity was responsible for the complication. Proper venting of the medullary canal would largely eliminate the occurrence of these emboli. Nevertheless, the monomer continues to be suspected as a toxic agent for various functions of the biological system. In high enough doses the esters of methacrylic acid not only exert an adverse influence on cardiac and respiratory functions but also on the immunological defence mechanism against bacterial attack and the development of the fetus during pregnancy. In general, studies indicate that acrylate esters are more toxic than methacrylate esters and that a longer alkyl chain of the ester group diminishes the acute toxicity.

The major question for biomedical usage is whether during clinical application of acrylic material, especially bone cements, the level of monomer in the human bloodstream can become high enough to produce the effects to an unacceptable extent. Maximum levels of methyl methacrylate in blood during hip arthroplasty, which is undoubtedly the operation where the patient is most exposed, have been reported to be of the order of 1 mg per 100 ml. This low figure in comparison with the levels at which significant effects were found in animal studies, together with the fact that methyl methacrylate is rapidly removed from the blood stream by the metabolism (citrate cycle) is the reason that this monomer (and thus most acrylic implants) have been given the benefit of the doubt as far as systemic effects in humans are concerned.

In tissues immediately adjacent to curing bone cement, maximum concentrations of methyl methacrylate of 0.01 wt% to about 3 wt% at some time after application of the cement have been measured. These values are, temporarily, higher than the toxicity limits indicated by cell culture tests. Clinically, this means that implantation of acrylic bone cement will be accompanied by at least some damage to the surrounding tissues.

A special effect of acrylic resins is that monomeric (meth)acrylates are capable of causing allergic reactions. The first reports identifying acrylic resin as a possible allergen came from the field of prosthetic dentistry. Patients who receive an acrylic denture sometimes develop sore mouths which may be diagnosed as allergic stomatitis. The differential diagnosis, however, is complicated by psychological factors concerning the acceptance of having to wear a denture and by mechanical irritation as a result of an ill-fitting prosthesis. In orthopedic surgery, surgeons may experience pruritus and erythema of the finger skin after handling acrylic bone cement during total hip replacements, whereas their patients sometimes develop complications associated with hypersensitivity to the monomer. Again, an unequivocal diagnosis of allergy for monomeric methacrylates is difficult to make because of the many other possible traumatic stimuli, the low incidence of significant cases and the uncertain residual monomer content of the various biomedical acrylic appliances.

In industrial applications of acrylic resins, the potential of a monomer to cause irritation of the skin is expressed by the Draize-value, an ordinal measure for the dermatological reactions of rabbit skin after intensive exposure to the compounds. Acrylic esters have higher Draize-values than methacrylic esters and the alcohol part of the ester has a strong influence on the irritating power.

(*b*) *Other low-molecular-weight constituents.* Depending on the polymerization conditions, residues of auxiliary compounds such as hydroquinone, dibenzoylperoxide or aromatic amines can be expected in the final implant. High-purity ophthalmological PMMA will obviously contain only traces of such compounds, but cured bone cement will necessarily contain larger concentrations. The toxic potency of these residuals depends, however, on the rate at which they leach out and diffuse into the surrounding tissues. The concentration of these auxiliaries in the implant varies from the part per million level for hydroquinone to about 1% for peroxide or for the amines. Although the rate of leaching is low by the nature of the polymeric material, the level of contamination of the biological system (especially with dimethyl-*p*-toluidine) has aroused concern. Leaching rates of $0.6\ \mu g\ cm^{-2}$ per day initially to $0.002\ \mu g\ cm^{-2}$ per day after several weeks have been measured in *in vitro* experiments. The systemic toxicity is probably irrelevant at these concentration levels but, as with methyl methacrylate, the cytotoxicity is high. Growth in cell cultures is retarded at concentrations as low as 10^{-4}%. In addition, dimethyl-*p*-toluidine is a suspected, but not proven, carcinogen.

Dibenzoylperoxide, hydroquinone and the aromatic amines should be considered to be allergenic. Reports of clinical cases, however, come from applications of these compounds other than acrylic implantations. It has, for instance, been shown that persons sensitized by benzoylperoxide may demonstrate allergic reactions after applications of an acne prescription containing 5% benzoylperoxide. Where such high concentrations of peroxide are not met in acrylic implants it is conceivable that the reported incidence of these cases will remain scarce.

3.2 Influence of Physical Properties

(*a*) *Reactions to wear particles.* When there is mobility of, or between, parts of PMMA implants, the low wear strength of this hard and glassy polymer may result in the formation of small wear particles or wear debris. The adverse effects of this loose particulate material which eventually settles in the surrounding tissues were first experienced in connection with the Plexiglass Judet femoral prosthesis. Chronic inflammation and foreign-body reactions (acrylosis) invariably observed in connection with this type of prosthesis were the main reasons for its being abandoned. Similar observations have been made to a lesser extent where acrylic cement had been used to anchor metallic prostheses. Small particles of cement, probably due to friction between the metal prosthesis stem and the cement or between cement and adjacent bone, are sometimes found in the new capsule around the artificial joint. Acrylic debris seems to be a major factor in the formation of a typical partly granulomatous foreign-body reaction in this capsule. In severe cases, the capsule can develop to a thickness that impairs joint mobility. The reactions seem to follow a pattern which is specific for reactions to particulate material but less specific for the material *per se*.

(b) *Reactions to curing temperature*. During *in situ* curing of acrylic resins such as bone cements, the exothermic nature of the polymerization reaction initiates a process of autoacceleration where the temperature increases sharply. The possibility of this rise in temperature causing thermal damage to the surrounding tissues (and, if so, to what extent) has received much attention. Laboratory experiments concerning the actual temperatures that occur in the cement mass and, more significantly, at the cement–bone interface, or even in the layers of adjacent bone, show considerable variation. In clinical situations the observed maximum temperature varies from 120 °C in the center of thick masses to values in the range 45–70 °C at the bone–cement interface. The uncertainty about the actual temperatures is inherent due to the multitude of parameters that will determine the temperature–time relationship at a certain distance from a heat source: geometry, thermal conductivities, heat capacities and boundary conditions varying from case to case and from individual to individual. This also explains the lack of agreement between the results of extensive histopathological investigations on thermal damage of the surrounding tissues after bone cement implantation.

Great emphasis is often attached in this respect to temperatures at which body proteins are observed to coagulate (56 °C) or at which bone collagen has been found to deteriorate (70–72 °C). However, it has been pointed out that not only the temperature but also the time for which a certain temperature is maintained should be taken into consideration. Thus, the probability of tissue damage is related to the product of time and temperature implying that temperatures much lower than 56 °C can be harmful provided they prevail long enough. Elaboration of such principles leads to the conclusion that regular PMMA bone cements are only marginal as to their safety with respect to thermal damage. Any attempt, therefore, to reduce the maximum occurring temperatures by proper design of materials and clinical conditions is relevant for the prevention of untoward tissue reactions.

(c) *Reactions to mechanical properties*. Owing to the anatomy of the eye, implantation of an intraocular lens (extensively performed in cataract surgery) results in the eye tissues being pressed against the implant. It has been shown in animal experiments that short, pressurized contact of the corneal endothelium results in loss of endothelial cells. The effect is seen to a much lesser extent, or not at all, in the case of lenses made of materials with much lower rigidity (e.g., hydrogels and silicones).

3.3 Long-Term Histological Reactions

(a) *Reactions of hard tissues*. The histological response of bone in contact with acrylic resin is very well documented in the field of joint replacement, especially that of total hip replacement. Unfortunately, this area represents applications of the material where many unspecific reactions have to be expected because of the complex surgical preparation. Charnley (1970) and Willert and Puls (1972) have described the basic histology in connection with total hip cementations. Three phases of reactions have been recognized: the initial phase, the repair phase and the stabilization phase.

The initial phase shows the reactions to the cement superimposed on the reactions to the operation trauma. The most prominent reaction is necrosis of bone and marrow in the immediate vicinity of the cement. Feith (1975) found that the cause of this reaction may be ascribed to monomeric methyl methacrylate, possibly in combination with the temporarily high temperatures during curing of the cement.

The repair phase starts after about three weeks. Bone and marrow necrosis have been cleared by phagocytosis, bone and collagenous connective tissue being formed in their place, the latter eventually encapsulating the implant.

The stabilization phase is characterized by bone growth in the direction of the implant and bone remodelling. New bone growth may grow up close to the implant surface, although this is thought to depend on stress distributions and levels. In hip joint cementations a 0.1–1.5 mm thick connective tissue membrane prevails in separating bone from implant, but when bone grows into pores of acrylic cement—where low stress levels are likely—direct contact between bone and cement can occur. When a membrane has formed it may contain foreign-body giant cells and small cement particles surrounded by giant cells. Although foci of chronic inflammation may also occur in this capsule, the situation is often stable for many years.

On the still longer term, from seven to 15 years, loosening of hip prostheses occurs in a relatively high percentage of cases. The cause of this is not fully understood: if it is not due to late infection by aerobic or anaerobic microorganisms it is probably a combination of the shortcomings of the material and the surgery, unfavorable stress distributions and gradual resorption of bone.

Another longer term histological reaction is the disturbance of mineralization of bone directly adjacent to the cement. Residuals of dimethyl-*p*-toluidine, the accelerator of the polymerization reaction, are suspected by some investigators to be responsible for such disturbances.

(*b*) *Reaction of soft tissues.* One of the incongruities in soft-tissue reactions to PMMA is the absence of capsule formation when intraocular lenses are placed in the anterior or posterior chamber of the eye. In all other parts of the body a solid implant of this type of polymer becomes encapsulated, at least to some extent. It is likely that this lack of reaction (which is fortunate from the ophthalmological standpoint) is a specific property of the eye tissues rather than a merit of the material.

The short-term reaction to the rigid PMMA lenses—the loss of corneal epithelium—has already been mentioned (see Sect. 3.2*c*). In the longer term, corneal dystrophy and inflammation of the uvea with resulting opacification of the vitreous body have been associated with the implanted material, although without proven specificity because other materials such as hydrogels and silicones give rise to the same observations. This holds for provoked precipitates and pigment dispersion which may occur in some cases but which is associated with technical considerations such as avoiding permanent contact of the lens with the iris.

A recent tendency in cataract surgery is an increased interest in so-called "soft" lenses made of polyhydroxyethylmethacrylate copolymers (as hydrogels) and silicones. By their low rigidity, these materials allow the fabrication of lenses which, in folded position or in a dehydrated shrunken form, can be introduced into the eye via a much smaller incision than is necessary for the rigid acrylic lenses.

When, for example, a neurocranial defect is repaired with a precured acrylic plate or with *in situ* curing cement, the implant will be in contact with the dura mater at one side and with muscle tissue at the other. Soft tissues appear to have the same or a greater tolerance than bone to acrylic resins. Provided that the influences of high temperature and monomer are minimized by suitable precautions, direct curing of acrylic cement on the dura mater has been reported not to offer problems in cranioplasty. After implantation, solid acrylic implants soon become encapsulated in fibrous connective tissue. The thickness of the capsule and the extent and duration of inflammatory cell infiltration largely depend on mechanical and geometrical factors, such as mobility or sharp edges. These factors can easily be controlled in the case of implants opposing soft tissue layers.

3.4 Carcinogenicity

There has been considerable doubt as to the capability of implanted plastics to induce neoplastic changes in the surrounding tissues. With the exception of the possible role in tumor induction of low-molecular-weight contaminants of polymers, there is no evidence that PMMA in the composition and shapes as used in implant surgery behaves differently from other plastic materials. The physical nature of the implant, and geometrical factors such as shape, dimensions and surface roughness are of primary importance rather than the chemical nature of the material. Moreover, tumor formations directly related to implants have largely been observed in animals and only rarely in humans. In animals, the probability is strongly dependent on the nature of the fibrous tissue capsule that is formed around the implant. It is possible, therefore, that wear particles, sometimes observed in the capsules around heavily loaded implants such as bone cements in hip arthroplasty, offer some risk in this respect.

Even when an induction period for the formation of tumors of 20–25 years is taken into consideration, the absence of reports of implant related malignancies after more than 40 years of clinical use supports the conclusion that PMMA is intrinsically safe in this respect.

4. Developments

4.1 Toxicity of Residuals

Little can be done to reduce levels of residual low-molecular-weight substances in cold-curing acrylic resins. Residual monomer, strongly associated with the condition of polymerization at low temperatures in a short time is a factor beyond the control of material designers. In practice, the toxicity of methyl methacrylate seems not to offer serious or irreversible tissue damage. More concern has recently been expressed about the influence of dimethyl-*p*-toluidine, the commonly used accelerator of the polymerization reaction, which has been associated with disturbances of bone mineralization. The suspicion has been intensified by observations that cements in which the polymerization was initiated by an experimental tri-*n*-butyl-borane–O_2 system did not seem to give rise to mineralization problems.

4.2 Reduction of Maximum Temperature in Cold-Curing Resins

There are several ways to reduce the maximum temperature during the exothermic polymerization of methyl methacrylate, but the method of adding a heat sink in the form of water or an aqueous gel is the only approach that has been proven to be of practical value. One bone cement has been developed in which 20 wt% water is dispersed. Owing to the heat capacity of water and the fact that it "dilutes" the system, the maximum temperature is thus significantly reduced. The mechanical properties of the cement are claimed to be unimpaired by the finely dispersed droplets of water, although the effect of the proposed improvement on clinical and histological reactions is still under investigation.

4.3 Porous Cement

A further development is the dispersion of an aqueous gel through the still fluid mixture of cold-curing acrylic resin. When the gel is added in amounts of 35–50 wt% the dispersed filaments form an interconnected network in the mixture. After curing of the acrylic phase, the aqueous phase can leach out of the material by virtue of its continuous geometry leaving a network of pores. The effect is twofold: during curing the maximum temperature is reduced significantly, while afterwards the network of pores makes it possible for the surrounding tissues to grow in, thus firmly anchoring the implant. Although the pores impair the mechanical properties to a level that the material is no longer suitable for cementing heavily loaded hip prostheses, histological evaluation has shown the benefits of admixing the aqueous gel. In rabbit experiments this type of cement, when introduced into the femoral cavity, did not provoke more cortical necrosis than when the cavity was only reamed and left unfilled. This was in contrast to the reaction on normal solid cements and most certainly the consequence of reducing the maximum temperature. After longer residence times, when implanted in bone, the pores of the cement become filled, first with healthy well-vascularized connective tissue, then with bone. Eventually, most of the pore volume is filled with bone and the implant is "integrated" into the host tissue without becoming encapsulated (the latter is invariably the case with dense cements). When implanted in soft tissue, the same differences between dense cement and porous cement occur. The porous implant is immobilized by the ingrowth of connective tissue without capsule formation while dense cements become encapsulated and remain mobile. Irritation of the surrounding tissues by the mobile implant may even lead to ectopic mineralization and cartilage formation around the implant.

4.4 Increasing Strength of Bone Cements

From clinical experience and follow-up studies it appears that, for the purpose of hip cementation, acrylic resins are only marginal in strength. Two paths are followed to strengthen the existing formulation. The first is to reduce microporosity in the cement caused by enclosure of air during mixing and by the volatility of the monomer. The second is to incorporate fibers or other fillers to reinforce the cement.

Reducing microporosity is accomplished by centrifuging techniques and procedures of preparing the cement under vacuum. Although these preparation techniques complicate the task of the surgeon in the theatre, the gain in strength that may be realized seems to warrant the efforts or costs of special equipment.

Incorporation of fillers and fibers in polymers is a well-known technology of reinforcing polymers. The same principles applied to bone cements give the expected improvement of mechanical properties. Carbon fibers improve strength and stiffness considerably when added in amounts as low as 1–2 wt%. Reinforced cements, however, are still in an experimental phase and the increase of the viscosity of the initial mixture by these fillers may conflict with a good adaptation to, or penetration of, the bone walls when injected in the femoral cavity. This aspect represents another field of recent developments, namely low-viscosity cements. Loosening of cemented hip prostheses without fracture of the cement is often associated with poor adaptation of the cement to the surrounding bone. When the still-uncured cement mixture interpenetrates the bone trabeculae of the cavity walls this might improve the longevity of the arthroplasty. Many manufacturers of bone cements now include a low-viscosity type of cement in their program to meet clinical demands in this respect.

Bibliography

Allarkhia L, Knoll R L, Lindstrom R L 1987 Soft intra-ocular lenses. *J. Cataract Refract. Surg.* 13: 607–70

Autian J 1975 Structure–toxicity relationships of acrylic monomers. *Environ. Health Perspect.* 11: 141–52

Charnley J 1960 Anchorage of the femoral head prosthesis to the shaft of the femur. *J. Bone Jt. Surg. Br.* Vol. 42: 28–30

Charnley J 1961 Arthroplasty of the hip. A new operation. *Lancet* 1: 1129–32

Charnley J 1970 *Acrylic Cement in Orthopedic Surgery.* Churchill-Livingstone, Edinburgh

Feith R 1975 Side effects of acrylic cement implanted into bone. *Acta. Orthop. Scand.* Suppl. 161

Galin M A, Turkish L, Chowchuvech E 1977 Detection, removal and effect of unpolymerized methylmethacrylate in intra-ocular lenses. *Am. J. Opthalmol.* 84: 153–9

Huiskes R 1980 Some fundamental aspects of the human joint replacement. *Acta. Orthop. Scand.* Suppl. 185

Lintner F, Bösch P, Brand G 1982 Histologische Untersuchungen über Umbauvorgänge an der Zement-Knochen-Grenze bei Endoprothesen nach 3–10 jähriger Implantation. *Pathol: Res. Pract.* 173: 376–89

Lundskog J 1972 Heat and bone tissue. *Scand. J. Plast. Reconstr. Surg.* Suppl. 9

Malten K E 1984 Dermatological problems with synthetic resins and plastics in glues, parts I, II. *Dermatosen* 32: 81–6; 118–25

Oppenheimer B S, Oppenheimer E T, Stout A P, Daniskefsky I 1953 Malignant tumors resulting from embedding plastics in rodents. *Science* 118: 305–6

Oppenheimer B S, Oppenheimer E T, Stout A P, Eirich F R 1955 Further studies of polymers as carcinogenic agents in animals. *Cancer Res.* 15: 333–40

Ridley H 1952 Intra-ocular lenses; Recent developments in surgery of cataract. *Br. J. Ophthalmol.* 36: 113–22

Ridley H 1970 Long-term results of acrylic lens surgery. *Proc. R. Soc. Med.* 63: 309–10 van der Walle H B, Klecak G, Geleick H, Beusink T 1982 Sensitizing potential of 14-mono(meth)acrylates in the guinea pig. *Contact Dermatitis* 8: 223–35

van der Walle H B, Waegemakers T H J M, Beusink T 1983 Sensitizing potential of 12-di(meth)acrylates in the guinea pig. *Contact Dermatitis* 9: 10–20

van Mullem P J, de Wijn J R 1988 Bone and soft connective tissue response to porous acrylic implants; A histokinetic study. *J. Cranio-Max.-Fac. Surg.* 16(3): 99–109

Willert H G, Buchhorn G (eds.) 1987 Knochenzement. *Aktuelle Probleme in Chirurgie und Orthopädie*, Vol. 31. Hans Huber, Bern–Stuttgart–Toronto

Willert H G, Puls P 1972 Die Reaktion der Knochens auf Knochenzement bei der Allo-Arthroplastik der Hüfte. *Arch. Orthop. Unfall-Chir.* 72: 33–71

J. R. de Wijn
[University of Leiden, Leiden, The Netherlands]

P. J. van Mullem
[University of Nijmegen, Nijmegen, The Netherlands]

Adhesives from Protein–Polymer Grafts

Proteins are remarkable molecules that present an optimal mix of intimate contact with a substrate and cooperative secondary force interactions with the groups found on those substrates. The former quality is due to their very flexible backbone while the latter is due to 20 different side chains that comprise acidic, basic, hydroxylated and even hydrophobic moieties. The molecules thus satisfy all of the theoretical requirements for adhesion.

Even though proteins have been used as carpenter's glue and as a component of early mortars throughout history, they have not been considered suitable for state-of-the-art applications. This has been because of their sensitivity to moisture; a major problem with proteins. This sensitivity leads to their plasticization (collagen has over 70 MPa tensile strength and over 21 MPa shear strength when dry but about 10% of these values when wet) and eventual degradation due to extracellular proteases produced by microorganisms in the moist state. There are, however, proteinaceous adhesives (≥99% of total weight) produced by marine organisms, such as barnacle cement and mussel adhesive, that not only set in the presence of water, but remain unaffected by it as well as by proteolytic enzymes. The resistance of barnacle cement to proteolytic enzymes, acids and alkali suggests a highly cross-linked structure. Thus it appears that three-dimensional cross-linking can overcome the problem of plasticization and degradation, although very few proteins besides these adhesives are known to have this property.

Starting with this premise, several studies have been undertaken to develop methods for the intermolecular cross-linking of proteins. Even though cross-linking can be accomplished by utilizing any multifunctional polymer, such as toluene diisocyanate, that can react with amino and hydroxyl groups found on amino acid side chains, it was decided to build and graft a synthetic polymer on the protein backbone. The polymer chosen was epoxy resin which meant that the phenolic side chains (i.e., tyrosines) of the protein would be utilized.

The selection of the protein was based on practical considerations such as water solubility, the availability of extensive chemical data in order to be able to follow and quantify each step of the reactions, and the availability of consistently good quality material at a reasonable price. These conditions inevitably led to the selection of 300 Bloom gelatin as the material of choice.

1. Industrial Applications: Protein–Polymer Grafts

1.1 Epoxy Grafts

Since collagen, the precursor of gelatin, contains only five tyrosine residues while there are 38 lysine and hydroxylysines, 53 arginines and 75 dicarboxylic acids on a backbone of over a thousand amino acids, it was necessary to modify the gelatin backbone to introduce additional phenolic groups in order to achieve higher cross-linking density. Although the literature is full of reactions developed for the modification of most amino acids, many are not suitable for the purpose at hand because reactions and reagents used for amino acid modification leading to the development of materials of construction have to fulfill the following requirements:

(a) to be able to form stable, irreversible covalent bonds with a particular amino acid; and

(b) to carry another reactive or potentially reactive moiety so that after modification is completed further manipulations required for grafting can take place.

Of all the procedures found in the literature, only reductive alkylation fulfills these requirements and utilizing phenolic aldehydes, such as 4-, 2,3-, 2,4- or 3,4-dihydroxybenzaldehyde, phenolic groups held covalently can be introduced. The reaction occurs smoothly, as depicted in Fig. 1, and converts about 60% of the available lysines. The failure to increase the proportion converted was traced to the reversibility of the first step in the reaction (the formation of the Schiff's base between the aldehyde and the ε-NH$_2$ of the lysines) because the unconverted lysines react with epichlorohydrin during the next step of the procedure. This reaction provides approximately 30 phenolic anchoring positions for grafting, corresponding to a 3% cross-linking density. Since all high-performance adhesives have a very high cross-linking density, it is necessary to

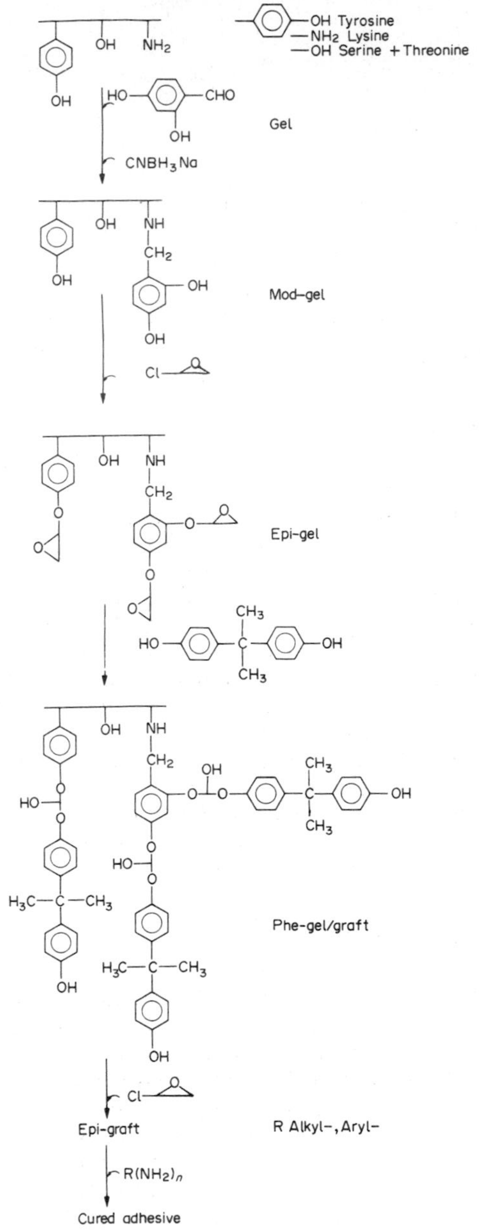

Figure 1
The grafting procedure: there are 5, 55 and 38
similar chains of tyrosine, lysine and serine +
threonine in the molecule, respectively (after Kaleem
et al. 1987d © *Nature (London)*, Macmillan Journals.
Reproduced with permission)

modify other amino acids with reactive moieties,
such as arginine, aspartic and glutamic acids, serine,
threonine and hydroxyproline. A method for the
nearly quantitative modification of arginine, which is

difficult to modify due to its very high pK (12.5), was
developed utilizing 4-hydroxybenzil, after the grafting studies had been completed (Kaleem et al.
1987b, c). The next step involves treating the modified gelatin with epichlorohydrin, where a slight
excess of the reagent results in a quantitative yield.

The epoxidized gelatin can be reacted with any
polyfunctional phenol such as bisphenol A and
resorcinol sulfide. Again the reaction occurs with
nearly quantitative yield. To convert the graft to a
state that can be cured, it is reacted with epichlorohydrin a second time, after which it can be cured
with any epoxide curing agent (e.g., amines, anhyd-
rides, Lewis acids, preformed novolaks). Up to the
point of curing, the graft is completely soluble in
water, dimethylsulfoxide and ethylene glycol. If
desired, instead of curing, a second or a third cycle
of grafting can be performed by reacting this product
with a polyfunctional phenol followed by epoxida-
tion; however, too much cross-linking may lead to
much more brittle products and may cause steric
problems during curing, leaving uncured loose ends.
All of these parameters require a systematic study
for optimization. A similar study is required to find
the optimal protein size to achieve maximum
strength. During reductive alkylation as well as arginine modification, a certain degradation of the backbone occurs (Kaleem et al. 1987c), making such a
systematic study desirable. Experience gained with
bioadhesive development, however, suggests that
even in the denatured state, full-sized molecules
may block certain reactive sites by coiling. Hence it
is desirable to have smaller molecules for optimal
cross-linking (Kaleem et al. 1987a). Since these
studies were completed before the development
of the arginine modification procedure, a
polyaziridine—XAMA-2 or 7—that reacts with carboxyl groups was also used for curing, in order to
increase the cross-linking density. No studies were
performed to find out whether the use of aziridine
did interfere with curing.

Using tetra-aminobenzene, aziridine and heating
to 150 °C for 5 h followed by slow cooling, such a
graft has achieved approximately 28 MPa tensile
strength with aluminum coupons that were prepared
according to a Forest Products Laboratory method.
With sand-blasted stainless steel 10.5 MPa was
obtained. The graft also bonds appropriately prepared polyethylene samples. The cured adhesive is
very brittle. Considering the graft has less than 5%
epoxy resin and about 8% cross-linking density
these results are quite impressive; however, owing
to low cross-linking density, moisture absorption is
quite high. By including higher fatty acid chlorides
during curing, moisture absorption can be reduced.
Using isocyanates with ethylene glycol during curing
also reduces water uptake. Even though the grafts
were prepared stepwise to be able to quantify each
step, there is no reason why similar grafts cannot be

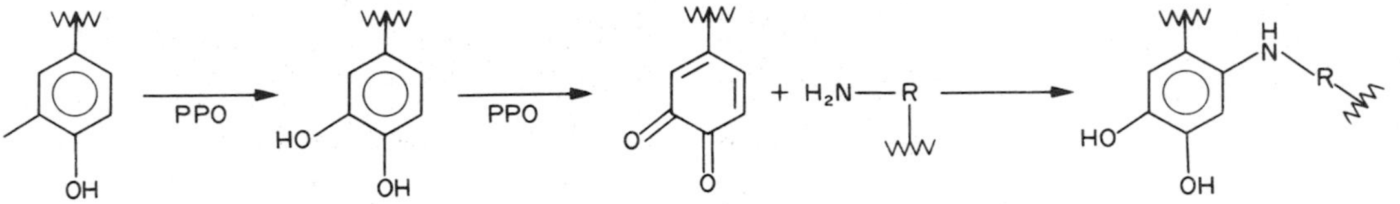

Figure 2
Scheme showing the hydroxylation of tyrosines to L-DOPA, the subsequent oxidation to dopaquinone and the spontaneous reaction with ε-NH$_2$ of lysines

formed between modified protein and preformed epoxy resins.

Epoxy grafts are discussed further in Kaleem et al. (1987d).

1.2 Phenolic Grafts

Modified gelatin can also be grafted with phenolic resins. In a typical case, modified gelatin is reacted with formaldehyde followed by resorcinol. The graft behaves like any phenolic resin and can be cured with substances such as formaldehyde or hexamethylene tetramine, but the more interesting observation made was that both the modified protein and the graft were thermoset resins that became insoluble at 65 °C. They can be molded and accept fillers.

1.3 Expectations and Potential

It has to be stated here that these studies have only demonstrated the feasibility of the idea that such grafts can be prepared and that they had some interesting properties. However, there is no reason to doubt that other polymers such as elastomers and heterocyclics can also be grafted onto protein backbones. Once the basic steps are established, such grafts can also be produced with other soluble proteins such as whey protein or insoluble proteins such as zein. These grafts become very interesting when it is considered that more than one polymer can be grafted onto a protein backbone, either successively or simultaneously, after the base lines for different grafts have been established. Since these grafts represent upgrading of a renewable resource, the major emphasis has to be directed to finding out where such grafts can best be utilized, while continuing to improve them. Since fossil fuel resources are decreasing, an ability to utilize renewable resources, either exclusively or together with only a small amount of raw materials derived from fossil fuel, is a highly desirable possibility.

Knowing that it takes considerable time to bring a concept from the laboratory to the market, it seems appropriate to begin such studies now. There are, however, certain problems that have to be taken into consideration. Since native proteins have a preponderance of amino acids with aliphatic side chains, they cannot be expected to perform much above 200 °C for long periods. Furthermore, it is highly doubtful whether it will ever be possible to find a protein that has a preponderance of easily modifiable amino acids uniformly distributed along the protein backbone, which is the important criterion to obtain optimum performance.

2. Biomedical Applications

Exposure to moisture is the biggest problem for all adhesive applications, with one exception: biomedical adhesives, where the adhesive has to be applied on a wet tissue because living tissues cannot be dried. For this reason barnacle cement and mussel adhesive, which set in the presence of water and are unaffected by proteolytic enzymes, have been considered ideal biomedical adhesives. None of the available polymers really adhere to soft or calcified tissues and the success with enamel bonding is because of mechanical interlocking made possible by a preceding acid etch; a procedure that also cannot be applied to live tissues such as dentin. The chemistry of barnacle cement is not well understood owing to its rapid setting and the inability to find an inhibitor that can keep it in the unpolymerized state. Histoenzymological studies have implicated the involvement of an enzyme, polyphenol oxidase, in the setting of the adhesive. More recently, a protein responsible for the adhesion of mussel byssus was purified and sequenced and it too relies on similar enzymes for its cross-linking. This enzyme hydroxylates tyrosines to L-DOPA and then oxidizes them to dopaquinone which spontaneously reacts with the free ε-NH$_2$ of lysines found on the same protein backbone according to the scheme shown in Fig. 2.

Consequently, modified gelatin with its increased phenolic content was reacted with polyphenol oxidase together with a polyamine-like spermine, with the expectation that the protein could function as a substrate for the enzyme. Spectrometric studies have shown that the enzyme accepted modified gelatin as a substrate. A concentrated solution of modified gelatin, together with the enzyme and a polyamine, was found to be able to bond slices of dry bone (Kaleem et al. 1987a).

See also: Adhesives in Medicine

Bibliography

Kaleem K, Chertok F, Erhan S 1987a Collagen-based bioadhesive barnacle cement mimic. *Angew. Makromol. Chem.* 155: 31

Kaleem K, Chertok F, Erhan S 1987b Modification of free arginine using 4-hydroxybenzil. *J. Biol. Phys.* 15: 63

Kaleem K, Chertok F, Erhan S 1987c Modification of protein-bound arginine with 4-hydroxybenzil. *J. Biol. Phys.* 15: 71

Kaleem K, Chertok F, Erhan S 1987d Novel materials from protein–polymer grafts. *Nature (London)* 325: 328

S. Erhan
[Albert Einstein Medical Center, Philadelphia, Pennsylvania, USA]

Adhesives in Medicine

At present, only a limited range of adhesives is used routinely in medical practice. Methyl silicones are employed as temporary bonding aids for the attachment of diagnostic or prosthetic devices to human skin, some cyanoacrylates are used in dental applications, and common acrylics such as poly-(methyl methacrylate) find applications as cements in both dental and orthopedic reconstructions. Moisture-curing and ultraviolet-curing polyurethanes are now achieving some success in their use as "pit and fissure" sealants for children's teeth, aiding in the prevention of dental cavities.

Problems abound with these and all other adhesive substances proposed for medical use. For example: adhesive bond strengths of the silicones are usually too low; the cyanoacrylates degrade slowly, releasing potentially toxic by-products; and the acrylics often contain migratable monomeric species that can be taken up by the body tissues and fluids with adverse secondary effects. In addition, the heats of polymerization of the acrylics are often so high as to kill adjacent living cells. Polyurethanes and many of the polycarboxylates used in dental applications can break down quickly due to the high-wear features of the oral environment.

1. Challenges of the Medical Environment

The fixation of biomedical or dental prosthetic devices in humans provides a severe challenge to adhesive science and technology. Custom-fitted strong (usually metallic) devices called subperiosteal implants provide a good example of the challenges to be met and the extreme complexity of the problems posed by contrasting adhesive requirements over very short distances along the same structures. These implants are expected to adhere to the remaining bone structures in a patient whose teeth have been lost as well as adhering to the remaining tissue flaps used to cover them. They must provide a post which protrudes with a bacterial-tight infection-free adhesive seal through the tissue into the nonsterile intraoral region to provide a mounting site for dentures or other load-bearing functional prosthetic restorations. The portion of the implant buried in the tissue must be firmly adherent in order to provide the immobile support required. The integrity of the tissue bond immediately adjacent to the permucosal post must be nearly perfect to prevent bacterial seepage, infection and resulting inflammatory responses. The structural parts residing in the saliva-bathed oral cavity should resist adhesion and colonization by biological materials, especially plaque-forming bacteria. The distance over which the adhesive properties must change so drastically can be as little as a few micrometers along the surface of such implants. Furthermore, the requisite conditions must be achieved and maintained in environments that begin as surgically prepared, often bloody and usually nonsterile, potentially contaminant-rich sites.

As may be expected, even though medical and dental implants of many configurations have been used clinically for over a decade, the results obtained have been at best ambiguous. Many failures have been noted as a result of poor tissue bonding to the implanted portions, serious cellular and debris accumulation on the exposed segments, and very poor healing around posts protruding through external tissue. It is not uncommon for the implants to extrude completely from their placement sites within only a few weeks. It is obvious that trials in animals, and premature applications of these otherwise promising devices in humans, have proceeded too rapidly in the absence of both necessary and sufficient knowledge of the critical properties of the materials that promote or inhibit biological adhesion.

One major difficulty with studies of biomedical adhesives for any use, be it in dental, orthopedic or cardiovascular environments, has been the tendency of investigators to focus on the longer-term behavior of the materials—from weeks to months to years after application—without proper characterization or even an elementary knowledge of the original surface states of the materials. Little knowledge has been gained of the interfacial conditions at the time of device placement or of the early bioadhesive sequela (over minutes, hours and days) which change the original surface properties of the adhesive and tissue in ways that promote or inhibit adhesion. In more traditional engineering studies, such as those dealing with the rate and consequences of biological fouling of heat-exchange materials (as in power-plant condenser tubes), the same area of ignorance of the early adhesive or "conditioning" events is usually dismissed with the label "induction period."

There is much to be learned from the adhesion of marine fouling organisms. The biofouling of ship hulls, propellers and cooling-water piping includes the spontaneous adhesion of microfouling films and macrofouling aggregations of barnacles, tube worms, algae, bryozoa, and hydroid communities that can diminish ship operating efficiency by 40% in as little as six months in tropical waters—an impressive demonstration of strong adhesion in wet, salty, biochemically active circumstances.

2. Requirements of Good Medical Adhesives

Assessing surface properties of adhesive materials in the biological environment must be the subject of extensive and continuing research and development. The magnitude and difficulty of the subject can be appreciated by considering the number and types of different requirements.

The problem at hand requires the synthesis and study of selected materials for use as encapsulants, sealants and lead insulators exposed to biological environments, especially *in vivo* environments. Topics of interest in this study are dimensional stability, adhesion, permeability, ion migration, possibly electrical and radio-frequency interactions and physical fatigue.

Useful candidate adhesive systems should have a minimum sterile shelf life of 18 months, have a minimum adhesive force to oily or wet biological surfaces of 35 MN m^{-2} or greater, cure effectively at temperatures from 5 °C to 40 °C, support a weight of at least a few kilograms within 30 s of application with an adhesive area of only 5 cm^2 or so when employed at 37 °C, and function effectively on rough or irregular surfaces with heterogeneities as large as 0.25 cm. In addition, they should be able to maintain their attachment for periods in excess of one year without compromising the viability of sensitive adjacent tissues.

Assessment of candidate adhesive systems requires knowledge of their characteristics and performance in various environments:

(a) response to water and water vapor with respect to dimensional stability and permeability;

(b) response to inorganic ions typically found in biological fluids;

(c) methods and conditions of application that minimize failures in bonding;

(d) adherence to typical substrates in a saline environment;

(e) fatigue resistance to repeated flexure, rotation and elongation;

(f) ability to protect the electrical characteristics of semiconductor and interconnect circuits;

(g) attenuating effects over the entire electromagnetic frequency range;

(h) relative compatibilities and stabilities as passive layers in contact with sensitive living tissues; and

(i) flex life in physiological surroundings.

Similar information on composite adhesives or layered materials (i.e., reinforced composites) must also be compiled.

3. Surface Effects on Bonding

Recognizing that the use of medical adhesives to bond living tissue to solid substrates must bear some relationship to the original surface properties of the solid and tissue phases, studies have been made of the influence of some relevant surface-energy parameters in the presence and absence of significant variations of surface charge and texture (roughness). At least one empirical parameter of surface free energy, the "critical surface tension" defined 40 years ago and determined from simple contact-angle measurements with a variety of pure liquids (Baier et al. 1968), provides a useful "label" for the apparently bioadhesive character of solid or semisolid (hydrogels, tissues) substrates. Materials with critical surface tensions between 20 mN m^{-1} and 30 mN m^{-1} exhibit minimal underwater adhesion, whereas materials with critical surface tensions above this range (including most common engineering polymers) or below this range (primarily fluorocarbons) support adhesion in saline media to greater degrees.

Examining the actual initial adhesive events at biological interfaces, it has been found that attachment is preceded by adsorption of macromolecular conditioning films on the nonbiological surfaces. These films are largely composed of amphipathic biopolymers, depositing from aqueous media at about the same rates but in different configurations, on substrates of differing surface properties. Initial rates of subsequent cellular or bacterial adhesion to such adsorbed films are also comparable, but the tenacity of adhesion varies with the degree of apparent "denaturation" of the conditioning film of macromolecules. The critical-surface-tension parameter is a useful descriptor of the bioadhesive properties of these conditioned substrates, revealing changes in surface properties as the initial biopolymer layer forms and as the cellular or bacterial material is deposited. Specific examples of adhesion to such devices as the artificial heart, substitute blood vessels and dental implants support the generalization of these findings to adhesion in most natural circumstances, and imply that practical and inexpensive approaches, such as contact-angle measurements, can be used in researching and controlling biological adhesion.

Particular attention has been given to the experimental requirements for making reliable determinations of surface-energy parameters from contact-angle measurements on living hydrated complex biological materials. These studies demonstrate the inadequacy of using water or water-

25

miscible fluids alone. Although water droplets will quickly penetrate, swell or adsorb surface-active components when placed on biological substrates, exhibiting low contact angles or complete spreading, this complicated event cannot be taken to indicate an operationally high surface free energy for those substrates in contact with other environmental agents. In fact, contact angles measured with water-immiscible fluids of a variety of types on biological surfaces in equilibrium with their surroundings, reveal the apparent surface free energy to be that of the "biologically stable" surface state, characterized by a critical surface tension of 20–30 mN m^{-1}, as opposed to greater than 70 mN m^{-1} (normally expected for an aqueous surface). Desiccated biological surfaces, as exemplified by isolated cell-wall preparations and by extracted proteins, polysaccharides and their complex conjugates, typically display critical surface tensions of above 30 mN m^{-1}. The practical result of these findings is that medical adhesives and/or priming agents must, to be efficacious on living substrates, have surface-active ingredients capable of depressing the interfacial tensions between adhesive and tissue to above 20 mN m^{-1}.

4. Role of Adsorbed Films

One of the most important observations about adhesion in relevant biological circumstances is that it never occurs without the involvement of preadsorbed films of biological macromolecules, primarily glycoproteins. These films are spontaneously adsorbed from the aqueous biological media in which the adhesive is placed. It has become obvious that all bulk materials allow spontaneous deposition and continuing buildup of proteinaceous material from the biological medium in the same relative amounts and at the same rates. The immediate supposition could be that differences in the surface chemistries of these various materials are thus effectively masked or obscured by an overcoating of essentially the same biological macromolecules, accumulated in the same amounts over the same exposure times. If this supposition were correct, subsequent cell contacts, obviously limited to the new exterior face of each protein-coated substrate, could no longer exhibit differential degrees of cell adhesion or cellular spreading. However, the empirical observation is that striking differences in adhesive behavior are noted.

Returning to the parameter of critical surface tension as one indicator of continuing substrate differences (even after conditioning-film coverage), the changes of the critical surface tension have been measured for different test materials vs incubation time at physiological conditions. Such tests revealed that the highest-surface-energy substrate used,

plasma-cleaned (glow-discharge-cleaned) glass, underwent dramatic decreases in critical surface tension within seconds of immersion into the living media.

On the other hand, various glass surfaces treated with different "siliconizing" compounds to lower their surface energy remained at their original critical surface tensions (i.e., 20–30 mN m^{-1} within experimental error) in spite of the independently demonstrated accumulation of about the same amount of proteinaceous material.

Using an improved discriminator, based on measures of the spread areas of adherent living cells, different spread areas were found for the same cell types in the same tissue culture media. The cells exhibited 50% greater spread areas on the high-energy glass (plasma-cleaned) than on the siliconized specimens. The differentiation was even more striking when assessing the number of pseudopods, or irregular projections, on cells adherent to substrates varying in surface energy. The initial substrate surface properties (e.g., surface energy) appear to influence cell adhesion at a considerable distance from the original interface; that is, through the conditioning layer. It is crucial to note that all such interactions occur through, or are mediated by, the initially adsorbed glycoproteinaceous layers. It is necessary to understand, predict and control the qualities of this biological coating in order to achieve the desired improvements in medical adhesives.

5. Observed Failure Mechanisms

There are at least three basic reasons why medical adhesives fail: rejection by the host tissue or some element of it; failure of the physical form as a result of structural or load-bearing forces; and chemical attack on the adhesive by some part of the biological environment. Each of these alone is a possible route to failure, but all three often act simultaneously.

The phenomenon known as biorejection has at its root the mismatched surface energetics of materials placed in intimate contact with a biological system. The manifestation of rejection may take the form of the biosystem rapidly encapsulating or "walling off" the adhesive itself, thereby rendering it ineffective. This may occur by the simple deposition of biological materials as in the formation of a lesion around an implanted device, thereby causing a widespread inflammatory or immunological reaction. Both of these reactions have been characterized by a number of investigators.

Interaction of biological fluids or other living systems with an adhesive phase may also be chemical in nature. Encapsulating metallic devices with adhesive polymers usually helps to extend the serviceable lifetime of the device, but problems often occur with the adhesion of the encapsulant to the metal, leading to blistering, loss of encapsulating film integrity and device leakage. The basis for this mode of attack lies in the loss in dimensional stability of the encapsulating agent due to swelling, the permeation of water and ions through the film and the lack of sufficient mechanical strength to withstand forces in the biological environment.

Water and water vapor are known to penetrate certain organic films to different degrees. This penetration can lead to failure of a protected device by promoting corrosion or other secondary effects. The rate of penetrant invasion, a thermally activated process, depends on the nature of the penetrant, the nature of the material to be penetrated and temperature.

Biological fluids contain several species of inorganic ions that are known to cause damaging effects to sensitive solid-state electronic devices. These ions are able to penetrate adhesive layers and either contribute to the breakdown of bonding between the adhesive and the substrate (thereby leading to rapid corrosion and possible catastrophic failure) or enter the electronic device and interfere with its function. This frequently occurred in early models of heart pacemakers.

6. Current Research Needs

At present, many aspects of medical adhesives and their use require investigation. These may be summarized as follows:

(a) An attempt should be made to resolve the relative influence of surface energy vs surface texture of the medical substrates as controlling factors in obtaining long-term adhesion. Further, resolution of the relative roles of the solid–liquid interfacial tension vs the critical surface tension of the cured adhesives should be attempted.

(b) A program of tensile testing should be developed to overcome the remaining difficulties in obtaining both valid and useful results. It should be aimed at defining the potential utility of a variety of surface treatments, including application of various "coupling agents," in establishing superior adhesion. These tests should also be addressed to determining the influence of water, of salt solutions and of protein solutions or biological fluid simulants on sealant and adhesive bond strengths.

(c) An attempt should be made to establish and validate routine and inexpensive quality control checks for adhesive formulations.

(d) The critical surface properties of human tissues *in vivo*, after various time lags from their initial exposure for bonding, should be determined on children and adults of both sexes. The measurement techniques must be safe and relatively simple.

(e) Investigations should be made of the reactivity of the candidate adhesive monomers and polymers with substrates precoated with different protein films, to model more closely situations encountered clinically.

(f) Studies should be made of the failure surfaces produced when medical adhesives are separated from substrates of different character. These should include consideration both before and after exposure to salt solutions, blood, saliva or tissue fluids.

(g) Experimentation with the use of tailor-made interface modifiers, also known as coupling agents, should be encouraged. This approach has given excellent results in other systems where strong adhesive bonds to moist surfaces were required, such as glass-fiber-reinforced plastics.

(h) Surface chemical characterization should be undertaken for all monomers, polymers and resins now being considered for medical-adhesive applications. Legislative mandates require knowledge of such properties before the approval of materials for human applications.

See also: Chemical Adhesion in Dental Restoratives

Bibliography

Baier R E (ed.) 1975 *Applied Chemistry at Protein Interfaces*, Advances in Chemistry, Vol. 145. American Chemical Society, Washington, DC

Baier R E 1984 The basics of biomedical polymers: Interfacial factors. In: Gebelein C G (ed.) 1984 *Polymeric Materials and Artificial Organs*, ACS Symposium Series 256. American Chemical Society, Washington, DC, pp. 39–44

Baier R E, Meyer A E 1988 Implant surface preparation. *Int. J. Oral Maxillofac. Implants* 3: 9–20

Baier R E, Meyer A E, Natiella J R, Natiella R R, Carter J M 1984 Surface properties determine bioadhesive outcomes: Methods and results. *J. Biomed. Mater. Res.* 18: 337–55

Baier R E, Shafrin E G, Zisman W A 1968 Adhesion: Mechanisms that assist or impede it. *Science* 162: 1360–8

Fowkes F M 1987 Role of acid–base interfacial bonding in adhesion. *J. Adhes. Sci. Technol.* 1: 7–27

Glantz P-O, Baier R E 1986 Recent studies on nonspecific aspects of intraoral adhesion. *J. Adhes.* 20: 227–44

Kronenthal R L, Oser Z, Martin E (eds.) 1975 *Polymers in Medicine and Surgery*, Polymer Science and Technology, Vol. 8. Plenum, New York

Manly R S (ed.) 1970 *Adhesion in Biological Systems*. Academic Press, New York

R. E. Baier
[State University of New York, Buffalo, New York, USA]

Aluminum Oxide

From the large number of ceramic materials generally available only a few have been found to possess the combination of properties from which improvements in prosthetic devices can be expected. Among the ceramics used in reconstructive surgery, the aluminum oxide ceramic (alumina ceramic) is the most chemically inactive. After its early evaluation, the alumina ceramic was soon regarded as the prototype of the so-called bioinert materials.

Due to its mechanical properties, especially its rigidity and hardness, the alumina ceramic has found its main uses in hard-tissue, structural applications in orthopedic surgery and dentistry. Most of these applications are based on new aspects of bone remodelling resulting from observations of the tissue reactions in the vicinity of such implants. As a consequence of its inertness, the remodelling of bony tissue adjacent to alumina implants is not disturbed biochemically (by ions or other matter going into solution) or by immune reactions. The interface reactions have, subsequently, been found to be controlled solely by the stress and strain fields created inside the bony tissue by the insertion of the implant. Using this knowledge, rules have been derived for the design of joint and dental implants so that the remodelling results in close bone contact and thus "osseo-integration." Many years of experience with these implant systems has allowed for some judgement about the validity of these rules and the information on which they are based.

In addition, the bioinertness of these ceramics has led to improved devices in nearly load-free applications in ear, nose and throat surgery and in some soft tissue replacements.

1. Alumina Ceramic

The alumina ceramic is the main representative of the oxide ceramics. Their definition and position within the realm of ceramic materials can be seen in Fig. 1.

From the structural point of view, ceramics are clearly different from metals and plastics (Heimke 1984). Generally, homopolar and heteropolar compounds and solids are distinguished because either they share some of their electrons which are delocalized from the parent atoms, or they have the electrons clearly bound to the atoms concerned; that is, they are ions. In metals, the freely moving electrons are an essential feature controlling most of their basic properties. In ceramics, and particularly in all oxide ceramics, the ionic nature of the bond is predominant. This implies that ceramics always contain more than one kind of atom (pure carbon is the exception to this), often with several types of both positive and negative species. Oxide ceramics usually consist of one or very few metal oxides.

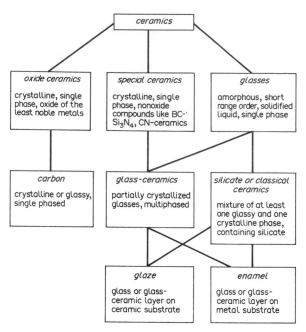

Figure 1
Survey of ceramic materials according to crystallinity and number of phases

The different groups of ceramic materials can also be distinguished by their manufacturing processes. All classical ceramics contain silicates, a proportion of which are introduced by the use of clays, which allows the shaping of the desired parts when mixed with the correct amount of water. The oxide and the special ceramics start from industrially prepared mostly pure powders to which some plastics are temporarily added to allow approximate shaping. However, in some ceramics prepared from industrially made raw materials the final compound may not be formed until the "firing" process. (Firing is the process for all ceramics in which the shaped or molded raw piece is transformed into a solid.)

There is also another difference between silicate and oxide ceramics. In the silicate-containing ceramics one or more melts are formed during the firing cycle that solidify and partially crystallize during the process. Thus, the final product is always a mixture of different phases. However, in oxide ceramics, the shaped agglomeration of powder particles is transformed into a polycrystalline solid by solid-state reactions only, mostly by diffusion.

The essential properties of alumina ceramics for all load-bearing applications are summarized in Table 1. Included here are the specifications according to the first generation of standards as well as the actual values for one of the most widely used commercial materials. The differences between these two sets of values result from experiences which

Table 1
Properties of medical-grade alumina ceramics[a]

Property	Ceramic requirements according to ISO 6474 ASTM F603-83 DIN 58 8353	Frialit bioceramic
Density (g cm^{-3})	>3.9	>3.98
Alumina content (%)	>99.5	>99.9
SiO$_2$ and alkali metal oxides (%)	<0.1	<0.05
Microstructure, average grain size (μm)	<7	<2.5
Microhardness (MPa)	23 000	23 000
Compressive strength (MPa)	4000	4000
Flexural strength (MPa)	>400	>450
Young's modulus (MPa)	380 000	380 000
Impact strength (cm MPa)	>40	>40
Wear resistance[b] (mm^3 h^{-1})	0.01	0.001
Corrosion resistance[b] (mg m^{-2} day^{-1})	<0.1	<0.1

a many of these values are not average but maximum or minimum requirements to be met by all test pieces   b the wear and corrosion resistance values refer to particular test arrangements

indicate the necessity for an extension of the safety margins in some high load-bearing applications (Griss and Heimke 1981a, Heimke and Griss 1981); these improvements will also be included in forthcoming revisions of the standards. For implants in load-free (or nearly load-free) situations, for example in ear, nose and throat surgery or ophthalmology, somewhat reduced values for the mechanical strength will be stated in the revised versions of some of these standards.

2. Reasons for Using Alumina Ceramics in Medicine

The first large scale clinical testing and subsequent applications of alumina ceramics were in total hip replacements and in dentistry.

In the 1960s, total hip replacement became a standard treatment in orthopedic surgery after the introduction of polymethylmethacrylate (PMMA) as the so-called bone cement for the fixation of the polyethylene acetabular socket and the metal stem in the medullary canal of the femur. By 1970, increasing numbers of patients requiring repeat

operations stimulated effort to find the causes for these failures. In the early 1970s, it was widely agreed that the soft tissue layer, which separates the surface of the PMMA along most parts of the interface from the normally proliferating bony tissue, was the major cause of the problem. It was also recognized that the polyethylene wear particles contributed, among other things, to a thickening of the soft tissue layer and, therefore, to implant loosening. In dentistry, a direct correlation exists between the thickness of the soft tissue interlayer, the mobility of the implant, the pocket depth of the gingiva surrounding the implant, and the probability of implant failure (Spiekermann 1980). At that time, it was generally assumed that a basic cause for the formation of the soft tissue interlayer was the chemical instability of the materials used, mainly the stainless steels, the cobalt-based alloys, and the PMMA. This instability resulted in products (metal ions or monomers) disturbing the highly sensitive differentiation processes necessary for normal bone formation. The particular type of biocompatibility of some ceramics raised hopes of avoiding such soft tissue interlayers and thereby achieving direct anchorage of implant devices. This soft-tissue-free implant fixation was termed osseo-integration. For joint replacements, the low friction and high wear resistance of alumina ceramics offered additional advantages.

The early compatibility studies of alumina ceramics, including sarcoma rate testing (Griss and Heimke 1981b), had confirmed the expected biological inertness of this material as well as its tribological advantages along the articulating surfaces. The results of other scientists contradicted these findings by showing that soft tissue interlayers between the alumina implants and the surrounding bone could be related to the mechanical effects (Heimke et al. 1981).

3. Osseo-Integration by Biomechanically Controlled Tissue Remodelling

Since a bioinert material by definition (summarized in Table 2) does not react with the surrounding tissue, it cannot form any bond with the adjacent bone. Thus, only a purely mechanical fixation could be anticipated.

In the initial experiments, in which the test pieces were placed in the bony tissue under nearly load-free conditions, it was recognized that the reorganization of the bony tissue adjacent to an alumina implant follows exactly the same sequence of reactions characteristic of fracture healing (Griss et al. 1975). Further experiments with fully functional total hip replacements in sheep and dogs made it possible to define the design criteria, allowing for a stable and reliable anchorage of implants in the adjacent bony tissue (Griss et al. 1976).

Table 2
The concept of a bioinert material

Requirement	Result
"Nothing goes into solution": leakage of ions or other matter from the implant into the surrounding tissue is below detectability by the cells and without any systemic effect	No biochemical influence on cell differentiation and proliferation. No biochemical information to the cells about the presence of the implant
Strong and fast adsorption of molecules contained in body fluid so that the surface of the implant is covered completely (coated) by the body's own matter	No enzyme reactions: the implant is "camouflaged" against the host's immune system. No foreign body reactions

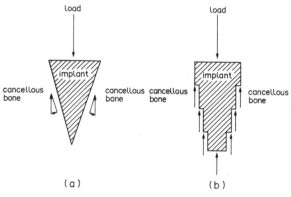

Figure 2
The influence of implant design (shape) on the interfacial stresses (small arrows): (a) large tangential component of force, resulting in shear movement and (b) mainly pressure along the surfaces of the steps, thus avoiding relative movements along these parts of the interface

The essential biomechanical requirement for fracture healing is to avoid any shear motion along the interface. In addition to fracture healing, however, it was found that, for maintaining the close bone contact with a bioinert prosthesis, it was essential to preserve this condition of motionlessness along the interface concerned for the entire lifetime of the prosthesis.

Basically, there are only two mechanical situations allowing for a motionless contact: the load-free situation and the situation where the interface is oriented perpendicular to the force acting along the interface. The first, trivial, situation is of no interest for load-bearing implants. Thus, all load-bearing implants must offer sufficiently large interfaces to the adjacent bony tissue so that they are transmitters of mostly pure pressure (forces perpendicular to the surface). Surfaces along which the forces are mainly transmitted parallel (resulting in shear along the interface) are separated by a soft tissue interlayer from the surrounding bond and therefore cannot contribute to osseo-integration and direct load transmission. Figure 2 summarizes these considerations. Clinical experience has shown that the requirement of perpendicularity of the forces meeting the interface allows for a deviation of up to 15°.

The validity of the concept of purely stress-and-strain-field controlled tissue reactions along the interface between bone and an implant of a bioinert material such as alumina ceramic was confirmed by the discovery of the "load-line shadow effect" (Heimke et al. 1982). The details of this effect are shown schematically in Fig. 3. Immediately after pressfitting the implant, the lacunae are filled with blood clots (Fig. 3a). If the condition of motionlessness is maintained during the healing-in period, the blood clot serves as a scaffold for the formation of new bone, filling the lacunae primarily homogeneously (Fig. 3b). After load application, the bony tissue reorganizes to adjust to the stresses acting, resulting in an area of reduced calcification where no

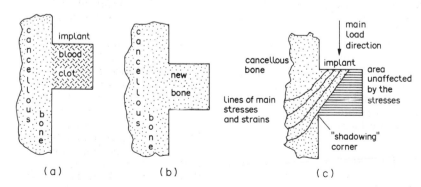

Figure 3
The load-line shadow effect: (a) immediately after pressfitting the implant; (b) the formation of new bone; and (c) the area shadowed from the load lines

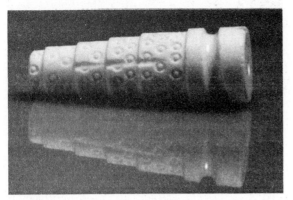

Figure 4
Alumina ceramic Frialit dental implant, Tübingen type: the implant body is cylindrical with decreasing diameters connected by steps; the surface of the cylinders carry lacunae for interlocking with the bony tissue; the groove in the coronal portion receives the gingiva

Figure 5
Frialit cylindrical sockets with asymmetric threads for biomechanically favorable load transmission, made of dense, high purity alumina ceramic

stresses are acting—the area shadowed from the load lines (Fig. 3c).

According to the previously described requirements for close bone contact, such contact could not be expected along mainly tangentially loaded interfaces such as the cylindrical surface of the Tübingen dental implant shown in Fig. 4. However, the evaluation of the tissue surrounding two implants, which became available for histology because of an accident suffered by a patient, revealed a close bone apposition not only along the pressure-transmitting interfaces (the surfaces of the steps connecting the cylinders of different diameters) but also along the cylindrical, mainly tangentially loaded areas. Due to this interlocking effect (Heimke et al. 1982), the relative movements created by all load changes resulting from the much higher stiffness of the implant compared with the surroundings (mostly cancellous bone) are realized by the deformation of bony structures some distance away from the surface of the implant rather than by shear movements along the interface. Thus, an interdependence exists between the shape of the implant and the reactions of the bony tissue surrounding it. To achieve a stable and reliable osseo-integration, the shape of the implant must be chosen so that the load pattern (the stress and strain field) created by its insertion allows for remodelling reactions resulting in a close bone contact along as many interfaces as possible.

4. Alumina Ceramic Implants in Orthopedic Surgery

The first implant design to follow the above conclusions was that of the cylindrical socket of Frialit total hip replacement (see Fig. 5). This is combined with a ball of the same ceramic fixed on a metal stem. Clinical trials started in September 1974 and the reports of many clinics, up to the late 1980s, show they have endured the test of clinical applications. The results of the first five years had been reported in detail and have led to some improvements which became possible because of further progress in the science of the alumina ceramic concerned. In the first generation the ceramic sockets and balls had been combined with femoral components designed for cement fixation. Results of further work led to the design of stems with specially shaped steps and lacunae made from a titanium alloy. This now completely cement-free implantable system stood the test of clinical applications throughout the late 1980s.

The other total hip replacement system using a bioinert alumina ceramic and based approximately on the biomechanical ideas described above was introduced into clinical trials by Mittelmaier in 1974 and this has also withstood widespread clinical tests.

Some years earlier Boutin had already started the clinical use of alumina ceramics in hip surgery. His design, however, closely followed the shape of plastic cups designed for cement implantation and did not take account of the aforementioned biomechanical stress and strain field considerations. Within the field of orthopedics, relatively early alumina ceramic implants have also been tested clinically in tumor surgery of the proximal femur and of the upper arm, but there have been no convincing results.

In knee surgery, early attempts to use alumina ceramic components have not led to large scale applications. Extended experimental work with ceramic-on-ceramic articulating total knee replacements has never entered the state of clinical trials

(Geduldig et al. 1976). It was not until the introduction of knee endoprostheses with anchoring portions, femoral condyles of alumina ceramic and an articulating tibial plateau of polyethylene, and of ceramic ankle joints with a similar combination, that ceramic components also found application within the lower extremities (Oonishi et al. 1983). However, the loosening rate of the tibial components of the knee prostheses appears to be too high.

In the late 1970s, considerable reduction in wear was reported when polyethylene sockets of hip prostheses were articulating against alumina ceramic balls rather than metal balls. After some exaggerated initial hopes, it is now well established that this combination results in a reduction of the polyethylene wear debris by a factor of approximately two.

5. Alumina Ceramic Dental Implants

The application of dense alumina ceramics for dental implants was suggested in 1964, but the first ceramic dental implant system fully accounting for the biomechanical rules discussed, the Tübingen dental implant (see Fig. 4), was not designed until 1975 and it was subsequently tested on a strictly experimental basis with a complete follow-up of each single implant for the next five years. Only after the results of this survey became available (Schulte 1984) were these implants made available to dental practitioners.

These implants can be used to replace teeth immediately or shortly after extraction (and after a reshaping of the alveoli to achieve the press-fit necessary for an undisturbed period of bone healing and osseo-integration). They are also used as so-called late implants in edentulous regions of the jaw. In 1983 a modification of the Tübingen implant was introduced after careful clinical tests (Nentwig 1985). Its thinner shape underneath the groove into which the gingiva is positioned allows for its implantation even in cases where the alveolar ridge is too small for the original Tübingen implants.

Scientifically, all the different studies (e.g., histology) on retrieved implants, the different types of mobility tests, and the detailed studies of the gingival attachment, either confirmed the initial design rules or, like the discovery of the load-line shadow effect, further contributed to a more detailed understanding of the load-pattern-controlled tissue reactions around bioinert implants. In addition, the observation of the full preservation of the shape of the alveolar ridge around such immediate implants after more than twelve years opens a new regime of dental care with the possibility of improving public health considerably.

A surprising discovery is also worth mentioning, amplifying the statement about dental care. From the definition of the term "bioinert" (see Table 1),

and considering the adsorption behavior with respect to macromolecules in tissue, one would have expected a tendency of increased plaque formation with such ceramic implants. However, experience has shown beyond any doubt that the opposite is true: these implants cause much less plaque formation than metal implants or natural teeth.

The follow-up of all implants inserted in the Tübingen clinic (now more than 1000) has yielded more than 1 000 000 sets of data which have been computerized and evaluated after five and ten years (d'Hoedt et al. 1987) from all relevant points of view. Preliminary results of the ten-year analysis are given in Table 3.

The comparison of the two different follow-up periods (ten years in the left-hand and three years in the right-hand column) is justified by the fact that implant losses of this system are confined nearly completely to the healing-in and early loading period; that is, to the first year. This had already been shown in the five-year statistical evaluation and is clearly confirmed by the ten-year statistics. From the more detailed evaluations of all the data, including those treatments which deviated from the standard, the following additional results are worth mentioning.

(a) The success rate does not depend on the location of the implant.

Table 3
Summary of ten years follow-up evaluation of all Frialit Tübingen implants inserted in the Tübingen Clinic by the standard treatment (including reimplantation)

	All standard treatments 1975–1985	Treatments commenced after reaching final routine, 1982–1985
Total number of treatments commenced	610	352
Disappeared in follow-up	18	5
Treatments remaining in follow-up	592 (100%)	347 (100%)
Failed treatments	92 (15.5%)	26 (7.5%)
Attempts of reimplantation contained in these failures	14	5
Successful treatments	500 (84.5%)	321 (92.5%)
Reimplantation contained in the successful treatments	28	10

(b) The success rate is reduced considerably for any deviation from the standard procedure, in particular if no initial press-fit can be achieved or if the implants are loaded within three months of the operation.

(c) The sulcus fluid flow rates are identical with those mentioned for teeth of the same patients, and they are constant over time as are the pocket depths.

(d) There are indications of a markedly reduced plaque adhesion to alumina ceramic implants as compared to metal implants or natural teeth.

(e) The observations indicate that the shape of the alveolar ridge is maintained around these implants as it is maintained around natural teeth.

Several other implant systems use either alumina ceramics or alumina single crystals (Kawahara et al. 1980). These do not, however, take account of the biomechanical requirements for the true osseo-integration of bioinert implants, but rather duplicate more or less closely metallic implant designs. Therefore, all of these implants are separated from the surrounding bony tissue by a soft tissue interlayer (Ehrl and Frenkel 1980, Kawahara 1983). This, of course, results in some mobility of the implant which, in turn, creates some mechanical irritation of the gingival attachment.

6. Alumina Ceramics in Ear, Nose and Throat

Extended animal experiments have shown that the middle ear mucosa proliferates normally on the surface of dense, pure alumina ceramic (Plester and Jahnke 1981). Total and partial oscicular replacements made of alumina ceramic (see Fig. 6) have, since the early 1980s, stood the test of clinical experience showing an improved success rate as compared to plastic parts. More recently, trachea-supporting rings have passed the clinical testing period and have been introduced into clinical application. Orbital support plates have also been tested clinically and have proved successful.

7. Alumina Keratoprostheses

Another completely soft tissue application of dense, pure alumina ceramics is the Frialit Kerato-prosthesis. It consists of a corundum single crystal as its optical part and an alumina ceramic holding ring. It has since the early 1980s stood the test of clinical application in the case of implantation without perforating the lid (Polack 1983). The attempts at using this implant in the "through the lid technique" have shown that in this application mechanical irritation prevents a close integration of the corneal tissue.

8. Outlook

The highly pure, dense alumina ceramic is the main bioinert material and has withstood long-term applications in hip surgery and dental implantology. For other joint replacements, the combination with polyethylene may enable certain improvements.

The interrelation between the implant shape, the stress and strain field created by the implant in the surrounding tissue, and the remodelling reactions has only recently been qualitatively understood. However, combining this knowledge with recent more detailed information on the basic process of bone remodelling can result in implant designs from which an improved osseo-integration can be expected. The success rates achieved and documented with the already well established systems must be regarded as the standard that any new system must surpass. Any judgement about the success of an intended improvement can only be given after at least five years of carefully documented clinical experience.

See also: Dental Implants

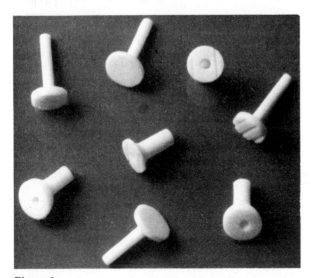

Figure 6
Frialit ossicular replacements, Tübingen type

Bibliography

Boutin P, Christel P, Dorlot J M, Meunier A, de Roquancourt A, Blanquaert D, Herman S, Sedel L, Witvoet J 1988 The use of dense alumina–alumina ceramic combination in total hip replacement. *J. Biomed. Mater. Res.* 22: 1203–32
d'Hoedt B, Heimke G, Schulte W 1987 Bioinert ceramics in dental implantology. In: Vincencini P (ed) 1987 *High Tech Ceramics*. Elsevier, Amsterdam, pp. 219–33

Ehrl P A, Frenkel G 1980 Experimental and clinical experiences with a blade vent-abutment of Al_2O_3-ceramic in the shortened dental row-situation of the mandible. In: Heimke G (ed.) 1980 *Dental Implants*. Hanser, Munich, pp. 63–7

Geduldig D, Lade R, Prüssner P, Willert H G, Zichner L, Dörre E 1976 Experimental investigations of dense alumina ceramic for hip and knee joint replacements. In: Schaldach M, Hohmann D (eds.) 1976 *Artificial Hip and Knee Joint Technology*. Springer, Berlin pp. 434–45

Griss P, Heimke G 1981a Five years' experience with ceramic-metal-composite hip endoprostheses, I. Clinical evaluation. *Arch. Orthop. Traumat. Surg.* 98: 157–63

Griss P, Heimke G 1981b Biocompatibility of high density alumina and its application in orthopedic surgery. In: Williams D F (ed.) 1981 *Biocompatibility of Clinical Implant Materials*. CRC Press, Boca Raton, FL, pp. 155–98

Griss P, Heimke G, von Adrian-Werburg H, Krempien B, Reipa S, Lauterbach H J, Wartung H J 1975 Morphological and biomechanical aspects of Al_2O_3 ceramic joint replacement. Experimental results and design considerations for human endoprostheses. *J. Biomed. Mater. Res. Symp.* 6: 177–88

Griss P, Heimke G, Krempien B, Jentschura G 1976 Ceramic hip joint replacement—Experimental results and early clinical experience. In: Schaldach M, Hohmann D (eds.) 1976 *Advances in Hip and Knee Joint Technology*, Engineering in Medicine, Vol. 2. Springer, Berlin, pp. 446–55

Heimke G 1984 Structural characteristics of metal and ceramics. In: Ducheyene P, Hastings G W (eds.) 1984 *Metal and Ceramic Biomaterials* Vol. 1. CRC Press, Boca Raton, FL, pp. 7–61

Heimke G, Griss P 1981 Five years' experience with ceramic-metal-composite hip endoprostheses, II. Mechanical evaluations and improvements. *Arch. Orthop. Traumat. Surg.* 98: 165–71

Heimke G, Griss P, Werner E, Jentschura G 1981 The effect of mechanical factors on biocompatibility test. *J. Biomed. Mater. Res.* 15: 209–13

Heimke G, Schulte W, d'Hoedt B, Griss P, Büsing C M, Stock D 1982 The influence of fine surface structures on the osseo-integration of implants. *J. Artif. Organs* 5: 207–12

Howie D W, Vernon-Roberts B 1988 Synovial macrophage response to aluminium oxide ceramic and cobalt–thromium alloy wear particles. *Biomaterials* 9: 442–8

Kawahara H 1983 Cellular responses to implant materials: biological, physical and chemical factors. *Int. Dent. J.* 33:(4) 350–75

Kawahara H, Hirabayashi M, Shikita T 1980 Single crystal alumina for dental implants and bone screws. *J. Biomed. Mater. Res.* 14: 597–605

Nentwig G H 1985 Late implantation with the Frialit implant. *Type Munich Quintessenz* 36: Report 6772

Oonishi H, Hamaguchi T, Okabe N, Nabeshima T, Hasegawa T, Kitamura Y 1983 Cementless alumina ceramic artificial ankle joint. In: Ducheyene P, van der Perre G, Aubert A E (eds.) 1983 *Biomaterials and Biomechanics*, Advances in Biomaterials, Vol. 5, Wiley, Chichester, UK pp. 85–90

Plester D, Jahnke K 1981 Ceramic implants in otologic surgery. *Am. J. Otology* 3: 104–8

Polack F M 1983 Clinical results with a ceramic kerato-prosthesis. *Cornea* 2: 185–96

Schulte W 1984 The intra-osseous Al_2O_3 (Frialit) Tübingen implant; Development status after eight years. *Quintessenz* 15: (I–III) Report 2267

Schulte W, Heimke G 1976 The Tübingen dental implant. *Quintessenz* 6: 17–23

Spiekermann H 1980 Clinical and animal experiences with endosseus implants. In: Heimke G (ed.) 1980 *Dental Implants*. Hanser, Munich, pp. 49–54

G. Heimke
[Clemson University, Clemson,
South Carolina, USA]

Arteries, Synthetic

Synthetic (substitute) arteries were not designed by engineers who used mathematics and logic to produce tubes that simulated the arteries they were intended to replace: the development was rather haphazard, with each step in the process (not necessarily forward) dependent on advances in medical science that were often not related to surgery of the vascular system. Reliable anesthesia, blood transfusion and antibiotics made vascular surgery possible; smoking and increased longevity made it necessary. Lessons were learned slowly; for example, since 1900 large arteries have been repaired successfully using sutures made of silk and when prostheses were first inserted in the 1950s, they were joined to arteries by silk sutures. Sutures made from natural fibers are slowly destroyed by the body, eventually allowing prostheses and artery to part, there being no permanent bond between them. In retrospect, this result seems predictable, but it took nearly ten years before natural fiber suturing of arterial prostheses was discontinued.

All manner of materials were tried as substitutes for arteries, including glass, silver and silk, but by the 1950s divergent lines of enquiry had narrowed to prostheses made from preserved human cadaver arteries and prostheses fashioned from synthetic fibers. Arteries taken from cadavers were preserved (the most successful method was freeze-drying) and stored in a "bank". Little, if anything, was known about antigenicity, tissue typing or rejection and not surprisingly the results were poor. Early occlusion was caused by fibrosis (the natural response of the body to foreign material is to surround it with fibrous tissue which forms a dense and constricting layer—a scar) or by dilatation of the tube as the wall was adsorbed with the formation of an aneurysm.

The biological inert and pliable prostheses made from the newly available synthetic fibers (nylon, vynon, Dacron and Teflon) worked well in certain circumstances, particularly when the prostheses were large-bore. Paradoxically, their success has had

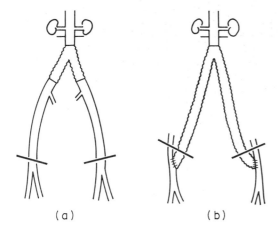

Figure 1
An aortic bifurcation graft may extend from the infrarenal aorta to (a) the common iliac arteries (aortoiliac graft) or (b) the common femoral arteries below the inguinal ligament (aortobifemoral graft)

an adverse effect on the development of small-bore prostheses. (While there is no strict definition as to what is large-bore and what is small-bore, any prostheses over 8 mm in diameter can be considered as large-bore and anything under 6 mm as small-bore.)

1. Criteria for Success

Success of prostheses may be claimed on the basis of either absence of complications that can be attributed to the prosthesis itself, or on the basis of long-term patency, that is, it continues to allow bloodflow (this does not necessarily mean good function). For example, an "aortofemoral" bypass (see Fig. 1) may

(a) remain patent and the patient remain symptom-free,

(b) remain patent but the symptoms persist,

(c) occlude in time, but without return of symptoms, or

(d) occlude with immediate return of symptoms.

Results (a) and (c) would both be classed as successes if symptoms were the only criteria considered, while (b) and (d) would be failures.

When a prosthesis is patent and symptoms persist, the prosthesis itself may be acting as a stenosis (a narrow section of artery) or, more commonly, there are further occlusions downstream which are the cause of a high impedance to flow. It is obvious from Fig. 1 that the function of the prosthesis will depend on an adequate (normal) inflow and outflow.

Long-term patency is calculated from life-table analysis, but the statistical method is not ideal (the end points are difficult to define). Regrettably, life tables are often miscalculated and even more often misinterpreted. Therefore, unless all the data are published, a check on an author's claims is impossible (Underwood et al. 1984).

2. Materials

Prostheses are made from two fundamentally different types of material: synthetic fibers (polymers of Dacron and Teflon) and tubes of animal origin which are modified (tanned) to make a durable tube.

2.1 Synthetic Fibers

Single fibers of a polymer (e.g., Dacron or Teflon) may be woven or knitted into a fabric and fashioned as a tube. It is important that in the manufacture, no residual products of the processes remain on the fibers and that the fibers are not damaged (Conway et al. 1979). It is also essential that the properties of the fibers are not altered by sterilization or by the passage of time. Rearrangement of the molecules within individual polymers may result in the formation of stable crystals or a cyclic oligomer.

The essential difference between woven and knitted prostheses is that woven ones are less porous. It is usual to use a velour knit to produce a fairly porous wall. In the late 1970s, both surgeons and manufacturers were persuaded that if a prosthesis were porous, it would act as a matrix through which small blood vessels and cells could pass. The advantage of this would be that a lining composed of living (and hence functioning) cells would form on the inside of a prosthesis and prolong its functional lifetime. There is, however, no evidence that this does occur and the adverse effect of a porous wall is that it leaks blood. It is normal practice to give an anticoagulant to patients when an operation is being carried out on iliac arteries (the practice is not universal and there is no evidence to suggest it is necessary) and in these circumstances the loss of blood can be enormous. Hence the prosthesis must be preclotted by immersion in the patient's blood before anticoagulants are given or by the incorporation of some substance within the wall of the prosthesis that is absorbed in time (e.g., albumen). Fashion and competition between manufacturers has more to do with these variations than surgical science.

It is essential that contact with body fluids and cyclical changes in pressure and mechanical trauma such as stitching or bending do not produce permanent damage to the wall. Prostheses made from polymers generally fulfill these requirements.

2.2 Natural Fibers

Several prostheses made from animal tissues have been tried, including:

(a) freeze-dried human cadaver arteries,

(b) chemically modified bovine carotid arteries,

(c) tanned human umbilical veins, and

(d) fibrous tubes formed round a mandrel.

It is reasonable to say that these have met with varying success. The walls, however, are "dead" and are rapidly replaced with scar tissue. The bypasses that remain patent often show localized dilatations (aneurysms) and some rupture.

3. Large-Bore Prostheses

Large-bore prostheses are used to replace the aorta and its immediate branches, the arteries arising from the aortic arch and the iliac arteries. The arteries that are most commonly replaced are the abdominal aorta and the iliacs. The most popular material for replacement is Dacron, woven or knitted into a fabric and fashioned into a tube. There is no evidence to suggest that woven tubes are superior to knitted ones or to support claims for the superiority of Dacron over Teflon.

Most large-bore prostheses are crimped in an attempt to resist kinking. The crimps effectively reduce the declared diameter by about 10% and secondary flow phenomena may produce areas of relative stasis within the crimps in which thromboses are initiated (Butler 1979). Since the wall of the prosthesis is stiff, there will be an impedance mismatch at both ends of an insert and, in consequence, a loss of pulsatile energy.

The difficulty is that it is not possible to control or to measure accurately all the variables that affect success. There is good evidence that an obstruction in the superficial femoral artery affects the success of an operation in which a prosthesis is inserted into the iliac artery. It affects success both in terms of function and in terms of patency. To compare one prosthesis with another it would be necessary to stratify the data to take into account the effect of the obstruction downstream, but in at least half the patients with such an obstruction the prosthesis would stay open and function well. When this problem is translocated further (e.g., down the leg for a femoropopliteal bypass), quantification of the variables becomes even more difficult.

Despite all problems, large-bore prostheses are very successful when used in appropriate circumstances. The reason for this success is probably that the velocity of flow is high and only a very small proportion of the blood flowing through the tube ever comes into contact with the wall. When run-off into the thigh is normal (i.e., the superficial and profunda branches of the femoral artery are patent), 95% of the prostheses can be expected to remain patent for at least five years (see Fig. 2). Extending the prosthesis by stretching alters neither its resistance to kinking *in vivo* nor the longitudinal impedance (see Fig. 3).

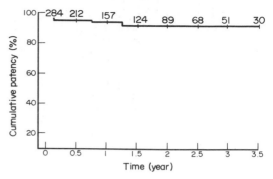

Figure 2
Cumulative patency curve obtained by life-table analysis of 284 aortofemoral bypasses followed for three and a half years. The number of prostheses available for analysis at a particular point in time are indicated on the curve

4. Small-Bore Prostheses

There is no generally accepted small-bore prosthesis that works well although there are several prototypes in the process of evaluation. The most promising materials are polyurethanes, from which small-bore elastic tubes (4 mm diameter) can be made. Small arteries are viscoelastic and because

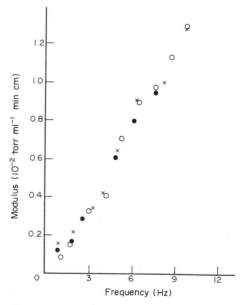

Figure 3
Graph of modulus of impedance against frequency for a single length of a knitted prosthesis (10 mm Knitted DeBakey Vasculour–D) undergoing stretching: (●) unstretched, (×) +50% stretch, and (○) +100% stretch

this property is expressed in real and imaginary terms it is difficult to quantify. An approximation to viscoelasticity can be made by measuring the speed of propagation of a pressure wave in arteries and veins and then calculating a propagation constant (Chapman and Charlesworth 1983a,b). Ideally, the wall of a small-bore prosthesis should be elastic with a wave speed of the order of 13 m s^{-1}. In addition, the wall must be microporous and resist kinking (e.g., across the knee joint a prosthesis may be bent at an angle of over 90°).

In the leg (below the inguinal ligament) the principal artery is the superficial femoral; branches of the profunda femoris supply the muscle of the thigh, while the superficial femoral conducts blood to the leg below the knee. The superficial femoral is approximately 4 mm in diameter, but the size varies according to the build and sex of the patient. It tapers slightly from its origin to the terminal portion of the popliteal artery where it divides into three branches. These branches (the tibial arteries) are about 1–2 mm in diameter. Their capacity to conduct flow and their combined input impedance are usually referred to as the "run-off". The quality of the run-off is usually estimated from arteriograms and it has so far not been possible to measure it accurately.

The input to the distal popliteal artery may vary either physiologically or pathologically. In the physiological case, an increase in flow to the muscles of the lower leg in exercise is brought about by reducing tone in circular muscles which control the diameter of the tibial arteries and their branches. This increases the pressure gradient between the heart and the muscles of the lower leg and so the flow increases. When one or all of the tibial arteries are affected by arteriosclerosis, the remaining vessels will be permanently diluted even at rest and no variation is possible. The run-off is then pathological, impedance is high and it is fixed.

It is known through long experience that a bypass from the common femoral to the popliteal artery will work well when the bypass is fashioned from the patient's own saphenous vein and the run-off is not compromised (see Fig. 4). It is also known that if a prosthesis is used for the bypass, the results will be less satisfactory. The expected patency is about 70% at three years for a bypass made from saphenous vein but only approximately 55% if the bypass is constructed using a synthetic prosthesis. This patency applies when run-off is good but is hopelessly worse when run-off is abnormal. To understand why this is, it is convenient to consider what are regarded as the essential properties of small-bore prostheses. These properties can be summarized as biocompatibility, a smooth low-friction nonthrombogenic flow surface, and dimensions and mechanical and elastic properties that approximate to those of the arteries that are being bypassed.

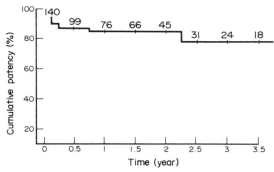

Figure 4
Cumulative patency curve obtained by life-table analysis of 140 femoropopliteal bypasses followed for three and a half years. The number of bypasses available for analysis at a particular point in time are indicated on the curve

4.1 Biocompatibility

The material from which a prosthesis is made must not provoke any adverse tissue response while *in vivo*; more specifically, it must be nondegradable, nontoxic and noncarcinogenic. It must also be fatigue resistant; that is, robust enough to maintain its function over an indefinite period of time. Fulfilling these requirements as well as having a nonthrombogenic internal surface is particularly demanding in practice. These requirements also apply, of course, to large-bore prostheses.

4.2 Nonthrombogenic Flow Surface

One prerequisite of a good internal surface is smoothness. If the surface is not smooth, frictional drag may raise local shear stresses to the point where perturbations in the flow pattern, and ultimately turbulence, can develop close to the wall. This in turn may lead to aggregation of platelets (small cells that initiate the clotting process) and thrombosis. This process, which once started is often self-propagating, is more of a problem in small-bore prostheses than in large-bore. There are two reasons for this: first, the layer of fluid closest to the wall (the boundary layer) is proportionately thicker in smaller tubes; and second, a lining of proteins and cells called the pseudointima forms on the wall, reducing the cross-sectional area of the lumen and, in some small-bore tubes, acting as a stenosis. This lining is usually about 1 mm thick and if this reduces the diameter of the lumen to below 4 mm then considerable pressure gradients will result because of the increased hydraulic resistance and the volume of flow may be reduced. High resistance in a bypass which reduces the velocity of flow predisposes the patient to thrombosis and is obviously detrimental.

There are several ways of trying to obtain a nonthrombogenic flow surface. One is to make the wall of the prosthesis porous. This, hopefully, will promote the formation of neointima, a new and wholly natural interface, the cells of which produce chemicals called prostaglandins which resist the formation of thromboses. In essence this is a good idea, but so far the formation of a complete neointima in prostheses implanted in a human has not been demonstrated. Another drawback is that once initiated, this natural process is difficult, if not impossible, to control, particularly around the anastomoses (junctions at which an artery is sutured to another blood vessel), and the lining can overgrow (intimal hyperplasia) and occlude the bypass.

An alternative approach is to present to the bloodstream a nonporous surface made from an inert synthetic material that will not stimulate any kind of reaction. This is, however, an extremely rigorous demand to make of a synthetic material. Attempts to alter the characteristics of such a surface (e.g., by bonding heparin to the wall) have met with little success.

Although the flow surface is difficult to simulate, several approaches have been tried or are still being developed, including the following methods.

(a) The production of an ultrasmooth surface, the critical surface tension of which is in the same order of magnitude as natural endothelium. Teflon is a fairly close approximation. The wall of a polyurethane tube may be lined with a different polymer with better surface properties, by copolymers or by binding the naturally occurring anticoagulant heparin to the surface.

(b) The use of a naturally occurring endothelium like that from the umbilical vein or from bovine arteries. These materials are dead when implanted and are soon replaced by fibrous tissue.

(c) The recreation of endothelium by "seeding" the surface of a prosthesis with living cells taken from the intended recipient of the graft. Endothelial cells can be harvested from veins (or mesothelial cells from omentum) and grown further on tissue culture. The cultured cells are then used to line small-bore tubes. There are the following considerable difficulties with this method.

  (i) It is not sure whether cultured endothelial cells will continue to behave biochemically as they do *in vivo*. (*In vivo* they produce prostacyclins which help to resist thrombosis.)

  (ii) Within hours of implantation, only a very small proportion (less than 20%) of the cells seeded on a prosthesis remain and within days the only endothelial cells that are adherent to the wall are probably derived from the patient's blood. The use of a chemical adjuvant to improve the adherence of

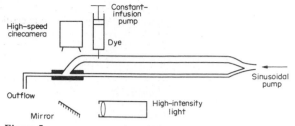

Figure 5
Diagram of a model used to investigate disturbances in flow that occur at the junction of a bypass to a small "artery." Dimensional analysis is used to simulate flows at comparable sites in patients

seeded cells to the wall of a prosthesis has had little success.

Despite careful tuning of porosity and cell-seeding techniques, there is no evidence that a confluent layer of functioning endothelial cells can be created within a prosthesis implanted in humans. It is possible to create a lattice from polyurethanes that has mechanical properties that closely mimic those of the saphenous vein, but copolymer bonding alters these properties and of the prototypes that have been implanted in humans, most have shown fairly rapid degradation with eventual dilatation and aneurysm formation.

4.3 Dimensions

First principles suggest that the perfect dimensions for a bypass graft are those of the host artery and scientific investigations confirm this. There is no advantage to be had from using a larger diameter graft—its lower hydraulic resistance will not increase the volume of flow. This is because, *in vivo*, the principal constraint on flow is the resistance of the arteries downstream from the graft (i.e., run-off) rather than the resistance of the bypass itself. Hence, for the same volume of flow in a graft of large diameter there will be a corresponding low-velocity flow. This is a disadvantage because high-velocity flow actively discourages thrombus formation. A further important consequence of not matching the size of the bypass to the recipient arteries is the creation of flow disturbances at the anastomoses. Figure 5 shows a model of an anastomosis constructed from a clear silicone rubber. The large tube represents the bypass, the smaller tube a tibial artery. Pulsatile flow is in the direction of the arrow. Dye injected upstream clearly demonstrates separation of flow during part of the cardiac cycle. The variations in shear stress at this point may cause damage to the wall. If the two tubes had been of the same diameter, there would still have been some degree of disturbance owing to angulation (the angle at which the prosthesis is joined to the artery), but

Table 1
Young's modulus and characteristic impedances for various arterial prostheses

Prosthesis	Young's modulus (dynamic) (10^5 N m^{-2})	Impedance (torr ml^{-1} min^{-1})
Human saphenous vein	11.39	4.0
Bovine artery	9.31	4.5
Umbilical vein	27.21	7.9
Woven Teflon	24.39	5.0
Velour Dacron	26.15	5.4
Knitted Dacron	51.77	8.0
Polytetrafluoroethylene	44.83	5.75

the effects would have been minimized (Klimach et al. 1984).

4.4 Mechanical and Elastic Properties

The importance of matching the mechanical and elastic properties is perhaps not so immediately obvious, but is nevertheless crucial for the proper function of a bypass. Arteries transmit pulsatile flow with considerable efficiency and minimal energy loss as, therefore, should a bypass graft. At an anastomosis the efficiency of energy transfer is governed by the impedance of the bypass (determined by the elastic properties of the wall and its dimensions) relative to that of the artery, the ideal situation being when they are equal. If they are not, then the anastomosis becomes a point of discontinuity and a site for reflection of the arterial pulse. As already shown, to minimize flow disturbances the dimensions of graft and artery should be the same, so it is clear that for the optimum transfer of pulsatile energy, their elastic properties should also be the same. A mismatched anastomosis is inefficient and *in vivo* inefficiency is magnified because, for every bypass, there are two anastomoses.

All available prostheses are incompliant and in no way simulate the mechanical behavior of human arteries or veins (see Table 1). However, the mechanical properties of arteries are difficult to measure, especially as the wall is viscoelastic and compliance is a crude measure of viscoelasticity. (The compliance is a measure of the distensibility of an artery and is given by $(\Delta D/DP) \times 100$ where D is the diameter and P is the pressure.) The mechanical behavior of a saphenous vein bypass has been characterized by measuring pulse-wave velocity and calculating a propagation constant (Chapman and Charlesworth 1983a,b). Assuming a vein to be 4 mm in diameter and the pulse pressure to be approximately 40 torr, the wave speed is 13.6 mm s^{-1} which corresponds to an expansion of the lumen of 28% at peak systole (maximum pressure). The pulse-wave speeds and the characteristic impedances of various prostheses have been measured and they differ widely from arteries and veins, as shown in Table 1.

The arterial tree matches the low output impedance of the heart to the high impedance of the peripheral circulation. This matching is difficult to duplicate when impedance is very high at the distal end of a bypass; that is, run-off is fixed and pathologically high. One way of achieving an impedance match at both the proximal and distal ends of a bypass would be to use a tapered bypass. Calculations suggest that there is no advantage to doing this other than when run-off is abnormal.

In summary, large-bore prostheses are very successful in spite of themselves. So far, the complex problem of producing a viscoelastic narrow tube that matches the impedance of the vascular tree and the properties of which are not altered after implantation remains unsolved.

See also: Biodegradation of Medical Polymers; Polyurethanes

Bibliography

Butler G R 1979 Blood flow in arteries with and without prosthetic inserts. Ph.D thesis, University of Manchester

Chapman B L W, Charlesworth D 1983a An *in vivo* method of measurement of the mechanical properties of vascular prostheses: The mechanical properties of saphenous vein bypass grafts. *Phys. Med. Biol.* 28: 1067–74

Chapman B L W, Charlesworth D 1983b The mechanical properties of glutaraldehyde stabilized human umbilical vein measured *in vivo*. *Br. J. Surg.* 70: 530–1

Conway P F, Simmens S C, Charlesworth D 1979 A study of arterial prostheses by scanning electron microscopy. *Br. J. Surg.* 66: 262–4

Klimach O, Chapman B L W, Underwood C J, Charlesworth D 1984 An investigation into how the geometry of an end-to-side arterial anastomosis affects its function. *Br. J. Surg.* 71: 43–5

Underwood C J, Faragher E B, Charlesworth D 1984 The uses and abuses of life-table methods in vascular surgery. *Br. J. Surg.* 71: 495–8

D. Charlesworth
[University Hospital of South Manchester, Manchester, UK]

B

Base-Metal Casting Alloys for Dental Use

Base-metal alloys find many applications in aspects of dentistry for which gold and precious-metal casting alloys have been used. Perhaps the first major intrusion of base-metal alloys into the previously nearly exclusive domain of dental golds was in the 1920s, for use in removable partial dentures (RPDs). RPDs consist of cast-metal frameworks with clasp arms for passive retention of the appliance by natural teeth. This removable framework supports replacement teeth, which are usually made of plastic. The cobalt–chromium alloys were and still are the primary replacement base metals for RPDs, but in the mid-1950s nickel–chromium alloys was also introduced.

Dental crowns (sometimes referred to as "caps" or "jackets") were fabricated from porcelain in the late 1800s. With the advent of gold-alloy casting techniques (1907), gold crowns began to be used, as did fixed partial dentures (FPDs, or bridges). An FPD consists of two or more crowns to which replacement teeth are connected to restore chewing function and maintain dental health.

In the mid-1960s, porcelain was fused to metal crowns and FPDs for aesthetic purposes. During the same period, nickel-base alloys with 12–20% chromium were introduced for cast crowns and FPDs, and by 1980 it was estimated that roughly 50% of such dental treatment used these alloys.

The 1970s saw the advent of severe political instability of some nations supplying cobalt and chromium. Concern over availability led to current investigations for replacement alloys, including the casting of titanium-base alloys for RPDs, crowns and FPDs. Titanium ore is available in western nations including the USA.

1. Cobalt–Chromium–Molybdenum Alloys

Cobalt–chromium–molybdenum alloys are used for many cast dental services (ADA/ANSI Specification MD-156-14) which require high strength, rigidity, hardness, corrosion resistance and biocompatibility. The most common application is for RPD frameworks. Dental technology for casting these alloys was developed in the late 1920s and early 1930s. This led to a rapid replacement of RPD gold alloys, as the new alloys were less expensive and RPDs fabricated from them were less massive.

Alloy compositions generally fall within the range of 28–30% chromium, 5–7% molybdenum, 0.5–1% silicon, 0.2–0.5% carbon and 0.1–1% manganese, and contain less than 1% nickel or iron, the balance being cobalt. Small additions of other elements have been employed, depending on the manufacturer. At least one alloy contains up to 15% nickel in place of cobalt in order to improve ductility.

As-cast microstructures of these alloys show precipitates of M_6C, M_7C_3 and $M_{23}C_6$ carbides in a cobalt-base matrix. High corrosion resistance is primarily due to the chromium. This forms a rapidly passivating surface layer, probably of chromium oxides, which stabilizes the alloy against further oxidation and corrosion. Some alloys of this type are highly biocompatible (ANSI/ASTM Specification F75-76) and are used as surgical implant materials. Molybdenum tends to compete with chromium and cobalt for the available carbon during solidification, thereby minimizing local depletion of chromium from the matrix. This renders the alloys more resistant to crevice and pitting corrosion, and enhances biocompatibility. Silicon and manganese act as deoxidizers, reduce the melting points and control the formation of oxides on melt surfaces, providing suitable emissivity for optically sensed temperature control during casting. For torch casting with a neutral to slightly reducing flame, these oxides break up, and a glassy surface exists just above the proper casting temperature. This provides a visual indication of proximity to the casting temperature. Carbon is the primary strengthening and hardening agent in the form of as-cast carbides and carbon retained in the cobalt-base solid-solution matrix.

As carbon content increases, precipitates change from a cored dendritic structure to a eutectic network. Carbides raise the hardness and improve wear resistance, to the extent that highly polished surfaces persist over many years of use. Plaque does not accumulate as easily on polished surfaces, hence devices stay cleaner, are easier to maintain, and therefore less likely to promote caries or other oral disease. Carbon also aids in casting by reducing melting temperatures.

Iron and nickel are generally not deliberate additions, but rather accompany the other ingredients. As cobalt becomes less available and/or more expensive, iron might be substituted in limited quantities in the future.

The mechanical properties of these alloys should meet the minimum requirements as set forth in ADA/ANSI Specification MD-156-14: 0.1% offset yield strength of 500 MPa, modulus of elasticity 172 000 MPa, and elongation 1.5%. The density of these alloys is 8400–8800 kg m^{-3}.

In addition to the primary use for RPDs, a variety of partial denture attachments are fabricated from these alloys. Bite raisers and serrated inserts for plastic teeth are also produced. Owing to the combination of outstanding biocompatibility and mechanical properties, some of the alloys have been used for making dental implants of various types (Cranin 1970). One of the earliest and most successful implants is the subperiosteal implant; others include endosseous endodontic pins to "lengthen" the root of a tooth, and endosseous blade implants. Blade implants have been used as single-tooth replacements and as abutments for fixed partial dentures. Controversy surrounds the use of all dental implants; the blade implant (a development of the 1960s) has been the most controversial (see *Dental Implants*).

Crown and bridge alloys that employ cobalt and chromium as major elemental constituents are also appearing on the market.

2. Nickel–Chromium Alloys

Nickel–chromium alloys are being used extensively in dentistry as alternatives to cobalt–chromium, gold and precious-metal alloys. Applications are primarily for porcelain-veneered crowns and FPDs, but also for unveneered crowns and bridgework (C and B), as well as for RPDs; "C and B" and "FPD" refer to the same type of appliances.

The mechanical-property requirements for RPDs are as given in ADA/ANSI Specification MD-156-14. Alloys used for FPDs must conform to guidelines of the Acceptance Program of the American Dental Association. The mechanical properties and compositions of the alloys used for crowns and FPDs vary considerably, as shown in Tables 1 and 2. The high rigidity and strength compared with gold alloys allow for some reduction in the size of a prosthetic device, thereby permitting fabrication and insertion where space would not otherwise be considered adequate for larger, more bulky appliances.

The wide variation in properties and composition makes generalizations on performance of the alloys difficult. For example, bonding of porcelain to metal depends on the oxides developed at high temperatures, and this depends very strongly on alloy chemistry; the same is true of melting and casting behavior and mechanical properties. Microstructures vary accordingly, leading to marked differences

in corrosion behavior and perhaps also to variations in toxic and/or allergic responses.

Microstructures range from that of an almost single-phase alloy, with isolated regions of precipitate, to distinctly multiphase alloys. Some contain eutectic-like structures.

Differences in laboratory corrosion resistance of one multiphase beryllium alloy and a corresponding multiphase beryllium-free alloy have been demonstrated. A beryllium-rich phase corroded preferentially at a lower corrosion potential in artificial saliva (von Weber and Fraker 1980). The beryllium-containing alloy was of a type that comprises a significant fraction of those available, and was of a relatively low chromium content (~12–13%).

The corrosion behavior of high-chromium-content alloys cannot be presupposed as unequivocally superior. An example of the strong dependence of corrosion resistance on specific alloy composition is illustrated by the fact that one alloy with approximately 12% chromium, but modified with gallium, has been shown to have superior laboratory corrosion resistance (Mayer 1977).

Nickel–chromium alloys evolved for C and B applications during the late 1960s and early 1970s. One crown and FPD alloy for porcelain application had already been on the market by then, as well as one of a similar composition for RPDs. Consequently, alloys of like composition were developed, as gold-alloy substitutes were earnestly sought in the 1970s.

2.1 Effect of Elemental Alloying Additions

Chromium is the primary addition for corrosion resistance in these alloys. Beryllium is added as a solid-solution strengthener, and acts to render the alloys self-fluxing at porcelain-veneering temperatures. Alloys with beryllium generally produce acceptably thin, transparent oxide boundaries between porcelain and alloy, an important aesthetic consideration for anterior restorations. Beryllium also acts in chemical bonding of porcelain to metal (see *Porcelain–Metal Bonding in Dentistry*). A highly important effect of beryllium addition is the lowering of the melting temperature and improved castability. Aluminum, also, promotes an acceptable oxide layer, aids in bonding of porcelain, improves corrosion resistance and can produce significant strengthening by precipitation of $AlNi_3$. Silicon lowers the melting temperature, can affect the

Table 1
Mechanical properties of nickel–chromium C and B alloys

	Yield strength at $E = 0.002$ (MPa)	Elongation (%)	Vickers hardness	E (MPa)	Density (kg m^{-3})
As cast	296–717	Up to 18	167–309	158 000–212 000	~8700
After fired porcelain veneer	310–572	2.3–28.3			

Table 2
Chemical composition range (wt%) for nickel–chromium dental alloys

	Ni	Cr	Mo	Fe	Si	Al	Nb	Mn	Be	Other
With beryllium	71–78	12–16	3.0–5.5	0–1.5	0–3.0	0–2.2		trace	0.4–1.9	Ti, Co, Zr, Cu
Without beryllium	58–78	13–23	2–10	0–8	0–4.3	0–6	0–3	0–5		Ta, Sn, B, La, Cu, Ti, Zr, Co, W, Ga

chemical bond of porcelain (improving or degrading, depending on the alloy) and, like manganese, acts as a deoxidizing agent.

Molybdenum and niobium are added primarily to inhibit pitting or crevice corrosion, and these elements also help control the thermal expansion. Iron affects expansion and helps chemical bonding with porcelain; gallium greatly improves corrosion resistance, lowers the melting temperature, assists in chemical bonding and tends to stabilize alloys against multiphase development; both gallium oxide and the stabilization appear to improve corrosion resistance. Boron improves chemical bonding and strengthens alloys.

3. Ferritic and Austenitic Steels

Although stainless-steel alloys are used for dentistry, they are primarily in the wrought form for partial-denture clasps and orthodontic wires and bands. Cast stainless steels are used in the eastern European countries, but they have not found wide acceptance elsewhere.

One stainless-steel casting alloy was introduced in the USA during the early 1970s for C and B for porcelain veneering or RPDs. This alloy had a weight-percent composition of 27% chromium, 8% cobalt, 5% nickel, 2.25% molybdenum, 0.1% carbon (max.), 0.1% manganese (max.), 1.0% silicon (max.) and a balance of iron. In the as-cast state it consisted of nearly 50% austenite in a slightly higher percentage ferrite base. Castings had to be quenched immediately into cold flowing water to prevent embrittlement due to the formation of σ phase. This presented some difficulty with dental castings, as the variable thickness of dental castings led to a poorly controlled cooling rate in the hot dental molds; embrittlement could occur at some sections. Although embrittlement might be eliminated by heat treatment, this constituted additional costly steps in processing. These factors, coupled with additional steps for porcelain veneering, led to low levels of acceptance by dental laboratories. The as-cast properties of the alloy are: yield strength, 536 MPa; tensile strength, 653 MPa; elongation 6.9%; Brinell hardness, 255–269; and density, 7700 kg m^{-3}.

Economic and aesthetic considerations have apparently acted to inhibit development and use of stainless-steel castings in the west, as opposed to in eastern Europe. Furthermore, in the USA during the 1920s, cobalt–chromium alloys were introduced for use in place of gold alloys for RPDs. The high strength, corrosion resistance and structural-stability requirements for RPDs were well met by these alloys, whereas cast stainless steel of that time and up to the 1950s had insufficient as-cast strength (Schoefer 1980) and required precisely controlled heat treatments (beyond the scope of the dental-laboratory technician of that time) for corrosion resistance and strengthening. Welding or soldering would also require post-heat treatments. Sensitization or embrittlement of castings could also occur during cooling.

Reports on the use of stainless-steel alloys for dentistry in eastern Europe continue to raise the question of suitability of stainless alloys. Tissue inflammation and ulceration are reported for stainless crowns; allergic responses, corrosion and poor-quality castings have also been reported. Unreproducible microstructures and properties have been noted for some castings fabricated using oxygen–acetylene melting. Stainless-steel alloys in eastern Europe are used without porcelain veneers, hence lower strength alloys for C and B applications would seem acceptable. However, the problems expected and apparently confirmed by experience in eastern Europe, coupled with the needs for porcelain veneering and/or high strength in western countries, appear to have inhibited development of stainless steels in the west.

Despite the apparent inadequacy of alloys of the 1930s and 1940s, and those currently in use in eastern Europe, new stainless steels have been developed for industrial castings that may have the potential to overcome those problems (Schoefer 1980). Additional alloy development could conceivably produce satisfactory iron-base casting alloys that would meet the needs of high-quality dental care.

4. Titanium-Base Alloys

Titanium possesses various properties, such as good strength, high ductility, excellent corrosion resistance and biocompatibility, which make it a promising candidate for use in dentistry (Table 3). Titanium alloys have never been used for clinical

Table 3
Comparison of the properties of pure metals used in dentistry

Property	Nickel	Titanium	Gold
Density ($kg\,m^{-3}$)	8900	4500	19 300
Tensile strength (MPa)	317	235	124
Yield strength (MPa)	59	138	12
Elongation (%)	30	54	30
Elastic modulus (MPa)	193 000	117 000	76 000
Thermal expansion at room temperature ($10^{-6}\,°C^{-1}$)	13.3	11.9	14.2
Corrosion resistance (body fluids)	poor	excellent	excellent
Biocompatibility (soft tissues)	moderate	excellent	excellent
Cost ($US\$\,cm^{-3}$)	0.082	0.084	400.00

dental castings, however, mainly because they become contaminated when conventional casting methods are used and this has detrimental effects on the mechanical behavior, surface quality and dimensional accuracy. Contamination is reduced in the lower-melting-point (1430 °C) binary alloy containing 13% copper. In 1978, this alloy was cast into the first titanium-base dental castings using conventional techniques. The castings had a satisfactory appearance and good dimensional accuracy, but they were embrittled by contamination during casting. However, castings made in an electric-arc furnace under argon meet the requirements in ADA/ANSI Specification MD-156-14 for partial-denture alloys, as shown in Table 4.

The microstructure of the Ti–13% Cu alloy consists of finely decomposed prior grains of β-titanium. Assuming a normal eutectoid decomposition, this structure would be expected to consist of α-titanium plus the compound Ti_2Cu, but metastable martensitic structures can be expected in such alloys. The Ti–13% Cu alloy has not met with commercial clinical use. Titanium dental castings are currently made from commercially pure (CP) titanium using a partial-vacuum–pressure-assisted method or centrifugal casting in an inert atmosphere. Electric-arc melting is employed in both cases. Certain commercial phosphate-bonded investment mold materials and some improved mold materials based on zirconia and/or rare-earth oxide coatings have produced

high quality titanium castings with improved dimensional accuracy. Although surface contamination does not appear to be a problem with these materials, further characterization of the surface microstructures is needed.

The laboratory-tested corrosion resistance of Ti–13% Cu–4.5% Ni (and other titanium alloys) is superior to that of a cast cobalt–chromium alloy. Minimal soft-tissue reactions have been reported. Bonding of porcelain to titanium has been reported without discoloration and there are various methods for brazing or soldering titanium alloys.

It appears that a major obstacle to the use of relatively inexpensive titanium-based alloys in dentistry is a lack of information about effective economical casting methods. Further progress will depend to a great extent on the adequacy of developmental research in this field.

Bibliography

Baran G R 1983 Oxidation kinetics of some Ni–Cr alloys. *J. Dent. Res.* 62 (1): 51–5

Cranin A N (ed.) 1970 *Oral Implantology*. Thomas, Springfield, IL

Hruska A R 1987 Intraoral welding of pure titanium. *Quintessence Int.* 18: 683–8

Lang B R, Morris H F, Razzoog U (eds.) 1985 *International Workshop: Biocompatibility, Toxicity and Hypersensitivity to Alloy Systems used in Dentistry*. Michigan Press, Michigan, MI

Mayer J-M 1977 Corrosion resistance of nickel–chromium dental casting alloys. *Corros. Sci.* 17: 971–82

Moser J B, Lin J H C, Taira M, Greener 1985 Development of Dental Pd–Ti Alloys. *Dent. Mater.* 1: 37–40

Schoefer E A 1980 Properties of cast stainless steels. *Metals Handbook*, 9th edn., Vol. 3. American Society for Metals, Metals Park, Ohio, pp. 104–24

Taggart W H 1907 A new and accurate method of making gold inlays. *Dent. Cosmos* 49: 1117

Togaya T, Suzuki M, Tsutsumi S, Ida K 1983 An application of pure titanium to the metal porcelain system. *Dent. Mater.* 2: 210–19

Valega T M (ed.) 1977 *Alternatives to Gold Alloys in Dentistry*, HEW Publication (NIH) 77–1227. US

Table 4
As-cast mechanical properties of Ti–13% Cu alloy

Property	Ti–13% Cu alloy	ADA/ANSI Specification MD-156-14
Tensile strength (MPa)	786	>618
Yield strength (MPa)	607	>500
Elongation (%)	3.6	>1.5
Hardness (Rockwell 30 N)	50	>50

Department of Health, Education and Welfare, Washington, DC, pp. 68–79, 81–93, 224–33

von Weber H, Fraker A C 1980 Anodisches Polarisationverhalten von Ungeglühten und Mehrfach Geglühten Ni–Cr Legierungen. *Dtsch. Zahnaerztl. Z.* 35: 942–6

Weber H, Geis-Gerstorfer 1987 *In vitro* corrosion behavior of four Ni–Cr dental alloys in lactic acid and sodium chloride solutions. *Dent. Mater.* 3: 195–289

J. A. Tesk and R. M. Waterstrat
[National Institute of Standards and Technology, Gaithersburg, Maryland, USA]

Beam Ion Implantations

The materials processing technique of beam ion implantation has its origins in nuclear physics studies and in the early nuclear industry, where ion production and magnetic mass spectrometry were techniques for isotopes separation. Ion implantation by use of directed ion beams is by far the most widely practiced form of high-voltage ion implantation. As a method for modifying the near-surface properties of materials, the technology finds wide use in the solid-state electronics industry. The technique is used for the introduction of dopants in carefully controlled quantities into elements of integrated circuits. The success of this enterprise has fostered interest in the use of the high-energy process for improvement of a variety of surface properties in a variety of materials. It was recognized early that a number of features commended the technique for possible use in biomaterials.

The most important single feature of ion implantation is that the high energy of the process means that significant quantities of almost any atomic species can be injected into the surface of almost any target regardless of chemical equilibrium propensities. This feature affords the possibility, in principle, of fabricating thermodynamically metastable surface products with enhanced properties—optical, electrochemical, corrosive or conductive, for example, on polymeric, ceramic, semiconducting or metallic substrates. These capabilities seem particularly attractive when the traditional relationship of bioengineering to the materials industry is considered. Biomaterials comprises such a small market that, in the past, materials for surgically implantable devices have been chosen largely from readily available multipurpose polymers, ceramics and metal alloys that have been developed for wide commercial use. Medical grades of materials have differed from commercial ones in impurity control, quality, certification, and so on, but not in basic properties. The biocompatibility of the material, together with the best optimization of properties for the given purpose consistent with biocompatibility, has provided the criteria for selection. Against this background, a surface modification technique with the broad capability and flexibility of ion implantation may offer many new choices for optimization of materials in the eclectic field of bioengineering.

Another factor of importance in biomaterials, especially in comparison with surface coating techniques, is the integrity of the ion-implanted layer. For ion implantation, the doping constituent is introduced one atom at a time into a substantial range of the target substrate, so that intermixing with the substrate is at the atomic level, at least initially. There is a continuous gradient of concentration between the treated layer and the rest of the substrate. Abrupt interfaces which can lead to cracking, delamination and spalling can be avoided.

Broad features of ion implantation seem compatible with biomaterials needs. Products are processed in small batches in high vacuum. Operations are carried out in a clean environment. Since the ion-treated layer is very thin, it is expected that ion implantation will be the last step in the manufacturing process for many products. This observation suggests that ion implantation might often serve as an add-on step for remedy of a problem or improvement of a product in situations where most of the manufacturing technology for the product is already set. Moreover, the integrity of the treated layer, together with the generally excellent protocols already provided by the ion implantation industry, means that a strong *a priori* argument that ion implantation at least produces no deleterious consequences can often be made. The ability to make such an argument should help expedite regulatory acceptance. Products could also be designed from the outset with the idea of using one or more ion implantation steps in the manufacture.

Ion implantation for wear inhibition of alloy components of artificial hips, knees and other orthopedic prosthetic components has reached commercial maturity. In 1989 approximately 100 000 such components, of the orthopedic Ti–6 wt% Al–4 wt% V alloy, were implanted with nitrogen ions in the USA and in the UK. In addition to the striking improvement in wear performance, the add-on character of the process, the excellent adhesion, the cosmetics of the product and the modest cost addition were factors that contributed to rapid acceptance.

From the materials science standpoint, the orthopedics application is simple and direct, and by no means exploits all of the capability ion implantation has to offer. Ion implantation of biomaterials should, therefore, be regarded as a young field. Due to the large number of possibilities and variations that the technique affords, however, each new application may require considerable research and development. Therefore, rapid advancement is not necessarily assured.

1. Science, Technology and Practice

The term beam ion implantation naturally implies that the ions impinge on the product essentially from one direction, although beam divergence may be as great as ±20° for uniform treatment of larger products or of an array of products. Variations in the angle of incidence will also be related to the shapes of products. The definition also includes the concepts that the beam is preformed, preanalyzed for atomic (usually isotopic) constituency and preaccelerated relative to the target. The ions are in free flight when they impinge on the target. Hence the product is not a field terminal in the process and therefore arcing, with possible attendant surface damage, will not occur. The product can be, and usually is, essentially at ground potential, except for small biases introduced by the need for on-line ion dosimetry. Current practice is to use equipment that provides accelerating voltages of up to 200 kV. Thus, large doses of singly charged ions are economically available at energies of up to 200 keV. By use of doubly charged ions, energies of up to 400 keV are available at some penalty in dose or economics.

Ionization of the particle serves two purposes: it allows acceleration of the atom to high velocity so that penetration will occur and it allows magnetic mass analysis. Ionization plays little or no role, however, in the resulting property changes of the target. For most target materials, excluding perhaps the best of polymeric insulators, all effects in terms of materials property changes produced by ion implantation are the same as would be produced by impingement of neutral atoms of the same atomic species and energy. Charging of targets during processing is not a problem for insulators of electrical conductivity no lower than that of Al_2O_3, for example. The fact that the beam is ionized affords the possibility of accurate dosimetry by measurement of the ion-beam current.

1.1 Physics

When an ion of about 100 keV in energy enters a solid, it loses energy by interaction with electrons of the target atoms and by elastic collisions with nuclei of the target atoms. The electronic losses excite electrons, but again charge separation is usually not a factor in the outcome of the process. The ion makes several nuclear collisions before coming to rest and in many of these the ion is scattered through large angles. The trajectory of the particle then consists, essentially, of reasonably straight path segments where electronic losses dominate and of occasional major changes in direction where a collision occurs. The parameters of these major events vary statistically from one collision to the next. Ions sometimes receive lateral deflections, others are backscattered and so on. As a result, a population of ions incident at a given energy comes to rest over a somewhat Gaussian distribution on depth. Target atoms, ejected from the solid matrix at points of major collisions, go on to eject others in similar collisions. This entire slowing-down process is called a collision cascade.

Figure 1 illustrates the cascade concept. Some 1000 target atoms are typically displaced in these collisions for one incident ion. The volume of material affected is approximately 100 000 atomic volumes. These large effects in the atomic structure, for the introduction of one foreign atom, are a unique characteristic of high-energy ion implantation as a materials processing technique. These processes have traditionally been termed "damage" because the lattice perfection of crystalline solids is disturbed. However, as far as practical properties are concerned, the effects of damage are as often beneficial as they are deleterious. The damage process often produces favorable results such as homogenization, refinement of microstructures, amorphization and improved adhesion of ion-treated layers to the substrate. Ultimately most of the vacancies and interstitials produced by the damage process recombine with each other to produce considerable healing.

For process design, accurate predictions of numbers of displaced atoms vs depth, numbers of sputtered atoms, distribution of implanted ions in depth and other results can often conveniently be made by computer calculation. Figure 2 compares an experimental determination of concentration vs depth with the concentration profile as calculated from the well-known TRIM code (Zeigler et al.

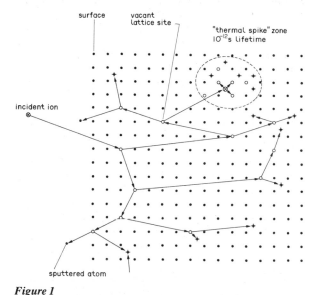

Figure 1

Collision cascade due to an ion incident on a crystalline solid. The array of black dots represents the atoms of the target material

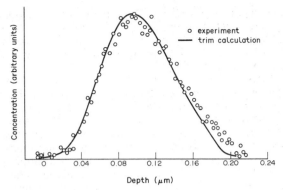

Figure 2
Concentration vs depth of ion-implanted 280 keV
zirconium in Ti–6 wt% Al–4 wt% V alloy as
determined by experiment and TRIM calculation.
The purpose of the implant treatment was to produce
a wear marker (Williams and Buchanan 1985)

1984). The case is for a small dose of zirconium ions
incident at 280 keV onto a Ti–6 wt% Al–4 wt% V
alloy. Experimental data were obtained by use of
Rutherford backscattering. Results are typical for
ion implantation at a single energy. A difference
between calculation and experiment is detected at
the larger depths, but it cannot be stated whether
the experiment or calculation is the more accurate in
this case. This result also illustrates the main disadvantage of ion implantation; that is, the rather
shallow layer of material affected. Such calculated
results are invaluable in helping to predict the outcomes for ion treatments, but nevertheless must not
be accepted uncritically. By varying the energies for
one or more implant treatments, the form of the
profile can be somewhat tailored.

Sputter ejection of atoms from the surface of the
target is part of the collision cascade (see Fig. 1).
The sputtering phenomenon, including angular
dependence of sputtering and selective sputtering of
substrate atoms, has a large influence on process
design for high-dose implantations, such as are often
used for prevention of wear or corrosion. One
important point is that the maximum concentration
of implanted dopant that can be introduced by direct
ion implantation is about $1/Y$, where Y is the sputtering yield; that is, the number of atoms ejected
from the surface per incident ion, usually having a
value in the range 0.3–10.0. This effect is the most
important limitation on the amount of dopant that
can be introduced by direct ion implantation.

Partly due to the sputtering limitation, there is
interest in making surface products by "coprocess-
ing" techniques. Such techniques employ two or
more matter sources, at least one of which is a high-
energy ion beam. The other(s) may be low-energy
sources such as evaporation sources. By virtue of

damage mechanisms, partly termed "ion mixing,"
adhesion and homogeneity of the ion-treated layer
are improved, in comparison with deposition techniques that make use of only low energies. It is
generally possible to add more matter by codeposition than by direct ion implantation and some
additional flexibility in choice of chemical makeup
may also be available. Ion mixing is thought to be
greatly assisted by the "thermal spike" phenomenon
(see Fig. 1), in which collision processes near end
of range start to simulate heating in character.

Aside from property changes produced by the
chemical effects of the ion-implanted dopant, most
other property changes result from the nuclear, or
collisional, losses rather than the electronic losses.
The total effect of ion implantation, then, consists of
effects related to the chemistry produced by the ion-
implanted dopant, plus the effects produced by the
large energy deposition. The two can act in synergistic ways.

Although the energy losses are termed nuclear by
the practitioner of ion implantation, atomic nuclei
are not penetrated in these elastic collisions at the
energies commonly used in the process (up to a few
hundred keV). Thus there are no nuclear reactions
and freedom from biological hazard due to radiation
can be absolutely ensured, both during processing
and for the final product.

1.2 Equipment, Practice and Quality

The equipment and practice of ion implantation are
logically discussed in relation to quality of result.
The possibility of excellent protocols is considered
to be an inherent advantage of ion implantation, but
quality does not automatically ensue in a process
without due attention.

Figure 3 is a representation of a beam-ion
implanter of the prevailing geometry and features.
Implanters with a variety of advantages and disadvantages exist, but this type of layout, having
emerged as the industry standard for use in doping
of semiconductors, is the most widely used. Such an
implanter can be regarded as having only two main
parts: the accelerator, which produces, accelerates
and manipulates the beam, and the endstation,
where the beam is directed to the product. In Fig. 3
the accelerator includes all components starting near
the lower left with the ion source and proceeding
around to the endstation. In this part, the ions are
generated, the beam is preaccelerated in the
extraction/preacceleration stage and is analyzed to
the isotopic mass level by use of the analyzing
magnet. The beam content up to the analyzing
magnet depends on the recipe of the feedstock,
which must include atomic constituents of the
desired type. Substantial beams of most chemical
elements can be derived. The chosen isotope is then
postaccelerated to the desired ultimate energy in the

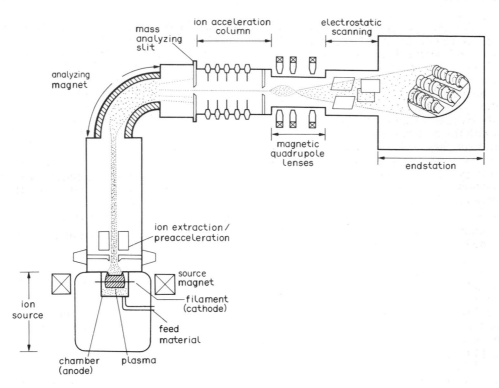

Figure 3
Ion implanter of the common "postacceleration" type. The target in the endstation is intended to represent an array of femoral components of artificial knee joints for implantation (courtesy of Spire Corporation)

acceleration column. Finally, the beam manipulation features (lenses and scanning) allow transport of the beam to the target and scanning of the beam over sizable areas in the endstation.

For large-scale production, as currently applies for solid-state electronics, the endstations are highly particularized to the product. For the foreseeable future, most biomaterials will be processed in large multipurpose vacuum chambers into which custom fixturing suitable for the given product has been fitted.

(a) Beam quality. This is not expected to be a serious issue for most biomedical applications. Sufficient beam purity is usually assured by the magnetic analysis. Nevertheless, undesired elements will often be in the beam up to the magnet due to the complexity of the recipe in the plasma and, possibly, due to previous use of the ion source. Questions of what happens in the event of a magnet current excursion or of whether there are undesired multiply charged ions at the same analysis position (magnet current) might need to be addressed for some cases.

(b) Ion dosimetry. This is usually achieved by measurement of the ion current, either directly from the product itself or by use of Faraday cups which intercept part of the beam during part of the cycle.

Uniformity of dose resulting from manipulation of beam and product is an issue. In addition to magnitude of dose delivered to a flat surface, another aspect of dosimetry is uniformity of treatment of curved surfaces in the directed beam. Manipulation of the product may be required to achieve uniformity. Needed accuracy of dosimetry depends on the product. Doping levels ranging from the parts per million level to 50 at.% are routinely achieved. For hips and knees of the titanium alloys a minimum of about 20 at.% of nitrogen in the implanted layer is needed. Fortunately moderate overdosing in regions of overlap (during manipulation) has no deleterious consequences in this case. Precise dosimetry is much easier for flat products; nevertheless, designs for uniform treatment of such products as ball bearings have been produced.

(c) Beam heat. The energy of the process eventually appears as by-product heat deposited directly in the target. One of the frequently claimed advantages of ion implantation is that it is a low-temperature process. The claim is true, in that high temperatures are not needed to achieve penetration. However, to maintain low temperatures, by-product heat must be managed. This is usually achievable and ion implantation treatments can be performed at a range of temperatures from cryogenic to quite high values.

(*d*) *Particle contamination*. For processing of silicon wafers in the semiconductor industry, contamination with particles (of about 1 μm in size) is a very important issue. These particles come from several sources such as microchipping of the brittle wafers during automatic manipulation or breaking of an entire wafer. Other particle sources may include mechanical devices, vacuum valves and the like associated with the chamber. Particle contamination is a problem for these silicon wafers only because demands are so stringent—the devices are so small that one such particle can ruin an entire chip. This type of contamination is not likely to be a problem for most biomedical products.

(*e*) *Chemical contamination*. Contamination with unwanted atoms can result from several sources. Generally the beam will impinge on chamber parts, apertures, product fixturing devices, Faraday cups and so on other than the product. Atoms sputtered from these components can contaminate the product. Atomic constituency of these accessory components and the possibility that they themselves also have undesirable ion-implanted contaminants retained from previous use should be examined. The sputtering process described previously normally helps prevent the accumulation of unwanted atoms deposited adventitiously at low energies. Nevertheless, a few such atoms might also be ion mixed into, as well as sputtered off, the treated surface. Appropriate remedies, depending on product and process, will generally be manageable.

Surfaces can become contaminated with carbon by a process involving beam-assisted decomposition of carbonaceous molecules in the residual vacuum, near the surface of the product. Residual gas content in high-vacuum systems typically consists of mainly water molecules, CO and CO_2, and hydrocarbons, some of which may originate in the pumping system. Carbonaceous molecules can decompose on the surface in the beam. Despite continuous sputter cleaning, for targets containing constituents that have a strong chemical affinity for carbon (e.g., titanium), carbon is retained on the surface and is ion mixed into the surface to form an alloy. The process is not necessarily deleterious in all cases; in fact, the effect has been design enhanced to produce favorable wear properties for some engineering materials. Discoloration, apparently due to this carburization, has been a quality factor for implantation of orthopedic appliances at the commercial scale. The simplest remedy is to operate the endstation with sufficiently good vacuum practice.

1.3 Materials Science

The materials science of ion implantation, in contrast to the particle physics, is not highly developed insofar as the ability to predict the detailed outcome of an ion implantation treatment is concerned. In recent years most basic research has concentrated on the materials science rather than the physics of ion implantation. A large body of literature regarding ion implantation treatments in all classes of condensed matter exists. This information provides important guidance in the use of ion implantation for materials surface engineering. Detailed product design and anticipation of side effects is not yet generally possible, however. Instead considerable testing is often required to bring products to full maturity.

(*a*) *Amorphization*. Ion bombardment amorphizes certain materials whose chemical makeup lends a susceptibility toward amorphization. The following chemical and other factors favor amorphization from an empirical point of view:

(i) chemical bonding (in order)—covalent, ionic, metallic;

(ii) complex equilibrium crystal structure for the chemical composition at the bombardment condition;

(iii) narrow composition range for equilibrium phase field (e.g., intermetallic "line" compounds);

(iv) more elements;

(v) two or more phases of differing crystal structures in chemical equilibrium at the bombardment temperature;

(vi) low melting point; and

(vii) low-temperature bombardment.

Amorphization appears to be a true critical phenomenon. For temperatures and chemical situations where amorphization occurs, bombardment with small ion doses that produce only 0.2–0.5 displacements per target atom, relatively independent of beam ion or target, produces amorphization. If amorphization does not occur for the low dose, then the material is not amorphization susceptible and amorphization never occurs, even for bombardments many times larger than the critical for amorphization-susceptible substances. The factors listed previously are interconnected. Covalently bonded SiC is readily amorphized at room temperature, but the more nearly ionically bonded Al_2O_3 can only be amorphized by bombardment at cryogenic temperatures. Alloys whose single-phase structures are in equilibrium over a wide composition range are usually not amorphized. An exception to this generality, item (iii) in the list, is the case of the shape-memory alloy nickel–titanium, for which both the austenite and martensite can be amorphized. Pure metals are not amorphized at any temperature unless the dose is so high as to bring the composition of the resulting surface alloy into a favorable range.

(*b*) *Effects in polymers*. At low doses ion bombardment produces increased cross linking, polymerization and increased ultraviolet absorption (for polystyrene). The effect is an exception to the observation that most ion-induced changes result from nuclear stopping; the effect correlates with electronic stopping. At higher doses hydrogen gas evolution, densification and, finally, carbonization occur. These latter effects apparently correlate with nuclear stopping.

(c) *Semiconductors*. Effects in semiconductors that are of possible special interest in biomaterials include production of fully coherent and insulating buried layers of SiO_2 or Si_3N_4, and production of cobalt silicide layers with full metallic conducting characteristics (White et al. 1989). Such phenomena might be of use in the design of miniature array electrode devices.

(d) *Metals and alloys*. Ion implantation effects in Ti–6 wt% Al–4 wt% V and in stainless steels have been studied extensively. Ion implantation of nitrogen into chromium, titanium or aluminum, and alloys high in these elements, often produces very favorable effects on wear and fatigue, presumably due to nitride formation. Nitrogen implantation usually does not reduce friction in wear couples. The amorphous TiC product, produced in part by beam carburization, can reduce friction. Choice of ion-implanted constituents for corrosion resistance depends on application. Although titanium and titanium alloys are relatively corrosion resistant in saline at 37 °C, dissolution rates are nevertheless measurable. Implantation of the noble metals platinum, iridium and rhodium greatly improves corrosion performance.

(e) *Ceramics*. For high-dose implantations, metallic dopants precipitate as metal particles in the Al_2O_3 matrix provided that the postimplantation anneal is in vacuum or in a reducing atmosphere. For annealing in oxidizing atmospheres, metallic dopants are incorporated as oxide alloys. Buried conducting thin metallic layers, analogous to the thin buried insulating or conducting layers in silicon, have not been formed in sapphire. Ion implantation of chromium into sapphire produces an approximate factor of eight increase in the stress for 1% failure (99% failsafety) for single-crystal bend bars. Ion bombardment of graphitic and glassy carbons produces dramatic improvements in wear; the effect correlates with damage effects alone, having little to do with either the chemical nature of the ion-implanted dopant or electronic effects (Kenny et al. 1989).

2. Products and Development

Ion implantation of orthopedic prostheses of Ti–6 wt% Al–4 wt% V alloy comprises the main part of ion implantation of biomaterials at commercial scale (see Fig. 4). In the past the orthopedic titanium alloy often exhibited severe wear against a wear partner of ultrahigh molecular-weight polyethylene (UHMWPE). Ion implantation of nitrogen to a level of about 20 at.% or greater on a depth of about 0.1 µm strongly inhibits the triggering of this severe wear (Williams and Buchanan 1985). The precise reason for the improvement has not been proved, because of the complexity of the corrosion–wear situation, but an argument can be made that ion implantation hardens the surface to the point that it is harder than the amorphous titanium oxide wear debris. Thus, in the three-body wear situation, the autocatalytic wear process is arrested.

Ion-implanted components include femoral components of artificial knees, heads of hip joints and shoulders. These applications illustrate the value of an acceptable surface treatment such as ion implantation in allowing a higher level of design optimization for devices. Inherent advantages of the Ti–6 wt% Al–4 wt% V alloy are its excellent biocompatibility, its low modulus of elasticity and its high fatigue strength. It was thought that these properties could be used in designs that would ameliorate loosening of prosthetic devices in the bone. The wear issue was perceived as a risk factor, however, that tended to inhibit use of the titanium alloy designs. Apparently, nitrogen ion implantation has totally eliminated this concern, and ion implantation is the prevailing industry practice in the USA for titanium alloy components whose use entails rubbing contact with UHMWPE.

Important humanitarian and economic benefits are possible as a result of incremental improvements in orthopedics materials technology, such as this ion implantation process. Each year approximately 250 000 patients in the USA receive either a total hip or a total knee reconstruction at a cost to the economy of about US$5 billion. Some 5% to 10% of

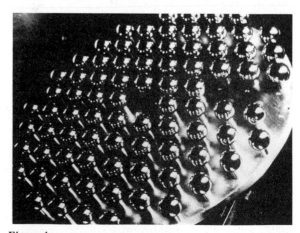

Figure 4
An array of hip joint heads of Ti–6Al–4V alloy, as mounted in preparation for ion implantation treatment (courtesy of Spire Corporation)

these surgeries are reconstructions of previous implants. The average lifetime of an artificial joint is considered to be about ten years. Actually, within ten years of surgery 30% have loosened in the bone to the extent that they are painful or dysfunctional to some degree. Moreover, many younger, heavier or otherwise more active patients are partially disabled and need an artificial joint. Most cannot have surgery because the projected lifetimes of devices are too short. It can be concluded that the ion implantation process may result in alleviation of much suffering, in addition to annual savings of hundreds of millions of US dollars because of reductions in revision surgery, in disability and in patient care.

Since wear of the alloy component has been eliminated, or at least greatly reduced, there is interest in the use of ion implantation for improving the UHMWPE side of the wear couple; wear of that component may emerge as a limitation on lifetimes of orthopedic prosthetics.

Nitrogen implantation of bone plates is also used as an economical and acceptable added insurance against fatigue failure.

The future is expected to bring expansion of ion processing of biomaterials, in terms of both available techniques and types of applications. Newly available ion processing equipment will improve the economics and utility of ion processing for various ionic species and process parameters. Useful energy ranges are likely to be extended in both directions, up and down. In general, the future of ion processing of biomaterials is likely to be closely linked to developments in ion processing of other advanced materials such as space-optical materials and microelectronic materials.

Emerging techniques include the ion cluster beam technique (Ina et al. 1989), which is able to deliver more material than direct ion implantation at energies of tens of electron volts per atom. At these energies, still much greater than chemical energies, the integrity and properties of coatings are much improved, but radiation damage can be avoided if desired. Controllability of energy is an important feature. Products include a variety of metal, oxide and compound coatings on a variety of substrates. Codeposition techniques, mentioned previously, will become more available and economical. These will be used to improve the thickness and the quality of surface compounds and alloys of all kinds (e.g., TiN).

As to future applications, interface problems in the joining of dissimilar materials of all combinations—paralene on various substrates, hydroxylapatite coatings on various substrates, metallic layers on or in polymers and ceramic insulators—are highly amenable to improvement by ion processing techniques. These types of materials problems play a large role in bioengineering. Improvement of biocompatibility itself is a closely related endeavor.

Surgically implantable biosensors, neural electrodes and other bioelectronic devices is one area in which ion processing would appear to have much to offer. Use of ion processing for improved performance of metals and ceramics surfaces for tribology, corrosion, fatigue, fracture and biocompatibility should see steady growth. In addition to orthopedics, possible areas of application include orthodonture, dental reconstruction and alloy sutures.

3. Acknowledgements
The work of the author is sponsored by the US Department of Energy under Contract No. DE-AC05-84OR21400 with Martin Marietta Energy Systems, Inc.

Bibliography
Dearnaley G 1985 Adhesive and abrasive wear mechanisms in ion implanted metals. *Nucl. Instrum. Methods Phys. Res. B* 7/8: 158–65
Hanker J S, Giammara B L (eds.) 1989 *Biomedical Materials and Devices*. Materials Research Society, Pittsburgh, PA, pp. 669–720
Ina T, Minowa Y, Koshirakawa N, Yamanish K 1989 Development of an ionized cluster beam system for large-area deposition. *Nucl. Instrum. Methods* A37/38: 779–82
Kenny M J, Pollock J T A, Wielunski L S 1989 Ion implanted graphitic carbons. *Nucl. Instrum. Methods Phys. Res. B* 39: 704–7
Licciardello A, Puglisi O, Calcagno L, Foti G 1989 UV absorption and sol–gel transition in ion-bombarded polystyrene. *Nucl. Instrum. Methods Phys. Res. B* 39: 769–72
Namba S, Itoh N, Iwaki M (eds.) 1989 *Nucl. Instrum. Methods Phys. Res. B* 39
Picraux S, Peercy P S 1985 Ion implantation of surfaces. *Scientific American* 252: 102–13
Ryssel H, Glawischnig H (eds.) 1982 *Ion Implantation Techniques*. Springer, Berlin
Takagi T (ed.) 1989 *Ion Implantation Technology*. North-Holland, Amsterdam
White A E, Short K C, Dynes R C, Hull R, Vandenberg J M 1989 Mesotaxy: Synthesis of buried single-crystal silicide layers by implantation. *Nucl. Instrum. Methods Phys. Res. B* 39: 253–8
Williams J M, Buchanan R A 1985 Ion implantation of surgical Ti–6Al–4V alloy. *Mater. Sci. Eng.* 69: 237–46
Zeigler J F, Biersack J P, Littmark U 1984 *The Stopping and Ranges of Ions in Solids*. Pergamon, New York

J. M. Williams
[Oak Ridge National Laboratory, Oak Ridge, Tennessee, USA]

Biocompatibility: An Overview

Biocompatibility has evolved over several years as a descriptive term concerned with the biological acceptability and biological performance of materials used in medicine and dentistry. Difficulties have arisen over the usage of this word, however, because of a poor understanding of the concepts and mechanisms involved and an unreasonable expectation that the phenomena can be classified, quantified and characterized. This article aims to put the phenomena of biocompatibility into the context of current thinking and the potential interactions between materials and body tissues.

1. Definition

Conventionally, biocompatibility has been equated with the lack of a significant interaction between a material and tissues. This has implied a combination of inertness and nontoxicity, and typical descriptions of an ideal "biocompatible" material would be a list of negatives, such as nondegradable, nonirritant, nontoxic, noncarcinogenic and nonallergenic. In view of the fact that so few materials even approach inertness in the body, and probably no material has zero influence on tissues, this concept of biocompatibility has been questioned. Perhaps even more importantly, the whole concept of inertness and lack of interaction as a prerequisite for biocompatibility has been challenged on the basis that total inactivity at the interface may equally be considered as a passive ignorance of the material by the tissues and that an active acceptance might be more appropriate. Biocompatibility is, therefore, now considered in a more positive light; the definition used is that "biocompatibility refers to the ability of a material to perform with an appropriate host response, in a specific application."

There are several points to be made about this definition.

(a) Biocompatibility is not a single event or a single phenomenon. It refers, instead, to a collection of processes involving different but interdependent mechanisms of interaction between the material and the tissue.

(b) Biocompatibility refers to the ability of the material to perform a function. This reflects the fact that all materials are intended to perform a specific function in the body rather than simply reside there. The ability to perform this function, and to continue to perform this function, depends not only on the intrinsic mechanical and physical properties of the material but also on its interaction with the tissues.

(c) The definition refers to the appropriate host responses. It does not stipulate that there should be no response, but rather that the response should be appropriate or acceptable in view of the function that has to be performed. It may be that the appropriate response is a minimal response. This is clearly allowed for in the definition, but so are any more extensive reactions

that are necessary for the continued safe and effective performance of the material or device.

(d) The definition also refers to the specific application. Biocompatibility of a material always has to be described with reference to the situation in which it is used. While one type of interaction, or indeed one type of response of the tissue, may be seen with respect to one material in one situation, a different reaction or a different response may be seen in another situation. For example, the same material in two different physical forms (e.g., a solid monolithic object and particulate matter) may elicit quite different responses. Thus, while it is perfectly satisfactory to deal with biocompatibility as a collection of phenomena with respect to defined conditions, there is no justification for using the adjectival counterpart biocompatible to describe a material. No material is unequivocally biocompatible; many materials may be biocompatible under one or more defined conditions but cannot be assumed to display biocompatibility under all conditions. Biocompatibility is not an intrinsic material property and cannot be considered as such.

2. Components

The above definition implies that biocompatibility is a multifactoral complex series of events which control the performance of a material in contact with tissues. It is convenient to divide these events into several categories:

(a) the initial physicochemical interactions between materials and tissue components at the interface;

(b) the effect of the tissue environment on the material (i.e., corrosion and degradation phenomena);

(c) The development of the host response to the material immediately adjacent to it (i.e., the local host response); and

(d) the transport of products of the interfacial reaction away from the site and the possibility of remote or systemic effects.

These four different components will be described in turn before reviewing the phenomena collectively.

2.1 Initial Interfacial Reactions

The implantation of any material into any tissue compartment immediately establishes a solid–liquid interface involving the material itself and blood, serum or an extracellular fluid. Since the fluid will, by definition, contain biological macromolecules, especially protein and glycoproteins, there will be a tendency for such molecules to adsorb onto the solid at the interface. Protein adsorption is, therefore, the first event in biocompatibility and the process will be initiated within seconds. It is an important feature of biocompatibility that, from the first few seconds onwards, the tissue no longer interfaces directly with the material but rather with the material covered by a layer—possibly of monomolecular thickness—of a protein.

A great deal has been learnt in recent years about the thermodynamics and kinetics of the adsorption process and the sequence of events that may occur with the complex protein solutions that are encountered. The following points have to be borne in mind when considering this phenomenon.

(a) When materials are exposed to single-protein solutions under experimental conditions, the amount of protein adsorbed will vary with time and with concentration (determined by adsorption isotherms). Typically, the amounts increase until a monomolecular layer is established; this takes place over a time period measured in minutes or a small number of hours and results in less than $1 \, \mu g \, cm^{-2}$ adsorbed protein.

(b) The process is not necessarily uniform over a surface and can be variable, particularly with surfaces of heterogeneous microstructure or composition.

(c) In solutions containing more than one protein, including fluids encountered *in vivo*, there will be competitive adsorption, with different proteins having different affinities for surfaces. This does not necessarily relate to the concentration of the proteins in the solutions, and certain proteins of high adhesiveness, such as fibronectin and vitronectin, can adsorb preferentially compared with the more quantitatively important but less surface-active proteins.

(d) The nature of the protein layer does not remain uniform with time. Once adsorbed, proteins may undergo conformational change (as molecules may fold and unfold over the surface) and they may desorb and be replaced. In complex solutions, and particularly in blood or serum, sequential adsorption processes can take place as layers are replaced or augmented.

(e) The character of the adsorbed protein layer will vary from material to material. In the majority of cases, it is assumed that the binding is reversible and mediated through hydrophobic interactions. Surface energy, surface chemistry and surface charge are assumed to be important variables, but clear unequivocal relationships are hard to establish. Generally, the more hydrophobic the surface the greater the amount of protein adsorbed, and the greater the hydrophobicity of the protein the greater the attraction.

(f) The significance of protein adsorption is not always clear, although it certainly is an important event in the overall interaction between blood and biomaterials. With soft and hard tissues, the role of the protein layer has not been elucidated, but it is likely to be important in determining the nature of cell–material interactions since it will be cell–protein rather than cell–synthetic-substrate interfaces that will be established.

2.2 Stability of Materials in Tissues

It is well established that the tissue environment is an extremely hostile and aggressive medium with respect to nonbiological synthetic substrates. Although physiological functions and biochemical reactions neither take place at high temperatures nor under aggressive radiation conditions, the combination of an electrolyte with active biological species, including catalytic enzymes, intermediate oxygen species and free radicals, constitutes a particularly reactive environment. Thus most polymers will suffer some degree of degradation, all metals will corrode to some extent and even many ostensibly stable ceramics will show some signs of aging.

It should be emphasized that a wide range of effects on the materials are included in this heading. All interactions between the material and the tissue resulting in a change in composition, structure or properties have to be considered, the following list providing an indication of the individual mechanisms which might be involved:

(a) depolymerisation of polymers,

(b) cross-linking of polymers,

(c) oxidative degradation of polymers,

(d) leaching of additives from plastics,

(e) hydrolysis of polymers,

(f) crazing and stress cracking in polymers,

(g) oxidation of metals,

(h) crevice, pitting and galvanic corrosion of metals,

(i) metal-ion release from passivated metals,

(j) stress corrosion cracking and corrosion fatigue,

(k) aging of oxide ceramics, and

(l) dissolution of ceramics.

It will be noted that some of these processes clearly release constituents or reaction products into the surrounding tissue. In some cases the degradation process results in a significant change to the structure of the material, either causing deleterious loss of properties or occasionally increasing some property, as with the increase in rigidity that often follows cross-linking. In some cases, the environment acts synergistically with mechanical stress to produce cracking or other accelerated mechanical damage.

Whatever the precise mechanism, this material degradation is a component of, and indeed is at the very heart of, biocompatibility. The consequences of the degradation are clearly two-fold with respect to the overall biocompatibility situation. First, the effects on the material may compromise its ability to perform the required function. Stress cracking in polyurethanes, corrosion fatigue in stainless steels, plasticizer leaching from poly(vinyl chloride), lipid absorption and swelling in silicones are all well-known examples of this. Second, and probably more important, the release of any component from the material, whatever it comprises and whether soluble or particulate, will act as a potential stimulus to inflammation in the tissue and, therefore, as a key determinant of the host response. The kinetics and mechanisms of these processes, therefore, are of the utmost importance, as are the various methods now becoming available for the surface treatment of biomaterials that aim to reduce these degradative interactions.

2.3 Local Host Response

At this stage there are no simple, well-understood and universally applicable mechanisms by which the tissues respond to the presence of an implanted device. Knowledge and understanding of these events which are collectively responsible for the development of this response are rapidly increasing, however, and a broad pattern is emerging. The response is clearly a dynamic phenomenon, a sequence of events that is influenced by the environmental features created by the implant. These events center around the activity of different cell types, this activity being mediated by a variety of biochemical substances which are present in, or released into, the tissue. The implant may be considered as a source of irritation, of either mechanical or chemical origin, which stimulates cellular and humoral activity in its vicinity. In many ways, the stimulus provided by the implant will be very similar to the stimulus provided by other "insults" to the tissue, including trauma and infection, and most features of the tissue response to an implant will bear a close similarity to the classical features of wound repair after trauma or of the cellular and humoral response to invading bacteria.

It is most instructive, therefore, when considering the mechanisms by which implanted biomaterials influence the localized tissues, to consider the sequence of events in normal wound healing and to consider how this sequence is modified by the presence of an implant.

The immediate response to injury, whatever its cause, is inflammation. This involves vascular, neurological, humoral and cellular response and is essentially the same whether it is induced by mechanical trauma, microbiological infection or electrical, chemical or radiological energy. This arises because the response is mediated by the same substances within the tissue in each case. The inflammatory process is aimed at eliminating, or at least containing, the causative agent, and removing damaged or dead tissue components, so that the tissue can be subsequently repaired. The process of repair involves the replacement of lost or destroyed cells by vital cells and of damaged tissue by new tissue. Although the functions of inflammation and repair processes are quite distinct and chronologically the containment of the injurious agent has to precede the repair, the repair process can begin during the inflammatory phase and the two are very much interwoven. This is of the greatest significance when considering the modification of this response in the presence of biomaterials and then degradation products, since the material may represent an injurious agent that cannot be eliminated and which will act as a persistent stimulus to inflammation.

Considering the wound healing process that takes place after tissue injury, the features of inflammation and repair are identifiable and in a simple case the inflammation will be of the acute transient type. Acute inflammation is the immediate response to injury and is associated with the two principal forces in the body's defense capability: the leucocytes and the antibodies. Since both of these are blood borne, changes in the vasculature are the main features of acute inflammation. As shown diagrammatically in Fig. 1, the microvasculature undergoes considerable change, the vessels becoming dilated and filling with excess blood, and the vessel walls becoming more permeable to allow components of the defense to diffuse into the extravascular tissue. The leucocytes are the more important cells in this context and of these the neutrophils and monocytes are the more mobile and active in the early phase. Neutrophils in particular dominate the extravascular spaces at this time. The cells migrate in the tissue under the chemotactic influence of substances released into and activated within the tissue. Within a short time, the monocytes—slightly less mobile than the neutrophils and present initially only in small numbers—accumulate in the tissue. Both neutrophils and monocytes (which are known as macrophages when released into extravascular spaces from the blood vessels) are phagocytic; that is, they are able to migrate towards, engulf and destroy debris within the tissue. The process of phagocytosis is initiated by the attachment of the particle to the surface of the cell, and this is facilitated if the particle becomes coated with one of the substances (e.g., the immunoglobulin IgG) for which the cell surface has receptors. Since bacteria may be coated with IgG antibodies from the serum (this receptor-mediated attachment), recognition may result in a far more effective clearance rate than with, for example,

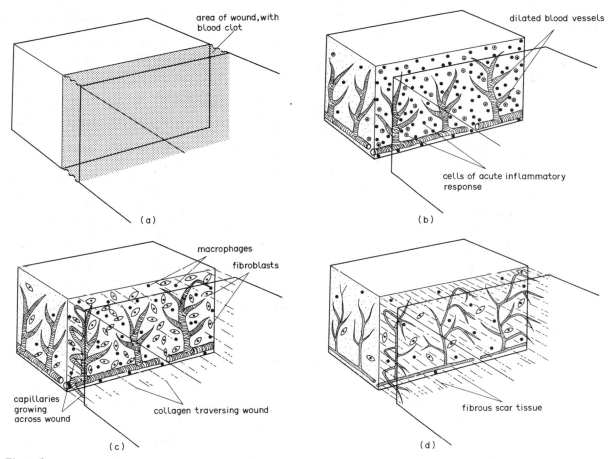

Figure 1
Sequence of events in tissue during normal wound healing: (a) incision in volume of soft tissue; (b) acute inflammatory response; (c) repair process; and (d) resolution with fibrous scar tissue

biologically unrecognisable degradation products of a biomaterial. Conversely, however, it demonstrates the importance of surface design of biomaterials in respect of cell recognition and attachment within the framework of biocompatibility.

After attachment, the cell engulfs the object. This may be easy in the case of a bacterium, which becomes enclosed within a membrane, that was hitherto part of a cell membrane, producing a phagosome. The cytoplasmic granules of the cell then fuse with the phagosome and shed their collection of powerful enzymes to kill and cause the disintegration of the trapped bacterium. The development of this inflammation is controlled by chemical mediators, derived either from the plasma or directly from the tissue. Some mediators control vascular permeability while others control chemotaxis. Examples are histamine, serotonin, complement and prostaglandins.

In the case of simple wounding, the vascular and exudative changes will soon subside and lead

directly to the repair stage. The extent and nature of tissue repair and regeneration is dependent on the ability of the cells within the tissue to replicate. Labile cells in the skin have a large regenerative capacity so that skin is easily replaced. Most often tissues have far less of a regenerative capacity and healing may take place solely through fibroblast activity and collagen formation. Therefore, in normal wound healing, the acute inflammation gives way to a period of fibrogenesis and revascularization, a zone of scar tissue being produced.

If the damage to the tissue is more extensive, the inflammatory response may not be so easily resolved and may develop into the chronic state. The chronic inflammation is a proliferative rather than exudative response and the tissue is characterized by fibroblasts associated with repair, and by an accumulation of white cells that attempt to carry on the defensive action. These consist largely of macrophages, plasma cells and lymphocytes. When the stimulus to inflammation is particularly severe or resistant to elimination, a specialized form of inflammatory tissue may develop, known as "granulation" tissue. The consequence of a chronic inflammation is that the tissue response may never be totally resolved and that the tissue may always contain inflammatory cells as well as new collagen and blood vessels.

The significance of wound healing in the presence of an implant is that the implant may act as a persistent stimulus to inflammation. This may arise through mechanical factors (including abrasion of the tissue) or chemically, through the release of degradation products or leachables, for example.

In the case of a monolithic solid consisting of a single material that is neither toxic to the host nor degraded by the tissues, the inflammatory response and repair processes may take place virtually unaffected. Exactly the same cellular activity and tissue regeneration may be seen, the implant serving only as a physical barrier to capillary regrowth and collagen bridging. The end result (see Fig. 2) is that a zone of scar tissue forms around the implant. This is a significant point, for the collagen will not normally be adherent to the implant surface but will be aligned parallel to the surface, effectively walling off the material from the tissues. In no sense is the implant incorporated into the tissues, but rather it remains in its own self-generated capsule, thereafter being ignored by the tissue. This response is the basis of the classical fibrous encapsulation of implants.

If the material is not totally inert, or if there is any other way in which it is able to irritate the tissue, then the process is likely to be more extensive. With a slightly greater degree of interaction, the acute phase may be a little more noticeable in terms of the number of cells present, but more importantly the chronic response will be more significant and prolonged. As noted above, a more extensive chronic inflammation, occasioned by the continued stimulus, will act as a persistent mediator of fibrosis so that the fibrous capsule may not reach equilibrium so readily and will progressively thicken. A slightly thicker capsule may not necessarily be undesirable, and indeed in some implant applications attempts are made to actively encourage this fibrosis to give a greater tissue mass. The problem is, however, that if a stable state is not achieved, the capsule may alter its characteristics, not only in the amount of collagen but also in the number and type of inflammatory cell. The granulation tissue referred to earlier may arise in response to the presence of an implant; this has been observed on many occasions.

There are at least two important consequences of this behavior in relation to the concept of biocompatibility.

The first consequence concerns the timescale of the events and the interplay between degradation and the host response. As noted above, the degradation of many materials is influenced by the precise nature of its environment and, specifically, with the characteristics of the biological environment. As a tissue reaction becomes more cellular, and the cells themselves become more active, so the chemical hostility of the tissue increases. This is not surprising since the inflammation is chemically attempting to remove exogenous debris. With a complexity of enzymes, peroxides and free radicals, it is quite possible that the implant materials are going to suffer greater degradation. The greater the degradation of the material, however, the greater the stimuli to inflammation; thus a self-perpetuating phenomenon is established (see Fig. 3). Once a material becomes unstable within the tissue environment, the host response is likely to become progressively more significant.

Second, while it has been stated that the minimal fibrous tissue is not ideal, the stimulation of extensive inflammation and fibrosis is hardly ideal either. What is required here is neither unaffected wound healing nor stimulated inflammation, but rather controlled cellular activity to give the adaptation of the most appropriate tissue to the surface. It is possible that this will be achieved by either control of the cellular response through the use of anti-inflammatory agents or promotion of the tissue regeneration through growth factors.

2.4 Systemic Distributions and Responses

If there is an interfacial reaction that leads to the generation and release of reaction products, the question arises as to the fate of these products in the tissue. It is possible for the products to be soluble and to be taken up rapidly in the extracellular fluid, to be transported away in either lymphatic or vascular systems. Thereafter, it is possible for the solubilized material to be metabolized and excreted, or to be stored in some tissue depot.

Alternatively, the products may be particulate. Depending on their size, such particles may again be taken up and circulated (e.g., by macrophages and then the lymph vessels) and subsequently deposited (e.g., in the lymph nodes).

If these products have the capacity to irritate tissue, which is highly likely, then problems could arise systemically (throughout the body) or at some remote site through which the products pass or in which they are deposited. Very little is known about these systemic effects as yet, but it is a matter of considerable debate and controversy.

3. Comments

The above analysis represents a relatively simplified approach to the components of biocompatibility. The details of biocompatibility in any one situation will naturally vary and there are very many mediators, or controllers, of the phenomena including characteristics of the materials and the host. It is believed, however, that the general sequence of events will follow the above pattern.

It should be noted, however, that there are also some special features of biocompatibility where an additional factor (or factors) come into play and where, therefore, quite different events are seen. If, for example, bacteria are introduced into the

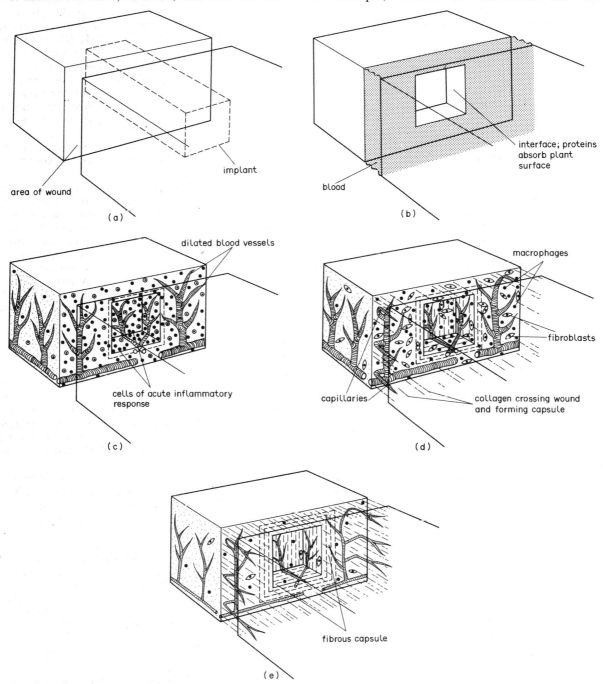

Figure 2
Sequence of events in tissue response to implantation of an inert material: (a) implantation in soft tissue; (b) initial protein adsorption at interface; (c) acute inflammatory response; (d) repair process—collagen formation and revascularisation influenced only by the physical presence of the implant; and (e) fibrous encapsulation of implant

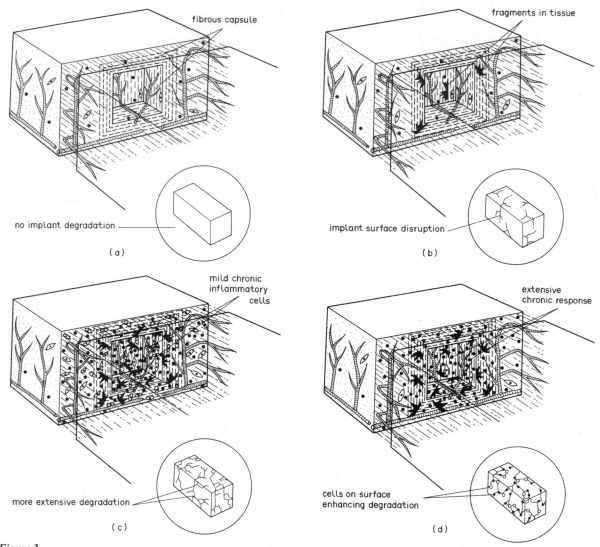

Figure 3
Interaction between degradation and host response: (a) fibrous capsule with no degradation of the implant material; (b) limited surface degradation, with products in the tissue, leading to more extensive response; (c) more extensive degradation, giving persistent chronic inflammatory response; and (d) extensive chronic response providing a more aggressive tissue environment and, in turn, a continued inflammatory response

implant site, there will be a significant contribution from biomaterial–bacterial interactions. If some feature of the material or devices causes cell mutation, then carcinogenesis may arise, a matter of considerable controversy and uncertainty. None of these additional features of biocompatibility is, as yet, fully understood.

Bibliography

Anderson J M, Miller K M 1984 Biomaterial biocompatibility and the macrophage. *Biomaterials* 5: 5–10
Bakker D, van Blitterswijk C S, Hesseling S C, Grote J J, Daers W T 1988 Effect of implantation site on phagocyte/polymer interaction and fibrous capsule formation. *Biomaterials* 9: 14–23
Christel P, Meunier A 1989 A histomorphometric comparison of the muscular tissue reactions to high density polyethylene in rats and rabbits. *J. Biomed. Mater. Res.* 23: 1169–82
Miller K M, Anderson J M 1988 Human monocyte/macrophage activation and interleukin 1 generation by biomedical polymers. *J. Biomed. Mater. Res.* 22: 713–32
Miller K M, Anderson J M 1989 *In vitro* stimulation of fibroblast activity by factors generated from human monocytes activated by biomedical polymers. *J. Biomed. Mater. Res.* 23: 911–30

Williams D F 1987a Tissue–material interactions. *J. Mater. Sci.* 22: 3421–44

Williams D F 1987b *Definitions in Biomaterials*. Elsevier, Amsterdam

Williams D F 1990 Biocompatibility: Performance in the surgical reconstruction of man. *Interdiscip. Sci. Rev.* (in press)

Ziats N P, Miller K M, Anderson J M 1988 *In vitro* and *in vivo* interactions of cells with biomaterials. *Biomaterials* 9: 5–13

D. F. Williams
[University of Liverpool, Liverpool, UK]

Biocompatibility of Dental Materials

The biocompatibility of dental materials is of ultimate importance for the longevity of dental restorations. The testing of these materials is complex and expensive. The consumer has to pay for this testing, but the alternative—a lack of testing—has for years led to the clinical use of toxic materials that cause pulpal and periapical damage. This has resulted in increased costs for repeated replacement of dental restorations, with increasing loss of tooth structure, and finally to the loss of teeth. Improvement in the biocompatibility of dental materials will prevent additional damage at each stage of the disease process.

Dental materials have been categorized as medical devices. Full implementation of the US Medical Device Bill (1976) should prevent the marketing of toxic materials. On a voluntary basis, methods and criteria for the evaluation of biologic properties of materials have been established by the Fédération Dentaire Internationale (FDI) (1980) and by the American National Standards Institute/American Dental Association (ANSI/ADA 1982).

1. Testing Procedures and Extent of Testing

If a law is to be efficient, methods and criteria for compliance must be available. Such recommendations have been worked out by the ANSI/ADA and the FDI, as indicated previously. The basic premise is that it remains the manufacturer's responsibility to test new material and to assure its safety and efficacy, according to available adequate methods. A survey of the methods recommended by the FDI—adopted in a slightly different form by the International Standard Organisation (ISO)—will exemplify the different levels of testing, as summarized in Table 1.

2. Initial Tests

Although the dental profession thinks of reactions to dental materials mainly in local terms, it should be

Table 1
Recommended levels of biological testing of dental materials[a]

Initial tests	Secondary tests
Short-term systemic toxicity test: oral route	Subcutaneous implant test
	Bone implant test
Acute systemic toxicity test: intravenous route	Sensitization test
	Oral mucous membrane irritation test
Inhalation toxicity test	
Hemolysis test	
Ames mutagenicity test	*Usage tests*
Styles cell transformation test	Oral mucous membrane irritation test
Dominant lethal test	Pulp and dentin test
In vitro cytotoxicity test (chromium release)	Pulp capping and pulpotomy test
Cytotoxicity test (millipore filter)	Endodontic usage test
	Bone implant usage test
Tissue culture agar overlay test	

a Fédération Dentaire Internationale 1980

realized that any material in contact with bleeding tissue may also have a systemic effect. These aspects and other aspects dealing with the general biocompatibility are addressed under the initial tests.

The short-term systemic toxicity test has the objective of assessing the toxicity of the material after short-term oral administration. Because of frequent misunderstandings, it should be emphasized that it is not the intent of this test to protect the patient from toxic effects of swallowing the material. It is not a usage test. One gram per kilogram of body weight would be the equivalent of a 70 kg person swallowing 70 g of a test material. That is hardly a practically occurring condition. The intention of this test is to establish a toxicity profile for the material.

The acute systemic toxicity test has the objective of assessing the acute toxicity of the material after intravenous administration.

The inhalation test is designed for dental remedies that have significant volatility under usage conditions.

The hemolysis test is designed for *in vitro* evaluation of the hemolytic activity of materials intended for prolonged tissue contact.

The Ames mutagenicity and the dominant lethal tests are all designed to assess the potential carcinogenic activity of a material.

It is the manufacturer's knowledge of the chemical composition of the material and the interaction between the material and the tissue that ultimately decides which test or tests will be chosen. For economic reasons, it is clear that tests will be chosen that are as inexpensive as possible while providing the necessary information for maximal protection.

The *in vitro* cytotoxicity tests—the chromium

release method (Spångberg 1973) (see Fig. 1), the millipore filter method (Wennberg et al. 1979) and the tissue culture agar overlay tests (Autian 1977)— are all designed to assess and rank the local toxicity of materials.

The chromium and the millipore filter methods have the advantage that they may be applied to materials in various stages from freshly mixed to totally cured. In this way, they relate to how the materials are used in dentistry. Both restorative materials and endodontic materials are applied to the dental tissues in their uncured form. Therefore, these tests have the ability to detect the initial toxic effect which other methods that depend upon the use of cured materials will not record. They also afford direct material–tissue contact. Moreover, extracts of materials can be tested against the cells

and indicate toxicity related to what may diffuse out from the materials over a longer period. However, these tests do not give proper information regarding long-term effects of materials.

Since the agar overlay test depends upon diffusion through agar, the diffusibility of materials under these conditions influences the toxicity. However, it should be realized that the objective of all tissue culture tests is to rank the toxicity of the material under this system and, although the ultimate goal is to replace the usage test with simpler tests, a dependable correlation must first be established. In that case, it seems that tests that depend upon a direct cell–material contact would be superior to those involving a dividing medium.

3. Secondary Tests

These tests give specific information about the local *in vivo* biocompatibility of dental materials and may, therefore, be of great importance for an understanding of these aspects of the materials.

The sensitization test may be necessary for materials that contain chemicals with high allergic potential, but is not to be considered a usage test for allergic patients in general.

The oral mucosa membrane irritation test may be used as a secondary test to rank the toxicity of materials against an epithelialized tissue, but may also serve as a usage test for materials that, in clinical use, intermittently or permanently contact oral mucosa.

The subcutaneous implant test has the objective of assessing the *in vivo* toxicity of materials that are intended for prolonged contact with subcutaneous tissue. It also gives information, in general, about what happens in the interface between material and soft tissue and the effect this has on the underlying tissue. The method may not be of particular interest for restorative material where the dentin modifies the soft tissue response; however, it is directly relevant to pulp capping, pulpotomy and endodontic materials.

The method involves the placement of the test material in 5–7 mm Teflon tubes with an outer diameter of 1.7 mm and an inner diameter of 1.3 mm. Utmost care should be exercised to prevent spillage on the outside of the tube which should be kept meticulously clean to be used as an inert control in each experiment. A needle implantation technique is employed, using an intravenous 14-gauge thin-wall needle accommodating the Teflon tube. This method (see Fig. 2) ensures passive placement of the tube with less damage than the incision technique. To eliminate differences and further standardize the method, the tube is closed at one end with nonirritating paraffin wax and filled with a syringe to avoid spillage of the material. The needle implantation technique is used to reduce initial tissue damage.

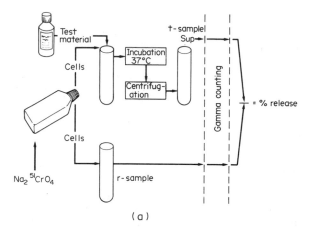

(a)

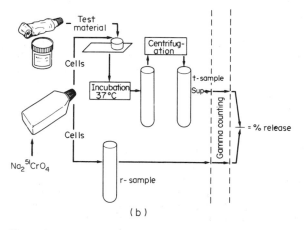

(b)

Figure 1
Chromium release method for testing cytotoxicity *in vitro*: (a) procedure for fluid or soluble materials; (b) modified procedure for solid or semisolid materials (t-sample, test sample; r-sample, reference sample; sup, supernatant)

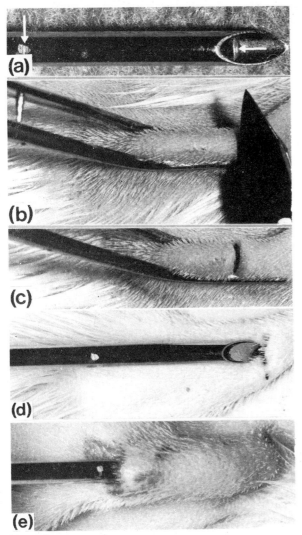

Figure 2
Subcutaneous implant test (modified needle implantation technique): (a) mark (vertical arrow) indicating how far the needle will be advanced into the subcutaneous tissue and end of stylus (horizontal arrow) as it will appear when needle is withdrawn (by which the tube is placed in the tissue without pressure); (b) following shaving, a cut through the skin only for easy introduction of the needle; (c) bloodless incision; (d) introduction of the needle with the original stylus to prevent entry of tissue into the needle; (e) advancement of the needle to the mark indicated in (a)

When a double-ended tube is used, different reactions may occur from the same material in both ends, related to different physical conditions (see Fig. 3).

The bone implant test is intended to assess the *in vivo* toxicity of materials intended for prolonged contact with bone. This test involves the placement of Teflon cups in the edentulous areas on both sides of the symphysis of guinea pig mandible with instrumentation developed by Spångberg (1969). The cup, with an inner diameter of 1.3 mm, filled with various materials or empty for control, is implanted in bone cavities. In each experiment the surface of the cup, except the uneven rough surface grooves, serves as a control. At short observation periods, the reaction to the material is modified by the reactions to drilling in bone (see Fig. 4a, b). At longer observation periods (e.g., 180 days), this initial reaction is eliminated and if no toxic materials are present, healthy bone appears inside the cup, though some remaining inflammatory cells are present between the bone and the inner surface of the cup (see Fig. 4c, d).

Although this method is somewhat more complex than the subcutaneous implant method, it has the advantage that the cup is immobile in the bone. It has the disadvantage that there is a more severe reaction to the implantation method *per se*. With increasing time of observation, this disadvantage is eliminated. In the final analysis, the two methods complement each other and the use of both has the further advantage that the materials are tested in two different animal species where reactions could be different (Olsson et al. 1981).

4. Usage Tests

4.1 Pulp and Dentin Test

The objective of this test is to assess the response of the dentin and the dental pulp to restorative procedures and materials. Manufacturers who want to compare materials with different chemical components may use one of the tissue culture or implant tests. However, after these initial comparative tests, the material must eventually be tried out in a cavity under the conditions in which it is intended to be used.

Since the intention is to record and compare reactions to the materials *per se*, it is important to control and reduce to a minimum experimental variables. Thus, the experimental teeth should be as equal as possible and free of caries and attrition, since these conditions cause pulpal changes and increases in severity with the depth and width of dentin involvement.

Cavity preparation must occur under sufficient water spray to prevent pulp reactions from the cavity preparations from adding to those from the material *per se*. Only surface drying of the cavity should take place, to prevent desiccation. These reactions have been known for many years and should be avoided if the effect of the material *per se* is to be evaluated (Langeland 1961).

For comparison from case to case, it should be realized that similarity of width of the cavities is as important as the depth. The wider the cavity, the more dentinal tubules are open to penetration of

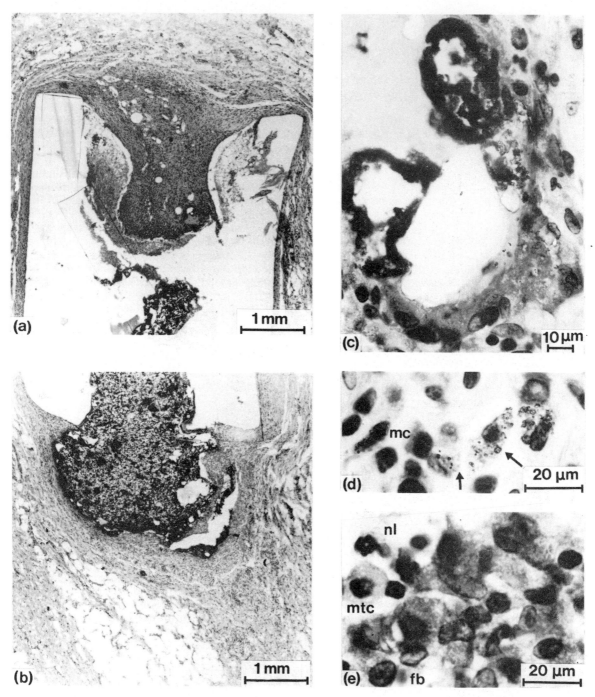

Figure 3
Tissue sections 14 days after subcutaneous implant (of freshly mixed AH26), showing the difficulty encountered in experiments introducing materials using a Teflon tube open at both ends: (a) material short of the tube end; (b) material beyond the tube end; (c) tissue next to the material, showing vessels and foreign body cells with material, adjacent lymphocytes and neutrophilic leukocytes, indicating a chemotactic reaction; (d) material particles in macrophages (mc) and in a capillary (arrows); (e) inflammation in tissue at a distance from the implant (nl, neutrophilic leukocytes; mtc, mast cells; ly, lymphocytes; fb, fibroblasts); the condition demonstrated in (a), (b) negates any quantitative comparison between reactions in opposite ends

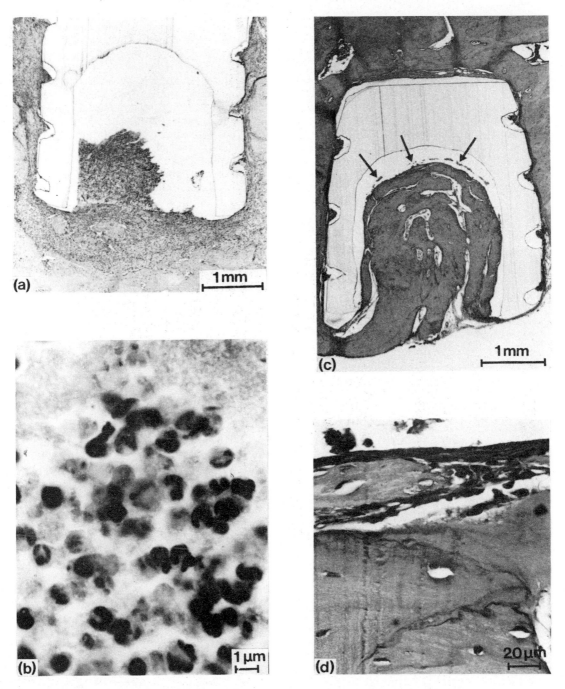

Figure 4
Results of bone implant test with Teflon cup containing Kloroperka-N.Ø. in guinea pig mandible for 30 days (a,b) and control test with empty cup for 180 days (c,d): (a) dense concentration of cells beyond the opening of and halfway into the cup (material in soft tissue beyond the cup opening); (b) at top of picture, necrosis in the area of immediate material contact and in the adjacent tissue dense concentration of neutrophilic leukocytes; (c) healthy bone filling the entire cup, with cells and cell remnants between bone and bottom of cup (arrows) (empty space is an artifact); (d) from area of middle arrow in (c), healthy bone with osteocytes covered by a fibrous tissue, with some inflammatory cells remaining in the adjacent space. In (a) and (b) the reactions are partly due to the operative trauma, partly to the toxicity of the material

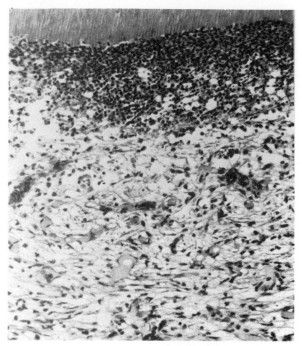

Figure 5
Experimental cavity drilled under water spray in premolar of young individual, with carious dentin applied to the cavity floor. (The cavity was obturated with amalgam which was perforated following its hardening to allow saliva access.) Observation period, 21 days; shows dense concentration of cells in the odontoblast layer and layer of Weil where the cut dentinal tubules terminate in the pulp. Density of cell concentration tapers off in central pulpal direction. Morphometric measurement by cell counting will give widely varying results dependent on location (Langeland and Langeland 1981)

irritants. In regard to controls, many investigators request homologous teeth. However, because histologic artifacts are common in pulps, a more dependable control is the pulp tissue surrounding that where the involved dentinal tubules from the experimental cavity terminate in the pulp in each case. This tissue has been exposed to exactly the same histologic procedures and would, therefore, respond with the same artifacts (e.g., see Figs. 5–7) (Langeland and Langeland 1981, Dowden and Langeland 1983).

If pretreatment of the cavity is a requirement for the use of the materials tested, the effect of this procedure must be tested. Thus, some materials require the use of acid etching and in this case the pulpal reaction would be the cumulative effect of the acid etching and the material to be used (Macko et al. 1978).

Since materials are intended to be used for the restoration of carious cavities, they should be tested under these conditions. However, the pulpal response varies greatly under apparently similar carious conditions. To control this variable, it has been suggested that a pulpitis should be induced experimentally and then the effect of, for example, anti-inflammatory materials tested (Mjör and Tronstad 1972). Unfortunately, the variables are no better controlled under these conditions than when using naturally carious teeth (see Fig. 5). It should particularly be emphasized that morphometric measurements may be highly deceiving, because the numbers vary with the location of the section and the histologic artifacts. These conditions are aggravated in electron microscopy.

4.2 Pulp Capping and Pulpotomy Test

The objective of this test is to assess the response of the dental pulp to pulp capping and pulpotomy materials. Since these materials are intended to contact connective tissue directly, the subcutaneous and intraosseous implant tests become more directly relevant. Adequate information may be derived from these secondary, less costly tests. However, ultimately the tests have to be performed in experimental teeth.

If the pulp capping material is introduced from a cavity in the cervical region and interferes with the circulation to the pulp tissue in the pulp horn, the severity of the reaction is aggravated by the fact that necrotic disintegration products seep back into the pulp (see Fig. 6). Therefore, the reaction in that pulp will be more severe than when the material protrudes only halfway into the pulp horn or when it does not cut off the pulp horn at all (see Fig. 7).

4.3 Endodontic Usage Test

The objective of this test is to assess the response of the pulp wound and the periapical tissue to endodontic materials. In these experiments, the results depend greatly on the level at which the pulp tissue is cut off and the total removal of all pulp tissue in the root canal system. Prognosis studies involving adequate observation periods and numbers of teeth indicate that the success of endodontic therapy is greatly dependent upon the termination of the root filling approximately 1 mm from the apical foramina. The total removal of all pulp tissue is not possible except in the straight part of the canal (where it is also circular). In irregularly shaped canals and in canals where pathologic changes have occurred, various amounts of tissues will remain.

Thus, in this test, the end result will be influenced by the method perhaps as much as by the material (Langeland and Walton 1983). Therefore, the subcutaneous implant test and the intraosseous test give information about the biocompatibility of the material. In these tests, for example with Hydron, it is evident that the material is transported from the area of original placement. This transport

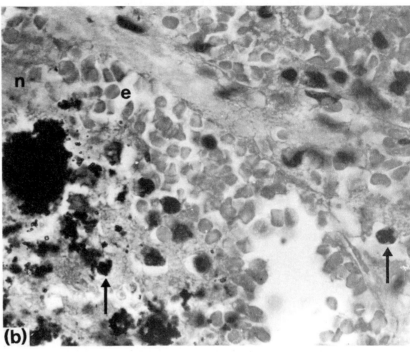

Figure 6
Upper canine of monkey capped with calcium–eugenol cement, after 8 days, with cervical perforation short of the pulp horn; (a) calcium–eugenol cement forced through pulp tissue to opposite pulpal wall; (b) tissue apical to the pulp capping with scattered material, showing necrotic tissue (n), scattered extravasated erythrocytes (e) and neutrophilic leukocytes (arrows) with material, the inflammatory reaction being due not solely to the material but also to the disintegration product from the necrotic pulp horn (Hørsted et al. 1981)

is confirmed following an endodontic procedure (see Fig. 8).

Methodological errors should be considered. Sections selected outside the reaction area may not show any material, whereas material is present in sections nearer to the foramen. Therefore, only sections passing through the foramen and the adjacent periapical tissue are acceptable to meet the requirements of the endodontic test. If material has been transported, it will be evident, as also will be the periapical lesion (Langeland et al. 1981).

4.4 Bone Implant Usage Test

The objective of this test is to evaluate all materials that, during their intended use, penetrate the oral mucosa and the subjacent bone. The secondary subcutaneous test and the bone implant test can be used to evaluate the tissue reaction to the material *per se* (see Figs. 3, 4). However, in the usage test, the requirements are that the implant material be placed in jaws of animals, protrude through the oral mucosa and be exposed to masticatory forces. Under these circumstances, oral plaque may gather on the implant surface, the amount depending upon the quality of the surface. Materials that, when

totally embedded in the tissue, cause a fairly minimal reaction except when they are highly corrosive will cause considerable reactions which may lead to the loss of the implant due to bone destruction. This condition is illustrated by the implant of a Teflon tube which, *per se*, does not cause tissue damage, but does so when protruding through the oral mucosa (see Fig. 9) although it is not exposed to masticatory forces in this test; the tissue destruction and adjacent inflammation are a response to the bacterial plague.

See also: Biocompatibility: An Overview; Dental Implants

Bibliography

American National Standard Association/American Dental Association 1982 ANSI/ADA Document 41. *J. Am. Dent. Assoc.* 104: 680

Autian A 1977 Toxicological evaluation of biomaterials: Primary acute toxicity screening program. *Artificial Organs* 1: 53

Browne R M 1985 *In vitro* cytotoxicity testing of dental restorative materials. *CRC Crit. Rev. Biocompat.* 1: 85–110

Bumgardner J D, Lucas L C, Tilden A B 1989 Toxicity of copper based dental alloys in cell culture. *J. Biomed. Mater. Res.* 23: 1103–14

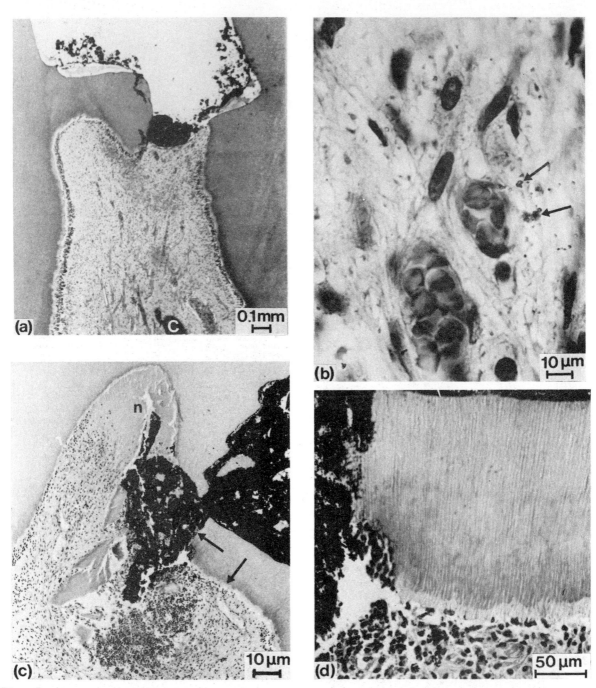

Figure 7
Lower molar of monkey capped with Dycal (a,b) or eugenol cement (c,d) after 8 days: (a) perforation to one pulp horn and application of the Dycal to the exposed pulp tissue, showing slight to moderate concentration of cells below the exposure and calcifications (C) in the adjacent pulp tissue; (b) scattered particles of Dycal in the adjacent pulp tissue (arrows); (c) cement pushed into the pulp occupying a third of the space, apical to the pulp horn, showing pulp horn necrotic (n), a dense concentration of cells apical to the cement; (d) from area of arrows in (c), showing remaining dentin with adjacent cement and a concentration of neutrophilic leukocytes. No conclusion can be drawn as to the toxicity of the material, because the clinical procedure in (a,b) has caused less damage than in (c). The necrosis of the pulp horn of (c) is due to the mechanical interference of the material with the circulation, not to the toxicity of the material (Hørsted et al. 1981)

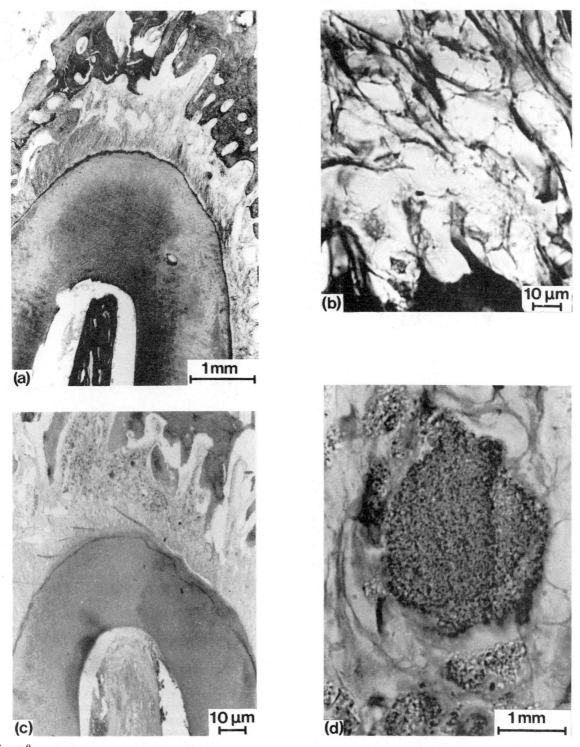

Figure 8
Root filling with Hydron following pulp extirpation, observed after 41 days: (a,b) no Hydron particles in the periapical tissue; (c,d) (~200 μm from (a,b)) hydron particles in periapical tissue, despite absence of a root canal exit also in this section (Langeland et al. 1981)

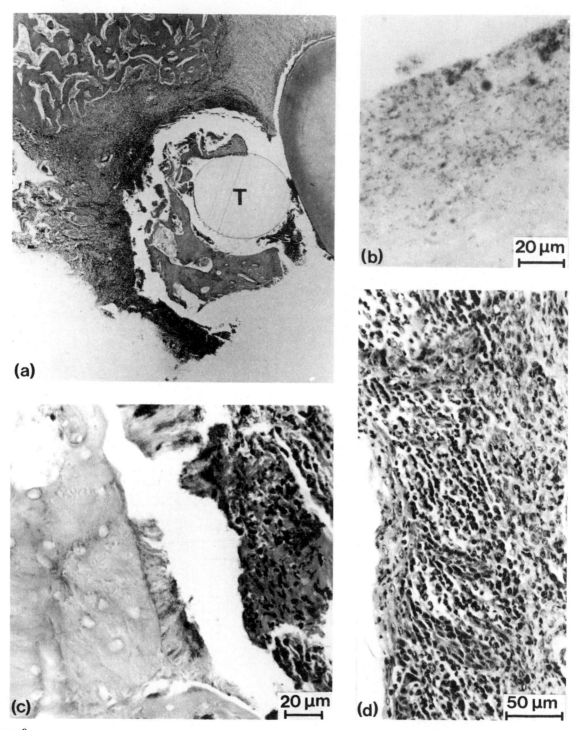

Figure 9
Teflon rod implanted in guinea pig maxillary bone exposed to the oral cavity for 12 weeks: (a) cross section of Teflon rod (T) near perforation through the oral mucosa: (b) Brown and Brenn stained sections from such an area, showing bacteria; (c) bone and soft tissue next to implant, showing bone without osteocytes indicating necrosis, and adjacent dense concentration of inflammatory cells; (d) dense concentration of inflammatory cells in tissue near implant and perforation to the oral cavity (Langeland and Spångberg 1975)

Dowden W E, Emmings F, Langeland K 1983 The pulpal effect of freezing temperature applied to monkey teeth. *Oral Surg.* 55: 408–18

Dowden W E, Langeland K 1983 An evaluation and comparison of the pulpal response to gold foil and indium alloy. *J. Prosthet. Dent.* 50: 497–504

Fédération Dentaire Internationale 1980 FDI Technical Report 9: Recommended standards for the biological evaluation of dental materials. *Int. Dent. J.* 30: 140

Hetem S, Jowett A K, Ferguson M W S 1989 Biocompatibility testing of a posterior composite and dental cement using a new organ culture method. *J. Dent.* 17: 155–61

Hørsted P, El Attar K, Langeland K 1981 Capping monkey pulps with Dycal and Ca-eugenol cement. *Oral Surg.* 52: 531

Langeland K 1961 Tissue changes incident to cavity preparation. An evaluation of some dental engines. *Acta Ondont. Scand.* 19: 397

Langeland K, Langeland L K 1981 Problems of intradental testing of restorative materials. *Int. Endod. J.* 14: 80

Langeland K, Olsson B, Pascon E A 1981 Biologic evaluation of Hydron. *J. Endod.* 7: 196

Langeland K, Spångberg L 1975 Methodology and criteria in evaluation of dental endosseous implants. *J. Dent. Res.* Special Issue B. 54: 158

Langeland K, Walton R 1983 Sargenti (N2) technique. In: Clark J (ed.) 1983 *Clinical Dentistry*, Vol. 4. Lippincott, Philadelphia, PA, Chap. 13, pp. 1–31

Lin L M, Langeland K 1981 Light and electron microscopic study of teeth with carious pulp exposures. *Oral Surg.* 51: 292

Macko D J, Rutberg M, Langeland K 1978 Pulpal response to the application of phosphoric acid to dentin. *Oral Surg.* 45: 930

Meryon S D 1987 The influence of surface area on the *in vitro* cytotoxicity of a range of dental materials. *J. Biomed. Mater. Res.* 21: 1179–86

Mjör I A, Tronstad L 1972 Experimentally induced pulpitis. *Oral Surg.* 34: 102

Olsson B, Sliwkowski A, Langeland K 1981 Intraosseous implantation for the biological evaluation of endodontic materials. *J. Endod.* 7: 253

Safavi K E, Pascon E A, Langeland K 1981 A simplified method for evaluation of biocompatibility of endodontic sealers. Proceedings, International Association for Dental Research. *J. Dent. Res.* 60A: 353

Safavi K E, Pascon E A, Langeland K 1983 Evaluation of tissue reaction to endodontic materials. *J. Endod.* 9: 421–9

Smith D C, Williams D F (eds.) 1984 *Biocompatibility of Dental Materials*, Vols. I–IV. CRC Press, Boca Raton, FL

Spångberg L 1969 Biologic effect of root canal filling materials, 7: Reaction of bony tissue to implanted root canal filling materials in guinea pigs. *Odont. Rev.* Suppl. 16, 20: 501

Spångberg L 1973 Kinetic quantitative evaluation of materials. Cytotoxicity in vitro. *Oral Surg.* 35: 389

Stanley H R 1985 *Toxicity Testing of Dental Materials*. CRC Press, Boca Raton, FL

Syrjanen S, Hentsten-Pettersen A, Kangasniemi K, Yli-Urpo A 1985 *In vitro* and *in vivo* biological responses to some dental alloys. *Biomaterials* 6: 169–76

US Congress (Senate) 1976 *Medical Device Amendments of 1976*. Public Law 94-295, 94th Congress, S 510. US Government Printing Office, Washington, DC

Wennberg A, Hasselgren G, Tronstad L 1979 A method for toxicity screening of biomaterials using cells cultured on millipore filters. *J. Biomed. Mater. Res.* 13: 109

K. Langeland
[University of Connecticut, Farmington, Connecticut, USA]

Biodegradation of Medical Polymers

The subject of polymer degradation is of the greatest importance in biomaterials science, but is also one of considerable confusion and controversy. Its importance lies in the fact that the effects of the physiological environment on a polymer will critically control both the functional performance of the material (and any device made from it) and the response of the tissue to the material. The confusion and controversy lie in the uncertainty relating to the susceptibility of polymers to degradation in this environment.

It has long been considered that polymers have a significant advantage over metals in the context of medical applications, as the isotonic saline solution that comprises the extracellular fluid is extremely hostile to metals, but is not normally associated with the degradation of many synthetic high-molecular-weight polymers. It should be possible to select polymers that will be stable under the conditions found in the body, taking into consideration that the circumstances under which polymers are most susceptible, including those of elevated temperature, electromagnetic radiation and atmospheric oxygen, are not operative here. The fact is, however, that polymers are a great deal more susceptible to degradation under these conditions than previously accepted and indeed it may now be assumed that virtually all will be altered to some extent by exposure to this particular environment.

The situation is made a little more confusing, and indeed urgent, by the fact that there are also several conditions where degradation is desirable. While it is clearly a fundamental requirement for most materials used in the replacement of tissues that they are inherently and permanently stable in the body, there are many applications where the function of the material is only transient and where there is sound logic in assuming that the materials should be removed once the function has been performed. Thus, intentionally degradable materials, which are spontaneously eliminated once the function has been performed, are of considerable relevance. Polymers provide a very suitable class of material for consideration for this purpose, although the kinetics of the processes and the toxicology of the breakdown products have to be considered very carefully.

It is, of course, necessary to define the terms, although this is not easy with degradation and biodegradation. Degradation is sometimes meant to denote molecular breakdown of a material resulting in a change in structure and properties. Others, however, use the term to denote a change in properties, whether or not this is caused by molecular breakdown. There is a fundamental difference between one process that causes main chain scission in a polymer and another that causes extraction of some leachable additives, even though both processes may result in the same type of property change. It is an unresolved issue as to whether both processes are to be described as degradation.

The term biodegradation is often used to denote degradation occurring in biological environments. There is again, however, a fundamental difference between a degradation process occurring in and a degradation process caused by a specific environment. In the former case it may be a coincidence that the environment happens to be of a biological nature. In the second case there has to be a specific function of the biological environment that is actively causing the degradation. For this article, biodegradation refers to the breakdown of a material that is mediated, or caused, by specific biological activity. It is necessary for there to be a specific mechanism associated with the biological environment that is involved with the degradation, such as cells, enzymes, bacteria or yeasts.

1. Mechanisms of Polymer Degradation In Vivo

As mentioned previously, most polymers are susceptible to degradation under certain conditions, but these conditions and the kinetics of the reactions are extremely variable. The degradation processes may be divided into two different types. First, there are those that involve the absorption of some kind of energy which then causes the propagation of molecular degradation by secondary reactions. As mentioned previously, these conditions are not conventionally thought to be operative in the body. Second, there are hydrolytic mechanisms, which can result in molecular fragmentation. This is usually seen in heterochain polymers where the process may be regarded as the reverse of polycondensation. This is clearly quite feasible in the aqueous extracellular fluid. A number of conditions have to be met for this to take place. First, the polymer has to contain hydrolytically unstable bonds, hence its prevalence in heterochain polymers, especially those with the

$$-\underset{\underset{O}{\|}}{C}-O- \quad \text{and} \quad -\underset{\underset{H}{|}}{N}-\underset{\underset{O}{\|}}{C}-$$

groups. Second, for any significant degradation to occur in a reasonable time, the polymer should be hydrophilic and preferably absorb a measurable amount of water. Third, the hydrolysis should take place at physiological pH and temperature.

On this basis, polymers can be placed in a ranking order of predicted susceptibility to *in vivo* degradation. The most stable should be the non-hydrolyzable, hydrophobic polymers (e.g., polytetrafluoroethylene (PTFE)). Those that are non-hydrolyzable, but with some degree of water permeability, may not suffer molecular degradation, but may display swelling and cracking or the leaching of components, thus altering properties. Hydrolyzable but hydrophobic polymers, with very limited water permeability, should suffer hydrolysis very slowly and directed from the surface (e.g., aromatic polyesters, polyamides) while hydrolyzable hydrophilic polymers should suffer bulk degradation (e.g., aliphatic polyesters).

The degradation seen clinically and experimentally follows this pattern for the most part. It may be described, therefore, as degradation rather than biodegradation, since it is only water and not active biological species that are involved. There are, however, many examples where the hydrolysis can be influenced by such species, especially enzymes, and also where nonhydrolytic mechanisms, such as oxidation, take place.

2. Hydrolysis

Heterochain polymers, particularly those containing oxygen or nitrogen, are likely to be susceptible to hydrolysis. Among those that have been shown to degrade by hydrolysis are polyamides, poly(amino acid)s, aliphatic and aromatic polyesters, and some polyurethanes.

While the hydrolytic degradation process is relatively easy to understand in these polymers, questions have arisen as to whether different rates of hydrolysis can be seen with variations in the tissue environment. Of particular importance here is the possibility of hydrolytic enzymes influencing the overall degradation process. This is of some importance semantically when deciding whether a material is suffering biodegradation or simply degradation, but more significantly, it is an important issue in the predictability and uniformity of polymer degradation *in vivo*.

According to the definition given in Sect. 1, if hydrolysis of a polymer takes place *in vivo* by a similar mechanism and with the same kinetics as that achieved in a simple *in vitro* aqueous environment, then it cannot strictly be biodegradation, for there is no specific role for biological components in the tissue. Enzymes are proteins that have a catalytic function, normally reserved for, and indeed very specific to, certain biochemical or physiological reactions. As potentially very reactive substances, however, it is possible for them to act somewhat less specifically and a variety of substrates may be affected. Some enzymes are now widely used in technological processes to exert their catalytic properties on nonphysiological reactions.

It seems fairly clear from the literature that enzymes are able to influence the hydrolytic degradation of certain polymers. From *in vitro* experiments it is known that a variety of hydrolytic enzymes, especially those with esterase activity, are able to influence the degradation of synthetic polyesters. It is somewhat more difficult to demonstrate this effect *in vivo*, but a certain amount of circumstantial evidence does exist to show that the hydrolysis is dependent upon the degree of cellular activity which is suggestive, but not demonstrative, that biologically active species are involved.

Susceptibility to this type of degradation will vary with the structural characteristics of the polymers. Crystallinity, molecular weight and polydispersity are among the important features here.

3. Oxidative Degradation

While hydrolysis is likely to be the most important degradation mechanism for the heterochain polymers, it will not be of significance with homochain polymers. These materials, including polyethylene, polypropylene and PTFE, are therefore much more stable, but it is known that they may suffer changes. The ultrahigh molecular-weight high-density polyethylene used in joint prostheses is slowly oxidized. Similarly, if an antioxidant-free polypropylene were implanted, it would suffer oxidative degradation, even at body temperature. There will be some materials that potentially suffer from both hydrolytic and oxidative degradation, including the poly(ether urethane)s.

It is worth noting that extensive oxidation of polymers within tissue fluids at ambient temperatures is not normally thought to be of significance, since this process usually takes place at elevated temperatures and in the presence of atmospheric oxygen. The fact that it does occur in the tissue suggests that a more active oxidation process is occurring and again the possibility of enzymatic attack and the presence of superoxides, peroxides and free radicals may have to be taken into account.

4. Cellular Degradation

It has been shown in Sect. 2 that degradation rates may be affected by various components of the tissue environment. Obviously, there are many different types of cell within this tissue and since those cells with phagocytic ability are normally able to remove debris from the tissue by engulfment and digestion, the possibility of cellular digestion of polymers must be considered. In general terms, it seems unlikely that cells could do much damage to monolithic smooth-surfaced polymers, because they do not have any specific mechanism for their recognition. The situation with particles may be a little different since a macrophage should be able to digest fragments up to 10 μm in size, thereby exposing the particle to concentrated biochemical attack. However, such cells only have a limited lifetime, measured in days or weeks and it may be impossible for sufficient degradation to have occurred before the cell itself dies. There is much evidence to show that cells are able to phagocytose this type of foreign particle (e.g., polymer wear debris, graphite fragments), eliminate them from that site and deposit them, on completion of their life cycle, elsewhere in the body. The lymph nodes or liver are common sites for deposition.

There is, however, very little evidence of polymer degradation taking place directly under the influence of cells. Macrophages have been observed to cause pitting on the surface of silicone polymers by virtue of the peroxides released onto the surface and some degradation of labelled polyester by macrophages *in vitro* has been seen.

5. Bacterial Degradation

If enzymes are able to cause polymer degradation, it might be expected that bacteria could also be involved in this process, especially as much of their activity is mediated by enzymes. It is certainly known that bacteria are capable of degrading some nonproteinaceous macromolecular structures through the action of intracellular enzymes. The question therefore arises as to whether bacteria in contaminated wounds might affect the rate of polymer degradation, this clearly being important with wound-closure materials, but also in other situations. It is quite probable that the degradation of the collagenous material catgut, used as a suture, is modified when it is used in infected wounds where enzymes of both hydrolase and peptidase activity are present and it is a clinical observation that the absorption of catgut takes place more rapidly under these contaminated conditions.

The situation regarding synthetic absorbable sutures is not so clear. Poly(glycolic acid) sutures are degraded faster in infected urine compared to sterile urine. However, the same material degrades much slower in the presence of *strep. mites, e. coli* and *staph. albus* compared to sterile broth conditions. Experiments with *staph. albus in vivo* confirm this finding and suggest that these synthetic absorbable sutures possess a significant advantage over catgut. Bacteria, both aerobic and anaerobic, have been shown to display growth patterns and, especially, far greater survival times, when they are in the presence of polyester substrates, possibly indicating utilization of carbon derived from the polymer in their metabolism.

6. Examples of Polymer Degradation

6.1 Poly(Lactic Acid) and Poly(Glycolic Acid)

These are the best known examples of the poly(α-hydroxy acid)s, of general formula $(OCHRCO)_n$, which are degradable in the body, where R is hydrogen in poly(glycolic acid) (PGA) and R is methyl in poly(lactic acid) (PLA). Since the 1960s these materials, either as homopolymers or copolymers, have been used in clinical situations by virtue of their degradability.

It is difficult to determine from the literature what is the precise mechanism of degradation of these polymers. It is widely believed that chain scission occurs through simple hydrolytic reactions, although species present other than water, such as the various anions and cations as well as enzymes, may influence the kinetics.

When PGA fibers are implanted (in the form of surgical sutures) they may lose strength completely within a short time (e.g., 20 days), but suffer insignificant mass loss during this time. Complete mass loss may not occur for periods of two to three times longer than this, during which time crystallinity changes occur. Typical commercial preparations will initially have a crystallinity of 80% with 7–8 nm ellipsoidal crystallites. Water absorption takes place with characteristics consistent with hydrophilic polymers.

Chu has reviewed the mechanisms of degradation of PGA and PLA in physiological environments (see *Suture Materials*). Also referring to PGA commercial sutures, they describe two distinctive phases of degradation, the first, taking place over a 20 day period, being associated with attacks on amorphous regions and tie-chain segments, free chain ends and chain folds. Some glycolic acid is released in this time, but it is only when all or most of the amorphous regions have been removed by hydrolysis that the water can begin to attack the crystalline regions. As the last of the amorphous regions are lost, the rate of glycolic acid production temporarily decreases (the degree of crystallinity is now at a maximum). The second phase of degradation starts more slowly than the first, because of the greater difficulty of hydrolyzing the crystalline regions, but this rate increases and glycolic acid is then released rapidly.

The rate of degradation is influenced by the number of structural features, environmental variables and testing conditions, including pH, annealing treatments and applied mechanical strain.

A further member of this group of polymers polydioxanone has also been introduced as a de-gradable biomaterial. This has the structure $[OCH_2OCH_2CO]_n$. Its physical properties are different from those of PLA and PGA and, possibly because of the presence of the ethylene oxide unit, the degradation mechanism and kinetics appear to be different as well.

6.2 Poly(β-Hydroxybutyrate)

A further interesting example of an aliphatic polyester that has been considered in the context of biodegradation is poly(β-hydroxybutyrate) (PHB). These polymers $(\text{+}OCH(Me)CH_2\text{+})_n$, where Me is methyl) prepared either as homopolymers or copolymerized with hydroxyvalerate (HV, $\text{+}OCH(Et)CH_2CO\text{+}_n$, where Et is ethyl) have been introduced as potentially useful degradable materials because of their putative degradability in various biological environments.

PHB is a naturally occurring polymer, being a principal energy and carbon storage compound that is synthesized by a range of prokaryotic cells. These energy reserve substances are produced within cells in the presence of excess nutrient, but are broken down during periods of starvation to prepare useful materials from the polymer. Since it is known that bacteria other than those that initially produced the PHB are able to affect this material under some circumstances (e.g., when the polymer is buried in the soil), it has been assumed that degradation will also occur in other biological environments. However, the degradability of PHB and its copolymers in animal or human tissues or related environments is questionable. High-molecular-weight fibers of PHB and PHB–PHV copolymers (up to 27% PHV) do not degrade in tissues or simulated environments over periods of up to six months. This stability is severely influenced if the polymer is predegraded by y-ray irradiation. Material fabricated in different ways may also show different behavior with respect to degradation. Cold-pressed tablets of PHB and PHB–PHV may be degraded, the kinetics of which varies with fabrication route, molecular weight and HV content.

6.3 Aromatic Polyesters: Poly(Ethylene Terephthalate)

Poly(ethylene terephthalate) (PET) is the major aromatic polyester used medically, with extensive application in vascular prostheses. Although apparently clinically successful, failures do occur because of degradation and extensive studies have been reported. The basic degradation mechanisms of PET in acid and alkaline media have been known for some time, with the hydrolytic route shown in this equation usually being observed:

$$-\overset{O}{\overset{\|}{R}}OC- \ + H^+ \longrightarrow -R^+\overset{O}{\overset{\|}{O}}\underset{H}{C} \longrightarrow -R^+ + HO\overset{O}{\overset{\|}{C}}-$$

Water diffusivity into PET is very slow, however, with a diffusivity rate of 10^{-11} m d^{-1} so that degradation is slow and, depending on the configuration, is a surface-dominated phenomenon, with a slow loss of strength of PET in tissues. After eight years, the strength may be reduced to nearly one-half of the original value, but with little loss of substance and no detectable change in molecular weight ($M_v = 2 \times 10^4$). The rate and mechanism of degradation does appear to be influenced by infection, the normally surface-dominated effect changing to one in which internal changes are seen, especially in amorphous regions. It was claimed that this was due to the change in pH and the release of lysosomal enzymes.

A series of failures of Dacron arterial prostheses, mostly arising because of false aneurysms or infection, has been reported. Fiber breakage and thinning were obvious in scanning electron micrographs of these prostheses removed from patients. The strength of these explanted materials varied from 10% to 98% of the original strength.

6.4 Polyamides: Nylons

It has been known for many years that nylon fabrics lose their tensile strength during implantation. Some nylons are both hydrophilic and hydrolyzable, although the extent of the water absorption is variable. Nylon 6 with a water content at saturation of 11% hydrolyzes faster than nylon 11, which has a value of 1.5%.

In the hydrolysis of these polyamides, the primary attack is on the oxygen atom of the carboxyl group:

$$\underset{\text{RCNHR}'}{\overset{\text{O}}{\|}} + H^+ \longrightarrow \underset{\text{RC}=\text{N}^+\text{HR}'}{\overset{\text{OH}}{\|}} \longrightarrow \underset{\text{RCOH}}{\overset{\text{O}}{\|}} + H_2NR'$$

Both acid and amine end groups are formed. The protonation reactions will vary depending on the pH of the environment.

It is also interesting to note the influence of enzymes on the rate of degradation. Smith et al. (1987a) have shown that a radiolabelled nylon 6,6 is degraded *in vitro* by the enzymes papain, trypsin and chymotrypsin, and this polymer is degraded faster in tissues that are inflamed (i.e., containing a significant infiltration of enzyme-releasing cells) than in quiescent tissue.

6.5 Polyurethanes

A wide variety of polyurethanes exists and many of these have been used or contemplated for use in medical applications. Urethane, urea, ester and ether groups may be present and different classifications based on the dominant groupings are available. In biomedical applications, polyurethanes are usually described as either poly(ester urethane)s or poly(ether urethane)s. The former group found the earliest use, but *in vivo* degradation and disintegration became apparent in applications such as bone adhesives and arterial prostheses, and generally little use has been made of them in recent years.

However, poly(ether urethane)s should be more stable and applications have been seen in artificial hearts, pacemaker leads, flexible leaflet heart valves and catheters. While success appears to have been achieved in several of these situations, it is clear that stability cannot be guaranteed and several problems have arisen.

The most significant observations have been with pacemaker lead encapsulation. High failure rates of these leads have been reported and two degradation mechanisms identified which appear to be responsible for fissuring or cracking of the poly(ether urethane) (PEU). First, there is a possibility of metal ion catalyzed oxidation of the PEU, the metal being derived from the lead. Second, the PEU is reported to be susceptible to environmental stress cracking.

In respect of the oxidation, Thoma and Phillips (1987) studied the effects of metal ions and did not find any significant oxidation under *in vitro* conditions and so the role of this mechanism is unclear, even though the literature demonstrates metal–ion complexation and stiffening of the soft segments of the PEU and related effects.

Stokes and Chem (1984) have reported on the phenomenon of environmental stress cracking and demonstrated the significance of residual stress. Material affected by this phenomenon is characterized by crazes with little or no reduction in strength or molecular weight. The lower the ether content of the PEU, the greater the resistance to this cracking; annealing of the materials to reduce residual stresses is also helpful.

Several detailed studies have defined some of the parameters concerning the biodegradation of PEUs. Smith et al. (1987b) prepared a series of radiolabelled PEUs and studied the effects of various hydrolytic and oxidative enzymes on these preparations. The PEU synthesized with a polyether of molecular weight 1000 was affected by esterase, papain, bromelain, ficin, chymotrypsin, trysin and cathepsin, but not by collagenase, xanthine oxidase or cytochrome oxidase. The characteristics of the degradation were very variable, however, and while it was suggested that ethylenediamine was among the degradation products, the mechanism of degradation was difficult to define. Anderson and colleagues (Marchant et al. 1984, Marchant et al. 1986) have made observations on similar PEUs in animal experiments and found specimens to become pitted, with the pit dimensions of 1–2 μm being compatible with the cytoplasmic structures of cells, especially macrophages, that can attach to polymer surfaces.

Ratner et al. (1988) have also studied the enzymatic and oxidative degradation of polyurethanes.

The PEUs and one poly(ester urethane) were studied *in vitro* and degradation was assessed by changes in molecular weight and polydispersity by gel permeation chromatography methods. Varying degrees of both enzymatic degradation (leucine aminopeptidase, papain and chymotrypsin) and oxidative degradation (hydrogen peroxide) were seen, although the amounts involved were low (see *Polyurethanes*).

7. Conclusions

It is clear that the environment of the body can be very aggressive towards polymers. Many suffer hydrolytic degradation, while oxidation may also occur in some cases. The wide variety of aggressive biological species including cells, enzymes, superoxides, peroxides, lipids and so on shows that biodegradation is an extremely important phenomenon in biomaterials science.

Bibliography

Chu C C, Browning A 1988 The study of thermal and gross morphologic properties of polyglycolic acid upon healing and degradation. *J. Biomed. Mater. Res.* 22: 699–712

Holland S J, Jolly A M, Yasin M, Tighe B J 1987 Polymers for biodegradable medical devices. *Biomaterials* 8: 289–95

Marchant R E, Anderson J M, Castillo E, Hiltner A 1986 The effects of an enhanced inflammatory reaction on the surface properties of cast Biomer. *J. Biomed. Mater. Res.* 20: 153–68

Marchant R E, Anderson J M, Phua K, Hiltner A 1984 *In vivo* biocompatibility studies. II. Biomer. *J. Biomed. Mater. Res.* 18: 309–15

Miller N D, Williams D F 1984 The *in vivo* and *in vitro* degradation of poly(glycolic acid) as a function of applied strain. *Biomaterials* 5: 365–8

Miller N D, Williams D F 1987 On the degradation of PHB and PHB–PHV copolymers. *Biomaterials* 8: 129–34

Paynter R W, Askill I N, Glick S U, Guidoin R 1988 The hydrolytic stability of Mitrathane. *J. Biomed. Mater. Res.* 22: 687–98

Ratner B D, Gladhill K W, Howbett T A 1988 Analysis of *in vitro* enzymatic and oxidative degradation of polyurethanes. *J. Biomed. Mater. Res.* 21: 525–30

Smith R, Oliver C, Williams D F 1987a The enzymatic degradation of polymers *in vitro*. *J. Biomed. Mater. Res.* 21: 991–1003

Smith R, Williams D F, Oliver C 1987b The degradation of poly(ether urethane)s. *J. Biomed. Mater. Res.* 21: 1149–66

Stokes K, Chem B 1984 Environmental stress cracking in implanted polyurethanes. In: Planck H, Egbers G, Syre I (eds.) 1984 *Polyurethanes in Biomedical Engineering.* Elsevier, Amsterdam, pp. 243–55

Thoma R J, Phillips R E 1987 Studies of poly(ether urethane) pacemaker lead insulation oxidation. *J. Biomed. Mater. Res.* 22: 509–27

Vinyard E, Eloy R, Descotes J, Buden J R, Guidiecelli H, Magne J L, Patra P, Berruet R, Huic A, Chauchard J 1988 Stability of performances of vascular prostheses: Retrospective study of 22 cases of human implanted prostheses. *J. Biomed. Mater. Res.* 22: 633–48

Williams D F 1985 Physiological and microbiological corrosion. *CRC Crit. Rev. Biocompat.* 1: 1–24

D. F. Williams
[University of Liverpool, Liverpool, UK]

Biomaterial–Blood Interactions

Since 1940, and chiefly since 1950, numerous materials and devices have been introduced into the circulatory system, such as artificial heart valves, circulatory assist devices, totally artificial hearts, oxygenators, liver support systems, renal dialysers, blood vessel substitutes and catheters.

Whenever materials are introduced, severe problems can arise, most of which are related to the interaction between the blood and the devices in use. When blood comes into contact with an artificial surface adsorption and/or degradation of blood components, consumption of coagulation factors, and adhesion and aggregation of platelets induce the generation of microthrombi and occasionally life-threatening thromboembolism. In addition, device dysfunction, impairment of blood flow to distal organs, embolization of thrombus and removal or modification of circulating blood elements are all possible and may be described as "artificial surfaces hemopathy" or "post-perfusion syndrome." Therefore, pharmacological inhibition of the activation of hemostasis must often be employed when cardiovascular biomaterials are used. After heart–lung bypass, for example, the result may be a prolonged incidence of post-perfusion bleeding by 5% to 25%, with a general blood cell count decrease sometimes associated with functional activation and/or impairment. The amount of the coagulation substrate fibrinogen decreases with initiation of extracorporeal oxygenation and levels of fibrinolytic activity are low.

After the insertion of mechanical heart valves, local thrombotic occlusion and thromboembolism caused by the prostheses, associated with the anticoagulant therapy intended to prevent these complications, account for the majority of prosthesis-related deaths. The incidence of thromboembolism with mitral prostheses ranges from 1–11% per patient year. The incidence of significant hemorrhage is 4% per patient year, and up to 10% of these will be fatal. "Blood compatibility is therefore of the utmost importance in determining the performance of devices within the cardiovascular system" (Williams 1987).

The sequence of events leading to thrombus formation when blood comes into contact with an artificial surface is complex due to the many plasma components involved, the nonspecific and competitive nature of protein adsorption, the infinite possibilities for cooperative interactions between proteins and cells at the interface, and the continuously transient state of the interface. Further, physicochemical reactions involved in the adsorption of a mixture of proteins, cells and other nonproteic plasma components remain difficult to analyze and to correlate with the known superficial physicochemical structures of biomaterials. Nevertheless, an understanding of the origin of the surface activity of proteins, the multiple states of adsorbed protein and the competitive adsorption behavior of proteins is a requisite for the description of interfacial phenomena.

1. The Blood–Material Interface

The surface of a biomaterial is an area of transition; chemical constituents of the superficial layer a few nanometers from the surface are responsible for the forces accounting for the interfacial phenomena of surface tension, adsorption and adhesion.

Surface energy can be used to classify materials approximately and correlates directly with the nature and concentration of superficial chemical groups. Contact angle measurement of a drop of fluid deposited on the surface is a classical method of determining surface energy (γ_s) which is, for example, 17×10^{-6} J cm^{-2} for glass, 100×10^{-6} J cm^{-2} for pure metals and $1.5–5 \times 10^{-6}$ J cm^{-2} for polymers. Values of γ_s may be enhanced by the introduction of polar groups (OH, Cl, CN). Relative wettability based on contact angle measurements has been used to predict blood compatibility of materials but such a single criterion is inadequate.

The first occurrence is the adsorption of inorganic ions and water molecules. On contact with water the superficial structure of hydrophobic materials is not changed: it is the structure of the water itself that is modified. An intermediary layer a few molecules thick in which the molecules have a more ordered structure than in pure liquid water is formed by a dynamic process with constant rearrangement of ordered microdomains. However, the surface of hydrophilic materials is greatly modified by water. Water molecules become enmeshed in the macromolecular network and break certain interactions between the polymer chains.

2. Protein Adsorption: An Early Event

Proteins adsorb to any surface rapidly and selectively to form a layer about 100 nm thick. Adsorption of proteins and cells on the surface of biomaterials occurs in three steps: transport to the interface, the reaction of adsorption, and the conformational rearrangement of proteins and changes in shapes of the adherent cells. The kinetics of adsorption are dependent on the rate at which the first two steps occur. When a wet biomaterial comes into contact with a flowing solution of protein an intermediary layer is formed, which does not contain protein. The transfer of proteins to the surface through the intermediary layer is due to molecular diffusion which may be supplemented by convective transport in flow situations.

The most obvious means of binding protein to a surface is via an electrostatic bond formation between charged groups on the protein and oppositely charged surface sites. On glass, electrostatic adsorption is high, shows considerable reversibility and desorption occurs in response to electrolytes. Other possible protein–surface interactions are hydrogen bonding, probably for relatively polar substrates, and β-turns. These areas of the protein where the polypeptide chain folds back on itself by nearly 180° lie in the surface region of the protein and have been incriminated in adsorption. However, the most important binding mechanism is probably the hydrophobic reaction defined as the interaction of nonpolar groups in aqueous media. Proteins that contain hydrophobic patches on their surface are able to interact with hydrophobic-contacting surfaces. Proteins probably bind in this way to surfaces such as polyethylene, silicon and certain segmented polyurethanes. On such surfaces compact monolayers are formed, there is no desorption and only a slow partial exchange between adsorbed and dissolved protein occurs. The character of the surface, whether it is hydrophobic or hydrophilic, charged or neutral, polar or nonpolar, determines the degree to which any specific process can occur. Quantities adsorbed are generally greater for hydrophobic than for hydrophilic substrates, where desorption or ion exchange may take place.

The spontaneity of protein adsorption to artificial surfaces is promoted by:

(a) their size, since larger molecules may have multiple contact points with a surface (although this is not a determining factor);

(b) their charge, since molecules nearer their iso-electric pH may adsorb more easily;

(c) their structure, since more stable molecules, and covalent and sulfur–sulfur bridging reduce surface activity; and

(d) their chemical properties, related to their amphipathic nature, hydrophobicity and low solubility resulting from their high molecular weight.

The composition of the protein layer which has a controlling influence on the subsequent biomaterial interactions depends on the concentration of proteins, on their relative mobility in plasma and on their affinity for the surface. The smoothness of the surface and the heterogeneity of the distribution of reactive groups are partly responsible for the selectivity of composition of the adsorbed plasma proteins.

Adsorption is relatively rapid, but variable, being as much as 75% complete in 5 min and it reaches an equilibrium in about 1 h. The amount of fibrinogen adsorbed is independent of surface shear rate from $0 s^{-1}$ to $2100 s^{-1}$ (transport is not the controlling step under these conditions). Real-time studies have shown that in adsorption to quartz from an equimolar solution of bovine serum albumin (BSA) and γ-globulin, BSA reaches an equilibrium in 8 s and γ-globulin in 40 min.

The isotherms of adsorption are generally of the Langmuir type. Most protein surfaces exhibit a limited surface concentration, and adsorption isotherms showing surface concentration as a function of solution concentration usually approach a plateau beginning at a moderate solution concentration and reaching the concentrations found in blood. Although many studies used this simple Langmuir model to describe protein adsorption, more recent models suggest that adsorbed proteins may exist in more than one state. The mechanisms that could lead to multiple states of adsorbed proteins include the "occupancy" effect owing to the unavailability for adsorption of surface sites that are already occupied, the conformational changes before, during or after adsorption and the multiplicity of binding models due to amphipathicity of the molecules and the multiple site nature of the surface.

Biological indicators suggest the existence of different states of adsorbed proteins; for example, enzymes such as thrombin may lose some of their activity and proteins such as fibronectin and fibrinogen will exhibit different reactivity towards subsequent cell events. Conformational changes of protein molecules during the adsorption–desorption process may be either positive (surface passivation with an irreversible adsorbed protein layer) or negative (initiation of blood-enzyme systems, in particular the activation of the coagulation and complement system because of the desorption of the surface-activated protein components).

Adsorption is most often described as being irreversible on an ordinary time scale on hydrophobic solid surfaces while for hydrophilic surfaces both reversibility and irreversibility have been found. While desorption is either nonexistent or extremely low, there is a measurable exchange of proteins between the surface and the solution as has been clearly demonstrated by the use of labelled proteins. The extent of such an exchange varies with the surface, being greater for hydrophilic than for hydrophobic surfaces, and depends on protein concentration and flow. The rates of exchanges also vary from days (polyethylene–albumin) to hours (glass–fibrinogen) exhibiting slow-exchanging and fast-exchanging population of proteic molecules.

Although the amounts adsorbed at the plateau vary from one surface to another, the surface concentrations at these plateaux are of the order of $0.1–1.0 \mu g \, cm^{-2}$ and correspond to monomolecular layers. Proof of the existence of multilayer adsorption has been obtained for a number of polymers. Apparent exceptions to the "monolayer rule" are hydrogel-like materials such as segmented polyurethanes which essentially adsorb no protein, but the rapid reversibility of adsorption on these materials has to be considered.

2.1 Adsorption from Solution of Single Proteins

Protein adsorption has been studied experimentally by measuring the adsorption isotherms of single purified proteins involved in hemostasis and thrombus formation (fibrinogen, fibrin, factor XII, high molecular weight kininogen (HMWK), factor VIII (Von Willebrand factor, VWF) and thrombin), in inflammatory and immunological reactions (immunoglobulin G, immunoglobulin M, immune complexes, C5a and C3a), and in cell adhesion (albumin, fibronection and collagens).

Measurements of equilibrium layer thickness, performed on a fibrinogen–glass system, support the fact that adsorption occurs at the monolayer level. The molecules in the layer do not seem to be drastically altered since their average dimensions are similar to their dimensions in solution.

For the proteins studied, as the isoelectric pH increases so does the adsorption of the proteins. Until the early 1990s, there have been no attempts to determine the properties of protein–surface systems that govern the relative surface affinity of different proteins. On a hydrophobic substrate, increased surface concentration of the four proteins tested (immunoglobulin G, immunoglobulin M, human serum albumin and α_2 macroglobulin) correspond . to increased protein hydrophobicity. For polyethylene, plasma protein affinity varies as follows:

hemoglobin $\gg$ fibrinogen $>$ albumin $\simeq \gamma$-globulin

Small proteins, by virtue of their higher diffusion coefficients, will initially adsorb faster than large proteins during competitive adsorption. The relative extent of protein adsorption onto polymer surfaces is influenced by the surface tensions of the substrate material, the suspension liquid and the proteins themselves. The more hydrophobic proteins will adsorb to the largest extent with one substrate material. Protein adsorption changes the surface properties of the "naked" surface markedly, and most polymers are rendered more hydrophilic as a result of protein adsorption.

Mixture studies using a two- or three-protein system on a variety of surfaces with various mixture compositions and various total concentrations demonstrate that fibrinogen is preferentially adsorbed when it is in solution with albumin and immunoglobulin G. For glass, and for several other surfaces, the surface enrichment of fibrinogen (between 10- and 200-fold) has been demonstrated. However, many of these measurements refer to equilibrium and involve long adsorption times; the relative quantities adsorbed in multiprotein systems are relatively time independent. One of the few studies showed that, in a one-to-one mixture of albumin and fibrinogen, albumin predominated in the first seven minutes and was then gradually displaced by fibrinogen. The faster diffusion of albumin will delay but will not halt the subsequent adsorption of fibrinogen.

In a multiprotein system the competitive adsorption behavior of proteins at interfaces is of great importance. The factors influencing competitive adsorption may be, in order of importance, the chemical surface properties (in terms of charge, hydrophobicity and chemical functional groups) and the protein properties (such as electrical charge, hydrophobicity, available chemical functional groups, stability of the protein structure, interactions between proteins in the adsorbed layer, relative concentration in the bulk phase and molecular size). The problem of how to combine all of the possible factors into a quantitative model of competitive adsorption is complex since data for such models are unavailable.

2.2 Adsorption of Proteins from Plasma

Since blood is a multicomponent system where the influence of one protein on another is an important factor, only studies using blood or plasma can provide definitive answers regarding blood–material interactions.

In general, it has been found that preferential or selective adsorption occurs so that certain proteins may be enriched in the surface and vice versa. Nevertheless, the currently available data concern only major plasma proteins (fibrinogen, albumin and immunoglobulin G) and proteins involved in hemostasis and cell adhesion. Surface chemistry, time of adsorption and protein type are major factors in determining the composition of the adsorbed layer. This layer is formed from the mixture and thus contains a rather complex and changeable combination of proteins.

Adsorption of proteins from plasma to various hydrophilic–hydrophobic copolymers showed that nine or more protein peaks can be observed after gel electrophoresis of the sodium dodecyl sulfate (SDS) eluates from the surfaces. The adsorption of each protein varies in a characteristic way with copolymer composition.

A typical hydrophilic surface like glass shows not only some albumin, immunoglobulin G and fibrinogen and some plasminogen, but also some fibrin

degradation products (FDP), suggesting surface activation of plasminogen, which is even reduced when plasminogen-deficient plasma or factor-XII-deficient plasma is used. Contact factors of blood coagulation are known activators of the fibrinolytic system but many other still unidentified species are also present in the glass eluates. Activation of fibrinolysis attested by the formation of FDP is a common feature of all hydrophilic materials.

On hydrophobic materials such as polystyrene, fibrinogen is the major constituent of the layer. FDP are also observed but in smaller amounts than for glass. The eluates are in general less complex than those for glass and this reflects the increased difficulty of eluting proteins from hydrophobic surfaces.

On polyethylene and siliconized glass, fibrinogen adsorption is greatest at 2 min and then decreases to zero. Immunoglobulin G is detected on all surfaces though in relatively low surface concentrations. More polar polymers have low initial adsorption *in vitro* which increases with time, but *in vivo* fibrinogen deposition is characterized by a second stage of greatly increased adsorption that can be inhibited by heparin. These polymers also cause enhanced platelet consumption. Gas discharge treatment on (Dacron) poly(ester terephthalate) reduces the initial adsorption of fibrinogen.

For hydrogels the most frequent observation is a quantitative reduction in the amount of a particular protein adsorbed to the hydrogel compared with the hydrophobic surface. Desorption and exchange studies indicate that proteins are less tightly bound to hydrogels than to nonhydrogels. A qualitative difference in protein adsorption is the apparent preference of certain proteins, especially albumin, for hydrogel surfaces.

Much work remains to be done to establish the exact composition of proteins adsorbed from plasma and blood, particularly for trace proteins. The major constituents of the adsorbed protein layer are fibrinogen, γ-globulin and albumin. Other proteins adsorbed in smaller quantities include VWF, fibronectin, α- and β-globulins, transferrin, caeruloplasmin, the coagulation factors XI and XII, and HMWK.

The spectra and amount of protein adsorbed vary with time. The adsorption of hemoglobin and immunoglobulin G to different polymers increases with time although the degree of adsorption depends on the polymer, whereas fibrinogen and albumin adsorption increase initially with time but then decrease.

Enrichment of the proteins, calculated as the ratio of the weight fraction of each protein in the surface to the weight fraction of the protein in the bulk, varies with the polymer composition and is for fibrinogen 0.8–3.4, for immunoglobulin G 0.8–1.5, for albumin 0.3–0.8 and for hemoglobin 150–400.

Thus, the surface enrichment relative to the bulk phase is not very great for immunoglobulin G, albumin or fibrinogen but hemoglobin, which is not considered as a normal plasma constituent (with a concentration of 0.003 mg ml^{-1}), is an exception to this rule.

Generally, however, plasma proteins appear to adsorb in proportion to their bulk concentration as might be expected from mass action. Since the major plasma proteins are only minimally adsorbed on surfaces that are known to adsorb compact monolayers of proteins from single solutions, it has been suggested that in the plasma there are trace proteins that are strongly surface active and that these are adsorbed in large quantities or that the plasma itself restricts adsorption. The ability of hemoglobin to inhibit fibrinogen adsorption depends on the ratio of the proteins but it is also strongly dependent on the total concentration at which the competition takes place. For fibrinogen the maximum adsorption occurs at concentrations in plasma of 0.1% for glass, 1% for polyethylene and about 10% for polytetrafluoroethylene (PTFE). Both time and concentration effects on competitive adsorption of fibrinogen from hemoglobin solution are consistent with a mass action model for competitive effectiveness. Dilution of the plasma reduces the rate of adsorption of any protein as well as its rate of replacement by another protein and this is referred to as the Vroman effect, manifested either as a maximum in fibrinogen adsorbed to a surface as a function of time or as a maximum in fibrinogen adsorbed as a function of plasma concentration. The residence time of fibrinogen on the surface varies with the material. Fibrinogen is displaced within 2 s of contact by HMWK and is also displaced to some degree by factor XII, often within 1 min. Since there is no change in the protein film thickness and since fibrinogen is no longer reactive to antifibrinogen antiserum, it is assumed that fibrinogen is not covered but replaced by other proteins.

The competitiveness of protein adsorptions in the context of blood led Vroman (1987) to postulate that a rapid sequence of adsorption and displacement occurs with time, by which more abundant proteins are displaced by less abundant ones. Albumin is adsorbed and displaced in a fraction of a second (it is not often observed on a surface after plasma contact), and the absence of the Vroman effect for hydrophilic polyurethanes may reflect a very rapid sequence. This competitive adsorption is probably controlled not only by concentration but also by protein affinity for the surface (free energy of adsorption), protein–protein interactions in the layer and kinetics factors. It must be stressed that all studies of the Vroman effect were performed under static conditions.

The replacing factors are HMWK, kallikrein, factor XIa and/or high density lipoproteins. On activating surfaces the sequence appears to be albumin followed by the immunoglobulins, fibrinogen and fibronectin, and lastly HMWK. Plasmin does not affect the Vroman effect which is, however, dependent on an unidentified protein that binds tightly to lysine-agarose. All the surfaces studied except for some hydrophilic polyurethanes displayed the Vroman effect. However, quantitative differences (peak height, peak position) were found among surfaces, and there is some evidence that the height of the peak in the adsorption versus plasma concentration relation is correlated with gross thrombogenicity. It is at present difficult to understand the Vroman effect in relation to surface thrombogenesis, since contact phase activation occurs where fibrinogen is displaced from the surface whereas platelets will adhere only where fibrinogen remains on the surface.

2.3 Consequences of Protein Adsorption on Blood–Material Interactions

It is now accepted that when albumin layering occurs in preference to fibrinogen layering, thromboresistance is favored since albumin minimizes platelet adhesion and thrombogenesis. On the contrary, if fibrinogen and γ-globulin are preferentially adsorbed thrombosis may ensue, causing enhanced platelet interactions and platelet activation. Fibrinogen is also believed to interact with leucocytes. γ-globulin coatings on the surface not only enhance platelet and granulocyte adhesion but they also stimulate platelet release. After the initial layer of plasma proteins is adsorbed onto a surface, a crosslinked fibrin network may develop over a period of minutes or hours. The formed blood elements will be entrapped within this fibrin network. If the rate of thrombus formation is slow, the process of fibrinolysis may prevent gross thrombus formation.

2.4 Fate of Adsorbed Proteins

The local concentration of adsorbed proteins is extremely high (about 1000 mg ml^{-1}) under a variety of adsorbed states (but the structure of adsorbed protein is not well understood, including whether proteins are denatured at the solid–liquid interface) thus modifying cellular interactions controlled by the presence of specific proteins on the foreign surface at a sufficiently high surface density and degree of reactivity.

The organization of adsorbed protein films suggests that on polyvinylchloride (PVC), polytetrafluoroethylene (PTFE) and polyvinyldifluoride (PVDF) bare areas of polymer remain after adsorption, and the degree of coverage of the polymer surface with adsorbed proteins increases with increasing critical surface tension of the polymer. Protein films form in "islands." Albumin exhibits a globular surface deposition pattern with low coverage of the surface, influenced by fluid shear rate. In contrast, the glycoproteins, fibrinogen, γ-globulin and plasma fibronectin exhibit a more extensive coating of the surface with a characteristic reticulated pattern. The extent of surface coverage depends on the roughness of the surface; preferential binding of fibrinogen and, possibly, fibrinogen clusters is observed in surface cracks of the order of 1 μm. These results obtained by partial gold decoration transmission electron microscopy support the finding that smooth surfaces are less thrombogenic than rough ones.

Fibrinogen molecules are visible, and side-on binding configurations as well as side-to-side and end-to-end alignments are observed. Since the recent understanding of the activity of the fibrinogen D region in the end-on configuration, which is more reactive and prone to interact with cell membrane receptors, the side-on configuration of fibrinogen observed for hydrophobic surfaces suggests less reactivity. Albumin pretreatment significantly reduces the total amount of protein visualized on the surface. The conformational changes of protein are observed to a greater extent on hydrophobic than on hydrophilic surfaces.

Denaturation of adsorbed proteins is another matter for debate. Some systems have shown evidence of denaturation and others have not. For example, factor XII adsorbed on quartz was shown to undergo a conformational change while thrombin adsorbed on cuprophan and PVC retained most of its biological activity. Denaturation should be evaluated on the surface, which would be the best method but a difficult aim to achieve, or on eluted proteins. Fibrinogen eluted from glass tubing was examined by circular dichroism and a loss of alpha-helix content of the order of 50% relative to the native fibrinogen was observed. However, a different behavior of fibrinogen molecules initially eluted (less firmly bonded) with considerable chain degradation and of later-eluting fractions hardly different from "native" fibrinogen has been reported. The gel band patterns of the degraded fractions resemble those of early plasmin-induced fibrinogen degradation products (FDP). Plasminogen activation and subsequent fibrinogen partial degradation are non-negligible components of blood–material interactions.

3. Coagulation Activation

When normal plasma comes into contact with negatively charged surfaces, such as glass, kaolin or connective tissue, the intrinsic pathway of blood coagulation is activated.

3.1 Proteins Involved in the Contact Phase

The proteins that have been identified as being involved in the contact activation reaction are factor XII, prekallikrein, HMWK and factor XI.

(*a*) *Factor XII*. Human factor XII is a single-chain glycoprotein with a molecular weight of about 80 000 D, and it is almost entirely in the zymogen form. Conversion of the zymogen into an active serine protease is accomplished by cleavage of a single Arg–Val bond, producing the so-called α-factor XIIa possessing amidolytic and esterase activity. The three domains recognizable in the molecule are the amino-terminal domain (40 000 D), which contains the structural information for the binding of factor XIIa to negatively charged surfaces, followed by a 12 000 D connecting region and by the 28 000 D light chain, which contains the active site.

A number of additional peptide bond cleavages can occur in the human α-factor XIIa, in which case the amino-terminal region is lost, and the resulting enzyme β-factor XIIa (or Hageman factor fragment) lacks surface-dependent coagulant activity but it does contain the active site.

(*b*) *Factor XI*. This is an α-glycoprotein that contains two identical glycoproteins (80 000 D each) that are held together by disulfide bonds. Factor XI is present in plasma in a complex with HMWK and it is activated to factor XIa by limited proteolysis by factor XIIa.

Factor XIa is capable of propagating the intrinsic coagulation pathway by activating factor IX and it is also capable of proteolytically activating factor XII, factor VII (involved in the extrinsic pathway) and plasminogen. Factor XIa is mainly inhibited by α_1-antitrypsin. It can also be inhibited by antithrombin III, and heparin stimulates this inhibition.

(*c*) *Plasma prekallikrein*. Prekallikrein is a glycoprotein that exists as a single polypeptide chain with a molecular weight of approximately 80 000 D. It is a serine protease zymogen that is activated by limited proteolysis by factor XIIa. Kallikrein exhibits homology to factor XIa. As a protease, kallikrein is capable of liberating kinins from kininogens, of activating factors XII (particularly surface-bound factor XII) and IX, and of activating plasminogen. Inactivation of plasma kallikrein is accomplished by the C1 inhibitor and α_2-macroglobulin; less potent in this inhibition is antithrombin III.

(*d*) *High molecular weight kininogen*. HMWK contains approximately one-fifth of the kinin content of plasma and it exists as a single protein chain with a molecular weight of approximately 105 000 D. It is a nonenzymatic cofactor that is central to contact activation reactions. It has been shown to contain a most unusual region of amino acid sequences that is rich in histidine, lysine and glycine. This region is essential for the contact activation cofactor activity, and it is tempting to speculate that this highly positively charged region of the molecule is responsible for the critical binding of the molecule to negatively charged surfaces.

3.2 The Contact Activation Reaction

Data have been accumulated during the 1980s to explain the contributions of negatively charged surfaces to the activation of factor XII and to the expression of the activities of factor XIIa. Available evidence indicates that the three major roles of negatively charged surfaces are:

(a) to induce a structural change in factor XII such that it becomes highly susceptible to proteolytic activation,

(b) to promote HMWK-dependent interactions between factor XII and prekallikrein that result in reciprocal proteolytic activations of each molecule, and

(c) to promote the HMWK-dependent activation of factor XI by surface-bound α-factor XIIa.

However, the actual trigger event that initiates contact activation, after the exposure of a suitable surface on which the activation of factor XII, prekallikrein and factor XI can take place, is a catalytic activity responsible for the initial conversion of these contact activation zymogens into active serine proteases. Four different hypotheses have been put forward for the initiation of contact activation reactions once a negatively charged surface is exposed.

(a) The zymogens, factor XII and prekallikrein may assemble on the negatively charged surface and slowly generate the first active enzymes that subsequently account for the burst of activity.

(b) The interaction of surface bound factor XII with its natural substrates (factor XII and/or prekallikrein) may result in the expression of full proteolytic activity toward these molecules without requiring the formation of two α-factor XIIa chains.

(c) Low levels of factor XIIa and/or kallikrein may be permanently circulating in the blood.

(d) The protease not involved in the extraneous enzyme (coagulation system) may be responsible for the initiation of contact activation reactions.

3.3 Interactions Between Coagulation Contact Activation and other Biological Systems

Many interactions have been demonstrated between coagulation, fibrinolysis and the kallikrein–kinin complement systems during *in vitro* studies. However, the *in vivo* significance of many of the these recorded observations is not established. Clearly, however, the potential for interactions between these and other systems in the control of physiological and pathological processes, such as thrombus formation on artificial surfaces, is considerable.

4. Complement Activation

The complement activation process is similar to the clotting system of blood in that restricted proteolysis of some of the factors is one of the main regulation mechanisms and activation is of the cascade type. The system consists of more than 19 components and regulatory plasma proteins. C3 is the quantitatively dominating complement protein and it has a central position in the complement activation cascade. Most of the methods used are based on the analysis of complement degrading factors but some of these factors, especially C3, are also deposited on the solid surface. Contact of blood with an activating surface triggers complement activation through one of two pathways: the classical pathway or the alternative pathway. These cleave the third component of complement C3 and generate the anaphylatoxin C3a and a major cleavage fragment C3b. An amplification pathway consisting of alternative pathway proteins augments C3 cleavage once C3b has been generated. A common effector sequence generates the vasoactive, chemotactic, leucocyte-stimulating immune regulatory and cytolytic activities of the activated complement. The classical pathway, initiated by the binding activation of C1 to the modified F_C part of the immunoglobulin G– or immunoglobulin M–antigen complex, can also be induced by the surfaces of viruses, by lipopolysaccharides, by certain artificial surfaces or by immunoglobulin G adsorbed on artificial surfaces. The alternative pathway represents a recognition mechanism for foreign surfaces in a nonimmune host.

The alternative pathway is activated by surfaces which exhibit specific biochemical characteristics allowing bound C3b to initiate the assembly of the amplification C3 convertase on that surface. The activation of the alternative pathway is as follows:

(a) binding to the activating surface of C3b molecules formed continuously in plasma,

(b) Mg^{2+}-dependent binding of factor B to surface-bound C3b,

(c) cleavage of B within the C3b, B complex by a serine protease factor, and

(d) formation of C3 by the surface-bound C3b, Bb and deposition of additional C3b.

Complex C3b, Bb rapidly loses its convertase activity by spontaneous dissociation (half-life of 3 min at 37 °C). This dissociation of Bb is retarded by P which binds to C3b and increases by five-fold the half-life of the convertase. However, the enhancing effect of P is counterbalanced by H which binds to C3b and impairs the binding of B to bound C3b, actively dissociates Bb from C3b, Bb and facilitates proteolytic inactivation of bound C3b by I, yielding C3bi, an inactive form of C3b that cannot bind B.

The competition between H and B for binding to surface-bound C3b discriminates between activating and nonactivating surfaces. On a nonactivating surface fixed C3b binds H with an affinity almost 100-fold greater than that with which it binds B. On an activating surface C3b binds H less effectively, and C3b is relatively resistant to inactivation by H and I. A protected formation and the expression of the amplification convertase on an activating surface result in enhanced C3 cleavage and the deposition of additional molecules of C3b.

Covalent coupling of heparin, and heparin in the fluid phase, inhibit alternative pathway activation. The site of the molecule which is responsible for the anticomplementary effect of heparin is not the same site involved in binding antithrombin III. With increasing amounts of C3b, molecules are deposited on a surface, binding of C5 leads to C5 cleavage by both the alternative and classical pathway convertases, and cleavage of C5 releases the anaphylatoxin C5a and a major fragment C5b which will give rise to the subsequent cytolytic complexes C5b–9. Complement activation also liberates the anaphylatoxins C3a, C5a and C4a which are of similar molecular structure and have similar biological properties, such as contraction of smooth muscle cells, chemotaxis for neutrophils, eosinophils and monocytes, histamin release, oedema, and granulocyte aggregating ability.

The interactions between C5a and specific high affinity receptors on neutrophils (2×10^5 receptors) play an important role in the deleterious effects of the activation of the complement system. The embolization of granulocyte aggregates leads to leucostasis, the increased expression of C3 receptors on the cells, lysosomal enzyme release, the generation of toxin oxygen products, damage to endothelial cells, and the enhancement of granulocyte adhesion to endothelial cells which results in increased capillary permeability. C3b on activating surfaces may induce the release of lysosomal enzymes or derivatives from arachidonic acid metabolism and trigger superoxide formation from phagocytic cells.

Both the deposition reaction from serum and the subsequent reaction with C3 are smaller on hydrophilic surfaces than on hydrophobic surfaces. It is likely that the interaction of C3 with the surface is one of the recognition mechanisms that leads to surface-induced complement activation. C3 undergoes conformational changes on the hydrophobic surface similar to SDS denaturation or biological activation. The affinity of H for surface-bound C3b is also influenced by biochemical characteristics of the particle surface. A nonactivating surface of the alternative pathway should either be incapable of covalently binding C3b, possess a surface with preferential binding of H over that of B to surface-bound C3b, or express few antigenic sites recognized by natural or acquired antibodies.

5. Blood Cell–Surface Interactions

5.1 Platelet–Surface Interactions

The adhesion and aggregation of platelets are an invariable accompaniment of the exposure of blood to an artificial surface (see Fig. 1). The adhesion is linked to protein adsorption and the interaction of platelets with adsorbed fibrinogen or γ-globulin has been attributed to the formation of a complex between glycosyltransferases located in the platelet membrane and incomplete heterosaccharides in the protein layer. The adhesive proteins for platelets should be present both in the plasma and in secretory pools within the platelets. In this light, the α-granule of the platelet contains several candidates: fibronectin, VWF, fibrinogen and thrombospondin.

It is accepted that if a polymer or glass surface is precoated with albumin platelet adhesion is less than if the platelets were exposed to the bare polymer or glass surface, and it has been claimed that platelet adhesion to surfaces preadsorbed with proteins parallels the saccharide content of the adsorbed proteins. Fibrinogen precoating even at low levels causes increased platelet adhesion of about 40-fold that of albumin coating and lies in the same range as γ-globulin and fibronectin. Greater platelet adhesion than with fibrinogen is obtained with preadsorption of VWF, collagen and thrombospondin. An α-globulin coating on an artificial surface not only enhances platelet adhesion but it stimulates the platelet release reaction as well.

However, platelet adhesion does not correlate with fibrinogen adsorption to glass pretreated with a series of plasma dilutions. Other plasma proteins binding fibrinogen or acting in concert with fibrinogen may be important in influencing platelet adhesion. Alternatively, the state of the adsorbed fibrinogen may be different at different plasma concentrations, that is, how fibrinogen is adsorbed rather than how much is adsorbed may be most important. Platelet and macrophage responses do not correlate with the amount of adsorption of proteins for which they have receptors. *In vitro* platelets do not adhere to "converted" (i.e., having lost its antigenicity) fibrinogen, so the shorter the time for transition of native to converted form, the lower the probability of platelet adhesion.

Following platelet adhesion, the platelet release reaction takes place in the adhering platelets, and platelet aggregation then occurs on the surface. Among the different platelet constituents released, serotonin is less dependent than platelet adhesion on surface properties and it is different from platelet adhesion in that the amount of serotonin released increases with hydrophilicity.

The release of B thromboglobulin, platelet factor IV and thromboxane B2 (TXB2) is related to the implantation of prosthetic valves used in cardiopulmonary bypasses. Both platelets and white cells

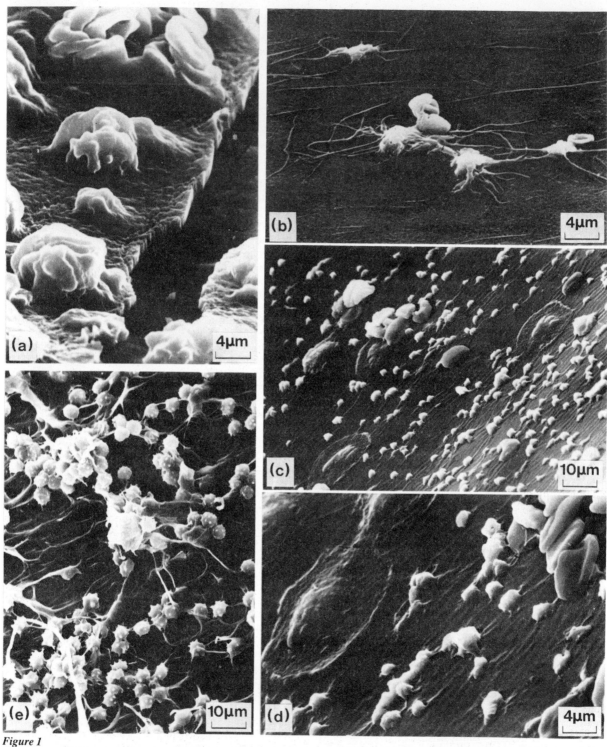

Figure 1
Blood–material interactions: (a) platelet adhesion on a proteic layer deposited on the material; (b) platelet spreading and pseudopodes emission in an early interaction; (c) and (d) large mononuclear cells spreading and platelet adhesion; and (e) thrombus formation with fibrin mesh development and erythrocytes engulfment (by permission of Editions Medicales Internationales, Paris)

(monocytes and granulocytes) liberate tissue factor III (FTIII), which will then activate the intrinsic coagulation system. As blood clotting on the artificial surface proceeds, an interaction between platelets and the intrinsic pathway is likely. Thus, the generation of thrombin causes the parallel platelet release reaction and aggregation. Platelet aggregation leads to the secretion from the alpha granules of thrombospondin which is then partially bound to the platelet membrane and may play an important role in mediating platelet aggregation on artificial surfaces.

Hemodynamic conditions are of major importance in the adhesion of platelets to the vessel wall and in determining localization, growth and fragmentation of thrombi. Platelet adhesion to a surface is governed by two independent mechanisms: the transport of platelets to the surface, which depends on the flow conditions, and the reaction of platelets with the surface, which depends on the nature of the surface and of the adsorbed protein layers.

Platelet response following the contact of blood with an artificial surface is also influenced by diffusion and shear forces. Single cells nonuniformly distributed on the surface appear in conditions of low flow rate (0.3 ml s^{-1}) while multiple cell aggregates form at higher shear rates. The adhesion and diffusion of platelets onto surfaces are markedly increased by the presence of red cells until the hematocrit reading reaches about 40%.

Taking into account the multitude of tests performed, no firm conclusions can be drawn as to which materials are the most passive to platelets although, in general, polyurethanes, polyethylene oxide, silicone rubber and PTFE tend to be "good" materials. Materials recognized to be reactive to platelets include polystyrene, PVC, polyethylene and polymethylmethacrylate. Platelet adhesion and activation by polyalkylacrylates have been found to increase with the length of the alkyl side chains and the resultant hydrophobicity. It has been suggested that clusters of surface-bound fibrinogen molecules with minimal conformational alterations may provide a stimulus for platelet activation by surfaces. The alteration of surface-bound protein through adherence and detachment of platelets is one way in which cells can alter a protein substrate.

5.2 Red Cell–Surface Interactions

Red cells, the most numerous cells in blood, do not adhere to endothelium, but they can bind weakly to some nonendothelial materials without spreading. They contribute significantly to blood–surface interactions in several ways. Rheological effects of red cells have been described. Collision with red cells increases the force of diffusion which brings other blood cells and plasma proteins towards the vessel surface. The addition of erythrocytes to protein solution reduces the amount of protein adsorbed.

Red cells do not interact directly with biomaterials to a great extent, but they do promote the adhesion of other blood cells, particularly platelets, due to the hemodynamic behavior of red cells, a reduction in the adsorption of platelet protective proteins and even the deposition of an adhesive substance by red cells.

The function and morphology of red cells can be affected by repeated contacts with artificial materials. First, the turbulence associated with prosthetic and extracorporeal devices can produce the shearing stresses necessary to damage red cells with the loss of ADP (a powerful vasoconstrictor and platelet aggregating agent) and of a procoagulant material. Both high shear stresses (300 Pa) and low shear stresses (150 Pa) cause overt hemolysis. The hemoglobin level released from red cells *in vitro* is related to shear stress, time and the ratio of the surface area of contact to the blood volume and varies with different plastic surfaces. Although controverted, the nature of the material may be most important in low shear stress conditions. Second, the impairment of binding sites of C3b and PGI$_2$ (a platelet antiaggregating agent), acting effectively as inhibitors of C3b and PGI$_2$, can lead to significant complications for complement activation and platelet deposition.

Exposure of blood to biomaterials does not usually have measurable effects on red cells but occasionally, during the use of cardiac assist devices, serious anemia and red cell fragmentation can occur. Low-grade hemolysis associated with cardiopulmonary bypass circuits may contribute to platelet fibrin deposition and to the post-perfusion syndrome.

5.3 Leucocyte–Surface Interactions

Leucocytes, particularly neutrophils and monocytes, have a strong tendency to adhere to surfaces; they react with platelets, complement, coagulation and fibrinolytic systems, and they are major mediators of the inflammatory response. The half-life of neutrophils in the blood is about 7 h, irrespective of the age of the cells. Many paths lead to white cell adhesion and aggregation; kallikrein formation and complement activation, such as the formation of C5a, can be expected to link the process of surface-activated clotting to that of immune adhesion. Immunoglobulin G deposited onto hydrophobic rather than onto hydrophilic surfaces will irreversibly bind one or more complement components, thus localizing chemotactic activity at the surface and causing granulocyte adhesion and spreading.

As a result of leucocyte adhesion, several reactions are initiated. These include platelet–platelet and platelet–leucocyte interactions, the detachment of adherent thrombi by the action of leucocyte proteases, the detachment of adherent platelets and adsorbed proteins by leucocytes, the release of leucocyte products (e.g., leukotriene B4) which may give rise to both local and systemic vascular reactions, inflammatory reactions which are leucocyte dependent, and the promotion of fibroblast ingrowths into prosthetic material.

Further, chemoattractants derived from platelets or released during blood coagulation contribute to the adherence of leucocytes to the edge of developing thrombi. Significant quantities of procoagulant activity are generated, at least by the provision of tissue thromboplastin. Leucocytic proteases can also initiate intrinsic coagulation and cleavage of C3. Sensitized basophils and macrophages can elaborate the platelet activating factor, which is a potent inflammatory agent.

Fibronectin and immunoglobulin G preadsorptions promote macrophage attachment while all other proteins (albumin, hemoglobin, fibrinogen) decrease or prevent attachment. Receptors for fibronectin and immunoglobulin G on macrophages are probably responsible for the effects of these proteins on attachment.

5.4 The Adhesion of Blood Cells to Artificial Substrates

It is now accepted that blood cell adhesion to physiological substrates (e.g., collagen, immunoglobulin G, endothelial cells and bacteria) is mediated by cytoadhesins, a family of molecules present in cell membranes or penetrating the cytoskeleton (integrins). These molecules are membrane receptors of glycoproteic nature (glycoprotein Ib, IIb/IIIa, Ia complexes) for plasma proteins (VWF, thrombospondin, fibronectin, vitronectin, immunoglobulins and activated components of the complement system: that is, C3b). They are present on platelets, leukocytes, monocytes and lymphocytes T and B.

These plasma proteins might, on an artificial surface, behave as substrate adhesion molecules (SAMs) and exhibit by adsorption on a surface the characteristic amino acid sequence recognized by cytoadhesins; that is, the arginine–glycine–aspartate–serine (RGDs) sequence. This sequence is localized in the NH^2 or COOH terminal domain for VWF, in the α-and γ-chain for fibrinogen, and in fragment IV for fibronectin. Vitronectin and human thrombin also contain RGDs sequences. The precise analysis of the behavior of SAMs on polymeric substrates is thus considered as an essential step in the understanding of cell–substrate interactions.

6. Conclusions

Large numbers of people have been treated with biomaterials and devices introduced within the cardiovascular system that are able to both prolong life and improve its quality. Progress is not so much hampered by the difficulties and constraints related to the functional requirements of the devices as by our lack of understanding of the interaction between the blood and the devices. In spite of a large body of knowledge relating material characteristics to blood compatibility, it remains unclear which materials are supportive of compatibility with blood. Further, specific materials do not display the same interactions in all situations, in particular, low and high blood flow conditions have different mechanisms of interaction.

In physiological conditions the blood interface is the vascular endothelium, a unique monocellular layer which by active and/or passive participation maintains a nonthrombotic state. Endothelial cells are significantly involved in all the major processes associated with hemostasis and thrombosis (vasoregulation, platelet reactivity, coagulation and fibrinolysis). Moreover, they exhibit highly specialized functions, according to their distribution in the vascular system, as required by the hemodynamic conditions and/or locoregional regulatory systems. These cells are able to synthesize in the pulmonary artery PGI_2, a potent vasodilator and platelet anti-aggregating agent, whereas at the venous level endothelial cells synthesize the tissue type plasminogen activator (tpA) thus regulating fibrin deposition in venous vessels. In addition, these polarized cells possess on their luminal membrane specialized microdomains and, in particular, receptors for thrombin. Clearance is in fact an essential step since this protease activates protein C in the presence of thrombomodulin, protein C being essential in the proteolytic inactivation of clotting factors, induces synthesis and excretion of tpA, and stimulates PGI_2 production.

Therefore, mimicking the normal endothelium will represent a major advance in blood-contacting materials research. Until this can be achieved, partial improvements have been found:

(a) by developing new materials of reduced surface energy with sequential hydrophobic and hydrophilic microdomains, or with relative mobility of chains bearing anionic sites at the interface;

(b) by improving the physical surface of existing materials or by modifying their chemical surface, by coating or grafting proteins, as well as by plasma surface treatment;

(c) by introducing pharmacological agents at the interface (e.g., heparin, antiaggregants or fibrinolytic molecules) covalent or ionic binding may be used and result in a short or long term efficacy; and

(d) by introducing into the polymeric backbone chains that mimic glycosaminoglycans of endothelial cells or of heparin.

See also: Drugs: Attachment to Polymers; Heparinized Materials

Bibliography

Anderson J M, Kottke-Marchant K 1987 Platelet interactions with biomaterials and artificial devices. In: Williams D F (ed.) 1987, pp. 103–50

Brash J L 1983 Mechanism of adsorption of proteins to solid surfaces and its relationship to blood compatibility. In: Szycher M (ed.) 1983 *Biocompatible Polymers, Metals and Composites*. Technomic, Lancaster, PA, pp. 35–52

Brash J L, Horbett T A 1987 *Proteins at Interfaces: Physicochemical and Biochemical Studies*. American Chemical Society, Washington, DC, pp. 706

Brash J L, Uniyal S 1979 Dependence of albumin–fibrinogen simple and competitive adsorption on surface properties of biomaterials. *J. Polym. Sci.* 66: 377–89

Cazenave J P, Davies J A, Kazatchkine M D, Van Aken W G (eds.) 1986 *Blood–surface Interactions*. Elsevier, Amsterdam

Forbes C D, Courtney J M 1987 Thrombosis and artificial surfaces. In: Bloom A L, Thomas D (eds.) 1987 *Haemostasis and Thrombosis*, 2nd edn. Churchill Livingstone, Edinburgh, pp. 902–21

Kazatchkine M D, Carreno M P 1988 Activation of the complement system at the interface between blood and artificial surfaces. *Biomaterials* 9: 30–5

Salzman E W, Merrill E W 1987 Interaction of blood with artificial surfaces. In: Colman R W, Hirsh J, Marder V J, Salzman E W (eds.) 1987 *Hemostasis and Thrombosis Basic Principles and Clinical Practice*, 2nd edn. Lippincott, Philadelphia, PA, pp. 1335–47

Sevastianov V I 1988 Role of protein adsorption in blood compatibility of polymers. *CRC Crit. Rev. Biocompatibility* 4(2): 109–54

Van Aken G G, Davies J A 1986 Interaction of leucocytes and red cells with surfaces. In: Cazenave J P, Davies J A, Kazatchkine M D, Van Aken W G (eds.) 1986, pp. 107–24

Van Mourik J A, Van Aken W G 1986 Initiation of blood coagulation and fibrinolysis. In: Cazenave J P, Davies J A, Kazatchkine M D, Van Aken W G (eds.) 1986, pp. 61–74

Vroman L 1987 The importance of surfaces in contact phase reactions. *Semin. Thromb. Hemostasis* 13(1): 79–85

Vroman L, Adams A L, Fischer G C, Munoz P C 1980 Interactions of high molecular weight kininogen factor XII and fibrinogen in plasma at interfaces. *Blood* 55: 156–9

Williams D F (ed.) 1987 *Blood Compatibility*, Vol. 1. CRC Press, Boca Raton, FL

R. Eloy and J. Belleville
[Inserm, Bron, France]

C

Calcium Phosphates and Apatites

Calcium phosphates are a class of materials which fulfill a wide range of uses and functions. For example, they are a major source for agricultural fertilizers, both natural and synthetic, and in the form of apatite-like systems, they are to be found in the skeletal tissues of all mammals. It is because of this diversity that the properties and behavior of the apatites are of interest.

1. Chemistry and Structure

Interest in calcium phosphates stems from the fact that bones and teeth contain a high percentage of mineralized calcium phosphate. DeJong first observed the similarity between the powder x ray diffraction pattern of the *in vivo* mineral and the basic calcium phosphate, hydroxyapatite

$$(Ca_{10}(PO_4)_6OH)_2$$

in 1926. Basic calcium phosphates, precipitated from aqueous solutions *in vitro*, form a nonstoichiometric series, partly because of substitutions by other ions and surface effects but also due to the extremely small crystallite sizes formed by this low-solubility mineral. These properties, coupled with the fact that the kinetics due to the large complex unit cell of apatite favor dissolution over accretion and the possible existence of one or more distinct precursor phases, make the exact chemistry and structure of apatitic calcium phosphates difficult to study. Conflicting hypotheses are best interpreted as the result of this complexity, with different authors focusing on particular aspects of the problem that are accessible to their experimental techniques. It is also fair to say that a combination of all the present experimental techniques will still provide an inadequate picture of the structure of these microcrystalline materials. Hence, much current thinking is influenced by analogies with calcium phosphates of large crystal size. These compounds are prepared at high pressures and/or temperatures in most cases, the exceptions being the acid phosphates which can be grown to large crystal size in solution at room temperature.

The major deviations from stoichiometry in apatitic calcium phosphates as observed by x-ray diffraction are calcium and hydroxyl deficiency, the substitution of carbonate ion for hydroxyl and/or phosphate ions, fluoride or chloride substitution for hydroxyl ions, acid phosphate ions substituting for orthophosphate ions and magnesium, sodium and strontium ions substituting for calcium ions. Since apatitic calcium phosphates can have crystallite dimensions ranging from a few nanometers to hundreds of nanometers, surface area effects can also be important factors in a chemical analysis of the material. Both surface absorption of these ions and distinct calcium phosphate surface phases may be present. Octacalcium phosphate $(Ca_4H(PO_4)_3.3H_2O)$ and dicalcium phosphate dihydrate $(CaHPO_4.2H_2O)$ have been proposed as surface phases as well as precursors of hydroxyapatite.

The final composition of apatitic calcium phosphates depends on the ion concentrations during the precipitation. At near physiological pH and high supersaturation, acid phosphate groups are incorporated (as well as magnesium, sodium, carbonate and fluoride if present) into platelike crystallites.

If slowly heated below 600 °C, apatitic calcium phosphates become dehydrated and acid phosphate groups form pyrophosphate groups:

$$2HPO_4^{2-} \rightarrow P_2O_7^{4-} + H_2O(g)$$

Between 600 °C and 700 °C, pyrophosphate and hydroxyl groups are lost:

$$P_2O_7^{4-} + 2OH^- \rightarrow 2PO_4^{3-} + H_2O$$

The resulting phase is β tricalcium phosphate.

However, hydroxyapatite prepared at high pH and heated rapidly to above 700 °C maintains its apatitic character and grows to a much larger crystal size. The ignition products include α and β $Ca_3(PO_4)_2$, $Ca_2P_2O_7$, $(CaMg)_3(PO_4)_2$ and $CaCO_3$, as found in biological calcium phosphates. Although apatitic calcium phosphates commercially prepared from high pH solutions tend to remain apatitic on heating, the presence of ammonium sulfate in small quantities can promote the formation of $Ca_3(PO_4)_2$.

Hydroxyapatite (Fig. 1) is a hexagonal structure of space group P6₃/m, with basal plane edge $a = 0.9432$ nm and the height $c = 0.6881$ nm. The cell contents are given by formula $Ca_{10}(PO_4)_6(OH)_2$. The hydroxyl ions lie in projection at the corners of the rhombic base of the unit cell forming columns of hydroxyls with a spacing of half the unit-cell height. Six of the calcium ions are associated with these hydroxyls, forming equilateral triangles centered on and perpendicular to the hydroxyl columns. Successive calcium triangles are rotated by 60 °C. Four calciums lie along two separate columns parallel to the hydroxyl columns at heights halfway between the calcium triangles. These calcium ions are coordinated by oxygens from the orthophosphate tetrahedra. The hydroxyl oxygen atoms are

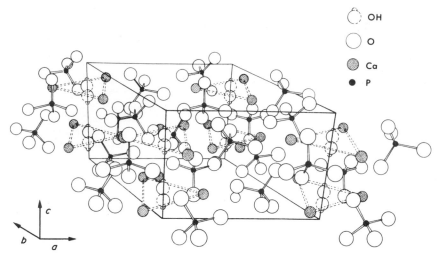

Figure 1
Hexagonal hydroxyapatite; each OH position is statistically only 50% occupied as indicated by the dotted outline of half of each OH group (courtesy of Centre National de la Recherche Scientifique, Paris)

displaced by 0.03 nm from the plane of the calcium triangles. The hydroxyls are oriented such that the oxygen–hydrogen bond is along the column axis but does not cross the plane of the calcium triangle. If fluorine is substituted for hydroxyl it lies at the center of the calcium triangles. In mixtures of fluoride and hydroxyl ions in this structure, the formation of hydrogen bonds between the two ions is thought to explain the lower solubility of such mixed apatites than either hydroxy- or fluorapatite. In fact, the lowest solubility occurs for a one-to-one mixture of hydroxyl and fluoride ions.

Although studies have determined the solubility of hydroxyapatite, these results apply only to large crystallite sizes of pure materials and have been corrected for the recrystallization phenomenon that is a concurrent process in such experiments. Apparent solubilities are higher and the presence of impurities will shift the observed values.

2. Biological Apatites

All mammalian skeletal tissues, bones and teeth, have as a major constituent an inorganic substance, the chemistry and crystallography of which have been identified by various techniques (wet chemistry, x-ray and neutron diffraction, electron microscopy and diffraction, and infrared spectroscopy) as being equivalent to the hydroxyapatite found naturally or produced synthetically in the laboratory. These tissues are often known as calcified or mineralized tissues because they contain apatite-like materials.

Bones consist of approximately 58 wt% apatite, 12 wt% water and 25 wt% organic substances, principally collagen. In bone, the apatite is present in the form of submicroscopic crystallites of about 5–20 nm by 60 nm, as measured by x-ray diffraction and transmission electron microscopy. The other mineralized tissues have differing compositions which are compared with stoichiometric apatites (hydroxyapatite, OHAp; chlorapatite, ClAp; and fluorapatite, FAp) in Table 1.

While the crystallite sizes in dentin and cementum are roughly comparable to those found in bone, the crystallites found in enamel grow much larger, over 100 nm in length. Also, the protein found in enamel is not collagen but is a material called enamelin. However, the crystallography of enamel OHAp is very similar to that of synthetic hydroxyapatite as seen in Table 2.

Investigations of the physical properties of mineralized tissues have indicated that the observed properties, especially the elastic and electromechanical properties, often depend on the hierarchical levels of structure found in the given system. For bone, it appears that there are at least four levels of structure that should be considered when trying to understand its properties, behavior and the role of the mineral phase.

The OHAp crystallites themselves (along with the collagen molecules) can be depicted as the molecular level. The intimate juxtaposition or infusion of these crystallites on and through the collagen fibrils constitutes the ultrastructural level. It is here that the concept of a two-phase material composite becomes apparent. The high-modulus OHAp crystallites are mechanically interfaced with the compliant energy-absorbing (tough) collagen, resulting in a composite system with properties quite disparate from either component. This collagen–OHAp composite may be organized randomly, as found in

Table 1
Comparison of composition of mineralized tissues and stoichiometric apatite[a]

Component (%)	Material						
	Bone[b]	Enamel	Dentine	Cementum	OHAp	ClAp	FAp
Water	12.2	2.3	13.2	32			
Organic constituent	24.6	1.7	17.5	22			
Mineral constituent	57.9	96.0	69.3	46			
calcium	22.5	36.3	35	35.5	39.9	38.5	39.7
phosphorus	10.3	17.3	17.1	17.1	18.5	17.9	18.4
carbon dioxide	3.5	2.8	4.0	4.4			
magnesium	0.26	0.40	1.2	0.9			
sodium	0.52	0.45	0.2	1.1			
potassium	0.09	0.20	0.07	0.1			
chlorine	0.11	0.25	0.03	0.1		6.81	
fluorine	0.05	0.02	0.02	0.02			3.77
hydroxyl					3.38		

a not including trace elements   b 5.3% not accounted for

woven bone or in certain regions of dentin; alternatively, it may form various organized lamellar structures, such as found either in young bovine cortical bone (known as plexiform), where long thin lamellae are found, or in mature human or bovine cortical bone (known as Haversian bone), where the collagen–OHAp composite is formed into densely packed concentric lamellae around a central blood canal. These secondary osteons, which are usually circular or elliptical in cross section, have their long cylindrical structures arranged approximately parallel to the long-bone axis. In the latter cases, the lamellar structures appear to be bonded to each other through a material known as the ground substance. This level of organization for long bones is the microstructural level. The microstructural organization may differ in some details from types of bone other than that found in the cortex of long bones. The whole bone is considered at the macroscopic level; size, shape and organization must all be considered at this level of structure.

While all the bones are of interest, the long bones of the lower body (the femur, tibia and fibula) are of most significance when considering material

properties. This is because of their role as the structural supports for the body. Thus, while the role of the apatites in all biological calcified tissues will be presented here, particular emphasis will be placed on their effects in long bones.

3. Physical Properties

The principal interest in the physical properties of the apatites is due to the important roles they play in the mechanical and structural functions of the skeletal system. It is the structural relationship between the apatitic and organic components that is considered responsible for the range of properties and behavior observed in the mineralized tissues. Attempts to analyze the properties of such composite systems or to understand how they arise require a complete description of the properties of each component. Thus a description of the elastic properties of bones or teeth requires, in part, a knowledge of the corresponding properties in biological apatites. Similarly, attempts to understand other physical properties of a composite system such as thermal or electrical conductivity require corresponding data on the specific components.

Collagen is a compliant material with elastic properties resembling rubber. A long low-slope linear stress–strain curve with a slight onset of plastic deformation prior to failure is characteristic. In comparison, the elastic behavior of most minerals may be categorized as stiff and brittle (i.e., a linear high-slope stress–strain curve with no onset of plasticity prior to failure). Either material by itself would not provide the mechanical performance required for weight bearing in mammals. The superposition of such disparate materials in the form of a composite system could provide the appropriate response, depending on the way in which the composite is

Table 2
Crystallographic parameters for dental enamel and synthetic hydroxyapatite

Material	Parameter (nm)	
	a	c
Enamel OHAp	0.944	0.688
Synthetic OHAp	0.942	0.688

formed. In any case, the elastic properties of the apatitic phase provide the starting point for calculations and attempts to model and understand the mechanical response of calcified tissues. Ultrasonic wave propagation measurements are ideal for obtaining the anisotropic elastic properties of crystalline materials in small specimens.

In many mineral systems, single crystals of suitable size for various physical experiments can be found in nature or be grown synthetically in the laboratory. OHAp appears to be one of the few exceptions to this. Relatively large single crystals of FAp and ClAp are found in several places; larger, gemlike synthetic single crystals of both have also been grown by several different researchers. However, no OHAp single crystals of appreciable size have either been found in nature or been grown in the laboratory. Usable single crystals of Durango apatite unfortunately have about 5% or more fluoride ions replacing hydroxyl ions, thus rendering them inappropriate for studies of OHAp properties. Similarly, since the apatitic crystallites in calcified tissues are so small, they are generally not available for use in experiments measuring physical properties; certainly they are too small for the single-crystal ultrasonic wave propagation studies.

Powdered compacts of both natural and synthetic apatites as well as of the inorganic component of calcified tissues can be used to obtain isotropic elastic properties. Only two elastic parameters are necessary to characterize an isotropic material. Thus, any pair of the four commonly defined technical elastic constants (Young's modulus E, shear modulus G, bulk modulus K and Poisson's ratio ν) may be used for this purpose.

Values of the four constants as obtained from ultrasonic wave propagation measurement in isotropic compacts of mineral and synthetic apatites, as well as in other calcium compounds, are given in Table 3.

Similar ultrasonic measurements have been carried out on powders prepared by grinding specimens of various calcified tissues. In addition, ultrasonic measurements have been made on intact oriented specimens of human and bovine calcified tissues. Both the powders and oriented specimens contain the organic components as well as the apatitic phase. Measurements are also available for preparations of powdered specimens of various marine life in which the organic components have been extracted with ethylenediamine leaving only the mineral or apatitic phase behind. Table 4 presents the elastic moduli for all these materials.

All the ultrasonic measurements on powdered specimens presented in the tables were obtained by back extrapolation from high-pressure measurements (up to 5×10^9 Pa) in order to obtain the properties at maximum density at atmospheric pressure.

Differences in the properties between powders and oriented whole specimens presented in Table 4 reflect the alignment of crystallites in the whole specimens. It is the anisotropic elastic properties of the apatitic crystallites that are responsible in part for the macroscopic properties in the bulk specimens. Thus, it is the anisotropic elastic constants rather than the isotropic constants presented in Tables 3 and 4 that are necessary for describing the mechanical behavior of calcified tissues.

The effects of the inorganic constituents on the ultrasonic measurements of elastic properties can be observed in Table 4. Here, data on elephant ivory, which is similar in structure and composition to dentin, is presented both for powdered specimens including the organic constituents as well as for the equivalent ivory mineral. It is clear that the organic constituents significantly affect the elastic properties of the tissue: the increase in stiffness is almost 250% when the organic component is removed. It is important to recognize that the actual effect on stiffening due to the apatitic phase will be different in whole specimens of many bones (especially the long weight-bearing bones) as compared with isotropic powders, because of the possibility of oriented crystallites in bulk specimens. This effect is also reflected in part by the fact that the elastic moduli of

Table 3

Elastic properties of mineral and synthetic apatites and various calcium-bearing minerals from ultrasonic measurements

Material	Bulk modulus (10^{10} Pa)	Shear modulus (10^{10} Pa)	Young's modulus (10^{10} Pa)	Poisson's ratio
FAp (mineral)	9.40	4.64	12.0	0.28
OHAp (mineral)	8.90	4.45	11.4	0.27
OHAp (synthetic)	8.80	4.55	11.7	0.28
ClAp (synthetic)	6.85	3.71	9.43	0.27
CO$_3$Ap (synthetic, type B)	8.17	4.26	10.9	0.28
CaF	7.74	4.70	11.8	0.26
Dicalcium phosphate dihydrate	5.50	2.40	6.33	0.31

Table 4
Elastic properties of biological calcified materials

Material	Bulk modulus (10^{10} Pa)	Shear modulus (10^{10} Pa)	Young's modulus (10^{10} Pa)	Poisson's ratio
Bovine dentin				
powdered	3.22	1.12	3.02	0.35
fresh intact longitudinal specimen	1.8	0.80	2.1	0.31
Bovine enamel				
powdered	6.31	2.93	7.69	0.32
fresh intact longitudinal specimen	4.6	3.0	7.4	0.23
Bovine femur				
fresh intact longitudinal specimen	3.05	1.22	3.23	0.34
Human femur				
powdered	2.01	0.800	2.11	0.33
Human tibia				
fresh intact longitudinal specimen			2.44	
Flounder vertebrae	5.26	2.42	6.39	0.30
Brown trout vertebrae	5.49	2.45	6.40	0.31
Brown trout rib	4.52	2.01	5.26	0.31
Sea bass rib	4.38	2.53	6.37	0.26
Mollusc shell				
(aragonite and calcite)	5.76	3.55	8.86	0.25
Ivory	1.92	0.730	1.96	0.34
Ivory mineral	4.72	1.86	4.84	0.33

long bones, such as the femur and tibia, are angularly dependent; that is, the Young's and shear moduli differ for different orientations of specimens relative to the long axis.

Anisotropic elastic constants based on ultrasonic measurement are available for FAp. By comparing these values with the results for powdered specimens of FAp and OHAp, anisotropic elastic constants have been modelled for OHAp. Similar measurements have been made on oriented specimens of bovine and human bone and dentin studied as if they were single crystals with the same hexagonal (axial) symmetry (Table 5).

The usual technical moduli (E, G, K and v) can be calculated as a function of specimen orientation with respect to the bone axis from the single-crystal data using standard transformations. These anisotropic technical moduli can also be calculated from experiments using standard mechanical testing to obtain five independent moduli.

Table 5
Anisotropic elastic constants of calcified tissues and apatites from ultrasonic measurements at room temperature

Material	Anisotropic elastic constants (10^{10} Pa)					
	C_{11}	C_{33}	C_{44}	C_{12}	C_{13}	C_{66}
Apatite						
FAp	15.05	18.5	4.251	4.88	6.22	5.087[a]
OHAp	13.7	17.2	3.96	4.25	5.49	4.73
Bovine phalanx						
fresh	1.97	3.2	0.54	1.21	1.26	0.38
dried	2.12	3.74	0.75	0.95	1.02	0.59
Femoral bone (dried)						
bovine	2.38	3.34	0.82	1.02	1.12	0.68
human	2.34	3.25	0.871	0.906	0.911	0.717[a]
Dentin	3.70	3.90	0.570	1.66	0.87	1.02
Enamel	11.5	12.5	2.28	4.24	3.00	3.63

a C_{66} measured independently; for all other specimens, C_{66} calculated from $C_{66} = \frac{1}{2}(C_{11} - C_{12})$

Table 6
Elastic stiffness constants and densities for wet bovine Haversian and plexiform bone

Physical property	Bone type	
	Haversian	Plexiform
Density ($kg\ m^{-3}$)	1920	2120
Elastic stiffness constants (10^{10} Pa)		
C_{11}	21.2	22.4
C_{22}	22.0	25.0
C_{33}	29.0	35.0
C_{44}	6.30	8.20
C_{55}	6.30	7.10
C_{66}	5.40	6.10
C_{12}	11.7	14.0
C_{13}	12.7	15.8
C_{23}	11.1	13.6

Although apatitic crystallite orientation in long bones does contribute to the anisotropic elastic properties in bone, it is not the only factor. The microstructural differences, for instance, between bovine Haversian and plexiform bone, described in Sect. 2, also play a significant role in the respective elastic properties. Indeed, while the Haversian bone symmetry may be described as hexagonal (or transverse isotropic) and requires only five elastic constants for its description, plexiform bone apparently is best described as having orthorhombic (or orthotropic) symmetry, which requires nine independent constants. The differences can readily be seen by comparing the measured ultrasonic velocities in specimens cut with orientations longitudinal, radial and transverse to the bone axis for both Haversian and plexiform bone obtained from different aspects of a single bovine femur: Young's moduli calculated from these velocities and the respective densities of the specimens also clearly show this significant difference in properties. While the organic–apatitic compositions and the crystallite–collagen orientations appear comparable in both structures, the

properties clearly appear to be dominated by the microstructural symmetry. Confirmation of the microstructural influence is provided by examining the full set of C_{ij} measured ultrasonically for both bovine Haversian and plexiform bone. In order to provide an appropriate comparison, all nine C_{ij} are presented for both structures in Table 6 even though there are redundancies in the former system since only five constants are required if hexagonal symmetry is assumed. The redundancies required in the C_{ij} for any system exhibiting hexagonal symmetry are $C_{11} = C_{22}$, $C_{23} = C_{13}$, $C_{44} = C_{55}$ and $2C_{66} = C_{11} - C_{12}$ (see Acoustic Measurements of Bone and Bone–Implant Systems).

Other physical properties of the apatites that are of interest are the thermodynamic, electrical and optical properties (Tables 7 and 8). Many of these can be interrelated. The elucidation of the electromechanical properties of the apatites, especially OHAp is particularly important with regard to understanding the electromechanical properties of bone, as these interactions may be the prime mechanisms whereby the signals controlling the resorption and remodelling of bone (so-called Wolff's Law) are established. These properties, while of basic interest in understanding the conditions determining the formation and growth of crystals in nature and in the laboratory, have provided little insight into the corresponding conditions for crystallization and growth of mineral in calcified tissues.

4. Applications

Applications of the calcium phosphates are found in a variety of fields. Bone char (the inorganic residue produced by heating bones) is used in sugar processing. Many of the calcium phosphates are important as fertilizers in agriculture. Others find use as catalysts in chemical processes.

A principal application of the calcium-bearing inorganic compounds as materials is in the biological system, as prosthetic and restorative materials for

Table 7
Electrical and optical properties of bone, OHAp, FAp and ClAp

Material	Refractive index		Dielectric constant		
	ω	ε	Resistivity (Ω m)	$\varepsilon_\perp$	$\varepsilon_\parallel$
Bone			63[a]	10(4% H_2O)	
OHAp	1.667	1.664	3×10^{8} [b]		
FAp	1.633	1.629	3.8×10^{10} [b]	10.02	7.62
ClAp	1.667	1.664	7.1×10^{10} [b]		

a wet sample   b powdered sample

Table 8
Thermodynamic properties of bone, OHAp, FAp and ClAp

Material	Melting point (K)	Heat of formation (kJ mol^{-1})	Specific heat capacity (J kg^{-1} K^{-1})	Entropy[a] (J mol^{-1} K^{-1})	Heat content[a] (kJ mol^{-1})
Bone			778[b]		
OHAp		13562	439[b]	781	128
FAp	1933	13843	765–803	776	127
ClAp	1803	13322			

a at 298 K   b at 250 K

the skeletal system. Since calcium phosphates are already present in the body as OHAp, it is possible that certain synthetic analogues may be introduced into the body without any untoward effects. Even more significant is the possibility that such materials may even be so biocompatible as to engender or favor an adhesive interface between the body's calcified tissues and the synthetic implant. Several applications are found in orthopedic and dental practice when function of a particular bony tissue is lost or destroyed due to pathology, accident, removal or other causes.

The earliest and still the most promising applications were made in dentistry, where the amount of material needed is generally small and the systems usually can be shielded from loading until healing and adhesive attachment have progressed to the level where normal loads may be applied. One of the first applications to be suggested for the calcium phosphates was the tooth root implant. After a tooth is removed, a shaped plug of the material is inserted in the extraction site. On healing, a metal post is introduced into the implant so that it projects above the gum line. This enables a crown to be attached, ideally providing an integrated replacement which maintains dental aesthetics while permitting biting and chewing comparable with the real root.

When the above is contraindicated, a second application is to place the appropriate material in the extraction site so that the teeth on either side do not displace because of the void between them, as well as to reduce the resorption of the alveolar ridge.

Similarly, if a portion of the mandible or maxillary is destroyed or removed, a suitable replacement may interface adhesively with the remaining bone to restore the function of the jaw.

A number of calcium phosphate and related materials have been suggested and have even been utilized clinically for one or more of these applications. Porous β tricalcium phosphate, one of the earliest materials to be tried as a scaffolding material in dentistry, has been used because of its dissolution in the biological environment. Thus as bone grows into the pores, the β tricalcium phosphate ideally disappears leaving only reformed bone. However, it is important in the manufacturing process that the conversion procedure be such that nearly 100% conversion from the starting material to β tricalcium phosphate occurs. Otherwise incomplete dissolution of the implant may occur with subsequent complications.

OHAp is an obvious choice because it is already present in the body. Several different forms have been implanted for all three purposes. A dense form of OHAp prepared at atmospheric pressure by sintering at high temperatures provides an improvement in elastic and mechanical properties over those obtained by pressure compacting alone, which generally results in higher porosity and lower density. Technical moduli obtained from ultrasonic measurements on such dense compacts are compared with the results for their intermediate precursor as well as with those for powdered synthetic OHAp and ivory mineral in Table 9.

Table 9
Elastic moduli and Poisson's ratio for various synthetic and biological apatites

Material	Bulk modulus (10^{10} Pa)	Shear modulus (10^{10} Pa)	Young's modulus (10^{10} Pa)	Poisson's ratio
OHAp (dense)	8.15	4.25	10.9	0.28
OHAp (powdered, high pressure)	8.80	4.55	11.7	0.28
Ivory mineral	4.72	1.86	4.84	0.33
Intermediate precursor (for dense OHAp ceramic)	2.09	1.49	3.61	0.21

Table 10
Elastic properties of sintered synthetic hydroxyapatites

Sintering temperature (°C)	Volumetric porosity (%)	Bulk modulus (10^{10} Pa)	Shear modulus (10^{10} Pa)	Young's modulus (10^{10} Pa)
1000	34.1	13.7	13.5	30.3
1200	0.040	90.0	41.5	108
1300	0.053	97.3	39.1	103
1350	0.203	94.1	40.2	106

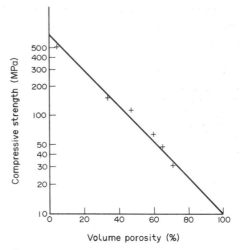

Figure 2
Graph of compressive strength (MPa) for OHAp compacts vs volume porosity (%)

Mechanical measurements have also been used to measure the properties of such dense OHAp compacts. These dense compacts appear to give the desired response when implanted (i.e., bone does appear to grow into them forming an adhesive interface).

Another form of OHAp under investigation as an implant is a preparation of porous OHAp produced by conversion from coral skeletons, so as to maintain the original interconnected porosity. It is proposed that, on implantation, connective tissues can grow into the pores, mechanically stabilizing the implant until mineralization occurs followed by subsequent adhesive attachment of the newly formed bone to the OHAp. Here the problems may be resorption, due in part to the possibility of incomplete conversion as well as to the large amount of surface area of the OHAp now accessible to the biological fluids, and the relatively poorer mechanical properties (especially strength) when compared with the dense ceramics. Although not a calcium phosphate, another calcium-containing material, aragonite, the high pressure form of calcite ($CaCO_3$), is obtainable in porous form directly from the unconverted coral skeletons. After proper cleaning and analyses assuring that there are no contaminants which might affect its biocompatibility, this material is also one to which bone is found to adhere when implanted. Since this material is resorbed in the body, its applications are dedicated to clinical situations in which this material is only to function temporarily and in which long term resorption is desired.

Porosity may also be achieved directly, starting with powdered OHAp, by adjusting the temperature and length of time of sintering the material. In Table 10 are data on the average elastic properties of OHAp compacts prepared at various sintering temperatures. The amount of porosity also affects the strength of the material. Classically, the relationship between strength and porosity is a log–linear one. This relationship holds reasonably well for the compressive strengths of various sintered OHAps, as measured by several investigators; Fig. 2 is the graph of the compressive strength (on a log scale) vs percentage of volume porosity for these data.

Improvements in the mechanical properties will be necessary before such porous materials find use as skeletal implants in weight-bearing situations. However, OHAp and FAp may be used to clad presently used weight-bearing implant metals such as titanium and titanium alloys. The cladding provides the adhesive interface between bone and the implant, with the metallic core providing the appropriate weight-bearing properties.

Calcite and its magnesium analogue dolomite have, in dense form, also proved successful in implant studies as regards bone adhesion. Thus, calcite–tricalcium phosphate (TCP) mixtures (in molar ratios of 1:3, respectively) have also been proposed as a potential replacement material. Here too, varying amounts of porosity may be achieved by controlling the temperature and length of time of sintering. In Table 11 are the average elastic properties of such calcite–TCP porous compacts as a function of sintering temperature. Poor mechanical properties at present relegate the potential use of

Table 11
Elastic properties of sintered synthetic calcite–TCP mixtures

Sintering temperature (°C)	Volumetric porosity (%)	Bulk modulus (10^{10} Pa)	Shear modulus (10^{10} Pa)	Young's modulus (10^{10} Pa)
1000	29.5	11.7	10.4	24.0
1200	21.8	23.3	17.5	40.0
1300	16.2	36.9	20.1	50.8
1350	11.9	36.6	27.1	65.0

these materials to non-load-bearing clinical applications, again where resorption is desired.

See also: Collagen; Dental Implants

Bibliography

Centre National de la Recherche Scientifique 1975 *Physico-Chimie et Crystallographie des Apatites d'Intérêt Biologique*, Colloques Internationaux du Centre National de la Recherche Scientifique, Vol. 230. Centre National de la Recherche Scientifique, Paris
de Groot K (ed.) 1983 *Bioceramics of Calcium Phosphate.* CRC Press, Boca Raton, FL
Ducheyne P, Lemons J E (eds.) 1988 Bioceramics: Material characteristics versus *in vivo* behavior. *Ann. N. Y. Acad. Sci.* 523
Frost H M 1963 *Bone Remodeling Dynamics.* Thomas, Springfield, IL
Gilmore R S, Katz J L 1982 Elastic properties of apatites. *J. Mater. Sci.* 17: 1131–41
Grenoble D E, Katz J L, Dunn K L, Gilmore R S, Linga Murty K 1972 The elastic properties of hard tissues and apatites. *J. Biomed. Mater. Res.* 6: 221–33
Guillemin G, Patat J-L, Fournie J, Chetail M 1987 Use of coral as a bone graft substitute. *J. Biomed. Mater. Res.* 21: 557–67
McConnell D 1973 *Apatite: Its Crystal Chemistry, Mineralogy, Utilization, and Geologic and Biologic Occurrences.* Springer, New York
Nancollas G H (ed.) 1982 *Biological Mineralization and Demineralization: Report on the Dahlem Workshop, Berlin 1981.* Springer, Berlin
Zipkin I (ed.) 1973 *Biomedical Mineralization.* Wiley, New York

J. L. Katz and R. A. Harper
[Rensselaer Polytechnic Institute, Troy, New York, USA]

Carbons

Carbon is the element most frequently found in bioorganic compounds. It is unique among the elements as a result of the vast number and variety of compounds it can form. There are over a million carbon compounds, many thousands of which are vital to organic and life processes. Carbon, occurring in many different atomic structural forms, is important in the biomedical field for yet another reason. Some carbon forms, because of their chemical and mechanical structure, have been found exceptionally suitable for use in partial, and at times total, replacement of damaged biological structures such as bones, cartilage, tendons, ligaments and vascular tissues. The acute and chronic responses of tissues to these special carbon forms are minimal. Equally important, these same elemental carbon structures are unaffected by the *in vivo* environment and are permanently incorporated into the living host. Hence, the element carbon is important both from a biochemical and from a biomechanical point of view.

Processes such as hydrocarbon gas pyrolysis, vapor deposition and polymer vitrification have made it possible to produce carbon structures in a variety of forms with a wide range of properties. Some selected forms of carbon produced by these methods have substantial clinical significance.

1. Clinically Relevant Carbons

Three types of carbon are used extensively in biomedical devices: the low-temperature isotropic (LTI) variety of pyrolytic carbon, glassy (vitreous) carbon and the ultralow-temperature isotropic (ULTI) form of vapor-deposited carbon. Table 1 is a compilation of the generic, chemical and trade names for these carbons.

At present, only these three materials are clinically significant, but other forms of carbon have been and are being considered for use in the reconstruction of living tissues. Braided carbon-fiber systems have been used in the reconstruction of tendons and ligaments. Carbon–carbon composites have received attention as possible bone substitutes and load-carrying members in joint replacement. Particulate forms of carbon have been investigated as adsorbents and for use in blood purification.

Difficulties encountered with these latter carbon materials are generally not a result of their chemical composition, but rather of their structure. The composite, particulate and fiber systems suffer from a lack of permanent integrity. Particles migrate and can be entrained in blood. Fibers and filaments can be shed from composites and migrate into the lymphatic system. Research is directed towards making these materials more stable structurally.

In contrast, the carbon materials in use are integral and monolithic materials (glassy carbon and LTI carbon) or impermeable thin coatings (ULTI carbon). These three forms do not suffer from the integrity problems typical of the other available carbon materials. With the exception of the LTI carbons codeposited with silicon, all the clinical carbon materials are pure elemental carbon. Up to 20 wt% silicon has been added to LTI carbon without significantly affecting the biocompatibility of the

Table 1
Isotropic carbons used in clinical devices

Generic name	Other designations	Trade names[a]
Pyrolitic carbon	LTI carbon	Pyrolite Carbon
Glassy carbon	vitreous carbon, polymeric carbon	bioCarbon
Vapor-deposited carbon	ULTI carbon	Biolite carbon

a Pyrolite and Biolite are trademarks of CarboMedics, bioCarbon is a trademark of Bentley Laboratories material. The composition, structure and fabrication of the three clinically relevant carbons are unique compared with the more common naturally occurring form of carbon (i.e., graphite) and other industrial forms produced from pure elemental carbon.

2. Composition and Structure

The broad range of properties of biomedical carbons is directly attributable to variations in their crystal structure. All have a structure based on that of graphite. In graphite (see Fig. 1a), the carbon atoms are arranged in highly ordered planar hexagonal arrays. Van der Waals forces weakly bond the planes together. The weak binding between adjacent planes explains the lubricity demonstrated by graphite when adjacent crystal planes are placed in shear. Since the binding between layer planes is very weak, they slip over one another.

The LTI, ULTI and glassy carbons are subcrystalline forms and represent a lower degree of crystal perfection (see Figs. 1b, c). There is no order between the layers as there is in graphite, so the crystal structure of these carbons is two-dimensional. Such a structure has been called turbostratic. In Figs. 1a, c the highly organized graphite crystal structure is compared with the wrinkled, disorganized structure of the subcrystalline turbostratic carbons. The random, hence isotropic, arrangement of the small crystallites (1–5 nm) of these carbons is depicted in Fig. 1d.

The densities of turbostratic carbons fall between about $1400 \, \text{kg m}^{-3}$ and $2100 \, \text{kg m}^{-3}$. High-density LTI carbons are the strongest bulk forms of carbon and their strength can further be increased by adding silicon. ULTI carbon can also be produced with high densities and strengths, but it is available only as a thin coating (0.1–1.0 μm) of pure carbon. Glassy carbon is inherently a low-density material and as such is weak. Its strength cannot be increased through processing.

3. Material Fabrication

The methods used to fabricate medical devices from carbon differ considerably from those used to fabricate them from metals and polymers. Very high temperatures are required to produce both the LTI and the glassy carbon structures. The extremely high melting or sublimation temperature (~3500 °C) of

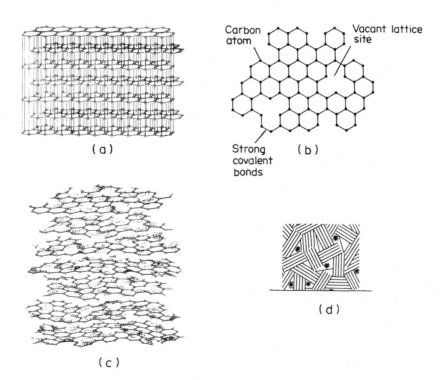

Figure 1
Possible atomic arrangements of graphitic carbon: (a) graphite (hexagonal polymorph), (b) single imperfect layer plane, (c) three-dimensional turbostratic structure (layers viewed edge-on), and (d) aggregate of crystallites together with some single layers and unassociated carbon

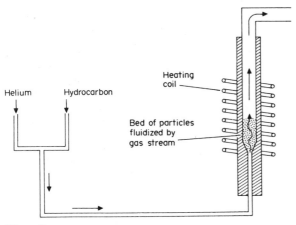

Figure 2
LTI carbon production

carbon makes it impossible to produce thick slabs or billets of dense turbostratic carbon from which parts can be machined. Consequently, the LTI carbons are generally applied as structural coatings of less than approximately 1 mm in thickness. The glassy carbons are monolithic structures limited to approximately 7 mm in thickness. Lower temperatures are involved in the formation of ULTI carbon, as will be explained.

LTI carbons are generated in a tube furnace with a fluidized bed in the hot zone (see Fig. 2). A hydrocarbon gas is passed through the furnace and is pyrolyzed in the fluidized bed containing the parts to be coated. The structure of the deposited carbon can be controlled and tailored as desired by adjusting the composition of the gas, the temperature of pyrolysis, the residence time of the hydrocarbon in the bed and the bed surface area. Typically, a coating of no more than 1 mm in thickness is deposited on a preshaped refractory substrate, transforming it into a strong biocompatible structure. The silicon-alloyed variety of LTI carbon is formed by mixing a silicon-containing carrier gas with the hydrocarbon. The alloyed carbons are stronger and have greater wear resistance than the unalloyed carbon.

Glassy carbons are produced by slowly heating a solid polymeric preform to drive off the volatile constituents. The heating rate is dependent on the size of the part (maximum thickness <7 mm) and the diffusion rate of volatiles through the polymeric mass. If the pyrolysis takes place too quickly, bubbles may form or the part may shatter like glass. The isotropic monolithic carbons formed in this process may therefore take as long as a month to produce. Unlike the LTI carbons, glassy carbon is process limited to a maximum density of approximately 1500 kg m^{-3}, which explains the low strength

of this carbon. A 50% volume shrinkage can be expected in transforming the polymeric form to the glassy carbon form.

The clinically important ULTI carbon is a turbostratic carbon formed in a hybrid vacuum process by using a catalyst to deposit carbon at high rates from a carbon-bearing gaseous precursor. The density, crystallite size and isotropy of the coating can be varied within wide limits as a result of flexibility in processing. Cooling the substrate before or during coating allows the coating of low-melting-point materials. Such a coating is often less than 1 μm thick and can be structured so that it does not significantly affect the surface topography or the mechanical properties of the substrate. The chemical, physical and structural properties of the ULTI carbons are virtually identical to those of the fluidized-bed-derived LTI carbons.

4. Mechanical Properties

The mechanical properties of the various carbons are intimately related to their microstructures. Within any particular structure, atoms are bonded either covalently or by a combination of covalent bonds and van der Waals forces. The bond energy for a graphite covalent bond is 477 kJ mol^{-1}, whereas the bonding energy of van der Waals forces is 17 kJ mol^{-1}. The highest strength and stiffness ever reported for any material are for graphite whiskers, which are predominantly covalently bonded.

However, the lubricity characteristically associated with graphite results from the weak van der Waals attraction between the three-dimensional graphite planes. Due to this weak attraction, large shear strains at low stresses are possible. The orientation and proportions of these two bonding mechanisms within a material significantly affect its properties. In an isotropic carbon, where there is a controlled random mixture of these two bonding mechanisms, it is possible to generate materials with low elastic moduli (20 GPa) and high flexural strength (275–620 MPa). There are many benefits as a result of this combination of properties. Large strains (~2%) are possible without fracture. Structures can be made that, because of similar moduli and strength, are mechanically compatible with bone.

The turbostratic carbons are extremely tough compared with ceramics such as aluminum oxide. The energy to fracture for LTI carbon is approximately 5.5 MJ m^{-3}, compared with 0.18 MJ m^{-3} for aluminum oxide; that is, the carbon is approximately 25 times as tough. The strain to fracture for the vapor-deposited carbons is greater than 5.0%, making it feasible to coat highly flexible polymeric materials such as polyethylene, polyesters and nylon without fear of fracturing the coating when the substrate is flexed. For comparison, the strain to failure of aluminum oxide is approximately 0.1%, about one-fiftieth that of the ULTI carbons.

These turbostratic carbon materials have extremely good wear resistance, some of which can be attributed to their toughness; that is, their capacity to sustain large local elastic strains under concentrated or point loading without galling or incurring surface damage. High abrasive forces caused by surface disparities are distributed and reduced. The wear resistance of LTI carbons can be increased further by increasing the hardness, reducing the crystallite size and adding alloying elements such as silicon.

Wear studies of spinning disks on plates have been performed at room temperature in distilled water. The disks were made of polished high-density LTI pyrolytic carbon, and various pure and silicon-alloyed LTI pyrolytic carbons and glassy carbon were used as the plate. Similar testing was also performed using LTI pyrolytic carbon and Stellite 21 disks on flat plates of pure and alloyed β III titanium and on ultrahigh-molecular-weight polyethylene (UHMWPE). The results indicate that the combination of LTI carbon on UHMWPE is slightly better than Stellite 21 on UHMWPE. LTI carbon on LTI carbon or pure titanium was significantly better than the Stellite 21–polyethylene combination. In other tests where the flat plate was LTI carbon and the disk was either Stellite 21 or pure titanium, the wear of the LTI carbon was approximately 30 times that for the LTI carbon on LTI carbon combination.

Since the vapor-deposited carbons are a relatively thin coating, their mechanical biocompatibility, including strength, modulus of elasticity, wear and bond strength, depends on the substrate material. It has been determined that the wear properties of vapor-deposited carbons of the ULTI variety, when coated on graphite, are quite similar to those of the LTI carbon.

The bond strength of the ULTI carbon to stainless steel and to Ti–6% Al–4% V exceeds 70 MPa as measured with a thin-film adhesion tester. This excellent bond is, in part, achieved through the formation of interfacial carbides. The ULTI carbon coating generally has a lower bond strength with materials that do not form carbides. The bond strength of ULTI carbon to a polyimide was measured as greater than 23 MPa with failure occurring in the polymer, suggesting that the carbon–polymer bond strength is greater than the polymer strength.

Another unique characteristic of the turbostratic carbons is that they do not fail in fatigue. The ultimate strength of turbostratic carbon, as opposed to metals, does not degrade with cyclical loading. The fatigue strength of these carbon structures is equal to the single cycle fracture strength. It appears that unlike other crystalline solids, these forms of carbon do not contain mobile defects which at normal temperatures can move and provide a mechanism for the initiation of a fatigue crack.

Some of the more important known properties of the turbostratic carbons are listed in Table 2. It should be noted that the properties listed for glassy carbon are a function of the density, which is dependent on the precursor material, whereas the vapor-deposited, pyrolytic and alloyed carbons exhibit a wide range of properties that can be preselected through the control of the fabrication parameters.

5. Biocompatibility

Charcoal and lampblack have been used since ancient times as tatooing materials, because of their inertness in tissue. This apparent chemical biocompatibility and advancements in methods of producing carbons with appealing biomechanical properties such as high strength, low wear and immunity to fatigue have led to extensive studies to determine the limits of their biological application.

Carbons found immediate applications in the cardiovascular area, such as in heart valves, the first of which was implanted in 1969. Since then, more than 600 000 pyrolytic carbon valve components have been produced for implantation. The cardiovascular application is particularly demanding. Early attempts at producing successful heart valves failed because the materials used were either thrombogenic or suffered from high wear and mechanical failure. Thrombus, wear, distortion and biodegradation have been virtually eliminated, because of the biocompatibility and mechanical durability of pyrolytic carbon, clearly establishing it as the material of choice for heart valves.

Prior to the use of pyrolytic carbon in such applications, only surfaces treated with a coating of graphite, benzalkonium chloride and heparin (GBH) could be exposed to blood for long periods of time without thrombus formation. Other materials would activate the clotting mechanism of the tissue. Blood compatibility tests to determine if blood is in any way hemolyzed by contact with pyrolytic carbon have shown it to be equivalent to siliconized glass, which is known to cause little damage to blood in comparison with most foreign surfaces.

The LTI and ULTI carbons have been tested extensively in the form of rings, the glassy carbons to a lesser extent. LTI and ULTI carbon rings, placed in either the inferior vena cava or the renal artery, exhibit good to excellent thromboresistance. Testing with carbon beads has shown that the carbon surfaces are not only thromboresistant, but also compatible with the cellular elements of blood. The materials do not influence plasma proteins or alter the activity of plasma enzymes. In fact, one of the proposed explanations for the blood compatibility of these materials is that they adsorb blood proteins on their surface without altering them.

Chronic reactions of soft and hard tissues are also important in determining the compatibility of implantable devices such as prosthetic joints. This compatibility has been tested, among other ways, by successfully cultivating normal Wish human amnion cells as a monolayer culture directly on LTI and silicon-alloyed LTI carbon. Five year clinical testing of glassy carbon dental implants in humans and three year testing of pyrolytic carbon implants in baboon mandibles yielded none of the foreign body reactions such as the fibrous tissue encapsulation found with incompatible materials.

Biomechanically, pyrolytic carbon is thought to be particularly well suited to hard tissue load-bearing applications, because it can be fabricated with an elastic modulus that is similar to that of bone. Problems associated with mismatch between implant material and supporting tissue elastic moduli such as stress bypass osteopenia and pressure necrosis can effectively be eliminated by proper design with prosthetic materials of compatible elastic moduli. The pyrolytic carbons are ideally suited for these applications, as shown by the response of bone to such implants.

Wear is an important factor to be considered in determining the biocompatibility of a material for certain prosthetic applications. This characteristic is of special concern in cardiovascular and orthopedic applications where moving parts articulate with one another, resulting in a wear process that may damage fluid flow characteristics or smoothness of closure in the heart valve application or lead to eventual destruction of joint surfaces in the orthopedic applications.

Often, the tissue reaction to a material in bulk form is benign and the particulate form triggers a foreign body reaction. Particulate forms of implant material can be introduced as the products of wear and are therefore a concern in prosthetic devices that articulate with one another. Although the wear resistance of pyrolytic carbon has been shown to be very good, studies have nevertheless been performed on the reaction to carbon particulates both in moving joints and in blood. The living host reaction to particulate carbon is generally passive. It has been observed that particulate carbon shed from braided fibers migrates from the implantation site into the lymph system and deposits passively in the lymph nodes. Rats injected intravenously with up to 20 mg of carbon particulates per kilogram body weight showed a survival rate and weight gain identical to that of control rats.

The environment of body fluids is highly corrosive and there is a potential for the generation of galvanic cells when using carbon in concert with other implant materials. Carbon and graphite have an inertness that is comparable with that of the noble

Table 2
Properties of biomedical carbons

	Vitreous carbon[a]	Low- to high-density LTI carbon	Silicon-alloyed LTI carbon	ULTI carbon
Optical constants (real, r; imaginary, i)		1.79(r), 0.45(i)	1.79(r), 0.45(i)	1.7(r), 0.4(i)
Density (kg m^{-3})	1400–1600	1500–2200	2000–2200	1500–2200
Crystallite size L_c (nm)	1–4	3–4	3–4	0.8–1.5
Flexural strength (MPa)	70–207	275–550	550–620	345–>690
Young's modulus (GPa)	24–31	17–28	28–41	14–21
Fatigue limit/flexural strength	1	1	1	1
Poisson ratio		0.2	0.2	
Hardness (DPH)	150–200	150–250	230–370	150–250
Thermal expansion coefficient (10^{-6} K^{-1})	2.0–5.8	4–6	5	
Electrical resistivity at 20 °C (10^{-3} Ω mm)	30–55	5–20	3–20	
Thermal conductivity at 20 °C (W m^{-1} K^{-1})	4–25	4	4	
Silicon content (wt%)	0	0	5–12	0
Strain energy to fracture (MJ m^{-3})	0.6–1.4	2.7–5.5	5.5	9.9
Strain to fracture (%)	0.8–1.3	1.6–2.1	2.0	>5.0
Impurity level (ppm)		<100	<100	<100

a values are dependent on precursor material

Table 3
Successful applications of glassy, LTI and vapor-deposited ULTI carbons

Application	Material	Cellular compatibility	Strength	Wear resist	Stiffness
		\multicolumn{4}{l}{Material characteristics contributing to success}
Mitral and aortic heart valves	LTI	×	×	×	
Dacron and Teflon heart valve sewing rings	ULTI	×			
Blood access device	LTI/titanium	×			
Dacron and Teflon vascul grafts	ULTI	×			
Dacron, Teflon and polypropylene septum and aneurism patches	ULTI	×			
Pacemaker electrodes	Porous glassy carbon–ULTI-coated porous titanium	×			
Blood oxygenator microporous membranes	ULTI	×			
Otologic vent tubes	LTI	×	×		
Subperiosteal dental implant frames	ULTI	×			
Dental endosseous root form and blade implants	LTI	×	×		×
Dacron-reinforced polyurethane aloplastic trays for alveolar ridge augmentation	ULTI	×			
Percutaneous electrical connectors	LTI	×	×		
Hand joints	LTI	×	×	×	×

metals and are therefore relatively nonreactive with the corrosive fluids of the body. However, when coupled with less noble materials, carbon may through galvanic action accelerate their corrosion. For example, carbon coupled with the common implant material 316L stainless steel may accelerate the corrosion of the steel, especially if the surface area ratio of carbon to steel is large. This, however, is not the case when carbon is coupled, even at a high surface ratio, with titanium, titanium alloy Ti–6% Al–4% V or the medical grade ASTM F-75 cobalt–chromium alloy.

6. Applications

Table 3 reviews the successful uses of glassy, LTI and ULTI carbons in various medical areas. The greatest advantage of carbon over other materials is the excellent cellular biocompatibility in concert with appealing mechanical properties. For example, the strength, wear resistance and fact that carbon structures are unaffected by the *in vivo* environment have all contributed to their successful application in mechanical heart valves. Over 1.5 million patient years have been accumulated with carbon components in mechanical heart valves with a very small number of component-related complications. In contrast, polymeric and tissue component failure is measurable and considerably greater.

Analysis of the *in vivo* wear of carbon articulating surfaces indicates that the results of the accelerated disk-on-plate testing are exaggerated. Thus, in the orthopedic area, the carbons offer an option in the choice of articulating materials for particular applications. Materials whose mechanical properties may be ideal, but have unacceptable tissue biocompatibility characteristics may be coated with the ULTI carbon to form an acceptable composite.

Considerable work is being done in the cardiovascular and orthopedic areas as a result of the mechanical and chemical properties of the various carbons. Due to the wide range of properties available with these carbons, many further applications are being explored in order to take advantage of these special biologically relevant materials.

See also: Biocompatibility: An Overview

Bibliography

Bokros J C 1969 Deposition, structure and properties of pyrolytic carbon. In: Walker P L Jr (ed.) 1969 *Chemistry and Physics of Carbon*, Vol. 5. Dekker, New York, Chap. 1

Bokros J C 1977 Carbon biomedical devices. *Carbon* 15: 355–71

Bokros J C, LaGrange L D, Schoen F T 1972 Control of structure of carbon for use in bioengineering. In: Walker P L Jr (ed.) 1972 *Chemistry and Physics of Carbon*, Vol. 9. Dekker, New York, p. 103

Christel P 1986 The applications of carbon fibre reinforced carbon composites in orthopaedic surgery. *CRC Crit. Rev. Biocompat.* 2: 189–218

Haubold A D 1977 Carbon in prosthetics. *Ann. N. Y. Acad. Sci.* 283: 383–95

Haubold A D 1983 Blood/carbon interactions. *J. Am. Soc. Artif. Internal Organs* 6: 88–92

Haubold A D, Shim H S, Bokros J C 1979 Carbon cardiovascular devices. In: Unger F (ed.) 1979 *Assisted Circulation*. Springer, New York, pp. 520–32

Haubold A D, Shim H S, Bokros J C 1981 Carbon in medical devices. In: Williams D F (ed.) 1981 *Biocompatibility of Clinical Implant Materials*, Vol. 2. CRC Press, Boca Raton, FL, Chap. 1

Krouskop T A, Brown H D, Gray K, Shively J, Romovacek G R, Spira M, Runyan R S 1988 Bacterial challenge study of a porous carbon percutaneous implant. *Biomaterials* 9: 398–404

More N, Baquey C, Barthe X, Rovais F, Rivel J 1988 Biocompatibility of carbon–carbon materials. *Biomaterials* 9: 328–34

Schoen F J 1983 Carbons in heart valve prostheses: Foundations and clinical performance. In: Zycher M (ed.) 1983 *Biocompatible Polymers, Metals and Composites*. Technomic, Lancaster, PA, pp. 240–61

Schoen F J, Titus J L, Lawrie G M 1982 Durability of pyrolytic carbon-containing heart valve prostheses. *J. Biomed. Mater. Res.* 16: 559–70

A. D. Haubold, R. A. Yapp and J. C. Bokros
[Carbomedics, Austin, Texas, USA]

Chemical Adhesion in Dental Restoratives

A tooth is a living organ, at equilibrium with a complex biological environment and set in the mouth close to sense organs, salivary glands, blood vessels and nerves. It is not surprising, therefore, that the first successful *in vivo* bonds to teeth were not made until 1955 and this was by acid etching. The first chemical bonds were not achieved until 1963. Enamel, though a fossilized shell of hydroxyapatite, is coated with an ever-changing gelatinous layer of protein, polysaccharides and microorganisms, a good repellant for any adhesive. The dentin that forms the tough inner core of the tooth is alive, made of a composite of collagen and hydroxyapatite, and containing tubules that radiate from the pulp cavity at the center of the tooth. The tubules exude fluid when cut and contain living cell processes as well as nerves. Chemical bonding to dentin has taken much longer to achieve and is probably the most researched sector of dental materials. The problems of bonding to enamel and dentin are difficult enough in the laboratory, but in the mouth time is of the essence; four minutes is probably the maximum time any practitioner would allow for any surface preparation process prior to bonding. The environment is hostile: patients breathe on the prepared surface, salivate on it and, if allowed, lick it. Once made, the bond is subjected not only to moisture, but also to attack by acidic bacterial endotoxins and hydrolytic enzymes present either in saliva or dentin. The bonds will be thermally cycled during either food ingestion or breathing during cold weather. The bonds are mechanically stressed by masticatory forces, traumatic events and habitual probing by the tongue or opposing teeth.

To solve the problem, surface treatments have ranged from simple nontoxic cleansing methods, through surface etching, to growing mechanically retentive crystals on surfaces. The hydroxyapatite has been chemically treated, much as fillers for composites are, and the collagen has been grafted with polymers just as fibers are treated in the wool and cotton industries. Clinical success has come with the evolution of fast-acting simple-to-use systems which, in both enamel and dentin bonding, have produced bonds with morphologies more akin to welds than to thin-layer adhesives interfacing two substrates.

1. The Clinical Need for Adhesive Restorations

Restoration of a tooth to its original strength and integrity prior to attack by disease, or to its original aesthetic appearance following either drug-related discoloration or trauma, usually requires some form of tooth preparation with a dental drill. In this way, clean-sided cavities are cut which will retain any cement, composite filling material or crown, while not deteriorating at the margin between the tooth and restoration. If a strong adhesive bond can be achieved between the tooth and the restorative, there is no need to cut mechanically retentive cavities; thus sound tooth tissue is saved. In addition, the restored tooth is a stronger unit since the outer enamel shell has been adhesively repaired. This is particularly useful in molar teeth whose cusp may fracture during mastication if a large cavity is drilled in the center of the tooth, separating the cusps of the tooth from each other's support.

In an intact tooth the enamel seals and protects the dentin beneath from the oral environment. When a cavity is cut through the enamel into dentin the tubules are exposed and a route exists from the mouth to the dental pulp for bacteria and toxins.

Adhesive restoratives can not only seal the dentin from infiltration, but can protect the caries-prone enamel–dentin interface from attack. The adhesive layer may also prevent toxins within the restorative from diffusing into the pulp and causing inflammation. There is one school of thought that considers bacterial colonization of the interface between restoratives and dentin the main cause of postoperative pain. Hence adhesives that prevent the bacteria colonizing the base of cavities should reduce the incidence of postoperative pain.

Aesthetically, chemical adhesives are permitting practitioners to veneer anterior teeth with either composite or porcelain to hide unsightly staining of the enamel, with very little need to remove healthy tooth substance. Artificial teeth supported by thin bridges of base metal made to adhere to the back of adjoining teeth offer a conservative alternative to fixed bridges (which require considerable tooth preparation of the healthy teeth either side of the space in the mouth where the false teeth are to fit).

To summarize, adhesive restorative dentistry allows more healthy tooth tissue to remain, less use of drills, and a better chance of protecting the dentin and pulp from bacterial attack due to infiltration at the tooth–restoration interface.

2. Biophysical Considerations

In order for an adhesive to be effective it must wet the surface of the substrate. When bonding restoratives, two substrates are of interest: enamel and dentin. Enamel is exposed to the proteinaceous solution saliva, a biofilm or pellicle adsorbs onto the hydroxyapatite surface and prevents wetting by adhesion promoters and resins alike. Cut dentin is covered in a smear layer consisting of heat-denatured collagen and mineral debris, usually minute crystals of hydroxyapatite. The success or failure of dental bonding systems is greatly dependent on their interaction with these surface films. Acids will remove all but the most mature cross-linked pellicle and any dentinal smear layer. However, since hydroxyapatite dissolves readily at a pH lower than 4.6, the underlying surfaces that have been cleaned by the acid may well be demineralized, more porous and less structurally sound. The enamel surface, when cleaned with strong acid (e.g., 37% phosphoric acid) will dissolve unevenly and if etching is continued long enough, will result in a mechanically retentive surface easily wetted by organic resins. The etch patterns arise from the heterogeneous nature of enamel and from the fact that hydroxyapatite crystals dissolve more rapidly along some crystal planes than others. Light curing a resin that has penetrated the etched enamel surface produces well-polymerized resin tags some $200 \, \mu m$ long. The apatite is probably grafted to the resin by

a light-induced radical reaction initiated at the crystal surface.

Etching dentin produces no such bond. Strong acids open up the tubules and demineralize the peritubular dentin. Though resin can flow into the tubules, it does not wet their sides and no improvement in bond strength has been demonstrated by these means. It has been demonstrated that high bond strength to dentin can be achieved, even when the smear layer is left intact. The fact that collagen can aid in the sealing of mineral within the bond is one reason for leaving the smear layer, another reason being that it occludes the tops of the cut tubules; therefore, if consolidated, the smear layer can act as a seal against bacterial and toxin infiltration of the dentin and pulp.

It has already been stated that dentin is heterogeneous, having tubules rather than being a monolithic composite of collagen and apatite. The dentin is also heterogeneous in terms of mineral distribution. Near the pulp there is less mineralization of the collagen matrix and near the enamel–dentin junction the tubules themselves may be occluded with mineral. The distribution of mineral varies with age, mineral density increasing with age, especially up to the age of about thirty. Hence, one bonding system may be successful in dentin in shallow cavities, while yet another may be more successful in more collagenous deep cavities. However, the deeper the cavity the more nerves and cell processes are encountered, and the nearer the pulp is, making blandness of system a great virtue.

3. Cement-Based Systems

In 1963, D. C. Smith produced the first chemically acting dental adhesive. It was a cement based on zinc oxide as the powder and a solution of poly(acrylic acid) as the binder, and was known as polycarboxylate cement. The cement bonded well to enamel and less well to the less mineralized dentin. It was assumed that the cement binds to the substrate by calcium–carboxylate ionic bonds, the correlation between bond strengths and mineral density of the substrate supporting this assumption. However, calcium polyacrylate is not stable in water, and poly(acrylic acid) is a mild etchant for both enamel and dentin. The reason for permanent bonding of the cement lies in the presence of zinc ions from the filler particles which not only cross-link the gel matrix of the cement, but stabilize the gel matrix at the bond interface, perhaps even modifying the apatite crystals at the enamel surface to zinc rather than to calcium phosphates which are known to be very stable in water.

The carboxylate cements, once mixed, must be used immediately, partly because the bonding ability of the cement relies on there being sufficient unreacted carboxylate groups available to etch and wet the substrate. The setting rheology of these cements is often criticized by practitioners, because of the rapid increase in viscosity and with it an increase in elastic memory as the amount of zinc–acrylates increases. This problem is particularly notable in the cementation of crowns and bridges where small film thicknesses of cement are essential ($<40\,\mu$m), and recovery of any elastic memory in cement during placement causes the crowns to rise off the support cut in the tooth. For this reason, complex crown and bridge work is still cemented in place using nonadhesive cements such as zinc phosphate, which have both a Newtonian rheology and a slower increase in viscosity during setting.

Derivatives of polycarboxylate cements are the glass ionomer cements known increasingly as glass polyalkenoate cements. These were originally made by mixing aluminosilicate glass and were traditionally used for silicate restorations with polyacrylic acid. The setting time of these cements proved too slow, and Wilson et al. (1980) improved the practicality of the system by speeding the set by increasing the alkalinity of the aluminosilicate glass. Further improvements included the copolymerization of the acrylic acid with unsaturated dicarboxylic acid capable of forming intramolecular ionic cross-links during setting, rather than the intermolecular cross-links that caused the rapid initial increase in viscosity of the original polycarboxylate cements. The acids in common use are itaconnic acid and malonic acid. Further delay of initial set of these cements can be achieved by chelating free ions with tartaric acid, thus denying them the chance to cross-link the polymer matrix.

The nature of the glass is very important in the successful formulation of a glass ionomer cement. The early glasses used contained inclusions of calcium fluoride which scattered light to such an extent that the cements looked dead compared to the translucency of the tooth that surrounded them. Later formulations did not contain these inclusions and blended much more into their surroundings. The dissolution of the glass is also critical. The first ions to leach from the glass are calcium and fluoride ions. Calcium, as has been stated previously, does not produce a very insoluble cement with poly(carboxylic acids). It is not until at least one hour after mixing that the aluminum ions leaching from the glass are present in a sufficient quantity to stabilize both the cement and the bond, making them resistant to dissolution by saliva. Hence these cements must be protected by waxes or varnishes from salivary attack if they are not to wash out.

Modern glass ionomers have sufficient translucency to be used as anterior restorative materials and lining cements. Their great advantage is the ability to adhere to both enamel and dentin in the same way as the previously mentioned polycarboxylate cements do. They also contain fluoride which is released during the setting process and diffuses from the gel matrix, which binds the cement, into the surrounding tooth structure, helping to either remineralize early carious lesions or forming acid-resistant crystals of fluorapatite. The glass ionomer cements are particularly useful in restoring erosion cavities near the margin with the gums, without the need to remove healthy tooth substances, and in modifying the shape of molar teeth to make them easier to keep clean. Glass ionomers have been shown to seal dentin from invasion by bacteria very well and this is discussed further in Sect. 4.

The traditional restoratives for molars have been silver–mercury amalgams. A glass ionomer has been produced commercially that challenges some of the traditional roles of amalgam. Silver cermets are produced using aluminosilicate glasses as the base. These are reacted with poly(acrylic acid–malonic acid) copolymer solution to give a cement of metallic appearance. The first silver cermet cement was used clinically for the restoration of molars, building up cores on badly damaged teeth to take a porcelain crown, and as an etchable lining material for composite restorations. Silver cermet cements bond to both enamel and dentin, and are claimed to have superior abrasion resistance to other glass ionomer cements by virtue of the metallic silver in the silver cermet phase.

It has been known for some time that the strength of glass ionomers can be increased by increasing the molecular weight of the polymeric phase in the gel matrix. Glass ionomers containing a copolymer capable of cross-linking by blue light have become available. The pendant cross-linking group contains a large alkyl chain surmounted by an hydrophilic methacrylate moiety. The cement combines the convenience of the command setting that light curing brings with the adhesive properties of conventional glass ionomers. However, since glass ionomer bonds to both enamel and dentin fail cohesively within the cement, the improved cohesive strength of the light-cure glass ionomers is translated into a threefold increase in bond strength to levels previously only achieved by the best dentin bonding agents. Finally, since the light curing increases the cross-link density of the gel matrix, the water and its attendant ions diffuse out of the set cements more rapidly. As a result the fluoride release from the light-cured cements is enhanced.

4. Adhesive Composite Restorations

4.1 Bonding to Enamel

Ever since Buonocore in 1953 demonstrated that phosphoric acid placed on human dental enamel produced a mechanically retentive etch pattern, the chemical treatment of enamel to produce bonds has largely stopped. Prior to the acceptance of acid etching, glues derived from secretions of molluscs, silanes and organophosphates were patented as enamel adhesives, but none compares for simplicity of action and reliability with the 17–20 MPa bond strengths in tension achieved by acid etching enamel. Since apatites exposed to ultraviolet light can initiate additional polymerization on their surface, there may be some chemical adhesion in the process. Most of the adhesive strength, however, derives from the increased wetting of the freshly etched surface by hydrophobic resins, the roughness of the surface itself acting as a fracture toughness enhancer; and the capillary action of the deeper pits which can compete for the resin in composite restoratives and compete with the interfiller particle capillary faces to improve wetting. The very highly filled composites intended for restoration of occluding surfaces in the rear of the mouth wet etched enamel less efficiently because of greater capillary forces within the uncured composite. These products require the application of an unfilled resin to the surface prior to the placement of composite for best results. The increased use of fluoride, as in toothpaste, has had an influence on the etchability of enamel since Buonocore's experiments in the 1950s; it is now increasingly necessary to abrade away the outer 50 μm of the tooth prior to etching in order to achieve a suitable etch pattern. If this is not done the only etching may be in the form of pinholes at the surface, opening out into bottlenecked inclusions in the less fluoridated enamel below.

Etching the enamel provides good bonds between enamel and composite restorations, but it does not necessarily seal the margin of a restoration from acid attack or from bacteriological infiltration. The primers being developed for dentin may be combined with the unfilled resin to improve this seal and with it that marginal integrity of restorations so important for their longevity. Bonding to enamel has made possible chair-side restoration of fractured edges, concealment of tetracycline-stained enamel, veneering of teeth with porcelain and the use of minimal cavity preparation of erosion cavities with subsequent retention of more mature tooth structure.

4.2 Bonding to Dentin

Since acid etching of enamel had produced strong resin bonds, the acid etching of dentin was also tried. It removed the gelatinous smear layer from the surface and allowed resin to penetrate many hundreds of micrometers into the dentin. However, there was no useful improvement in bond strength, largely because the hydrophobic resin failed to wet the gelatinous hydrophilic lining of the tubules, penetration being by means of applied pressure rather than by capillary action. Early researchers in the field viewed the dentin substrate as a mixture of protein (collagen) and mineral (hydroxyapatite),

and the adhesion problem as one of conditioning the surface with a traditional difunctional molecule. One functional group of the molecule should have an affinity either for collagen or hydroxyapatite, and this group should be joined via a spacer group of the correct length and rigidity to a monomer moiety capable of radical addition with the growing resin matrix of the composite restorative. This approach assumes that the cut dentin surface can be treated as a monolithic structure and that the physiology of dentin will allow a thin layer at its surface to remain as part of that living structure. The time allowed for such treatments is very short (a few minutes at most) and this is to be achieved using chemicals bland to the pulp.

The materials that were based on this approach used, as their surface reactive groups, either organophosphates or *n*-phenyl glycine when reacting with hydroxyapatite, and isocyanates when reacting with collagen. These systems were capable of producing high maximum bond strength (17 MPa in shear). However, they were not consistent in their action and wide variations in bond strengths were noted, between 10–20% of bonds failing owing to stresses arising from the uptake of water by the overlying composite. The hydrolytic stability of bonds was also questioned, especially those based on organophosphates. These bonding systems, relying on binding to the mineral phase, can be improved by applying mineralizing solutions to the dentin surface prior to bonding in the same way as proved effective with polycarboxylate-type cements.

However, applying mineralizing agents adds another step to the surface preparation technique and adds to the cost of the restoration. Two types of mineralizing solution have been used: isotonic calcium phosphate producing systems, and nonisotonic systems based on transition metals, of which one based on ferric oxalate is in commercial production. The mineralizing systems have one advantage in that they occlude the dentinal tubules with mineral and therefore aid in sealing the dentin from bacterial attack. However, the isotonic solutions require two minutes of exposure to be effective, and the more rapid transition-metal-based systems, though twice as fast, can be shown to remove calcium from the peritubular dentin and only partially replace it with ferric ions. Even with the use of mineralizing solutions, the monomeric surface treatment methods are not as reliable as the acid etch bonds achieved on enamel and the dentistry profession has never trusted them enough to abandon cutting cavities with mechanical keying.

The first bonding systems for dentin that were reproducible enough to eliminate the need for mechanically retentive cavities began to be reported in the late 1970s. Nakabayashi et al. (1982) reported the first system based on methyl methacrylate, tributyl borane as initiator and 4-META as bonding comonomer. Methyl methacrylate is placed on the cut dentin surface with its smear layer intact. The monomer diffuses through the smear layer and into the dentin. The tributyl borane, in the presence of water, splits into butyl radicals which graft onto the collagen molecules and initiate the polymerization of the methyl methacrylate. The 4-META is a methacryl-substituted melittic anhydride which is hydrolyzed to melittic acid and chelates calcium. This chelation adds to the binding between the growing methyl methacrylate chains, increasing both grafting and cross-link density of the final acrylic–dentin composite layer. The dentin and its smear layer are embedded in hydrophobic resins and can now be bonded to by the curing composite restorative material by anaerobic copolymerization. The bond is very strong (~20 MPa in tension) and very resistant to hydrolysis.

The embedding of the dentin is taken further in the GLUMA system. Here an equimolar mixture of gluteraldehyde and hydroxyethyl methacrylate is placed on dentin that has been cleansed of its smear layer by EDTA solution. The glute–HEMA complex penetrates the tubules to depths greater than 300 μm. There is some reaction between the collagen and the gluteraldehyde, and between the hydroxy group on the HEMA and the gluteraldehyde. The result is a dentin surface well wetted by a hydrophilic monomer that is light cured. An intermediate unfilled resin is then placed on the treated surface, which forms a graded bridge between the hydrophilic HEMA and the hydrophobic composite resin.

The composite restoration, bridging resin and HEMA–glute complex are then finally light cured together using an α-diketone–amine photoinitiator. Once again, strong bonds are achievable and the bonds are very resistant to acids. It can be demonstrated that this system protects the vulnerable interface between the enamel and dentin very well by diffusing along the gap between the two, and then outwards along dentinal tubules and the proteinaceous sheaths surrounding the enamel prisms.

A variant of the GLUMA system has been introduced in which the gluteraldehyde is replaced by malaic acid. The bonds of both systems fail in the unfilled resin layer. The GLUMA system has shown a tendency to produce poor bond strengths if insufficient exposure to the blue light is given during the curing of the composite layer. The malaic-acid-based system, however, has shown a tendency to poor bond strengths if the unfilled resin layer is applied too sparingly. In each case low bond strengths occur because of too low a degree of conversion of monomer in the intermediate unfilled layer.

The commercial derivative of the ferric oxalate work mentioned previously first cleanses the dentin with aluminum oxalate solution containing nitric acid. This removes all traces of mineral from the surface layer of the dentin and considerably enlarges the tubules. The treated surface is then reinfiltrated with surface-active monomers dissolved in acetone, followed by an application of unfilled resin. The result is very similar to the HEMA-based materials described previously, bond failure being in the infilled resin layer once again. Since failure is focused on this infilled resin layer in these last three systems, higher bond strengths will only be possible when methods of guaranteeing high rates of conversion of this layer can be found.

Other systems infiltrate the intact dentin smear layer, cross-linking it by the reaction of isocyanate groups with collagen. The smear layer is first defatted and dehydrated with acetone to maximize this reaction. The cross-linking extends into the very outer part of the intact dentin, but no further. The isocyanate also bears a methacryl group which is capable of copolymerizing with the composite restorative. This is a very quick technique, marred only by the very rapid reaction of the isocyanate which can cause the applicator brush to drag in the cavity. This system seals the dentin well, provided the cut dentin is covered in an intact smear layer. If the layer has been removed for any reason, protection from acid attack is less in that area.

The simplest bonding agent to use consists of the resin used to bind composite restorative materials, a diacrylate of bisphenol A known as bisGMA, that has been reacted with phosphorus oxychloride. Pendant hydroxyl groups on the bisGMA react to form a mixture of mono-, di- and triphosphate esters. This mixture, when mixed with alcohol, organic sulfinate and a suitable curing agent, will react with and partially solubilize the dentinal smear layer. When viscous composite restorative is packed onto the dentin, the reacted smear layer mixes with the composite and the whole is cured to form a weld-like structure which, though it contains organophosphorus linkages, stands up well to hydrolysis and acid attack.

The successful dentin bonding agents have one thing in common: they are more like welds than the simple bonds between two substrates in which the adhesive is present only as an almost monomolecular layer. The bond region between composite restorative and dentin may be 50–300 μm thick, having gradients of intimately mixed components that have originated in either substrate or the bonding agent, extending throughout the bond. Composite components can be detected deep in the dentin and calcium from the dentin can be detected in the composite.

5. Future Developments

The retentive properties of modern enamel and dentin adhesion promoters appear adequate for all applications. However, two aspects of their use require further development: bonds break down *in vivo*, and curing contraction stresses are transmitted to the tooth causing postoperative pain and, occasionally, enamel fracture. Clinical trials report marginal breakdown rates of about 20%. This may be because of operator error; however, hydrolytic stability of the resins used in these bonds and greater bond strengths that allow greater margins of operator error are being developed. The contraction stresses of composite restorative materials on setting varies between 2% and 4% depending on filler content. In a bonded composite these stresses are transmitted to the tooth and can either distort cusps or pull bonds apart. One method of reducing these stresses at the bond interface is to make the bonding agent elastomeric. This has been shown to be a viable approach and has resulted in a 50% reduction in strain in cusps adjacent to bonds.

See also: Adhesives in Medicine

Bibliography

Bowen R L, Cobb E N, Rapson J E 1982 Adhesive bonding of various materials to hard tooth tissues: Improvement in bond strength to dentin. *J. Dent. Res.* 61: 1070–6

Causton B E 1984 Improved bonding of composite restorative to dentine. A study *in vitro* of the use of a commercial halogenated phosphate ester. *Br. Dent. J.* 156: 93–5

Causton B E, Johnson N W 1979 The role of diffusible ionic species in the bonding of polycarboxylate cements to dentine: An *in vitro* study. *J. Dent. Res.* 58: 1383–93

Davidson C L 1986 Resisting the curing contraction with adhesive composites. *J. Prosthet. Dent.* 55: 446–7

Fan P L 1987 Dentin bonding systems: An update. *J. Am. Dent. Assoc.* 114: 91–4

McLean J W, Powis D R, Prosser H J, Wilson A D 1985 The use of glass–ionomer cements in bonding composite resins to dentine. *Br. Dent. J.* 158: 410–14

Munksgaard E C, Irie M, Asmussen E 1985 Dentin–polymer bond promoted by gluma and various resins. *J. Dent. Res.* 64: 1409–11

Nakabayashi N, Kojima K, Masuhara E 1982 The promotion of adhesive by the infiltration of monomers into tooth substrates. *J. Biomed. Mat. Res.* 16: 265–73

Prosser H J, Powis D R, Wilson A D 1986 Glass–ionomer cements of improved flexural strength. *J. Dent. Res.* 65: 146–8

Smith D C 1971 A review of the zinc polycarboxylate cements. *J. Can. Dent. Assoc.* 37: 22–9

Wilson A D, Crisp S, Prosser H J, Lewis B G, Merson S A 1980 Aluminosilicate glasses for polyelectrolyte cements. *Ind. Eng. Chem., Prod. Res. Dev.* 19: 263–70

B. E. Causton
[King's College School of Medicine and Dentistry, London, UK]

Cobalt-Based Alloys

The use of cobalt-based alloys in medicine can be said to have started in 1929 with the investigation by Erdle and Prange of different cobalt–chromium alloys for use as removable dental restorations (Lane 1949). This led to the patenting of one particular alloy, Vitallium, which demonstrated good mechanical properties and the ability to be cast into the required intricate shapes. As a result, the alloy began to be used extensively in dentistry and by 1937 it came to the attention of Venable and Stuck (1941) for possible use in surgery. At that time, the only surgical alloys available were comparatively crude stainless steels possessing marginal corrosion resistance to body fluids; consequently it was possible with Vitallium to demonstrate a clear superiority. This led to the use of Vitallium for selected surgical applications such as bone plates.

By 1940, Vitallium had been chosen as a turbine bucket material for the early designs of jet engines, due to its good oxidation resistance at high temperatures. However, it was quickly realized that the alloy possessed insufficient fatigue strength. A period of extensive research was begun which resulted in the development of a whole new group of cobalt-based alloys. This research was unfortunately curtailed by the shortage of cobalt during the Korean War, prompting a switch to the nickel-based alloys and leaving cobalt-based alloys to be used only for extremely high-temperature applications.

Shortly before this time, interest began to develop in the possibility of treating arthritic degenerative changes to the hip joint by prosthetic replacement. In 1938, Smith-Petersen (1939) introduced a cup arthroplasty for interposition between the femoral head and natural acetabulum. Since this required an extremely corrosion-resistant material, Vitallium was chosen. Approximately 20 years later, Moore developed a hemiarthroplasty for replacement of the entire femoral head, again fabricated from Vitallium. The good casting properties of this alloy allowed for easy fabrication of the intricate shape required for this application. These two devices clearly demonstrated the feasibility of using metallic implants to treat orthopedic disorders and led directly to the development of the total hip prosthesis by both McKee and Watson-Farrar (1966), who used a metal-upon-metal sliding configuration for the femoral and acetabular components, and Charnley (1961), who used a metal-upon-plastic combination.

Even though it was subsequently found that the McKee–Farrar design tended to loosen (Evans et al. 1974), the use of Vitallium for these devices increased as the operation gained in acceptance. By the early 1970s, however, and with an increasing number of total hip prostheses being inserted into younger and more active patients, it became clear

Table 1
Composition (wt%) of surgical grade cobalt–chromium alloys[a]

Component	Cast F-75	Forged F-75	Hot isostatically pressed F-75	Wrought F-90	Wrought F-562
Chromium	27–30	26–28	27–30	19–21	19–21
Carbon	≤0.35	≤0.05	0.23	0.05–0.15	≤0.025
Molybdenum	5–7	5–7	5.81		9.0–10.5
Tungsten				14–16	
Nickel	≤2.5	≤1.0	0.14	9–11	33–37
Iron	≤0.75	≤0.75	0.15	≤3	≤1
Manganese	≤1.0	≤1.0	0.40	≤2.0	≤0.15
Silicon	≤1.0	≤1.0			≤0.15

a Balance Co; ASTM specifications except for hot isostatically pressed F-75, for which a typical melt composition is given

'that fatigue failures of the femoral stems were constituting a problem (Galante et al. 1975) and necessitated an improvement in mechanical properties. This caused a resurgence of interest in the metallurgy of the alloy and led to the introduction of the current range of cobalt-based surgical alloys and also to the fabrication techniques currently available. These alloys are now employed in a wide variety of applications within medicine and possess mechanical properties that have allowed manufacturers to produce substantially improved devices.

1. Surgical Grade Cobalt–Chromium Alloys: Composition and Structure

Within the overall classification of cobalt–chromium alloys, there are four which are currently used for surgical applications. Of these four, only one, F-75, is fabricated by casting; the others are fabricated by some form of thermomechanical processing to produce the required configuration. Obviously these differences in fabrication technique imply a different physical metallurgy for each alloy and this is reflected in their respective compositions as shown in Table 1.

1.1 Casting Alloy

The structure of F-75 cobalt–chromium casting alloy was traditionally determined more by concern for good corrosion resistance and ease of casting rather than by considerations of mechanical properties. As a result, a chemical composition was set to ensure that these parameters were optimized. The two most important additions to the cobalt base are chromium and carbon, the former being used to enhance corrosion resistance and the latter to improve the castability of the alloy.

In the unalloyed condition, cobalt possesses insufficient corrosion resistance for any surgical application. With the addition of chromium, however,

the corrosion rate is reduced, due to the formation of chromic oxide on the surface of the alloy. This serves to inhibit further interaction between the environment and the alloy and by so doing reduces corrosion. Even though many cobalt-, nickel- and iron-based alloys, all of which utilize the same phenomenon, contain only 10–20% chromium, F-75 has always contained 27–30% chromium. This provides the best corrosion resistance of any of the cobalt–chromium surgical alloys.

The use of carbon as an alloying element for cobalt was originally intended to lower the melting point while providing for solid-solution hardening of the matrix. This lowering of the melting point was important, in that it enabled the alloy to be cast using simple foundry equipment and inexpensive materials. For this reason, the dental grade Vitallium (Phillips 1973), which is generally cast in a small laboratory, still maintains a substantially higher carbon content (up to 0.50%) than surgical grade F-75, which contains a maximum of 0.35%. Even this level of carbon addition, however, is sufficient to lower the melting point to about 1350 °C, compared with approximately 1450–1500 °C for a binary cobalt–chromium alloy.

In addition to chromium and carbon, both molybdenum and nickel are present in F-75. The role of these elements remains uncertain; however, both are likely to improve mechanical properties by solution hardening of the matrix.

For many years, F-75 was used solely in the as-cast condition with little attention to the ways in which the mechanical properties could be improved. More recently, however, the need for these improved properties has stimulated research interest in the use of heat-treatment programs designed to modify the microstructure.

In the as-cast condition, F-75 consists of a highly cored dendritic matrix within which are dispersed chromium-rich carbides of the type $M_{23}C_6$, M_7C_3 and M_6C (M = Cr, Mo or Co) (Clemow and Daniell 1979). This coring causes the alloy composition to vary across the microstructure and therefore also results in small differences in the mechanical properties between the cobalt-rich dendrites and the chromium-rich interdendritic regions. A further feature of F-75 is the extremely large grain size frequently observed in the original castings, which reduces both the strength and ductility of the alloy. This problem is particularly severe in castings having a large cross section, since here any chilling effect from the mold wall is quickly lost and the remaining liquid solidifies slowly, resulting in a coarse grain size.

Modification of this microstructure by heat treatment may be conducted in two ways (Clemow and Daniell 1979). In the first, the alloy is given a simple homogenization treatment by heating it to above 1150 °C and allowing the compositional inhomogeneities to be removed by diffusion. At this temperature, however, the carbide particles remain unchanged and the only gain in mechanical properties is an increase in the ductility. The second and more common treatment involves a full solution treatment of the alloy, in which not only is the coring removed but the carbide particles are dissolved in the matrix. This allows the carbide-forming elements to harden the alloy by solid-solution hardening and so causes an improvement in mechanical properties. This treatment is normally conducted at 1210–1250 °C, the optimum temperature depending upon the carbon content of the particular alloy.

However, certain adverse effects may still be caused by such a treatment. One such effect is that the carbides may reprecipitate on cooling unless a sufficiently rapid cooling rate is used, resulting in a loss of ductility. More seriously, excessive grain growth may occur at the solution treatment temperature once the carbides have been dissolved, since this removes any barrier to grain-boundary motion. This problem is normally circumvented by ensuring that full carbide dissolution does not take place and that a few small particles remain to pin the grain boundaries. For this reason, the normal solution-treated microstructure of F-75 consists of a homogeneous solid solution with a dispersion of small residual carbides.

1.2 Thermomechanically Processed Alloys

At present, there exist three different cobalt–chromium alloys which are processed by thermomechanical treatment and used for surgical applications. Like the iron- and nickel-based alloys (to which they are metallurgically very similar), F-75, F-90 and F-562 are highly dependent for their final microstructures on the composition and thermomechanical processing history. This is due in part to the allotropic transformation from the hexagonal-close-packed (hcp) crystal structure to the face-centered-cubic (fcc) structure; this occurs in cobalt on heating at approximately 430 °C. However, since only a very small energy difference separates these two crystallographic forms, both may be made to coexist at room temperature by careful alloying and/or mechanical working. In this way, a single-phase microstructure may be converted to one having two phases and a substantial improvement in mechanical properties obtained. The hcp phase is promoted by additions of chromium, molybdenum, tungsten and silicon, while the fcc phase is stabilized by carbon, nickel, iron and manganese.

The importance of good corrosion resistance for cobalt-based alloys in surgery dictates that chromium be added within the range 19–30% (Acharya et al. 1970). As a result, the cobalt matrix is strengthened by solid-solution hardening and the relative stability of the hcp phase is increased. This stability is further increased by the addition of either molybdenum in forged F-75 or tungsten in F-90, so that during cold working it is comparatively easy to induce the allotropic transformation. For the wrought F-75 alloy, the chemical composition is maintained identical to that of the casting alloy except that the percentage of carbon is reduced to almost a minimum, to inhibit carbide formation.

For both F-90 and F-562 alloys, a slightly lower chromium concentration of 19–21% is used, although with the further addition of either molybdenum or tungsten. To ensure adequate fcc stability, F-90 contains a small percentage of carbon, 9–11% nickel and small additions of both iron and manganese, while F-562 relies on the addition of 33–37% nickel. For all three alloys, an annealing treatment before any mechanical deformation causes the fcc structure to form. This phase may be retained at room temperature by cooling, in which condition the alloys exhibit poor mechanical properties (Graham and Youngblood 1970). When mechanical energy is applied, however, the hcp transformation is induced and a significant improvement in mechanical properties is obtained. Obviously the final properties depend greatly on the degree of plastic deformation and generally a compromise between sufficient strength and adequate ductility is chosen. The microstructure of the alloys in this condition consists of an fcc matrix containing a dispersion of hcp platelets, the spacing between these platelets decreasing with increasing deformation.

For F-562 alloy, additional hardening may be obtained by aging cold-worked specimens at temperatures of 500–600 °C for 1–4 h (Graham 1969). A two-stage mechanism for this hardening has been proposed, whereby initially molybdenum is forced out of solution and to the interface between the fcc and hcp phases where, on further aging, it is precipitated out as Co_3Mo. This reaction is generally sufficient to cause a 10–20% increase in hardness, making the alloy one of the strongest available at present.

2. Fabrication

2.1 Casting

Of the four cobalt–chromium surgical alloys currently used, only F-75 is fabricated by casting. For the majority of applications, investment casting is used, since this process has been found to allow for easy design changes while ensuring that close dimensional tolerances are achieved. Recent developments in casting practice have concentrated on improving the quality of the final casting, in particular, controlling the inclusion content and grain size. To achieve this, most major manufacturers have installed either vacuum melting or vacuum casting systems to minimize the level of oxide inclusions. At the same time, a lower melt temperature can be used, with a subsequent refinement in final grain size. Typically this reduction in temperature is in the region of 150 °C from the normal 2750–2800 °C for air-melted material. In the same way, the vacuum-processed alloy allows a lower mold temperature of 900 °C to be used as opposed to the usual 950–1000 °C for air-melted alloy. Both of these practices induce a greater homogeneous nucleation rate within the alloy, resulting in finer grain size within the casting.

An alternative approach to vacuum melting and casting is the practice of inoculating the mold with complex silicate compounds. This serves to increase the heterogeneous nucleation rate of the alloy and causes a reduction in grain size.

2.2 Hot Isostatic Pressing

As an alternative to conventional casting practices, powder metallurgical techniques have been used successfully to achieve extremely good mechanical properties in F-75 (Bardos 1979). In this process, near-final shapes are produced by compacting alloy powder under the influence of isostatic pressure (~100 MPa) and a high sintering temperature (1100 °C). In this way, the individual powder particles bond together to form a fully dense material having a very fine grain size and hence good mechanical properties. To achieve these properties, slight modifications to the original chemical composition have been found to be necessary, and because hot isostatic pressing involves greater production costs than investment casting, it is currently used only for components subjected to high levels of stress, most notably the femoral stems of total hip prostheses. More recently, however, it has been found that substantial improvements in the mechanical properties of standard castings may be obtained by hot isostatically pressing them in the as-cast condition. In this way, any gas porosity or shrinkage voids can be closed, while a slight refinement in grain size is achieved.

2.3 Forging

To achieve the maximum strength in cobalt–chromium alloys, F-78, F-90 and F-562 may all be forged to near final shape from rough billets. In this process, it is common to use several steps, during which the alloy is successively brought closer to the required shape. The temperature at which these operations are effected varies, since even though a higher temperature allows a greater deformation to be achieved, it results in less strengthening. Consequently, it is usual for the early forging steps to be conducted at a comparatively high temperature to allow easy deformation and the later steps at a lower temperature to induce as much cold working as possible. In this way the finished product may exhibit extremely high strength levels, while there are minimal quantities of metal to be removed subsequently by machining.

3. Mechanical Properties

3.1 F-75

The F-75 alloys used in surgical devices are capable of exhibiting a wide range of mechanical properties, as shown in Table 2. Generally, adequate properties are obtained in the as-cast condition to satisfy most applications and therefore the majority of items manufactured from this alloy are produced in this condition. In certain instances, however, and particularly when a higher ductility is required, a solution treatment program may be used, which serves to increase the percentage elongation while slightly improving the tensile strength (Devine and Wulff 1975). Since this treatment increases the likelihood that the alloy will pass the required American Society for the Testing of Materials (ASTM) specification for ductility while also providing a slight increase in strength, its use is becoming more widespread for orthopedic devices.

For the hot isostatically pressed alloy, a substantial improvement in mechanical properties can be achieved—a fatigue strength more than double that of the as-cast alloy being attainable (Bardos 1979). This processing technique serves to illustrate the importance of a fine grain size and carbide distribution in the F-75 alloy system to achieve optimum mechanical properties. In the same way, the forged F-75 is capable of exhibiting extremely high strength values through a combination of a fine grain size and the cold-work-induced allotropic transformation. In this case, a high alloy ductility can be maintained, even with a large degree of cold work, since the carbon content is reduced to very low levels. Such processing treatment as hot isostatic pressing or forging can only be justified for specific applications where extremely high mechanical properties are important, since the associated cost is substantially greater.

3.2 F-90

Since in the annealed conditions F-90 consists of a single-phase fcc microstructure, only very poor mechanical properties can be expected. With cold work, however, the allotropic transformation can be induced and similar or even superior properties to those of the forged F-75 can be obtained. This difference in properties between F-90 in the annealed and work-hardened conditions means that great care must be taken to ensure even and thorough deformation of the component. Without this, the variation in properties may allow unexpected failure to occur.

3.3 F-562

In the same way as for F-90, F-562 requires cold working to achieve its excellent mechanical properties (Graham 1969). In fact, in the annealed condition F-562 exhibits the poorest properties of all of the cobalt-based alloys, due to the high nickel content. This element is very similar to cobalt and so is unable to provide solid-solution hardening of the matrix. Cold working of this matrix by itself is sufficient to improve the properties to levels similar to those attained in the other forged or wrought alloys. However, for F-562 there exists the additional possibility of aging at approximately 550 °C, which causes an intermetallic compound to precipitate and produces the highest strengths of any of the surgical alloys. It is interesting to note that even in this condition, when a yield strength of 1600 MPa may be expected, the alloy is still capable of exhibiting elongations in excess of 8%.

4. Corrosion Resistance

Because of the environment within which they must operate, cobalt–chromium alloys must exhibit good corrosion resistance, and numerous investigators have demonstrated this property (Kuhn 1981). Certainly it was this property that first distinguished these alloys from those available in the 1930s, and even today cobalt–chromium alloys are superior in this respect to almost every other group of elements available.

In any surgical application, a number of different types of corrosion can occur, including uniform, galvanic, crevice and stress-corrosion cracking.

4.1 Uniform Corrosion

The overall level of corrosion from uniform surface attack on the cobalt–chromium alloys can be considered to be minimal, only marginal differences being found between the various types. Recent reports have indicated, however, that even though no corrosion may be detectable, a certain amount of ionic dissolution can occur, leading to an increase in metallic ion concentration in the blood (Woodman et al. 1981). This increase indicates that even apparently inert materials interact with their body environment, although the significance of this interaction remains uncertain.

4.2 Galvanic Corrosion

The coupling of cobalt–chromium alloy components with those manufactured from other alloys is generally avoided, for fear of galvanic corrosion. This is particularly true for the stainless steels with their lower resistance to corrosion, which may lead to severe attack. For the cobalt–chromium alloys, however, this danger is less pronounced, so even when they are coupled with carbon or titanium, both of which are cathodic to them, no galvanic corrosion effects are observed. Similarly, combinations of the various cobalt–chromium alloys have been found to be unaffected by their slightly differing potentials and no galvanic corrosion attack has been reported (Kuhn 1981).

Table 2
Typical mechanical properties of surgical grade cobalt–chromium alloys

Alloy	Condition	0.2% yield strength (MPa)	Ultimate tensile strength (MPa)	Elongation (%)	Fatigue strength (MPa)
F-75	As-cast	515	725	9	250
F-75	Solution treated	533	1143	15	280
F-75	Hot isostatically pressed	840	1275	16	765
F-75	ASTM minimum	450	655	8	
F-75	Forged	962	1507	28	897
F-90	Annealed	350	862	60	345
F-90	Cold worked	1310	1510	12	586
F-90	ASTM minimum	310	860	10	
F-562	Annealed	240	795	50	333
F-562	Cold worked	1206	1276	10	555
F-562	Cold worked and aged	1586	1793	8	850

Table 3
Summary of surgical uses of cobalt–chromium alloys

Alloy	Use
F-75 (cast)	Prosthetic replacements of hips, knees, elbows, shoulders, ankles, fingers
	Bone plates and screws
	Bone rods and staples
	Heart valves
F-75 (forged)	Prosthetic replacement of hips
F-90 (forged)	Prosthetic replacement of hips
	Heart valves
	Wire
F-562 (forged)	Prosthetic replacement of hips

4.3 Crevice Corrosion

Crevice corrosion is the principal form of corrosion observed around multicomponent surgical implants and is found when two or more components meet to form an interface. Due to the frequency with which such corrosion is encountered, a great deal of research has been conducted in this area and it is generally agreed that the surgical alloys exhibit decreasing crevice-corrosion resistance in the order: titanium and its alloys; cobalt and its alloys; and finally the stainless steels (Kuhn 1981). Within the cobalt-based alloys themselves, slight differences in resistance have been found, F-75 exhibiting the highest resistance followed by F-90 and finally F-562 (Devine and Wulff 1976).

4.4 Stress-Corrosion Cracking

The cobalt–chromium alloys are apparently immune to stress-corrosion cracking, even though the body environment contains a high level of chloride ions (Kuhn 1981).

5. Surgical Uses

Because the environment within which surgical alloys must operate is extremely corrosive, only four types of metal or alloy are currently used. These include 316 stainless steel, titanium and its alloy Ti–6% Al–4% V, tantalum, and the cobalt–chromium alloys. Of these, stainless steel and the cobalt-based alloys are the most widely used, the former being chosen where the device is intended to be temporary and the latter where it is likely that the implant will be *in situ* for an extended period of time. The two major areas of use for the cobalt–chromium alloys are the fields of orthopedic and cardiovascular surgery (see Table 3).

For orthopedic applications, the alloy is used for either the replacement of natural joints or the internal fixation of bone fractures. For most designs, F-75 alloy has been found to be perfectly satisfactory, combining good mechanical properties with excellent corrosion resistance and the ability to be cast into near-final intricate shapes. An additional advantage to the implant manufacturer is that investment casting requires only that a wax pattern be produced, so new or modified designs can be introduced without the need for expensive retooling.

To date, the list of joints which have been successfully replaced includes the hip, knee, elbow, shoulder, finger and ankle, with designs generally imitating the natural configuration. For the vast majority of these, it is sufficient only that the implant be manufactured to close tolerances while exhibiting good corrosion resistance, and for these applications, as-cast F-75 has proved ideal. The only exceptions are total hip prostheses, where it has been found that greater fatigue strengths than those obtainable from as-cast F-75 are required, to minimize the danger of stem failure caused by inadequate fixation to bone. For this reason, the new generation of cobalt–chromium alloys, such as the forged or hot isostatically pressed F-75 and the wrought F-90 and F-562 alloys, has been introduced; with their substantially higher fatigue strengths, the incidence of stem failure has been reduced. With all these alloys, however, it is usual for an F-75 head to be welded onto the high-fatigue-strength femoral stem, since the fabrication of a polished and spherical part from such alloys is not technically and economically feasible. These welds have been found to be structurally sound and there have been no reports of galvanic corrosion effects between any of the dissimilar metals.

For the fixation of bone fractures, it is common practice to use a plate secured across the fracture by screws. Even though many of these devices are made from 316 stainless steel, the incidence of both crevice and fretting corrosion at the interface between plate countersink and screw chamfer has caused sufficient concern to warrant the manufacture of certain implants from F-75 alloy. In these cases the ability of this alloy to be cast into intricate shapes becomes paramount, since the difficulty of machining screws from such a hard material would otherwise render the operation infeasible. Even with the excellent castability of F-75, the use of cobalt–chromium internal fixation devices is usually only warranted when it is expected that the device will be required to remain *in situ* for long periods of time.

Aside from the orthopedic applications previously described, cobalt–chromium alloys have also found considerable use in the manufacture of cardiovascular implants. Here they are employed as either support rings or confinement cages in artificial heart valves. In early designs of these devices, a small amount of metal-upon-metal sliding contact was common and it was because of this that F-75 was originally chosen. Today, however, the designs have changed sufficiently, so even though excellent corrosion resistance remains vital, the major concern is a more easily machinable and formable alloy. Consequently, a number of manufacturers have chosen to use the F-90 alloy instead; this machines more easily and to closer tolerances than F-75. An added advantage of this material is believed to be that a polished F-90 surface possesses a greater resistance to thrombus formation than the equivalent F-75 surface and as a result improves the long-term performance of these implants.

In summary, the different cobalt-based surgical alloys are used in applications requiring their combination of high strength, excellent corrosion resistance and ease of fabrication. For those devices of intricate shape, such as finger, elbow or knee joint prostheses, F-75 alloy allows for the necessary close dimensional tolerances to be achieved through investment casting. However, when high strengths or easier machining ability are required, one of the forged or thermomechanically processed alloys is generally adequate.

Bibliography

Acharya A, Freise E, Greener E H 1970 Open circuit potentials and microstructure of some binary Co–Cr alloys. *Cobalt (Engl. Ed.)* 47: 75–80

Bardos D 1979 High strength Co–Cr Mo alloy by hot isostatic pressing of powder. *Biomater. Med. Devices Artif. Organs* 7: 73

Bartoluzzi A, Black J 1985 Chromium concentrations in serum, blood clot and urine from patients following total hip arthroplasty. *Biomaterials* 6: 2–7

Charnley J 1961 Arthroplasty of the hip. *Lancet* 1: 1129

Clemow A J T, Daniell B L 1979 The influence of heat treatment upon the microstructure of Co–Cr–Mo alloy. *J. Biomed. Mater. Res.* 13: 265–79

DePalma V A, Baier R E, Ford J W, Gott V L, Furnse A 1972 Investigation of three surface properties of several metals and their relation to blood biocompatibility. *Biomed. Mater. Symp.* 3: 37–75

Devine T M, Wulff J 1975 Cast vs wrought cobalt–chromium surgical implant alloys. *J. Biomed. Mater. Res.* 9: 151–67

Devine T M, Wulff J 1976 The comparative crevice corrosion resistance of cobalt–chromium base surgical implant alloys. *J. Electrochem. Soc.* 123: 1433

Evans E M, Freeman M A R, Muller A J, Vernon-Roberts B 1974 Metal sensitivity as a cause of bone necrosis and loosening of the prosthesis in total joint replacement. *J. Bone Jt. Surg.* 56B: 626

Galante J O, Rostoker W, Doyle J M 1975 Failed femoral stems in total hip prostheses. *J. Bone Jt. Surg.* 57A: 230–6

Graham A H 1969 Strengthening of "MP-alloys" during aging at elevated temperatures. *Trans. Am. Soc. Met.* 62: 930–5

Graham A H, Youngblood J L 1970 Work strengthening by a deformation-induced phase transformation in MP-alloys. *Metall. Trans.* 1: 424–30

Koegel A, Black J 1984 Release of corrosion products by F-75 cobalt base alloy in the rat. *J. Biomed. Mater. Res.* 18: 513–22

Kuhn A T 1981 Corrosion of Co–Cr alloys in aqueous environments. *Biomaterials* 2: 68–77

Lane J R 1949 A survey of dental alloys. *J. Am. Dent. Assoc.* 37: 414

McKee G K, Watson-Farrar J 1966 Replacement of arthritic hips by the McKee–Farrar prosthesis. *J. Bone Jt. Surg.* 48B: 245

Phillips R W 1973 *Skinner's Science of Dental Materials*. Saunders, Philadelphia, PA

Pilliar R M, Weatherly G C 1985 Developments in implant alloys. *CRC Crit. Rev. Biocompat.* 1: 371–402

Smith-Petersen N M 1939 Arthroplasty of the hip: A new method. *J. Bone Jt. Surg.* 21: 269

Thomas I T, Evans E J 1986 The effect of cobalt–chromium–molybdenum powder on collagen formation by fibroblasts in vitro. *Biomaterials* 7: 301–4

Venable C S, Stuck W G 1941 Three years experience with Vitallium in bone surgery. *Ann. Surg.* 114: 390

Williams D F 1981 The properties and clinical uses of cobalt-based alloys. In: Williams D F (ed.) 1981 *Biocompatibility of Clinical Implant Materials*, Vol. 1, CRC Press, Boca Raton, FL, pp. 99–127

Woodman J L, Black J, Nunamaker D M 1983 Release of cobalt and nickel from a new finger joint prosthesis made of cobalt–chromium alloy. *J. Biomed. Mater. Res.* 17: 655–8

A. M. Weinstein and A. J. T. Clemow
[IatroMed, Phoenix, Arizona, USA]

Collagen

Collagen is the main protein component of all connective tissues; that is, the tissues of the body that provide load carrying and protective functions, including bones, tendons, ligaments, cartilage and skin. Because of its structure and properties, it can be considered as a useful material and for many years it has been a focus of attention in the biomaterials field.

Collagen was, in fact, one of the first biomaterials to be used in surgery, since it is the basis of the catgut suture. In this particular form it is derived from animal tissue and processed, rather like the tanning process in the leather industry, into the reasonably strong but degradable filaments that are needed for absorbable sutures. In more recent years, a wide variety of other possibilities has arisen. There are several ways of preparing collagen and several types of structure can be generated; thus the material is potentially quite versatile. In this article, the chemistry and structure of collagen will be considered first, followed by the preparation and properties of collagenous biomaterials and, finally, the examples of medical applications.

1. Chemical Structure

Collagen is the name given to a series of structural proteins which have the same basic molecular arrangement but some differences in amino acid sequences. All collagens consist of three chains which are arranged in a triple helix configuration. They are characterized by every third residue in the chains being the amino acid glycine, as determined by the configuration. The variety of collagens is determined by the nature of the other residues and the number of repeat units in the molecule. The different collagens are referred to as "types" and currently 11 (Types I–XI) genetically different types have been identified and characterized. Collagen is synthesized in and secreted from cells (fibroblasts). The precursor chains, as synthesized, are polypeptides with the appropriate sequence, but with different domains at the nitrogen and carbon terminals. The transcription and translation processes that take place cause the polypeptide chain to undergo a series of changes, including hydroxylation and glycosylation of some residues such that the molecules attain the structure necessary for helix formation. Three precursor chains combine and fold from the carbon terminal, the resulting structure being known as procollagen. At each end of the helix are telopeptides, which are short nonhelical domains that, being reactive, are able to provide the molecules with cross-linking capability.

Individual collagen molecules are about 300 nm long and these are arranged with a staggered overlap such that the resulting fibrils have a characteristic banding pattern, on electron microscopic examination, with a periodicity of 67 nm. The cross-linking takes place initially by the action of an enzyme on the lysine or hydroxylysine residues in the telopeptides. A variety of other intramolecular and intermolecular cross-linking reactions can then take place, the extent of which controls, to a large extent, the resulting properties of the collagen.

Type I collagen is the most abundant in connective tissue, especially in tendon and skin. Types II and III can also be found in relatively large quantities, particularly in cartilage and blood vessels, respectively. These three types are quite similar structurally. The remaining types are less frequently encountered and possess differences in their structural features appropriate to their particular function.

2. Properties

If collagen is to be used as a biomaterial, it has to possess certain functional properties and display appropriate biocompatibility. Properties most important with respect to its use as a structural or space-filling material are the mechanical characteristics. Associated with this are the properties of stability and/or controlled degradation or turnover. As far as biocompatibility is concerned, collagen, as a natural biopolymer, should be extremely well accepted from the toxicological and conventional foreign-body reaction point of view, but equally could be capable of eliciting immunological reactions because of its proteinaceous nature.

Before describing these properties it is necessary to comment on the form of the collagen that is most appropriate to biomedical applications. There are, in fact, two different forms that may be used. The first is the collagen obtained directly from natural tissues and used as a structural material after appropriate processing. The second is reconstituted collagen, which is prepared as a soluble purified form and then reprecipitated. The significance of these two approaches is that, as a tissue-derived material, the former structure possesses good mechanical properties, but is quite highly immunogenic. Solubilizing and reprecipitating collagen, on the other hand, tends to detract from the mechanical characteristics as the highly organized molecular arrangement is disrupted, but the very derangement of the protein reduces the immunogenicity.

The strength of collagen is determined by the preferred orientation of the molecules and the extent of cross-linking, as well as by the primary bond strengths within the molecules and fibrils. In tendon, the Type I collagen fibrils are arranged coaxial with the tendon so that a highly anisotropic structure is produced, with axial strength and relative inelasticity, but lateral flexibility. In skin, on the other hand, a far more uniform arrangement is produced with quasi-isotropy and a tensile strain approaching 50%. This almost elastomeric behavior is the result of macromolecular structuring where a crimp or waviness allows for extensive strain at low stress in a manner similar to the coiled structure found in rubbery materials.

As a natural tissue, collagen undergoes a constant turnover, the rate of which may vary from days to years. The stability is largely related to the degree of cross-linking, the most heavily cross-linked materials being the most stable. It follows that collagenous biomaterials will have varying degrees of stability in the body. This, of course, is a useful and widely used variable.

With respect to immunological properties, it is now widely recognized that collagen proteins are able to generate antibodies when placed in a human host, but that, in comparison with other proteins, collagen is a poor immunogen. There are, in fact, at least three antigenic domains on collagen molecules, with most evidence being associated with the nonhelical telopeptide determinants. Different patterns of immune response are seen with different collagen types and species. Type I collagen is known to induce both humoral and cell-mediated, or delayed hypersensitivity, responses. In practice, however, the extent of the immunological reaction to collagenous biomaterials is unclear. It is certainly true that the collagen-based xenografts widely used in cardiac surgery (see Sect. 3.5) are capable of producing such responses, but that the cross-linking process usually used with these tissues is able to alter this capability radically. While chemical modification of the collagen within the tissue by glutaraldehyde can either expose new epitopes (increasing antigenicity) or mask them (reduce antigenicity), it is usually accepted that the treatment is beneficial in this respect. Solubilization and reconstitution, especially if it involves treatment with pepsin to remove the more reactive telopeptides, effectively renders the collagen nonimmunogenic.

It should be expected that collagen, when used as an implant material, should induce a host response a little different to that seen with totally synthetic biomaterials. The host response to an implanted material is best considered as a modification to a wound healing process, which itself involves inflammatory and repair components (see *Biocompatibility: An Overview*). Since collagen synthesis is the central point in wound repair, it might be expected that the implantation of collagen into the wound area would have an influence on the process. This is particularly important when the various reactions that take place here are considered as cell–extracellular matrix reactions, since it is known that collagen actively enhances cell activity. Many cells show good adhesion to collagen, sometimes necessitating the presence of adhesive proteins such as fibronectin or chondronectin. Collagen also promotes cell growth and, perhaps not surprisingly, fibroblasts are affected the most. This is clearly an important consideration in the use of collagenous biomaterials in wound repair.

Since some of the clinical applications of collagen involve the vascular system, it is also important to consider blood compatibility. Collagen is, in fact, a potent mediator of platelet aggregation. Damage to the endothelial lining of a blood vessel, which exposes collagen to the blood, results in rapid hemostasis following platelet adhesion and activation. The evidence would suggest that adhesive proteins, such as fibronectin, in the plasma are collagen receptors. Collagen also directly activates the Hageman factor in the coagulation cascade.

Because of these properties, collagen acts as a hemostat to produce clotting in areas of extensive bleeding. On the other hand, great care has to be exercised when contemplating the use of collagen or collagen-based products in vascular applications.

3. Medical Applications

3.1 Preparation

The starting point of many purified collagen preparations is minced bovine tissue such as skin.

Treatment with an enzyme such as pepsin cleaves the molecules in the telopeptide cross-linking regions and causes swelling. The enzyme is then inactivated and the solubilized collagen purified by one of a variety of methods. This enzymatic treatment of the tissue results in molecules without the potentially immunogenic ends, making the materials very suitable for implantation. However, this elimination of the cross-linking also substantially reduces the strength and other physical properties. The reconstituted collagen, therefore, requires additional treatment to regain these properties. Cross-linking induced by a suitable agent is most commonly used. This is also an important stage in the process of preparing and adapting native collagenous material for implantation since immunogenicity has to be minimized and this is best achieved by cross-linking. At the present time, the most widely used method of collagen cross-linking involves treatment with glutaraldehyde. Used as a solution, this is a complex substance and acts in a complex way. While the cross-linking action is very effective, problems do exist over the residues, especially polymeric residues of the aldehyde. Also, there is a tendency for aldehyde cross-linked collagens to calcify *in vivo*. A number of other dialdehydes have been investigated, although none yet has proved useful.

3.2 Hemostatic Agents

Because of its influence on the blood clotting process, collagen is used in a variety of proprietary hemostatic agents. The material used is typically collagen derived from bovine tendon, enzymatically treated to remove noncollagenous proteins, dispersed and homogenized, lyophilized to give a sponge and finally cross-linked. The products are available in a variety of forms, including sponges, fleeces and powders. The collagen provides a structure that allows fibrin to rapidly form and the whole mass is resorbed within a few weeks. These agents are used to control bleeding in a variety of situations, especially in cellular organs such as the liver, where there is normally little structural tissue to support a blood clot.

3.3 Wound and Burn Dressings

Collagenous products have been used widely as wound or burn dressings, either in the form of sheets or sponges. They possess the properties of a hemostatic agent to control bleeding, an excellent substrate for cell adhesion, migration and growth, and flexibility to accommodate surface strain and degradability for eventual elimination from the site. As well as providing a dressing material in its own right, collagen is under intensive investigation as a substrate or matrix for other biological products. For example, a so-called "synthetic skin" has been produced which comprises a porous layer of cross-linked bovine collagen reconstituted with the glycosaminoglycans, chondroitin-6-sulfate, covered by an impermeable layer of silicone rubber. These membranes can be seeded with autologous epidermal cells which facilitate the regeneration of dermal tissue. These products may also be prepared with active agents such as heparin and antibacterial substances.

3.4 Injectable Collagen

If collagen is prepared as a solution or a gel it may precipitate as a fibrous mass under the right conditions. It is possible to arrange for this transition in tissues after injection of such a gel. Perhaps more importantly, a suspension of collagen fibers can be prepared with a viscosity suitable for injection from which an insoluble fiber mass can form *in vivo*. This is the basis of the injectable collagens (which are able to stimulate a fibrotic response on implantation) for tissue regeneration, with subsequent gradual replacement by host tissue. This approach is most widely used for the augmentation or build-up of dermal tissue for cosmetic reasons. The material is typically a sterile suspension of bovine collagen fibrils, pepsin solubilized and purified, suspended in a sodium phosphate, sodium chloride solution. At 37 °C, after injection, the fibrils tend to entangle to form a network. This is invaded by fibroblasts and fibrous tissue forms, thereby increasing tissue volume or altering skin morphology. There does appear to be some contraction in the longer term. Although generally quite safe, a small percentage of recipients do show a reaction and preliminary sensitivity testing is usually carried out.

3.5 Bioprosthetic Heart Valves

One of the most notable biomedical uses of collagen-based products is that of heart valves constructed from xenograft tissue, especially bovine pericardium and porcine aortic valve tissue. The use of this tissue allows trileaflet valves to be produced, providing more normal hemodynamic conditions compared with the mechanical valves, and better blood compatibility. The latter property is obtained because the collagenous material is cross-linked with glutaraldehyde, eliminating the potential for platelet and factor XII activation described in Sect. 2. These tissues do, however, have their problems, particularly a slow degradation and a tendency to calcify. The latter process, most marked in younger patients, tends to originate in the cuspal areas subjected to greatest mechanical stress, the process of calcification tending to increase surface stresses and render the tissue liable to tearing.

3.6 Vascular Prostheses

Replacement of segments of blood vessels is an extremely important procedure. It is most often achieved in the case of medium-sized arteries through the use of autologous vein transposition, but a significant number of patients do not have lengths of vein (usually the saphenous vein) of appropriate quality and length available for this purpose and there is a desire for synthetic artificial arteries for use in this situation. With larger diameter arteries, particularly the aorta, synthetic materials such as Daron are used successfully, but with anything less than 10 mm the problems tend to increase. Expanded polytetrafluoroethylene is reasonably successful in some situations, but the replacement of arteries of less than 6 mm diameter remains elusive.

Collagenous materials have been used in this situation, including modified and stabilized bovine ureter or human umbilical vein and some fabricated collagen tubes. Although these do have the advantage of more closely matched mechanical characteristics with respect to the natural artery, these are as yet, in common with their synthetic counterparts, far from satisfactory.

Bibliography

Bajpai P K 1985 Immunologic aspects of treated natural tissue prostheses. In: Williams D F (ed.) 1985 *Biocompatibility of Tissue Analogs*, Vol. 1. CRC Press, Boca Raton, FL, pp. 5–25

Cooperman L, MacKinnon V, Bechler G, Pharris B B 1985 Injectable collagen. *Anesth. Plastic Surg.* 9: 145–52

DeLusto F, Condell R A, Nguyen M A, McPherson J M 1986 A comparative study of the biologic and immunologic response to medical devices derived from dermal collagen. *J. Biomed. Mater. Res.* 20: 109–20

Nimni M E 1988 *Collagen*. CRC Press, Boca Raton, FL

Nimni M E, Cheung D, Strates B, Kodama M, Sheikh K 1987 Chemically modified collagen: A natural biomaterial for tissue replacement. *J. Biomed. Mater. Res.* 21: 741–71

Oliver R F, Barker H, Cooke A, Grant R A 1982 Dermal collagen implants. *Biomaterials* 3: 38–40

Ramshaw J A M 1990 Collagen as a biomaterial. In: Williams D F (ed.) 1990 *Current Perspectives on Implantable Devices*, Vol. 2. JAI Press, London

Yannas I V, Lee E, Orgill D P, Skrabut E M, Murphy G F 1989 Synthesis and characterisation of extracellular matrix that induces partial regeneration of adult mammalian skin. *Proc. Natl. Acad. Sci. U.S.A.* 86: 933–7

D. F. Williams
[University of Liverpool, Liverpool, UK]

Composite Materials

Composite materials may be defined as those materials that consist of two or more fundamentally different components that are able to act synergistically to give properties superior to those provided by either component alone. It is usual, although not mandatory, for the components to be of quite different structures and quite distinct types. For example, the most commonly used composites consist of resins and glasses, or carbons and thermoplastics. It is particularly desirable that the most attractive properties of each component are preserved in the composite and the less attractive properties are obviated. Thus a composite of a glass and a resin should reflect the modulus and strength of individual glass fibers and the ductility and resilience of the resin to give a strong tough material, in which the brittleness of the glass, and the low yield and creep resistance of the resin are of no consequence.

In practice, composites are often seen as a compromise between metals and polymers. Metals may be considered as strong rigid tough materials that are, however, susceptible to corrosion, and are often costly and heavy. Alternatively, polymers are often considered to be degradation resistant, light and relatively inexpensive, but of low elastic modulus, high viscoelasticity and poor strength. A material that is strong, tough, of average modulus, environmentally stable and of high strength-to-weight ratio as well as being relatively inexpensive can be considered to approach the ideal compromise between metals and polymers. Composites involving either glass, carbon or ceramic fibers distributed in a thermosetting or thermoplastic matrix offer suitable properties in this respect and are widely used in many industrial applications.

This general principle has been followed within the field of biomaterials and composite materials are used in this classical sense. However, there are at least three additional reasons for using composites in medical devices that, while consistent with this general philosophy, are unique to this area.

The first reason concerns the requirements of a material intended to replace or augment bone. Since bone is a living tissue that is consistently being remodelled, the properties and structure of which reflect this dynamic state, it is inappropriate to replace or augment it with substances that have fundamentally different characteristics that could interfere with this dynamic state. One of the principal factors here is the elastic modulus of the material and the use of composites in orthopedics is largely based upon the perceived need to match the elasticity of replacement or augmentation material with that of the bone itself.

The second reason concerns the frequently found conflict between bulk properties and surface properties of materials intended for tissue replacement. Whatever the precise function of a prosthesis, it must possess appropriate mechanical and physical properties to perform, and continue to perform, this function. For this reason traditional biomaterials have largely included the mechanically robust alloys, ceramics and thermoplastics. Such synthetic materials, however, do not necessarily possess the most appropriate characteristics to interface with viable biological tissues. An increasing trend in biomaterials science is to incorporate, into one composite structure, components that provide engineering performance and components that provide biological performance. In most cases the engineering material tends to form the bulk component and the biological material is used as a coating, giving a composite structure rather than a composite material. There are some situations, however, where true composite materials have been developed on the basis of dispersed components of this type, including a material consisting of hydroxyapatite particles dispersed in a polymer matrix intended for replacement of, or interface with, bone.

The third reason concerns the use of materials to replace tooth structures. The principle described previously for replacing bone with a material of equivalent modulus can also apply to the replacement of tooth substance, although the significance here is far from clear. There are certainly moves towards the development of dental materials with structures and properties analogous to tooth enamel and dentin, but this has not provided the driving force for the development of composites in dentistry. The real reasons here may be seen quite easily by a consideration of the performance of the main tooth filling substance amalgam. Although in use before the 1900s, amalgam is unaesthetic, of questionable biological safety and of high operator variability. For many years attempts were made to find alternatives to amalgam that would be equally strong and abrasion resistant, but also tooth colored, nontoxic and technique tolerant. Polymers capable of curing under intraoral conditions became the focus of attention, but these were insufficiently strong and wear resistant, and suffered considerable shrinkage on curing. Very significant developments have been made with composites which provide both improved strength and wear resistance while reducing the shrinkage to clinically acceptable levels.

1. Composites in Orthopedic Surgery

1.1 The Concept of Modulus Matching

Bone has to be considered as both a composite material and a composite structure. As a material, bone is heterogeneous, consisting on the microstructural scale of hydroxyapatite crystals in a proteinaceous matrix (largely collagen). The collagen, as a protein, has properties similar to those of polymers while the hydroxyapatite is a ceramic, with classical hardness and brittleness. The hydroxyapatite is present as very fine crystallites within the collagen matrix. This composite material is then arranged with a highly complex architecture, the nature of which depends on location and a number of other conditions. Within the structure are blood vessels, cells and various interconnecting channels. In cortical bone, the solid material that constitutes the cylindrical shaft of the long bones, there are concentric rings of the collagen–hydroxyapatite composite together with the cells and vessels. In other parts of the bony skeleton the material, known here as cancellous bone, is far more porous, with an open architecture.

In both types of bone, the structure is highly dependent on the stress. The cells in the bone are of two types: the osteoclasts, which are responsible for slow but constant removal of bone, and the osteoblasts, which produce new bone. These cells, in a healthy patient, are constantly working together to remodel the bone, but this process is responsive to mechanical stress. The reasons for this are not entirely understood, but both piezoelectric and streaming potential phenomena may be involved. In particular, if stress is applied to the bone under normal physiological conditions, the bone responds by a continual turnover of both collagen and hydroxyapatite such that the orientation and spatial distribution of the components are optimized with respect to load transmission. Most of the features relating to shapes, sizes and cross sections of bones are a consequence of the need to transmit forces. If, however, the stress is reduced, there is less stimulus to the cells and the bone remodelling will not be in equilibrium. Indeed, if the stress is substantially reduced, the bone mass and apatite volume fraction decrease, without substantial new bone formation. The bone becomes portic, weaker and less tough as a consequence.

Clearly, load transmission through a bone such as the femur or tibia is dependent upon the structure and properties of the whole of the bone. Within a femur, for example, the stress and strain distribution will depend on the forces applied (obviously dynamic), the geometry of the whole bone and the elastic constants of the bone material. If part of the bone is replaced with a synthetic material, then the geometry and elastic properties of that replacement will significantly influence the stress and strain fields.

The elastic constants of bone vary with location, species, age and other factors. Typical values are in the range 10–20 GPa. If a segment were replaced with exactly the same modulus and the same geometry, there would be no change in the stress distribution. If, however, a bone segment is replaced with a metal of modulus 200 GPa then, depending on the geometry and the method of fixation, the stress distribution is likely to be radically altered. This may be seen in practice with a rigid metal used for the replacement of the upper end of the femur, where the metal stem is placed in the cylindrical shaft of the bone or with a metal plate placed on the outside of the bone to assist in fracture healing. In either case, and perhaps most dramatically in the latter, it is possible for the stress in the surrounding bone to be reduced to below physiological levels. With the fracture plate, for example, the plate will transmit a very large proportion of the load, because of the tenfold differences in modulus between stainless steel and bone. This phenomenon is known as stress (or strain) protection, disuse atrophy or disuse osteoporosis, and the effects on the bone may be considerable loss of bone substance and eventual fracture.

There are several potential solutions to this problem. With the fracture plate, whose function is temporary, the plate may be removed after healing takes place. This, of course, is an expensive procedure and exposes the patient to a second operation. Alternatively, it could be made of a degradable material that is resorbed or eliminated after healing, but before osteoporosis becomes established. Another alternative would be to use a material with a lower elastic modulus. Composite materials may have a role in both of the latter two solutions. With joint replacement, elimination is not appropriate and it is necessary to consider the possibility of lower modulus materials, where again composites are obvious candidates.

1.2 Degradable Composites

Some materials are designed to be degradable in the body. This property is utilized in sutures, drug delivery systems and a variety of other applications. Unfortunately, the requirements for degradability, of easily broken interatomic or intermolecular bonds, are not usually compatible with the requirements for good strength and toughness, of strong stable bonds. Generally, therefore, degradable materials have not been used in highly stressed situations.

The materials most widely used as intentionally degradable biomaterials are polymers such as the aliphatic polyesters (e.g., polyglycolic and polylactic acids). These are clearly too weak for bone plates, but the possibility does exist for them to be reinforced for this purpose. It is an absolute requirement here, of course, for the reinforcement, as well as the polymer matrix, to be degradable. One possibility is to use resorbable calcium phosphates, but it is extremely difficult to produce such composites with compatible degradation kinetics of matrix and filler. A more likely alternative is the reinforcement of polymer matrices with polymer fibers and certain "self-reinforcing composites" have been developed. The most important here is a polylactic acid fiber-reinforced polylactic acid matrix. This type of composite does not have sufficient strength to use in major long bones, but it has been used in smaller bones and in the bones of the facial skeleton.

1.3 Thermosetting Resin Composites

Most of the composites used in engineering and construction industries utilize a thermosetting resin as the matrix. Typically this will be an epoxy or phenolic resin. The filler will usually be glass although higher performance, and usually higher cost, composites may utilize Kevlar (i.e., poly(paraphenylene terephthalamide)) or carbon fibers. These composites offer tremendous advantages in strength-to-weight ratios. They are typically produced with an anisotropic structure in which the fibers can be arranged to suit the requirements, including unidirectional or cross-ply formats. The matrix is usually added to the fibers, arranged in a prepreg and cured in the standard way for the polymer.

While these materials have revolutionized composite technology and the selection of materials in general, they have not proved very successful as biomaterials. Composite epoxy–carbon fiber materials have been employed, but without consistent success. The main problem is that thermosetting resins are water sensitive and their properties may deteriorate upon implantation, with water penetration and loss of interfacial coupling. In general, the biocompatibility of these thermosetting resins is not good and it is unlikely that the future of composites in medicine rests with these matrices.

1.4 Thermoplastic Composites

Traditionally, thermoplastic matrices have been less useful in fiber-reinforced composites because of the greater difficulty in fabrication. The best results in mechanical performance in composites are normally achieved with long fibers and it is very difficult to fabricate devices out of long-fiber-reinforced thermoplastics by normal thermoplastic processing routes. Due to the greater stability and general biocompatibility of thermoplastics compared with thermosetting resins, however, there has been a considerable impetus to the development of fiber, especially carbon-fiber-reinforced thermoplastics.

Some attempts have been made to incorporate carbon fibers into high-density polyethylene and poly(methyl methacrylate), the two standard materials used in joint replacement, in order to confer superior properties, but these attempts have not met with much success. Of greater relevance are the attempts to utilize the thermoplastics polysulfone and polyetheretherketone (PEEK) (see Scheme 1). These are both stable polymers apparently showing good biocompatibility and, largely because of extensive work on aerospace applications, they have been incorporated into glass and carbon fiber composition. These two polymers, whose structures are shown below, have strengths approaching 100 MPa and tensile moduli between 2 GPa and 4 GPa. PEEK is particularly interesting and is incorporated into a variety of commercially available composites,

Table 1
Mechanical properties of PEEK and CFPEEK

	Fraction of fiber (wt%)	Flexual modulus (GPa)	Flexual strength (MPa)
PEEK	0	3.8	156
Chopped CFPEEK	30	21	343
Chopped CFPEEK	50	41	372
Continuous CFPEEK, quasi-isotropic	68	47	803
Continuous CFPEEK, uniaxial	68	125	1940

including chopped carbon fiber (CF) injection and pultrusion forms and continuous carbon fiber pre-impregnated forms. With the short-chopped-fiber-reinforced PEEK a relatively isotropic composite is produced while highly oriented materials or quasi-isotropic structures can be produced with the prepregs. The range of mechanical properties achievable with these materials is shown in Table 1. The materials appear to have good stability and studies have been published that demonstrate good *in vitro* and *in vivo* biocompatibility. The composites are reasonably versatile in fabrication technologies and it is quite possible to prepare bone fracture plates, as shown in Fig. 1.

1.5 Hydroxyapatite-Based Composites

It has been noted previously that bone itself is a composite material, consisting of a dispersion of hydroxyapatite in collagen. Attempts have been made to reproduce this type of structure with hydroxyapatite dispersed in synthetic polymers. It has proved possible to incorporate particles of hydroxyapatite in polyethylene with quite high volume fractions. Young's moduli of between 1 GPa and 9 GPa have been produced in these composites the modulus increasing with increasing volume fraction, although there is a limit to this because of the brittleness of such composites at volume fractions above 40%.

Obviously the elastic properties obtained with these composites are in the region of those of bone, which is highly desirable. As far as bulk properties are concerned, however, the fact that the filler is hydroxyapatite is irrelevant, since it cannot reinforce polyethylene in the same way as it reinforces the quite different collagen and other particulate fillers could probably perform this function better.

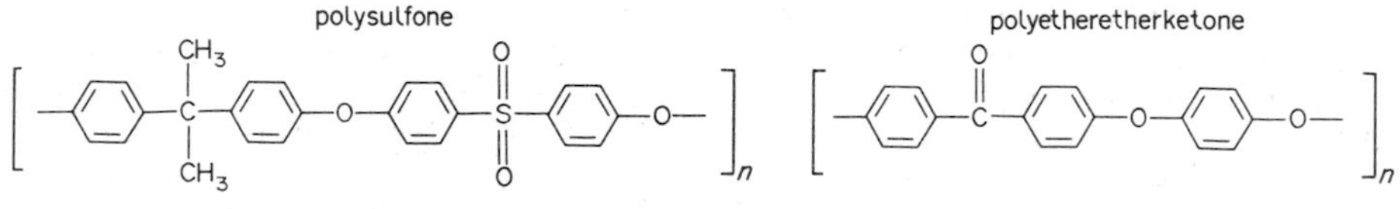

Scheme 1

Figure 1
Sample of bone plate manufactured from carbon-fiber-reinforced polyetheretherketone

There is some suggestion, however, that the hydroxyapatite present at the surface could favorably influence the response of bone to the material.

1.6 Carbon–Carbon Composites

One further type of composite suggested for use in orthopedic surgery is the carbon–carbon composite. Again with origins in aerospace, where high strength-to-weight ratios and especially high-temperature strengths are required, these composites are attractive for medical devices because they consist only of carbon. Carbon has many advantages from the biocompatibility point of view, but it is not a particularly good material for all-round mechanical performance. Carbon fibers can be made exceptionally strong and rigid, properties that are used in carbon-fiber-reinforced polymers. Under some circumstances, it is possible to pyrolyze a carbon-fiber-polymer composite such that the polymer is converted to carbon, any hydrogen, nitrogen or other atoms being driven off. Naturally this results in the development of porosity so that it is possible to have a carbon-fiber-reinforced porous carbon matrix. It is also possible to densify this composite by the addition of further polymer and repyrolysis.

Carbon–carbon composites have been investigated as potential implantable materials. Although some good characteristics may be observed, they have yet to display a combination of mechanical and biological properties appropriate to these applications.

2. Composite Materials in Restorative Dentistry

The composite materials used in restorative dentistry represent a totally different approach to that seen in orthopedic surgery, and quite different properties and structures have emerged. The early rationale for the development of composites related to the need for improvements in the performance of materials used for the restoration of anterior teeth. As soon as synthetic polymers became available around the 1940s, it was found that poly(methyl methacrylate) could be polymerized under intraoral conditions and the material was used for filling cavities in front teeth. It had the advantage of being chemically resistant and, with appropriate pigmentation, was tooth colored. It did, however, have three major disadvantages. First, it was neither hard nor strong and suffered mechanical failure, including abrasion through tooth brushing. Second, the polymerization process involves a considerable amount of contraction which led to difficulties in filling the cavity exactly. Third, the material has a large coefficient of thermal expansion, which resulted in differential expansion and contraction with respect to the tooth and the opening up of gaps between the tooth and the filling.

The solution to these three problems was seen to lie with composites using a filler such as quartz. It proved possible to improve the hardness and strength, and to reduce both the polymerization contraction and coefficient of expansion, since the quartz does not participate in the polymerization and has a much lower expansion coefficient. Composites have been used for these fillings for a considerable time, with generally very good success. They have not employed poly(methyl methacrylate) however, and nor have they used simple quartz fillers, as described later.

While these materials proved successful in the anterior part of the mouth, attempts to use them for posterior cavities met with little initial success. This largely arose from the much higher stresses in these large cavities, which usually involved occlusal surfaces and, therefore, large biting forces. The composites were insufficiently strong and abrasion resistant to withstand this harsh mechanical environment.

During the 1980s, however, considerable improvements have been made to these materials, both with respect to resins, fillers and interfacial couplings, and there are very many formulations available.

2.1 Resins

Several criteria have directed the choice of new and improved resins. It became obvious at a very early stage that poly(methy methacrylate) could not be used in these composites, because of its own polymerization shrinkage, water absorption and inferior mechanical properties (see *Acrylic Dental Polymers*). However, the readily polymerized vinyl double bond was ideal for intraoral polymerization and it was necessary to produce resin that retained this capacity while providing better properties. For many years, the major resin used was an aromatic dimethacrylate, known as BIS–GMA, prepared as an adduct of bisphenol A and glycidyl methacrylate (see Scheme 2). More recently, various urethane dimethacrylates have been employed. There are several varieties of these resins, including the following (see Scheme 3). These resins, where the

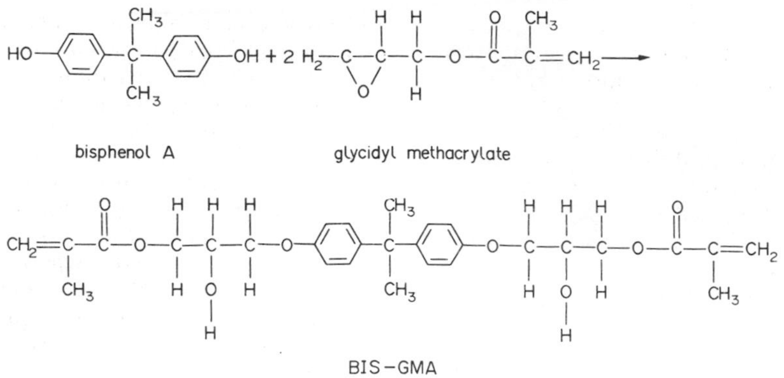

bisphenol A   glycidyl methacrylate

BIS—GMA

Scheme 2

monomers terminate in methacrylate groups, can be polymerized by conventional acrylic free-radical processes. At one stage, ultraviolet light was used to initiate the polymerization, but visible light with a camphoroquinone activator is used more frequently. Most resins will, in addition be cross-linked with ethylene glycol dimethacrylate (EGDMA) or tri-ethylene glycol dimethacrylate (TEGDMA) (see Scheme 4).

2.2 Fillers

There are several features of the filler that are important with respect to the properties of the composite. In contrast to the fiber-reinforced composites in orthopedics, these are particulate composites, where the variables are filler chemistry, particle size, and size distribution and filler shape. Among the fillers that have been used are quartz, aluminum silicate, borosilicate glass, barium glass and strontium glasses. The choice is governed by a need for hardness and radiopacity (to allow the

fillings to be distinguishable from the teeth on x-ray). It is generally agreed that the size and size distribution of the particles should be adjusted to give a maximum volume fraction, while retaining an appropriate viscosity. Usually a multimodal distribution and a variable geometry are employed for this purpose, with size ranging from less than 0.1 µm to 50–70 µm. The variation in filler content has led to a classification of dental composites, based on filler size. Traditional composites tend to have a reasonably uniform size, between 5 µm and 20 µm. Hybrid composites contain some large particles (50 µm) with the matrix reinforced by very fine distributions, possibly of pyrolytic or fumed silica. Microfilled composites contain only very small particles, usually of the fumed silica, which is of the size 0.04 µm.

2.3 Coupling Agents

It is well known that optimal performance of composites can only be obtained if there is an adequate bond between the filler and the resin. Most dental

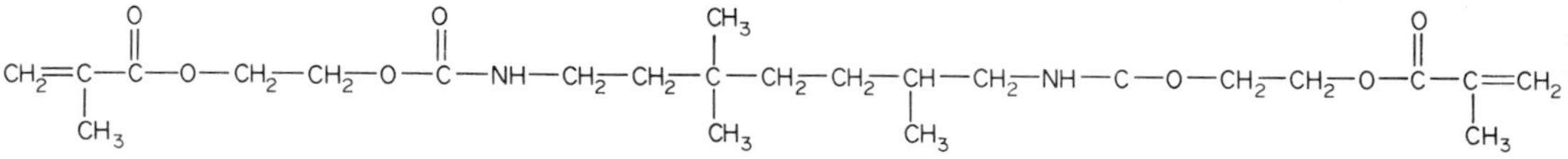

Scheme 3

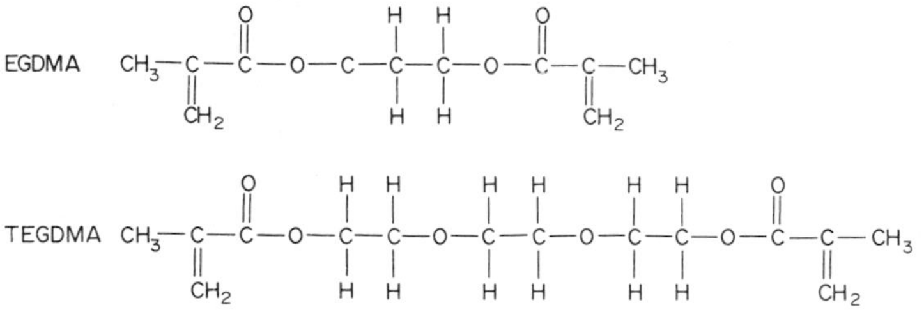

EGDMA

TEGDMA

Scheme 4

composites use a coupling agent that is of the silane type, such as vinyl triethoxysilane, although other agents such as titanates may be employed.

2.4 Properties

The properties of these dental composites will vary quite considerably, but it is generally agreed that their mechanical performance is very good. The weight percentage of filler is often in the region of ·70–80%, and compressive strengths in excess of 300 MPa, transverse strengths of 100 MPa and fracture toughnesses of $2\,MN\,m^{-3/2}$ can be achieved. Clinical experience shows that some of these materials perform as well as amalgams, at least for up to four or five years in posterior cavities.

Bibliography

Bonfield W 1987 Materials for the replacement of the osteoarthritic hip. *Met. Mater.* 3: 712–16

Christel P 1986 The application of carbon fibre-reinforced carbon composites in orthopaedic surgery. *CRC Crit. Rev. Biocompat.* 2: 189–218

Gillett N, Brown S A, Dumbleton J H, Pool R P 1985 The use of short carbon fibre reinforced thermoplastic plates for fracture fixation. *Biomaterials* 6: 113–21

Lloyd C H, Adamson M 1987 The development of fracture toughness and fracture strength in posterior restorative materials. *Dent. Mater.* 3: 225–31

Pilliar R M, Vowles R W, Williams D F 1987 The effect of environmental ageing in the fracture toughness of dental composites. *J. Dent. Res.* 66: 72–6

Roulet J F 1987 *Degradation of Dental Polymers*. Karger, Basel

Shirandami R, Esat I I 1990 New design of hip prosthesis using carbon fibre reinforced composite. *J. Biomed. Eng.* 12: 19–22

Skinner H B 1988 Composite technology for total hip arthroplasty. *Clin. Orthop. Relat. Res.* 235: 224–36

Wenz L M, Merritt K, Brown S A, Moet A, Steffee A D 1990 In vitro biocompatibility of polyetheretherketone and polysulfone composites. *J. Biomed. Mater. Res.* 24: 207–15

Wiliams D F, McNamara A, Turner R M 1987 Potential of polyetheretherketone and carbon-fibre reinforced PEEK in medical applications. *J. Mater. Sci. Lett.* 6: 188–90

D. F. Williams
[University of Liverpool, Liverpool, UK]

R. M. Turner
[ICI Materials Research Centre, Wilton, UK]

Corrosion of Dental Materials

Synthetic materials placed in the oral cavity react with the environment and deteriorate. The two main forms of attack on metallic materials are sulfide tarnishing and chloride corrosion. The reactions involved in both processes are electrochemical in nature. Chloride corrosion causes deterioration of less noble metals; the attack is usually in the form of pitting and sometimes penetrates deep into the structure. The effect can range from degradation of appearance to loss of mechanical strength. Corrosion can also release metallic ions into the digestive tract or directly into the tissues.

Electrochemical processes in the oral cavity can be accelerated if different metals come into contact, forming a galvanic cell. The galvanic currents have been reported to cause pain and pathological changes; more recently, attention has been given to the intensification of the corrosion effects.

1. Environment in the Oral Cavity

Dental restorations and orthodontic appliances are exposed to an ever-changing oral environment. The principal component in the environment is saliva, but the chemistry and temperature of the environment is frequently modified by the intake of food and drink.

Saliva is a moderately corrosive complex liquid containing both inorganic and organic components. It contains substantial amounts of chlorides, which are the most corrosive species, and phosphates, which inhibit corrosion. With the exception of proteins, the role of other inorganic and organic components in the corrosion processes has not been established. Proteins have been reported to slow down the corrosion processes without actually taking part in the electrochemical reactions (Finkelstein and Greener 1978). Other studies have shown an element-specific effect of proteins on the corrosion rate (Clark and Williams 1982). Dissolved oxygen is the primary cathodic reactant, while CO_2 mainly affects the acidity of saliva.

Saliva is well buffered, mainly by the $NaHCO_3$–CO_2 system, phosphate being the only other buffer of significance. It is usually nearly neutral (pH 6–7), although substantial variations in acidity of the oral environment may occur as a result of the intake of food and drink.

Biofilms of organic matter form on solid surfaces exposed to saliva. Pretreatment of some dental alloys in human saliva has been shown to effect the corrosion reactions (Holland 1984).

In laboratory corrosion tests, the oral environment has been simulated by synthetic salivas of various compositions, Ringer's solution or simple saline. The corrosion behavior of dental materials in synthetic saliva approximates to their behavior in natural saliva: Ringer's solution and saline are more corrosive because of their higher chloride contents, the absence of inhibiting components and their low buffer capacity (Marek and Topfl 1986).

2. Electrochemical Reactions in the Oral Environment

Corrosion of dental materials is caused by electrochemical reactions. The liquid environment acts as an electrolyte and the dental alloy as an electrode. The alloy attains a corrosion potential E_{corr}, which can be measured with respect to a suitable reference electrode. The value of E_{corr} is the result of the interaction of at least one cathodic reaction, which has an equilibrium potential more positive than E_{corr} and proceeds at E_{corr} in the direction of reduction, and at least one anodic reaction, which has an equilibrium potential more negative than E_{corr} and proceeds at E_{corr} in the direction of oxidation. Corrosion is directly caused by the anodic reactions.

The practical upper limit for E_{corr} is the redox potential of the electrolyte. The oral environment usually has a redox potential in the range of $-0.05\,V$ to $+0.20\,V$ with respect to a saturated calomel electrode (SCE) (Ewers and Greener 1985).

The most important cathodic reaction in the oral environment is reduction of dissolved oxygen:

$$O_2 + 2H_2O + 4e^- \rightarrow 4(OH)^- \qquad (1)$$

The reaction has a very high equilibrium potential (about $+0.55\,V$ SCE in neutral electrolytes), but is relatively sluggish. Another common cathodic reaction is the reduction of hydrogen ions

$$2H^+ + 2e^- \rightarrow H_2 \qquad (2)$$

which is less likely in the oral environment, but may be important locally in crevices and pores on base metals and alloys, where it is favored by acidic conditions and the lack of oxygen. Other possible cathodic reactions are probably of minor importance.

The important anodic reactions on dental materials result in the oxidation of the metal. Oxidation may take the form of dissolution or of the formation of solid corrosion products. In the case of dissolution, or when the solid corrosion products are nonprotective, the result is deterioration of the dental material. If the corrosion products are protective, they cover the surface and lower the rate of the electrochemical reactions. The most effective protection is provided if the metal passivates, forming a thin self-healing highly protective film, which slows down the reactions to a rate that can be considered insignificant.

The behavior of electrodes is best described by polarization diagrams, which show the rate of the electrochemical reaction, expressed as the electrical current, as a function of the electrode potential. Figure 1 shows the anodic behavior of an active electrode (dissolving or forming nonprotective products) and a passivating electrode. In the presence of

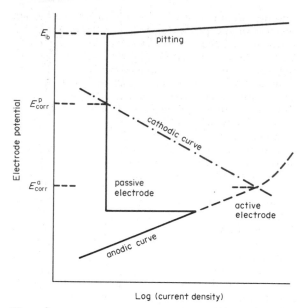

Figure 1
Schematic polarization diagram for active and passive electrodes

chlorides in the electrolyte, passivity ceases when the breakdown potential E_b is reached; at more positive potentials rapid localized corrosion takes place in the form of pitting. Figure 1 also shows that the corrosion potentials of both the active electrode E_{corr}^a and the passive electrode E_{corr}^p are determined by the intersection of the anodic and cathodic polarization curves.

Tarnishing of dental materials is mainly caused by the electrochemical formation of silver and copper sulfides:

$$2Ag + S^{2-} \rightarrow Ag_2S + 2e^- \qquad (3)$$
$$2Cu + S^{2-} \rightarrow Cu_2S + 2e^- \qquad (4)$$
$$Cu + S^{2-} \rightarrow CuS + 2e^- \qquad (5)$$

The equilibrium potentials of these reactions are much more negative than the corrosion potentials of dental materials; consequently, the reactions are anodically polarized at E_{corr}.

Many electrochemical reactions involve hydrogen or hydroxyl ions, and are thus affected by the acidity of the electrolyte. The thermodynamic stability of various phases, such as metal ions and oxides, can be illustrated as a function of the electrode potential and pH of the electrolyte in a Pourbaix diagram. Pourbaix diagrams can be used to predict the corrosion tendencies, such as immunity, active dissolution and passivation, of the metal at a given pH and E_{corr}, but they do not predict the rates of the reactions.

Many dental alloys are heterogeneous either because of the presence of different phases in the structure or because of segregation of alloying elements. Anodic reactions on heterogeneous alloy electrodes are concentrated on the corrosion susceptible phases or regions, and macroanodes and macrocathodes can be identified. There is then a galvanic relationship between the different phases or regions, in which the presence of less susceptible cathodic areas accelerates corrosion of the more susceptible anodic areas. True galvanic corrosion, however, occurs when an electrode is electrically connected with another electrode that has a different E_{corr} (e.g., when two different dental restorative materials are in contact). Corrosion of the more active electrode (i.e., the one with a more negative E_{corr}) is then accelerated, while the more noble electrode is partially or completely protected. Galvanic interaction requires two separate conductive paths between the electrodes. In the oral cavity only a negligible interaction takes place between restorations that are not in direct contact.

When electrochemical reactions take place in a confined area, such as a crevice between restorations, the changes in electrolyte chemistry caused by the reactions cannot be easily eliminated by mass transport. The chemistry in such occluded cells then often becomes substantially different from the chemistry of the bulk environment. The most important changes include the depletion of dissolved oxygen, which is consumed according to the cathodic reaction in Eqn. (1), an increase in the concentration of anions (e.g., Cl^-) to preserve electrical neutrality and an increase in acidity, which results from the precipitation of solid corrosion products, such as oxides, in a reaction of the type

$$M + H_2O \rightarrow MO + 2H^+ + 2e^- \qquad (6)$$

The difference in the concentration of dissolved oxygen at different regions of the electrode surface, called differential aeration, contributes to the acceleration of corrosion within the occluded cell, because the absence of the reaction in Eqn. (1) allows the acidity to increase, while the corrosion potential is maintained at a high level due to oxygen reduction outside the cell. At high acidity and high Cl^- concentration, many metals lose the protection provided by passive films, the surfaces within the occluded cell become active and the material suffers severe localized corrosion.

3. Corrosion Characteristics of Dental Amalgams

Dental amalgams are more susceptible to chloride corrosion than most other dental materials. The conventional γ_2-containing amalgams corrode along the networks of the thermodynamically least stable

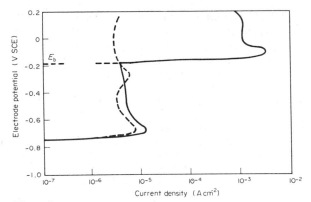

Figure 2
Anodic polarization curves of two dental amalgams in deaerated synthetic saliva at 37 °C: ——— low-copper amalgam; – – – high-copper amalgam

γ_2 (tin–mercury) phase. The high-copper γ_2-free amalgams are more corrosion resistant, but they also suffer deterioration, the least stable phase being the η (copper–tin) phase.

In nearly neutral saliva, both types of dental amalgam exhibit passivity, as shown by the polarization curves in Fig. 2. The γ_2-containing amalgams exhibit a clearly defined breakdown potential E_b, related to the presence of the γ_2 phase. In human saliva the breakdown occurs at about -0.15 V SCE. At potentials more positive than E_b, the protective film of tin oxide breaks down and the γ_2 phase is rapidly attacked; the main corrosion products are tin oxide and tin chloride hydroxide. Since the corrosion potential of the amalgam in the oral environment is close to E_b, some corrosion damage in the form of pitting occurs. The attack is accelerated by deformation and abrasion, which disturb the protective passive film (Averette and Marek 1983, Marek 1984). The most severe attack occurs in occluded cells. In these the combined effects of differential aeration, increased acidity and increased concentration of chlorides cause loss of protection and active dissolution of the γ_2 phase, as indicated by the Pourbaix diagram for tin (see Fig. 3). Since porosity is often associated with the γ_2 network, occluded cells are easily established and the attack follows the γ_2 network through the structure. Depth of penetration of the attack as high as 0.2 mm after 2.5–3 years *in vivo* has been observed (Espevik and Mjör 1979). Corrosion resistance of all amalgams is decreased by porosity.

The corrosion damage causes loss of strength. The most intensive attack takes place in the areas of the margins, where deformation disturbs the passive film and occluded cells form in the crevice between the restoration and the tooth. Corrosion is thus thought to be one of the major causes of the breakdown of restoration margins. Another form of

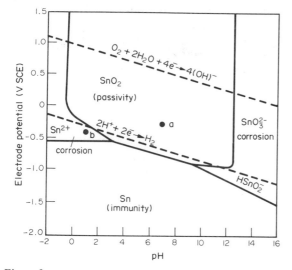

Figure 3
Simplified Pourbaix diagram for the system tin–water at 25 °C: (a) free corrosion conditions of a low-copper dental amalgam; (b) corrosion conditions in an occluded cell

attack, observed on occlusal surfaces, is grain boundary corrosion of the γ_1 (silver–mercury) phase (Espevik and Mjör 1979).

The high-copper γ_2-free amalgams are more resistant to both corrosion and breakdown of margins. The better corrosion behavior is evident from the anodic polarization curve (see Fig. 2), which does not show the sharp current increase associated with the breakdown of passivity. The corrosion attack takes the form of deterioration of the individual grains of the η (copper–tin) reaction phase. Copper is released by dissolution into the electrolyte and tin remains in the structure in the form of solid corrosion products. Since the η phase does not form an interconnected network, the attack is less destructive than corrosion of γ_2-containing amalgams.

4. Corrosion Characteristics of Noble Dental Alloys

Gold alloys conforming to American Dental Association Specification No. 5 (more than 75% gold and platinum-group metals) are highly resistant in the oral environment to electrochemical forms of deterioration other than sulfide tarnishing. Their resistance is due to the thermodynamic stability of the noble elements, which dominates the behavior of the essentially single-phase materials. The corrosion potential is usually in the range −0.15 V SCE to +0.05 V SCE.

The resistance to chloride corrosion decreases as the gold and platinum-group metal content of the alloy decreases. The corrosion potential is usually in the range of −0.2–0.0 V SCE for medium- and low-gold alloys. Silver and copper have been found to corrode selectively from regions in which segregation of these elements occurred (Sarkar et al. 1979a). Preferential grain boundary attack, also attributed to segregation of silver and copper, was observed in some cases.

Silver is attacked by the oral environment, but its resistance to chloride corrosion can be substantially improved by alloying with palladium, which induces passivation (Vaidyanathan and Prasad 1981, Sastri et al. 1982). Silver–palladium alloys containing little or no gold are more susceptible to chloride corrosion than both high- and low-gold alloys. Application of sensitive electrochemical techniques, such as cyclic polarization (Sarkar et al. 1979b), has shown the beneficial effect of gold and possibly detrimental effect of copper. In the cored dendritic structure of the silver–palladium alloys, the matrix is richer in silver and more susceptible to corrosion than the palladium-rich dendrites. High-palladium alloys for porcelain fused to metal (PFM) applications, based on the palladium–copper–gallium and palladium–cobalt–gallium systems, have shown passive behavior and satisfactory corrosion resistance (Sumithra et al. 1983, Mezger et al. 1985).

Noble dental alloys are relatively resistant to occluded cell corrosion because the electrochemical reactions that take place do not result in an increase in acidity within the cell. The seriousness of *in vivo* chloride corrosion of noble alloys has not been established.

5. Corrosion Characteristics of Base Dental Alloys

Cobalt, nickel, iron and titanium as well as the major elements used in their alloys, such as chromium, molybdenum, vanadium and aluminum, are base metals that are not thermodynamically stable in the oral environment. The corrosion resistance of these alloys is because of the formation of a protective passive film. Due to the presence of chlorides in the oral environment, passivating alloys are potentially susceptible to pitting corrosion. The schematic anodic polarization curve in Fig. 1 shows that protection ceases when the breakdown potential E_b is reached. The more positive E_b, the more resistant is the alloy to the initiation of pitting corrosion.

The corrosion resistance of nickel-based dental alloys is mainly a function of the chromium content (Meyer 1977). Figure 4 shows anodic polarization curves of a cobalt–chromium alloy, a high-chromium nickel alloy and a low-chromium nickel alloy in synthetic saliva (Marek 1983). The limited region of passivity exhibited by the low-chromium nickel alloy indicates high susceptibility to pitting. Nickel ions also were found in the electrolyte in which a low-

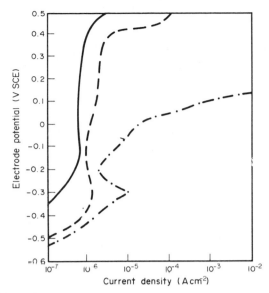

Figure 4
Anodic polarization curves of three base dental-casting alloys in deaerated synthetic saliva at 37 °C: ——— 60% Co–30% Cr–5% Mo, – – – 67% Ni–20% Cr–5% Mo, –·–· 80% Ni–12.5% Cr–2% Mo

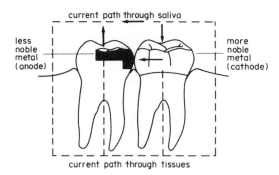

Figure 5
Galvanic interaction between two dissimilar dental restorations

chromium nickel alloy was immersed *in vitro* (Espevik 1978). Chromium-containing alloys can suffer grain boundary attack if the boundaries become depleted in chromium (e.g., after improper heat treatment). Titanium and its alloys are highly resistant to chloride corrosion (Sarkar and Greener 1973, Mueller and Greener 1970), because of the superior properties of the passive film. Base alloys are generally highly susceptible to occluded cell corrosion in chloride environments. Alloying with molybdenum generally increases the resistance to both pitting and crevice corrosion. Titanium and its alloys show best resistance to localized corrosion. Base alloys are not susceptible to sulfide tarnishing.

Since the barrier to corrosion for base alloys is a thin surface film, abrasion may affect the corrosion behavior of passivating alloys. In laboratory tests, including ultrasonic agitation of abrasive particles suspended in the electrolyte, abrasion has been shown to increase the corrosion current density and activate the electrode potential, indicating destruction of the passive film (Wright et al. 1980). The severity of the abrasive effect of mastication on the corrosion behavior *in vivo* has not been determined.

6. Galvanism in the Oral Cavity

When dissimilar metals come in contact in the oral cavity the flow of the galvanic current can cause pain as well as intensified corrosion of the more active metal. The contact can be intermittent or permanent. On contact, the current quickly rises to a maximum and then decays as the electrodes adjust to the new potential by surface changes. A schematic illustration of the galvanic interaction between two dental restorations is shown in Fig. 5.

The galvanic couple that has received most attention is dental amalgam in contact with a noble dental alloy. In this cell, the amalgam becomes the anode and suffers accelerated corrosion. The noble alloy becomes the cathode and the galvanic effect is stronger for higher cathode–anode area ratios. The intensified corrosion of dental amalgams has been shown to cause a significant decrease in strength (Wang Chen and Greener 1977). In γ_2-containing amalgams, corrosion of the γ_2 phase releases mercury which can react with gold and cause discoloration of some noble alloys. Substantial galvanic effects have also been reported for an interaction between high- and low-copper amalgams. Galvanic interaction between noble and base alloys can cause breakdown of passivity of the base alloy as the corrosion potential changes to a more positive value.

Although the driving force for galvanic corrosion is the difference between the electrode potentials, there is little correlation between the open-circuit potential-difference and the intensity of the galvanic effect, in which the polarization characteristics of both electrodes are of major importance (Marek 1983).

See also: Dental Amalgams

Bibliography

Averette D F, Marek M 1983 The effect of tensile strain on corrosion of dental amalgam. *J. Dent. Res.* 62: 842–5
Clark G C F, Williams D F 1982 The effects of proteins on metallic passivation. *J. Biomed. Mater. Res.* 16: 125–34
Espevik S 1977 *In vitro* corrosion of dental amalgams with different Cu content. *Scand. J. Dent. Res.* 85: 631–6

Espevik S 1978 Corrosion of base metal alloys *in vitro*. *Acta Odontol. Scand.* 36: 113–17

Espevik S, Mjör I A 1979 Degradation of amalgam restorations *in vivo*. In: Syrett B C, Acharya A (eds.) 1979 *Corrosion and Degradation of Implant Materials*, ASTM Special Technical Publication 684, American Society for Testing and Materials, Philadelphia, PA, pp. 316–27

Ewers G J, Greener E H 1985 The electrochemical activity of the oral cavity—A new approach. *J. Oral Rehabil.* 12: 469–76

Finkelstein F G, Greener E H 1978 Role of mucin and albumin in saline polarization of dental amalgam. *J. Oral Rehabil.* 5: 95–100

Hodges R J 1977 The corrosion resistance of gold and base metal alloys. In: Valega T M (ed.) 1977 *Alternatives to Gold Alloys in Dentistry*, US Department of Health, Education and Welfare Publication No. DHEW 77-1227. US Government Printing Office, Washington, DC, pp. 106–38

Holland R I 1984 Effect of pellicle on galvanic corrosion of amalgam. *Scand. J. Dent. Res.* 92: 93–6

Jørgensen K D 1965 The mechanism of marginal fracture of amalgam fillings. *Acta Odontol. Scand.* 28: 347–89

Jørgensen K D, Saito T 1970 Structure and corrosion of dental amalgams. *Acta Odontol. Scand.* 28: 129–42

Marek M 1983 The corrosion of dental materials. In: Scully J C (ed.) 1983 *Corrosion: Aqueous Processes and Passive Films*, Treatise on Materials Science and Technology, Vol. 23. Academic Press, New York, pp. 331–94

Marek M 1984 Acceleration of corrosion of dental amalgam by abrasion. *J. Dent. Res.* 63: 1010–13

Marek M, Topfl E 1986 Electrolytes for corrosion testing of dental alloys. *J. Dent. Res.* 65 (SI): 301

Meyer J M 1977 Corrosion resistance of Ni–Cr dental alloys. *Corros. Sci.* 17: 971–82

Mezger P R, Vrijhoef M M A, Greener E H 1985 Corrosion resistance of three high-palladium alloys. *Dent. Mater.* 1: 177–80

Mueller H J, Greener E H 1970 Polarization resistance of surgical materials in Ringer's solution. *J. Biomed. Mater. Res.* 4: 29–41

Sarkar N K, Fuys R A, Stanford J W 1979a The chloride corrosion of low-gold casting alloys. *J. Dent. Res.* 58: 568–75

Sarkar N K, Fuys R A, Stanford J W 1979b The chloride corrosion of silver-base casting alloys. *J. Dent. Res.* 58: 1572–7

Sarkar N K, Greener E H 1973 *In vitro* corrosion resistance of new dental alloys. *Biomater., Med. Devices, Artif. Organs* 1: 121–9

Sarkar N K, Greener E H 1975 Electrochemistry of saline corrosion of conventional dental amalgams. *J. Oral Rehabil.* 2: 49–62

Sastri S, Vaidyanathan T K, Mukherjee K 1982 Potentiodynamic polarization analysis of silver–palladium alloys in chloride solutions. *Met. Trans. A* 13: 313–7

Sumithra N, Vaidyanathan T K, Sastri S, Prasad A 1983 Chloride corrosion of recent commercial Pd-based alloys. *J. Dent. Res.* 62 1983 IADR No. 346

Vaidyanathan T K, Prasad A 1981 *In vitro* corrosion and tarnish analysis of Ag–Pd binary system. *J. Dent. Res.* 60: 707–15

Wang Chen C P, Greener E H 1977 Galvanic study of different amalgams. *J. Oral Rehabil.* 4: 23–7

Wright S R, Cocks F H, Pearsall G W, Gettleman L 1980 An ultrasonic abrasion simulation test method applied to dental alloy evaluation. *Corrosion* 36: 101–3

M. Marek
[Georgia Institute of Technology, Atlanta, Georgia, USA]

D

Dental Amalgams

Dental amalgam is prepared by mixing an alloy powder with mercury to produce a plastic mixture which is condensed into a prepared cavity. It has maintained its popularity as a filling material over the years since it is durable, inexpensive and requires only one dental appointment for placement.

1. History of Dental Amalgams

During the T'ang Dynasty, *c.* AD 700, Chinese physicians used a silver–tin amalgam as a filling material (Hsi-T'ao 1958). Historical accounts of amalgams (Craig 1985, Roggenkamp 1986) state that amalgams were not used in the West until much later. Following the development of a silver–mercury paste by Traveau in France in 1826, amalgams for the restoration of teeth were introduced into the USA by the Crawcour brothers in 1833. Although the use of amalgams has always been controversial, 150 years of research in both fabrication technology and alloy chemistry has greatly improved the quality of the amalgams currently available.

The first systematic investigations of silver–tin alloys and their amalgams were started in the nineteenth century by J. Foster Flagg and were continued by G. V. Black. In 1929, an alloy similar to Black's ternary silver–tin–copper alloy was approved by the American Dental Association (ADA) in the American Dental Association Tentative Specification for Dental Amalgam Alloys. This was revised several times until the current ADA specification was adopted in 1977 and formed the basis for the original International Organization for Standardization (ISO) Standard 1559 adopted in 1978.

The required limits of chemical composition (in wt%) outlined in the previous 1969 ADA specification were similar to those of Black's alloy: a minimum of 65% silver and maximums of 29% tin, 6% copper, 2% zinc and 3% mercury. It was not until the 1960s that formulations that differed from the Black alloy were recognized. The first divergence was high-copper amalgam, produced by mixing a small amount of plasticized copper amalgam into low-copper amalgams. The second clinically significant modification of the amalgam was the development of a high-copper content amalgam by Innes and Youdelis (1963). This discovery did not receive much attention until 1970, when it was shown that amalgam restorations prepared from this high-copper powder were clinically superior to low-copper amalgams, especially in resisting marginal breakdown (Mahler et al. 1970). The work of Innes and Youdelis has since led to the development of numerous high-copper alloys.

2. Alloy Powder

The current ADA Specification No. 1 (1977) requires that the chemical composition of alloy powders consists essentially of silver and tin, but it allows for the inclusion of copper, zinc, gold and/or mercury in amounts smaller than the silver or tin content. Alloy powders containing zinc in excess of 0.01 wt% are designated as "zinc-containing". Those powders containing zinc less than or equal to 0.01 wt% are designated as "nonzinc". The present ADA specification does not list compositional ranges for a low- or high-copper alloy powder. In contrast, the most recently adopted ISO standard (1986) specifies a chemical composition (in wt%) of a minimum of 40% silver and maximums of 32% tin, 30% copper, 2% zinc and 3% mercury. Usually, an alloy powder containing less than 6% copper is classified as a low-copper alloy or traditional powder, whereas an alloy containing greater than 8–10% copper is classified as a high-copper alloy powder. The amalgams made from these two types of alloy powder are called low-copper and high-copper amalgams, respectively; usage of the former is becoming less common in developed countries.

As mentioned in Sect. 1, the first major change in alloy composition occurred in the 1960s when the silver–copper binary eutectic powder was mixed with a low-copper powder. The silver–copper eutectic particles were intended to "dispersion" harden the amalgams produced, though it is now known that, in strict metallurgical terms, this does not occur. The resulting copper-enriched powder was called "high-copper admixed" or "dispersant" powder.

The next major change occurred in 1974 when Asgar developed a new type of alloy for amalgams by preparing a powder from a single silver–copper–tin ternary alloy, rather than by mixing two kinds of powder together, and this was called high-copper single composition powder. The ranges of chemical compositions for commercially available copper alloys are given in Table 1.

2.1 Alloy Phases

Studies of the silver–tin–copper ternary equilibrium phase diagram date back as far as the 1920s. When an alloy is produced by normal melting and heat treatment, the major phase in alloys within the compositional ranges listed in Table 1 is always an intermetallic compound in the silver–tin binary system, Ag_3Sn, called the γ phase (rhombically deformed hexagonal-close-packed). A solid-solution phase of the silver–tin alloy system, the β phase (hexagonal-close-packed), is also sometimes present. Since copper dissolves in Ag_3Sn to approximately 1%, any excess copper will form copper-rich phases, most commonly copper–tin. Depending on the composition of the alloy and its thermal history, other phases which may be present are ε copper–tin (Cu_3Sn, hexagonal), η' copper–tin (Cu_6Sn_5, NiAs-type hexagonal) and/or α tin. Admixed powders contain a silver–copper eutectic (α silver plus α copper) powder and, in some cases, include a hypoeutectic silver–copper alloy powder. In the low-copper powder range, zinc forms Cu_5Zn_8 when the amount of zinc exceeds its solubility in the γ and/or β silver–tin (1.6% and 5.9%, respectively) (Jensen et al. 1973).

2.2 Shape of Particles in the Alloy Powder

The particles, usually 10–30 µm in size, used to prepare dental amalgams may be irregularly shaped powders fabricated by lathe-cutting or ball-milling the alloy ingot, or spherical particles made by atomizing molten metal. The shape of spherical particles depends on the cooling rate during fabrication, ranging from nearly perfect spheres to spheroidal shapes with irregular surfaces. In either case, particles are annealed under inert or reducing gas conditions, followed by acid washing in order to alter the reaction rate with mercury. Powders composed of irregular particles were used until spherical powders were introduced in 1962 (Demaree and Taylor 1962). Alloy powders currently in use may be composed entirely of irregularly shaped particles, entirely of spherical particles, or be a mixture of irregular and spherical particles. Although particle shape does not affect the reaction of the powder with mercury, dramatic differences in shape between lathe-cut and spherical powders can affect the handling characteristics of the dental amalgam.

Table 1
Ranges of chemical compositions (wt%) for commercially available copper alloys

	Low-copper alloys	High-copper alloys
Silver	66.7–71.5	39.9–70.1
Tin	24.3–27.6	17.0–30.2
Copper	1.2–5.5	9.5–29.9
Zinc	0–1.5	0–0.7
Mercury	0–4.7	0–0.25

3. Amalgamation Reaction

Modern dental amalgams are prepared by a process called "trituration": this entails vigorously mixing a powder or tablets with mercury for a short period of time (10–20 s) using a mechanical mixer.

3.1 Low-Copper Alloy

When trituration begins, the silver and tin in the outer surface layers of the particles dissolve into the mercury. Owing to the limited solubility of silver (0.035 wt%) and tin (0.06 wt%) in mercury at room temperature, two kinds of intermetallic compounds, Ag_2Hg_3 (body-centered-cubic) and Sn_8Hg (simple hexagonal) (Fairhurst and Ryge 1962, Fairhurst and Cohen 1972) start to precipitate soon after mixing. These compounds are called γ_1 and γ_2, respectively. In a dissolution–precipitation process, a species must be supersaturated before precipitation can begin. Silver reaches its saturation concentration in mercury, before tin, so silver supersaturation results in the precipitation of γ_1 before the precipitation of γ_2. While the reaction takes place, the alloy powder and the precipitates coexist with the mercury, keeping the mix at a plastic consistency for approximately 10–15 min.

The alloy is usually mixed with 40–50 wt% mercury, which is insufficient to consume the alloy particles completely. Consequently, unconsumed particles are present in the hardened amalgam and these become surrounded and bound together by minute equiaxed γ_1 grains (1–3 µm in size) and irregularly shaped, discrete γ_2 crystals. The amalgam is thus a composite in which alloy particles of approximately 30 vol.% are embedded in γ_1 (60%) and γ_2 (10%) phases (Mahler 1988). The structure also includes ε-Cu_3Sn particles and voids in the γ_1 matrix (Phillips 1982).

The copper included in the powder, in the form of either dissolved atoms in the γ-Ag_3Sn lattice or atoms in the ε-Cu_3Sn when copper exceeds its solubility in γ-Ag_3Sn, participates in the reaction in a similar manner to that described in Sect. 3.2 for high-copper alloy. However, copper does not affect the amalgamation reaction to a noticeable level since it is of a low concentration in the alloy.

3.2 High-Copper Alloy

In the case of high-copper admixed alloys, silver and tin enter the mercury from the γ and/or β silver–tin powder, silver also entering the mercury from silver–copper alloy particles. Silver from both sources precipitates as γ_1-Ag_2Hg_3 grains. If sufficient copper is available, the dissolved tin combines with copper to form η'-Cu_6Sn_5 crystals rather than forming γ_2-Sn_8Hg. In most high-copper amalgams, formation of the γ_2 phase is totally suppressed, which is important since γ_2 is the weakest, softest phase of the amalgam phases (Craig 1985). In the admixed powder reaction, the η' crystals form at two different sites: at the surface of the silver–copper eutectic particles and at the γ_1 matrix (Okabe et al. 1979b). Apparently, before supersaturation, the dissolved tin diffuses to the surface of the eutectic particles to combine with copper to form η' crystals. Copper in the high-copper powder acts as a sink for tin migration. As the amalgamation reaction proceeds, the reaction layer (usually 1–2 μm thick) forms on the particle surface. This layer is a mixture of ultra-fine η' crystals (approximately 30 nm) (Okabe et al. 1977) and a lesser amount of γ_1 crystals. Saturation of silver occurs after nucleation and growth of η' crystals. The structure of high-copper admixed amalgams includes the unconsumed low-copper alloy particles, unconsumed silver–copper particles, and, sometimes, ε-Cu_3Sn particles embedded in the γ_1 matrix. In addition, individual thin η' rod crystals, which are believed to have formed by the dissolution and precipitation process, are embedded within γ_1 crystals in the high-copper admixed amalgam.

Similarly, in the amalgamation reaction of the high-copper single composition alloy powder (Okabe et al. 1978a, b), silver and tin dissolve from the γ and/or β silver–tin phase in the particle. Concurrently, copper dissolves mainly from the ε-Cu_3Sn phase, which coexists with the silver–tin phases in the silver–tin–copper alloy particles. Tin atoms dissolved in the mercury are attracted to the ε-Cu_3Sn areas located on the alloy powder surfaces where meshes of η'-Cu_6Sn_5 crystals form. The η' crystals in these amalgams are much larger than those found in high-copper admixed amalgams. Included in the structures are unconsumed alloy particles, η' copper–tin crystals and the γ_1 matrix. An important finding is that unconsumed alloy particles are covered by a number of η'-Cu_6Sn_5 rod crystals. Other η' crystals are seen embedded within the γ_1 grains.

4. Mechanical and Physical Properties

4.1 Mechanical Properties

Sufficient strength to resist clinical fracture is a prerequisite for any restorative material. Extensive investigations of mechanical and physical properties have been made with amalgams prepared from experimental and commercial alloy powders, producing a variety of microstructures and phase distributions. Specimens made from these different amalgams have been subjected to many mechanical tests including measurements of tensile, shear, compressive and bending strengths, as well as being subjected to fatigue, creep and, more recently, fracture toughness tests. Thus far, only creep tests have shown a significant correlation with clinical performance (Mahler et al. 1970). The extent of marginal fracture of amalgam restorations increases with creep rate and, consequently, national and international specifications for amalgams now include creep values. For example, the ISO Standard 1559 for creep measures the percentage change in the length of week-old specimens over a three-hour period under a constant compressive stress (36 MPa) at 37 °C. Recent emphasis has been placed on studying amalgam deformation at slow-strain rates because of the established relationship between creep and clinical performance.

Table 2 consists of a summary of the mechanical properties of types of amalgam. The compressive strength ranges given for 30 min, 1 h and 24 h after specimen preparation demonstrate how amalgams strengthen as amalgamation progresses; most commercial amalgams reach their final strength within 24 h. The compressive strength ranges for the three types of amalgam overlap each other. However, the values indicate that high-copper alloys could have higher compressive strengths. Both the diametral tensile strength and elastic modulus for each type of amalgam fall into a similar range. The much lower values in tensile strength ranges as compared with

Table 2
Ranges of strengths and creep values of low-copper and high-copper amalgams

Amalgam	Compressive strength (MPa)[a]			Diametral tensile strength (MPa)	Elastic modulus[b] (GPa)	ISO Creep (%)
	30 min	1 h	24 h			
Low-copper		124–200	310–390	53–62	20–40	0.70–3.70
High-copper admixed	55–110	80–180	330–500	42–62	35–50	0.05–2.10
High-copper single composition	70–210	100–320	320–510	34–64	30–60	0.03–0.040

a loading rate $= 4.2 \times 10^{-3}$ mm s^{-1} b loading rate $= 2.0 \times 10^{-3}$ mm s^{-1}
Table compiled from data of Osborne et al. (1978), Duke et al. (1982), Bryant (1979, 1980), Bryant and Mahler (1986)

those of compressive strength reveal the brittleness of these amalgams. On the other hand, there is a dramatic difference in creep rates between low-copper and high-copper amalgams, indicating that the low-copper amalgam is much more plastic.

Published studies (e.g., Mitchell et al. 1987) demonstrate that the fracture mode and strength of the amalgam are strongly sensitive to strain rate. At all strain rates, the mechanical properties of dental amalgams are controlled by the γ_1-based matrix. At a high strain rate, amalgam is very brittle and γ_1 grains undergo intergranular fracture. The plasticity of γ_1 and γ_2 increases as the strain rate is lowered; at extremely low strain rates both γ_1 and γ_2 grains exhibit a high degree of plastic deformation, producing many fine "needles". In addition, significant grain boundary sliding of the γ_1 grains becomes evident (Okabe et al. 1983). The sliding rate of γ_1 grains increases when the structure contains γ_2, the softest component among the amalgam phases. Therefore, low-copper amalgams, all of which contain γ_2, have lower creep resistance than high-copper amalgams.

A restraint on sliding is operative in high-copper amalgams. When η' crystals span γ_1 grain boundaries, the γ_1 grains must move plastically around much harder η' crystals for sliding to occur; such grain boundaries slide very slowly. Another example of a durable interface is that between the high-copper alloy particles, which are covered by a dense mesh of minute η' crystals, and the γ_1 matrix. The presence of η' crystals, therefore, increases resistance to creep deformation.

Crack nucleation and propagation in amalgams at various strain rates has been studied: at high ($\sim 10^{-1}$ mm s^{-1}) and moderate ($\sim 10^{-3}$ mm s^{-1}) loading rates, cracks have been found to nucleate at different interfaces between γ_1 and other components and to propagate through the γ_1 matrix. Failure is either by intergranular separation of γ_1 or by cleavage. The tendency for cleavage to occur increases with strain rate and with the percentage of η' crystals.

At low strain rates (10^{-6} mm s^{-1} or less), γ_1 grains fracture transgranularly in a ductile manner. Cracks nucleate near grain boundaries; however, cracks remain within the grain so that individual γ_1 grains fracture by microvoid coalescence and dimpled "cup and cone" rupture surfaces are observed. At higher strain rates, the activation of multiple slip systems required for plastic deformation is blocked by the inability of superdislocations within the γ_1 intermetallic to cross-slip. At low strain rates, dislocation climb reactivates the required slip systems and plastic deformation becomes possible.

4.2 Dimensional Change

The amalgamation reaction continues long after the amalgam is condensed into the prepared cavity. Since the reaction includes the dissolution of alloy particles, the precipitation of crystals and the growth of crystals, dimensional changes occur during hardening. An ideal amalgam for use as a restorative material would neither expand nor contract. In theory, an unconstrained amalgam will first contract due to dissolution and then expand towards the end of the reaction as impinging crystals push one another apart. The dimensional change of the amalgam is the result of the total effects of the type of alloy used and, more importantly, manipulative variables such as mercury percentages, trituration conditions and condensation technique. The ISO specification states that the dimensional change should be measured at 37 °C, between 5 min and 24 h after the preparation of the specimen. Currently, its value is regulated to be within $0 \pm 0.20\%$ or $0 \pm 20\,\mu\mathrm{m\,cm}^{-1}$.

Dimensional change is an important clinical problem, since it is thought to relate to microleakage at the interface which results in postoperative sensitivity and possibly secondary caries. A recent *in vitro* study recommended that the specification be revised to reflect the fact that microleakage involves not only dimensional changes but also the surface texture of the amalgam at the cavity wall. In addition to the evaluations of the dimensional changes of various amalgams *per se*, much research has been devoted to investigating the effects, for example, of cavity varnishes and thermocycling on microleakage. Mahler (1988) summarizes the present limited information.

4.3 Corrosion Resistance

Deterioration of metals in the mouth may be caused by direct chemical action and/or electrolytic processes. Dental amalgam is subject to tarnishing and electrochemical corrosion. Over a period of time, tarnish or surface discoloration may develop into corrosion which reduces the structural integrity of the amalgam. Electrochemical evaluation of various amalgam phases reveals that the γ_2 phase is considerably more anodic than γ_1 and γ-Ag_3Sn. The absence of γ_2 in high-copper amalgams results in a dramatic improvement in their corrosion resistance (see *Corrosion of Dental Materials*). High-copper amalgams also suffer some corrosion, however, since η' is anodic with respect to the surrounding mercury-containing phases.

In most oral environments both types of amalgams are initially in a state of passivity and little corrosion takes place. However, high acidity and a high concentration of chloride ions lead to the development of crevices and pores which destroy passivity and allow corrosion to occur. The surface of amalgam restorations should thus be as free of pores and crevices as possible.

Crevice corrosion can be beneficial, as corrosion products at the interface between the tooth structure and amalgam appear to contribute to the seal that forms between the restorative and the tooth. This seal reduces microleakage. In studies, the following insoluble corrosion products have been identified: SnO, SnO_2, $Sn_4(OH)Cl_2$, $CuCl_2 \cdot 3Cu(OH)_2$, Cu_2O, $CaSn(OH)_6$, $ZnSn(OH)_6$, $Zn_5(OH)_8Cl_2 \cdot H_2O$ and $Zn(OH)Cl$.

A contact between dissimilar metals in the same electrolyte, in this case saliva, results in galvanic corrosion in which the less noble metal suffers more intensive deterioration. Since dental amalgams are less noble than most other dental alloys, amalgam restorations should not be placed in continuous contact with cast restorations, especially alloys of gold and palladium. Even intermittent contacts of dissimilar restorations in opposing teeth will result in the flow of a galvanic current which may cause a discomfort to the patient.

5. Marginal Fracture and Bulk Fracture

Many clinical studies show that high-copper amalgams have a significantly lower rate of marginal breakdown than low-copper amalgams. Figure 1 shows surfaces of both low- and high-copper amalgam restorations after four years of service. The photograph clearly shows the difference in their clinical performance: the margins of the low-copper amalgam restoration (right) are severely ditched. In addition, the surface near the interproximal area of the low-copper amalgam has been indented because of the action of the opposing tooth cusp. This is an indication of its low creep resistance. Some studies show that restorations with less breakdown need to be replaced as a result of recurrent caries less often, though not all studies agree with this. Continued research on marginal fractures should clarify the situation.

Research has focused on understanding the mechanism of marginal breakdown. Mahler et al. (1970) demonstrated a correlation between the

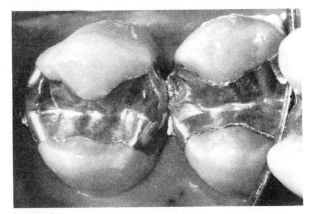

Figure 1
Four-year-old low-copper (right) and high-copper (left) amalgam restorations (courtesy of D. B. Mahler)

creep of γ_2-containing amalgams and marginal fracture. There have been many attempts to explain this relationship, a large number placing primary emphasis on the role of corrosion (e.g., Sarkar 1978). It has been reported that amalgam restorations with the largest quantities of corrosion products had the worst margins; however, other studies have shown that marginal cracks develop in restorations even in the absence of significant corrosion (Hanawa et al. 1988). A recent study of amalgam restorations in denture teeth correlated increasing marginal fracture with the length of time in service (Marker et al. 1987). The mode in the fracturing process appeared to be similar to that reported for natural teeth. Examination of cross sections of these specimens revealed that marginal breakdown occurred even when no defect was apparent near the margins of freshly placed amalgam restorations. Evidence of plastic deformation was observed near crack tips, suggesting that crack nucleation occurred at low strain rates since only at such rates can this plasticity in amalgams occur (Hanawa et al. 1988). Once nucleated, cracks propagate and final fracture occurs at high strain rates.

Explanations of marginal breakdown that rely primarily on mechanical deformation have also been offered. It has been proposed that phase changes in the amalgam or corrosion products in the interface might produce hydrostatic stresses tending to extrude an amalgam, and it has also been suggested that long-term linear expansion might result in extrusion and lead to marginal breakdown. Tooth deflection is another factor that might explain marginal breakdown. Calculations showed that creep resulting from compressive stresses at the margins tended to produce a gap there (Derand 1977). Any change in cavity design tending to increase the deflection of the cusp will increase this gap; for example, deeper or wider preparations would increase the deflection. In this light, it is interesting that significantly greater marginal breakdown has been reported in large restorations. Moreover, it has been found that marginal breakdown increases with biting force.

Several studies have shown that high-copper amalgams undergo less bulk fracture than low-copper amalgams. Since low- and high-copper amalgams are weak in tension, both types of amalgam would be expected to be equally brittle. It is possible that easier crack nucleation at low strain rates could account for the brittleness of γ_2-containing amalgams.

6. Mercury Release

Controversy has existed over the use of dental amalgams for the restoration of teeth ever since the technique was introduced. The reason is apparent: mercury without oxide film has a high vapor pressure at room temperature (1.20×10^{-3} torr at $20\,°C$)

and evaporates freely (Okabe 1987). Thus, there has always been a belief that amalgam restorations could cause mercury poisoning by chronically exposing the dental personnel and patient to mercury vapor. In many countries, there has been an increased public awareness and professional concern over the release of mercury from amalgam. This concern has been fuelled by the introduction of highly sensitive mercury vapor detectors with detection limits as low as $1 \mu g\, m^{-3}$ in air or $0.2 \mu g\, l^{-1}$ in solution. Measurable amounts of mercury vapor have been reported in the intraoral air of patients with amalgam restorations, and there have been a number of reports of mercury in breath, saliva, blood, nails and hair, resulting from both mercury release and various clinical procedures. Many of these studies have been reviewed in a recent article by Langan et al. (1987). However, current reports reveal that far less mercury is actually released from amalgam restorations than was previously suggested. It is known that mercury is liberated *in vivo* and that the release level temporarily rises when the amalgam is abraded. In *in vitro* studies, mercury release increases when the amalgam is heated in an oxygen-free atmosphere or after it has been corroded. However, in air, processes such as aging, surface oxidation and water film formation reduce mercury vaporization from the amalgam.

An x-ray diffractometry study showed that the rate of reaction of mercury during amalgamation is dependent on the alloy used and that, in a normal situation, all mercury used for trituration is consumed within several hours (Ferracane et al. 1986). However, formation of the solid phases and the consumption of the mercury does not mean that amalgamation has ended. The peritectic isotherms of some of the amalgam phases are near the highest service temperature of the amalgam (60–80 °C), that is, 127 °C for the γ_1-Ag_2Hg_3 phase and 216 °C for the γ_2-Sn_8Hg phase. The homologous temperature (the ratio of the ambient temperature to the melting point temperature, T/T_m) could thus be as high as 0.8. Body temperature is high enough to initiate the solid-state diffusion process necessary to change an amalgam from an initial nonequilibrium structure to a structure that is closer to equilibrium. These diffusional processes result in the decomposition of both the γ_1 and γ_2 phases and the transformation of the γ_1 into the β_1 phase, which contains less mercury than γ_1. While these structural changes are taking place, it is probable that some mercury is liberated by evaporating into mouth air, by dissolving into oral fluids or by diffusing into the tooth structure. However, at the same time, unconsumed alloy particles are likely to be a sink for the mercury produced as a result of such changes.

None of the *in vivo* and *in vitro* corrosion products listed above include mercury in their structure and, consequently, mercury liberated during the corrosion process must either go into solution or reamalgamate with unconsumed alloy particles. Studies of amalgam dissolution into various electrolytes give conflicting results: some investigators report more release of mercury from high-copper amalgams than from low-copper amalgams, but others show the opposite tendency (Okabe 1987). However, almost all of the data are consistent in that the rate of dissolution of mercury decreases drastically with time of immersion. The dissolution rate slows after releasing atoms located near the surface, dissolution becoming a solid-state diffusion control process. Oxide film formation (Hanawa et al. 1987) is another factor which slows down the release of metallic ions to the media. There is no doubt that mercury dissolves at different rates from various amalgam phases, dissoluton being found to be greatest from γ_1 though this is reduced in the case of dental amalgams where γ_1 is galvanically coupled with the less noble phases γ and γ_2. This coupling reduces the driving force for mercury release since it keeps the corrosion potential of the system at a low level.

As for the toxicity of mercury and the biocompatibility of amalgams, recent reviews (Craig 1986, Langan et al. 1987, Mahler 1988) summarize the literature published thus far. Although occupational exposure to mercury is a potential hazard for dental personnel, mercury can be handled in the dental office without exceeding limiting values. There is no evidence in the published data that the minute amount of mercury released from amalgam restorations into oral air, saliva or tooth structure can cause mercurialism to patients, except for those who have allergies to mercury, and, furthermore, such allergies are rare.

It has been estimated that the amount of mercury vapor inhaled during the first three-hour period after placement of an amalgam restoration with a surface area of $50\, mm^2$ is $200\, ng$, approximately 10% of the normal daily human exposure to mercury through air, water and food intake. Amalgam restorations are unlikely to expose a patient to mercury vapor levels near the threshold limit value (TLV) of $0.05\, mg\, m^{-3}$ per 40 h working week as set by the US National Institute for Occupational Safety and Health. If several large restorations are releasing enough mercury to keep the mercury vapor concentration at the TLV level, all the mercury in the restoration would evaporate in a few years. Clearly, the life span of amalgam restorations is much longer than this; hence, mercury vapor from restorations cannot exceed the TLV, except very intermittently (Okabe 1987).

7. Selection of Alloys

Since numerous alloys for dental amalgams are available, selecting the best amalgam for use can be confusing. Creep and compressive strength are two important factors to consider when making a selection. An amalgam with an ISO creep value of less than 1.0% is likely to show good marginal integrity for a long time span.

The compressive strength one hour after preparation of the amalgam, sometimes called "early strength," is another important factor. Most high-copper single composition amalgams have much higher early strengths than high-copper admixed amalgams. It is generally recommended that impressions of newly carved amalgam posts are not taken; nevertheless, such clinical practices are becoming more widespread. If crown preparations are to be made on freshly placed amalgam, one with high early strength should be selected. Another advantage of high early strength in terms of amalgam polishing has been suggested—that some such high-copper amalgams can be polished 10 min after the end of trituration without harmful effects.

Some generalizations can be made about single composition and admixed high-copper amalgams. Both can have low creep, mainly due to the absence of γ_2 in both types. Since low creep is important, either type of high-copper amalgam will be an improvement over low-copper amalgams. Compressive strengths of single composition amalgams tend to be higher. However, both types of high-copper amalgam are more than strong enough in compression. Admixed amalgams resist condensation well since most admixed amalgams contain irregular particles whereas the totally spherical, single composition amalgams are mushy. Amalgams which resist condensation are generally easier to adapt to cavity walls and may exhibit less leakage. If it is necessary to work with the restoration immediately after placement, single composition amalgams should be used owing to their high early strength. If strength properties are less important, admixed amalgams may be preferred because of their higher resistance to condensation.

Finally, it should be noted that the performance of even the best of these amalgams is influenced by operator technique. High-copper amalgams have been shown to be less technique-sensitive than low-copper amalgams. Nevertheless, careful cavity designing and preparation, condensing, carving and polishing are still essential. High-copper amalgams, placed by skilled dental personnel, allow patients to receive higher quality, possibly longer lasting restorations.

Bibliography

Bryant R W 1979 The strength of fifteen amalgam alloys. *Aust. Dent. J.* 24: 244–52

Bryant R W 1980 The static creep of amalgams from fifteen alloys. *Aust. Dent. J.* 25: 7–11

Bryant R W, Mahler D B 1986 Modulus of elasticity in bending of composites and amalgams. *J. Pros. Dent.* 56: 243–8

Craig R G 1985 *Restorative Dental Materials*. Mosby, St Louis, MO, pp. 198–200

Craig R G 1986 Biocompatibility of mercury derivatives. *Dent. Mater.* 2: 91–6

Demaree N C, Taylor D F 1962 Properties of dental amalgams made from spherical alloy particles. *J. Dent. Res.* 41: 890–906

Derand T 1977 Marginal failure of amalgam class II restorations. *J. Dent. Res.* 56: 481–5

Duke E S, Cochran M A, Moore B K, Clark H E 1982 Laboratory profiles of 30 high-copper amalgam alloys. *J. Amer. Dent. Assoc.* 105: 636–40

Fairhurst C W, Cohen J B 1972 The crystal structures of two compounds found in dental amalgam: Ag_2Hg_3 and Ag_3Sn. *Acta Crystallogr., Sect. B* 28: 371–8

Fairhurst C W, Ryge G 1962 X-ray diffraction investigation of the Sn–Hg phase in dental amalgam. In: Mueller W M (ed.) 1962 *Advances in X-Ray Analysis* Pergamon, New York, pp. 64–70

Ferracane J, Mafiana P, Spears R, Okabe T 1986 Rate of liquid mercury consumption in freshly made amalgams. *J. Dent. Res.* 65: 192

Hanawa T, Ota M, Marker V A, Mitchell R J, Okabe T 1988 Crack initiation and propagation in retrieved amalgams. *J. Dent. Res.* 67: 118 (abstract 42)

Hanawa T, Takahashi H, Ota M, Pinizzotto P F, Ferracane J L, Okabe T 1987 Surface characterization of amalgams using x-ray photoelectron spectroscopy. *J. Dent. Res.* 66: 1470–8

Hsi-T'ao C 1958 The use of amalgam as filling material in dentistry in ancient China. *Chinese Med. J.* 76: 553–5

Innes D B K, Youdelis W V 1963 Dispersion strengthened amalgams. *Canad. Dent. Assoc. J.* 29: 587–93

Jensen S J, Olsen K B, Utoft L 1973 A zinc-copper phase in dental silver amalgam alloy. *Scand. J. Dent. Res.* 81: 572–6

Langan D A, Fan P L, Hoos A A 1987 The use of mercury in dentistry: A critical review of the recent literature. *J. Amer. Dent. Assoc.* 115: 867–80

Mahler D B 1988 Research on dental amalgam: 1982–1986. In: Reese J A (ed.) 1988 *Advances in Dental Research*, Vol. 2. International Association of Dental Research, Washington, DC, pp. 71–82

Mahler D B, Terkla L G, Van Eysdan J, Reisbick M H 1970 Marginal fracture vs mechanical properties of amalgam. *J. Dent. Res.* 49: 1452–7

Marker V A, McKinney T W, Filler W H, Miller B H, Mitchell R J, Okabe T 1987 A study design for an *in vivo* investigation of marginal fracture in amalgam restorations. *Dent. Mater.* 3: 322–30

Mitchell R J, Ogura H, Nakamura K, Hanawa T, Marker V A, Okabe T 1987 Characterization of fractured surfaces of dental amalgams. *Int. Symp. Testing and Failure-Analysis*. American Society for Metals, Metals Park, OH, pp. 179–87

Okabe T 1987 Mercury in the structure of dental amalgam. *Dent. Mater.* 3: 1–8

Okabe T, Butts M B, Mitchell R J 1983 Changes in the microstructures of silver-tin and admixed high-copper amalgams during creep. *J. Dent. Res.* 62: 37–43

Okabe T, Mitchell R J, Butts M B 1977 Analysis of Asgar–Mahler reaction zone in dispersalloy amalgam by electron diffraction. *J. Dent. Res.* 56: 1037–43

Okabe T, Mitchell R J, Butts M B, Fairhurst C W 1978a A study of high-copper amalgams, III. SEM observation of amalgamation of high-copper powders. *J. Dent. Res.* 57: 975–82

Okabe T, Mitchell R J, Butts M B, Fairhurst C W 1979a A study of high-copper dental amalgams by scanning electron microscopy. In: LeMay I, Fallon P A, McCall J L (eds.) 1979 *Microstructural Science*, Vol. 7. Elsevier, New York, pp. 165–74

Okabe T, Mitchell R J, Butts M B, Wright A H, Fairhurst C W 1978b A study of high-copper amalgams, I: A comparison of amalgamation on high-copper alloy tablets. *J. Dent. Res.* 57: 759–67

Okabe T, Mitchell R J, Fairhurst C W 1979b A study of high-copper amalgams, IV. Formation of η'Cu–Sn (Cu$_6$Sn$_5$) crystals in high-copper dispersant amalgam matrix. *J. Dent. Res.* 58: 1087–92

Osborne J W, Gale E N, Chew C L, Rhodes B F, Phillips R W 1978 Clinical performance and physical properties of twelve amalgam alloys. *J. Dent. Res.* 57: 983–8

Phillips R W 1982 *Skinner's Science of Dental Materials*. Saunders, Philadelphia, PA, pp. 311–5

Roggenkamp C L 1986 A history of copper in amalgam and an overview of setting reaction phases. *Quintessence Int.* 17: 129–33

Sarkar N K 1978 Creep, corrosion and marginal fracture of dental amalgams. *J. Oral Rehab.* 5: 413–23

T. Okabe
[Baylor College of Dentistry, Dallas, Texas, USA]

Dental Implants

The use of implantable devices and materials in oral surgery can be classified into three categories. The first category comprises implants used in prosthodontics for the replacement of missing teeth. This category can be subdivided into: (a) those used directly, by providing a replacement for the missing dentition, possibly by anatomically shaped replacements for teeth themselves, or more commonly by providing a structure below the tissue surface (but with protruding posts) which is able to support crowns, bridges or dentures; and (b) those used indirectly, by procedures such as ridge augmentation which is designed to alter the shape of the bone to facilitate the wearing and retention of dentures.

The second category includes implants used in periodontal surgery, such as in the restoration of bony defects around the teeth and in the stabilization of periodontally involved teeth.

The last category comprises implants used in facial reconstruction in cases of acquired defects (from trauma or tumor resection), or congenital or developmental defects and abnormalities.

All of these implants have been described as dental implants, although this particular term is most usually confined to implants that are used in direct dental prosthetics. In this article all of the above devices will be considered.

1. Direct Replacement of Missing Teeth

Several methods for the replacement of missing teeth have been available for many years. In particular, the fixed-bridge, removable-partial-denture and full-denture techniques are well practised and entirely adequate for the majority of patients. There is a sizable minority, however, for whom these methods are less than satisfactory and alternative methods become desirable. During the 1970s and 1980s, several alternative methods were devised:

(a) the subperiosteal implant, used to provide a permanent means of support for lower and, occasionally, upper dentures (Sect. 1.1);

(b) the blade-vent endosseous implant, used singly or in small numbers to provide posts for attachment of individual crowns or dentures (Sect. 1.2);

(c) other metallic endosseous implant systems (Sect. 1.3);

(d) alumina endosseous implant systems (Sect. 1.4);

(e) the tooth-root replacement, implanted within alveolar bone and subsequently used to provide a base for transgingival posts, again for single or multiple teeth (Sect. 1.5) (included here are the so-called osseointegrated dental implants); and

(f) the transosteal implant, also used to provide a permanent means of denture support, applicable in this case only to the mandible (Sect. 1.6).

Each of these different types of implant is associated with its own indications, contraindications, clinical methodology and performance. A full review of the status of certain of these implants was published in 1981, as the report of a National Institute of Health (NIH)–Harvard Consensus Conference (Schnitman and Shulman 1980).

1.1 Subperiosteal Implants

These devices are specifically intended for the patient without teeth who, usually because of the resorption of the bony ridge, cannot achieve acceptable stability with a conventional lower denture. There has been considerable discussion on the merits of subperiosteal implants, but under the right conditions they do provide a reasonable prognosis (Mentag 1980).

Although it is feasible to construct and use maxillary (upper jaw) subperiosteal implants, the mandibular (lower jaw) implant is far more commonly encountered. The implant is most frequently used for the patient who, with a history of difficulty in wearing dentures, suffers from chronic soreness of the tissue in contact with the denture and who has a loose lower denture with poor masticatory function. Ridge resorption is the most common cause, although poor neuromuscular coordination and a psychological intolerance of dentures may also be involved.

Since the implant has to adapt to a bony surface of variable geometry, it has to be fabricated individually for each patient. The implant consists of an open framework that sits on top of the bony ridge but under the soft tissue (mucosa). Care has to be taken, of course, to avoid the main nerve (the mental neurovascular bundle). The implants may or may not have receptacles for retaining screws, depending on the operator's preference. Several different designs of post, which protrude through the mucosa into the mouth, have been used. These are most commonly made of a round section with either a neck (1–2 mm thick) or a continuous taper. The former provides better tissue adaptability, while the latter allows better hygiene.

The primary requirements of the materials used in the preparation of a subperiosteal implant are biocompatibility, mechanical strength and rigidity, and castability in the form outlined above. Considerations of the mechanical properties dictate that the device is metallic and the cobalt–chromium alloys are universally used because of their castability. Usually the struts are of a satin finish and the posts are of a polished finish. This is to facilitate anchorage of the overlying fibrous tissue in the former case, and a good seal between the post and the tissue at the point of entry in the latter. There have been some attempts to improve the biocompatibility of the implants by using surface coatings (e.g., vapor-deposited pyrolytic carbon) but these techniques are not yet in widespread use. Hydroxylapatite-coated subperiosteal implants, disclosed in the mid-1980s, may prove more beneficial.

1.2 Blade-Vent Endosseous Implant

The endosseous dental implant is one which both penetrates the alveolar bone and protrudes through the mucosa into the oral cavity. Although some clinicians undoubtedly have reasonable success rates, these are by no means guaranteed, and there is considerable variation in the conditions under which the implants are used and the results obtained (Babbush 1986).

There is no uniformity of opinion regarding indications and contraindications for these implants. Some have argued that the best results are achieved with single tooth replacements rather than with the use of multiple implants in the edentulous (toothless) arch (jaw arch), while others have argued the converse. Generally, these implants are used as the abutment for a fixed bridge, but not as an abutment for a removable prosthesis. A single interdental implant may be a freestanding single implant supporting a crown between two natural adjacent teeth, or a single tooth implant that is splinted to at least

one natural tooth abutment. Although some success can be achieved with the former type, there is considerable evidence to show that splinting to adjacent teeth gives better results. Between four and six implants may be used in a jaw.

Implants can be used in both mandible and maxilla (lower and upper jaws), although it is important that there is sufficient depth of bone to avoid the sinus and the nasal cavity in the maxilla and the inferior alveolar canal and mental foramen in the mandible. Generally, the implants should only be used in adults; splinting blade implants to natural teeth in the young could lead to a restricted lateral jaw development.

In contrast to the subperiosteal implant, the endosseous implant is a commercially available standardized product. Historically, the blade-vent endosseous implant has been most widely used. With this type of implant there is an abutment head (a neck and a body) which will contain vents of varying size and shape. Many features of the design are based on the necessity to give a large implant–bone interfacial area and to maximize resistance to lateral forces. Titanium is widely used for this type of implant, a surface-textured coating frequently being used in the buried area to increase attachment to the bone.

There is widespread disagreement over the performance of blade implants. In cases where problems have arisen clinically, it is the mobility of the implant, giving pain and discomfort, that is involved. It is possible for implants to be unacceptable radiographically, but clinically asymptomatic. Once an implant is loose, it should be removed.

Clearly, the assessment of the success of these implants depends on the criteria used, and the NIH–Harvard Consensus Conference included the following factors: adequate function, absence of discomfort, the patient's belief that aesthetic and psychological attitudes are improved, good geometrical relationship between opposing sets of teeth, bone loss no greater than a third of the vertical height of the implant with functional stability for five years, gingival inflammation amenable to treatment, mobility less than 1 mm, and no infection. Kapur (1980) presented cumulative survival rates for free-end blade implants (see Fig. 1), the results suggesting that five-year survival is approximately 50%. Full-arch applications in maxilla and mandible are less favored than free-end or interdental applications.

1.3 Other Metallic Endosseous Systems

The blade-vent implants described in Sect. 1.2 were designed to maximize the interactions with the bone that lead to stabilization of the implant. Variations on this theme were introduced in the mid-1980s, in which different implant geometries and surface structures play a role in increasing this mechanical

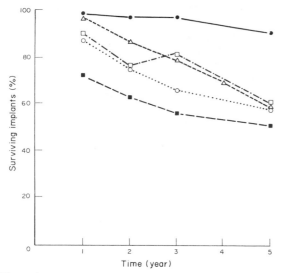

Figure 1
Cumulative survival rates for a series of clinical
studies of endosseous implants used for free end
supports (after Kapur 1980)

stability. The rationale for several of these developments can be found in the work of Schroeder during
the mid-1970s (Schroeder et al. 1981), where
increased attachment to bone was demonstrated
with titanium plasma-sprayed onto the surface.

1.4 Alumina Endosseous Implant Systems

Although considerable use has been made of titanium in endosseous implants, there has been an
increasing amount of interest shown in nonmetallic
systems, and especially the alumina ceramics. The
material used is usually a high-purity fine-grained
alumina (Al_2O_3).

The implants may be shaped as tapered cylinders,
with a series of circumferential stops providing the
taper. This design is aimed at achieving optimal load
transfer to the implant. The coronal end of the
implant contains a polished circular groove, into
which the gingival tissue will be adapted; the anticipated contraction of the circumferential fibers
providing a seal between the surface tissue (the
epithelium) and the alumina. The external surfaces
of the stepped cylinder are polished but contain
regularly spaced circular recesses of approximately
900 μm diameter, which are described as lucunae
and allow some ingrowth of bone. A longitudinal
groove also gives stability against rotation after bone
ingrowth.

1.5 Tooth-Root Replacement Implants

It is a widely held view that endosseous dental
implants have a better prognosis if they are not
stressed during the immediate postoperative phase;
thus temporary crowns are placed out of contact

with opposing teeth or splinted to adjacent teeth.
This idea is taken one stage further with tooth-root
implants in which only the root is implanted at the
initial stage. Healing takes place (and bone grows
around the implant to achieve stabilization) during
the time when the implant is buried and protected
from the oral environment and from functional
loads. Only when this stage is complete is a post
placed on the root, and this may be used to support a
crown, bridgework or full dentures, depending on
the implant and clinical condition.

Several different designs and materials have
emerged within this approach. Of particular relevance is the "osseointegrated" implant system of
Branemark. These implants were first placed in
patients in 1965. A detailed description of the
implants and the technique is given by Albrektsson
et al. (1986).

The implants are principally used in the treatment
of the edentulous patient, and usually four to six
implants will be placed in each jaw. Generally, only
patients who have been completely edentulous for a
minimum of one year and who have had problems
wearing dentures have been regarded as suitable
candidates.

The implanted tooth roots are usually referred
to as "fixtures" in this system. These consist of
threaded cylinders with a number of perforations
and a removable cover screw at the upper end. This
cover screw, which is located within the internally
threaded upper part of the fixture, is removed once
healing has taken place and replaced with an abutment, to which can be attached the bridgework. The
implants are made of commercially pure titanium,
and considerable attention is paid to the perfection
of the implant surface.

Published figures indicate that more than 6000
fixtures had been inserted in over 1000 patients in
numerous centers, with 19 years as the longest time
with continuous bridge function. An analysis of
performance after five years showed 95% and 85%
fixture survival rates in lower and upper jaws, respectively; figures not very different from those
after one year.

1.6 Transosteal Implants

These involve a device that penetrates the bone of
the mandible through the inferior border, where
rigid fixation is achieved; it is solely used in the
edentulous mandible, and especially in those that
are atrophic (reduced) or have suffered from bone
loss due to trauma or tumor surgery.

The implant consists of an inferior horizontal
plate, which acts as a base for a number of short
vertically attached retentive pins. Two threaded
vertical components are much longer and penetrate
the alveolar crest and the mucoperiosteal tissue in
the cuspid region. Fasteners and lock nuts are used
to secure the dentures to these pins. Different size

implants are available, ranging from four pin to seven pin. The implant is available in titanium–aluminum–vanadium alloy.

2. *Response of the Periodontium to Implants*

The success or otherwise of the tooth-replacement implants described in Sect. 1 will depend on many factors. The most important of these is the response of the tissues to the device. This in turn, will depend on variables such as material chemistry, surface finish or texture, implant design, the implant–bone geometrical relationship, functional loading during and subsequent to the healing phase and patient health and pharmacological status.

It is well known that a surgically created defect in bone may repair spontaneously and completely if the defect is not too large. Depending on the conditions, this osteogenesis may occur by bone growth emanating from existing bony walls (osteoconduction) or from foci within the reorganizing blood clot in the defect (osteoinduction). If a biomaterial is placed within the bony cavity the healing will take place in the space between the implant and bone—again osteogenesis will occur following blood clot reorganization. As this process takes place, the gap will become filled partly with new bone and partly with fibrous unmineralized tissue. It is a matter of crucial importance whether this process of new bone growth can proceed to the point where the gap is completely filled with bone, implying direct bone–material contact. Since fibrogenesis is taking place within this space as part of the normal response to an implant, it is likely that a layer of fibrous tissue adjacent to the implant surface will develop before the slower process of osteogenesis is able to generate new bone at that interface. It should be expected, therefore, that the normal response of bone to an implanted material is some bone regeneration, but with a layer of unmineralized tissue at the interface itself. It is extremely important to note that these processes are governed not only by the material and design features outlined above, but also by the mechanical stresses applied to the system.

Using this information (shown diagrammatically in Fig. 2), at least four types of response to endosseous implants may be suggested, taking into account epithelial and bony components (Williams 1981).

First, let us assume that the epithelium migrates a short distance during the healing phase such that it grows up to, and directly interfaces with, the implant. Under these circumstances the first possibility is that the deeper tissue will respond by the development of the classical fibrous encapsulation of the implant, this tissue preventing any attachment between bone and the surface, and demonstrating the usual collagen orientation with respect to that

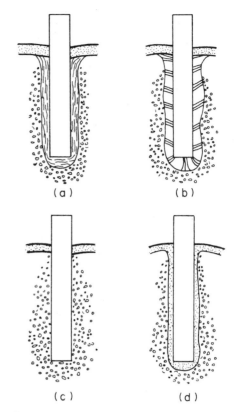

Figure 2
Variations in the host response to permucosal dental implants: (a) an epithelial seal with the development of a classic fibrous tissue encapsulation of the implant; (b) the development of a pseudo-periodontal ligament; (c) direct bone–implant bonding; and (d) internal migration of the epithelium

surface. The second possibility is that this fibrous tissue forms, but instead of the collagen fibers being aligned with the surface, they are oriented obliquely to it. Third, we may suggest that under some circumstances osteogenesis is dominant in this peri-implant tissue such that direct implant–bone bonding occurs, thereby essentially producing ankylosis, or fusion to the bone. Finally, the assumption that an epithelial seal will form may not be valid. Indeed, it is highly probable that the epithelial migration will not cease when it interfaces with the implant superficially, but rather that it will migrate internally, thus producing an epithelial barrier along the implant, preventing either direct bone contact or a pseudoperiodontal ligament. It is partly for this reason that so much emphasis has been placed on the two-stage procedure, involving the totally buried root at the first stage.

At this time there is ample evidence to suggest that, unless specific attempts are made to control the events by virtue of material selection and mechanical factors, the likely tissue response to a permucosal endosseous implant is a migrating epithelium and proliferating fibrous connective tissue without direct implant–bone attachment.

3. Implant–Bone Attachment

As mentioned in Sect. 2, the normal response of bone to an implant will be the formation of new bone in the vicinity of the interface, but with a thin layer of fibrous tissue preventing absolute implant–bone contact. It is claimed, however, that certain materials are able to encourage direct contact. This concept has been developed most extensively in the context of the Branemark implant system, where it is claimed that these implants are fully incorporated into the bone without an intervening fibrous layer, at least to the limits of resolution of electron microscopy. This process is referred to as osseointegration. The evidence of Branemark and his colleagues suggests that this property is limited to a few materials, and is controlled to a very considerable degree by the extent of trauma caused during bone preparation (e.g., thermal trauma during drilling), the cleanliness of the implant surface and the mechanical environment.

The Branemark system utilizes titanium and much of the experimental evidence and theoretical considerations relate to osseointegration with respect to this metal. It is important to point out that on the basis of some ultrastructural analyses, Albrektsson et al. (1986) have stated that a key stage in the development of the response is the adsorption of a layer of glycoprotein on the implant surface. It is claimed that, on the basis of certain characteristics of the oxide layer on the titanium surface, a very thin proteoglycan layer forms and collagen filaments can be seen only 20 nm from this oxide surface within the proteoglycans. Since it is normal for there to be proteoglycan layers of 20 nm thickness between collagen filaments anyway, it is assumed that titanium is acting as if it were a natural tissue component. All other surfaces studied appeared to develop much thicker collagen-free proteoglycan layers.

It is clear that an exceptional degree of implant–bone attachment takes place with the Branemark system and that this osseointegration, defined by Branemark as the direct structural and functional connection between ordered living bone and the surface of a load-carrying implant, is a real phenomenon.

It remains to be seen whether this phenomenon is related specifically to titanium or is more attributable to other factors inherent in the technique. There is evidence, for example, that sound bony anchorage may take place with alumina. There is also the increasing interest in materials of more natural origin, such as the calcium phosphate ceramics that are analogous to the apatite structures of bone mineral (Williams 1985). Work with calcium hydroxylapatite, for example, has shown epitactic deposition of new bone mineral on the surface of the synthetic substrate (Jarcho 1981). This material has so far attracted much attention as a particulate bone substitute material, and as such is discussed below, but attempts have also been made to prepare implants with hydroxylapatite-coated surfaces.

Two, somewhat different, approaches to bone–biomaterial bonding should be mentioned here. The first concerns the use of porous-surfaced implants in which bony ingrowth into the surface porosity may take place. The use of this concept in the development of porous plasma-sprayed titanium dental implants was mentioned in Sect. 1.3. The second concerns the use of the ceramics or glass ceramics, such as Bioglass and Ceravital, that have a controlled surface reactivity that allows bonding to take place to mineralized tissue. Since these materials have relatively poor mechanical properties, their use in functional implants has so far been restricted. However, in view of the likelihood of their use as surface coatings on tougher substrates, applications in dental implants are highly probable (Gross et al. 1988).

4. Implants in Surgery of the Alveolar Ridge

Extensive ridge resorption is a common occurrence after tooth loss and frequently leads to problems of denture retention. Many patients probably tolerate a poorly retained denture, compensating for the loss of intrinsic stability associated with an atrophic ridge by the use, possibly subconsciously, of the lips, cheek and tongue to retain the denture. As the ridge resorption progresses, however, speech, mastication and general comfort are severely compromised. Several proprietary substances are available to assist in denture retention, but these are no more than palliative and many patients will require some form of treatment.

4.1 Ridge Augmentation

Preprosthetic surgical techniques used in patients with an atrophic mandible are of several types and include those designed to add height to the ridge by augmentation.

Augmentation has been performed for many years using bone grafts. While excellent initial results may be achieved, especially with a substantial increase in ridge height, there are two major disadvantages of autogenous grafting in this situation. First, there is the risk of morbidity associated with the donor site. Second, the reconstructed mandible undergoes resorption just as the original bone does, and in four to five years, typically, the graft will have resorbed completely leaving the ridge at its presurgical deficient height. In view of this, several different synthetic materials have been used for ridge augmentation—acrylics, silicone rubber and the polytetrafluoroethylene–carbon composite, Proplast, being the most prominent until recently. None of these materials was particularly successful, especially because of chronic tissue reactions. With Proplast, the overlying alveolar mucosa was unable to withstand the forces transmitted by dentures, leading to tissue breakdown and infection. The use of these materials has now largely been discontinued and current interest is focused on the use of calcium phosphate ceramics.

Calcium phosphate ceramics can be prepared in a variety of forms and with a variety of characteristics. Nonresorbable calcium phosphates appear to offer greater potential, and dense hydroxylapatite has received much interest in this situation. It is still a little early to define success rates and long-term benefits from this type of augmentation, but the signs are that, with careful patient selection, good results may be achieved. A review of the use of particulate hydroxylapatite, by itself or combined with bone, has been published by Kent (1986). Complications seem to be rare, although some form of delayed healing or tissue breakdown may be observed. A significant number of patients do experience paresthesia, a localized numbness (15% in Kent's study), but the vast majority return to normal.

4.2 Alveolar Ridge Preservation

Augmentation of the alveolar ridge is a reconstructive procedure that, by definition, is performed after resorption takes place. A more attractive concept is to prevent the resorption taking place at all, or at least to maintain substantial height and width of the ridge, and calcium phosphate ceramics have been used in this situation. This involves the placement of the implant material as an immediate root substitute after tooth extraction. The implants may be particulate or root-shaped solids, to match the socket.

5. Implants in Periodontal Surgery

Where there is a simple defect in periodontal disease (the disease of the gums), its elimination may be readily achieved by soft-tissue surgery and improved oral hygiene. The treatment of a defect in the underlying bone is not so easy and it may be decided that some reconstructive material is necessary. There are two basic choices, involving either bone grafts or synthetic materials. There are many reports in the literature on experimental and clinical analysis of bone-grafting procedures in periodontal surgery. The evidence indicates a considerable unpredictability and variability about bone grafting. Fresh bone from the patient's own hip and decalcified freeze-dried donated bone both seem to give reasonable results, but there is no certainty about true reattachments and resistance to root resorption and ankylosis.

A little unusually, only one class of synthetic material has been used in the surgical treatment of infrabony defects, and this involves calcium phosphate ceramics. Certain of the arguments concerning these ceramics have been rehearsed in the discussion of mandibular ridge augmentation.

The first uses of calcium phosphate ceramics in this situation involved the resorbable tricalcium phosphate. Animal studies showed that the material was well tolerated and that bony pockets regenerated with new bone. Although some clinical success has been reported, there is evidence that the pockets can fill just as fast, or even faster, with bone grafts. At least one of these materials has been made commercially available.

6. Prostheses in Facial Reconstruction

There are numerous reasons why the maxillofacial or plastic surgeon should want to reconstruct or augment tissues in the lower parts of the face, including congenital and developmental defects, tumor resection and degenerative disease. The tissue requiring reconstruction may be soft or hard and the need for reconstruction may be functional or cosmetic.

6.1 Mandibular Reconstruction

The successful surgical treatment of many cases of tumors of the head and neck will include the removal of all of the mandible, together with the regional lymph nodes and other soft tissues. However, resection of the anterior part of the mandible presents a very severe reconstructive problem that has significant implications in appearance and serious functional consequences.

There is no doubt that the best method for replacing or restoring the mandible is a bone graft. Generally, however, this is not recommended as a primary procedure in the case of tumor resection. This is because the removal of a considerable amount of soft tissue (giving poor quantitative coverage) and prior irradiation therapy (giving poor qualitative coverage) yield an inferior prognosis. Additionally, when scar contraction causes displacement of the tissue, it is often difficult to prepare bone grafts to follow and retain the normal contour of the mandible at this stage. The most significant requirement of mandibular reconstruction at the primary operation is the provision of immediate stability of the remaining bone fragments and muscles. If these can be maintained in position for a few months, possibly by some fixation device, the postoperative fibrosis may be sufficient to ensure long-term stability and maintenance of contours. Bone grafts, suitably restrained, may then be used as a secondary procedure to complete the reconstruction. This has led to the use of implanted devices under different conditions with different objectives.

(a) Temporary mandibular reconstruction implants. Among the polymeric materials that have been employed in this capacity are celluloid, acrylics, polytetrafluoroethylene and silicone rubber (Kent, 1984). Generally, these polymers are not sufficiently rigid. At one stage, acrylic polymers were thought promising, but the fixation tended to break down at the area of attachment of the acrylic to the host bone. The acrylic produced a substantial capsule which tended to prevent vascular ingrowth and reorganization of the graft.

Wires made of either stainless steel or cobalt–chromium alloy, in the form of Kirschner wires or Steinmann pins, are often used in this temporary situation, but they tend to loosen and have difficulty in restraining the bone in the desired position. Metal plates appear to fare a little better, but there is often inadequate fixation to the bone and the overlying soft tissue may break down.

Custom-made cast metallic implants have been described. More common, however, are the prefabricated implants available in kit form. The Bowerman–Conroy system, for example, has body, ramus and angle sections in various sizes. Manufactured from titanium, these may be riveted together in a form suitable for individual patients.

(b) Implants used with bone grafts. Stainless-steel or cobalt–chromium alloy plates can be used to attach solid bone grafts to residual fragments of bone. There are several disadvantages in this technique, and of greater significance are the developments in tray prostheses in which particulate cancellous bone chips can be contained in a mesh. This prevents the displacement of the bone chips and their dispersion into the surrounding soft tissue. Stainless-steel, cobalt–chromium alloy, titanium and tantalum meshes have all been used, but more recently a poly(ether urethane)–Dacron mesh has proved popular. This can be fabricated to suit individuals and host connective tissues grow through the mesh. It may be used for small defects without the need for additional fixation, but some difficulties arise with larger defects since interdental or extraskeletal stabilization is required and the material may lack sufficient strength.

6.2 Facial Reconstruction and Augmentation

Deformities may be either congenital or acquired and their reconstruction may be for either functional or cosmetic reasons, or both. In addition to the mandible already discussed, the ears, lips, eyes and eyelids, forehead, cheek, tongue, salivary glands and nose may all require surgery under some circumstances. Very occasionally some implantable device is necessary to complete the reconstruction or augmentation.

These have utilized a variety of polymers, including acrylics, polyethylene and, especially, silicone rubber and Proplast. Probably the greatest number of procedures have been performed with silicone rubber which can be easily custom-made to fit individual shapes or as preformed devices. With large defects, the best cosmetic results are achieved with custom-made implants, using a facial moulage, plaster impressions and wax models. Although good reconstruction can be achieved, migration of the implants is a problem, especially in exposed areas in chin and nose augmentation.

See also: Biocompatibility of Dental Materials; Corrosion of Dental Materials; Dental Materials: Clinical Evaluation

Bibliography

Albrektsson T, Jansson T, Lekholm U 1986 Osseo-integrated dental implants. *Dent. Clin. North Am.* 30: 151–74

Babbush C A 1986 Endosteal blade-vent implants. *Dent. Clin. North Am.* 30: 97–115

Cawood J I, Howell R A 1989 Anatomical considerations in the selection of patients for preprosthetic surgery of the edentulous jaw. In: Williams D F (ed.) 1989 *Current Perspectives on Dental Implants.* JAI Press, London, pp. 139–80

d'Hoedt B, Schulte W 1989 A comparative study of results with various endosseous implant systems. *Int. J. Oral Maxillofac. Implants* 4: 95–105

Gross U, Kinne R, Schmitz H J, Strunz V 1988 The response of bone to surface active glasses/glass-ceramics. *CRC Crit. Rev. Biocompat.* 8: 155–79

Jarcho M 1981 Calcium phosphate ceramics as hard tissue prosthetics. *Clin. Orthop. Relat. Res.* 157: 259–78

Kapur K 1980 The literature on blade implants. In: Schnitman and Shulman 1980, pp. 240–60

Kent J N 1984 Clinical experience with biomaterials in oral and maxillofacial surgery. In: Boretos J W, Eden M (eds.) 1984 *Contemporary Biomaterials.* Noyes, Park Ridge, NJ, pp. 254–303

Kent J N 1986 Reconstruction of the alveolar ridge with hydroxyapatite. *Dent. Clin. North Am.* 30: 231–57

Mentag P J 1980 Current status of the mandibular subperiosteal implant prosthesis. *Dent. Clin. North Am.* 24: 553–63

Schepers E, Ducheyne P, de Clerq M 1989 Interfacial analysis of fiber-reinforced bioactive glass dental root implants. *J. Biomed. Mater. Res.* 23: 735–52

Schnitman P A, Shulman L B (eds.) 1980 *Dental Implants: Benefits and Risk.* National Institutes of Health, Bethesda, MD

Schroeder A, Van der Zypen E, Stich H, Sutter F 1981 The reactions of bone, connective tissue and epithelium to endosteal implants with titanium-sprayed surfaces. *J. Maxillofac. Surg.* 9: 15–25

Steflik D E, McKinney R V, Koth D L 1989 Ultrastructural comparisons of ceramic and titanium dental implants *in vivo. J. Biomed. Mater. Res.* 23: 895–910

Williams D F 1981 Implants in dental and maxillofacial surgery. *Biomaterials* 2: 133–46
Williams D F 1985 The biocompatibility and clinical uses of calcium phosphate ceramics. In: Williams D F (ed.) 1985 *Biocompatibility of Tissue Analogs*, Vol. 2. CRC Press, Boca Raton, FL, pp. 43–70

D. F. Williams
[University of Liverpool, Liverpool, UK]

Dental Investment Materials

Dental investment materials may be classified as silicate-bonded, phosphate-bonded or gypsum-bonded. Silicate- and phosphate-bonded investments are commercially available for casting dental prostheses such as partial dental frameworks and crowns used for porcelain–metal restorations. The chief property of these investments is their ability to undergo high-temperature burnout (1100–1300 °C) without harmful decomposition while expanding sufficiently to compensate for the alloy shrinkage. Gypsum-bonded investments are used for the casting of alloys whose melting temperature is below 1000 °C—principally gold casting alloys.

1. Silicate-Bonded Investments

Silicate-bonded investments are complex and their chemical transformations are not fully understood. In general, ethyl silicate is hydrolyzed into colloidal silicic acid upon mixing. Equation (1) shows a simplified version of this reaction:

$$Si(OC_2H_5)_4 + 4H_2O \rightarrow Si(OH)_4 + 4C_2H_5OH \quad (1)$$

Elevating the temperature of the mold to 150 °C causes the excess water and alcohol to be driven off. Further heating to 1100 °C causes the thermal expansion of the silica necessary to produce accurately dimensioned castings.

Silicate-bonded investments have been largely supplanted by phosphate-bonded materials owing to the ease of manipulation and greater dimensional control inherent in the phosphate-bonded materials.

2. Phosphate-Bonded Investments

Phosphate-bonded investments consist of a refractory filler and a binder. Like gypsum-bonded investments, the filler is a blend of silica of the quartz and cristobalite polymorphs, the actual percentage of each polymorph depending on the amount of thermal expansion desired by the manufacturer. The particles range in size from several micrometers to coarse sand.

The binder is usually a binary chemical system consisting of an acidic phosphate such as monoammonium dihydrogen phosphate and a basic alkaline-

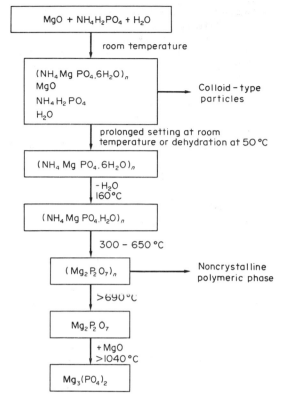

Figure 1
Summary of the setting and thermal reactions of the binding system of phosphate-bonded investments

earth oxide such as magnesium oxide. They react in the presence of water to initiate the setting mechanism. During the setting reaction, the reactants, reaction products and medium of reaction interact physicochemically forming a colloidal-type gel of multimolecular $(NH_4MgPO_4.6H_2O)_n$ by coagulation around excess MgO and grains of refractory fillers. A typical room-temperature setting reaction and subsequent thermal reactions are shown in Fig. 1. The magnitude and course of the reactions, however, are greatly affected by a number of variables, such as environmental factors (temperature and humidity), operational factors (type and duration of mixing) and inherent properties of the refractory system.

The early setting mechanism controls the performance criteria of the investment including time of setting, setting expansion, strength of the set investment, thermal expansion and nature of the surface. These physical properties vary from mix to mix according to the raw materials used and are prone to reflect any change in environmental or operational factors.

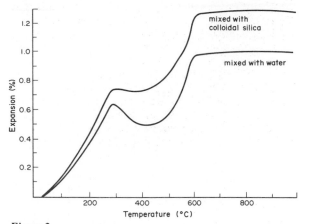

Figure 2
Typical expansion curves for a phosphate-bonded
investment

Figure 2 shows typical expansion curves for phosphate investments. The observed shrinkage in the region from 300 °C to 500 °C can be explained by the decomposition of the reaction products. This decomposition is accompanied by the evolution of ammonia vapor whose characteristic smell is clearly detectable.

Most phosphate-bonded investments use colloidal silica in aqueous suspension as one of the reactants rather than water. This substitution causes an additional setting and thermal expansion. Figure 2 shows how the use of colloidal silica greatly reduces the shrinkage between 300 °C and 500 °C, and increases the overall expansion. Dilution of the colloidal solution allows the user to match the expansion of the investment to the shrinkage of the casting alloy.

The effect that the concentration of the colloidal silica solution has on setting and thermal expansion is shown in Fig. 3. The maximum thermal expansion obtainable is shown to be 1.3%, while the maximum setting expansion is 1.1%. Therefore, it would be expected that the total expansion (setting plus thermal) would be 2.4%. However, the setting expansion is measured with a V trough and the thermal expansion with a quartz dilatometer, and the values obtained may not be proportionately applicable to the dimensional relationship of the wax pattern to the casting. There are limiting forces on the investment as it undergoes setting and thermal expansion which may cause effective total expansion other than that expected from Fig. 3.

When being stored prior to use, phosphate-bonded investments should be protected from high humidity or large temperature changes to prevent deterioration. Also, the manufacturer's directions regarding manipulation of the material should be carefully followed.

3. Gypsum-Bonded Investments

Gypsum-bonded investments are usually composed of calcium sulfate hemihydrate and some form of silica. Sufficient hemihydrate (usually the α form) is used to act as a low-temperature binder and the remainder of the investment (60–75%) is silica. The silica overcomes contraction of the gypsum on heating and provides sufficient mold expansion (1.2–1.7%) to compensate for contraction of gold casting alloys during solidification and cooling.

Cristobalite is the silica allotrope of choice for dental investments, because it has the largest thermal expansion coefficient. However, because of its cost, investments containing the quartz allotrope are used and additional expansion is obtained from hygroscopic expansion. Hygroscopic expansion of gypsum can provide a large portion of the mold enlargement needed to give accurate dental gold castings. The presence of silica weakens the mix and allows a high setting expansion to occur by outgrowth of gypsum crystals before the mix reaches sufficient rigidity to provide its own restraint. An asbestos sheet is usually employed to line the metal casting ring to produce minimal restraint during investment setting and hygroscopic expansion. The factors controlling setting time of gypsum investments are similar to those for gypsum model materials.

Some investments also contain small quantities of carbon or copper to help give a reducing atmosphere under the mold at casting temperature.

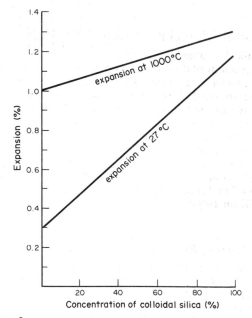

Figure 3
Effect of colloidal silica concentration on the
dimensional change of a phosphate-bonded investment

Casting investments can also be used to maintain the relationship of parts of bridges or partial dentures during soldering operations. In such procedures the investment expansion is useful as it helps to compensate for shrinkage of solders during their solidification.

See also: Dental Materials: Clinical Evaluation

Bibliography

Abu Hassan M I, Glyn Jones J C, Hallsworth A J 1989 Porosity determination of cast investments by a wax infiltration technique. *J. Dent.* 17: 195–8

Marsaw F A, Rijk W G, Hesby R A, Hinman R W, Pellev G B 1984 Internal volumetric expansion of casting investments. *J. Prosthet. Dent.* 52: 361–6

Mueller H J 1986 Particle and pore size distribution of investments. *J. Oral Rehab.* 13: 383–93

Neiman R, Sarma A C 1980 Setting and thermal reactions of phosphate investments. *J. Dent. Res.* 59: 1478–85

Phillips R W 1973 *Skinner's Science of Dental Materials*. Saunders, Philadelphia, PA

Santos J F, Ballester R Y 1987 Delayed hygroscopic expansion of phosphate bonded investments. *Dent. Mater.* 3: 165–70

US Council on Dental Materials, Instruments and Equipment 1981 *Dentist's Desk Reference: Materials, Instruments and Equipment*, 1st edn. American Dental Association, Chicago, IL

A. F. Steinbock
[Whip Mix Corporation, Louisville, Kentucky, USA]

Dental Materials: Clinical Evaluation

Clinical evaluation is an essential element of the assessment of dental materials intended for intraoral use; no single laboratory test or series of such tests is able to predict the clinical behavior and performance of such materials reliably. At a time when there is an ever increasing range and diversity of new materials and an increasing awareness of the limitations of laboratory evaluations in terms of providing meaningful data of clinical relevance, clinical evaluation is assuming new importance in the field of dental materials science and related clinical research (Wilson 1990).

1. Types of Evaluation

Clinical evaluation may be classified as "explanatory", a form of evaluation whereby materials are tested under strictly controlled conditions that favor success and optimum effectiveness, or as "pragmatic", the evaluation of the efficiency of materials in "real-world" situations (Jacobsen 1988). In general, pragmatic evaluations tend to follow explanatory studies once the value of a material has been established. However, many dental materials now have a limited commercial life expectancy. Consequently, most of the clinical evaluations of dental materials undertaken in recent years have been limited to forms of explanatory trials. Other forms of clinical studies such as observational studies have, on occasion, been used to advantage in the evaluation of some materials; however, the findings from such studies tend to compliment rather than provide an alternative to results from, in particular, explanatory evaluations.

2. Explanatory Evaluations

Given the predominance of explanatory evaluations in the field of clinical assessments of dental materials, some consideration of the different forms of this type of evaluation are pertinent. Explanatory evaluation may be single centered or multicentered, and of paired, unpaired or unbalanced design (Jacobsen 1988). It is widely accepted that, where possible, the best design is the paired type in which, for example, in the evaluation of a restorative, a pair of equivalent preparations are restored in each patient, one being filled with the material under test, the other with a suitable control material. Pairing and the involvement of at least two independent centers (i.e., multicenter, paired design evaluation) greatly increases the power of the test. As in all forms of clinical evaluation, good statistical practices are of paramount importance with regard to the interpretation of the findings; the statistical methods employed in explanatory evaluations are those of hypothesis testing and estimation theory rather than decision theory, as used in pragmatic evaluations of suitable design.

3. Protocols

Recommendations concerning clinical research protocols for dental materials have been published by a number of professional organisations, notably by the Federation Dentaire Internationale (1977a, b, 1982) and the American Dental Association (e.g., American Dental Association (Council on Dental Materials, Instruments and Equipment) 1989). These recommendations are either specific to certain generic types of materials such as posterior composites (the latter) or more general in terms of format (Federation Dentaire Internationale 1977b) and factors to be considered in the planning, initation and execution of evaluations (Federation Dentaire Internationale 1982). These recommendations, in particular those concerning format and the factors to be considered in evaluations, are considered essential reference for all those involved or contemplating participation in clinical assessments of dental materials. In essence, a protocol must be specially prepared for each evaluation and should include detailed information concerning: the identification of the problem to be addressed, the relevant background information, the materials and methods to be employed, and the administrative aspects of the programme (Federation Dentaire Internationale 1982).

4. Assessments

The outcome of clinical evaluations is generally dependent on subjective assessments of quality and decisions concerning failure made both directly in the clinical situation and indirectly as part of retrospective analyses of, for example, replicas of teeth included in evaluations of restoratives. With the application of new technology, systems have been developed to complete objective assessments during principally retrospective analyses, and these systems are now being used increasingly in programmes of evaluation.

4.1 Direct Clinical Assessments

Much of the original work in this field by workers such as Ryge and Snyder (1973) is embodied in guidelines such as those published by the California Dental Association (1977). Most clinical evaluations systems such as the one originally described by Ryge and Snyder (1973), now widely referred to as the United States Public Health Service (USPHS) system, are typically employed as a consequence of their ease of use, reproducibility and validity. Such systems employ criteria which describe the condition of, for example, restorations; the underlying principle is that the categories described for various characteristics relate to decisions that would be made clinically, providing findings that, as a consequence, should be relevant to clinical practice and the application of materials (Ryge and Mjor 1988).

Direct clinical evaluations may also be completed by reference to standards which may take the form of models, photographs or specially prepared aids to facilitate assessments of, for example, surface roughness or the size of steps or deficiencies at margins. However, even with such systems, statistically significant differences between findings may not be clinically significant.

4.2 Indirect Clinical Assessments

Indirect clinical assessments may be completed using techniques involving the use of replicas (Ryge and Mjor 1988) and clinical photographs together with the systems for categorization against standards (Mahler and Marantz 1980); the use of clinical photographs and replicas is well suited for use in longitudinal studies. These assessments may be equivalent to, and therefore as subjective as, most forms of direct clinical assessments, or objective when measurements of, for example, wear are recorded. Indirect clinical assessments involving subjective decisions may be more reliable than equivalent direct clinical assessments; however, such techniques still suffer many limitations, many of which relate to reproducibility and sensitivity. By contrast, objective indirect clinical assessments can, given appropriate techniques, be completed with relatively little error and provide a means of quantifying changes in surfaces and margins in clinical service. Such changes are generally reported in terms of deterioration.

Objective indirect clinical assessments using microscopic techniques (e.g., quantitative margin analysis of restorations using scanning electron microscopy (Roulet et al. 1989)) provide a valuable means of assessing the initial adequacy and subsequent pattern of deterioration of structures such as restorations. Furthermore, objective indirect clinical assessments of this type provide a means for the early detection of the limitations of materials. As such, objective indirect clinical assessments are of increasing importance in the clinical evaluation of dental materials, and will therefore have an important role in the new or at least modified forms of clinical testing that are required to bridge the gap between laboratory-based research and development and the existing understanding and knowledge of the performance and behavior of materials in clinical service (Wilson 1990).

5. The Future

As it remains the case that no single laboratory test or series of such tests is able to predict reliably the long-term clinical behavior and performance of a generic group of materials of similar composition and construction, let alone the long-term in-service expectations for specific materials, clinical evaluation will remain the conclusive test of dental materials. However, contemporary forms of clinical evaluation only tend to yield meaningful data after three to five years of investigation, and the findings of such studies typically fail to explain all aspects of the influences of the many factors which determine the clinical behavior and performance of dental materials in clinical service. Consequently, to complete conclusive (clinical) tests on new materials which increasingly have a market life expectancy of only a few (about five) years, it is believed that new more reliable and predictive forms of short-term clinical evaluation are required (Wilson 1990). Indirect objective clinical assessments and relatively sophisticated forms of data handling and analysis will inevitably form part of the protocol for such studies. However, the validity of such new forms of clinical evaluation may only be established by long-term (i.e., five-year) findings from more traditional forms of evaluation.

Bibliography

American Dental Association (Council on Dental Materials, Instruments and Equipment) 1989 *Revised (1989) Guidelines for Submission of Composite Resin Materials for Posterior Restorations.* ADA, Chicago, IL

California Dental Association 1977 *Guidelines for the Assessment of Clinical Quality and Professional Performance.* CDA, Los Angeles, CA

Federation Dentaire Internationale 1977a Recommended outline for a research protocol. Technical Report No. 5, FDI, London

Federation Dentaire Internationale 1977b Recommended format for protocol for clinical research programme: Clinical comparison of several anterior and posterior restorative materials. Technical Report No. 6, FDI, London

Federation Dentaire Internationale 1982 Recommendations for clinical research protocols for dental materials. Technical Report No. 16, FDI, London

Jacobsen P H 1988 Design and analysis of clinical trials. *J. Dent.* 16: 215–18

Mahler D B, Marantz R L 1980 Clinical assessments of dental amalgam restorations. *Int. Dent. J.* 30: 327–34

Roulet J F, Reich T, Blunck U, Noack M 1989 Quantitative margin analysis in the scanning electron microscopy. *Scan. Micros.* 3(1): 147–59

Ryge G, Mjor I A 1988 Quality assessments in operative dentistry. In: Horsted-Bindslev P, Mjor I A (eds.) 1988 *Modern Concepts in Operative Dentistry.* Munksgaard, Copenhagen, pp. 287–301

Ryge G, Snyder M 1973 Evaluating the clinical quality of restorations. *J. Am. Dent. Assoc.* 87: 369–77

Wilson N H F 1990 Buonocore Memorial Lecture 1990. The evaluation of materials: Relationships between laboratory investigations and clinical studies. *Oper. Dent.* (in press)

N. H. F. Wilson
[University of Manchester, Manchester, UK]

Dental Plaster and Stone

Dental plaster and stone are produced by the partial dehydration of gypsum. Dental types are impression and laboratory plasters, based on ordinary gypsum plaster (plaster of Paris), and dental stones and die stones, based on high-strength gypsum plasters. All are supplied as fine powders: laboratory plaster is white, while the other three types are color identified (see *Dental Investment Materials*).

All four types consist essentially of calcium sulfate hemihydrate ($CaSO_4.\frac{1}{2}H_2O$). The physical characteristics of the powder particles differ as a result of the different manufacturing methods. When mixed with water in suitable proportions they form a slurry or paste which rapidly sets to form a rigid mass (cast gypsum). In dental technology the mixing proportions (gauge) are given as a water-to-powder (W/P) ratio, expressed as a decimal fraction.

1. Clinical Uses

Impression plaster is used for making impressions of oral structures (particularly in the edentulous mouth) for the location of components in crown and bridge construction and for recording jaw relations. Laboratory plaster is used for pouring casts of oral structures when high strength and abrasion resistance are not needed, for articulating casts and for making molds for processing dentures from dental polymers.

Ordinary dental stone (artificial stone, laboratory stone) yields a set mass much harder and stronger than cast plaster and is used for pouring working casts on which dentures can be constructed. Die stone (improved stone, high-strength stone) gives the hardest and strongest set mass available from dental gypsum products, and is used for pouring dies (accurate replicas of individual teeth which have been prepared for inlays and crowns).

2. Chemistry

In the temperature range 20–100 °C two phase changes are observed in the $CaSO_4$–H_2O system:

$$CaSO_4.2H_2O \underset{}{\overset{40-45\,°C}{\rightleftharpoons}} CaSO_4.\tfrac{1}{2}H_2O\ (+\text{water})$$

calcium sulfate dihydrate (gypsum)   calcium sulfate hemihydrate

$$\overset{95-100\,°C}{\rightleftharpoons} CaSO_4(+\text{water}) \qquad (1)$$

hexagonal calcium sulfate

It is not possible to give unequivocal transformation temperatures; the ranges shown above are based on isothermal experiments and represent equilibrium conditions. Even at 90 °C, substantially complete conversion of gypsum to hemihydrate takes about 12 h. In calcining gypsum, higher temperatures are used for shorter times; the final product, gypsum plaster, is a white powder consisting very largely of calcium sulfate hemihydrate, crystallographically identical with the mineral bassanite.

Theoretically, this is the only stable hydrate of calcium sulfate in the temperature range 45–95 °C; however, in the absence of water it exists as a metastable compound at atmospheric temperature. It readily combines with water at temperatures below 40 °C to form the stable hydrate gypsum. Hexagonal calcium sulfate (soluble anhydrite) is very unstable at temperatures below 95 °C; even in the presence of atmospheric moisture it rapidly reverts to the hemihydrate.

2.1 Manufacture

Ordinary gypsum plaster is the product of dry calcination of ground gypsum in open pans, kettles or rotary kilns. In the absence of liquid water, the phase change to hemihydrate occurs without the opportunity for reorganization of crystal morphology; the resulting hemihydrate crystals are rough irregular pseudomorphs with porosity caused by loss of water of crystallization. The powder has a low apparent density, a high relative surface area and a high bulk volume.

High-strength gypsum plasters are prepared by wet calcination. Many dental stones are based on raw hemihydrates prepared by autoclaving lump gypsum in superheated steam; sufficient liquid water is present for through-solution conversion with hemihydrate recrystallizating as dense prismatic crystals. The resulting powder, therefore, has a smaller relative surface area and a lower bulk volume than dry calcined hemihydrate. A controlled amount of grinding rounds off the crystals and provides some fines, both factors improving the packing ability of the powder and thus further reducing the bulk volume.

Other methods of wet calcination have been developed that produce hemihydrate crystals which are predominantly short and squat. These methods include boiling lump gypsum in a 30% calcium chloride solution and autoclaving ground gypsum in a dilute solution ($<1\%$) of a crystal habit modifier, such as sodium succinate. After controlled grinding, these products yield a hemihydrate powder with a bulk volume even lower than that of ordinary autoclaved hemihydrate. This powder is used as the base material for formulating die stones.

2.2 Setting

When mixed with water, calcium sulfate hemihydrate combines with it to form the dihydrate by the exothermic reaction

$$2CaSO_4.\tfrac{1}{2}H_2O + 3H_2O \rightarrow 2CaSO_4.2H_2O \qquad (2)$$

Hemihydrate is sparingly soluble in water ($7\,g\,l^{-1}$), so the initial mix is a slurry of hemihydrate particles suspended in a saturated aqueous solution. The stable hydrate at normal temperatures is the dihydrate, which is even less soluble than the hemihydrate ($2\,g\,l^{-1}$). Therefore, the aqueous phase is supersaturated with respect to the dihydrate, which crystallizes out. This allows more hemihydrate to go into solution and, in turn, precipitate as gypsum. Thus the setting process is a through-solution reaction involving the progressive dissolution of hemihydrate, diffusion of Ca^{2+} and SO_4^{2-} ions to nucleation sites, and the precipitation of fine crystals of gypsum. The final set mass is a felted aggregate of interlocking acicular gypsum crystals.

The setting process is continuous from the beginning of the reaction until it ceases, by which time the cast material has attained its full wet strength. During this time, important physical changes can be recognized. Initially there is a continuous aqueous phase present, so the mix is a viscous liquid exhibiting pseudoplasticity. As the reaction proceeds, gypsum crystals grow at the expense of the aqueous phase and the viscosity of the mix increases. When the clumps of growing gypsum crystals begin to interact the mix becomes truly plastic and will not flow, even under vibration; casting is then no longer possible but the mix can be readily molded. As crystal growth continues the setting material becomes a rigid solid, initially weak and friable but gaining rapidly in strength as the reaction nears completion.

The setting reaction causes a decrease in the true volume of the reactants; however, once the mix attains rigidity an isotropic expansion is observed, the result of growth pressure of gypsum crystals. Thus, there is a decrease in apparent density, with formation of microscopic voids separating individual crystals in the aggregate.

From Eqn. (2), it can be calculated that in setting, 0.1 kg hemihydrate combines with 0.0186 kg water. Water in excess of this, added to the mix to produce a workable viscosity, remains uncombined in the set mass as a saturated solution of gypsum. This excess water weakens the cast gypsum and, if removed by drying, leaves porosity which reduces the dry strength. Strength properties of cast gypsum mainly depend on the amount of unreacted water remaining after setting and thus on the W/P ratio of the original mix; low W/P ratios, therefore, produce the strongest casts.

Within wide limits, the rate of the hydration reaction is independent of the W/P ratio. However, the rate at which the associated physical changes occur is highly dependent because these changes result from the interaction of clumps of gypsum crystals growing from nucleation centers in the slurry. Thick mixes (low W/P ratios) show more rapid hardening and higher setting expansions because the available nucleation sites are concentrated in a smaller volume; interaction of the solid phase occurs earlier and is more effective in promoting expansion.

2.3 Effect of Additives

Many salts and colloids accelerate or retard the setting of gypsum products by their effect on the rate of gypsum formation. They have been used on an empirical basis for many years, but their modes of action are still not fully understood. In low concentrations soluble sulfates and chlorides are effective accelerators, apparently by increasing the rate of solution of hemihydrate. Powdered gypsum accelerates by providing extra sites for heterogeneous

Figure 1
Scanning electron micrographs of fracture surface of cast gypsum: (a) set plaster, W/P = 0.50; (b) set die stone, W/P = 0.25

nucleation. Acetates, borates and citrates are effective retarders, which may act by nuclei poisoning, by reducing the rate of solution of hemihydrate or by inhibiting the growth of dihydrate crystals. Colloids also retard the set, presumably by nuclei poisoning.

Many accelerators and retarders reduce the setting expansion, in some cases by changing the crystal habit of precipitating gypsum from acicular to squat. At the same time strength properties are reduced.

3. Microstructure

The set mass consists of a tangled aggregate of monoclinic gypsum crystals, acicular in habit, usually ranging from 5 µm to 20 µm in length. Two types of inherent microporosity are evident:

(a) porosity associated with setting expansion—these voids result from crystal growth, and appear as very small angular spaces, separating individual crystals; and

(b) porosity caused by the removal of unreacted water—the voids are larger, occurring between clumps of crystals in the aggregate.

The relative amounts of both types are affected by the W/P ratio, but in opposite ways. The first type is increased by reducing the W/P ratio, because the effect of crystal growth is greater and the setting expansion higher. On the other hand, the second type is decreased by reducing the W/P ratio, because less uncombined water remains in the set mass.

At any W/P ratio, the total porosity is the sum of these two types. The second type predominates, so the net result is always a decrease in inherent porosity (i.e., an increase in apparent density) as the W/P ratio of the mix is decreased. Porosity represents about 40% of the total cast volume at a W/P of 0.50 and about 20% at a W/P of 0.25.

Typical microstructures are illustrated in Fig. 1. Both types of porosity can be seen, and the higher apparent density in the cast mixed at a W/P of 0.25 is readily apparent.

4. Physical Properties

Measured physical properties of four widely used dental gypsum products are shown in Table 1.

4.1 Water Requirement

The amount of water needed for a workable mix primarily depends on the bulk volume of the particular hemihydrate powder. Dry calcined plasters have a high bulk volume and rough porous particles, and need a relatively high proportion of mixing water to wet and "float" the powder. W/P ratios of 0.5–0.6 are usual. Wet calcined plasters have lower bulk volumes and dense regular particles; typical W/P ratios are 0.30–0.33 for dental stones and 0.22–0.25 for die stones.

4.2 Setting Time

During the setting process, various setting times can be arbitrarily defined and measured. Dental technology relies on an "initial set" measured by

Table 1
Physical properties of typical dental gypsum products

Type	W/P ratio	Vicat initial set (min)	Linear setting expansion (%)	Tensile strength		Compressive strength	
				Wet	Dry	Wet	Dry
Impression plaster	0.60	4	0.13	1.3		5.9	
Laboratory plaster	0.50	22	0.30	2.3	4.1	12.4	24.9
Dental stone	0.30	17	0.12	3.5	7.6	25.5	63.5
Die stone	0.23	19	0.10	4.3	9.9	40.7	80.7

penetrometers, such as Gillmore or Vicat initial needles. Results are similar and give an indication of the time when the solidified material can first be safely handled.

Impression plasters need a rapid set, while with laboratory materials a longer manipulation time is desirable. Dental manufacturers adjust the setting rates of raw hemihydrates by adding suitable accelerators or accelerator–retarder blends.

4.3 Setting Expansion

The observed setting expansion is highest with low W/P ratios; wet calcined hemihydrates, with their low water demand, have a high inherent setting expansion. Typical values for the linear setting expansion of raw hemihydrates are: dry calcined (W/P = 0.50), 0.3%; and wet calcined (W/P = 0.24), 0.5%. Impression plasters, stones and die stones are used in applications requiring high dimensional accuracy; accelerator–retarder blends added to control setting time also reduce the setting expansion very effectively. Typical combinations are potassium sulfate–sodium tetraborate (borax), and potassium sodium tartrate (Rochelle salts)–sodium citrate. These additives also reduce the strength of the set material. This is not a disadvantage in impression plasters, but in dental stones and die stones, strength as well as accuracy is important; formulation of the latter materials involves reaching a compromise between a desirable reduction in setting expansion and an undesirable reduction in strength.

4.4 Strength

Cast gypsum is a typically brittle material; tensile strengths range from about 20% of the compressive strength for the weaker materials to about 10% for die stones. In most cases drying produces about a 100% increase in both tensile and compressive strengths; almost all of this increase occurs during the removal of the last 2% of uncombined water.

The effect of drying is reversible; soaking a dried cast in water reduces its strength to the original level.

The temperature of drying should be carefully controlled. Gypsum is stable below 40 °C, but loss of water of crystallization begins at higher temperatures and causes a significant reduction in dimensions and strength properties.

Bibliography

Andrews H 1951 The production, properties and uses of calcium sulfate plasters. *Building Research Congress, Division 2, Building Materials*, Pt. F, *Paints and Plaster*. BRC, London, pp. 135–44

Combe E C 1975 Recent developments in model and die materials. In: von Fraunhofer J A (ed.) 1975 *Scientific Aspects of Dental Materials*. Butterworths, London, pp. 401–24

Combe E C, Smith D C 1964 Some properties of gypsum plasters. *Br. Dent. J.* 117: 237–45

Fairhurst C W 1960 Compressive properties of dental gypsum. *J. Dent. Res.* 39: 812–24

Greener F H, Harcourt J K, Lautenschlager E P 1972 Gypsum Products. *Materials Science in Dentistry*. Williams and Wilkins, Baltimore, MD, pp. 261–72

Harcourt J K, Lautenschlager E P 1970 Accelerated and retarded dental plaster setting investigated by x-ray diffraction. *J. Dent. Res.* 49: 502–7

Khalil A A, Hussein A T, Gad G M 1971 On the thermochemistry of gypsum. *J. Appl. Chem. Biotechnol.* 21: 314–16

Lautenschlager E P, Corbin F 1969 Investigation on the expansion of dental stone. *J. Dent. Res.* 48: 206–10

Lautenschlager E P, Harcourt J K, Ploszaj L C 1969 Setting reactions of gypsum materials investigated by x-ray diffraction. *J. Dent. Res.* 48: 43–8

Mahler D B 1955 Plaster of Paris and stone materials. *Int. Dent. J.* 5: 241–53

Phillips R W (ed.) 1982 Gypsum products: Chemistry of setting. Basic principles (Chap. 4); Technical considerations (Chap. 5). *Skinner's Science of Dental Materials*. Saunders, Philadelphia, PA, pp. 63–89

Ridge M J, Beretka J 1969 Calcium sulfate hemihydrate and its hydration. *Rev. Pure Appl. Chem.* 19: 17–44

Weiser H B, Milligan W O, Ekholm W C 1936 The mechanism of the dehydration of calcium sulfate hemihydrate. *J. Am. Chem. Soc.* 58: 1261–5

R. Earnshaw
[University of Sydney, Sydney, Australia]

Dental Porcelain

Dental porcelains have the general composition of vitrified feldspar, along with additions of metallic oxide pigments that simulate natural tooth enamel shades. There are two main types of dental porcelain: the denture-teeth porcelains, which are reacted and vitrified in metal molds to form the shape of denture teeth, and porcelains that are reacted and ground to fine powders by the manufacturer and supplied to dental laboratories for the fabrication of crown and bridgework by hand.

Nicholas de Chemant, a French dentist, is credited with producing the first denture teeth, around 1790. Denture-teeth raw materials are based on feldspar with additions of around 15% quartz, 4% kaolin and the pigment oxides. The kaolin improves the molding qualities of the mixture. This general composition in relation to other triaxial whiteware bodies is shown in Fig. 1. The product of the fusion is self-glazing and nonporous. Since the 1950s, denture teeth have been fired under vacuum to further reduce the porosity. Denture teeth are used in full dentures which replace all of the teeth of the upper or lower arch. Porcelain denture teeth show a natural translucency which simulates that of natural teeth; in fact, the main aesthetic problem of denture teeth is that they have the appearance of perfect teeth without the minor defects and nonalignment of natural teeth. This is a problem that some dentists attempt to solve by staining of the denture teeth and even simulating artificial restoration in some of the teeth. The main mechanical problems with porcelain denture teeth are their brittleness and the clicking sound they make during biting. Acrylic resin teeth have been gradually replacing porcelain denture teeth, as they are tougher and do not produce the clicking sound. Also, the acrylic denture teeth can be chemically bonded to an acrylic denture base, eliminating the gap between porcelain denture teeth and an acrylic denture base. Although acrylic denture teeth have aesthetics equal to that of porcelain teeth, they have lower long-term wear and stain resistance.

The second type of dental porcelain is used by dental technicians in the fabrication of inlays, and in crown and bridgework. All-porcelain crowns (jacket crowns) have been used in dentistry since the 1890s to restore the entire visible portion of a tooth that has been damaged by fracture or advanced caries. The cross section of a jacket crown is shown in Fig. 2. The crowns are cemented with zinc oxide cements over the existing stump of the patient's tooth and serve as excellent restorations. These porcelains were used to fabricate inlays, but most inlays are made from silver amalgam or cast gold alloys. The older porcelains used for jacket crowns and inlays have a composition similar to those of the denture-teeth porcelains. However, the raw material is first reacted in the factory, ground to a fine powder and colored with pigments before being supplied to the dental technician. Porcelains for this purpose are classified according to their maturing temperature as follows:

(a) high-fusing, 1290–1370 °C,

(b) medium-fusing, 1090–1260 °C, and

(c) low-fusing, 870–1060 °C.

The coefficient of thermal expansion of these porcelains is around $9\text{--}10 \times 10^{-6}\,°C^{-1}$. The main source of service failure with these porcelains is fracture, since they have low tensile strengths of around $35\,MN\,m^{-2}$. Recent developments in this area

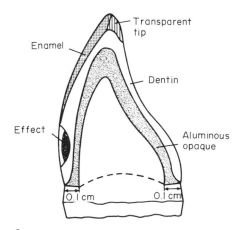

Figure 2
Cross section of aluminous porcelain crown (courtesy of Vita Zahnfabrik, Säckingen, FRG)

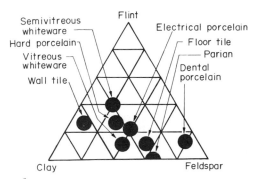

Figure 1
Dental porcelain raw-material composition in relation to other triaxial materials (after Norton 1952)

are the formulation of alumina-reinforced porcelains and porcelain enamels for fusing to alloy substructures.

Alumina-reinforced porcelains replaced the older jacket-crown porcelains during the 1960s. The porcelain for the inner core of the crown is essentially composed of the traditional porcelain, with 40–50% of crystalline alumina with an average diameter of 25 μm added as a mill addition. The alumina acts as a reinforcement and the transverse strength is almost doubled. As shown in Fig. 2, only the inner core of the jacket crown contains the alumina porcelain. The outer layers are made with the more translucent traditional porcelain. Although the alumina results in a stronger porcelain, there are no data available on the clinical breakage rate of these materials. Porcelain enamels were introduced during the 1950s for application to crown and bridgework. The cross section of a porcelain-alloy crown is shown in Fig. 3, which indicates the use of three porcelains to produce natural aesthetics. This type of restoration was made possible by the development of a porcelain enamel with a high coefficient of thermal expansion (slightly lower than that of dental gold alloys, with an average value of $13.5 \times 10^{-6}\,°C^{-1}$). These porcelain enamels are made from orthoclase feldspar or nepheline syenite with boron oxide and sodium, and lithium and potassium carbo-

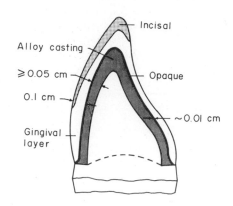

Figure 3
Cross section of ceramic–metal crown with full porcelain coverage (after O'Brien and Ryge 1978)

nates. The reacted product contains 5–15% of leucite, a crystalline phase with a high coefficient of thermal expansion formed either during the fritting or added as a mill addition. An opacifier, usually tin oxide or titania, is present in opaque porcelains which are applied to mask the color of the metal oxide layer. More translucent layers are applied

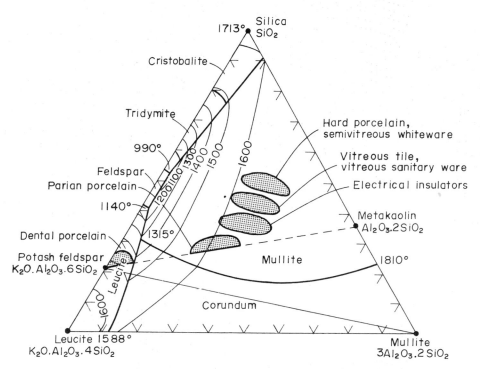

Figure 4
Composition of dental porcelains in relation to phases present in other triaxial whiteware materials

Table 1
Compositional ranges of dental porcelain enamels by analysis

Component	Range (%)
SiO_2	50–60
Al_2O_3	12–16
CaO or BaO	0–4
K_2O	9–10
Na_2O	5–8
TiO_2	12–15
ZrO_2	12–15
SnO_2	12–15

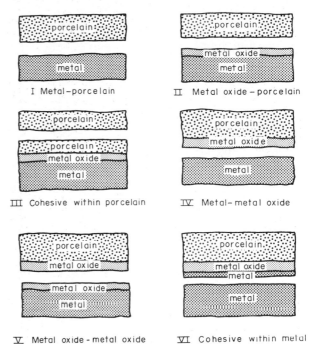

Figure 5
Classification of porcelain enamel failures according to interfaces formed. Type III represents cohesive failure indicative of a proper bond (after O'Brien and Ryge 1978)

above this layer. However, due to the opaque layer, the crown lacks the natural translucency of a natural tooth. The compositional ranges of porcelain enamels are given in Table 1 and the general compositions shown in Fig. 4. Porcelain–alloy restorations are used widely in dentistry for crown and bridgework, due to the greater strength of the overall structure. (For a discussion of the alloys used in porcelain–alloy restorations, see *Porcelain–Metal Bonding in Dentistry*.)

It is generally accepted that the bond strength should be at least equal to the strength of the porcelain. Failure through a weak boundary layer or adhesive failure indicates a weak bond. A classification of porcelain–enamel failures according to the interfaces formed is shown in Fig. 5. A type III failure would be indicative of a proper bond, since cohesive failure through the porcelain takes place. A type I failure is observed between porcelain and alloys coated with pure gold. Type V failures are observed with nichrome alloys when a thick chromium oxide layer is formed. Porcelain–alloy bond failures in service are due mainly to flexure of the metal substrate and impact. Therefore, impact and flexure laboratory tests are more representative than tensile or pull-through shear tests. The strengths of porcelain under different loading conditions are given in Table 2.

The porcelains supplied to the dental technician for fabricating crown and bridgework are fully reacted vitreous powders. The dental technician mixes the powder with distilled water to form a paste. The anatomy of the crown is formed by application of the paste to a metal crown using a small brush, and removing the excess water by vibration and blotting with paper in a process known as condensation. The porcelain in the wet state is slowly dried in front of the porcelain firing furnace and then introduced for baking. The process of densification is essentially a sintering process, involving vitreous flow of the particles. The entire restoration usually takes three firings for completion. The opaque layer is fired first, and is followed by the body and incisal portions of the crown. The porcelain may be brought to a glaze on the final

Table 2
Strengths of dental porcelains $(MN\,m^{-2})$[a]

Type	Transverse strength	Compressive strength	Diametral tensile strength	Shear strength
Porcelain enamel	90	345	38	111
Aluminous core	138			147

a After O'Brien and Ryge 1978

bake, or a low-fusing glaze may be applied and fired along with surface staining. The total shrinkage during firing depends upon the initial particle size distribution and the condensation method, but may be as high as 30%. After firing, the porosity is as low as 0.1% and as high as 7% for vacuum and air firing, respectively.

Three new ceramics have been introduced for the construction of all-ceramic jacket crowns. The first is a high-expansion core material which can be used with the same translucent porcelains applied to metal substrates. This core material has a flexure strength equal to that of the aluminous core material.

The second new material is another core material for jacket crowns that has a low firing shrinkage. This is achieved by adding magnesia to an aluminous core material which reacts to form a magnesia–alumina spinel during a heat treatment to 1300 °C. The reaction results in a volume expansion that compensates for the sintering shrinkage. Additional accuracy is achieved by injection molding the core material in the form of a plastic silicone resin composite. The flexure strength of the spinel core material is equal to that of the aluminous core material. After the spinel core is fired, outer layers of translucent porcelains are applied by hand and fired.

The third innovation is a mica glass ceramic that is used to fabricate crowns by casting. The mold is formed by the lost wax process with a phosphate bonded investment. The glass is cast at 1380 °C in a centrifugal machine similar to that used in jewelry casting. After casting, the glass is heat treated at 1075 °C for 6 h to produce crystallization. The resulting ceramic is then stained and glazed to the desired tooth shade. The flexure strength of the crystallized ceramic is 152 MPa which is slightly higher than that of the other dental porcelains.

Bibliography

Barreiro M M, Reisgo O, Vicente E E 1989 Phase identification in dental porcelains for ceramo-metallic restorations. *Dent. Mater.* 5: 51–7

McLean J W 1979 *The Science and Art of Dental Ceramics.* Quintessence Publishing, Chicago, IL

Norton F H 1952 *Elements of Ceramics.* Addison-Wesley, Reading, MA

O'Brien W J 1985 Ceramics. In: O'Brien W J (ed.) 1985 *Dental Clinics of North America.* Saunders, Philadelphia, PA

O'Brien W J, Craig R G (eds.) 1985 *Proc. Conf. New Developments in Dental Ceramics.* American Ceramics Society, Columbus, OH

O'Brien W J, Ryge G 1978 *Outline of Dental Materials and their Selection.* Saunders, Philadelphia, PA

Preston J D 1988 *Perspectives in Dental Ceramics.* Quintessence Publishing, Chicago, IL

Twiggs S W, Hashinger D T, Morena R, Fairhurst C W 1986 Glass transition temperatures of dental porcelains. *J. Biomed. Mater. Res.* 20: 293–300

Yamada H N (ed.) 1977 *Dental Porcelain: The State of the Art.* University of Southern California, School of Dentistry, Los Angeles, CA

W. J. O'Brien
[University of Michigan, Ann Arbor, Michigan, USA]

Denture Base Resins

Acrylic denture base materials have been widely available since the 1940s. Dentures are constructed using acrylic resin in a two-phase polymer–monomer dough-molding technique. The dough is packed into a mold of the denture that has been prepared in gypsum by a "lost wax" process.

The denture teeth are usually polymeric or, less frequently, porcelain. The polymeric teeth bond chemically to the denture base whereas the porcelain teeth rely upon the mechanical interlocking of the base polymer into holes or grooves, or around pins.

The polymer is essentially poly(methyl methacrylate) (PMMA) and since its introduction many different materials have been introduced to overcome its strength deficiencies. These materials have included such polymers as the polyamides, epoxy resin, polystyrene, vinyl-acrylic, rubber graft polymers and polycarbonate. Various modifications of PMMA and of the use of fillers have also been attempted. Dentures are constructed individually for each person to the requirements defined by an appropriate impression of the mouth and the various dimensions recorded. The geometry and size of each denture will vary. Each denture base will lie against tissue—some hard (such as the enamel of a tooth) and some soft (such as the mucosa and soft tissues that cover the bone of the jaws). In addition, each denture is subject to loading in both chewing and nonchewing contacts.

A complete denture is entirely supported by soft tissues and when loaded will flex in a complex manner that is partially determined by the uneven thickness of the mucosal base. It is difficult to estimate the number of flexures that a denture might undergo, but it is possible to suggest that a denture might flex about one million times per year. The amount of flexure will be limited by the compressibility of the mucosa and could be up to about 1 mm across the posterior dimension of a complete upper denture. Due to the low fracture strength of PMMA, these dentures may fracture easily in use in high strain rate situations. It has been proposed that PMMA dentures made for the upper jaw are liable to fracture in a fatigue mode (Smith 1961), but the evidence is not entirely conclusive. Lamb et al. (1985) are of the opinion that fatigue failure is not a

Table 1
Some requirements of a denture base material

Mechanical properties
  impact strength
  hardness
  fatigue strength
  stiffness
  creep/recovery
  flexural strength
Physical properties
  low water sorption and high diffusion coefficient
  radiopaque
  low specific gravity
  tasteless and odorless
  insoluble and no loss of constituents
Aesthetics
  lifelike
Technique
  ease and safety of manipulation in construction
    and repair
  technique of use appropriate to a dental
    laboratory
Cost
  reasonable cost

The polymerization of denture base polymers is generally promoted or activated by either heat or chemical activation (usually dimethyl-*p*-toluidine (DMPT)), or by a combination of these two. The initiator for the system is benzoyl peroxide.

Trevalon is an example of a standard heat-cured denture base resin. The powder component is PMMA (~97%) and a small amount of poly(butyl methacrylate) (PBMA) and about 0.2% benzoyl peroxide. The monomer is methyl methacrylate (MMA) which contains a cross-linking agent (~6% ethylene glycol dimethacrylate (EGDMA)) which is added to reduce crazing and solvent action. About 0.025% DMPT is added to the monomer and acts as the activator. Polymerization thus begins at a slow rate after mixing and, therefore, allows rapid curing cycles to be used (~20 min).

Lucitone 199 is an example of the microdispersed rubber phase polymers, probably formed in a similar way to that described in US patent No. 3,427,214. The MMA and the butadiene styrene are described as being copolymerized in an emulsion, a second coating of MMA being added to cover the bead. These polymer beads are then mixed with monomer in the usual way.

De Trey Self Cure is an example of a chemically activated resin and is similar to standard heat-cured materials, but contains approximately 1% of a tertiary amine in the monomer. The dough is packed into a warm mold (~30 °C) and the flask containing the mold is closed and held tightly in a dental press. The flask remains in the press for 15 min to allow a complete cure to take place.

Pour-n-Cure was developed as a resin to be used in a fluid state. The powder consists of PMMA and a small amount of ethyl methacrylate (EMA). The monomer is MMA with 8% EGDMA and 2% DMPT.

Nylon has been used occasionally as a denture base resin since 1950. Early studies used 66 and 610 nylon, but while these materials showed good mechanical advantages, various disadvantages were highlighted. The problems included the tendency of the material to lose color, stain, have high water absorption and develop surface roughness after a

cause of fracture in these dentures and generally these materials are used within their property limits. It is possible, however, that the bone of the jaws will resorb and if this should happen in the upper jaw then flexure could be great. In the upper jaw, bone loss will occur on the ridges, that is, on the outer aspect, but not centrally in the palate. The denture will then flex about its center line more when the resorption is greater. This is the clinical situation that could lead to clinical fatigue failure, particularly when a centrally placed incisal notch is present at the front of the denture. Some requirements of a denture base resin are set out in Table 1.

1. Current Materials

There are many varieties of denture base materials on the market. Examples of the basic types are set out in Table 2 with a summary of some of their properties.

Table 2
Denture base materials and properties

Material	Young's modulus (MPa)	Flexural strength (MPa)	Impact energy (10^{-4} J)	Hardness (10^{-4} mm)	Creep (%)	Water absorption (%)	Diffusion coefficient (10^{-8} cm^2 s^{-1})
Trevalon	2376	66	350	295	15	2.5	1.10
Lucitone 199	2220	55	1063	296	15	2.5	2.45
De Trey SOS	2398	48	485	297	18	2.36	1.62
Pour-n-Cure	2246	48	454	305	17	2.0	1.20
Nylon	1543		1130	299	4	0.8	0.69
Triad	2711	57	160	246	3	2.7	1.21

short period of clinical use. The problems of water absorption were overcome when nylon 12 was researched and used stiffened with glass fibers (Hargreaves 1971). Nylon has value as a denture base material in the rare cases of patients who show a proven allergy or when higher strength characteristics are required (MacGregor et al. 1984, Stafford et al. 1986). The nylons used have been either nylon 12 with small amounts of colorant and plasticizer or nylon 12 with 50 wt% glass spheres. These spheres have an average diameter of approximately 0.25 µm and are coated to improve adhesion. The nylon is injection-molded at 250 °C into a modified dental flask containing a hard gypsum mold preheated to 80 °C. Injection-molding pressure is 0.6 MPa and flasks are bench cooled before deflasking.

2. Physical Properties of the Resins

Particle sizes of the powders vary. A small particle size is useful to aid the wetting of the beads so that a smoother mix may be obtained and so that there will be a greater solution of the particles reducing the doughing time. The particle sizes of the powders of the resins listed in Table 2 are similar except for Lucitone 199 which has slightly larger particles.

Water sorption behavior for the acrylic polymers is similar with a water uptake of approximately 2% and the rate of water uptake is similar as defined by the diffusion coefficient. Nylon 12 absorbs less water than the acrylic polymers.

The molecular weights of the acrylate polymer powders, as determined by gel permeation chromatography, show that they all have an average molecular weight ($\bar{M}_w$) greater than 10^5. Above a molecular weight of 10^5, the material behaves as a typical long-chain polymer with optimum physical and mechanical properties. Thus molecular weight is not a factor affecting the properties, although Trevalon has a much greater average molecular weight than the other polymers. The molecular weight of the powder component will also affect doughing and working times, and high molecular weight powders will reduce the Trommsdorf gel effect and allow the use of rapid curing cycles. Thus, curing methods involving application of rapid heat may be used, reducing the production of a high exotherm. Etching of the prepared polymer surfaces shows that some of the larger beads undergo solution at their surfaces and the beads may be flattened (see Fig. 1). Fracture will cleave a bead (see Fig. 2) or may, in weaker situations, pass around beads (see Fig. 3). This has also been shown by Kusy and Turner (1977). Thus, there should be a good bond between the bead and the matrix.

3. Mechanical Properties

The mechanical properties of the standard resins in use only vary slightly (see Table 2). The modulus of

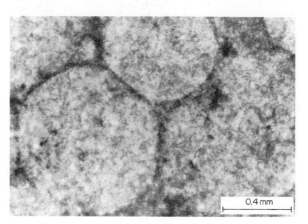

Figure 1
Polymer beads in a prepared etched specimen

elasticity and flexural strength are not very different in any of the usual polymers as measured in a three-point bend test. Nylon does not break in this test.

High-strain testing using the Hounsfield Plastics Impact Test shows that the rubber-toughened polymer is substantially better in impact resistance than the other resins, while nylon 12 has a similar impact resistance. The creep of the acrylates is usually about 15%.

4. Residual Monomer

Generally, heat-polymerized acrylates, processed at 70 °C for 7–14 h, show values of residual monomer in the resin of the denture of 1.5%. A terminal boil in the curing cycle will reduce this to about 0.3%. A cure with a final period at 100 °C for longer than one hour does not reduce the residual monomer in the resin any further.

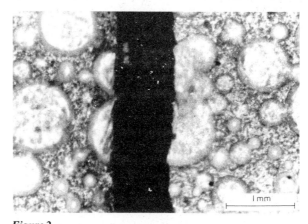

Figure 2
Etched beads of a polymer showing a fracture cleaving the beads

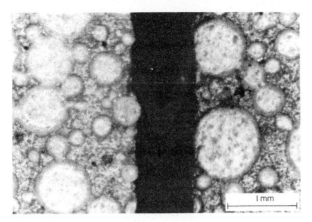

Figure 3
Etched specimens of polymer showing the fracture passing around the beads

A high level of residual monomer can occur in acrylates that have been modified for use in plates for orthodontic purposes. The use of a mixing technique that employs high levels of initial monomer is common in this situation and leads to a residual monomer content in the appliance of 1.5–4.5%. This reduces to 0.6–3.3% after soaking for 14 days in water. A further reduction in residual monomer levels may occur by continuing polymerization of the monomer. High residual levels are a potential danger to the patient in that sensitization may occur. It will also affect the mechanical properties and if continued polymerization occurs, it may cause dimensional instability (Stafford and Brooks 1985).

5. Future Developments

Further innovations that improve the existing range of PMMA resins and their modification using conventional approaches seem unlikely. Most of the possibilities have been researched.

5.1 Light Activation

A denture base material that may be cured by light activation has been reported. The material has the great advantage that there is no monomer present. A denture base material of this type is marketed under the trade name of Triad. The material is supplied in sheet and rope form, and in its unpolymerized form is soft and pliable. It contains a camphoroquinone-amine photoinitiator which is activated most efficiently by light in the 400–500 nm wavelength range. The polymerized material is a hard rigid cross-linked polymer consisting of urethane dimethacrylate and a silica filler. Its basic properties are similar to the other materials available (see Table 2), apart from the high strain rate

resistance property which is extremely poor. As can be seen, the impact strength is half that of the conventional denture base material Trevalon.

5.2 Microwave Processing

In this process, the microwave energy penetrates the special nonmetallic flask and the investing plaster of the denture and thus polymerizes the denture. The energy for the polymerization is generated from the monomer and not from the transmission of heat from the outside. The plaster investment may act as a heat sink for some of the exotherm generated. The process is rapid and the conversion of the monomer to polymer is similar to that obtained by a terminal boil.

Microwave curing at 70 W for 25 min minimizes porosity problems associated with rapid curing of acrylic dough. However, it has been shown that porosity-free material can only be guaranteed in sections of the denture not greater than 3 mm thick. It appears that microwave curing offers no advantage in terms of time saved over rapid water-curing systems.

5.3 Closed-Mold Processing Systems

Attempts to improve the impact properties and the dimensional accuracy have been made using a closed-mold pressure system. A standard method of heat delivery to the denture flask is used with either wet or dry heat. The polymer mix (in the stringy stage) is placed in a closed mold under a pressure of approximately 600 kPa and the material is held under pressure during the polymerization period. This ensures a compensating flow of polymer during the polymer contraction phase.

See also: Acrylic Dental Polymers; Dental Materials: Clinical Evaluation

Bibliography

Al-Mulla M A S, Murphy W M, Huggett R, Brooks S C 1989 Effect of water and artificial saliva on mechanical properties of some denture-base materials. *Dent. Mater.* 5: 399–402

Bates J F, Stafford G D, Huggett R, Handley R W 1977 Current status of pour type denture base resins. *J. Dent.* 5: 177–89

Chen J C, Lacefield W R, Castleberry D J 1988 Effect of denture thickness and curing cycle on dimensional stability of acrylic resin denture bases. *Dent. Mater.* 4: 20–4

Hargreaves A S 1971 Nylon as a denture-base material. *Dent. Pract. Dent. Rec.* 22: 122–8

Hargreaves A S 1983a The effects of cyclic stress on dental polymethylmethacrylate. I. Thermal and environmental fluctuation. *J. Oral Rehabil.* 10 (1): 75–85

Hargreaves A S 1983b The effects of cyclic stress on dental polymethylmethacrylate. II. Flexural fatigue. *J. Oral. Rehabil.* 10 (2): 137–51

Hill R G, Bates J F, Lewis T T, Rees N 1984 The fracture of acrylic polymers in water. *J. Mater. Sci.* 19: 1904–16

Jerolimov V, Brooks S C, Huggett R, Bates J 1989 Rapid curing of acrylic denture-base materials. *Dent. Mater.* 5: 18–22

Kusy R P, Turner D T 1977 Influence of the molecular weight of poly(methyl methacrylate) on fracture morphology in notched tension. *Polymer* 18: 391–402

Lamb D J, Ellis B, van Noort R 1985 The fracture topography of acrylic dentures broken in service. *Biomaterials* 6 (2): 110–12

MacGregor A R, Graham J, Stafford G D, Huggett R 1984 Recent experiences with denture polymers. *J. Dent.* 12 (2): 146–57

Ruyter I E 1982 Methacrylate-based polymeric dental materials: Conversion and related properties. Summary and review. *Acta Odontol. Scand.* 40: 359–76

Smith D C 1961 The acrylic denture. *Br. Dent. J.* 110: 257–67

Stafford G D, Bates J F, Huggett R, Handley R W 1980 A review of the properties of some denture-base polymers. *J. Dent.* 8: 292–306

Stafford G D, Brooks S C 1985 The loss of residual monomer from acrylic orthodontic resins. *Dent. Mater.* 1: 135–8

Stafford G D, Huggett R 1978 Creep and hardness testing of some denture-base polymers. *J. Prosthet. Dent.* 39: 682–7

Stafford G D, Huggett R, Causton B E 1980 Fracture toughness of denture base acrylics. *J. Biomed. Mater. Res.* 14: 359–71

Stafford G D, Huggett R, MacGregor A R, Graham J 1986 The use of nylon as a denture-base material. *J. Dent.* 14: 18–22

Stafford G D, Lewis T T, Huggett R 1982 Fatigue testing of denture base polymers. *J. Oral Rehabil.* 9: 139–54

G. D. Stafford
[University of Wales College of Medicine, Cardiff, UK]

Drugs: Attachment to Polymers

It has long been realized that the therapeutic benefits of a drug can often be enhanced by its association with a synthetic polymer or natural macromolecule. The earliest applications involved polymer gels or networks which were used as depots for controlled slow drug release (Donaruma and Vogl 1978). In such systems, a physically incorporated drug may be released by diffusion processes, or a chemically attached species may be liberated by the dissociative effect of enzymic hydrolysis or other types of reaction. An interesting account of the design and function of some slow-release systems has been given by Harris (1984). More recently, attention has been paid to soluble polymer–drug adducts and especially the opportunities they provide for constructing targetable drug-delivery systems. Aside from the possibilities of targeting, these soluble adducts are interesting in their own right; for example, their interactions with cell membranes are different from those of relatively small molecules, so that the adducts have different properties which may be of

therapeutic or mechanistic interest (Duncan and Kopecek 1984, Bamford et al. 1987a).

The use of polymeric implants to replace damaged or diseased parts of the body is also of great current interest. An approximate match between the mechanical properties of the implant and those of the natural tissue is generally thought to be desirable, so that synthetic elastomers find many applications, a large proportion of which involve contact with blood (Bamford and Middleton 1987). While the hemocompatibility of some poly(ether-urethane) elastomers is relatively good, it is not perfect, and attempts have been made to improve blood compatibility by chemical modification of the polymer; for example, by grafting or coupling antiplatelet drugs or other bioactive agents that inhibit clotting. A similar kind of treatment may also be useful for reducing the sensitivity of polymers of this type towards attack by body fluids.

The synthesis of polymer–drug adducts of types which may be generally useful in applications such as those outlined above are described here. Other important fields of application exist, based on adducts of polymers and bioactive molecules (e.g., affinity chromatography and polymer–enzyme systems) which will not be referred to specifically, but it is hoped that some of the procedures discussed may have relevance in these fields. As a rule, the distinction between the synthesis of a polymer–drug adduct in bulk and the coupling of a drug to a polymer surface shall not be made in this article, since the chemical reactions in the two cases are similar. A few special cases in which unusual features arise will, however, be mentioned.

1. Chemistry of Coupling Reactions

1.1 Coupling Agents

Many drugs carry OH, COOH or NH_2 groups in their molecules and these can be coupled directly to appropriate functional groups in the polymers (if such are present) or to monomers which may subsequently be polymerized. Coupling involves the formation of ester or amide linkages, and to bring this about under mild conditions it is generally necessary to use a "coupling agent." Two types of these are particularly widely employed. Carbodiimides probably have the longest history (Khorana 1953, Kurzer and Douraghi-Zadeh 1967). Dicyclohexylcarbodiimide (DCCI) functions well in nonaqueous media, especially dichloromethane, a base (often 4-dimethylaminopyridine) being added as a catalyst (e.g., Ziegler and Berger 1979). The essential reactions occurring during the esterification of a carboxylic acid RCOOH with an alcohol R'OH in the presence of DCCI are given in Scheme 1.

In addition to the major products, the desired ester RCOOR' and *N,N'*-dicyclohexylurea, the

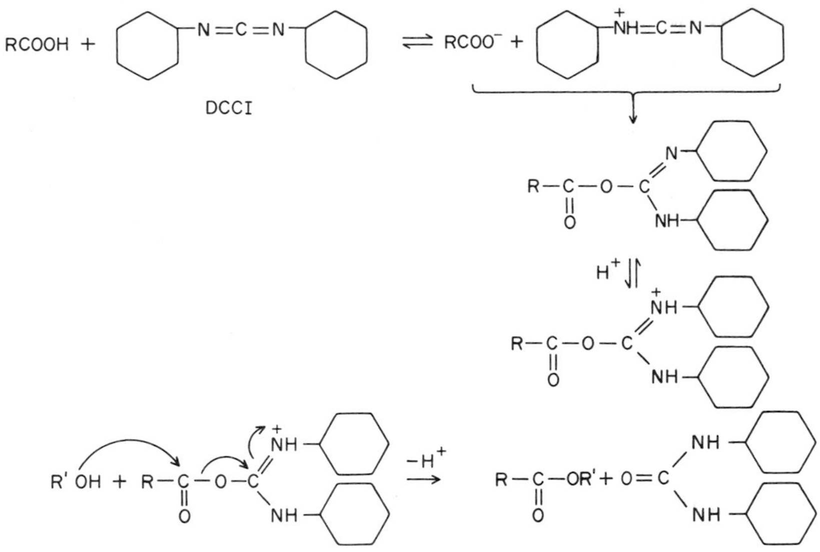

Scheme 1

N-acylurea is formed as by-product to an extent depending on the nature of the acid. Other carbodiimides find applications under special conditions. For example, the diisopropyl derivative may be used (in nonaqueous solution) when it is desirable that the urea formed (diisopropyl urea) should be soluble (Epton et al. 1987). In aqueous solution coupling agents frequently employed include 1-ethyl-3-(3-dimethyl aminopropyl) carbodiimide and 1-cyclohexyl-3-(2-morpholinoethyl) carbodiimide metho-*p*-toluene sulfonate.

Another valuable coupling agent is carbonyl diimidazole (CDI) (Staab 1962). CDI reacts with alcohols and carboxylic acids, as illustrated in Scheme 2 (Bamford et al. 1986) and esters may be synthesized by either of the routes indicated (i.e., starting with CDI + R'OH or CDI + RCOOH). However, for esterifying polyhydroxylic macromolecules when cross-linking must be avoided, route 2 is preferable since cross-linking of the macromolecules can occur through carbonate formation in route 1. In practice, route 2 is a two-stage process; CDI is first reacted with the carboxylic acid and when evolution of carbon dioxide has ceased the alcohol is added. Similarly, amides may be synthesized by use of amines instead of alcohols in route 2.

A few illustrative examples may now be quoted. The prostaglandin analogue BW245C, theophylline acetic acid and dipyridamole have been converted to polymerizable vinyl monomers or attached directly to polymers with the aid of the coupling agents mentioned. BW245C was esterified with hydroxymethacrylate esters and the monomers so obtained were subsequently copolymerized (see Fig. 1) (Bamford et al. 1987c). Since reaction of the hydroxyl on C15 of BW245C results in complete loss of

(antiplatelet) activity, care must be taken to avoid attack on this group. (For this reason CDI is not a suitable coupling agent.) A large excess of the esterifying alcohol ensures minimal attack at C15; this

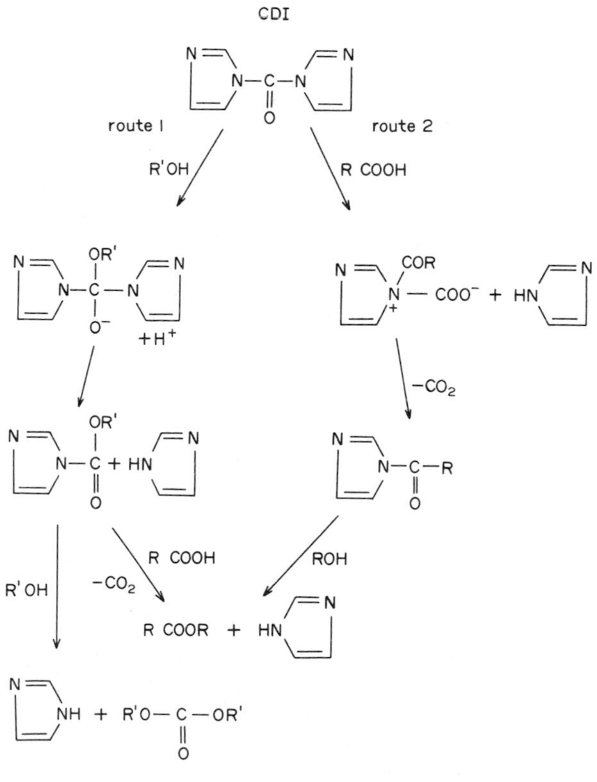

Scheme 2

157

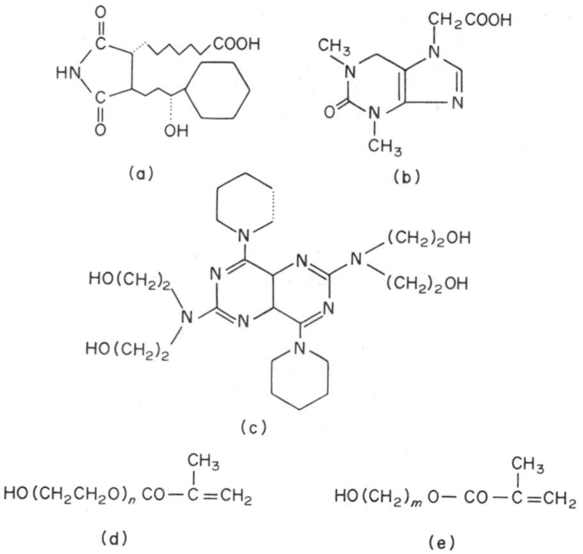

Figure 1
Structural representations of: (a) BW245C, (b) theophylline acetic acid, (c) dipyridamole, (d) hydroxymethacrylate esters ($n = 1,2,4$) and (e) hydroxymethacrylate esters ($m = 2,5,12$)

example illustrates the value of kinetic control in synthesis. DCCI in dichloromethane solution was used as coupling agent. BW245C was also attached to dextran and the terminal hydroxyls in poly(ethylene glycol) by use of DCCI. In the former case, the solvent was a mixture of formamide, N,N-dimethyl formamide and dichloromethane, with 4-dimethylaminopyridine as catalyst. A portion of the dextran chain with attached BW245C is shown in Fig. 2.

A wide variety of esters was synthesized from theophylline acetic acid by esterification of the carboxyl with low-molecular-weight alcohols, both

polymerizable and nonpolymerizable; theophylline acetic acid was also attached to hydroxylic polymers. In all cases, CDI was employed in the route 2 procedure of Scheme 2. Isolation and purification of low-molecular-weight products were carried out by flash-column chromatography, while polymer derivatives were subjected to extensive reprecipitation.

Dipyridamole was converted to the monomethacrylate ester by reaction of one hydroxyl with methacrylic acid (note that all four hydroxyls are equivalent). DCCI was used in chloroform solution and products were separated by thin-layer chromatography. The ester was copolymerized with N-vinyl pyrrolidone.

Activating groups; activated species. An examination of the mechanisms outlined above shows that the action of DCCI and CDI depends on the introduction into the carboxylic acid molecule of a good leaving group (e.g., imidazolyl in route 2 of Scheme 2). Many other readily displaceable groups are known, including those set out in Fig. 3.

The groups shown in Fig. 3a–d can be readily displaced from their acyl derivatives by reaction with nucleophiles (such as hydroxylic or amino compounds) which thereby become attached to the acyl moieties (through ester or amide linkages in the cases mentioned). Typical reactions are

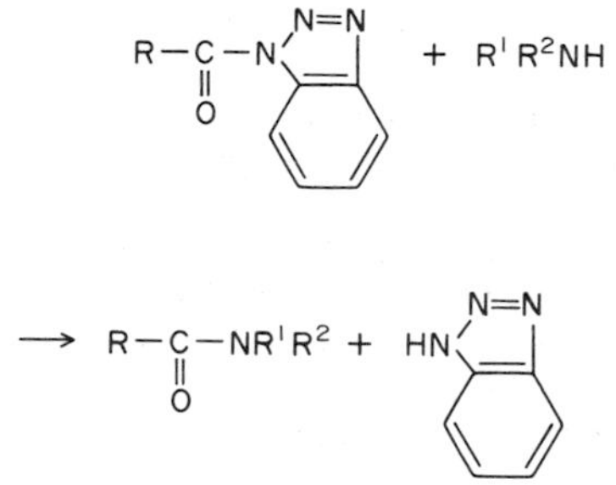

$$R-\underset{\underset{O}{\|}}{C}-N\diagdown + R^1 R^2 NH$$

$$\longrightarrow R-\underset{\underset{O}{\|}}{C}-NR^1R^2 + HN\diagdown \qquad (1)$$

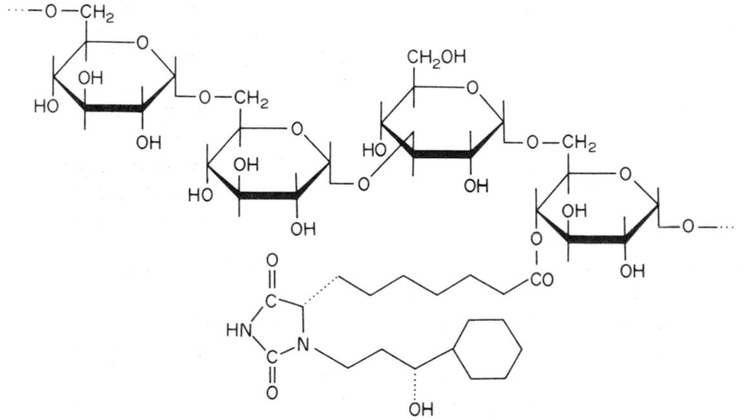

Figure 2
Portion of dextran chain with attached BW245C

and

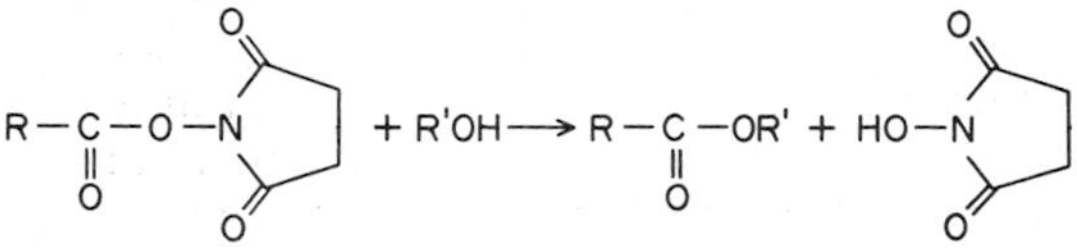

$$R-\overset{O}{\underset{O}{C}}-O-N\overset{O}{\underset{O}{\bigcirc}} + R'OH \longrightarrow R-\overset{O}{\underset{O}{C}}-OR' + HO-N\overset{O}{\underset{O}{\bigcirc}} \quad (2)$$

In these equations, R may be part of a polymer chain, so that these reactions provide a convenient method for attaching polymers to drugs carrying nucleophilic residues. Alternatively, if R embodies a polymerizable double bond, the reactions are useful for converting drugs to monomers which may be polymerized or copolymerized.

Tosyl sulfonic esters obtainable from the group shown in Fig. 3e via the sulfonyl chloride, also undergo facile nucleophilic substitution:

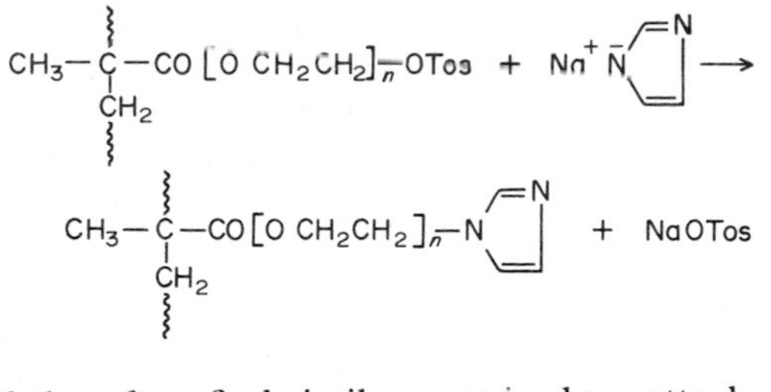

$$CH_3-\underset{CH_2}{\overset{|}{C}}-CO[O\,CH_2CH_2]_{\overline{n}}OTos + Na^+\,\bar{N}\overset{\frown}{\underset{\smile}{\bigvee}}^{N} \longrightarrow$$

$$CH_3-\underset{CH_2}{\overset{|}{C}}-CO[O\,CH_2CH_2]_{\overline{n}}N\overset{\frown}{\underset{\smile}{\bigvee}}^{N} + NaOTos \quad (3)$$

and therefore find similar uses in drug attachment. A related reagent, which also functions under mild conditions, is tresyl chloride (2,2,2-trifluoroethyl-sulfonyl chloride) which forms reactive sulfonate esters with hydroxylic compounds (Nilson and Mosbach 1981).

The alternative procedure of first coupling the drug to the activating group, then reacting with polymer or monomer, has also been used. It is normally restricted to those drugs which possess only one type of group capable of reaction with the activating molecule.

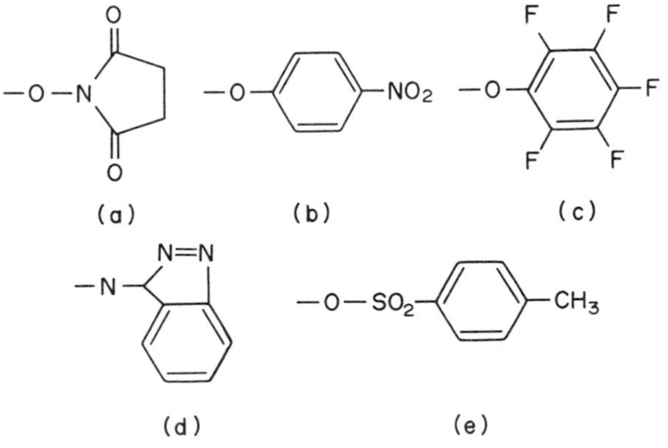

Figure 3
Structural representations of five readily displaceable groups useful in coupling reactions

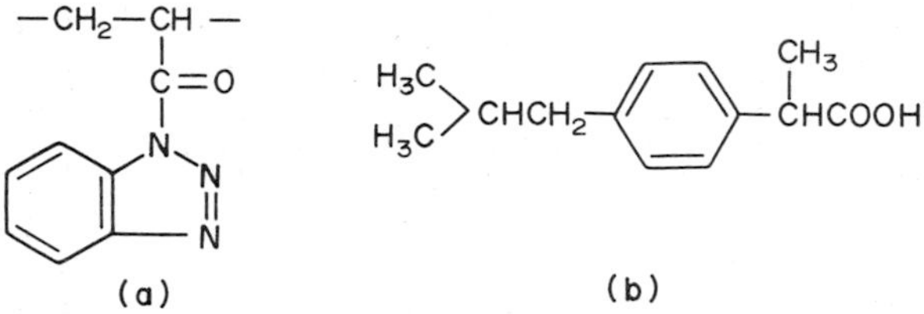

Figure 4
Structural representation of: (a) the activated polymer of acrylic acid and benzotriazole, and (b) Ibuprofen

Species such as those described above—whether polymer, monomer or drug—which are attached to an activating group are generally designated "activated" and they naturally find many applications, notably in preparing matrices for affinity separations and in synthesizing polymer–drug conjugates generally. The following are some representative examples.

Ferruti and Tanzi (1979) synthesized 1-acryloyl benzotriazole by coupling acrylic acid and benzotriazole with the aid of DCCI, and polymerized this monomer to form the activated polymer shown in Fig. 4a. The reactions of this and a series of analogous polymers with amines, hydroxylic compounds were studied and the expected results were reported and Ferruti and Tanzi gave the following order of reactivity of the activating groups:

$$\text{hydroxysuccinimides} < \text{benzotriazolides}$$
$$< \text{imidazolides}$$

The attachment of Ibuprofen (see Fig. 4b) and ursodeoxycholic acid to poly(ethylene glycol) provides examples of prior drug activation, the carboxyl groups of the drug being converted to imidazolides by reaction with CDI (route 2 of Scheme 2).

Duncan et al. (1983) have used the activated polymer technique to couple cytotoxic drugs such as daunomycin ($D-NH_2$) to polymer chains which will subsequently release the drug on attack by lysosomal enzymes. Formation of an activated precursor polymer with a backbone of poly N-(2-hydroxypropyl) methyacrylamide is illustrated in Scheme 3, in which the activating p-nitrophenyl group is shown attached to N-methacryloyl glycine. Subsequent coupling of daunomycin then occurs through amide link formation, as indicated.

The synthesis of matrices containing imidazole is an example of the use of sulfonic esters (Eqn. (3)).

1.2 Coupling Ligands

The coupling agents so far discussed owe their effectiveness to the ease by which activating groups may be introduced and removed from the species under consideration. No fragments characteristic of the coupling agents remain after the completion of the

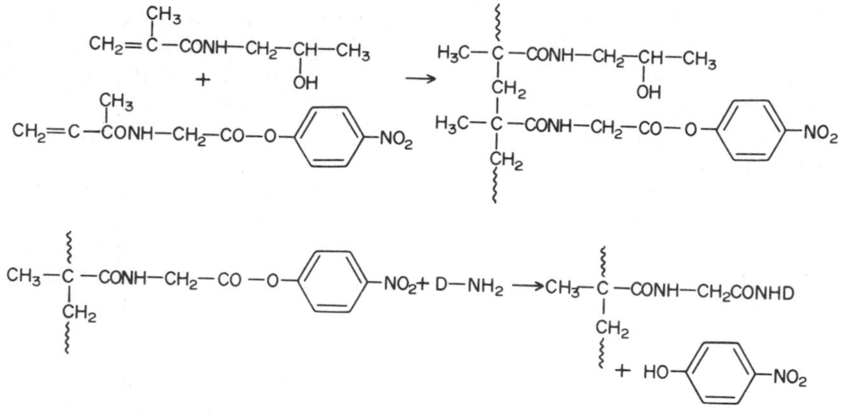

Scheme 3

reaction. There is a different category of reagent that brings about coupling by forming an adduct with the two groups to be coupled; these are called coupling ligands.

Coupling ligands have been much used in heterogeneous systems in which the introduction of additional cross-links is not of prime concern (e.g., in the activation of matrices for affinity chromatography) but they also find other applications. Two reagents commonly employed are cyanogen bromide and glutaraldehyde. Activation with cyanogen bromide proceeds in alkaline solution through the formation of an imino carbonate from neighboring OH groups (Scheme 4) (Porath and Axen 1976). This may react with species carrying amino groups in three ways, as shown in Scheme 4. Kohn and Wilchek (1983) recommend addition of triethylamine to BrCN, presumably to form a more reactive cyanotriethyl ammonium complex.

Glutaraldehyde has been much used as a coupling ligand, but the mechanism of its action is uncertain. One suggestion is that coupling is brought about by the addition of amino groups to 1,2-unsaturated groups of trimers present in aqueous glutaraldehyde:

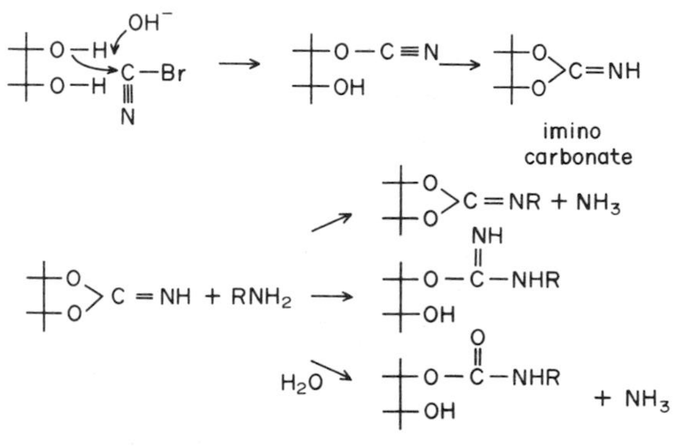

$$3 \, [OHC-(CH_2)_3-CHO] \longrightarrow$$

$$OHC-(CH_2)_3-CH=C-(CH_2)_2-CH=C-(CH_2)_2-CHO + 2H_2O$$

$$\overset{|}{CHO} \qquad \overset{|}{CHO}$$

$$(4)$$

The application of these coupling ligands can be illustrated by an example taken from the work of Margel et al. (1979), who pioneered use of functionalized colloidal particles as antibody carriers. Emulsion polymerization of mixtures of methyl methacrylate, methyacrylic acid, 2-hydroxyethyl methacrylate and ethylene glycoldimethacrylate (typically in the ratios 57:10:30:3 wt%, respectively) was used to prepare spherical particles with diameters in the range 30–300 nm. The particle surfaces carried relatively high concentrations of OH and COOH groups, since the polymerization took place in an aqueous medium. Smaller spheres could also be prepared by initiation with ionizing γ radiation (^{60}Co). A diamine (e.g., diaminoheptane) was first coupled to the spheres with the aid of a water-soluble carbodiimide or cyanogen bromide and the spheres were then activated with glutaraldehyde. Antibodies to be coupled were added after removal of excess glutaraldehyde. It is claimed that this technique has an advantage over carbodiimide coupling in that intramolecular and intermolecular cross-linking of antibody molecules is avoided.

Nucleophilic groups such as OH, NH, SH may conveniently be coupled by ligands bearing oxirane (epoxide) or epihalohydrin moieties. Thus epoxides enter into familiar reactions of the type

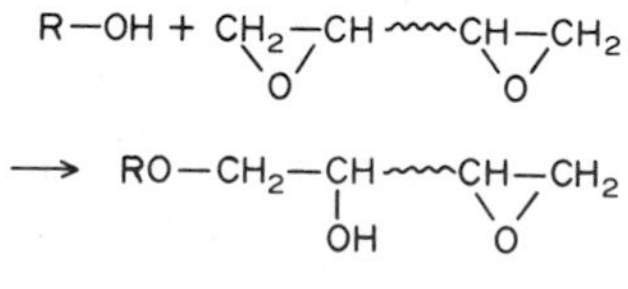

$$(5)$$

Scheme 4

160

Divinyl sulfone interacts in a similar manner:

$$R\text{—}OH + CH_2\text{=}CH\text{—}SO_2\text{—}CH\text{=}CH_2$$
$$\rightarrow RO\text{—}CH_2\text{—}CH_2\text{—}SO_2\text{—}CH\text{=}CH_2 \quad (6)$$

but is more reactive and lower pHs may be used. Rates of nucleophilic attack in both instances are in the order SH greater than NH which in turn is greater than OH.

Benzoquinone and sym-trichlorotriazine are active coupling ligands, reacting as follows:

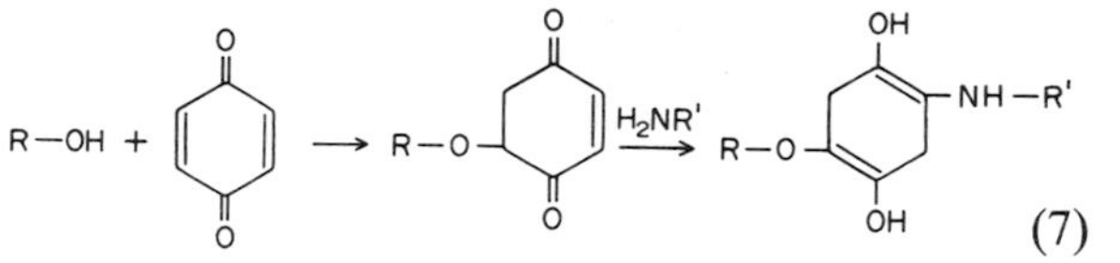

$$(7)$$

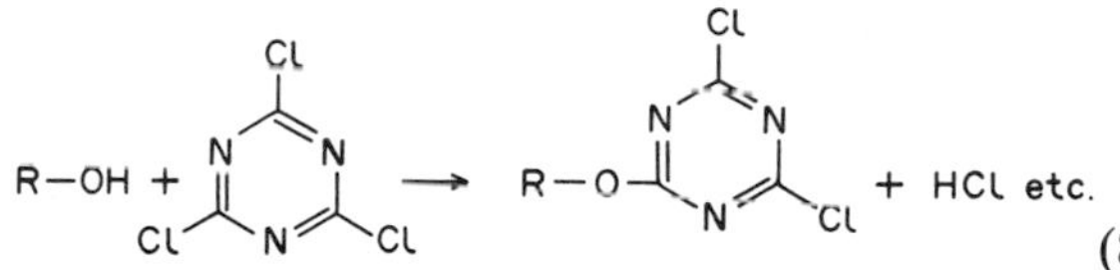

$$(8)$$

Some practical details may be found in Osterman (1986).

2. Free-Radical Methods in Drug Attachment

This category includes procedures in which a bioactive species is linked to a polymer by a free-radical reaction, generally a copolymerization. The technique therefore depends on the bioactive molecule or a derivative being able to participate in one of the component reactions of free-radical polymerization, viz initiation, propagation or chain transfer. Likewise, a preformed polymer molecule must function as a macroinitiator, a macromer or a (macro) transfer agent. Obviously the drug should not be susceptible to reactions with free radicals which cause inactivation.

The simplest examples are copolymerization of "drug monomers" such as the methacrylate esters of BW245C, theophylline acetic acid or dipyridamole (see Sect. 1.1). These have been copolymerized with styrene and *N*-vinyl pyrrolidone, to give, in the latter instance, water-soluble polymers. All copolymerizations were carried out photochemically (365 nm) in solution at 25 °C, with azo-bis-isobutyronitrile as photoinitiator; the polymers were isolated by appropriate precipitation and purified by multiple reprecipitations. Copolymers prepared in this way carry the bioactive moieties in side chains (e.g., as illustrated in Fig. 5).

Initiation of free-radical polymerization by transition metal derivatives (usually carbonyls) in which the metal is in a low (preferably zeroth) oxidation state, in association with organic halides as coinitiators, has been used in synthesising many polymer–

drug adducts in this class. The radical-forming reaction, which may be thermal or photochemical, is an electron transfer from metal to halide, generating a free-radical from the latter (Eqn. (9) in which M^0 represents the metal atom).

$$\overset{\frown}{M^0 + Cl}\text{—}R\cdot \rightarrow M^1 + Cl^- + R\cdot \quad (9)$$

In the presence of a monomer, the radical R· initiates polymerization so that a drug molecule linked to R· becomes attached to the polymer chain as a terminal unit. The mechanisms of these initiation reactions have been extensively studied and turn out to follow a common pattern (Bamford 1974, 1988). General principles determining the relative activities of halide coinitiators have been elucidated and the activities of many metal carbonyls are known. The process may fail if R· is strongly electron attracting, in which case the electron transfer yields $R^- + Cl\cdot$, so that R· does not become part of the polymer chain. Convenient metal carbonyls are $Mo(CO)_6$ (thermal initiation at 60 °C) and $Mn_2(CO)_{10}$ and $Re_2(CO)_{10}$ (photoinitiation at 436 nm and 365 nm, respectively). In suitable circumstances these photoinitiators give rise to very large photo-after-effects, during which grafting occurs (in the dark); this property is very useful, in practice, since it enables objects of irregular shape (carrying halogen atoms) to be grafted uniformly, although uniform irradiation is impossible (Bamford and Middleton 1983, Bamford et al. 1985). No metal becomes attached to the polymers in any of these reactions.

As an example of this technique the prostaglandin analogue BW245C can be considered again. Esterification with a halo-alcohol (e.g., CCl_3CH_2OH or $BrCH_2CH_2OH$ with DCCI as coupling agent) yields coinitiators which may be used in the reactions described above to synthesize polymers of the type shown in Fig. 6 containing terminally attached units of BW245C. Comparison of their properties

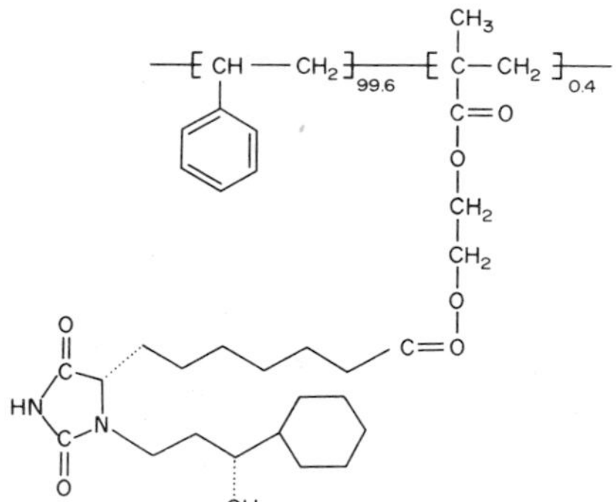

Figure 5
Copolymer carrying the active moieties in side chains

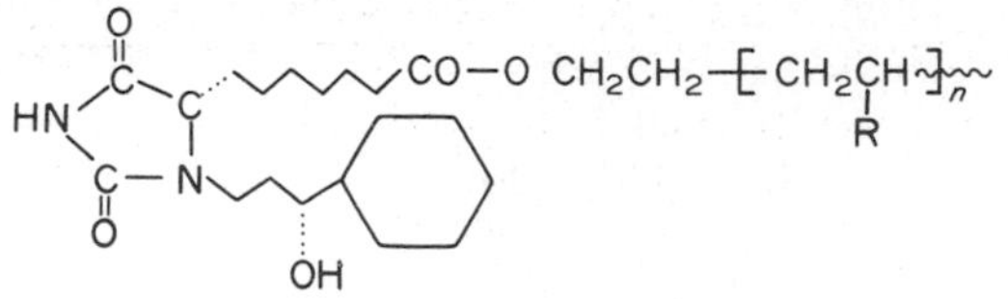

Figure 6
Polymer containing terminally attached units of
BW245C

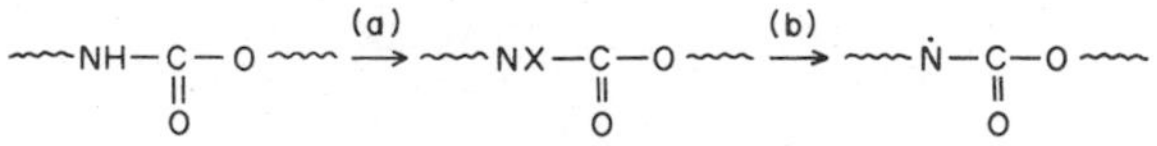

Scheme 5

with those of the polymers carrying the drug in side chains, shown in Fig. 5, thus becomes possible.

Polymers containing appropriately situated chlorine or bromine atoms are well-known coinitiators and so enter readily into grafting reactions which are often suitable for drug attachment. Several methods are available for introducing the required halogen functionality into the polymer. Direct halogenation is convenient for polyamides, polyurethanes and related materials. It has been shown that *N*-chloro- and *N*-bromo-imides are highly reactive in reactions of the type shown in Eqn. (9), forming nitrogen radicals which can initiate polymerization. Advantage has previously been taken of this in attaching grafts of vinyl polymers to polyamides, polypeptides and proteins and a similar technique is successful with polymers containing urethane linkages. The polymers may be halogenated either heterogeneously in dilute aqueous sodium hypochlorite or hypobromite solution, or homogeneously in *N,N*-dimethylformamide solution by reaction with bromine for a short time (~1 min). Both processes cause some degradation of the polymer, but this is not significant under mild conditions.

Heterogeneous halogenation is diffusion controlled and so by adjustment of conditions may be confined to a specified depth below the surface of the polymer; grafting and drug attachment are, of course, necessarily confined to the halogenated region.

The overall reactions are summarized in Scheme 5 in which (a) and (b) represent the halogenation and radical-forming steps (with transition metal derivative), respectively, and X is Cl or Br. If the material is required for biomedical applications it is necessary to remove excess halogen; this is easily affected by reaction with potassium iodide under conditions such that the polymer is adequately swollen (e.g., in a solution of KI in acetone containing 10% water).

Grafting or drug attachment (like halogenation) can be carried out heterogeneously or homogeneously. The former is particularly useful when fabrication must precede grafting; for example, it has been used successfully with (halogenated) arterial prostheses. Generally, solution grafting gives more uniform products and higher degrees of grafting.

Attachment of the HEMA ester of BW245C to the poly(ether-urethane-urea) Biomer and to the poly(ether-urethane) Pellethane by this technique has been achieved by the authors. Figure 7 shows that platelet adhesion to Biomer is effectively eliminated by grafting the HEMA ester of BW245C, 3 wt%.

A method of functionalizing polymers that appears to be of rather wide application is based on haloisocyanates, more correctly termed haloalkyl or haloacyl isocyanates. A number of these compounds are commercially available, and although they have been used since before the 1970s as derivatizing

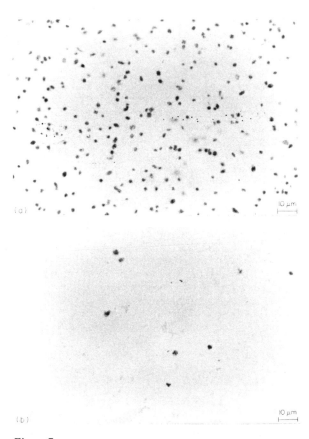

Figure 7
Blood-platelet adhesion to polymer films (the films cast from dimethylacetamide were immersed in sheep platelet-rich plasma for 4 h at ambient temperature, washed and stained): (a) pure Biomer and (b) Biomer grafted with HEMA–BW245C monomer (BW245C 3 wt%)

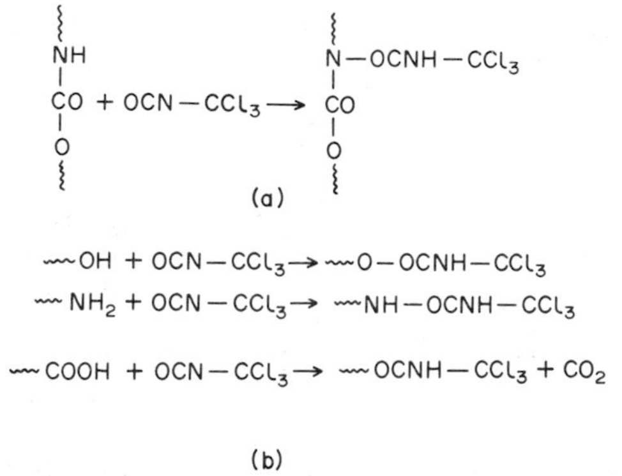

(a)

$$\text{\textasciitilde OH} + OCN-CCl_3 \rightarrow \text{\textasciitilde O}-OCNH-CCl_3$$
$$\text{\textasciitilde NH}_2 + OCN-CCl_3 \rightarrow \text{\textasciitilde NH}-OCNH-CCl_3$$
$$\text{\textasciitilde COOH} + OCN-CCl_3 \rightarrow \text{\textasciitilde OCNH}-CCl_3 + CO_2$$

(b)

Scheme 6

agents, they have received little attention from polymer chemists. As would be expected, their dual functionality and high reactivity makes them valuable in the synthesis of block and graft copolymers, and consequently in drug attachment. Among the common haloisocyanates are $ClCH_2NCO$, Cl_3CNCO, $ClCH_2CH_2NCO$ and $Cl_3CCONCO$. The isocyanate moieties of these molecules readily couple with groups carrying labile hydrogen atoms, including primary and secondary amines, hydroxyl, carboxyl, amide and urethane, all of which are frequently encountered in polymers as main-chain, side-chain or terminal residues. After coupling of the isocyanate, the halogen moieties are available for coinitiation of free-radical polymerization by interaction with a transition-metal derivative. The isocyanate reactions are summarized in Scheme 6, (a) showing main-chain isocyanation and (b) showing terminal-group or side-chain isocyanation. These reactions have been used in preparing polymer grafts to Biomer, dextran, poly(ethylene glycol) and poly(vinyl alcohol). In a number of instances the grafts incorporated monomer units derived from drugs; for example, BW245C, dipyridamole and indomethacin (shown in Fig. 8a).

These techniques have the advantage of allowing relatively high concentrations of drugs to be covalently coupled to the parent polymer: for example,

no difficulty was encountered in attaching 20 wt% indomethacin to Biomer.

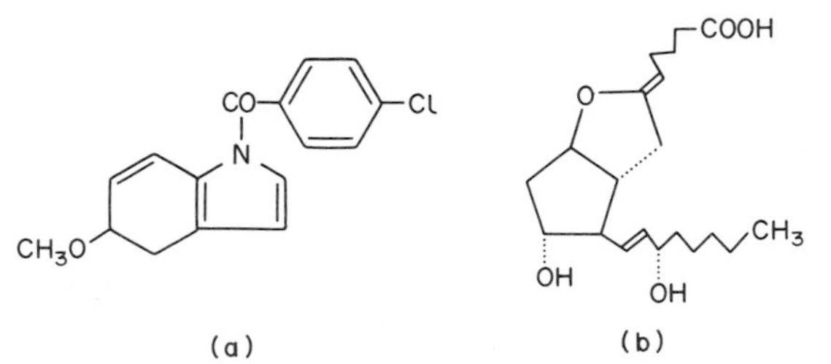

Single-electron transfer processes between an oxidizing metal ion and organic species, resulting in free-radical generation from the latter, have long been recognized. They provide a technique for the synthesis of graft copolymers and for this purpose use of Ce^{4+} was pioneered by Mino and Kaizerman (1958). The typical reactions shown in Eqn. (10a,b) have been employed for preparing grafts of cellulose (Hebeish and Guthrie 1981).

$$RCH_2OH + Ce^{4+} \rightleftharpoons complex \rightleftharpoons R\dot{C}HOH \text{ or } RCH_2O\cdot + H^+ + Ce^{3+}$$

$$(10a)$$

$$\underset{\substack{| \\ OH}}{R-CH} - \underset{\substack{| \\ OH}}{CH} - R' + Ce^{4+} \rightleftharpoons complex \longrightarrow \underset{\substack{\| \\ O}}{RCH} + \underset{\substack{| \\ OH}}{R'\dot{C}H} + H^+ + Ce^{3+} \quad (10b)$$

Some chain termination by Ce^{4+} also occurs in these systems. Clearly, these grafting reactions could be used for attaching drugs to polymer chains by methods similar to those already outlined, although little work of this kind has been reported. The covalent binding of heparin to poly(methyl methacrylate) chains by use of Ce^{4+} in acidic aqueous solutions has been achieved (Labarre et al. 1974); presumably initiation occurred from radicals formed by oxidation of hydroxyl groups in heparin (cf. Eqn. (10)).

3. Some Special Cases

It is important to realize that the general techniques for covalently binding drugs to polymer molecules, discussed above, may fail in special cases; for example, if suitable functionality cannot be incorporated into the drug without loss of activity, or if sensitive groups in the drug cannot be protected from unwanted reactions. To some extent each system constitutes a special case to be considered on its merits, as illustrated by the following examples.

3.1 Prostacyclin

Prostacyclin (PGI_2) (see Fig. 8b) is an extremely potent endogenous antiplatelet agent, secreted from endothelial cells of vascular intima. Interaction with platelets stimulates adenylate cyclase which liberates cyclic adenosine monophosphate (cAMP) and thus inhibits the platelet aggregation process initiated by thromboxane. Prostacyclin is very unstable on account of the presence of the vinyl ether grouping and it is virtually impossible to bind it directly to a polymer molecule.

Ebert et al. (1982) immobilized prostacyclin onto polystyrene surfaces by first coupling prostaglandin $PGF_{2\alpha}$ to the polymer and then converting the $PGF_{2\alpha}$ to PGI_2 by methods described by Tomoskozi et al. (1977) and by Whittaker (1977). Their procedure is illustrated in Scheme 7.

(a)   (b)

Figure 8
Structural representation of: (a) indomethacin and (b) prostacyclin

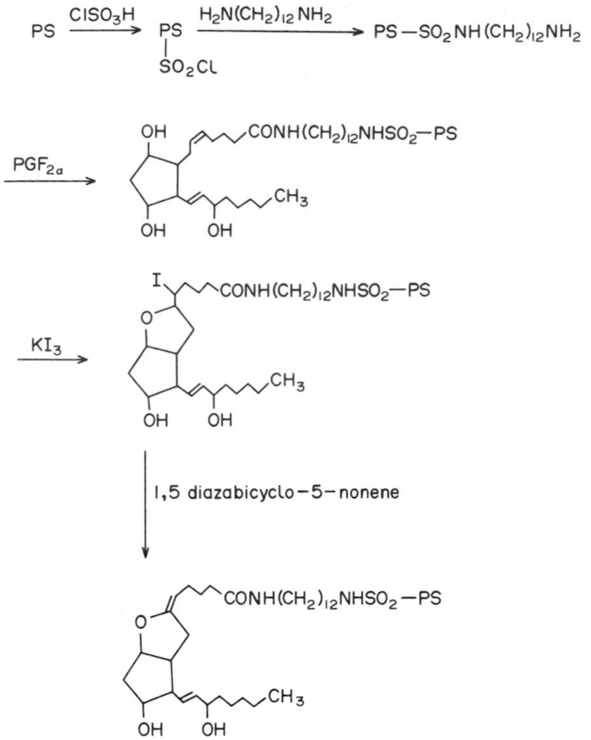

Scheme 7

The reaction was carried out on cross-linked polystyrene (PS) beads, which after swelling in dichloromethane were chlorosulfonated by addition of nitromethane and chlorosulfonic acid. The beads were then derivatized by reaction with diaminododecane, which served as a spacer, and subsequently coupled with $PGF_{2\alpha}$ in water–dioxan with the aid of 1-cyclohexyl-3-(2-morpholinoethyl) carbodiimide. Addition of KI_3 and $NaHCO_3$ resulted in the conversion of (coupled) $PGF_{2\alpha}$ to (coupled) 5-ξ-iodo-9-deoxy-6-ξ-9α-epoxyprostaglandin $F_{1\alpha}$, effectively through addition of iodine to the double bond followed by ring closure by elimination of HI. On addition of 1,5-diazabicyclo-5-nonene to the dried beads the $\Delta_{5,6}$ double bond was established by abstraction of hydrogen and iodine, with formation of immobilized PGI_2. Practical details for these procedures have been given by Ebert et al. (1982).

3.2 Heparin

Heparin is a linear polysaccharide with alternating uronic acid and glucosamine units; the predominating residues are L-iduronic and D-glucuronic acids with N-acetylated and N-sulfonated glucosamines. It is a strongly acidic polysaccharide carrying carboxyl, sulfonic acid and sulfonamide groups with rather low proportions of amino. Commercial heparin is polydisperse with a molecular-weight range of 6000–

30 000; its composition and activity vary according to the tissue of origin and mode of extraction. Heparin is a powerful naturally occurring anticoagulant, widely distributed in the tissues, particularly the lung and liver. The anticoagulant activity of heparin arises from its powerful potentiation of complex formation between antithrombin III (AT III) and the activated serine protease coagulation factors thrombin, XIIa, Xa, IXa, and VIIa, resulting in deactivation of these factors and suppression of blood coagulation. A type of mechanism that has been widely accepted assumes that heparin complexes with AT III forming a species which enters into complex formation with the coagulation factors much more rapidly than does AT III. Heparin is split off in this latter process, and so, according to this view, behaves in a purely catalytic fashion. An interesting account of the heparin–antithrombin system has been given by Rosenberg (1987).

Many attempts to synthesize thromboresistant surfaces have been based on the coupling of heparin to macromolecules. In view of the acidic nature of heparin, it is not surprising that early work made use of ionic binding. The products were effective in reducing blood coagulation, but their antithrombogenicity persisted for relatively short times since heparin was slowly eluted into the blood.

Heparin, unlike the bioactive species discussed so far, is a polyfunctional macromolecule so that the domains of the molecule responsible for interaction with AT III which extend over a number of carbohydrate units may be rendered inaccessible to AT III by multipoint covalent attachment of the carbohydrate to a synthetic polymer molecule (Larsson et al. 1987). In these circumstances anticoagulant activity would be diminished or lost, as indeed appeared to be the case from several early reports. The activity of immobilized heparin would therefore be expected to be strongly influenced by steric hindrance and the diffusivities of reacting species, so that the physical properties of the support become significant. These matters have been discussed in detail by Miura et al. (1980) who made a careful study of heparin covalently bound to Sepharose 4B, poly(2-hydroxyethyl methacrylate) and poly(vinyl alcohol) with the aid of the cyanogen bromide technique (Scheme 4). They paid special attention to the ability of the immobilized heparin to complex thrombin and factor Xa and to the properties of the resulting adducts. The prolongation of plasma recalcification time by immobilized heparin was found to be in the order of substrates:

$$\text{Sepharose } 4B > \text{poly(HEMA)} \gg \text{PVA}$$

A method for heparin immobilization that appears to be the most successful so far reported has been described by Larm et al. (1983). To avoid the problems outlined above these workers used one-point (terminal) coupling of heparin to the support, which was a synthetic material carying amino groups. An example of the latter was polyethylene, of which the surface had been partially oxidized with concentrated sulfuric acid and potassium permanganate to provide COOH and SO_2OH groups, which were subsequently used to bind a polybase (e.g., poly(ethylene imine)). Heparin was partially degraded with sodium nitrite at 0 °C and pH 7, yielding chains of molecular weight 5000–8000 with reactive aldehyde groups at the reducing terminal residues (2,5-anhydro-D-mannose). These were coupled to the derivatized (basic) support to give C=N links which were reduced by sodium cyanoborohydride. This process of reductive amination leads to CH—NH bonds which are not readily subject to enzymic attack. *In vivo* stability was reported to be very high, no significant loss in activity of treated polyethylene tubes being observed after implantation for four months in pigs' aorta. This finding obviously supports the view that heparin exerts a purely catalytic role and does not become passivated during its inhibition of blood coagulation.

Heparin immobilized on agarose is commercially available and finds uses as an affinity sorbent. Park et al. (1988) report the synthesis of an adduct in which heparin is linked to Biomer through chains of poly(ethylene glycol). The latter are more effective as spacers than hydrocarbon chains.

Bibliography

Bamford C H 1974 In: Jenkins A D, Ledwith A (eds.) 1974 *Reactivity, Mechanism and Structure in Polymer Chemistry*. Wiley, New York

Bamford C H 1988 *Encyclopedia of Polymer Science and Engineering*, 2nd edn, Vol. 13. Pergamon, Oxford, p. 767

Bamford C H, Middleton I P 1983 Studies on functionalizing and grating to poly(ether-urethanes). *Eur. Polym. J.* 19: 1027–35

Bamford C H, Middleton I P 1987 Rubbers as biomaterials. *Plast. Rubber Process. Appl.* 7: 137

Bamford C H, Middleton I P, Al-Lamee K G 1986 Studies in the esterification of dextran: Routes to bioactive polymers and graft copolymers. *Polymer* 27: 1981–5

Bamford C H, Middleton I P, Al-Lamee K G 1987a Polymeric inhibitors of platelet aggression. II: Copolymers of dipyridamole and related drugs with N-vinylpyrrolidone. *Biochim. Biophys. Acta* 924: 38–44

Bamford C H, Middleton I P, Al-Lamee K G, Paprotny J 1987b Haloisocyanates as transformation reagents. *Br. Polym. J.* 19: 269–74

Bamford C H, Middleton I P, Al-Lamee K G, Paprotny J, Satake Y 1987c Routes to bioactive hydrophilic polymers. *Polym. J.* 19: 475–83

Bamford C H, Middleton I P, Satake Y, Al-Lamee K G 1985 In: Culbertson B M, McGrath J E (eds.) 1985 *Advances in Polymer Synthesis*. Plenum, New York, p. 291

Donamura L G, Vogl O (eds.) 1978 *Polymeric Drugs*. Academic Press, New York

Duncan R, Kopecek J 1984 Soluble synthetic polymers as potential drug carriers. *Adv. Polym. Sci.* 57: 51–101

Duncan R, Kopecek J, Lloyd J B 1983 Development of N-(2-hydroxypropyl)methyacrylamide copolymers as therapeutic agents. In: Chiellini E, Giusti P (eds.) 1983 *Polymers in Medicine, Biomedical and Pharmacological Applications*. Plenum, New York, p. 97

Ebert C D, Lee E S, Kim S W 1982 The anti-platelet activity of immobilized prostacyclin. *J. Biomed. Mater. Res.* 16: 624

Epton R, Wellings D A, Williams A 1987 Perspectives in ultra-high load solid (gel) phase peptide synthesis. *Reactive Polym.* 6: 143–57

Ferruti P, Tanzi M C 1979 Synthesis and exchange reactions of some polymeric benzotriazolides. *J. Polym. Sci., Polym. Chem. Ed.* 17: 277

Harris F W 1984 Controlled release from polymers containing bioactive substituents. In: Langer R S, Wise D L (eds.) 1984 *Medical Applications of Controlled Release*, Vol. 1. CRC Press, Boca Raton, FL

Hebeish A, Guthrie J T 1981 *The Chemistry and Technology of Cellulose Chemistry*. Springer, Berlin

Khorana H G 1953 The chemistry of carbon diimides. *Chem. Rev.* 53: 145

Kohn J, Wilchek M 1983 1-Cyano-4-dimethylamino pyridimium tetrafluoroborate as a cyanylating agent for the covalent attachment of ligand to polysaccharide resins. *FEBS Lett.* 154: 209–10

Kurzer F E, Douraghi-Zadeh K 1967 Advances in the chemistry of carbon diimides. *Chem. Rev.* 67: 107

Labarre D, Boffa M C, Jozefowicz M 1974 Transformations of functional groups on polymers. *J. Polym. Sci., Polym. Symp.* 47: 131

Larm O, Larsson R, Olsson P 1983 A new nonthrombogenic surface prepared by selective covalent binding of heparin via a modified reducing terminal residue. *Biomater. Med. Dev. Artif. Organs* 11: 161

Larsson R, Larm O, Olsson P 1987 The search for thromboresistance using immobilized heparin. *Ann. N. Y. Acad. Sci.* 516: 102

Margel S, Zisblatt S, Rembaum A 1979 Polyglutaraldehyde: A new reagent for coupling proteins to microspheres and for labeling cell-surface receptors. II: Simplified labeling method by means of non-magnetic microspheres. *J. Immunol. Methods* 28: 341

Mino G, Kaizerman S 1958 A new method for the preparation of graft copolymers. Polymerization initiated by ceric ion redox systems. *J. Polym. Sci.* 31: 242

Miura Y, Aoyagi S, Kusada Y, Miyamoto K 1980 The characteristic of anti-coagulation by covalently immobilized heparin. *J. Biomed. Mater. Res.* 14: 619

Nilson K, Mosbach K 1981 Immobilization of enzymes and affinity ligands to various hydroxyl group-carrying supports using highly reactive sulfonyl chlorides. *Biochem. Biophys. Res. Commun.* 102: 449

Osterman L A (ed.) 1986 *Methods of Protein and Nucleic Acid Research: Chromatography*, Vol. 3. Springer, New York

Park K D, Okano T, Nojiri C, Kim S W 1988 Heparin immobilization onto segmented polyurethaneurea surfaces—Effect of hydrophilic spacers. *J. Biomed. Mater. Res.* 22: 977–92

Porath J, Axen R 1976 Immobilisation of enzymes to agar, agarose and Sephadex supports. *Methods Enzymol.* 44: 19

Rosenberg R D 1987 The heparin–antithrombin system: A natural anticoagulant mechanism. In: Colman R W, Hirsh J, Marder V J, Salzman E W (eds.) 1987 *Hemostasis and Thrombosis*, 2nd edn. Lippincott, New York

Staab H A 1962 Synthesen mit heterocyclischen Amiden (Azoliden). *Angew. Chem.* 74: 407

Tomoskozi I, Galombos G, Simonidez V, Kovacs G 1977 A simple synthesis of PGI$_2$. *Tetrahedron Lett.* 30: 2627

Whittaker N 1977 A synthesis of prostacyclin sodium salt. *Tetrahedron Lett.* 32: 2805

Ziegler F E, Berger G D 1979 A mild method for the esterification of fatty acids. *Synth. Commun.* 9: 539

C. H. Bamford and K. G. Al-Lamee
[University of Liverpool, Liverpool, UK]

E

Elastomers for Dental Use

The term elastomer embraces those polymers which have the rubberlike long-range extensibility at low forces. They are polymers where intermolecular forces are so low that the energy for extension is used preponderantly to uncoil molecules; in thermodynamic language the energy increase is mostly expended in entropy changes.

Natural rubber was the precursor to the wide and chemically diverse range of synthetic elastomers now available. Until the early 1950s synthetic elastomers were, like natural rubber, thermoplastic solids in the uncross-linked (unvulcanized) state. When thermally softened, such materials are still extremely viscous ($\sim 10^{11}$ Pa s), requiring heavy machinery to mix in compounding ingredients and to shape by extrusion, calendering and molding. Then elevated temperatures (>100 °C) are necessary to effect vulcanization. Indeed, this is still true of the bulk of the rubber industry. However, from about 1950, polymers became available that were mobile, pourable fluids in the uncross-linked state, and which could be cross-linked at room temperature in relatively short times. Major examples are room temperature vulcanizing (RTV) silicone rubbers and fluid polysulfides (Jorczak and Fettes 1950, Nitzsche and Wick 1957). These systems are widely used in the sealants industry, but have also come to be the basis of the major group of dental impression materials, as compliant linings for dentures and as maxillofacial prosthesis materials.

It has already been stated that elastomers owe their characteristics to low intermolecular forces. However, otherwise glassy rigid polymers can be rendered elastomeric by the addition of a plasticizer (a mobile liquid which reduces intermolecular forces to convert the polymer to an elastomer). Such materials are widely used in dentistry as compliant prostheses.

1. Dental Impression Materials

A dentist will need to take an impression of the oral structures in three general circumstances:

(a) an edentulous mouth, for the subsequent fitting of full dentures;

(b) a partially edentulous mouth, including standing teeth, for a partial denture; and

(c) prepared standing teeth in a dentate patient for supplying a crown, a bridge or an inlay.

Impressions of children's teeth are also taken in the course of orthodontic treatment.

From impressions, models are then cast, usually from plaster of Paris. When impressions of a number of teeth and the surrounding oral structures are involved, it is clear that there are very intricate undercuts engaged by the impression material. Hence a fluid RTV elastomer is useful in that it will both flow into the required structures and will have the necessary compliance and elasticity to facilitate withdrawal and faithfully reproduce the structures involved.

These materials are supplied to the dentist as a two-pack system, usually in two tubes (Fig. 1a); one tube contains the polymer, incorporating fillers and other additives, and the other contains the cross-linking (vulcanizing) agent. Equal lengths of each are squeezed out onto a mixing pad and mixed with a spatula. (The example shown is a polysulfide; a brown paste is the cross-linking system which is lead dioxide in a liquid vehicle.) The mixed material is loaded onto an impression tray (Fig. 1a), inserted in the patient's mouth (Fig. 1b), withdrawn after setting (Fig. 1c) and a plaster mold cast.

Before describing such materials in detail, it is instructive to consider briefly the alternatives. Historically, a form of plaster of Paris was used; when set, this had to be broken in the patient's mouth, and subsequently reassembled jigsaw fashion! The first quasielastomeric material used in dentistry, and still in use, is a thermally reversible Agar–Agar gel (Phillips 1982). This material is an elastic gel at either room or mouth temperature; when heated to about 50 °C it is fluid, and in this state is placed in the mouth in a cored impression tray. When in position, cold water is run through the tray, and the material gels so it can be withdrawn. Such materials are weak, and shrink rapidly due to water evaporation. Hence immediate casting-up of the plaster mold is necessary. A much more common hydrolloid is based on sodium or a related salt of alginic acid (Phillips 1982). This is cross-linked (set) by a divalent salt in the formulation. It is widely used and is suitable for nonprecision work. However, for partial denture and more critical crown and bridge work, elastomers have the necessary strength to avoid tearing on removal from intricate undercuts, the elasticity to recover well and the necessary dimensional stability not to change significantly over several days. A multicomponent bridge is a very expensive item of treatment, and it is clearly necessary that it must fit. This in turn places stringent demands on the impression material that is used in its fabrication.

Elastomeric impression materials fall into three main classes: silicone, polysulfide, and an imine-

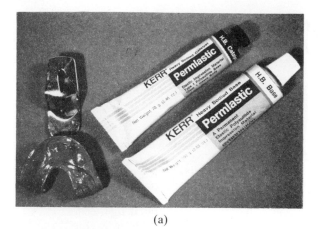

(a)

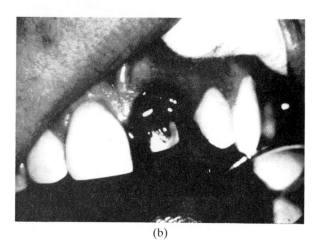

(b)

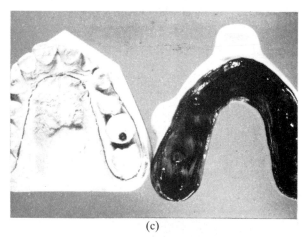

(c)

Figure 1
Polysulfide impression material showing: (a) tubes
of the impression material, and the mixed material
loaded in the impression tray; (b) the impression tray
in the mouth; and (c) a set impression and the mold
cast from it

terminated polyether. The first two are industrially
used polymers adapted for dentistry; the third was
developed especially for dentistry.

1.1 Silicone Impression Materials

There are a number of RTV silicone rubbers (Watt
1970), two of which are used in dentistry; namely the
so-called "condensation" and "addition" types.

(*a*) *Condensation silicones*. These are α, ω-hydroxy
terminated poly(dimethyl siloxanes) which are
cross-linked by an alkoxy ortho-silicate or related
reagent, in conjunction with an organotin activator
(Fig. 2). Stannous octoate and dibutyl tin dilaurates
are commonly used (Braden and Elliott 1966). The
alcohol eliminated will slowly evaporate, giving rise
to shrinkage (Fig. 3).

This type of silicone is commonly manufactured in
what is termed the "putty" and "wash" system. The
putty is a very highly filled system, which is used to
make a first impression. The lightly filled very fluid
system is applied to the surface of the first impression, and reinserted in the mouth. This fills in the
very fine detail (Fig. 4).

One disadvantage of this type of material is a
limited shelf life. If, as is common, the cross-linking
paste contains both the alkoxy ortho-silicate and
organotin compound, they react together very
slowly, particularly in the presence of tin. The net
result is that the impression material will not set.

(*b*) *Addition silicones*. These are vinyl terminated
poly(dimethyl siloxanes) and are historically the
latest of the impression materials, introduced at the
start of the 1980s. They are cross-linked with hydro-
silanes as in Eqn. (1).

$$-\left(\begin{array}{c} CH_3 \\ | \\ Si-O \\ | \\ CH_3 \end{array}\right)_n SiCH=CH_2 \quad \begin{array}{c} | \\ H-Si-CH_3 \\ | \\ O \\ | \\ H-Si-CH_3 \\ | \end{array} \; \rightarrow \; -\left(\begin{array}{c} CH_3 \\ | \\ Si-O \\ | \\ CH_3 \end{array}\right)_n \begin{array}{c} | \\ SiCH_2-CH_2-Si \\ | \\ O \\ | \\ \sim\!Si-CH_2-CH_2-Si \\ | \end{array} \quad (1)$$

$$\sim \; CH=CH_2 \qquad \qquad \text{Pt-salt catalyst}$$

This type of material, having no by-products eliminated on cross-linking, is dimensionally extremely
stable. Shrinkage of the order of only approximately
0.05% over several days is usually found. A number
of features seem to have limited their widespread
use.

(i) They are expensive.

(ii) If a plaster model is cast immediately with the
set impression, residual Si—H groups react with
water, with the evolution of hydrogen and the
generation of a porous surface to the model. (At
the time of writing this has apparently been
overcome by incorporating finely divided palladium in the formulation.)

Figure 2
Formation of condensation silicones

(iii) If the material is handled with rubber gloves, the sulfur in the rubber "poisons" the platinum salt catalyst, and the material does not set. (With the proliferation of the AIDS virus, the use of rubber gloves has become standard clinical practice. In this context it should be noted that sterilization of impressions is also standard practice, usually in glutaraldehyde solution.)

(c) General properties of silicone impression materials. All set silicone elastomers, with the exception of the highly filled "putty" materials, are extremely elastic, giving rapid and almost complete recovery from deformation. Residual strains of less than 1% are typical.

One disadvantage of silicone elastomers as impression materials is their hydrophobicity; this means that the unset impression wets the oral mucosa with difficulty, particularly if there is any

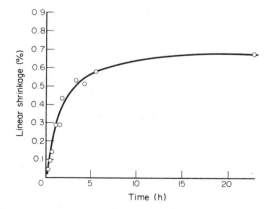

Figure 3
Plot of the linear shrinkage of silicone impression rubber as a function of time

Figure 4
Putty and wash silicone impression: the putty (first impression) is the lighter and the wash (second impression) is the darker

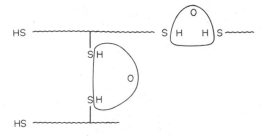

Figure 5
Cross-linking accompanied by chain-lengthening to produce polysulfides

accumulated blood or saliva. It is therefore necessary for the dentist to dry the appropriate area before taking the impression. However, silicone impression materials are now available that have a degree of hydrophobicity, conferred by the incorporation of a detergent in their formulation.

1.2 Polysulfides
These are invariably based on Thiokol LP-2, a fluid polysulfide (Braden 1966) having the general structure shown in Eqn. (2).

$$HS—(CH_2—CH_2—O—CH_2—CH_2—S—S)_n—CH_2$$
$$\mid \quad (2)$$
$$HS—CH_2—CH_2—O—CH_2$$

This also contains 2 mol% of pendant —SH groups as sites for cross-linking. This polymer, a yellow viscous liquid, together with a filler, plasticizer and a small quantity of sulfur (~0.5%) constitutes one of the pastes. The cross-linking paste is usually lead dioxide (PbO$_2$) in a liquid vehicle, often with a small quantity of oleic acid to control the reactivity of the lead dioxide.

Cross-linking accompanied by chain-lengthening is an oxidation process (Fig. 5). This impression material is extremely strong when set and has good dimensional stability that is only bettered by the addition silicones. Its disadvantages are that it has an unpleasant odor (characteristic of mercaptans), is messy to handle and is slow setting. Unaccountably, this class of material can be capricious in its setting behavior; occasionally a batch that takes a very long time to set will be obtained from the manufacturer. For these reasons, what is fundamentally an excellent impression material has lost popularity in favor of other materials.

Polysulfides are usually sold in three consistency grades. The least viscous grade ("light bodied") is mixed, placed in a syringe and injected round the oral structures of interest; this is then backed up by the most viscous grade ("heavy bodied") in an impression tray.

An obvious question concerns the use of lead dioxide. However, this is a very insoluble lead compound, and there is no evidence to suggest that either patient or dentist is at risk. Nevertheless, the lead dioxide results in an impression which is brown in color, the aesthetics of which combine unfavorably with the odor of the material. Hence various alternatives have been tried. The obvious alternatives are other oxidizing agents, and organic hydroperoxides have been used. Unfortunately, these materials are somewhat volatile, resulting in poor dimensional stability (Braden 1976).

One interesting proprietary material uses cupric hydroxide as a setting agent. The chemical mechanism is not clear.

A further alternative is to use an organic sulfide, which works by an interchange reaction:

$$\text{———— SH} \qquad\qquad \text{HS ————}$$
$$X—S—S—X'$$
$$\downarrow \qquad\qquad\qquad (3)$$
$$\text{———— S—S ————}$$
$$+ X—SH$$
$$+ X'—SH$$

The reverse reaction is quenched by the incorporation of zinc oxide or zinc carbonate.

1.3 Imine-Terminated Polyether
The silicone and polysulfide fluid RTV polymers so far described are industrial systems, adapted for use in dentistry. However, the material now to be described was developed as a dental impression material, and there is probably no industrial equivalent. Because the chemistry is ingenious, it is instructive to review the synthesis of the polymer used (Braden et al. 1972).

The starting material is a polyether comprising a random copolymer of ethylene oxide and tetrahydrofuran:

$$—\{[(CH_2)_2—O—]_n—[(CH_2)_4—O—]_m\}_p \qquad (4)$$

A copolymer is used to avoid the crystallinity which would result from a homopolymer.

This polymer has terminal hydroxy groups, which are esterified with an unsaturated acid, and an aziridine (ethylene imine) ring added to the double bond as in Fig. 6. Hence, the polymer paste of the impression material contains this imine-terminated polyether, together with a filler and plasticizer. The cross-linking paste contains a material capable of donating carbonium ions, and cross-linking is at chain ends via a ring-operating polymerization (Fig. 7). This impression material has the following advantages:

(a) it is rapid setting;

(b) it is clean and easy to handle;

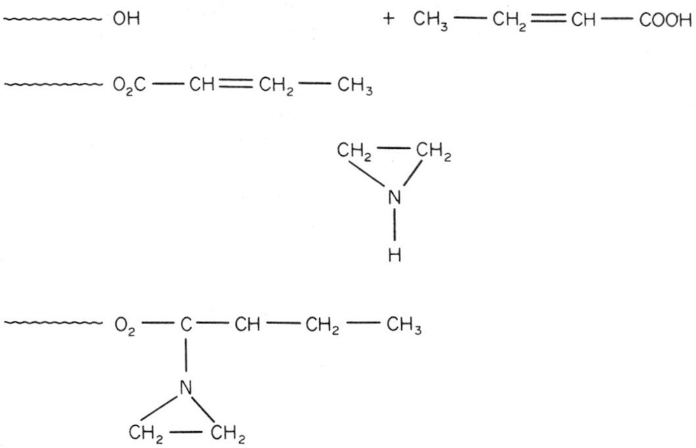

Figure 6
Esterification with an unsaturated acid and addition of aziridine to the double bond of the copolymer of ethylene oxide and tetrahydrofuran

(c) it is hydrophilic, hence readily wets the oral tissues; and

(d) it has excellent dimensional stability.

Indeed, the earlier versions of this material were too hydrophilic, resulting in unwanted dimensional changes if left wet for too long. The disadvantages are that it is very hard when set, necessitating an extra bulk of material in the impression tray to facilitate removal. More serious is the propensity of the cross-linking agent to produce hypersensitivity, notably on the hands of the operator (Nally and Storrs 1973). In the late 1980s a new version of this material appeared, but there is as yet no information on this important aspect.

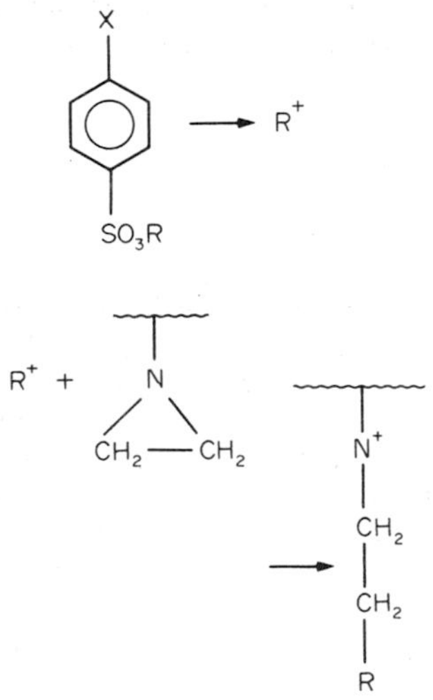

1.4 The Future

As stated in Sect. 1, the successful production of highly complex and expensive restorations rests on the accuracy and ease of use. Therefore, the question arises as to whether or not there are any other fluid polymer systems that can combine the advantages of the materials described and eliminate their disadvantages. There is one experimental system that shows some promise in this direction. Maleinized polybutadiene is a fluid polymer that can be cross-linked very rapidly by diols (Fig. 8). This is an addition reaction, in a hydrophilic polymer, and the set material is elastic and compliant.

2. Soft Prosthesis Materials

Some patients cannot tolerate a hard acrylic denture, and for this and a number of other clinical reasons, a denture is prescribed a "soft liner." This means that the palatal surface of the denture is to be lined with a compliant and/or elastomeric material. The requirements for a soft lining material are that it:

(a) is nontoxic and a nonirritant;

(b) has low water absorption;

(c) must not support the growth of microorganisms (e.g., *Candida Albicans*);

(d) must bond to the poly(methyl methacrylate) denture base; and

(e) must be amenable to processing by available dental technology techniques.

Figure 7
Cross-linking to produce imine-terminated polyether

This extremely demanding list disqualifies many, if not most, of the main classes of industrial elastomer. In fact, the materials currently used are predominantly RTV silicones and plasticized acrylic systems.

2.1 RTV Silicones

These are usually condensation silicones, as described in Sect. 1.1*a*, and are far from ideal. Not surprisingly, it is difficult to form a lasting bond to a poly(methyl methacrylate) denture. An adhesive, usually an α,ω-hydroxy terminated poly(dimethyl siloxane) in a solvent, is necessary. Silicone polymers seem particularly prone to the formation of *Candida Albicans* (Wright 1980), and surprisingly can absorb large quantities of water. The subject of water uptake is discussed in Sect. 2.4. The one advantage (cf. plasticized acrylics) is that they retain their compliance.

2.2 Plasticized Acrylics

Poly(methyl methacrylate) dentures are fabricated from a dough comprising a mixture of methyl methacrylate monomer and poly(methyl methacrylate) powder (see *Acrylic Dental Polymers*). Plasticized acrylics follow the same method of fabrication, and Table 1 gives a typical formulation. It will be noted that the polymer powder is poly(ethyl methacrylate) which has a lower glass transition temperature than poly(methyl methacrylate) and so needs less plasticizer to render it compliant. The 2-ethoxyethyl methacrylate, when polymerized, is compliant; the ethylene glycol dimethacrylate is a cross-linking agent; and the phthalate is a plasticizer. The dough

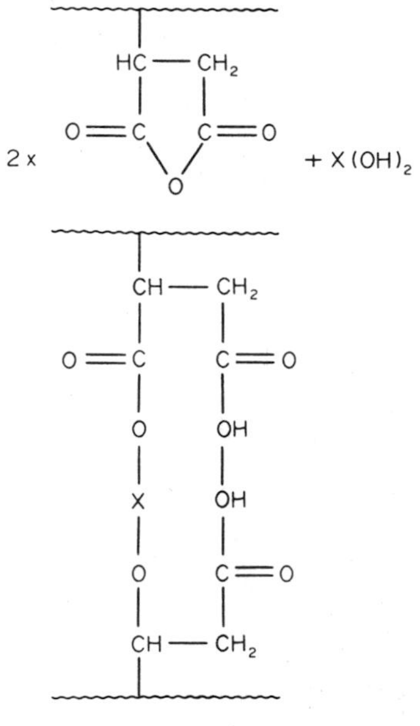

Table 1
Typical formulation of a plasticized acrylic

Polymer powder	Monomer liquid	Percentage composition
Poly(ethyl methacrylate)	2-ethoxyethyl methacrylate	70
	Ethylene glycol dimethacrylate	5
	Butyl phthalyl butyl glycolate	25

made from this is polymerized in contact with the poly(methyl methacrylate) dough, thus ensuring a good bond to the denture base. However, the good bond to the denture is probably the only advantage of this type of soft liner. The final material, while having an elastomeric compliance, is usually very sluggish in elastic recovery. More seriously, plasticizer leaching out in oral fluids results in hardening of the material in a few months.

Parker and Braden (1982) have developed materials where the plasticizer does not leach out, but these have other problems referred to in Sect. 2.4.

2.3 A Silicone–Acrylic Copolymer

It is obvious from Sects. 2.1, 2.2 that neither type of material is really suitable for its intended purpose. However, one material offers a reasonable compromise between the above two classes of material.

This material is an adduct of an α,ω-hydroxy terminated poly(dimethyl siloxane) and γ-methacryloxy propyl trimethoxy silane (Eqn. (5)).

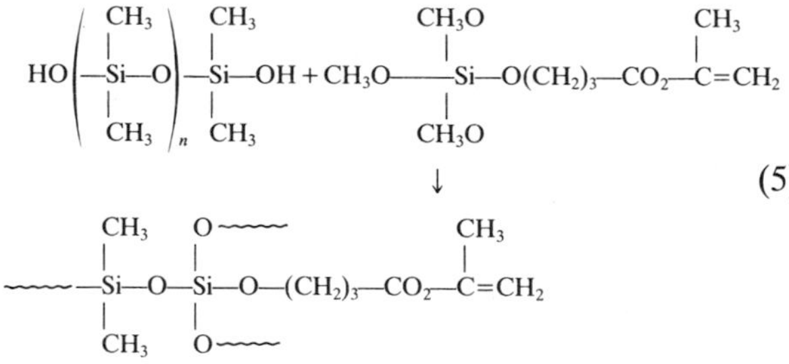

$$(5)$$

This is obviously a highly branched silicone polymer, but containing methacrylate groups to achieve bonding to poly(methyl methacrylate). It is cross-linked with a peroxide at 100 °C, usually being processed in contact with the acrylic dough when the denture is made.

2.4 Long-Term Water Absorption of Elastomeric Polymers

When Parker and Braden (1982) developed soft acrylics without extractable plasticizer, the resulting clinical materials failed mechanically after some

Figure 8
Cross-linking of maleinized polybutadiene by diols

time. This proved to be because, contrary to expectations, immersion in water resulted in very high prolonged uptake, without equilibrium after many years. This transpired to be a feature of elastomers in general, and in fact Muniandy and Thomas (1984) showed that this was due to the presence in most polymers of water soluble impurities. When water diffuses into an elastomer the water droplets form impurity sites. These grow in size until osmotic and elastic forces balance or, in weak polymers, until fracture occurs.

This almost unavoidable phenomenon makes the development of a long-term soft-lining material extremely difficult.

2.5 Other Materials

A promising group of materials has been described (Gettleman et al. 1987) comprising a polyphosphazine polymer, compounded with a methacrylate monomer, and heat polymerized in contact with the denture. It remains to be seen how such materials perform, particularly with respect to long-term water absorption.

2.6 Maxillofacial Prosthesis Materials

These materials are for the replacement of missing facial tissue associated with a serious accident or the surgical removal of a tumor. Clearly, apart from the physicochemical and biological requirements, the material needs to simulate facial tissue visually to minimize the obviously distressing situation involved. Most of the materials already discussed are used, but much depends on the skill of the surgeon and maxillofacial technician.

See also: Maxillofacial Prostheses

Bibliography

Braden M 1966 Characterisation of the setting process in dental polysulphide rubbers. *J. Dent. Res.* 45: 1065–71
Braden M 1976 The quest for a new impression rubber. *J. Dent.* 4: 1–4
Braden M, Causton B E, Clarke R L 1972 A polyether impression rubber. *J. Dent. Res.* 51: 889–96
Braden M, Elliott J C 1966 Characterisation of the setting process in dental silicone rubbers. *J. Dent. Res.* 45: 1016–23
Gettleman L, Vargo J M, Gebert P H, Farris C L, LeBoeuf R J Jr, Rawls H R 1987 PNF as a permanent soft liner for removable dentures. In: Gebelein C G (ed.) 1987 *Advances in Biomedical Polymers*. Plenum, New York, pp. 55–61
Jorczak J S, Fettes E M 1951 Polysulphide polymers. *Ind. Eng. Chem.* 43: 324–8
Muniandy K, Thomas A G 1984 Water absorption in rubbers: Polymers in a marine environment. *Proc. Conf. Institute of Physics and Institute of Marine Engineers*. Institute of Physics, London.
Nally F N, Storrs J 1973 Hypersensitivity to a dental impression material. *Br. Dent. J.* 134: 244–6
Nitzsche S, Wick M 1957 Vulcanization system for silicone rubber (in German). *Kunstoffe* 47: 431–4
Parker S, Braden M 1982 New soft lining materials. *J. Dent.* 10: 149–53
Phillips R W 1982 *Science of Dental Materials*, 7th edn. Saunders, Philadelphia, PA
Watt J A C 1970 RTV silicone rubbers. *Chem. Br.* 6: 519–24
Wright P S 1980 The effect of soft lining materials on the growth of *Candida Albicans*. *J. Dent.* 8: 144–51

M. Braden
[University of London, London, UK]

Endodontic Materials

The term "endodontic" literally means "within the tooth" but is used customarily to refer to therapy of the tooth when the dental pulp is involved directly. In the most conservative of these treatments, the pulp is treated to keep it alive and the interface between the therapeutic material and the pulp may be only micrometers across. In the most radical, not only the pulp but a substantial portion of the tooth root itself may be replaced. The material involved in the former procedure falls very much into the category of pharmacological materials, while those used in the latter are distinctly biomaterials. For materials used in all other endodontic procedures, it may be a point of argument whether the nature of the material is pharmacological or whether it falls into the category of biomaterials science. For completeness, the whole range of endodontic materials will be described in this article. First, however, it is appropriate to look at the structure of the system being treated: a diagram of the pulpal structure of a typical tooth is shown in Fig. 1.

1. Structural Elements of the Tooth

1.1 Enamel

Dental enamel is composed of large crystals of hydroxyapatite. It is permeated by micropores and there is a movement of fluid through it from the underlying dentin. Although relatively brittle, it is supported by a firm bond to the underlying dentin at the amelo–dentinal junction. The integrity of the surface enamel is related, therefore, to the mechanical properties of the underlying dentin which may change for a number of reasons, described later. Cracks are frequently observed at the enamel surface but these rarely propagate into the dentin, although when this occurs it can cause extreme pain. The enamel forms a relatively impermeable barrier which protects the underlying dentin.

1.2 Dentin

Dentin consists of microtubules of mineralized tissue within an organic matrix. In the newly erupted tooth, these tubules are occupied by the long cellular

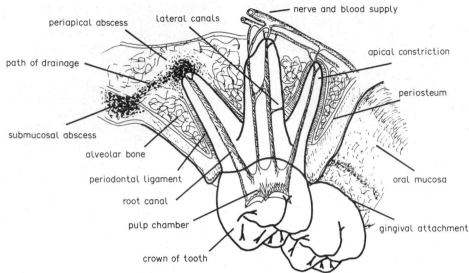

Figure 1
Pulpal structure of a typical tooth

processes of the odontoblasts, the cells which form the dentin. Although there are very few nerve endings within the dentin, the whole structure right up to the enamel is sensitive to pain. The fluid content of the dentin is maintained to a degree by the odontoblastic processes, and this will be lost in favor of simple diffusion when the pulp is removed. This may severely affect the mechanical properties of the dentin. Young dentin is relatively elastic but this diminishes as the tooth ages or when the pulp is lost. As the tooth ages, further deposition of mineral within the tubules may occur and occlude them to a greater or lesser degree. These combined effects will alter the physical properties of the dentin and in consequence the prognosis for the tooth after it has been treated endodontically.

Permeability. Because of its microtubular nature, dentin is extremely permeable. Furthermore, the large surface area within the tubules is relatively active chemically. Damage can occur to the pulp via the permeable dentin. When the pulp is removed and replaced with a chemical substance to obturate the space, irritant substances may penetrate the dentin in a reverse direction and affect adversely the cells of the periodontal ligament which is attached to the cementum surface of the tooth.

The permeability of the dentin can be affected in a number of ways. Most significantly, the permeability can be reduced greatly when a smear layer is produced by the dentist's instruments at the cut surface of the dentin. Chemicals used later may remove this smear layer and again increase the permeability. The presence of a smear layer may partly protect

the dentin and associated cellular elements from the effects of applied chemicals such as sealers or adhesives.

1.3 Cementum

Cementum is a thin coat of mineralized tissue that covers the outer surface of the root dentin. The gingival tissue is attached to it by hemidesmosomal junctions and by the penetration of fibers from the periodontal ligament. Cementum is produced continuously but slowly throughout life and this can be an advantage. After root treatment, it is possible for cementum to be laid down over the apertures into the pulp space provided these have been sealed with a biocompatible material.

When the pulp space is filled with obturating materials, toxic components can diffuse through the dentin and into the cementum, and the periodontal attachment of the tooth can be damaged in consequence.

1.4 The Pulp Space

The morphology of the pulp space may be highly variable from patient to patient, but there are certain common patterns for each tooth of the dental arch. Determination of the actual architecture of any particular pulp space is one of the greatest problems of pulp treatment.

2. Pulpal Damage

The primary reason for the pulp to become damaged is as part of the disease process of dental caries, although it may be damaged directly as a result of trauma or via a range of other mechanisms including direct infection via communicating links through the dentin. These links include lateral canals and furcation communications. These may become exposed to the oral environment by the loss of attachment at the neck of the tooth due to periodontal disease. Where this attachment is lost, pockets form and fill with oral bacteria or dental plaque. These bacteria may penetrate into the pulp directly through these channels and lead to pulp death.

The tooth has a number of defence reactions to the carious process. To what extent each of these may be activated depends on the speed and severity of the carious attack. These defences include the laying down of a reactive calcific barrier along the wall of the pulp chamber adjacent to the carious lesion. Such a barrier can make access to the pulp for endodontic treatment very difficult.

3. Endodontic Treatment

From the most conservative to the most radical, the range of endodontic procedures performed are: direct pulp-capping, pulpotomy, pulpectomy, apicectomy, hemisection, radicectomy and diadontic implant. The most commonly performed is the pulpectomy, with the others combined making up only a few percent of the total.

3.1 Pulp-Capping

The objective of pulp-capping is to maintain the viability of a dental pulp which has become exposed by loss or removal of dentin during an operative intervention or due to traumatic breakage of the crown of a tooth in an accident. The key to success lies with the status of the pulp prior to the exposure and the level of infection with microorganisms following the exposure. Existing inflammatory change and infection with pathogenic organisms compromise the prognosis severely. After cleansing, the cut surface is coated with a layer of calcium hydroxide in a carrier resin, and then the tooth is restored over it. There is an initial inflammatory response within the pulp and the cells in immediate contact with the calcium hydroxide become necrotic. Healing then occurs and a fibrous repair follows a period of localized inflammation. Finally, the pulp separates itself from the calcium hydroxide by laying down a stratum of amorphous calcification which is known as a "calcific barrier." This is in direct contact with the dentin surrounding the break and will usually seal the pulp off completely from calcium hydroxide.

Alternative materials have been used as pulp-capping agents. These include cyanoacrylates, glass polyalkanoates and others. Calcium hydroxide remains the first choice material.

3.2 Pulpotomy

The pulpotomy procedure is carried out when the coronal pulp is beyond rescue but the pulp in the root may be kept. This is usually in the case of damage to an immature tooth where the root apex has not fully formed. An open apex prevents a build-up of intracoronal pressure due to inflammation of the pulpal tissue, and pulp death due to ischemic necrosis is less likely.

Access is cut into the pulp chamber and the coronal pulp is cut out using a sharp rotary bur, a dental excavator or a diathermy instrument. The severed end of the radicular pulp is treated in the manner of a direct pulp cap; after cleansing, it is sealed with a layer of calcium hydroxide and the tooth is restored. Success of the procedure is judged primarily by radiographic evidence of the continuation of apex formation. A calcific barrier should also form between the calcium hydroxide layer and the dental pulp in the manner of healing after a direct pulp cap.

3.3 Pulpectomy

When the pulp is damaged more severely or is already necrotic, it is necessary to remove it in its entirety. This leaves a space which has to be filled with a suitable obturating material to prevent microorganisms from propagating in the avascular void of the root canal system and leading to chronic sepsis around the apical opening of the root. The obturating material should be biocompatible, because it may come into contact with the tissues around the tooth root. This contact may be direct where the material closes off apertures into the tooth at the apical foramen, lateral canals and furcation communications. Indirect contact occurs due to the tubular nature of dentin. Dentin permeability allows chemical constituents of the obturating material to diffuse through the tooth root and affect the supporting periodontium and its attachment to the root.

A further requirement of the obturating material is that it should form a total seal with the walls of the pulp canal against the ingress of microorganisms. The formation of this seal requires the material to take on the shape of the prepared canal space. This shape may be both complex and compound, so the material must deform easily in order to adapt to it. It is preferable that the obturating material is radiopaque so that its presence and position can be confirmed radiographically after, or sometimes during, placement.

The conventional method of sealing a tooth after removal of the pulp is to shape the canal to remove as much of the complexity of shape as possible and open and widen it to make insertion of the sealing material easier. This process also cleans the canal if it is infected and reduces considerably the microorganism count. The process of filing the walls of the canal system and the removal of debris by irrigation is known as the "biomechanical preparation" of the root canal. Irrigation of the canal during preparation is an essential part of the process and serves both to lubricate the cutting edge of the files in the canal as well as cleaning out dentin fillings. Sterile water will achieve these objectives, although sodium hypochlorite solution is used by many practitioners because of its ability to dissolve organic debris and its bleaching action on the dentin.

The canal system is then sealed temporarily to allow the resolution of apical inflammation prior to the placement of the definitive canal seal. During this period, which should be about three to seven days, the growth of microorganisms in the canal must be suppressed. This is achieved by incorporating an antiseptic into the dressing. A pledget of cotton wool is lightly moistened with a solution of *p*-monochlorophenol and placed into the canal opening before sealing temporarily with a zinc oxide–eugenol cement. Antibiotic pastes have been used but have little advantage over the method described and may in fact be irritant to the apical tissues. A further problem with such pastes is that they have to be completely removed from the canal structure prior to the placement of the definitive root filling, and this may be very difficult to achieve in practice due to the insoluble nature of many of the paste vehicles.

The most satisfactory method of sealing the canal system definitively is known as "lateral condensation." In this technique, a deformable plastic point is coated with a semifluid sealer and inserted to the full length of the canal. With the forcible insertion of pointed instruments called "spreaders," this point is squashed laterally against the sides of the canal. A further point is inserted into the space created and the process repeated until the canal is filled. The repeated squashing of the plastic points forces the sealer between the points and the canal walls, and between the points themselves, thus creating a tight-sealing laminate within the canal. This system of lateral condensation is the standard against which all other systems of canal obturation are compared. The most successful material used for constructing the plastic point is a composite of gutta percha and zinc oxide powder. Gutta percha is a natural product derived from the coagulated exudate of mazer trees, which are native to Malaysia. It is an isomer of natural rubber and has the same basic unit, isoprene. Other additives include plasticizing waxes, radiopacifiers such as barium sulfate and colorants. Gutta percha is remarkably low in cytotoxicity, even in tissue culture studies, and this contributes to its universal acceptability in this role. However, it should be remembered that a sealer paste will usually intervene between the surface of the gutta percha and the tissues and it is this that will determine to a greater extent the biocompatibility of the obturating system.

In the past, solvents such as chloroform have been used during the placement of gutta percha root fillings to help adapt the material to the shape of the canal. Several methods were used, ranging from injecting directly a paste of gutta percha dissolved in chloroform, coating a gutta percha point with such a paste as a sealer, or just dipping the gutta percha into the solvent before placing it into the canal. However, as the solvent evaporates, shrinkage occurs and the vital seal between the gutta percha point and the dentin of the root canal walls is lost. There is also a safety aspect in that chloroform is no longer considered acceptable as a material for direct clinical use. A recent innovation has been to melt the gutta percha in a specially designed heat gun and then inject it into the canal in the molten state. While this system works acceptably in skilled hands, the depth of penetration of the molten gutta percha can be impossible to determine during placement.

Synthetic polymers such as polystyrene have been used, but these are generally less deformable and therefore rely on the sealer paste to guarantee a seal. The use of silicone points was examined more than a decade ago, but has not proved advantageous over gutta percha and in consequence has not been developed.

Nondeformable points, such as those made of silver, were popular in the past but are now not much used. Titanium points are now available commercially and may have limited application in canals which can be shaped to a circular cross section, but such cases are few. A further disadvantage of metal points is that they are easily displaced with loss of the required seal during the subsequent restoration of the crown of the tooth.

The importance of the sealer paste is obvious. The success of root canal treatment is probably dictated more by this than by the obturating points that are used. Ideally, this material should be biocompatible, adhesive and highly thixotropic. The most popular sealers used today are based on zinc oxide–eugenol systems. These are acceptably biocompatible although they show no adhesive behavior. Various formulations of zinc oxide sealers have been promoted. One of the most popular is Grossman's formula which contains barium sulfate as a radio-opacifier. Anhydrous sodium borate in the powder removes water from the system and this slows down the setting reaction to a more workable level.

Epoxide resins have been used successfully and adhesion has been claimed even against wet canal walls. One such product uses a base paste of bisphenol diglycidyl ether which is mixed with a hardener containing hexamethylene tetramine. The material has an extremely long working time, measured in hours rather than minutes, and this may be advantageous in the time-consuming lateral condensation of multiple complex canals in molar teeth.

Several sealers based on calcium hydroxide have appeared on the market in the late 1980s. Excellent results from the use of these materials have been reported in the literature.

Glass polyalkanoate materials are being developed and show great promise. Sealers made of this material should adhere well to the dentin by the formation of chemical adhesive bonds.

Paste-fillers alone have been used and there are many permutations that have been marketed. These range from thermoplastic resins that are injected into the tooth in the molten state to two-part resins which set after injection. Many of the paste-fillers contain anti-inflammatory drugs such as the steroids dexamethasone or triamcinolone in order to damp down the response of the apical tissues to the material. Antiseptics are sometimes incorporated as well and may be highly irritant to the tissues, although the full force of the inflammatory response may be damped down and delayed by the presence of the anti-inflammatory steroids.

The major drawback with paste-fillers is the difficulty in establishing the seal at the correct place in the tooth root. By its very nature, the canal is an open-ended cylinder. A constriction at the apex may block successfully the passage of a semirigid point, but it may not limit the flow of a paste. The possibility of the paste being extruded into the tissues beyond the apex of the tooth limits both the range of materials which can be used and the technique and circumstances of placement.

Occasions arise when a tooth has to be root-treated when its apex is not fully formed or has been opened up by root resorption. In such cases, it is inevitable that a large area of root-sealing material is going to come into contact directly with the apical bone. The most desirable technique in such cases is to pack the apical aperture with calcium hydroxide powder to form an interface against which an insoluble and nonresorbable sealer may be packed. There are several commercial preparations based on calcium hydroxide that may be selected from with caution but offer little advantage over plain pharmaceutical grade calcium hydroxide powder. Such preparations contain a range of additives such as radioopacifiers, antiseptics and even antibiotics.

Whichever technique is used for obturating a root canal, it should be obvious that the primary criterion of success is the integrity of seal that has been achieved at every opening of the pulp space on the root surface. A wide range of scientific techniques has been developed to examine this seal, but there is no agreement on the best method. Without a standard assessment it is difficult to compare results for different materials. *In vitro* leakage techniques that have been reported have used dyes, radioisotopes, bacteria, scanning electron microscopic examination and electrical impedance measurement.

3.4 Apicectomy

When other, more conservative methods of sealing the apical tissues from the pulp space of the tooth have failed, a direct approach is sometimes undertaken and is known as an apicectomy. This procedure involves gaining direct access to the apex of the tooth by removing the overlying bone after a mucoperiosteal flap has been lifted. The apex of the tooth contains the often troublesome delta of apical canals and is resected to expose the main trunk of the root canal. This enables the direct placement of an apical seal. Traditionally, dental amalgam has been the chosen material for this seal and, perhaps surprisingly, it is still the most commonly used. When set, dental amalgam is well tolerated by the tissues. However, great care has to be taken to restrict the material to the prepared root canal and to avoid overspill of amalgam into the surrounding bone.

Plastic resins and composite materials have been used in this application, but generally they do not form an acceptable seal with the dentin walls in the compromised conditions of a surgical procedure. Polymerization shrinkage may serve to open the margins of the retrograde seal to microleakage.

Of all other materials that may be used to achieve the apical seal, probably the most promising are the glass polyalkanoates. These materials adhere to dentin and are relatively biocompatible, but few have the desired radioopacity necessary for postoperative radiographic examination. Lanthanum is added to some glass polyalkanoate cements to make them radioopaque, but in the volume used to fill a deftly prepared apical cavity they may be barely visible. For radioopacity, amalgam has no competitor.

An apicectomy causes great damage to the mechanical and structural integrity of a tooth and is a procedure to be avoided whenever possible. By the time an apicectomy is completed, the root has been hollowed out, had the apex cut off and is embrittled by the original removal of the pulp anyway. It is obvious that its strength must suffer considerably and fracture of the remaining root mass is not unusual in the long term.

3.5 Diadontic Implant

For a number of reasons, the area of root surface supporting a tooth may become reduced to the point where the tooth becomes nonviable. For example, the apical part of the root may be fractured away from the coronal part due to a blow. Root resorption can occur and reduce root support, or the level of the periodontal attachment can be reduced severely due to periodontal disease. In all these cases it may become necessary to supplement the support which the root provides. This can be achieved with a diadontic implant. The diadontic implant is a rod which can be cemented into the pulp space of the crown of the tooth and extended into the alveolar bone through what remains of the tooth root to provide increased radicular support. The implant is usually about 1–2 mm in diameter and made of surgical-grade cobalt–chromium alloy. The length of implant that may be employed depends on the anatomical structure in the vicinity of the tooth root.

Provided the implant remains sealed from the oral environment, its prognosis remains good. However, if contact with the mouth is established by the formation or extension of a periodontal pocket, then failure follows rapidly.

4. Restoration of Root-Filled Teeth

One common feature associated with pulp death and subsequent endodontic treatment is staining of the enamel and dentin by blood and tissue breakdown products, or even by endodontic materials themselves. Staining in anterior teeth is unacceptable and various techniques have been devised for bleaching the dentin and enamel. Bleaching agents include strong solutions of hydrogen peroxide, sodium hypochlorite and various acids. The decomposition of hydrogen peroxide to produce nascent oxygen is often accelerated by the application of heat or strong light to the outer surface of the tooth being bleached. Often, several lengthy visits are required to bleach a tooth to an acceptable color.

The vast majority of root-filled teeth have crowns that are damaged very severely. This is due in part to the original damage which was deep enough to destroy the dental pulp and in part to the clinician having to cut down into the pulp space to place the root filling. This means that, frequently, there is insufficient remaining crown mass to support a restoration. Furthermore, the mechanical properties of the remaining tissue are so compromised that fracture under loads that are normal in the mouth may be expected. In the few cases where a substantial bulk of coronal tissue remains it is possible to place an amalgam filling or a gold inlay, but the design of the restoration must be modified to take into account the reduced structural properties of the tooth. The entire load-bearing surface of the tooth is covered by an extension of the restoration known as "capping the cusps" so that occlusal loads are not concentrated in one weakened area such as a cusp tip. It has been demonstrated recently that a bracing effect can be achieved by employing an adhesive-bonded composite restoration. Instead of capping the cusps, the restoration is bonded firmly to acid-etched enamel and with dentin adhesives to the underlying dentin. However, the long-term strength of such bonds remains to be established.

When the loss of coronal tissue is greater it is necessary to provide a replacement crown of metal or porcelain. In cases of moderate loss it is possible to build up a core of replacement dentin using adhesively bonded composites or glass ionomer materials. Where loss is severe, gaining retention for a replacement crown is achieved by enlarging the radicular pulp space and inserting a metal post which is held in place with a cement. The prognosis for such restorations is not good and loss of the restoration due to fracture of the remaining root, breakage of the cement seal or recurrence of dental caries may occur.

The distribution of stress in post crowns has been examined by polarimetric analysis and by finite-element computer modelling methods. These methods have highlighted the critical importance of the cement layer in the system. Both the retention of the post and resistance of the root to fracture are markedly affected by modelling a different boundary layer.

Recent work has shown that adhesion to dentin can be gained by etching the dentin surface in the post hole and then washing out the dentinal tubules before permeating the tissue with a plastic resin. The usual resin used is bisphenol-A glycidyl dimethacrylate. The supporting post is constructed from an alloy that can be etched electrolytically in an acid bath to create a porous surface that can be permeated with the same resin. A filled composite resin may be used in the cement layer between the treated post and the treated dentin. In this way, a strong adhesive bond can be achieved between the post and the supporting dentin. This not only increases the retention of the system substantially, but also distributes stress better and reduces the possibility of root fracture. Furthermore, it has been used successfully to repair fractured roots.

Bibliography

Anon 1988. In vitro assessment of the biocompatibility of dental materials. Report of an international conference 1987. *International Endodontic Journal* 21: 49–187
Cohen S, Burns R C 1987 *Pathways of the Pulp*. Mosby, St Louis, MN
Mumford J M, Jedynakiewicz N M 1988 *Principles of Endodontics*. Quintessence, London
Schilder H 1967 Filling root canals in three dimensions. *Dent. Clin. North Am.* 11: 723–44

N. M. Jedynakiewicz
[University of Liverpool, Liverpool, UK]

F

Fracture Toughness

Biomaterials are used to replace or augment damaged or missing tissue. As yet it has not been possible to reproduce the complex biochemical functions of tissues and biomaterials are limited to relatively basic applications dependent upon mechanical properties. (Those materials that release chemicals with a pharmaceutical effect are an exception.) A crude division can be made between soft and hard (mineralized) tissues, and synthetic replacement materials are available for both. Soft tissues are elastomeric, being required to operate elastically over large strains. Brittle fracture of soft tissue does not take place and fracture toughness is not a restriction upon it, nor any suitable replacement. The function of mineralized tissue is to carry body weight, aid locomotion, protect delicate organs or provide biting surfaces. Mineralized tissue is stiff and hard, but it is also brittle. For this, fracture toughness is a critical property. To be successful, its replacement must be acceptable on a number of criteria; compromises are inevitable and the fracture toughness may be less than desirable. Indeed, many brittle biomaterials have been introduced by surgeons without reference to fracture toughness, but such an empirical approach relies upon the adequacy of the surgical technique reducing the incidence of fracture to an acceptable level. Often research has been undertaken in response to the occurrence of fractures. When few fractures are reported for any particular brittle material used in a (clinically) tried and tested design, its fracture toughness may be assumed adequate. As a consequence, the fracture toughnesses of a number of such biomaterials have not been determined as such, though values may exist from general engineering use.

A knowledge of the resistance to fracture of mineralized tissue has fundamental importance and defines a property requirement for any prosthetic material. Dentistry and orthopedic surgery are different in most respects and their restorative materials are generally equally distinct, though in some cases dental and surgical materials are related.

1. Mineralized Tissue

1.1 Bone

For research, bovine cortical bone has been used in place of human material for reasons of availability and size, though this has only been justified recently (Bonfield 1987). Fracture toughness as a concept to explain the fracture of bone was first used in 1973 (Melvin and Evans 1973). Before then, stress and total deformation energy were used as measures for fracture resistance, a practice that still continues. The applicability of linear elastic fracture mechanics to bone has been proven by a number of criteria in several investigations (Wright and Hayes 1977, Margel-Robertson et al. 1978, Behiri and Bonfield 1984).

Cortical bone has a directional structure and the value of the plane strain critical stress intensification factor K_{1C} for crack propagation parallel to the long axis of the bone ($2.0–5.6$ MN m$^{-1.5}$) (Bonfield and Datta 1976, Wright and Hayes 1977, Behiri and Bonfield 1984) is less than that perpendicular to the axis ($3–8$ MN m$^{-1.5}$) (Margel-Robertson et al. 1978). Bonfield (1987) has shown that the fracture toughness changes progressively, from 3.2 MN m$^{-1.5}$ to 6.3 MN m$^{-1.5}$, as the direction of crack propagation is changed from parallel to perpendicular to the bone axis.

The fracture toughness depends upon the density of the bone. Fracture toughness increases of 30% (Wright and Hayes 1977) and 40% (Behiri and Bonfield 1984) have been recorded for 5% increases in density. Such a finding has important implications for osteoporotic individuals. This is an extreme age-related condition associated with an increased incidence of bone fractures and subsequent protracted recovery. It is part of the natural aging process that humans become "more brittle" as they grow older. The change is progressive. The bones of a healthy 25 year old are much tougher than those of an individual of 90 years (approximately 4.5 MN m$^{-1.5}$ and 2.5 MN m$^{-1.5}$, respectively) (Bonfield 1987).

A strain rate dependency for K_{1C} has been demonstrated (Margel-Robertson et al. 1978). Fracture toughness decreases by about 8% with each order of magnitude change (decrease) in test machine displacement rate. This has been refined by relating the plane strain stress intensification factor K_1 to the crack velocity (Behiri and Bonfield 1984). Stable and controlled growth of cracks is possible at low crack velocities. As the velocity increases, K_1 increases from 2.8 MN m$^{-1.5}$ to 6.3 MN m$^{-1.5}$. At 1.2×10^{-3} m s^{-1} there is a transition, propagation becomes unstable as the crack accelerates to velocities typically in the range $10–40$ m s^{-1} and the value of K_1 falls. The fracture toughness decreases from 5.6 MN m$^{-1.5}$ to 2.0 MN m$^{-1.5}$ with increases in crack velocity over this unstable range. A change in the interaction between crack and bone structure takes place at this transition. At slower controlled velocities the crack passes around and intersects osteons (or lamellae) creating a rough fracture surface, while during fast unstable growth the crack cuts through the microstructural constituents indiscriminately. These microstructural and strain rate dependencies of fracture toughness can explain some of the differences (even apparent conflicts) between results from various research programs; even so, the variability of living material leaves a range of values greater than those to which materials scientists are accustomed.

1.2 Teeth

The structure of the tooth presents a practical problem. Teeth possess a thin (<1 mm) surface layer of highly mineralized (90%) dental enamel, arranged as fine prisms running perpendicular to the surface. Underlying and supporting this is dentin, a less mineralized (70%) tissue that contains fine liquid-filled tubules running through to the pulp chamber. The pulp chamber carries the nerve and extends up through the tooth roots to the center of the tooth. In place of enamel, the submucosal surface of the tooth is covered by a layer of cementum, a mineralized tissue similar to bone.

The problems encountered when testing such small volumes of material have deterred materials scientists, who have preferred to measure the fracture toughnesses of filling materials. This pragmatic approach suggests that materials can be developed (whereas teeth cannot), but it has lacked a reference point to indicate what is adequate. Fortunately two papers have appeared, one on enamel (Hassan et al. 1981) the other on dentin (El Mowafy and Watts 1986). The variability of values measured for bone gives strength to the argument that much work remains to be done for teeth.

The indentation technique developed by Palmqvist overcomes the limitations of thickness, size and contour of enamel (which rule out the production of conventional specimens). The fracture toughness of human enamel varies from $0.7 \, \text{MN m}^{-1.5}$ to $1.3 \, \text{MN m}^{-1.5}$ and depends upon the degree of mineralization and structure of enamel over the surface, though this has not been proven conclusively as the proposition depends upon the results from just four teeth! However, cracks do propagate along planes of weakness between hydroxyapatite prisms and the resistance to fracture varies with orientation (with the cervical/incisal direction being weakest).

It is possible to cut compact tension geometry specimens from human molar dentin, though the orientation of the specimen with respect to the dentin tubules is restricted and anisotropy determination is not possible. This gives a crack front that runs parallel to the axes of the dentin tubules. The scale of these specimens (4.5 mm × 4.5 mm × 1.6 mm) represents the extreme for test pieces, yet valid results have been produced. The fracture toughness of dentin ($3.08 \, \text{MN m}^{-1.5}$) is greater than that for enamel, explained by the blunt-

ing of the crack front as it passes through a succession of dentin tubules. The fracture toughness of dentin is invariant over the range of temperatures (0–60 °C) met in the mouth.

2. Surgical Biomaterials

2.1 Surgical Acrylic Bone Cement

Developed from dental autopolymerizing material, surgical acrylic bone cement fills the gap between the bone and the implant to locate and fix the prosthesis. The "cement" does not bond to the bone; it locks the implant in place and transfers load from the implant to the bone. Unfortunately, it has been estimated that 10% of hip joint failures are due to failure of this cement by fracture and loosening of the femoral stem at the bone–cement interface (Weber and Bargar 1983).

Although bone cements are primarily poly(methyl methacrylate) (PMMA), and hence may be called acrylic bone cement, variations in composition exist. Most products contain PMMA powder with between 5% and 13% $BaSO_4$ or ZrO_2 as an x-ray opacifier. One product has a copolymer (PMMA–2%polystyrene) powder. Another contains 0.5% Gentamicin antibiotic. MMA monomer is added to this powder in a one-to-two ratio to produce the paste. In a surgically acceptable time this paste sets by free-radical-induced autopolymerization (Freitag and Cannon 1976, Kusy 1978). Though small, these differences appear to affect the fracture toughness and have led to confusion when commercial products have been used to investigate clinically relevant variables. Confusion may also arise from inevitable batch variations and changes in composition in the course of time.

The presence of the $BaSO_4$ x-ray opacifying addition is a source of weakness (Freitag and Cannon 1976, Beaumont 1979, Sih and Berman 1980). The bond between the polymer and the 1 μm particles is weak, allowing voids to nucleate and open up at the interface during deformation. At any K_1 above that to effect craze initiation these particles increase the crack velocity and then decrease K_{1C} in proportion to the volume fraction present. In commercial products the price paid for inclusion of the x-ray opacifier is a reduction in the region of 18%. Reduced toughness has also been attributed to this dispersion inhibiting densification when pressure is applied during curing.

The first results gave an increased fracture toughness with the inclusion of antibiotic, though that improvement was lost when the Gentamicin dissolved in water (Kusy 1978). However, more recent research has revealed that the presence of antibiotic neither improves nor leads to deterioration in fracture toughness (Wright et al. 1984, Rimnac et al. 1986). This indifference continues when the antibiotic is eluted, whether the material is stored in

Ringer's solution or is implanted subcutaneously in dogs.

A lower viscosity mix improves intrusion of the cement into the bone. The effect of the modification of the composition to produce this viscosity is not clear. It has been said to reduce fracture toughness (Robinson et al. 1981) and to not affect it (Weber and Bargar 1983). A higher-molecular-weight polymer powder produces superior fracture toughness in the set cement and accounts for some of the variation between products (Rimnac et al. 1986). There is a widely held belief that centrifuging the cement before application improves strength by removing air incorporated during mixing, but such treatment has no effect on fracture toughness (Rimnac et al. 1986).

Clinical conditions are not constant when implants are placed. The prosthesis is at room temperature (23 °C) whereas the body is at 37 °C, and the surgeon may or may not maintain pressure on the prosthesis as the cement cures. These differences are significant; at higher temperatures a porous weaker material is produced, with a 25% loss in fracture toughness. Fortunately, maintaining a force as low as 12 N prevents this inferior structure from developing. Consequently, surgeons should maintain pressure on the implant until the cement has cured to maximize the fracture toughness (Sih and Berman 1980).

Slow stable growth of cracks in bone cement and Perspex (bulk polymerized PMMA sheet) obeys the relationship

$$v = AK_1^n \qquad (1)$$

where v is the crack velocity. From this it is possible to predict failure times at given stress levels. The exponent n is not affected by environment (air or water), but the constant A is. Extrapolation to essentially no growth (i.e., 10^{-10} m s^{-1}) gives a value for K_1 equal to that for craze initiation in Perspex, but this K_1 value should not be taken as a material property since it depends upon environment, strain rate and temperature. While changing the environment (air, saline solution, blood serum) has no effect on crack velocity at any K_1, it can increase K_{1C} (Beaumont 1979). Water induces a plasticized zone of crazing which absorbs energy and increases K_{1C} from 1.8 MN m$^{-1.5}$ (in air) to 2.1 MN m$^{-1.5}$ (in water). Though the presence of water appears to eliminate slow growth (i.e., at $K_1 < K_{1C}$) in Perspex, it does increase K_{1C} from 1.6 MN m$^{-1.5}$ to 1.7 MN m$^{-1.5}$. The higher values for bone cement are attributed to its heterogeneous structure (Beaumont and Young 1975). Storage in Ringer's solution or implantation also brings about an increase in fracture toughness through water absorption (Wright et al. 1984).

To improve the fracture toughness of any material is desirable, no less so for acrylic bone cement. The addition of aramid (Kevlar 29) fibers to dental auto-polymerizing PMMA (used as a substitute for bone cement) increases K_{1C} in proportion to the volume added. Fiber pullout, shearing and splitting absorbs energy to improve toughness. At the limit set by the flow characteristics of the mix (7% fiber content) the toughness is increased from 1.5 MN m$^{-1.5}$ to 4.4 MN m$^{-1.5}$ (Wright and Trent 1977) or to 2.9 MN m$^{-1.5}$ (Wright and Trent 1979), or from 0.96 MN m$^{-1.5}$ to 1.96 MN m$^{-1.5}$ (Pourdeyhimi et al. 1986). The effectiveness of the dispersion of the fibers is the most likely reason for the differences, as bunching of the fibers has been acknowledged to be a problem. Carbon fibers also increase fracture toughness. An addition of 2% increases K_{1C} by 30%. This poorer showing by carbon fibers has been attributed to poor interfacial adhesion and pullout of the very short (1.5 mm) fibers (Robinson et al. 1981).

2.2 Alumina

Polycrystalline fused alumina possesses properties that suggest use as an implant material—inertness, wear resistance, high strength and modulus—but it is a ceramic subject to brittle fracture. It is available with a purity between 95.0% and 99.7% but contains up to 6% of disconnected porosity. Purity and volume of porosity (within these ranges) do not affect the fracture toughness, but the morphology of the porosity does. (Angular porosity decreases K_{1C} by 10%.) The introduction of water to the crack tip has a corrosive effect reducing K_{1C} by between 15% and 20%. This is a crack tip phenomenon caused by the presence of water and no further decline takes place with prolonged immersion. However, body fluids are water-based solutions and are not pure water. The presence of dissociated salts in water contributes to the degradation whereas proteins do not (Dalgleish and Rawlings 1981). Coating alumina with a surface-active glass (bioglass) to promote bonding to bone provides an effective diffusion barrier to the biological environment to substantially reduce its effect (Ritter et al. 1979).

The reliability of ceramic implants can be improved by the application of fracture mechanics, to allow a reduction in empirical safety factors. In alumina, stable crack growth obeys Eqn. (1) until a condition of instability is reached (when $K_1 = K_{1C}$). The time to failure at a particular applied stress can be derived from this and the analysis extended to include proof testing to "guarantee" a service life (Ritter et al. 1979).

3. Dental Biomaterials

3.1 Denture Base Acrylic Resins

Comparison of dentally processed PMMA with Perspex is inevitable. Following processing the fracture toughness of dental heat-cured PMMA

(1.29 MN m$^{-1.5}$) is comparable to that of Perspex (1.15 MN m$^{-1.5}$). However, dentures operate in a wet environment and will absorb up to 2% water. Storage in water (followed by fracture in air) raises K_{1C} (to 1.50). As this effect is not produced in Perspex, the hypothesis advanced is that pores in the dentally processed material fill with water to maintain a wet environment at the advancing crack tip. The cross-linking comonomer ethylene glycol dimethacrylate (present to the extent of 10% in all dental products) is not very efficient and forms pendant groups which act to plasticize and increase K_{1C} (to 1.56). A few commercial products possess a fine dispersed rubber phase within the polymer beads to improve impact resistance; an effective addition, it doubles K_{1C} (to 3.00) (Hill et al. 1983).

Water promotes crazing in dental heat-cured PMMA and its presence extends the craze zone ahead of the crack tip. When the test environment is changed from air to water and if the crack growth rate is slow enough, crazing can develop fully and propagation proceeds at a higher value for K_1 (2.36 MN m$^{-1.5}$). At higher rates propagation changes from stable to unstable slip/stick, with the difference between (higher) initiation and (lower) arrest values decreasing with increasing rate (as the extent of crazing is reduced). The supply of water to the crack tip is rate dependent and when the crack growth is fast enough, testing is effectively being done in air (at which point $K_{1C} = 1.56$). In the rubber-toughened material the craze size is already large and the effect of water is small by comparison—growth is stable regardless of environment. Most testing has been conducted at room temperature (~19–25 °C), which is acceptable for comparisons between environments, but dentures are expected to operate in wet conditions between 0 °C and 60 °C. Raising the water temperature to 37 °C produces a fall in the toughness of heat-cured PMMA (Hill et al. 1984).

Earlier research (Stafford et al. 1980) suggested that the fracture toughness of autopolymerized dental PMMA is inferior to heat-cured material; this was attributed to a weaker interface between the prepolymerized beads and the autopolymerizing matrix as penetration of the monomer is restricted, and also to a less effectively polymerized matrix. However, more recent research (Hill et al. 1983) has given comparable values, as the higher level of residual monomer in the autopolymerized matrix offsets its lower molecular weight.

3.2 Dental Ceramics

A failure to appreciate fracture mechanics and the fact that it applies to all brittle biomaterials can lead to a false sense of security. This is clearly illustrated by the marketing of ceramic orthodontic brackets. The use of stainless steel brackets cemented to buccal teeth surfaces, and connected with steel arch-wires is a widely practised and effective treatment. However, their prominence and unattractive appearance is a major problem; most patients are young adults acutely aware of this. More recently introduced sapphire brackets are less visible and a reassuringly favorable tensile strength comparison has been given—1400 MN m^{-2} for sapphire, 350 MN m^{-2} for steel. Brackets possess sharp features, are often scratched, and on insertion and during service can be subject to high stresses. Sapphire is brittle and when the comparison is made on the correct basis of fracture toughness— 2.4–4.5 MN m$^{-1.5}$ and 80–95 MN m$^{-1.5}$ for sapphire and steel, respectively—the conclusion is less reassuring (Scott 1988). Clearly sapphire brackets are less abuse tolerant and a higher failure rate might be anticipated.

The composition of dental porcelain has evolved to meet specific dental requirements, in particular to improve translucency and to reproduce individual crowns, bridges and veneers with a high degree of accuracy. Dental felspathic porcelain contains 0.25 to 0.35 volume fraction of micrometer size crystalline leucite particles dispersed in a siliceous glassy matrix. In dental aluminous porcelain α-alumina replaces the leucite. The fracture toughnesses of these are 0.90–1.06 MN m$^{-1.5}$ and 1.48–1.56 MN m$^{-1.5}$, respectively. This difference results from interfacial stresses present in felspathic porcelain and not in aluminous porcelain, which alters the crack path to reduce the effectiveness of the dispersion (Morena et al. 1986).

3.3 Dental Filling Materials

(*a*) *Dental silver amalgam.* This is an established restorative material for load-bearing cavities in posterior teeth. A powdered 70 wt% Ag–30 wt% Sn alloy is titurated with the minimum amount of mercury necessary to form a firm paste. Constituent metals then react peritectically to form solid phases which cement together the superficially dissolved alloy particles. For many decades copper in amalgam alloy was restricted to less than 6% (now referred to as low-copper amalgam). However, material with increased copper showed improved performance *in vivo* and since the 1960s the level has been increased to as high as 30%. Initially this was brought about by the addition of a 50 wt% Ag–50 wt% Cu eutectic powder to conventional alloy powder (i.e., admixed high-copper amalgam), though currently most products contain a single high-copper powder (i.e., single-component high-copper amalgam). All three types are available. The application of fracture mechanics to dental materials comes at a time when an alternative to amalgam is actively being sought. Consequently, scientific interest in amalgam is declining and research on its fracture toughness has been minimal (in relation to its extensive use).

All publications (Roberts et al. 1978, Cruickshanks-Boyd and Lock 1983, Lloyd and Adamson 1985) have shown that high-copper amalgam has inferior fracture toughness to the original low-copper products. A significant (negative) correlation exists between copper content and fracture toughness. (This is unique amongst critical properties as higher-copper levels improve all others.) The fracture path is through the matrix avoiding the partly dissolved alloy particles. Differences in the relative proportions of the reaction products γ_1, γ_2 and η', their relative resistances to fracture and their distribution in the matrix lead to different fracture toughnesses for low-copper ($1.3–1.6 \text{ MN m}^{-1.5}$), admixed ($1.3 \text{ MN m}^{-1.5}$) and single-component high-copper ($1.0–1.3 \text{ MN m}^{-1.5}$) amalgams (Lloyd and Adamson 1985). Early results showed no difference between one day and one month (Roberts et al. 1978). More recent research has confirmed this, but in addition has shown that the full aging curve peaks at times between these and that the fracture toughness is lower at both shorter and longer times (Lloyd and Adamson 1987). This time dependence is a consequence of a peritectic setting reaction, and of the changes in composition and resistance to fracture of the matrix phases during setting. Whether there is a strain rate sensitivity is not clear, since the only two investigations have yielded conflicting conclusions (Roberts et al. 1978, Cruickshanks-Boyd and Lock 1983). However, the weight of evidence would suggest that there is a strain rate dependency due to the presence and activity of the same deformation processes responsible for creep in amalgam.

It is likely that many other composition and manipulation variables affect the fracture toughness of amalgam, but none has been investigated systematically. For example, surface contamination of the alloy powder or a dry mix, would lead to inadequate amalgamation, weak interfaces and, consequently, a lower fracture toughness. Such an explanation can (and has been used to) account for a lower fracture toughness and the accompanying appearance of intact alloy particles on the fracture surface when different batches of the same product are evaluated (Lloyd and Adamson 1987). Whether such findings are investigated further will depend upon how rapidly the successor to amalgam, posterior composite, is accepted by dentists and, in particular, by regulatory agencies.

(b) *Glass ionomer cement*. This is a direct development from the now obsolete silicate cement. Setting of this cement paste is brought about by a reaction between polyacrylic acid and an aluminosilicate glass powder. The attractive features of glass ionomer are aesthetic, pharmacological and adhesive: it is tooth coloured, it releases fluoride which arrests further decay, and it bonds to enamel and dentin to seal the interface between the filling and the tooth. However, its poor tensile strength has long been recognized and investigation of its fracture toughness has been limited to confirming an inferiority in this respect. Very low values in the range $0.25–0.55 \text{ MN m}^{-1.5}$ (Lloyd and Mitchell 1985, Goldman 1985) which change little with time (Lloyd and Adamson 1987) have been reported. To reduce surface friction and extend use to small (well-protected) cavities on occlusal surfaces one manufacturer has added a silver dispersion to the glass. Unfortunately this addition does not improve fracture toughness (Lloyd and Adamson 1987).

(c) *Composite filling material*. This began evolving in the early 1960s in response to the failure of pure resins to satisfy requirements. Composite contains a dispersion of fine inorganic particles coated with a silane compound to promote bonding to the glassy diacrylate resin matrix. Initially introduced as an aesthetic material for the visible anterior part of the mouth, development has reached the point at which it can replace amalgam in the more heavily loaded posterior teeth. This progressive development has resulted in diversity within the range of products available. Barium glass, strontium glass, zinc glass and silica powders are found with a variety of comonomers. Setting is brought about either by an amine/peroxide free radical polymerization (two paste presentation) or when the action of blue light upon another initiator yields the free radicles (single paste presentation). Early products contained coarse particles ($\sim 20 \text{ μm}$), the size of which has become finer in the course of time ($\sim 5 \text{ μm}$). In the late 1970s microfine (40 nm) silica was introduced, but at the expense of filler loading. More recently, both species have been mixed to achieve the higher filling loading required for posterior composite. Interest in the fracture toughness of these brittle composites postdates by a considerable time their widespread use; but it is now recognized that not only is fracture toughness a measure of resistance to bulk fracture, but also to edge chipping and to some aspects of wear.

The fracture toughness of composite increases as the volume fraction of the filler is increased (Lloyd and Iannetta 1982, Lloyd and Mitchell 1984, Pilliar et al. 1986, Davis and Waters 1987). A number of energy absorbing mechanisms directly attributable to the dispersion have been proposed. These include contributions from increased surface roughness (Roberts et al. 1977, Lloyd and Mitchell 1984, Davis and Waters 1987), pinning and bowing of the crack front (Lloyd and Iannetta 1982, Lloyd and Mitchell 1984, Ferracane et al. 1987, Davis and Waters 1987), and mismatch of the resin and particle moduli (Lloyd and Mitchell 1984). The pinning and bowing mechanism is analogous to the Orowan dispersion strengthening model and is more appropriate to widely spaced particles as found in less heavily filled composites containing microfine filler. Unfortunately such fine particles are less effective at crack pinning than coarse particles and this has been used to explain the lower fracture toughening of the former (Goldman 1985, Ferracane et al. 1987). However, the difference in volume fraction has been convincingly shown to be the root cause of this difference (Davis and Waters 1987). Most authors have noted that the fracture toughness is proportional to the product of strain energy release rate and elastic modulus. The modulus increases substantially with filler content and more than offsets a deterioration in the energy term at higher filler loadings; consequently, fracture toughness is improved by the presence of the filler.

The resistance to fracture of the resin matrix has long been recognized as a factor. Its capacity to absorb energy by plastic deformation is important (Lloyd and Iannetta 1982, Lloyd and Mitchell 1984, Pilliar et al. 1986, Cook and Johannson 1987, Davis and Waters 1987, Ferracane et al. 1987, Montes and Draughn 1987). Unstable stick/slip crack propagation during the fracture of double torsion specimens has been seen in composites containing up to 62 vol.% filler (Davis and Waters 1987, Montes and Draughn 1987). It is attributed to the existence of a plastic (viscoelastic) zone at the crack tip. This zone requires a higher energy (hence higher K_1 (initiation)) for the crack to cut through the zone, releasing this energy in the process until the crack arrests (at a lower K_1 (arrest)) as the zone builds up once more. Stable (continuous) growth results from a decreased deformation capacity at the crack tip during the fast fracture, or from when the zone formation is continuous ahead of a very slow-growing crack. Though the localization of deformation is essential, Davis and Waters (1987) have been able to identify a number of factors that enhance plasticity to improve fracture toughness and also increase general yielding.

The rise in fracture toughness which takes place in the first few days after curing has been interpreted as a post-cure phenomenon resulting from a strengthening of the resin. Peaking and a slight decline in the long term have been attributed to the continued loss of plasticizing monomer as the conversion to polymer approaches a limit (Lloyd and Iannetta 1982). This is supported by a correlation between the extent of conversion and the fracture toughness. Though cross-linking does reduce the plastic deformation (thus dissipation) at the crack tip, it increases the modulus and, consequently, the toughness (Cook and Johannson 1987).

The autopolymerizing chemistry necessitates mixing two pastes which leads to the incorporation of air and the presence of pores in the set material, not present when the light curing chemistry is used.

There is no apparent difference in the toughness between composites using these two initiation systems (Lloyd and Mitchell 1984, Goldman 1985). Even the deliberate introduction of porosity at levels of up to 2.5 vol.% does not affect fracture toughness (Yoshimoto 1984).

Within the mouth, composite is in a wet environment and the resin matrix will absorb water. After a week this produces a significant increase in toughness (averaging 17%) relative to composite held in air (Lloyd 1982, Lloyd and Iannetta 1982). However, after a further 29 weeks the difference becomes insignificant (Lloyd 1984). This has been attributed to the presence of plasticizers (absorption of water and loss of monomer—their relative effectiveness and the rate at which levels change). The plasticizing effect of absorbed water has been confirmed by measuring fracture toughness as a function of crack velocity, for which Eqn. (1) is obeyed. The exponent n is inversely related to the degree of time dependence for deformation—water absorption decreases n, implying more extensive viscoelastic deformation processes (Montes and Draughn 1987).

The corrosive nature of water should not be ignored. The silane coupling agent on the filler surface may be attacked in some composites leading to debonding and to a lower fracture toughness (Pilliar et al. 1986, Lloyd and Adamson 1987, Montes and Draughn 1987). The crack then takes an interfacial path and "clean" particles and pits appear on the fracture surface. However, it must be emphasized that many composites are more resistant and for a number of products experimenters have measured little difference between air- and water-stored material (Lloyd 1984, Pilliar et al. 1987), or substantially greater deterioration in affected material (Lloyd and Adamson 1987), or a preference to a path through the matrix at a distance from the interface (indicating cohesion) (Lloyd and Adamson 1987, Montes and Draughn 1987). The correlation between fractography and toughness is consistent and persuasive. However, it is difficult to reconcile this evidence with the finding of a single study in which the silane treatment was omitted; the effect though minor could be beneficial (Davis and Waters 1987). This should be the subject of further research.

Undoubtedly water absorption takes place, but composites are in pure water rarely; they are bathed in saliva and have intermittent contact with food products. Simple aqueous solutions such as 0.1% saline or 0.05M sucrose bring about the same effect as pure water (Lloyd 1984). Ethanol is a more effective solvent with a solubility parameter closer to that of the resin. Extended storage of composite in pure ethanol produces a significant increase in fracture toughness. It is argued that ethanol produces an environmental crazing at the crack tip which reduces the stress intensification and consequently increases the fracture toughness (Pilliar et al. 1987).

Composite fillings are also subjected to a range of temperatures; 0 °C and 60 °C are usually taken as the extremes met in the mouth. (Temperatures outside the range 32–37 °C are normally transient and associated with eating and drinking.) The temperature dependence for the fracture toughness of composite is small over the range 20–40 °C. The organic resin is more affected than the inorganic filler; consequently, a composite containing a microfine filler shows a greater change than another containing a higher loading of coarser filler ($+0.6\%$ °C^{-1} and $+0.1\%$ °C^{-1}, respectively) (Lloyd 1982).

3.4 Application of Fracture Mechanics to Restorative Dentistry

In structural engineering fracture mechanics is invaluable not only to define the resistance to fracture within the material used for fabrication, but also to predict the failure of loaded structures. Dental application of the former has been dealt with previously. The customized nature of dental reconstructions means that specific examples will probably never be analyzed, though general design guidelines may be refined by the application of fracture mechanics.

When decay leads to loss of a considerable amount of the tooth, it is not possible to effect a repair by preparing an undercut cavity to mechanically retain the filling. In this situation it is necessary to place very fine self-tapping pins in the dentin to anchor the filling. (The development of adhesive resin offers an alternative to this established treatment, but is unlikely to lead to the demise of these "dentin pins.") Overloading and mechanical failure of this restoration may occur by extraction of the pin from the filling material, by tensile failure of the pin or by fracture (initiated from the pin) through the body of the filling. It is possible to raise the force at which failure takes place and, consequently, improve the effectiveness of the treatment. In particular, an effective thread geometry prevents pin extraction and leads to fracture through the body of the filling. The force to produce fracture is linearly related to the fracture toughness of the embedding filling material, whether that is amalgam, glass ionomer or composite (Butchart and Lloyd 1986, 1987, 1988). Limits can be set; at the lower end a minimum force (hence fracture toughness) equal to the force to extract the pin from the dentin and at the upper end by the tensile strength of the pin. Thus pin retention is optimized simply by selecting a material that possesses a high fracture toughness!

It is often necessary to repair a fractured or worn filling, or to modify its form by an addition. It is usual to measure this interfacial strength by its fracture stress, but that is a function of both the bond strength and the quality of the repair. The interfacial defect size distribution is operator dependent and not readily controlled: published repair stresses show considerable scatter and comparisons are equivocal. However, fracture toughness specimens can be used to introduce a massive controlled defect and the interfacial bond strength can be measured by the fracture toughness of repaired specimens. This technique has been used to determine the effect of variations in resin composition, cleanliness and the presence of a dentin bonding agent (Dhuru and Lloyd 1985, Lloyd and Dhuru 1985). Replicative (of the clinical situation), practical and interpretive advantages commend this approach.

4. Conclusion

The application of fracture mechanics to both surgical and dental biomaterials is a recent phenomenon, even though many are brittle and the foundation for the science had long been established. The growth has been rapid and the subject matter is broad, fragmented and sometimes incomplete or inconclusive. The fragmentation of research is inevitable given the *de facto* division between medical and dental schools, and the establishment of biomaterials research groups with different objectives in each. Some dental biomaterials, the filling materials and cements, are quite distinct from surgical implant biomaterials and findings on these do not have a direct bearing upon each other. However, denture base acrylic resin and surgical acrylic bone cement share a common ancestry (though each has been developed to best suit its own usage) and effects noted for one are likely to apply to the other.

Finally, quantitative comparisons using data from different sources should be made with caution. Testing environment and specimen conditioning should be identical, as should be the method of determination of fracture toughness. Systematic differences have been noted between results obtained using different test piece geometries. The effect may be small and more often than not is left unexplained. Often absolute numerical values are less important than changes or trends brought about by altering variables. Identical conclusions may be drawn from separate pieces of work; however, taking numbers out of context from each for comparison can be meaningless. When comparisons are required data produced by a consistent procedure would be ideal; often this is not possible, in which case the presence of one (preferably more) common material(s) (giving similar results) could justify usage.

Bibliography

Beaumont P W R 1979 Fracture processes in acrylic bone cement containing barium sulphate dispersions. *J. Biomed. Eng.* 1: 147–52

Beaumont P W R, Young R J 1975 Failure of brittle polymers by slow crack growth. *J. Mater. Sci.* 10: 1334–42

Behiri J C, Bonfield W 1984 Fracture mechanics of bone—The effects of density, specimen thickness and crack velocity on longitudinal fracture. *J. Biomechanics* 17: 25–34

Bonfield W 1987 Advances in the fracture mechanics of cortical bone. *J. Biomechanics* 20: 1071–81

Bonfield W, Datta P K 1976 Fracture toughness of compact bone. *J. Biomechanics* 9: 131–4

Butchart D G M, Lloyd C H 1986 The retention of self-threading pins embedded in various restorative materials. *Dent. Mater.* 2: 125–9

Butchart D G M, Lloyd C H 1987 The retention of self-threading pins embedded in visible light-cured composites. *J. Dent.* 15: 253–6

Butchart D G M, Lloyd C H 1988 The retention of core forming materials by dentine pins. *J. Dent. Res.* 67: 673

Cook W D, Johannson M 1987 The influence of photocuring on the fracture properties of photo-cured dimethacrylate based dental composite resin. *J. Biomed. Mater. Res.* 21: 979–89

Cruickshanks-Boyd D W, Lock W R 1983 Fracture toughness of dental amalgams. *Biomaterials* 4: 234–42

Dalgleish B J, Rawlings R D 1981 A comparison of the mechanical behaviour of aluminas in air and simulated body environments. *J. Biomed. Mater. Res.* 15: 527–42

Davis D M, Waters N E 1987 An investigation into the fracture behaviour of a particulate-filled bis-GMA resin. *J. Dent. Res.* 1128–33

Dhuru V B, Lloyd C H 1985 The fracture toughness of repaired composite. *J. Oral Rehab.* 12: 413–21

El Mowafy O M, Watts D C 1986 Fracture toughness of human dentine. J. Dent. Res. 65: 677–81

Ferracane J L, Antonio R C, Matsumoto H 1987 Variables affecting the fracture toughness of dental composites. *J. Dent. Res.* 66: 1140–5

Freitag T A, Cannon S L 1976 Fracture of acrylic bone cements. I Fracture toughness. *J. Biomed. Mater. Res.* 10: 805–28

Goldman M 1985 Fracture properties of composite and glass ionomer dental restorative materials. *J. Biomed. Mater. Res.* 19: 771–83

Hassan R, Caputo A A, Bunshah R F 1981 Fracture toughness of human enamel. *J. Dent. Res.* 60: 820–7

Hill R G, Bates J F, Lewis T T, Rees N 1983 Fracture toughness of acrylic denture base. *Biomaterials* 4: 112–20

Hill R G, Bates J F, Lewis T T, Rees N 1984 The fracture of acrylic polymers in water. *J. Mater. Sci.* 19: 1904–16

Kusy R P 1978 Characterization of self-curing acrylic bone cements. *J. Biomed. Mater. Res.* 12: 271–305

Lloyd C H 1982 The fracture toughness of dental composites. II The environment and temperature dependence of the stress intensification factor (K_{IC}). *J. Oral Rehab.* 9: 133–8

Lloyd C H 1984 The fracture toughness of dental composites. III The effect of environment upon the stress intensification factor (K_{IC}) after extended storage. *J. Oral Rehab.* 11: 393–8

Lloyd C H, Adamson M 1985 The fracture toughness of (K_{IC}) of amalgam. *J. Oral Rehab.* 12: 59–68

Lloyd C H, Adamson M 1987 The development of fracture toughness and fracture strength in posterior restorative materials. *Dent. Mater.* 3: 225–31

Lloyd C H, Dhuru V B 1985 Effect of a commercial bonding agent upon the fracture toughness (K'_{IC}) of repaired heavily filled composite. *Dent. Mater.* 1: 83–5

Lloyd C H, Iannetta R V 1982 The fracture toughness of dental composites. I The development of strength and fracture toughness. *J. Oral Rehab.* 9: 55–66

Lloyd C H, Mitchell L 1984 The fracture toughness of tooth coloured restorative materials. *J. Oral Rehab.* 11: 257–72

Margel-Robertson D, Robertson D, Barrett C R 1978 Fracture toughness, critical crack length and plastic zone size in bone. *J. Biomechanics* 11: 359–64

Melvin J, Evans F G 1973 Crack propagation. In: Fung Y C, Brighton J A (eds.) 1973 *Biomaterials Symp. AMD2.* American Society of Mechanical Engineers, New York

Montes G G M, Draughn R A 1987 Slow crack propagation in composite restorative materials. *J. Biomed. Mater. Res.* 21: 629–42

Morena R, Lockwood P E, Fairhurst C W 1986 Fracture toughness of commercial dental porcelains. *Dent. Mater.* 2: 58–62

Pilliar R M, Smith D C, Maric B 1986 Fracture toughness of dental composites determined using the short-rod fracture toughness test. *J. Dent. Res.* 65: 1308–14

Pilliar R M, Vowles R, Williams D F 1987 The effect of environmental aging on the fracture toughness of dental composites. *J. Dent. Res.* 66: 722–6

Pourdeyhimi B, Robinson H H, Schwartz P, Wagner H D 1986 Fracture toughness of Kevlar 29/polymethyl methacrylate composite materials for surgical implantations. *Ann. Biomed. Eng.* 14: 277–94

Rimnac C M, Wright T M, McGill D L 1986 Effect of centrifugation on the fracture properties of acrylic bone cements. *J. Bone Joint Surg.* 68A: 281–7

Ritter J E, Greenspan D C, Palmer R A, Hench L L 1979 Use of fracture mechanics theory in lifetime predictions for alumina and bio-glass coated alumina. *J. Biomed. Mater. Res.* 13: 251–63

Roberts J C, Powers J M, Craig R G 1977 Fracture toughness of composite and unfilled restorative resins. *J. Dent. Res.* 56: 748–53

Roberts J C, Powers J M, Craig R G 1978 Fracture toughness and critical strain energy release rate of dental amalgam. *J. Mater. Sci.* 13: 965–71

Robinson R P, Wright T M, Burstein A H 1981 Mechanical properties of poly(methyl methacrylate) bone cements. *J. Biomed. Mater. Res.* 15: 203–8

Scott G E 1988 Fracture toughness and surface cracks. *Angle Orthod.* 58: 5–8

Sih G C, Berman A T 1980 Fracture toughness concept applied to methyl methacrylate. *J. Biomed. Mater. Res.* 14: 311–24

Stafford G D, Huggett R, Causton B E 1980 The fracture toughness of denture base acrylics. *J. Biomed. Mater. Res.* 14: 359–71

Weber S C, Bargar W L 1983 A comparison of the mechanical properties of Simplex, Zimmer and Zimmer Low Viscosity bone cements. *Biomat. Med. Dev. Art. Org.* 11: 3–12

Wright T M, Hayes W C 1977 Fracture mechanics parameters for compact bone—Effects of density and specimen thickness. *J. Biomechanics* 10: 419–30

Wright T M, Sullivan D J, Arnoczky S P 1984 The effect of antibiotic additions on the fracture properties of bone cements. *Acta Orthop. Scand.* 55: 414–18

Wright T M, Trent P S 1977 The strength and fracture properties of fibre-reinforced polymethyl methacrylate bone cement. *IRCS Med. Sci.* 5: 7–8

Wright T M, Trent P S 1979 Mechanical properties of aramid fibre-reinforced acrylic bone cement. *J. Mater. Sci.* 14: 503–5

Yoshimoto S 1984 Study on the fracture toughness of posterior composite resin restorative material. *Kanagawa Shigaku* 19: 19–30

C. H. Lloyd
[University of Dundee, Dundee, UK]

G

Glasses: Agricultural and Vetinary Applications

The chemical durability of glass has received attention for many years (Newton 1985) and the work has culminated in a level of knowledge that has led to improved glass applications. The most notable result of this research has been the development of glasses that are designed to contain highly dangerous radioactive waste materials (Hench 1985); these can last for 10^4–10^6 years in storage.

In contrast to these durable glasses, it is well known (Newton 1982) that problems exist, particularly in Europe, with regard to the preservation of historical stained-glass windows. The windows react with the atmosphere and slowly dissolve as a result of the effects of moisture and atmospheric pollution (Knott 1985). The problems associated with these glasses relate to their composition; the durability of glasses is a function of their glass-former content (see *Ormosils: Organically Modified Silicates*).

Little interest was shown in exploiting soluble glasses until the 1970s, when Drake (1974, 1978) proposed that soluble controlled-release glasses could be put to use; for example, as corrosion inhibitors and as medical and agricultural aids. Since this time, increasing interest has been shown in controlled-release glasses involving the systematic dissolution of a glass and the release of products from the glass into various systems.

The development of such systems has been most successful in their application to agriculture. Glasses are currently on sale that can cure trace-element deficiencies in animals (Telfer et al. 1983) and further products will become available that will slowly release active ingredients to benefit plants.

1. Controlled-Release Glasses Based on Phosphate Glass Systems

In order to understand the mechanisms involved in controlled-release glasses, it is necessary to consider the properties of phosphate-based glasses (Knott 1983, Knott et al. 1985). These glasses form the basis of all the applications to be discussed, although other systems such as borates and silicates display similar potentialities.

Very little work has been reported on phosphate glasses containing 50 mol% or less P_2O_5 (Ray et al. 1973). The bulk of the studies of this system have been concerned with phosphorus-rich glasses, which are insoluble. Telfer et al. (1983) showed that glasses could be produced with as little as 30 mol%

Table 1
Solubility of a copper-containing glass. Compound quantities are expressed in mol%[a]

P_2O_5	Na_2O	CuO	Release rate (mg cm^{-2} per day)
50	50	0	4120
46	46	8	3550
42	42	16	1550
38	38	24	163
30	30	40	15.6

a after Telfer et al. 1983

glass former in this system and glass systems containing 30–50 mol% P_2O_5 are the basis for most of the commercial exploitation that has followed.

Table 1 illustrates how copper (in ionic form) can be released in a controlled manner from simple phosphate glasses. Table 2 illustrates how the quantity of copper ions dissolved from a glass containing 24 mol% CuO can vary as the quantity of P_2O_5 and Na_2O is reduced from a total of 76 mol% to 64 mol%. Quantities of CaO, MgO or both are added to control solubility. The results of both investigations were obtained from experiments carried out on the insertion of the glasses into the reticulorumen of live sheep (after Telfer et al. 1973).

In the examples quoted, the whole glass dissolves, releasing all its components in direct proportion to their composition. The release rate is proportional to the square root of time even after periods of one year and release is therefore deduced to be controlled by diffusion.

Table 2
Solubility control using CaO and MgO. Compound quantities are expressed in mol%[a]

P_2O_5	Na_2O	MgO	CaO	CuO	Release rate (mg cm^{-2} per day)
38	38	0	0	24	163
36	36	0	4	24	40.5
34	34	4	4	24	11.0
34	34	0	8	24	6.5
32	32	0	12	24	3.2
32	32	4	8	24	1.3

a after Telfer et al. 1983

Table 3
Mineral requirements for animals

Large requirement (g kg^{-1} food per day)		Trace elements (mg kg^{-1} food per day)	
Calcium	15	Iron	20–80
Phosphorus	10	Iodine[a]	0.25–0.8
Potassium	2	Copper[a]	5–10
Sodium	1.6	Cobalt[a]	0.11
Chlorine	1.1	Selenium[a]	0.1–0.8
Sulfur	1.5	Zinc	<1
Magnesium	0.4	Manganese	<1

a most-common deficiencies in sheep and cattle

2. Trace-Element Deficiencies in Ruminant Animals

Ruminants are animals that have more than one stomach and the group includes cattle, sheep, goats and deer. All animals need carbohydrates, fats, proteins, vitamins and minerals. The mineral group can be divided into those minerals essential in large amounts and those needed in trace amounts, which are often referred to as trace elements (see Table 3). The microdeficiencies in ruminants are expressed in milligrams per kilogram of food per day.

The absence of trace elements in grazing pasture is a worldwide effect and Fig. 1 illustrates the areas of deficiency in the UK. Obviously, different countries have their own particular deficiencies; however, the problem is growing worldwide since, in general, the application of artificial fertilizers seems to have a detrimental effect on trace-element concentrations.

Copper deficiency is the most widely occurring deficiency in the world; it is, however, unusual for only one deficiency to occur in isolation. The most usual deficiencies needing treatment are combinations of two or more of copper, selenium, cobalt, zinc and iodine. The most common combination of deficiencies that cannot be dealt with by a single treatment is that of copper, selenium and cobalt (Telfer et al. 1983).

2.1 Treatment

Treatment for a trace-element problem such as copper deficiency is readily available by adding copper salts to drinking water or by giving injections. However, in the case of animals that are grazing over large areas, any individual treatment requires that they must be gathered together. There is a danger of exceeding the toxic level when animals are injected and, in the case of deficiency treatment, twice the required daily dosage in cattle may be lethal. Other remedies, such as animal "licks", assume that animals know best what is essential in a diet. The feasibility of supplementing the diet of ruminant animals with trace elements by means of soluble glasses has been shown by Allen et al. (1979), and Telfer and Zervas (1982). The method of implanting the glass subcutaneously, as used by Allen et al. (1979), eliminates cobalt supplementation of the animal, whereas the use of an oral rumen-bolus method of administration, as devised by Dewey et al. (1958), is a method that allows all trace elements to be made available to the animal.

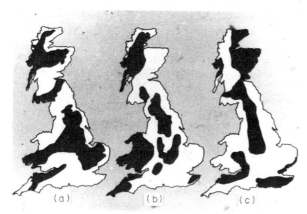

Figure 1
Areas of deficiency of (a) copper, (b) cobalt and (c) selenium. Trace-element deficiency is widespread, but the shaded areas illustrate the regions of highest incidence

3. Glass Compositions

Combined work at the University of Leeds between the department of Animal Physiology and the department of Ceramics has succeeded in producing a range of soluble glasses suitable for providing trace elements to ruminants (Telfer et al. 1983). As animals need phosphorus, calcium, sodium and magnesium in large amounts, the most obvious glass compositions are based on a range of phosphate glasses. The glasses have a general composition

28–50 ml% P_2O_5, 28–50 mol% Na_2O, 0–28 mol% CaO, 0–28 mol% MgO and 0.1–20 mol% trace elements.

The glasses are melted at temperatures between 1000 °C and 1150 °C, and are cast into boluses of variable dimensions dependent on the size of the animal and the life expectancy of the glass. The glasses are annealed after casting at 350–530 °C, depending on composition, the temperature being determined by differential thermal analysis (DTA). The desired release rate of trace elements is less than 25 mg cm^{-2} per day.

3.1 Administration and Operation

The dimensions of the bolus are designed so that it can be administered orally and so that it lodges in the reticulorumen (the first stomach) of the animal. For lambs, the bolus is approximately 1 cm in diameter and 4 cm in length with a weight of 17 g. For sheep, it is approximately 1.4 cm in diameter and 4.5 cm in length with a weight of 27 g, while for cattle it is approximately 2.5 cm in diameter and 8 cm in length with a weight of 115 g.

When lodged in the reticulorumen of the animal the bolus dissolves over a period of time, releasing rectifying elements into the animal's system. A release rate of 8 mg cm^{-2} per day will provide treatment for about one year.

4. Other Applications of Soluble Glasses in Treatment of Ruminant Animals

The most common deficiency in ruminant animals after trace-element deficiency is magnesia deficiency. Acute and chronic hypomagnesemic tetany are found in adult ruminant animals as a result of low levels of blood magnesium.

Acute hypomagnesemia is a result of blood-magnesium levels falling rapidly over 1–2 days, probably because of a variety of factors, but mainly owing to transport, cold stress or change in diet. Chronic hypomagnesemia develops over long periods of time and is caused by malnutrition.

In acute hypomagnesemia the magnesium levels in the blood plasma (red blood cells) drop from 2 mmol l^{-1} to 0.25 mmol l^{-1} very rapidly. There are no obvious signs that this is occurring until it has happened, at which point the animal will go into tetanic convulsions and will usually die. This type of death is responsible for 3% of the cattle losses in the world and in herds where it occurs the incidence is probably 10–15%, so that any one farmer may be severely affected.

Although the complete story is not known, the major consistent cause of acute hypomagnesemia is a change of diet from winter "stall" feed to lush spring grass. Although there may be a change in magnesium content of the diet, it is actually the high potassium content of spring grass (especially when fertilizers have been applied) that alters the magnesium metabolism.

The ruminant absorbs its daily dose of magnesium from the reticulorumen. This process is an active one, a small amount being passively absorbed. The passive component becomes more relevant as the concentration of magnesium in the rumen increases, so if this happens more magnesium will be absorbed.

There is no direct hormonal control over magnesium in the body and there are very few body reserves of the element. The ruminant is therefore heavily dependent on an adequate supply of magnesium in the rumen. Normally, about 30% of dietary magnesium intake is absorbed, but when there are high potassium levels in the rumen this can be reduced to 15–20%.

The problem is greatest in cattle and is seen mainly in pregnant and lactating animals, which have higher demands for magnesium. Even though the major risk is in the first few days, a long-acting supplement to the diet is advisable. Present methods of supplementation involve "dusting" the pastures with magnesium salts, but this does not ensure that each animal receives adequate magnesium. Soluble phosphate glasses based on the work of Telfer et al. (1983) have been used experimentally with most encouraging results (Beardsworth 1987). The glasses containing 15–30 mol% MgO have been shown to release 0.6–53 mg cm^{-2} per day in the rumen and samples of 3.5 cm × 2 cm diameter have dissolved over a six-week period, providing an adequate supply of magnesium to an animal during the critical period.

5. Soluble Glasses as Fertilizers

Drake (1978) has shown that it is possible to produce a particulate fertilizer composition which comprises a glassy matrix based on a phosphate glass that incorporates nitrogen in the form of calcium cyanamide. Conventional fertilizers suffer from many problems. In dry soils, plants are susceptible to damage by excessive nutrient concentrations, while in wet conditions the nutrients are liable to be wasted as a result of being removed from the soil in the neighborhood of the plant roots. In wet conditions, there is the further disadvantage that leaching is liable to produce pollution of adjacent water courses.

These risks may be partly ameliorated by recourse to lighter and more frequent applications of fertilizer, but this increases the application costs. A particular feature of a slow-release fertilizer is that relatively large concentrations can be applied at, or before, the seeding stage despite the susceptibility of seedlings to high concentrations of normal immediately assimilable fertilizer.

A vitreous matrix can be readily made comprising all the commonly known nutrients and trace elements, with the exception of nitrogen. Such a matrix is based on P_2O_5 as a glass former and typically uses calcium oxide as a modifier. The vitreous phase may contain all the normal major fertilizer elements such as potassium and magnesium and also trace elements such as iron, boron, manganese, sulfur, vanadium, copper, cobalt, zinc and molybdenum.

To incorporate nitrogen in the fertilizer, calcium cyanamide is added to the molten glass after the glass has been fused; that is, at temperatures of about 800 °C. About 20 mol% can be added, but stirring is essential. The glass is then cast and ground to a suitable particle size; the estimated amount of nitrogen in the glass is 10 wt%.

It is possible to increase the nitrogen content by incorporating the cyanamide into the base glass at a higher temperature and under a pressurized nitrogen atmosphere. A further method proposed by Drake (1978) is to coat cyanamide particles with a layer of metaphosphate glass which has a lower softening point than the cyanamide. The metaphosphate glass acts as a buffer protecting the cyanamide from chemical attack at higher temperatures. By incorporating the coated cyanamide powder into the base glass it is claimed that 15 wt% nitrogen can be fixed in the fertilizer. In the latter application, the cyanamide was first incorporated into a melt of metaphosphate glass at 800 °C; this was cooled, pulverized and added to the fertilizer base glass at 900–920 °C.

Solution rates quoted over a 30 hour period when leached at room temperature in 10 ml water range from 2600 µg ml^{-1} to 8 µg ml^{-1}. Commercial interest in these controlled-release fertilizers seems to be limited because of the problem of nitrogen fixation.

Bibliography

Allen W M, Drake C F, Sansom B F, Taylor R J 1979 Trace-element supplementation with soluble glasses. *Ann. Rech. Vet.* 10: 356–8

Beardsworth L 1987 Dietary factors affecting magnesium, calcium and phosphate absorption from the reticulo rumen of sheep. Ph.D. thesis, University of Leeds

Dewey D W, Lee H J, Marston H R 1958 Provision of cobalt to ruminants by means of heavy pellets. *Nature (London)* 181: 1367–71

Drake C F 1974 Vitreous controlled release fertiliser compositions. UK Patent No. GB 1,512,637

Drake C F 1978 Glass composition. UK Patent No. GB 2,037,735B

Hench L L 1985 Leaching of nuclear waste glasses. *Glass ... Current Issues.* Nijhoff, Dordrecht, Netherlands, pp. 631–7

Knott P 1983 The glassy state. Ceramic monographs. *Handb. Keram.* 32(6): 1–5

Knott P 1985 *Protection of Historical Stained Glass*, NATO Scientific Affairs Report. North Atlantic Treaty Organization, Brussels

Knott P, Algar B, Zervas G, Telfer S B 1985 Glass—A medium for providing animals with supplementary trace elements. In: Mills C F, Bremner I, Chesters J K (eds.) 1985 *Trace Elements in Man and Elements—TEMA 5. Proc. 5th Int. Symp.* Commonwealth Agricultural Bureaux, Slough, pp. 708–14

Newton R G 1982 *The Deterioration and Conservation of Painted Glass: A Critical Bibliography*, Corpus Vitrearum Medii Aevi, Occasional Papers. Oxford University Press, Oxford

Newton R G 1985 The durability of glass—A review. *Glass Technol.* 26: 21–36

Ray N H, Lewis C J, Laycock J N C, Robinson W D 1973 Oxide glasses of very low softening point. *Glass Technol.* 14: 50–9

Telfer S B, Zervas G 1982 The release of copper from soluble glass bullets in the reticulo-rumen of sheep. *Anim. Prod.* 34: 379–80

Telfer S B, Zervas G, Knott P 1983 Water soluble glass articles, their manufacture and their use in the treatment of ruminant animals. UK Patent No. GB 2,116,424B

P. Knott
[University of Leeds, Leeds, UK]

Glasses: Medical Applications

There are two major categories of medical applications of glasses: *in vitro*, where glass is used as a passive substrate for growth of cell cultures or immobilization of enzymes and antigen–antibodies; and *in vivo*, where surface-active glasses or glass ceramics are used as implants to replace living tissues. Glasses used *in vitro* are usually alkali borosilicates or porous silica, both of which are very resistant to surface chemical attack. In contrast, glasses or glass ceramics used as implants are designed to have controlled rates of surface reaction with body fluids, resulting in a chemical bond to living tissues.

1. In Vitro Applications

1.1 Tissue Culture Substrates

Glasses have been used in the development and production of vaccines and therapeutic hormones since the 1930s, when it was discovered that cell cultures could be made to grow on glass surfaces if the proper nutrients were supplied. Alkali borosilicate glasses, such as Pyrex, are used because the glass surface can be cleaned and sterilized repeatedly with little chemical damage. A thin protective dense SiO_2-rich film forms on the glass during cleaning, allowing cells of various types to attach, spread rapidly, grow and multiply. Limitations, however, of tissues grown *in vitro* are:

(a) inability to maintain differentiated cells,

(b) failure to establish cell aggregation and physiological three-dimensional structuring, and

(c) alteration of cell morphology and function with time.

Use of glasses with bioactive surfaces may solve some of these problems.

1.2 Enzyme Carriers

By chemically adhering enzymes to porous glass, such as Corning 96% SiO_2 with 90 nm pores, it is possible to make the enzymes retain activity for many days. Either direct enzyme–glass binding, or covalent coupling of the enzyme to the glass by amino-functional silane coupling agents is possible, but the coupling agents offer more versatility. Effective enzyme activity has been retained in this manner for trypsin, papain, ficin, urease, glucose oxidase, peroxidase, L-amino acid oxidase and alkaline phosphatase.

1.3 Antigen–Antibody Carriers

Similar techniques are used to bond antigens and antibodies irreversibly to insoluble porous-glass carriers without loss of specific combining capacity. Insoluble antigens or antibodies are termed immunoadsorbents: they are used for antibody and antigen isolation, purification, and study; enzyme removal or isolation; virus and microorganism detection; and radioimmunoassay. Human γ-globulin, covalently coupled to porous-glass particles, has yielded recovered antibodies that were 90–100% precipitable.

2. In Vivo Applications

2.1 Compositions

Surface-active glasses, also termed Bioglasses, based on a certain range of Na_2O–CaO–P_2O_5–SiO_2 compositions, have been developed for use as medical and dental implants. Partially or fully crystallized glasses are termed either Bioglass-ceramics, surface-reactive glass ceramics, or bioactive glass ceramics. These materials are unique in that they form a bond with living tissues by means of controlled chemical reaction at the surface of the implant. Thus, the implant becomes quickly and permanently fixed at the interface with the host, before fibrous encapsulation can occur. Loosening of the implant is prevented because of the interfacial bond.

For Bioglasses, controlled biological reactions occur at the interface only within specific limits of composition and surface reactivity. Within these limits the material bonds to tissues (region A in Fig. 1). When these limits are exceeded, the material behaves in the physiological environment either as a nearly inert material (region B) eliciting fibrous capsule formation, or resorbs in the tissue (region C). Compositional modifications, such as addition or

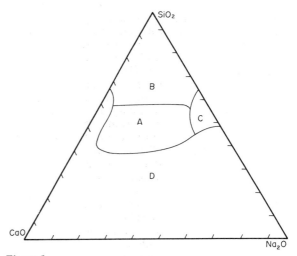

Figure 1
Bioglass compositional range for bonding to rat bone: A, bonding at less than 31 days; B, nonbonding, reactivity is too low, C, nonbonding, reactivity is too high; D, nonbonding, non-glass-forming

substitution of K_2O, MgO, CaF_2, B_2O_3 or Al_2O_3, generally contract region B.

2.2 Mechanism of Bioglass–Tissue Bonding

A comparison of the surface chemistry of glasses within region A with those of region B in Fig. 1, show that all bone-bonding Bioglasses develop a large increase in surface area when exposed to physiological solutions. The ultrapores of 3–30 nm size that result from formation of an active silica-rich gel on the implant surface, nucleate hydroxylapatite crystals on that surface. The hydroxylapatite agglomerates provide a means of incorporating collagen, mucopolysaccharides and glycoproteins within the surface-active layer of the implant. The resulting composite organic–inorganic interface is responsible for the mechanically strong bond at the interface exhibited by Bioglasses within region A, the bone-bonding boundary. Glass compositions that fall outside the boundary (region B) have too little reactivity to form a stable hydroxylapatite surface-layer and active bond sites for the organic molecules. Glasses in region C are too reactive, surface layers breakdown and total dissolution occurs.

Examination by transmission electron microscopy of well-mineralized rat bone bonded to 45S5 Bioglass shows a gradual change in electron contrast across the interface (Fig. 2). This is due to a gradation of hydroxylapatite concentration across the glass–bone bond. The inserts of electron diffraction patterns of bone (Fig. 2a) and glass reaction zone (Fig. 2b) show the continuity of the bonded interface for this type of biomaterial. The high strength of the bond is indicated by the fracture artifact (due to microtoming) in the glass reaction layer, which has failed to propagate into the bone.

2.3 Toxicology and Biocompatibility

Evidence for a lack of toxicity of various Bioglass and surface-active glass-ceramic formulations has been established from results of a variety of *in vivo* and *in vitro* test models (Table 1). The surface activity critical in bone adhesion is also seen in soft connective tissues, and without accompanying toxic effects in either type of tissue. Adherence to muscle and other soft tissues occurs if the implant is immobilized. Even if motion and wear should occur and produce particulate Bioglass, the material is quickly eliminated without adverse effect.

2.4 Orthopedic Applications

Because of low flexural strength, surface-active glasses must be used as coatings on high-strength

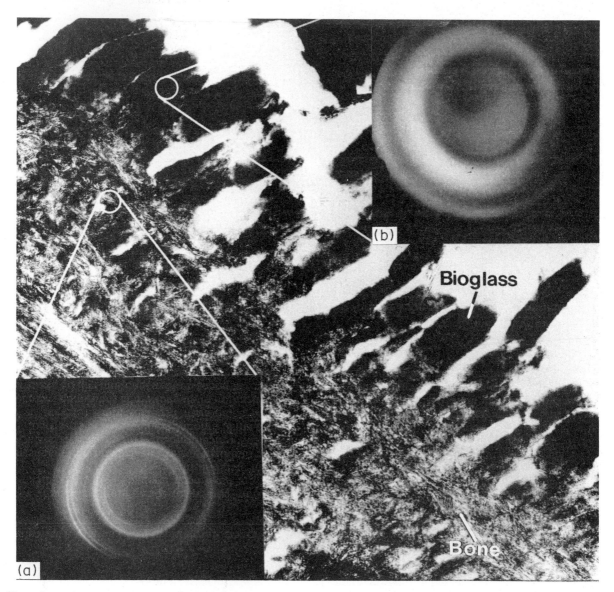

Figure 2
Transmission electron micrograph of a bonded bone–Bioglass interface: (a) electron diffraction pattern of the bone, (b) electron diffraction pattern of the glass reaction layer (Courtesy of T K Greenlee)

Table 1
Toxicology and biocompatibility tests for Bioglass

Type of test	Species	Material form	Results
In vitro			
bone cells	rat	solid	cells grew and divided normally (30 days)
fibroblasts	rat	solid	cells grew and divided normally (30 days)
fibroblasts	chick	solid	cells grew and divided normally (21 days)
fibroblasts	human	solid	cells grew and divided normally (21 days)
fibroblasts	hamster (CHO)	solid	grew slowly for 7 days, resumed normal rate when removed to more usual substrates
fibroblasts	hamster (NIL)	solid	grew slowly for 7 days, resumed normal rate when removed to more usual substrates
lymphocytes	human	solid	RNA synthesis remained normal
lymphocytes	human	extract	RNA synthesis remained normal
macrophages	mouse	solid	phagocytic activity remained normal
macrophages	mouse	extract	phagocytic activity remained normal
macrophages	rat	powder	cells with ingested material continued to behave normally in culture
In vivo			
tibia	rat	solid	bonded to bone with no toxic effects (up to 2.5 years)
femur	rat	solid	bonded to bone with no toxic effects (up to 2.5 years)
cervical spine	dog	solid	bonded to bone with no toxic effects (3 months)
femur	dog	solid	bonded to bone with no toxic effects (6 months)
femur	primates	solid	bonded to bone with no toxic effects (up to 1.5 years)
mandible	primates	solid	bonded to bone with no toxic effects (up to 6.5 years)
ossicles	mouse	solid	bonded to periosteum with no toxic effects (up to 1 year)
subcutis	rat	solid, powder	adhesion to collagen, no toxic effect
muscle	rat	solid	no toxic effect
muscle	rabbit	fibers	no toxic effect, some adhesion
peritoneal cavity	rat	powder	eliminated without effect, no fibrosis
peritoneal cavity	rat	solid	thin capsule, no fibrosis, normal cell growth on surface (up to 4 weeks)
pulmonary embolus	mouse	powder	no granuloma or thrombosis

ceramics, such as Al_2O_3, or metals, such as 316L stainless steel or cobalt–chromium alloys, for load-bearing devices such as segmental bone replacements and joint prostheses. The main motivation for the use of surface-active glass coatings is the replacement of poly(methyl methacrylate) (PMMA), commonly used as a self-polymerizing grouting agent between bone and the orthopedic device. Although excellent short-term stabilization of devices is achieved with PMMA, there are complications: low blood pressure due to release of monomer into the blood stream, necrosis of bone due to the heat generated in the exothermic polymerization reaction, and eventual loosening of devices due to lack of interfacial bonds. These problems are eliminated by the use of surface-active glass coatings. However, improved mechanical designs and surgical techniques are needed to achieve the glass–bone bond within the first few weeks after surgery. If motion at the interface between glass and bone occurs, mineralization of the bond is prevented, and a non-adherent fibrous capsule results. Figure 3 is an x radiograph of a successful 45S5 Bioglass-coated hip

prosthesis in a monkey. Mechanical tests of the femoral stem–bone bond showed that the distal condyles were torn off the bone with no loosening of the implant; three-point bend-tests causing fracture of the bone did not loosen the interface either. Use of a square cross section for the stem, and a tight fit at the time of surgery, gave a functional load-bearing prosthesis within days of surgery. Bonding of the coating to bone occurs as early as eight weeks. Bulk 45S5 Bioglass implants in rats develop a strong bond as early as two weeks. Less reactive Ceravital surface-active glass ceramics achieve a strong bone bond by the end of four weeks. Maturation of the bone bond continues for several months for most compositions.

Surface-active glass and glass-ceramic coatings can be applied to surgical metal alloys by several techniques. Flame-spray coatings are easily applied to simple or complex shapes, but the process is difficult to control. Rapid immersion of a preoxidized metal device in molten glass is the easiest and most acceptable procedure for simple geometries, but is difficult to apply to complex devices. Modified

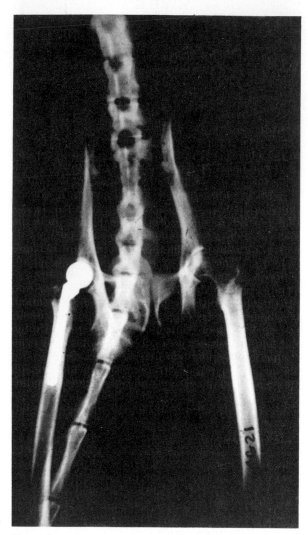

Figure 3
Radiograph of a Bioglass-coated hip prosthesis in a monkey

procedures. Although the glass-ceramic coating can decrease metallic corrosion, and perhaps wear, elimination of the cementing media is not possible with this composition because it lies within region B of Fig. 1 and therefore behaves as a nearly inert biomaterial.

2.5 Dental Applications

Several formulations of surface-active glasses and glass ceramics have performed successfully as implants for tooth replacements. Although breakage of bulk glass implants is a problem, it has been dealt with by using Bioglass coatings on dense Al_2O_3 or metal substrates. Even when gingival level breakage was observed in animal studies, the root portion of the implant remained solid. The mandibular ridge was restored distal to the fracture, and healed mucosa and healthy gingival tissues were present after two-year trials. A mandibular ridge augmentation in a baboon using 45S5 Bioglass remained solidly bonded and fully functional up to six years, when the experiment was terminated.

Orthodontic anchors in monkeys, composed of 45S5 Bioglass coatings on dense Al_2O_3-blade implants, show no movement of the implants in the bone and no resorption of bone adjacent to the implant, even with forces of 4.2–9.3 N applied by both lingual and maxillary arch supports. The direct bond of bone to the surface-active glass coating apparently allows force to be applied to the bone without initiating osteoclast activity. Finite-element and histological analysis of bonded Bioglass tooth implants, show evidence of a 100–200 μm thick, elastically compliant inorganic interface which dissipates the applied load. Electron microprobe and microhardness measurements across the bonded tooth implants confirm this conclusion. These favorable dental responses to surface-active glasses require firm fixation in the implant site and appropriate steps to prevent mobility as the bond develops. When motion occurs, the bond is disrupted, fibrous encapsulation follows and complications leading to loosening and implant removal are almost certain. However, if the initial placing of the implant in its host tissue is correct and immobilization during the 2–4 week bonding period is complete, then the surface-active glass-coated dental implant will function as an ankylosed tooth and may be treated as such.

Bioglass implants are being used in patients to maintain the alveolar ridge after multiple extractions. These are placed in the socket after extraction, bond to the surrounding bone and help support the dentures. In this way the resorption of alveolar bone under dentures is prevented.

2.6 Other Applications

Polymer and metal implants used in repair of the ossicular chain of the middle ear are often unsatisfactory due to development of excessive scar tissue,

enamelling methods are also used with a top coat of either Bioglass powder or Ceravital glass-ceramic granules fused to an undercoat.

An intermediate approach to eliminate PMMA, but still retain the surgical ease of the grouting technique, is the use of Ceravital granules in PMMA, to form a bioactive cement. Use of less than 70 vol.% of the bone-bonding active glass-ceramic phase significantly reduces the temperature rise during *in situ* polymerization and apparently also reduces free monomer transport in the circulatory system.

Total joint components and segmental bone replacements made of Nucerite glass-ceramic coated stainless steel are also used with PMMA cementing breakup or migration of the implant, chronic inflammation or infection, or extrusion of the prosthesis from the ear. Middle-ear implants made of 45S5 Bioglass and Ceravital show excellent adherence, absence of mobility, restoration of function, and little or no tendency for extrusion or exacerbation of infection in extensive animal tests and human trials. The bonding interface of the Bioglass implants consists of a very thin, tightly adherent, collagen layer similar to the bond between normal articulating ossicles. Ceravital bonds are osseous in character and these implants exhibit osteogenesis in this application. Surface-active glasses and glass ceramics appear especially suited to repair in the head and neck region, because of excellent interfacial adherence, absence of color to show through skin tissues, and low strength requirements. Successful animal results have been obtained in cranial repair, augmentation of the zygomatic arch, and mandibular and maxillary reconstruction.

Cervical spinal fusion is another area where standard metallic and polymeric biomaterials have not been successful, and autogenous bone grafts from the iliac crest of the pelvis (which do succeed) have painful side effects. Animal experiments using 45S5 Bioglass plugs produce fusion of the cervical vertebrae and therefore could provide an alternative to autogenous grafts.

One of the most intriguing potential applications of surface-active glasses is in the field of cell and tissue culture, and genetic engineering. Cell-culture studies show that cell types are sensitive to the active surfaces. Cells with very few membrane-attachment complexes are very slow to attach, spread and divide. Thus they can be maintained for long times in a resting state without modification of cellular morphology and without contact or confluence with other cells. Subsequent removal of the cells from the bioactive surface restores their normal behavior. Therefore, manipulation of primary cell lines in this stabilized resting state may result in unique modification of the cells.

See also: Biocompatibility: An Overview; Glasses: Agricultural and Vetinary Applications

Bibliography

Blencke B A, Bromer H, Deutscher K K 1978 Compatibility and long-term stability of glass-ceramic implants. *J. Biomed. Mater. Res.* 12: 307–16

Blumenthal N C, Posner A S, Cosma V, Gross U 1988 The effect of glass-ceramic bone implant material on the *in vitro* formation of hydroxyapatite. *J. Biomed. Mater. Res.* 22: 1033–41

Boretos J, Eden M (eds.) 1984 *Contemporary Biomaterials*. Noyes Data, Park Ridge, NJ

Gross U, Brandes J, Strunz V, Bab I, Sela J 1981 The ultrastructure of the interface between a glass ceramic and bone. *J. Biomed. Mater. Res.* 15: 291–305

Grote J J (ed.) 1984 *Biomaterials in Otology* 1984 Martinus Nijhoff, Boston, MA

Hench L L 1980 Biomaterials. *Science* 208: 826–31

Hench L L 1989 Bioceramics and the origin of life. *J. Biomed. Mater. Res.* 2: 685–704

Hench L L, Ethridge E 1982 *Biomaterials: An Interfacial Approach*. Academic Press, New York

Hench L L, Paschall H A 1973 Direct chemical bond of bioactive glass-ceramic materials to bone and muscle. *J. Biomed. Mater. Res. Symp.* 4: 25–42

Hench L L, Paschall H A, Allen W C, Piotrowski G 1975 Interfacial behavior of ceramic implants. In: Horowitz E, Torgesen J L (eds.) 1975 *Biomaterials*. NBS Special Publication 415. National Bureau of Standards, Washington, DC, pp. 19–35

Hench L L, Splinter R H, Greenlee T K, Allen W C 1971 Bonding mechanisms at the interface of ceramic prosthetic materials. *J. Biomed. Mater. Res. Symp.* 2: 117–41

Hench L L, Wilson J 1984 Surface active biomaterials. *Science* 226: 630–6

Merwin G, Atkins J, Wilson J, Hench L L 1981 Comparison of ossicular replacement material in a mouse ear model. *J. Am. Acad. Otol. (Head and Neck Surgery)* 90: 461–9

Pigott G H, Ishmael J 1970 A comparison between *in vitro* toxicity of PVC powders and their tissue reactions *in vivo*. *Ann. Occ. Hyg.* 22: 111–26

Piotrowski G, Hench L L, Allen W C, Miller G J 1975 Mechanical studies of the bone Bioglass interfacial bond. *J. Biomed. Mater. Res. Symp.* 6: 47–61

Reck R 1981 Tissue reactions to glass ceramics in the middle ear. *Clin. Otolarynogol.* 6: 63–5

Smith J R 1977 Bone dynamics associated with the controlled loading of Bioglass coated aluminum oxide endosteal implants. Masters Thesis, University of Washington, Seattle, WA

Stanley H, Hench L L, Bennett C G Jr, Chellemi S J, King C J III, Going R E, Ingersoll N J, Ethridge E C, Kreutziger K L, Loeb L, Clark A E 1982 The implantation of natural tooth form Bioglass in baboons—Long term results. *Int. J. Oral Implantol.* 2(2): 26–36

Stanley H, Hench L L, Going R, Bennett C, Chellemi S J, King C, Ingersoll N, Ethridge E, Kreutzinger K 1976 The implantation of natural tooth form Bioglasses in baboons. A preliminary report. *Oral. Surg., Oral Med., Oral Path.* 42: 339–56

Weetall H H 1972 Insolubilized antigens and antibodies. In: Hair M L (ed.) 1972 *The Chemistry of Biosurfaces*, Vol. 2. Dekker, New York, pp. 597–631

Weetall H H, Messing R A 1972 Insolubilized enzymes on inorganic materials. In: Hair M L (ed.) 1972 *The Chemistry of Biosurfaces*, Vol. 2. Dekker, New York, pp. 563–95

Wilson J, Pigott G H, Schoen F J, Hench L L 1981 Toxicology and biocompatibility of Bioglasses. *J. Biomed. Mater. Res.* 15: 805–17

L. L. Hench and J. Wilson
[University of Florida, Alachua, Florida, USA]

Gold Alloys for Dental Use

Gold alloys have been used in dentistry, not only because their gold color is preferred, but also because they have extremely high chemical stability in the mouth and several desirable mechanical properties such as high strength, ductility and elasticity.

According to American Dental Association Specification No. 5 and International Standard (ISO 1562), dental gold alloys for casting are classified as Type I, II, III and IV depending on their gold and platinum metals contents and their mechanical properties. Among them, the Type IV alloys are designed to be age-hardenable by an appropriate heat treatment. However, the cause of the age-hardening and the nature of the related phase transformations have only recently been been extensively studied. This is not surprising because it is difficult to elucidate the hardening mechanisms of alloys as complex as dental gold alloys. Frequently, they contain five or more elements, the essential components of the alloys being gold, copper and silver. There is a consensus among researchers in this field that a more fundamental understanding of phase transformations is necessary for developing new alloys for dentistry, and for explaining the mechanical, chemical and biological properties of dental alloys.

Progress in metallurgy and materials science has been rapid during the 1980s, because transmission electron microscopy (TEM) has made it possible to study directly the relationship between microstructure and phase transformations in alloys. In particular, high-resolution electron microscopic (HREM) observation, coupled with the selected area electron diffraction (SAED) technique, has readily enabled the analysis of age-hardening mechanisms in dental gold alloys.

1. Phase Diagram of the Au–Cu–Ag Ternary System

Although a knowledge of the equilibrium phase diagram is an important aid in predicting the phase transformations resulting from the aging of alloys, the ordering regions in the Au–Cu–Ag ternary system have not yet been established with sufficient accuracy. In 1925, Sterner-Rainer showed that in the Au–Cu–Ag phase diagram, there is a region where a disordered phase coexists with an Au–Cu ordered phase. His phase diagram, however, was too sketchy to be used for quantitative description of phase equilibria in this system.

The Au–Cu–Ag ternary system is characterized by ordering regions and a miscibility gap in which copper-rich and silver-rich face-centered-cubic (fcc) phases coexist. The boundaries of the two-phase region were outlined at 673 K and 987 K using an x-ray diffraction method (Masing and Kloiber 1940). Thereafter, McMullin and Norton (1949) reported the limits of the two-phase region for five different temperatures.

In contrast, information on the stable region of AuCu ordering in the ternary system is much more fragmentary. Raub (1949) found superlattice reflections in the x-ray diffraction patterns of Au–Cu–Ag alloys containing up to 30% silver, while Hultgren and Tarnopol (1939) indicated that the first 5 at.% of silver, when substituted for a corresponding amount of gold in the equiatomic Au–Cu binary alloy, lowered the critical temperature for ordering by as much as 65 K. Hultgren and Tarnopol also reported that the AuCu II orthorhombic ordered phase was stable over a wider range of temperature in the Au–Cu–Ag ternary alloys than in the binary Au–Cu alloys. The occurrence of spinodal decomposition was described by Murakami et al. (1975).

Although the studies by Hultgren and Tarnopol (1939) and by Raub (1949) showed in detail the existing regions and structures of the ordered phases in the Au–Cu–Ag ternary system, Uzuka et al. (1981) were the first to draw phase boundaries involving ordering in the phase diagram. Their experimental results indicated that with increasing silver content the transformation zone of the AuCu I superlattice in the ternary alloys was pushed towards lower temperatures. Moreover, the coexisting region of the AuCu II superlattice and silver-rich disordered phase (designated α_2) was extended towards the side of higher silver concentration. However, their phase diagram was actually a map of their experimental data and was not, apparently, intended to do justice to the phase rule.

Kikuchi et al. (1980) calculated a "coherent" phase diagram for this ternary system, using the interaction-energy parameter determined to fit only the binary phase diagrams of Au–Cu, Cu–Ag and Ag–Au systems. A single parent lattice, (i.e., the fcc lattice) was assumed for all possible phases. Consequently, the AuCu II long period super-structure ($L1_{0-s}$) could not be taken into account, and Cu_3Au ($L1_2$) and AuCu I ($L1_0$) superstructures were considered as well as disordered fcc phases. Eventually, Yamauchi et al. (1980) constructed a plausible "incoherent" phase diagram by superimposing the theoretical "coherent" phase diagram on experimental data of incoherent phases. An isothermal section of their plausible incoherent phase diagram is shown in Fig. 1. This plausible phase diagram explained TEM observations of the AuCu ordering and two-phase decomposition processes in Au–Cu–Ag ternary alloys rationally. However, more experimental work is required to determine the three-phase region topology with certainty in this ternary system. Recently, a coherent phase diagram of the $Au_x(Ag_{0.24}Cu_{0.76})_{1-x}$ pseudobinary system was depicted by Nakagawa and Yasuda (1988) on the basis of TEM and SAED examinations as shown in Fig. 2. Three-phase regions of AuCu II, α_1 (copper-rich fcc) and α_2 phases, AuCu I, AuCu II

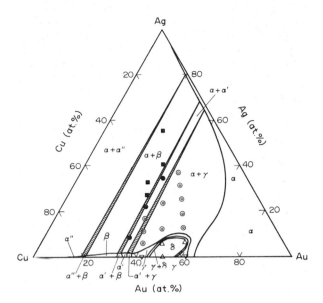

Figure 1
Isothermal section at 573 K of plausible "incoherent" phase diagram of the Au–Cu–Ag ternary system. α: Au- or Ag-rich disordered phase; α', α'': Cu-rich disordered phase; β: Au_3Cu type ordered phase; γ: AuCu II type ordered phase; δ: AuCu I type ordered phase (Yamauchi et al. 1980)

and α (disordered solid solution of fcc) phases, and Cu_3Au, AuCu II and α_2 phases in the SAED patterns were confirmed. Microstructural features of these regions were also studied in association with phase identifications.

2. Application of TEM to Studies of Age-Hardening Mechanisms in Au–Cu–Ag Ternary Alloys

TEM has important advantages over other methods of studying the hardening mechanisms in alloys in that:

(a) the results obtained are visual;

(b) it provides high resolution;

(c) additional information about structure and orientation can be obtained by using the SAED technique; and

(d) the dark field image formed using superlattice reflections readily enables ordered regions to be distinguished from disordered regions, since only ordered regions appear bright.

Notwithstanding these advantages, TEM studies of hardening mechanisms in dental gold alloys are relatively recent. The earliest TEM study was car-

ried out by Yasuda's group (Kanzawa et al. 1975) with the object of elucidating the correlation between microstructure and phase transformations in an 18 karat gold dental alloy, Au–35.7 at.% Cu–11.2 at.% Ag. They found that age-hardening in the alloy is brought about by a two-stage process, with initial hardening being due to the formation of coherent AuCu I ordered platelets, and secondary hardening resulting from twinning, as will be shown later. Shortly afterwards, Prasad et al. (1976) studied the age-hardening of a commercial Type III dental gold alloy, using TEM in addition to x-ray diffraction and hardness measurements. They reported that hardening was predominantly due to precipitation. The precipitates, which were formed on the {100} and {111} planes of the matrix, were homogeneously nucleated and coherent with it; ordering also played a role in hardening, but its contribution appeared to be very slight. The above results, however, were not in agreement with those of Yasuda's group. The data may still be insufficient for a definite conclusion to be made as to whether age-hardening in dental gold alloys results from ordering or from precipitation.

As can be seen in Fig. 1, the Au–Cu–Ag ternary system is characterized by ordering regions and a two-phase decomposition region. If the alloy has a composition falling in the ordering region, age-hardening will be due to an order–disorder transformation mechanism. Alternatively, if the alloy is located in the two-phase region, precipitation hardening will occur. Thus, it is supposed that the age-hardening characteristics of these alloys are

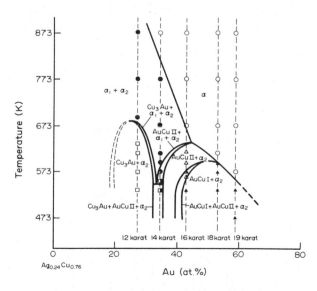

Figure 2
Experimental "coherent" phase diagram of the $Au_x(Ag_{0.24}Cu_{0.76})_{1-x}$ pseudobinary system (Nakagawa and Yasuda 1988)

affected by their composition, especially by the atomic ratio of gold to copper. Because the tendency for two-phase decomposition may markedly increase with decreasing gold content, the effect of the ordered phase on hardening may diminish or disappear altogether. However, it has not been clearly demonstrated that a difference in the age-hardening mechanism of ternary dental Au–Cu–Ag alloys occurs with changes in gold content. Therefore, Yasuda and his co-workers conducted studies to elucidate the differences in hardening mechanisms for 18, 16 and 14 karat gold alloys, in which the ratio of copper to silver was maintained at 65:35 by weight, using SAED and TEM in addition to x-ray diffraction, electrical resistivity measurements and hardness tests.

2.1 The 18 Karat Gold Alloy

Experimental results for the 18 karat gold alloy (Kanzawa et al. 1975), showed that age-hardening was due to the formation of platelets of AuCu I superlattice on the matrix {100} planes. These ordered platelets were arranged in a stepwise fashion that roughly followed ⟨110⟩ trace on the {100} planes, and their c-axes were distributed in the three cube edge directions of the parent disordered matrix. Thus, it was observed that at a later stage of aging, twinning took place on the $(101)_{tet}$ plane to relieve a considerable amount of strain induced in the matrix through changes in crystal symmetry.

2.2 The 16 Karat Gold Alloy

Age-hardening in the 16 karat gold alloy Au–43.2 at.% Cu–13.8 at.% Ag was, however, found to be due to the development of a long period antiphase domain structure in the AuCu II superlattice at lower aging temperatures (Yasuda et al. 1978). The unidirectional, alternating fine lamellar structure, consisting of the AuCu I superlattice, was also found to be present in several different areas of the same specimen. The age-hardening in this temperature range was thought to be due to the development of antiphase domain boundaries in addition to the strain field induced by the structural change. In a higher temperature range, an alternating coarse lamellar structure, consisting of an fcc lattice of copper-rich α_1 and silver-rich α_2 phases, was formed by discontinuous decomposition. These two phases were oriented parallel to one another. No age-hardening was observed in this temperature range. However, considerable age-hardening was found to occur in the middle temperature range, in spite of the similarity between the microstructures after aging at these temperatures and at higher ones. In this respect, it was thought that the ordering and the decomposition of supersaturated solid solution took place simultaneously during the aging process in this temperature range.

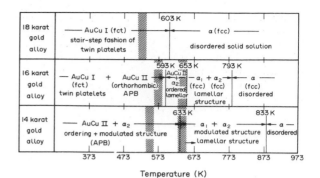

Figure 3
Schematic representation of age-hardening mechanisms and the associated microstructures in the 18, 16 and 14 karat ternary Au–Cu–Ag alloys: hatched bands indicate the temperature at which hardness peaks appear

2.3 The 14 Karat Gold Alloy

In the 14 karat gold alloy Au–49.7 at.% Cu–15.8 at.% Ag studied by Yasuda and Ohta (1979) which possessed an off-stoichiometric composition of AuCu, it was expected that age-hardening would also proceed from disordered solid solution by AuCu ordering accompanied by the two-phase decomposition. The microstructure of this alloy at lower aging temperature range showed a fine mottled contrast parallel to the ⟨100⟩ directions in the early stages of aging. This fine contrast was due to the formation of a modulated structure induced by spinodal decomposition, which was confirmed by the appearance of sidebands in x-ray diffraction patterns and satellite reflections in SAED patterns. With longer aging periods, the modulated structure changed to a "tweedlike" structure consisting of copper-rich α_1 and silver-rich α_2 phases. The α_1 phase gradually transformed to the long period superlattice AuCu II through the introduction of gold atoms from the alloy matrix into the α_1 phase. In the higher aging temperature range, a bright field TEM micrograph showed a large number of alternating dark and bright striations. These striations ran parallel to the ⟨100⟩ directions and were arranged almost at right angles to each other. The appearance of the striations suggested the presence of periodic variations of the lattice parameter and in the atomic scattering factor, caused by the periodic composition fluctuation along ⟨100⟩ (i.e., the formation of a modulated structure). The structural and morphological changes which give rise to age-hardening in the 18, 16 and 14 karat gold alloys are summarized and collated in Fig. 3.

2.4 Effect of Silver Content

The effect of changes in silver content on the age-hardening mechanism AuCu–Ag pseudobinary alloys was also studied by means of hardness tests, x-ray and SAED as well as TEM (Ohta et al. 1983). Hardening in a low silver-content alloy AuCu–6 at.% Ag aged at 573 K was caused by the formation of AuCu I ordered platelets formed along ⟨110⟩ directions of the matrix and the consequent increase in the elastic strain field. Growth of the ordered platelets caused twinning, which resulted in a softening of the alloy by releasing the elastic strain. The twinning of the AuCu I ordered phase was suppressed by the addition of more silver to the alloy.

The long period antiphase domain configuration was also observed in the alloy aged at 603 K (Yasuda et al. 1987). Fig. 4a shows the dark field image formed using the 110_z superlattice reflection. Figs. 4b,c show the corresponding SAED pattern and its schematic representation, respectively. It can be seen that a large number of the fringes abut each other at an angle which slightly deviates from a right angle, as was reported by Watanabe and Takashima (1975). HREM observation has proved to be an indispensable technique in unravelling the configuration of the long period antiphase domain boundary of AuCu II. Figure 5 shows an HREM image along [001] of periodic antiphase domain boundaries in AuCu–6 at.% Ag alloy. It is clear from this image that the antiphase domain boundaries have a wavy character; they are not strictly confined to {100} planes and their width also varies locally.

In contrast, age-hardening in the high silver-content alloys was attributed to the dual mechanisms of phase decomposition and ordering. In the AuCu–31.1 at.% Ag alloy, for example, aged at 573 K, phase decomposition was thought to proceed by spinodal decomposition, followed by the formation of a modulated structure of increasing periodicity which grows into a characteristic blocklike structure consisting of alternating ordered AuCu orthorhombic phase and silver-rich α_2 phase. In the initial stages of aging, most of the hardening was caused by the development of the modulated structure. The hardening that occurred upon further aging was attributed to ordering in the copper-rich α_1 phase.

3. Age-Hardening Mechanisms in Au–Cu–Ag–Pd Quarternary Alloys and Commercial Dental Gold Alloys

Since the works of Wise et al. (1932) and Wise and Eash (1933) concerning quarternary gold alloys, it has been known that the addition of platinum and/or palladium to ternary Au–Cu–Ag alloys gives rise to conspicuous age-hardening characteristics. To clarify the crystallography and morphology of the hardening phase in these alloys, Yasuda and Kanzawa (1977) carried out TEM studies on a quarternary gold alloy Au–35.4 at.% Cu–17.8 at.% Ag–9.7 at.% Pd which was prepared so as to contain the equivalent of 2 karat of palladium substituted for a corresponding amount of gold in a 16 karat Au–Cu–Ag

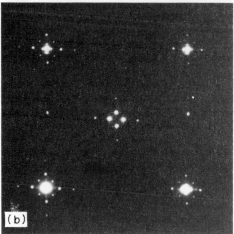

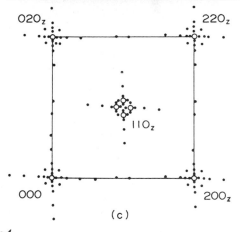

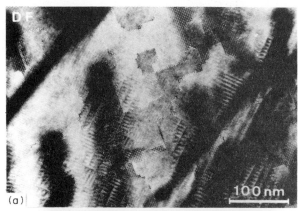

Figure 4
Dark field TEM micrograph (a) produced by using the 110_z superlattice spot and SAED pattern (b) of the AuCu–6 at.% Ag alloy after aging at 613 K for 100 ks

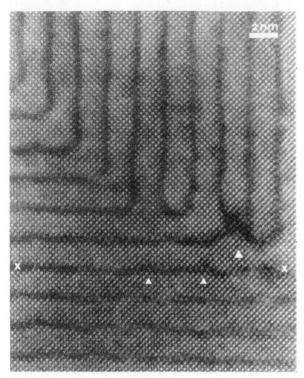

Figure 5
HREM image along [001] of periodic antiphase
domain boundaries in AuCu–6 at.% Ag alloy aged at
613 K for 100 ks

alloy. The age-hardening observed in this quarternary alloy was due to the formation of AuCu I ordered nuclei on the disordered matrix {100} planes. In adjacent ordered platelets of AuCu I, the c-axes were mutually perpendicular to compensate for the strain induced by their tetragonality, as was also observed in the 18 karat gold ternary alloy. To relieve the strain caused by the tetragonal distortion in the ordered domains, twinning took place on the $(101)_{tet}$ plane at a later stage of aging. The nodular precipitates that also formed at grain boundaries did not play an important role in age-hardening. Thus, it became clear that the hardening mechanism observed in the quarternary gold alloy was analogous to that previously found in the 18 karat gold ternary alloy.

Commercial palladium-containing dental gold alloys were also studied to make clear their hardening mechanisms and the associated phase transformations, since it was expected that the age-hardening mechanism varied greatly depending on the composition of the alloy. The two alloys studied were Au–26.5 at.% Cu–18.2 at.% Ag–7.9 at.% Pd (16 K–S) and Au–25.3 at.% Cu–27.4 at.% Ag–7.6 at.% Pd (14 K–S) (Yasuda et al. 1983). In these alloys, two stages of ordering could be distinguished

by TEM and SAED studies of isochronal aging of a supersaturated solid solution. In the first stage, ordering of a metastable AuCu I′ superlattice occurred as a stepwise arrangement of platelets situated on the {100} planes. This structural feature was also analogous to that found in the 18 karat gold ternary alloy. Prolongation of the aging period caused the formation of a lamellar structure consisting of the equilibrium AuCu I ordered phase and the silver-rich α_2 phase. The metastable AuCu I′ ordered platelets were formed prior to the equilibrium AuCu I ordered phase. Platelet formation was accompanied, in the 16 K–S alloy, by a contraction of about 7% in the surrounding matrix in a direction normal to the plane of the AuCu I′ ordered platelets (i.e., this direction coincided with the c-axis of the AuCu I′ superlattice). In the 14 K–S alloy, the contraction was about 6%. The strain introduced by the tetragonality of the AuCu I′ structure led to substantial hardening in the early stages of aging.

Isothermal age-hardening curves for the two alloys showed that, after a rapid increase in the early stages of aging, the hardness dropped drastically as aging was prolonged (Udoh et al. 1984). This marked decrease in hardness on overaging was due to the formation of the equilibrium AuCu I ordered phase and of the α_2 phase at grain boundaries as a lamellar structure by a heterogeneous mechanism. It was thought that, during lengthy aging, these lamellae would continue to grow at the expense of the metastable AuCu I′ ordered platelets formed inside the grains. The strain field generated by the formation of the AuCu I′ ordered platelets was therefore relieved as a result of the growth of the lamellar structure. Thus, loss of coherency at the interface between the ordered platelets and the matrix caused the hardness decrease.

Recently, a different type of age-hardening mechanism has been found to occur upon appropriate aging in a commercial 18 karat gold dental alloy Au–31.7 at.% Cu–8.1 at.% Pd–5.3 at.% Ag (18 K–S) (Udoh et al. 1986). X-ray and SAED studies as well as TEM showed that a Au_3Cu ($L1_2$) type superlattice and a disordered fcc phase were formed by aging at 673 K. Hardening occurred through the mechanism of the antiphase domain size effect. The $L1_0$ type AuCu I superlattice was also formed in regions adjacent to the Au_3Cu ordered phase in the later stages of aging below 623 K. Thus, it was concluded that the hardening should be attributed not only to the antiphase domain size effect, but also to the elastic strain field induced by the formation of the AuCu I ordered platelets in the aging temperature range below 623 K.

Although the hardening mechanism in dental gold alloys was thought to be attributable to a coherency strain field at the interface between the AuCu I platelets and the surrounding matrix, the configuration of the interface, which could provide direct evidence of the presence of a coherency strain field, could not be deduced by conventional TEM. However, HREM observation coupled with SAED provided direct information on crystal structures down to the scale of interatomic distances, and proved to be extremely effective for elucidating the structural characteristics of interfaces between different phases under suitable imaging conditions. These techniques were therefore employed for studying the configuration of the antiphase domain boundaries and the strain field generated at the interface between the AuCu I and Au_3Cu ordered phases in the 18 K–S commercial dental gold alloy (Yasuda et al. 1986).

Figure 6 shows the dark field image taken from the 18 K–S alloy aged at 673 K for 1.8 Ms by using a 110 superlattice reflection, with the incident electron beam parallel to the [100] direction using conventional TEM techniques. In the micrograph, the dark stripes which run approximately parallel to the ⟨100⟩ directions are antiphase domain boundaries that contribute to the hardening of the alloy. However, the atomic arrangement at the antiphase domain boundary could not be found in the micrographs obtained by the conventional dark field imaging technique. Figure 7 is an HREM image taken from the same region of the crystal as in Fig. 6. In this configuration it is clearly evident that the copper columns are at the correct positions with respect to the basic fcc lattice (i.e., at each corner of the square), and have the same configuration and scale as the projection of the copper columns in the Au_3Cu ordered structure. A shift in the arrays of the bright dots occurs across the interfaces (i.e., at the antiphase domain boundaries as indicated by the arrows

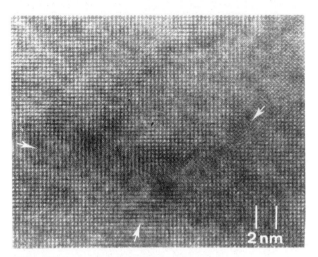

Figure 7
HREM micrograph along the [100] direction of the Au_3Cu superlattice; a triple junction of antiphase domain boundaries (indicated by arrows) is observed (the separation of the bright dots, which represent copper atoms, is 0.4 nm)

in Fig. 7). The presence of antiphase domain boundaries introduces a local change in chemical composition of constituents (i.e., excess atoms with respect to the stoichiometric composition will be segregated along the antiphase domain boundaries). Thus, it is thought that this segregation contributed to the hardening as well as to the higher energy at the antiphase domain boundaries.

When aging was carried out below 623 K, the AuCu I ordered platelet was formed in regions adjacent to the Au_3Cu ordered matrix. Figure 8a is a bright field high resolution image produced using the 001, 100, 110 and equivalent superlattice reflections, the fundamental reflections being excluded, and contains an interface between an AuCu I ordered platelet and the ordered Au_3Cu matrix. In the micrograph, the direction of the c-axis of the AuCu I ordered platelet is indicated by an arrow, while the arrowhead points to an antiphase domain boundary. Figure 8b is also a high resolution image which shows in detail the arrangement of the columns of atoms along the interface between the AuCu I and Au_3Cu ordered phases. In spite of the disturbance caused by the presence of strain and strain contrast, the interface is clearly visible owing to the small difference in interatomic distance between the AuCu I and Au_3Cu phases. These phases were further identified from optical diffraction patterns. In Fig. 8a, it can be seen that the amount of elastic strain increases with the thickness of the AuCu I ordered platelet (i.e., the length along the c-direction within the limited range of the elastic limit). The width, or thickness, of the AuCu I

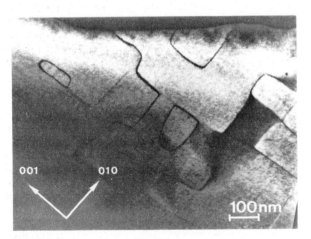

Figure 6
Dark field image of a commercial 18 karat gold dental alloy (18 K–S) aged at 673 K for 1.8 Ms taken by conventional electron microscopy using the 110 superlattice spot

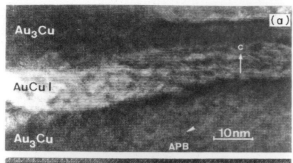

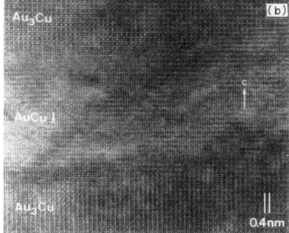

Figure 8
Bright field HREM image of a commercial 18 karat gold dental alloy (18 K–S) aged at 573 K for 1.8 Ms: (a) only the superlattice spots were selected for imaging, and (b) all reflections up to 220_{fcc} were selected for imaging

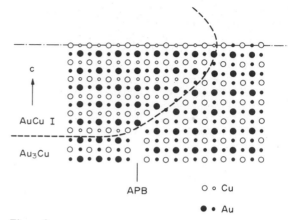

Figure 9
Schematic representation of a AuCu I platelet in an Au_3Cu matrix for axial ratio c/a of 0.9; an antiphase domain boundary (APB) is generated automatically if copper continuity is required along the center of the platelet (large circles are at level 0; small circles are at level 1/2)

platelet image is certainly limited by the strain contrast associated with the difference in lattice parameters along the c-axis for AuCu I and Au_3Cu ordered phases. If, as shown schematically in Fig. 9, it is assumed that at the top of such a platelet, the pure copper (001) planes in the AuCu I superlattice, coincide with the mixed gold–copper planes in the Au_3Cu superlattice, then with an axial ratio (c/a) of 0.90 for a platelet with a half-width of 5 AuCu I unit cells, the pure copper planes will now coincide with the pure gold planes. For a c/a ratio of 0.945, which was the measured value from the diffraction patterns, the width of the AuCu I platelet will be 18–19 units thick. This explanation is in good agreement with the experimental observations; indeed, such a mismatch of the antiphase domain boundary is observed in Fig. 8a. Thus, it was concluded that the AuCu I ordered platelets, being coherent with the Au_3Cu ordered phase as matrix, gave rise to considerable amounts of elastic strain which could be a major contribution to the age-hardening of the alloy.

Bibliography

Hultgren R, Tarnopol L 1939 Effect of silver on the gold–copper superlattice AuCu. *Trans. Metall. Soc. AIME* 133: 228–38

Kanzawa Y, Yasuda K, Metahi M 1975 Structural changes caused by age-hardening in a dental gold alloy. *J. Less-Common Met.* 43: 121–8

Kikuchi R, Sanchez J, de Fontaine D, Yamauchi H 1980 Theoretical calculation of the Cu–Ag–Au coherent phase diagram. *Acta Metall.* 28: 651–62

McMullin J, Norton J 1949 On the structure of gold–silver–copper alloys. *Metals Trans.* 1: 46–8

Masing G, Kloiber K 1940 Precipitation processes in the copper–silver–gold system (in German). *Z. Metallkd.* 32: 125–32

Murakami M, de Fontaine D, Sanchez J, Fodor J 1975 Ternary diffusion in multilayer Ag–Au–Cu thin films. *Thin Solid Films* 25: 465–82

Nakagawa M, Yasuda K 1988 A coherent phase diagram of the $Au_x(Ag_{0.24}Cu_{0.76})_{1-x}$ section of the Au–Cu–Ag ternary system. *J. Less-Common. Met.* 138: 95–106

Ohta M, Shiraishi T, Yamane M, Yasuda K 1983 Age-hardening mechanism of equiatomic AuCu and AuCu–Ag pseudobinary alloys. *Dent. Mater. J.* 2: 10–7

Prasad A, Eng T, Mukherjee K 1976 Electron microscopic studies of hardening in Type III dental alloy. *Mater. Sci. Eng.* 43: 179–85

Raub E 1949 Effect of silver on transformations in the gold–copper system (in German). *Z. Metallkd.* 40: 46–54

Udoh K, Hisatsune K, Yasuda K, Ohta M 1984 Isothermal age-hardening behaviour in commercial dental gold alloys containing palladium. *Dent. Mater. J.* 3: 253–61

Udoh K, Yasuda K, Ohta M 1986 Age-hardening characteristics in an 18 carat gold commercial dental alloy containing palladium. *J. Less-Common Met.* 118: 249–59

Uzuka T, Kanzawa Y, Yasuda K 1981 Determination of the AuCu superlattice formation region in gold–copper–silver ternary system. *J. Dent. Res.* 60: 883–9

Watanabe D, Takashima K 1975 Periodic antiphase domain structure in the off-stoichiometric CuAu II phase. *J. Appl. Crystallogr.* 8: 598–602

Wise E M, Crowell W, Eash J T 1932 The role of the platinum metals in dental alloys. *Trans. Metall. Soc. AIME* 99: 363–412

Wise E M, Eash J T 1933 The role of the platinum metals in dental alloys. *Trans. Metall. Soc. AIME* 104: 276–307

Yamauchi H, Yoshimatsu H A, Forouhi A R, de Fontaine D 1981 Phase relation in Cu–Ag–Au ternary alloys. In: McGachie R O, Bradley A G (eds.) 1981 *Precious Metals*, Pergamon, Willowdale, Canada, pp. 241–9

Yasuda K 1987 Age-hardening and related phase transformations in dental gold alloys. *Gold Bull.* 20: 90–103

Yasuda K, Kanzawa Y 1977 Electron microscope observation in an age-hardenable dental gold alloy. *Trans. Jpn. Inst. Met.* 18: 46–54

Yasuda K, Metahi H, Kanzawa Y 1978 Structure and morphology of an age-hardened gold–copper–silver dental alloy. *J. Less-Common Met.* 60: 65–78

Yasuda K, Nakagawa M, Van Tendeloo G, Amelinckx S 1987 A high resolution electron microscopy study of pseudobinary AuCu–6 at.% Ag alloy. *J. Less-Common Met.* 135: 169–83

Yasuda K, Ohta M 1979 Age-hardening in a 14 carat dental gold alloy. In: Browning M E, Edson G I, Bubes I S (eds.) 1979 *Proc. 3rd Int. Conf. Precious Metals* International Precious Metals Institute, Chicago, IL, pp. 137–64

Yasuda K, Udoh K, Hisatsune K, Ohta M 1983 Structural changes induced by ageing in commercial dental gold alloys containing palladium. *Dent. Mater. J.* 2: 48–58

Yasuda K, Van Tendeloo G, Van Landuyt J, Amelinckx S 1986 High-resolution electron microscopy study of age-hardening in a commercial dental gold alloy. *J. Dent. Res.* 65: 1179–85

K. Yasuda
[Nagasaki University, Nagasaki, Japan]

H

Heart-Valve Replacement Materials

The four natural cardiac valves (i.e., the tricuspid, pulmonic, mitral and aortic) allow unobstructed unidirectional blood flow through the heart. These delicate but complex and highly specialized structures can become distorted by diseases (e.g., rheumatic heart disease or infection) or degeneration ("wear and tear") thereby either obstructing forward motion of blood (stenosis) or allowing reverse flow (regurgitation or insufficiency). Damaged valves can contribute to impaired cardiac function, and in many cases, long-term survival is impossible unless the valve is repaired or replaced. Since valve repair is usually difficult and often not possible, frequently the only option available is surgical replacement by an artificial valve (a prosthesis).

A successful heart valve replacement should

(a) be easily implantable,

(b) evoke no thrombus (blood clot) formation,

(c) be quickly and permanently healed by host tissues,

(d) present minimal resistance to forward flow,

(e) allow insignificant retrograde flow,

(f) not damage cellular or molecular blood elements,

(g) undergo minimal wear or degenerative changes over extended intervals, and

(h) not generate excessive noise or other discomfort to the patient.

No available prosthetic heart valve completely satisfies these criteria.

1. Types of Replacement Valves

Heart valve prostheses are of two general types—mechanical and tissue. Several models of each are available (see Fig. 1). Mechanical valves generally have three essential components that are fabricated from rigid, nonphysiologic biomaterials:

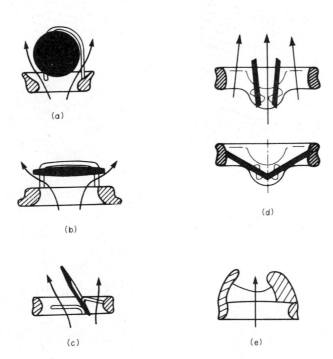

Figure 1

Designs and flow patterns of major types of prosthetic heart valves: (a) caged-ball, (b) caged-disk, (c) tilting-disk, (d) bileaflet tilting-disk, and (e) bioprosthetic (tissue) valves (in part © 1982, Pergamon Press Ltd; reproduced by permission from Schoen F J et al. 1982 Bioengineering aspects of heart valve replacement. *Ann. Biomed. Eng.* 10: 97–128 and in part © 1985, Springer-Verlag; reproduced by permission from Schoen F J et al. 1985 *Guide to Prosthetic Heart Valves*. Springer, New York, pp. 209–38)

(a) the flow occluder around both sides of which the blood must flow,

(b) the cagelike superstructure that guides and restricts occluder motion, and

(c) the valve body or base.

The poppet occluder moves passively, responding to pressure and flow changes within the heart. Caged-ball, caged-disk and tilting-disk mechanical valves have been used most widely.

Tissue valves have a flexible trileaflet anatomy, somewhat similar to that of natural valves; they also function passively. Historically, tissue valves have included homografts/allografts (i.e., human cadaver valves), heterografts/xenografts (i.e., porcine (pig) aortic valve or bovine (cow) pericardial tissue) or autografts (fabricated, for example, from the patient's own pulmonary valve, thigh connective tissue (fascia lata) or pericardium). The major advantages of tissue valves relative to mechanical prostheses are their pseudoanatomic central flow and relative freedom from surface-induced thrombus formation, usually without anticoagulation therapy. However, most tissue valves used today are glutaraldehyde-preserved (cross-linked/tanned) and are thereby nonviable altered biological (bioprosthetic) materials, which undergo durability-limiting progressive degenerative processes.

All prostheses, mechanical and tissue alike, have a fabric (Teflon, Dacron or polypropene) sewing ring that surrounds the valve orifice at the base and is used for suturing the device into the surgically prepared implantation site, the anulus. Since the anular anatomy is different for mitral and aortic valves (the valves most frequently diseased) sewing ring configurations vary slightly for prostheses which are to be used at each of these sites. The junction of the sewing ring with the anular tissues is intended to be the major area of direct patient–prosthesis interaction.

2. Hemodynamic Performance

All prosthetic valves present some degree of obstruction to forward flow, and thus the effective orifice area of almost all types of devices is less than that of properly functioning natural heart valves. Some functional prosthetic valve orifice areas are sufficiently low that they approach those measured in patients with moderate-to-severe valve stenosis who have not had surgery. Hemodynamic obstruction is accentuated in small sizes, even in the most efficient designs. Caged-ball prostheses are the most obstructive valves, while tilting-disk valves, especially the bileaflet tilting-disk designs, are the least obstructive. The central flow pattern of bioprosthetic heart valves generally yields enhanced hemodynamic function relative to most mechnical

prostheses, but even properly functioning bioprostheses can cause significant obstruction, particularly in small sizes, when the bulk of the struts supporting the tissue component is not reduced proportionally to the size of the anulus.

Due in large part to turbulent flow, some destruction of red blood cells by prosthetic heart valves is common. Nevertheless, hemolysis is generally slight and hemolytic anemia is unusual. However, materials degradation or paravalvular leak (i.e., a partial separation of the suture line anchoring the valve, leading to regurgitation of blood around the prosthesis) can cause turbulent blood flow through irregular small spaces, and may precipitate significant hemolysis.

3. Complications of Artificial Valves

3.1 Thromboembolism

In most clinical series using mechanical prostheses, the most frequent valve-related complications are thromboembolic problems, including gross thrombosis, causing local occlusion of the prosthesis (Fig. 2) or distant thromboemboli that may compromise blood flow to the heart, brain or kidneys. Thrombosis on artificial valves is potentiated by surface thrombogenicity, blood hypercoagulability and locally static blood flow. Sites of thrombus formation are predicted at areas of turbulence (e.g., the cage apex of caged-ball prostheses and the minor orifice of the outflow region of a tilting-disk

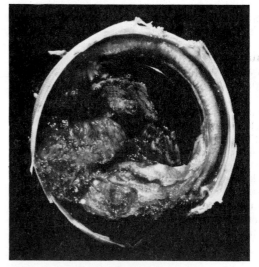

Figure 2
Thrombotic occlusion of tilting-disk prosthesis: the thrombus, initiated in a region of flow stasis, caused near-total occluder immobility (© 1987 W. B. Saunders Inc.; reproduced by permission from Schoen F J 1987 Surgical pathology of removed natural and prosthetic heart valves. *Hum. Pathol.* 18: 558–68)

prosthesis). The rate of thromboembolic complications with contemporary porcine bioprosthetic valves is approximately 1–2% per patient year, without anticoagulant therapy; this is less frequent than that with most mechanical valves, which is up to 4% per patient year, despite anticoagulant therapy. Thus, tissue valves, with low intrinsic thrombogenicity, are an attractive alternative to mechanical prosthetic valves when anticoagulation is undesirable. Thrombotic occlusion of bioprosthetic valves occurs rarely, and thromboemboli are unusual.

3.2 Infection

Infection associated with a valve prosthesis (endocarditis) is a devastating and often fatal complication. Bioprosthetic valve infective endocarditis occurs with approximately the same incidence as that with mechanical valves. However, although endocarditis associated with a mechanical prosthetic valve is localized to the prosthesis–tissue interface at the sewing ring (since the valve biomaterials cannot themselves support the growth of microorganisms) bioprosthetic valve infections can involve, and indeed can be limited to, the cuspal tissue. Cusp-limited infections are probably more readily treated by antibiotic therapy than most prosthetic valve endocarditis.

Infection is one of the most frequent clinically important complications of implanted biomaterials in general, occurring in approximately 1–10% of patients with a variety of prosthetic devices. Several potential mechanisms may be important. Microorganisms can be introduced inadvertently into a patient by supposedly sterile medical devices contaminated during manufacture or implantation. In addition, microorganisms are provided access to the circulation and to deeper tissue by damage to natural infection barriers during implantation surgery or subsequent device function. Moreover, implanted biomaterials and medical devices impede the free migration of inflammatory cells to the infected area and possibly interfere with the local bacteriocidal mechanisms of inflammatory cells.

Thus, the presence of any foreign material in tissues impairs the local resistance to infection, and the frequency and pathology of prosthetic valve endocarditis are almost independent of specific valve materials and design considerations. In contrast, thromboembolic considerations and durability considerations (discussed later) are clearly both materials- and design-dependent.

3.3 Structural Failure

Limitations to valve durability have been an important deterrent to the long-term success of cardiac valve replacement. Among patients who require reoperation for prosthetic valve failure or who have autopsy following prosthesis-associated death, materials degradation is frequently observed. Dura-

bility considerations vary widely among mechanical valves and bioprostheses, among different prosthesis configurations, among different models of a specific prosthesis type incorporating different design features and materials, and in some cases for the same model prosthesis placed in the aortic versus the mitral site. Since materials properties play a critical role in valve durability, degenerative dysfunction is discussed in detail below.

4. Degeneration of Mechanical and Tissue Valves

4.1 Mechanical Prostheses

The silicone ball occluder of early caged-ball aortic prostheses absorbed blood lipids, which caused distortion, cracking, embolization of poppet material or abnormal poppet movement, a spectrum of damage known as "ball variance." Mitral ball variance was distinctly less common than aortic, probably due to lower stresses incurred during mitral valve function, and changes in elastomer fabrication in 1964 have virtually eliminated lipid-related ball variance in subsequently implanted valves. Ball variance, although infrequently encountered today, exemplifies the complex interactions among mechanical, chemical and materials factors in degenerative valve failure.

In order to minimize thromboembolism, the cages of some caged-ball valves were covered with cloth to provide a scaffold on which recipient tissue could grow to yield a smooth, natural blood-contacting surface. However, cloth-covered caged-ball valves with metal or polymeric poppets suffered cloth abrasion and fragmentation by the occluder, and fibrous overgrowth with ball entrapment and resultant stenosis. Early tilting-disk designs had polymeric disks in which free rotation of the disk was not possible. These valves failed due to concentrated abrasive wear. Caged-disk valves with plastic disks developed severe disk wear with notching and reduction in diameter. Abrasive wear of mechanical components can interfere with local valve function, or also can cause distal migration (embolization) of fragments of material.

Most mechanical heart-valve prostheses currently in use have pyrolytic carbon occluders, and some have both carbon occluders and carbon cage components. Pyrolytic carbon, an advantageous material for construction of heart-valve components owing to its thromboresistance, high strength, wear resistance and ability to be fabricated into a wide variety of shapes, is applied to a preshaped graphite or metallic substrate using a fluidized bed coating process. Tilting-disk designs with pyrolytic carbon disks have good durability, with minimal abrasion of metallic cage components by carbon occluders. The wide use of pyrolytic carbon as an occluder and strut-covering material for mechanical valve prostheses has virtually eliminated abrasive wear as a long-term

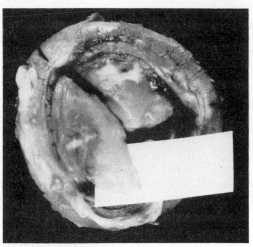

Figure 3
Porcine valve primary tissue failure due to calcification with secondary cuspal tear leading to severe regurgitation (© 1985 W. B. Saunders Inc.; reproduced by permission from Schoen F J and Hobson C E 1985 Anatomic analysis of removed bioprosthetic heart valves: Causes of failure of 33 mechanical valves and 58 bioprostheses, 1980–1983. *Hum. Pathol.* 16: 549–59)

complication of valve replacement. Nevertheless, catastrophic fractures of both metallic and carbon-coated valve components of some designs have been reported. Fractures are a particularly troublesome feature of mechanical valve prostheses used in circulatory support devices, probably due to the nonphysiological mechanical environment of such valves.

4.2 Bioprostheses

Deterioration of glutaraldehyde-treated porcine aortic valve bioprostheses by calcification or cuspal tearing (primary tissue failure), which causes stenosis or regurgitation, or both, now commonly necessitates reoperation following several years of function. The failure rate in adults is approximately 25% or higher, ten years after implantation, and recent evidence suggests that failure occurs in over 50% of cases at 15 years. Thus, deterioration of bioprosthetic valves is frequent and strongly time dependent; the low rate of valve failure earlier than five years following surgery (1% per year) accelerates to approximately 5% per year thereafter.

Calcification of the cusps is the most important pathologic process contributing to bioprosthetic valve failure. Regurgitation through tears, secondary to calcification, accounts for approximately three-quarters of porcine valve primary tissue failures (Fig. 3). Pure stenosis due to calcific cuspal stiffening and cuspal tears or perforations unrelated to calcification are less frequent. Noncalcific cuspal

tears reflect direct mechanical destruction of the valve structure during function, with fraying and disruption of collagen.

Bioprosthetic valve calcification increases with time after implantation, but patients vary widely in their propensity to calcify valves. Calcific deposits accumulate insidiously, and are present in the majority of bioprostheses examined three or more years after implantation. Nevertheless, some patients with bioprostheses functioning for ten years or more have no detectable calcific deposits. Specific factors which control which patients and which valves will have problems are largely unknown, except for age effects. Calcification is more likely to become severe and clinically significant in children and young adults than in older patients. Some data suggest that the increased risk of calcific failure may continue to age 35. The work on Hancock and Carpentier-Edwards has shown that there are no apparent differences in durability between these two most popular porcine bioprostheses.

Degenerative intrinsic calcification begins primarily at the cuspal attachments, which are the sites of greatest cuspal mechanical stresses. Mineral deposits are usually related to connective tissue cells and collagen of the transplanted valve (intrinsic calcification). Surface thrombi or vegetations also calcify (extrinsic calcification) but extrinsic mineralization is not usually clinically important. In contrast with the slow accumulation of intrinsic calcific deposits over several years, extrinsic calcification can occur within several days following implantation.

Bioprostheses fabricated from bovine pericardium may have a greater propensity to late primary tissue failure than porcine aortic valves. Calcification of pericardial valves, with or without cuspal tears, and large noncalcific cuspal disruptions are frequent. Cuspal tears often originate at the most peripheral portion of the cusps (i.e., the commissures).

5. Additional Considerations for Tissue Heart Valves

The natural aortic valve has optimal hemodynamics (no obstruction or regurgitation) and does not cause areas of localized stress concentration within the tissue, trauma to molecular or formed blood elements or thrombotic complications. The aortic valve achieves the basic requirements of valvular function by means of a highly specialized and complex structure. Although the pressure differential across the closed valve (approximately 80 mm Hg in the aorta, but less than 10 mm Hg in the left ventricle) induces a large load on the leaflets, the fibrous arrangement within the cusps assists in transferring the resultant stresses to the aortic anulus. In particular, aortic valve cusps have highly anisotropic material properties in the plane of the tissue; this reflects an oriented leaflet tissue architecture in which collagenous struts transfer stresses to the anulus.

Pericardial tissue, used as the material for the construction of some bioprosthetic heart valve leaflets, has approximately isotropic elastic properties. Since highly anisotropic elasticity is the major structural property of valve leaflets which promotes uniform distribution of stress during valve function, stresses in pericardial leaflet valves will be directed toward the natural mechanical focal points, the commissures, leading to high stress concentrations which are not encountered in the natural valve. Therefore, it is not surprising that cuspal tearing at the top of the stent posts is a common mode of failure of pericardial and other isotropic tissue valves. Also, repeated loading of pericardium causes dynamic creep of this viscoelastic material, progressively stretching the tissue and leading to permanent leaflet deformation.

The goals of aldehyde pretreatment of bioprosthetic tissue include enhanced material stability and decreased immunological reactivity, with maintained thromboresistance and preserved antimicrobial sterility. Glutaraldehyde is a dialdehyde which has been used since antiquity for the tanning of leather, and more recently as a fixative in the preparation of tissues for electron microscopy. Glutaraldehyde treatment cross-links proteins, especially collagen, the most abundant structural protein of bioprosthetic tissue. Interestingly, glutaraldehyde pretreatment of bioprosthetic tissue also appears to potentiate degenerative calcification. The specific alterations in the functional/mechanical characteristics of the natural aortic valve and other tissues used in bioprostheses by glutaraldehyde cross-linking procedures are controversial. Moreover, tissue mechanical properties may be further altered during valve function, but the exact nature and significance of the changes have yet to be elucidated.

Because of the problems encountered with mechanical prosthetic and bioprosthetic valves, there is considerable enthusiasm for use of sterilized aortic valve homografts derived from human cadavers. Like other tissue valves, homografts have central, unimpeded blood flow. Patients with these valves have almost complete freedom from thromboembolism without anticoagulant therapy. The disadvantages of homograft valves are their relatively more difficult technique of insertion, their limited availability, that their use is restricted to aortic (and not mitral) valve replacement, and that they exhibit a high incidence of late deterioration by tearing. Nevertheless, functional deterioration of the valve due to degeneration occurs slowly and usually can be managed successfully. Moreover, with the development of techniques for frozen storage of such valves

to minimize logistical constraints, and novel procedures to use homografts as mitral replacements, it is likely that the clinical use of valve homografts will broaden in the future.

6. Considerations in Valve Selection

With all mechanical valve prostheses, lifetime anticoagulation is necessary to minimize the risk of thromboembolic complications. Nevertheless, anticoagulation therapy imposes a significant risk of hemorrhage. Bioprostheses have the important advantage that chronic anticoagulant therapy is generally not required, but their long-term function is often compromised by degeneration. Therefore, the major consideration surgeons use to select the type of prosthesis to implant is the trade-off between risks of thromboembolic complications and hemorrhage related to anticoagulation use with mechanical prostheses (favoring bioprostheses) and the limited durability of bioprostheses (favoring mechnical prostheses). However, extraordinary clinical circumstances may prevail in selected patients. For example, bioprostheses are generally used in young women wishing to become pregnant, in order to avoid chronic anticoagulant drugs, which cross the placental circulation and potentially cause serious problems for the fetus. In contrast, mechanical valves may be preferable in children and young adults, despite the need for anticoagulation, because failure of bioprostheses within several years is almost certain in this population. Moreover, primary tissue degeneration of bioprosthetic valves, the overwhelming mode of valve failure, usually leads to slow progressive development of cardiac dysfunction, which can allow recognition of the problem and elective reoperation. This contrasts with the frequently catastrophic valve failure and subsequent rapid deterioration of patients with mechanical valves who have thromboembolic complications or structural valve failure. The most important advantages and disadvantages of the various types of replacement valves are summarized in Table 1.

Table 1
The most important advantages and disadvantages of contemporary valve substitutes

Valve type	Advantages	Disadvantages
Mechanical	Durable	Thrombogenic
Bioprosthetic	Thromboresistant	Propensity to deteriorate
Homograft	Near-normal hemodynamics	Limited availability
	Thromboresistant	Technically demanding insertion
		Late tearing

7. Future Directions

Problems encountered with the clinical use of bio-prostheses have stimulated several lines of investigation. Alterations in tissue preparation techniques for bioprosthetic valves and radically altered design configurations are under consideration. Since calcification is the major pathologic feature associated with bioprosthetic valve failure, considerable work is being directed toward elucidation of mechanisms of bioprosthetic valve mineralization and strategies to obviate this problem. The most promising approaches to effective prevention of bioprosthetic valve calcification are empirical and include pretreatment of valve leaflets to remove a structural component which initiates calcification (such as cell membranes) and the use of specific inhibitors of calcification (such as the diphosphonate drugs, which are potent inhibitors of minimal crystal-growth used in the therapy of metabolic bone disease). Drugs could be administered either systemically, by valve pretreatment or through sustained local controlled release. The best approach seems to include pretreatment of the valve cusps, either by diphosphonates or some treatment which removes crystal nucleation sites, in conjunction with controlled drug delivery into bioprosthetic valves that have a modified sewing-ring configuration that contains a diphosphonate-containing polymeric matrix.

Although modest development of mechanical poppet prostheses with new designs, particularly those using pyrolytic carbon occluders, has continued, it is unlikely that radical design changes will be forthcoming. Modest modification of tilting-disk configurations has concentrated on enhancing opening to improve hemodynamics and eliminating metallic struts from regions of stasis to reduce thromboembolic risk. Fixed pivot-point bileaflet tilting-disk configurations have achieved wide acceptability over the past several years with generally favorable clinical results. Major developments in the technology of polymeric materials have allowed recent progress in trileaflet flexible prostheses fabricated from polytetrafluoroethylene, polyurethane or silicones to mimic the natural aortic valve anatomy and function. However, long-term durability limitations remain the major obstacle to clinical application.

Bibliography

Ferrans V J, Tomita Y, Hilbert S L, Jones M, Roberts W C 1987 Pathology of bioprosthetic cardiac valves. *Hum. Pathol.* 18: 586–95

Jamieson W R E, Rosado L J, Munro A I, Gerein A N, Burr L H, Miyagishima R T, Janusz M T, Tyers G F O 1988 Carpentier-Edwards standard porcine bioprosthesis: Primary tissue failure (structural valve deterioration) by age groups. *Ann. Thorac. Surg.* 46: 155–62

Matsuki O, Robles A, Gibbs S, Bodnar E, Ross D N 1988 Long-term performance of 555 aortic homografts in the aortic position. *Ann. Thorac. Surg.* 46: 187–91

Milano A D, Bortolotti U, Mazzucco A, Guerra F, Stellin G, Talenti E, Thiene G, Gallucci V 1988 Performance of the Hancock porcine bioprosthesis following aortic valve replacement: Considerations based on a 15-year experience. *Ann. Thorac. Surg.* 46: 216–22

Morse D, Steiner R M, Fernandez J 1985 *Guide to Prosthetic Cardiac Valves*. Springer, New York

Rose A G 1987 *Pathology of Heart Valve Replacement*. MTP Press, Lancaster

Schoen F J 1987a Cardiac valve prostheses: Pathological and bioengineering considerations. *J. Cardiac Surg.*, 2: 65–108

Schoen F J 1987b Cardiac valve prostheses: Clinical status and contemporary biomaterials issues. *J. Biomed. Mater. Res.* 21(A1): 91–117

Schoen F J 1989 *Interventional and Surgical Cardiovacular Pathology: Clinical Correlations and Basic Principles*. Saunders, Philadelphia, PA

Schoen F J 1990 Mode of failure and other pathology of mechanical prosthetic and bioprosthetic heart valves. In: Bodnar E, Frater R (eds.) 1990 *Replacement Cardiac Valves*. Pergamon, Oxford

Schoen F J, Fernandez J, Gonzalez-Lavin L, Cernaianu A 1987 Causes of failure and pathologic finding in surgically-removed Ionescu–Shiley standard bovine pericardial heart valve bioprostheses: Emphasis on progressive structural deterioration. *Circulation* 76: 618–27

Schoen F J, Harasaki H, Kim K M, Anderson H C, Levy R J 1988 Biomaterial-associated calcification: Pathology, mechanisms, and strategies for prevention. *J. Biomed. Mater. Res.* 22(A1): 11–36

Schoen F J, Kujovich J L, Levy R J, St John Sutton M 1988 Bioprosthetic heart valve pathology: Clinicopathologic features of valve failure and pathobiology of calcification. *Cardiovasc. Clin.* 18(2): 289–317

Schoen F J, Levy R J, Ratner B D, Lelah M D, Christie G 1990 Materials considerations for improved cardiac valve prostheses. In: Bodnar E, Frater R (eds.) 1990 *Replacement Cardiac Valves*. Pergamon, Oxford

Silver M D, Butany J 1987 Mechanical heart valves: Methods of examination, complications, and modes of failure. *Hum. Pathol.* 18: 577–85

F. J. Schoen
[Brigham and Women's Hospital and Harvard Medical School, Boston, Massachusetts, USA]

Hemodialysis Membranes

The membrane-based procedure of hemodialysis, or the artificial kidney, is now established as a clinical treatment of high importance. In addition, the extensive use of this treatment means that hemodialysis membranes form a significant proportion of the overall clinical and industrial application of membranes.

The clinical role of hemodialysis imposes permeability requirements on the membrane, with respect to the removal from blood of solutes and water. The desire to enhance permeability has led to the application of the alternative procedures hemofiltration and high-flux hemodialysis. Contact of hemodialysis membranes with blood necessitates the maintenance of an acceptable level of compatibility, which is normally achieved by performing hemodialysis in association with the anticoagulant heparin. The most widely used hemodialysis membrane is regenerated cellulose, which has dominated the field for many years. Attempts to replace cellulose are based on the production of membranes with superior permeability characteristics or improved blood compatibility. Such membranes are obtained by cellulose modification or from synthetic polymers.

1. Clinical Use of Extracorporeal Blood Purification

Hemodialysis is an example of an extracorporeal blood purification procedure used principally in the treatment of uremia. The purpose of such extracorporeal blood purification is to replace, in renal failure, functions performed by the human kidney. These functions include maintenance of the normal composition of blood, regulation of the acid–base balance and participation in the control of blood pressure.

Extracorporeal blood purification does not completely replace the human kidney, but is intended to compensate partly for renal insufficiency by ensuring the controlled removal from the blood of solutes and water. The need to accomplish this removal under conditions appropriate to the nature and properties of blood, has led to extracorporeal blood purification being based on membrane separation processes.

2. Extracorporeal Blood-Purification Procedures

The relevant membrane separation processes for extracorporeal blood purification are dialysis, ultrafiltration and osmosis. Dialysis takes place when a permeable solute passes through a membrane, separating solutions of different concentration, under the action of a concentration-gradient driving force. In a multisolute system, each solute diffuses across the membrane under the influence of its particular concentration gradient. Ultrafiltration occurs when a solvent is transferred through a membrane under the action of a pressure-gradient driving force. If the passage of fluid through the membrane is associated with the transfer of solute, ultrafiltration will contribute to solute removal. Osmosis refers to the process when a membrane separates a solvent from solution containing impermeable solute, and solvent passes through the membrane into the solution under the action of a concentration-gradient driving force. Osmotic pressure is defined as the excess pressure which must be applied to the solution to prevent the entry of solvent, and being

dependent on membrane and solute characteristics it is not a simple solution property.

Hemodialysis is based on a combination of dialysis and ultrafiltration. To perform hemodialysis, blood from a patient is arranged to flow within a semipermeable membrane envelope, on the other side of which is passed a balanced isotonic solution, termed the dialyzate. Permeable solutes present in higher concentration in the blood, transfer through the membrane under the action of a concentration gradient by the diffusive process of dialysis and enter the dialyzate. The elimination of particular blood constituents can be prevented by the addition of these constituents to the dialyzate in approximately physiological concentrations. Ultrafiltration removes water from the blood under the action of a pressure gradient between blood and dialyzate. Osmosis is generally avoided, but can be used to achieve water removal by the addition of certain solutes to the dialyzate. Solute removal in hemodialysis is predominantly by diffusion with a minor contribution coming from convective solute transfer during ultrafiltration.

At present, there are two basic dialyzer design configurations in clinical use. These are the flat-bed or parallel-plate dialyzer and the capillary or hollow-fiber dialyzer.

In the parallel-plate dialyzer, the flow of blood is within a membrane envelope formed by two sheets of polymer, with the dialyzate normally flowing countercurrent to the blood. Different versions of the parallel-plate concept can be constructed by varying the number and dimensions of the membrane envelope, the method of introducing blood into the envelope and the configuration of membrane support structures.

The construction of hollow-fiber dialyzers is based on the shell and tube heat-exchanger principle. A typical dialyzer contains more than 10 000 fibers of about 200 µm internal diameter. Hollow-fiber dialyzers offer the highest ratio of effective membrane area to machine size and provide a means of reducing blood priming volume for a given membrane area. The need for membrane backing support is eliminated by the selection of an appropriate ratio of fiber diameter to wall thickness. Entry of blood into the fibers is achieved by embedding in a suitable polymer, usually polyurethane, and cutting to expose the fiber ends.

Hemodialysis is the most important clinical procedure for extracorporeal blood purification. However, alternative procedures are used in order to improve performance in terms of solute removal and to enhance clinical efficiency. In some cases, these procedures are based on a more effective use of ultrafiltration, either for the transport of water or for the convective transport of solute. Alternative procedures to hemodialysis include hemofiltration and high-flux hemodialysis.

An advantage of hemofiltration over hemodialysis is in the elimination of solutes which, because of their size, diffuse slowly and are removed inefficiently from the blood by dialysis. Hemofiltration utilizes membranes with hydraulic permeability and solute-molecular-weight cutoff greater than those of conventional hemodialysis membranes in order to purify blood by a combination of ultrafiltration and convective solute transport. Hemofiltration involves a much stronger ultrafiltration than hemodialysis and it is necessary to compensate for excessive fluid removal by the addition of a physiological saline solution, either before or after the ultrafiltration stage. Since it is based on ultrafiltration and partial restoration of fluid, hemofiltration approximates more closely than hemodialysis to the operation of the human kidney.

High-flux hemodialysis describes the procedure where membrane solute and water permeability exceed those of conventional hemodialysis membranes but are not comparable to those of hemofiltration membranes. Therefore, high-flux hemodialysis offers a more efficient blood purification without drastically altering the nature of the procedure.

3. Membrane Permeability

In considering the permeability of hemodialysis membranes, it is necessary to realize and accept that there can be no precise stipulation of permeability requirements. This is due to the fact that knowledge of the solutes occurring in uremia is incomplete.

In hemodialysis, the net solute flux of species A, J_A and the net volume flux J_V are given by

$$J_A = p_m \Delta^\circ C_A + \bar{C}_A (1 - \sigma) J_V \tag{1}$$

and

$$J_V = L_p (\Delta P - \sigma \Delta \Pi) \tag{2}$$

where p_m is the membrane permeability factor, $\Delta^\circ C_A$ and ΔP are the respective solute concentration and hydraulic pressure differences across the membrane, $\bar{C}_A$ is the average solute concentration in the membrane, L_p is the hydraulic conductivity coefficient, σ is the Staverman reflection coefficient and $\Delta \Pi$ is the osmotic pressure difference that would be exerted across the membrane if there were complete impermeability to the solute.

An important feature of hemodialysis is that solute diffusion is determined by factors additional to membrane permeability. The presence of concentration boundary layers associated with low-velocity regions on either side of the membrane gives rise, in the absence of ultrafiltration, to the following relationship between p_m and the overall permeability p_o

$$\frac{1}{p_o} = \frac{1}{p_m} + \frac{1}{p_b} + \frac{1}{p_d} \tag{3}$$

where b and d denote blood and dialyzate, respectively. In terms of mass transfer resistances, Eqn. (3) can be restated as

$$R_o = R_m + R_b + R_d \tag{4}$$

As indicated, there is uncertainty over the identity of solutes which should be removed in the treatment of uremia. It is generally believed that, in addition to the removal of low-molecular-weight solutes, it is important to remove solutes with molecular weights above 1500. With the low-molecular-weight solutes, R_m is of the same order as $R_b + R_d$. As solute size increases, R_m increases in comparison to $R_b + R_d$ and a stage is reached for higher-molecular-weight solutes when R_o and R_m can be considered equal. Therefore, if treatment effectiveness is related to the removal of higher-molecular-weight solutes, termed middle molecules, it should be possible, for a given membrane, to reduce treatment time and compensate by increasing the membrane area. In fact, this approach has been adopted clinically, despite lack of verification of the role of higher-molecular-weight solutes in renal failure.

Volume flux in hemodialysis is obtained from Eqn. (2). In the absence of osmotic effects

$$L_p = J_V / \Delta P \tag{5}$$

Eqn. (5) is an acceptable approximation for dilute solutions or when $\Delta P \gg \sigma \Delta \Pi$.

The rate of ultrafiltration is dependent on the hydrostatic and osmotic pressure gradients across the membrane and the ultrafiltration characteristics of the membrane. The applied pressure gradient is limited by the wet strength of the membrane and the design of any membrane support structure. The ultrafiltration characteristics are determined by membrane structure, thickness and swelling characteristics. Solutes up to a critical size, which is determined by the properties of the membrane, pass through the membrane at the same rate.

4. Membrane Compatibility

A consideration of the compatibility of a hemodialysis membrane must take into account not only the possible influence of the membrane on blood constituents but also the influence of the hemodialysis process on the clinical status of the patient. While the clinical status may be altered by the effect of the membrane on the levels of particular blood constituents, it could also be affected by other components of the hemodialysis circuit or by an alteration in the rate of solute removal.

The complex nature of blood coagulation can be considered in terms of contributions from different aspects, such as platelet activation, intrinsic and extrinsic coagulation pathways, complement activation, and the control systems involved in thrombosis inhibition and fibrinolysis. The events following contact of blood with a hemodialysis membrane can be considered in a similar manner provided two important distinctions are borne in mind. First, hemodialysis membranes cannot undertake the active role in the resistance to thrombus formation of which the natural blood vessel surface, the endothelium, is capable. Consequently, the use of hemodialysis membranes requires simultaneous therapy with an antithrombotic agent, usually the anticoagulant heparin. Second, while there is little evidence that the endothelium adsorbs proteins under physiological conditions, protein adsorption is a critical step in the interaction of blood with an artificial surface and must be expected during hemodialysis.

Protein adsorption to artificial surfaces is rapid from blood and protein deposition on membranes after hemodialysis has been demonstrated. On the basis of existing knowledge of protein adsorption and artificial surfaces, it can be accepted that the hemodialysis membrane structure will influence the nature of the adsorbed protein layer and this layer will, in turn, influence subsequent events. However, additional information on the relationship between membrane structure and the adsorbed protein layer is required. There is uncertainty over the extent to which proteins such as fibrinogen undergo degradation and over the possible influence of even trace proteins. Another important question is the effect of hemodialysis membranes on factor XII, prekallikrein and high-molecular-weight kininogen; proteins involved in the contact activation phase of blood coagulation.

A platelet response is an inevitable consequence of hemodialysis. Membrane structure influences platelet adhesion, although platelet loss during hemodialysis is not high. Hemodialysis membrane structure can also alter platelet function, as represented by induced platelet aggregation, and the release from platelets of substances such as platelet factor 4 and β-thromboglobulin.

An important consideration for hemodialysis is the relationship between the procedure and the leucocytes. A regular feature of hemodialysis is the occurrence of a sharp transient fall in leucocyte count. This fall is not attributed to the attachment of leucocytes to the membrane surface but to a temporary sequestration within the pulmonary microcirculation. The leucocytes involved are the nucleophilic granulocytes and the neutropenia is associated with granulocyte aggregation and embolization.

The leucocyte response to hemodialysis may include changes in cell function and a reduced ability to combat infection. Hemodialysis membranes may also induce the release of granulocyte components such as lactoferrin and granulocyte elastase. An important aspect of leucocyte changes during hemodialysis is the association between leucocytes and activation of the complement system.

The human complement system consists of series of plasma proteins, designated C1, C2, C3, C4 and so on, which are normally present as inactive molecules. When blood comes into contact with an appropriate stimulus, the inactive molecules are cleaved enzymatically to yield active proteins and polypeptides and activation of early components induces activation of later components in a cascading mechanism. There are two primary pathways of complement activation, the classical and the alternative, and a critical step is the cleavage of the C3 molecule. In the classical pathway, C3 activation follows activation of C1 and C2. This contrasts with the alternative pathway in which C3 activation does not involve earlier complement components. When complement activation takes place during hemodialysis, the time scale is similar to that for the observed neutropenia. Components of particular interest to hemodialysis are the split products C3a, C4a and C5a, which are released into the blood. These anaphylatoxin molecules function as inflammatory mediators. There is evidence that complement activation is responsible for granulocyte deviation, with anaphylatoxin molecules implicated.

The demonstration of neutropenia and complement activation has directed attention towards alterations to the immune system induced by hemodialysis. The contact of circulating cells with hemodialysis membranes may stimulate the release from mononuclear phagocytes (monocytes) of mediators with acute and chronic metabolic, immunological and inflammatory consequences. An important role in this mediation has been assigned to interleukin-1.

Another topic in membrane compatibility is that of hypersensitivity reactions. Although hypersensitivity reactions during hemodialysis are infrequent, they can have a serious outcome. Symptoms range from minor dermatological manifestations to pronounced cardiopulmonary disorder. These symptoms could be caused by an acute allergic reaction or by complement activation. Factors other than membrane structure include the sterilizing agent ethylene oxide and its breakdown products.

Present interest in the compatibility of hemodialysis membranes extends beyond monitoring of the alterations in the levels of blood constituents. Attempts are being made to achieve an understanding of the complex reactions which occur during hemodialysis and their relationship to the symptoms of the patient.

Finally, to the influence of the chemical nature of the membrane must be added the influence of membrane permeability. The disease state of uremia has an effect on cell function and any changes resulting from improved membrane permeability will have consequences for compatibility. Compatibility enhancement by improved membrane permeability is also exemplified by the removal of β2-microglobulin, a protein implicated in the dialysis-related amyloidosis bone disorder.

5. Membrane Design

The three fundamental membrane properties are permeability, mechanical strength and compatibility. Permeability is essential for the transfer of solutes and water, mechanical strength for membrane handling, and compatibility for clinical use. To achieve clinical acceptability, the membrane is required to meet appropriate levels of these properties.

Since knowledge of the solutes to be removed is limited, membrane permeability is normally directed towards a molecular-weight range. Transport of solutes and water can be accomplished by a combination of diffusion and convection, as in hemodialysis, or by convection alone, as in hemofiltration.

Approaches available for the design of membranes include the preparation of membranes from polymer structures providing a balance of hydrophilicity and hydrophobicity, the leaching of substances incorporated into the membrane, the stretching of wet membranes, and the manufacture of membranes with an asymmetric structure.

A balance of hydrophilicity and hydrophobicity can be achieved by cellulose regeneration, copolymerization of selected monomers, chemical modification and insolubilization of water-soluble polymers. In membrane manufacture, membranes based on regenerated cellulose are obtained by processes established for the production of packaging films, while other polymer membranes are generally produced by solvent casting techniques.

Hollow-fiber membranes are produced by extruding or spinning the polymer through a small annular die or spinneret. Production techniques include melt spinning, in which molten polymer is spun and cooled, wet spinning, in which a polymer solution is spun and coagulated in a nonsolvent, and dry spinning, in which a polymer in a volatile solvent is spun into an evaporative column.

Asymmetric membranes have a dense thin skin and an underlying porous substructure. The membranes are solvent cast, with the skin forming during a short drying period and the porous substructure during gelation. The structure of an asymmetric membrane is a combination of the dense skin for selectivity and the highly permeable porous substrate for support. The structure enables high rates of ultrafiltration for a given solute molecular-weight cutoff. Asymmetric membranes can be produced in flat-sheet or hollow-fiber form.

6. Membrane Systems

Hemodialysis membranes are dominated by the widespread clinical use of regenerated cellulose and cellulose membranes generally act as a basis of comparison for other systems. Membrane development is normally directed towards achieving permeability or compatibility properties superior to those of cellulose.

6.1 Cellulose Membranes

Regenerated-cellulose membranes have a long association with hemodialysis and remain the most widely used hemodialysis material. The structure of cellulose consists of a large number of anhydroglucose units joined together by 1–4, 1–6 β-glucosidic linkages and a feature of great importance to membrane formation and utilization is the strong intermolecular attraction resulting from hydrogen bonding. This bonding, which is promoted by the regular arrangement of hydroxyl groups, ensures that cellulose is insoluble in organic solvents, and since thermal degradation takes place prior to flow, cellulose cannot be melted. Therefore, the manufacture of cellulose membranes requires the production of a soluble or thermoplastic derivative as an intermediate. Another important consequence of the intermolecular bonding is that cellulose membranes, while swelling in an aqueous medium, remain insoluble, and are suitable for application in contact with blood.

The manufacture of cellulose membranes involves chemical modification of cellulose to produce a soluble or thermoplastic cellulose derivative, membrane formation by solvent casting or melting, and treatment of the cellulose derivate to achieve regeneration into cellulose. Since there is a reduction in molecular weight, regeneration is incomplete. A key factor is the production of the cellulose derivative. Cellulose membranes for hemodialysis are mostly produced by the cuprammonium process, in which a soluble complex is formed by dissolution of cellulose in cuprammonium solution and regeneration is by reaction with acid. Membranes can also be produced by the viscose process, in which a soluble xanthate is formed by reaction of cellulose with alkali and carbon disulfide, followed by acidic regeneration. Another technique is to produce thermoplastic cellulose acetate and hydrolyze to cellulose by reaction with alkali.

Cellulose films are generally classified by a system of figures and letters derived from a code used in the manufacture of regenerated cellulose films as wrapping materials. The figures refer to the nominal weight in grams per 10 m² of film and, therefore, are indicative of thickness. The letters denote the condition of the film by indicating whether it is coated or uncoated, since cellulose films may be coated with cellulose nitrate in order to reduce moisture sensitivity. Regenerated-cellulose films used as hemodialysis membranes do not require treatment for reducing moisture sensitivity and were formerly described by the letters PT, signifying that the membranes were uncoated or "plain, transparent." However, it is now more common to use the letters PM, denoting that the membrane is uncoated and suitable for medical application.

Regenerated cellulose membranes are too brittle to be handled satisfactorily in the dry state. The addition of a plasticizer is essential and glycerol is normally used. Displacement of the plasticizer occurs when the cellulose membrane contacts an aqueous environment.

With respect to solute-transport characteristics, cellulose membranes appear to act as microporous barriers or sieves. Therefore, solute permeability, for a given membrane, is principally inversely proportional to the molecular volume of the diffusing solute. Since the influence on solute transport of charge or adsorption properties can be ignored, it follows that a cellulose membrane is only selective to closely related solutes when the molecular volumes of the solutes approximate to the membrane pore size.

The most important regenerated-cellulose membranes for hemodialysis are manufactured by Akzo (Enka) under the tradename Cuprophan. These membranes are produced in flat-sheet and hollow-fiber form by the cuprammonium process and are available generally to device manufactures. Cuprophan hollow-fiber membranes are filled with a suitable liquid, such as isopropyl myristate, to minimize deformation during membrane handling and dialyzer assembly. The availabiity of Cuprophan membranes and their clinical use in a number of devices have led to these membranes serving as reference standards, with Cuprophan 150PM the most common reference-standard membrane.

Attempts to introduce alternative membrane systems to those of cellulose are generally based on achieving enhanced solute and water transport and improved compatibility. With respect to permeability, it should be noted that high-flux hemodialysis Cuprophan membranes, superior in permeability to Cuprophan 150PM, have been produced. However, in recent years the focus of attention regarding cellulose membranes has been on compatibility.

During hemodialysis with regenerated cellulose membranes, leucopenia and complement activation are consistent features. This marked immune response is generally much less pronounced with noncellulose membranes and this supports the view that the condition of a patient would benefit from the replacement of cellulose membranes. Despite this, the long clinical experience of regenerated-cellulose membranes and advantages of availability and cost, mean that dominance of cellulose membranes in hemodialysis will continue.

6.2 Modified-Cellulose Membranes

The strong influence of cellulose membranes on the complement system has been attributed to activation of the alternative pathway by hydroxyl groups. Hence, modification of cellulose by partial replacement of hydroxyl groups is a possible route to the enhancement of compatibility. However, modification can also lead to membranes with different permeability characteristics.

Partial replacement of the hydroxyl groups of cellulose by acetyl groups reduces the effect of hydrogen bonding, increases interchain separation, and makes the polymer less polar. The properties of a cellulose acetate polymer depend on the degree of acetylation, the nature and proportion of plasticizer, and the chain length of the cellulose molecule. Acetylation of cellulose introduces the possibilities for producing membranes by melting or solvent casting. Interest in cellulose acetate membranes for extracorporeal blood purification has been influenced by the success of cellulose acetate membranes in industrial ultrafiltration and reverse osmosis and the existence of the technology for producing asymmetric membranes.

Utilization of cellulose acetate ranges from cellulose triacetate membranes used in hemofiltration to polymers with a lower degree of substitution used in hemofiltration and hemodialysis. Cellulose acetate hemodialysis membranes induce less leucopenia and complement activation than cellulose membranes, although the response is greater than that obtained with other noncellulose membranes.

A more recent example of cellulose modification has been the manufacture by Akzo of Hemophan, a membrane in which diethylaminoethyl tertiary amino groups are introduced into some cellobiosic units of the cellulose molecule by means of stable ether bonds. In the case of Hemophan, the degree of substitution is less than 5% and much lower than the degree of substitution in cellulose acetate hemodialysis membranes. However, this low degree of modification results in significantly less leucopenia and complement activation than that found with cellulose acetate membranes and, with respect to the immune response, Hemophan appears markedly better than cellulose. The performance of Hemophan suggests that the nature of the substitution group is more important than the degree of substitution.

While the marked improvement with respect to leucopenia and complement activation brought about by hydroxyl substitution may be offset by a poorer performance in terms of other parameters, modified-cellulose membranes offer the considerable advantage of production under conditions similar to those of the well-established regenerated-cellulose membranes. Therefore, interest in modified-cellulose membranes is certain to continue.

6.3 Polyacrylonitrile-Based Membranes

Acrylonitrile polymers and copolymers have the ability to form excellent films and fibers. The characteristic properties of polyacrylonitrile result from the small size of the CN side group, the high polarity of this group, and hydrogen-bond formation between polymer chains. Polyacrylonitrile-based membranes for extracorporeal blood purification are favored by the fact that polymer purification benefits from the solubility of the acrylonitrile monomer in water and the incompatibility between monomer and polymer. A disadvantage is the restricted range of suitable solvents.

In order to function satisfactorily in extracorporeal blood purification, the normal structure of a polyacrylonitrile-based membrane must be altered to enable solute and water permeability. Suitable techniques include copolymerization, membrane stretching, chemical modification and the formation of asymmetric membranes.

A polyacrylonitrile-based membrane of particular interest is one produced from a copolymer of acrylonitrile and sodium methallyl sulfonate. This membrane was developed by Rhone-Poulenc SA and the production involves a combination of solvent casting, treatment of the cast membrane with water, and stretching. The membrane is plasticized with glycerol.

The acrylonitrile–sodium methallyl sulfonate copolymer membrane was introduced as AN69, but later a thinner membrane of the same chemical composition appeared as AN69S and this membrane is used clinically in flat-sheet and hollow-fiber form. Although incorrect, AN69 and AN69S are often referred to in the literature as polyacrylonitrile. The acrylonitrile copolymer membrane has a significantly higher permeability to solutes in the molecular-weight range 1000–2000 than Cuprophan 150PT and an ultrafiltration rate several times greater than that of Cuprophan 150PT.

Acrylonitrile–sodium methallyl sulfonate copolymer membranes are generally considered to be the most compatible and the influence on leucocytes and the complement system is certainly much less than that of regenerated cellulose. However, widespread clinical application has been hindered because membranes are not generally available but are restricted to particular dialyzers and there is a need to control the ultrafiltration rate with specialized monitoring equipment.

AN69S is presently by far the most important polyacrylonitrile-based membrane in clinical use.

Polyacrylonitrile hollow-fiber membranes have been produced in asymmetric form for hemofiltration and other polyacrylonitrile-based membranes may emerge.

6.4 Polycarbonate-Based Membranes

The polycarbonate membranes developed for extracorporeal blood purification utilize polycarbonate derived from bisphenol A. The development objective was to combine the mechanical properties of aromatic polycarbonates with appropriate permeability to solutes and water. Membranes for clinical use were produced by C. R. Bard Inc. by synthesizing polycarbonate–polyether block copolymers in order to achieve a balance of hydrophilic and hydrophobic properties. The members are asymmetric.

Polycarbonate–polyether block copolymer membranes offer potential advantages over cellulose membranes with respect to permeability and influence on the immune response. However, the fact that these advantages are less pronounced than those of AN69S may limit the clinical use of the polycarbonate-based membrane.

6.5 Methacrylate-Based Membranes

The approaches available for the production of methacrylate-based membranes range from the use of a hydrophobic polymer such as poly(methyl methacrylate) (PMMA) to provide wet strength to the use of a hydrophilic polymer such as poly(hydroxyethyl methacrylate) to provide permeability.

The methacrylate-based membrane to undergo clinical evaluation is that of PMMA hollow fiber. Since the membrane has been reported to be based on "pure" PMMA, it must be assumed that membrane permeability is accomplished by introducing a degree of porosity.

PMMA membranes can provide superior water and solute transport properties to those of cellulose hemodialysis membranes. In common with other synthetic membranes, the PMMA membrane is superior to regenerated cellulose with respect to leucopenia and complement activation, but, unlike other synthetic membranes, the PMMA membrane has been reported to induce a greater release of granulocyte elastase than regenerated cellulose. This is interesting not only from the membrane structure point of view but because the gradual increase in the release of granulocyte elastase during hemodialysis with both cellulose and noncellulose membranes contrasts sharply with the rapid transient decrease in leucocytes and increase in C3a generation.

6.6 Polysulfone Membranes

The first polysulfone membranes to be produced for hemodialysis were asymmetric hollow-fiber membranes developed by the Amicon Corporation. More recently, Fresenius have developed polysulfone membranes suitable for high-flux hemodialysis.

The Fresenius polysulfone membranes, reported to have a uniform pore size and even distribution, are produced in hollow-fiber form. Clinical data indicate that these polysulfone membranes have superior water and solute permeability characteristics than standard cellulose membranes. The compatibility of the polysulfone membranes, although apparently not quite reaching the standard of AN69S, is markedly better than cellulose. If the desirable properties become allied to commercial advantages, the Fresenius polysulfone membranes could achieve considerable clinical usage.

6.7 Poly(Vinyl Alcohol)-Based Membranes

The possession of water-solubility and film-forming properties makes poly(vinyl alcohol) (PVA) an obvious candidate material for hemodialysis membrane development. A fundamental step is insolubilization of the water-soluble polymer. This can be achieved by copolymerization or by a combination of copolymerization and cross-linking.

Membranes have been prepared by casting emulsions obtained from graft copolymerization of various monomers onto PVA. Comonomer selection and treatment of cast membranes offer the potential for useful products. However, the PVA-based membranes introduced clinically have been ethylene–vinyl alcohol copolymer membranes developed by the Kuraray Co. As with other synthetic polymer membranes, advantages over cellulose with respect to permeability and compatibility have been reported.

6.8 Membrane Development and Utilization

The development of a hemodialysis membrane requires considerable effort and cost. Therefore, the present options of cellulose, modified cellulose or synthetic polymers seem certain to continue. However, the strong focus on compatibility may encourage the utilization of membranes in association with antithrombotic agents other than heparin or the attachment of antithrombotic agents to membranes in order to reduce the requirement for systemic application of these agents.

Bibliography

Courtney J M, Gaylor J D S, Klinkmann H, Holtz M, 1984 Polymer membranes. In: Hastings G W, Ducheyne P (eds.) 1984 *Macromolecular Biomaterials*. CRC Press, Boca Raton, FL, pp. 143–80

Drukker W, Parsons F M, Maher J T (eds.) 1979 *Replacement of Renal Function of Dialysis*. Martinus Nijhoff, The Hague, The Netherlands

Forbes C D, Courtney J M 1987 Thrombosis and artificial surfaces. In: Bloom A L, Thomas D P (eds.) 1987 *Haemostasis and Thrombosis*, 2nd edn. Churchill Livingstone, Edinburgh, pp. 902–21

Gurland H J (ed.) 1987 *Uremia Therapy*. Springer, Berlin

Henderson L W, Cheung A K, Chenoweth D E 1983 Choosing a membrane. *Am. J. Kidney Diseases* 3: 5–20

Henderson L W, Mion C, Man N K (eds.) 1988 *Contemporary Management of End Stage Renal Failure*. Springer, New York

Klinkmann H, Bërgstrom J, Dzúrik R, Funck-Brentano J L (eds.) 1981 *Middle Molecules in Uremia and Other Diseases: Analytical Techniques, Metabolic Toxicity, and Clinical Aspects*. International Society for Artificial Organs, Cleveland, OH

Koch K M, Streicher E (eds.) 1987 *Biological Reactions within the Extracorporeal Blood Circuit during Hemodialysis*. Karger, Basel

J. M. Courtney and L. Irvine
[University of Strathclyde, Glasgow, UK]

M. Travers
[Akzo (Enka), Wuppertal, FRG]

Heparinized Materials

Materials that contain the anticoagulant heparin are known as heparinized materials. Heparin is a natural sulfated polysaccharide (a glycosaminoglycan) consisting largely of alternating *O*- or *N*-sulfated hexuronic acid (D-glucuronic or L-iduronic), and D-glucosamine residues (see Fig. 1). Certain *N*-sulfate groups are essential for biological activity. Commercial heparins are extracted typically from the intestinal mucosa or lungs of cows or pigs. Heparin exerts its anticoagulant activity by accelerating (in some cases by a thousandfold) the inactivation by antithrombin III of the activated serine proteases involved in the coagulation cascade: factors XIIa, XIa, IXa, Xa and, most importantly, thrombin. With thrombin inactivated or its formation prevented, fibrinogen is not converted to fibrin and the fibrin gel or "clot" does not form. Hence, incorporation of heparin is expected to take advantage of this activity and interfere with the thrombus

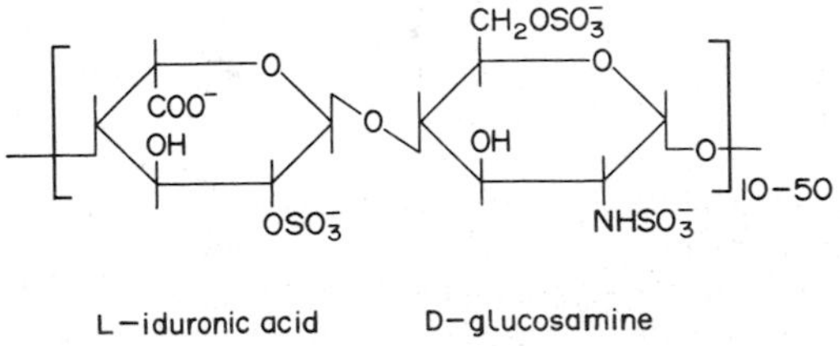

L—iduronic acid   D—glucosamine

Figure 1
Typical repeat structure of heparin

that would otherwise form on the material. Heparinization is potentially of particular benefit for devices used in low-flow areas of the circulation, or in situations dominated by thrombosis in stagnation zones or other flow irregularities. In these cases, red thrombus and fibrin formation predominate and the flow is unable to disperse locally concentrated activated clotting factors.

Heparin is either physically blended with the material or chemically bonded. The former method releases heparin from the material at a rate controlled by diffusion through the matrix. Such controlled-release systems provide a local heparinization near the point of release, thereby creating a microenvironment of dissolved heparin where coagulation is prevented; dilution with the bulk of the blood reduces the systemic heparin concentration to below effective levels. Heparin is a relatively large and heterogeneous molecule (mean molecular weight of 15 000 D), so its diffusivity and rate of release may be low. One approach used for certain polyurethanes has been to fractionate the heparin and incorporate only the low-molecular-weight fraction ($<10\,000$ D) into the polymer. This has the advantage that the low-molecular-weight fraction of heparin has a less adverse effect on blood platelets in many assay systems.

Most heparinized materials involve the use of chemical bonding, either covalently or ionically. Ionic bonding involves the sulfate or carboxylate groups in heparin, while covalent attachment may be through one of several functional groups: hydroxyls, carboxylic acid, aldehyde or amine. The amine may be from de N-sulfated residues along the chain or from the terminal serine in those chains retaining the linkage region from the original proteoglycan. The latter is particularly interesting since the resulting end-linked heparin may retain its native conformation and biological activity. Aldehyde groups at the end of heparin may be created by partial nitrous acid digestion and these can also be used to produce an end-linked heparin.

Ionically immobilized heparinized materials are functionally similar to the controlled-release systems using physically blended heparin. Due to ion exchange processes with the multitude of other ionic species in blood, ionically bound heparin is lost from the surface at a slow but steady rate. This released heparin then creates a microenvironment similar to that created by physically blended heparin. Experimental results confirmed by theoretical calculations indicate that a heparin release rate of 4×10^{-2} µg cm^{-2} min^{-1} is sufficient to create a microenvironment in which the heparin concentration at the blood–material interface is greater than 0.2 µg ml^{-1}, which is sufficient to prevent fibrin formation. Covalently bonded heparin in which the heparin–substrate bond is hydrolytically unstable acts in the same fashion. Heparin bonding to cyanogen-bromide-activated polymers such as agarose (heparin–Sepharose) suffers from this limitation, because of an unstable isourea bond, although heparin–Sepharose is a standard material for affinity chromatography and for certain mechanistic studies of immobilized heparin activity.

Ionically linked heparin typically involves the use of quaternary ammonium groups or similar cationic groups in the polymer. The first heparinized material discovered by Gott and co-workers at Johns Hopkins University involved the use of a quaternary ammonium salt of heparin (using benzalkonium chloride) that was adsorbed to graphite-coated surfaces. Other quaternary ammonium salts such as tridodecylmethylammonium chloride (TDMAC) were used, because of their nonpolar character and the absence of a need for graphite coating.

Rather than using a low-molecular-weight quaternary salt or amine, many investigators have incorporated the quaternizable amine directly into the polymer usually by copolymerization. N,N-dimethylaminoethyl or N,N-diethylaminoethyl groups have been the most popular. This approach avoids the problem of toxicity associated with leakage of the amine, but the heparin is still bound ionically. For example, Toray Industries in Japan market a polyurethane catheter which is coated with a hydrophilic heparinized N,N-dimethylaminoethyl methacrylate (DMAEM) graft copolymer (H–RSD). The graft copolymer is prepared by ultraviolet-initiated copolymerization of DMAEM and methoxypolyethylene glycol methacylate (MPEG) monomers on an appropriate substrate. The N,N-dimethylaminoethyl groups of the grafted copolymer are quaternized by ethyl bromide and the quaternized polymer in its final form is heparinized at 60 °C or 80 °C for 1–3 days. H–RSD technology has been used to prepare heparinized membranes, vascular grafts and blood bags. Polyamidoamines and diethylaminoethyl modified cellulose membranes have been treated in similar ways.

Covalent bonding (without appreciable heparin leakage) has been achieved, for example, using water soluble carbodiimides or gluturaldehyde. Carbodiimides are used to activate the carboxylic acid groups on the substrate or on heparin and are used to bind to amine groups on heparin or substrate, respectively. A diamine spacer (e.g., diaminohexane) is used typically to maximize biological activity.

Substrates have included an aminated polyhydroxyethyl methacrylate graft copolymer, hydrolyzed polymethyl acrylate beads, various agarose derivatives and even albumin to prepare albumin–heparin conjugates. Carbodiimide adducts that involve the hydroxyl groups of heparin or use acidic pH are presumed to result in the loss of anticoagulant activity, so that careful control of the experimental conditions is required. Glutaraldehyde has been used extensively in the preparation of a heparin–polyvinyl-alcohol hydrogel (heparin–PVA) in which the heparin is end-linked and retains its biological activity. The heparin–PVA is prepared from an aqueous solution of polyvinyl alcohol, glutaraldehyde, magnesium chloride (a catalyst) and heparin, which is allowed to dry and then cured at 70 °C for 2 h. Glutaraldehyde acts to both cross-link the polyvinyl alcohol and immobilize the heparin; the cross-link is presumed to consist of a pair of cyclic acetals. The hydrogel is used as a coating on an appropriately prepared substrate (e.g., polyethylene or segmented polyurethanes). Glutaraldehyde has also been used to stabilize ionically bound heparin and to reduce the rate of release.

Covalently bonded heparin surfaces, from which the heparin does not leak, are potentially useful for long-term applications. Previous concerns that such permanently immobilized heparin surfaces lose biological activity as a consequence of immobilization have been shown to be invalid. The immobilized heparin appears to act, without release, like heparin in solution to accelerate catalytically the inactivation of thrombin (and other serine proleases). Whether heparin augments the platelet reactivity of the substrate is more controversial, but studies suggest that this is not a significant problem. Clinically useful devices with such permanently heparinized coatings are in widespread development.

Bibliography

Kim S W, Ebert C D, Lin J Y, McRea J C 1983 Nonthrombogenic polymers: Pharmaceutical approaches. *ASAIO J.* 6: 76–8
Kim S W, Ebert C D, McRea J C, Briggs C, Byrn S M, Kim H P 1983 The biological activity of antithrombotic agents immobilized on polymer surfaces. *Ann. NY Acad. Sci.* 416: 513–24
Sefton M V, Cholakis C H, Llanos G 1987 Preparation of nonthrombogenic materials by chemical modification. In: Williams D F (ed.) 1987 *Blood Compatibility*, Vol. 1. CRC Press, Boca Raton, FL, pp. 151–8

M. V. Sefton
[University of Toronto, Toronto, Canada]

Hydrogels

Although the principles involved in the formation of hydrogel polymers cannot be regarded as being of recent origin, it is only since the late 1970s that the potential of these materials has begun to be realized. There is no precise and limiting definition of the term hydrogel and problems always arise when attempts are made to apply such definitions to the range of materials that may be encompassed by the term. Possibly the most useful description that may be given is that hydrogels are water-swollen polymer networks, of either natural or synthetic origin. Of these it is the cross-linked, covalently bonded, synthetic hydrogels whose biomedical use has grown most dramatically in recent years. It must be recognized, however, that composite structures involving both natural and synthetic hydrophilic materials have also begun to be exploited.

The vast majority of work in this field can be traced back to Wichterle and his co-workers who first indicated the usefulness, in biomedical applications, of highly cross-linked polymers of 2-hydroxyethyl methacrylate (HEMA) (Wichterle 1971). The great advantage of this material over most other hydrophilic gels (such as the synthetic acrylamide gels that have been known for many years) is its stability to varying conditions of pH, temperature and tonicity (osmotic concentration of salts) such as are commonly encountered in biomedical use. Any review of hydrogels in biomedical applications is bound, therefore, to take poly(HEMA) as a yardstick. When the polymer is prepared in the absence of water it is glassy and similar in many ways to poly(methyl methacrylate). The difference between poly(HEMA) and poly(methyl methacrylate) becomes quite apparent, however, when the materials are immersed in water. Whereas poly(methyl methacrylate) is affected relatively little by water, poly(HEMA) absorbs some two thirds of its own weight to form an elastic, water-containing gel. The amount of water absorbed is usually expressed as the equilibrium water content (EWC):

$$\text{EWC} = \frac{\text{weight of water in the gel}}{\text{total weight of hydrated gel}} \times 100\% \quad (1)$$

The EWC is the most significant single property of the gel since it is the water, held within the polymer substrate, that gives hydrogels their unique properties. Thus, the permeability of the membranes, their mechanical properties, their surface properties and the resultant behavior at biological interfaces are all a direct consequence of the amount and nature of water held in this way.

The EWC of hydrogels is governed by a range of factors. These include the nature of the hydrophilic monomer used in preparing the gel (of which HEMA is only one); the nature and density of cross-linking (the most common cross-linking agent being ethylene glycol dimethacrylate); and external factors such as the temperature, the tonicity (together with the nature of the constituent ions) and the pH of the hydrating medium. It is, therefore, the fact that the nature of hydrogels is dominated by the effect of water within the matrix that gives them their unique role in the field of biomedical applications. In simplistic terms the water conveys a degree of surface hydrophilicity that is absent in more conventional polymers. It also facilitates a degree of transport of water-soluble species that is difficult to achieve in other materials. Although it may be regarded as an advantageous influence, the role of water in controlling mechanical properties, while equally marked, is not unambiguously beneficial. The fact that the water acts as a plasticizer within the system certainly enables physical properties akin in many ways to those of soft tissue to be achieved. However, it normally carries the inherent disadvantage of reducing the tear strength to relatively low levels. The way in which permeability, surface properties and mechanical properties are variously employed in biomedical applications becomes apparent when a selection of the many potential applications for these materials is examined in more detail.

1. Contact Lenses

It is perhaps not immediately obvious that the use of polymers for contact lenses represents an example of the biomedical application of synthetic materials. The use of quite similar materials in joint replacement, heart valves, membrane oxygenators and hemodialysis membranes presents specific problems associated with, for example, their biocompatibility, strength and permeability that might seem to be absent in contact lenses. This is certainly not the case, however, and the general biomedical principle of designing the material to give a balance of properties appropriate to the particular environment is of prime importance. The situation is obviously less critical in the case of lenses intended for daily wear only than it is in the case of lenses worn for successive day and night periods (frequently referred to as extended wear). Nonetheless, in both cases very similar properties to those mentioned above, in connection with other biomedical applications, are involved.

The three important aspects of the ocular environment, in terms of materials design, are the cornea (permeability), the eyelid (mechanical properties) and the tears (surface properties). The cornea is avascular (has no blood vessels) and the need to ensure oxygen transport to the corneal surface governs the permeability requirement of the hydrogel which, in turn, is controlled by its equilibrium water content. The eyelid dictates the range of acceptable mechanical properties; comfort and retention of visual stability during the blink cycle dominate acceptable upper and lower limits of elasticity. The water content, although influential, is not the sole structural factor of importance here since the water structure within the gel, together with chain stiffness and interchain forces within the polymer matrix, is capable of exerting a dominating influence. The interaction of the hydrogel at the lens–tear interface is an important example of the interaction of a biomaterial with a complex biological environment. This behavior has many similar features to that observed at other body sites and is a function of the surface properties of the hydrogel.

The contact lens does have certain unique features that set it apart from other areas of biomedicine. The design and fitting of the lens can play an over-riding part in governing the patient's response to the given material, although this is to a large extent offset by the relative ease of insertion and removal of the device. This means that the clinician can optimize the "fit" of the lens. Thus, it is much easier in this field than in most others to compare the response of reasonably large numbers of patients to different materials, under conditions in which variables related to design and fitting have been isolated. For this reason, the research carried out in recent years into the use of polymers as contact lenses has provided information on a range of materials that will greatly assist future work on their use in other biomedical applications.

The history of the development of contact-lens materials is interesting. Glass was used exclusively for some years and it was usual for the lenses to be ground individually. Poly(methyl methacrylate) began to replace glass in the 1940s because of its toughness, optical properties and physiological inactivity, coupled with ease of processing by existing turning techniques. There have been many attempts to find alternative materials, but it was not until hydrogels appeared on the scene that any serious competitor emerged. Interestingly enough, it was the early recognition of the potential use of these materials in the manufacture of soft contact lenses that provided the commercial incentive responsible for much of the continued effort in the field of hydrogel chemistry.

An appropriate demonstration of the way in which soft contact lenses have influenced the development of hydrogel chemistry is found in the patent literature. Since the early work of Wichterle, principally on poly(HEMA) (Wichterle 1971), the contact-lens field has provided the basis for the examination of many more such polymers. Over one hundred patents appeared in the 1970s, covering a wide range of molecular structures and compositions. Although only a small proportion of these have reached commercial status, the necessary toxicology and clinical work has produced a useful introductory evaluation of several hydrogels of widely differing chemistry. There are many different hydrogel contact-lens materials, having EWCs ranging between 30% and 80% and based on various combinations of monomers including HEMA, vinyl pyrrolidone, glycidyl methacrylate, glyceryl methacrylate, methoxyethyl methacrylate, cyclohexyl methacrylate, methyl methacrylate, methacrylic acid and substituted acrylamides. This contrasts with a longstanding dependence on poly(HEMA) in most other biomedical applications, a situation that has, advantageously, begun to change.

2. Implant Materials

It has been appreciated for some time that synthetic hydrogels, such as poly(HEMA), may be suitable for general prosthetic application. Hydrogels are employed in a number of forms, such as optically transparent films, nonporous gels, porous sponges and as thromboresistant coatings bound to a less biotolerant polymer substrate. One of the early uses of poly(HEMA), in the form of a spongy gel, was in breast augmentation. The consistency of such gels can be modified to give an elasticity that is comparable with the original tissue. In addition, poly-(HEMA) has a potential advantage over alternative materials that have been used, such as poly(methyl methacrylate), polyethylene and silicones, in that, unlike these impermeable hydrophobic substrates, it does not provide a barrier to the transfer of body fluids from the surrounding tissue. This permeability to metabolites and ions of low molecular weight is believed to be advantageous to *in vivo* performance, although long-term implantation has produced problems, particularly the so-called calcification phenomenon.

A detached retina can be reattached to the underlying choroid, its source of nutrition, by means of a procedure known as scleral buckling. This is normally achieved through the use of a synthetic implant that causes the indentation of the sclera (the firm white fibrous membrane that forms the outer covering of the eyeball). The implant material has traditionally been made of silicone rubber (polysiloxane), but the potential advantages of a poly-(methyl acrylate-cohydroxyethyl acrylate) hydrogel implant are now recognized. Not only does such a material possess the desirable mechanical properties of softness and elasticity, but unlike its silicone rubber analogue it can be impregnated with a hydrophilic antibiotic which is then locally released. This can help prevent infection during the immediate postoperative period.

The swelling characteristics of the hydrogel can be exploited in a number of ways, depending on how it is preconditioned before insertion. It may be conditioned in hypertonic saline so that on absorbing body fluids it increases in size, thus achieving a higher buckle than is possible with a nonswelling implant. Alternatively, by conditioning in a physiologically isotonic saline, it can remain at a constant size. Finally, it can be inserted in a swollen state higher than its equilibrium value in the body such that it gradually decreases in size, thus reducing the degree of scleral buckling after choroid–retinal adhesion has occurred. This ability of hydrogels to swell on absorbing liquid has been exploited in the design of a nondislodgeable endoprosthesis for the treatment of patients with obstructions of the bile duct. Rings of hydrogel, located in grooves around the endoprosthesis, are placed on either side of the obstruction. The swelling of the rings prevents dislodgement of the device, thus allowing internal drainage to be maintained.

It was pointed out in Sect. 1 that the optical, mechanical and permeability characteristics of synthetic hydrogels have led to them being widely employed as contact-lens materials. In addition, they can also be used in keratorefractive surgery where they are implanted within the corneal stroma. Poly(HEMA) has been found to be well tolerated after implantation into rabbit eyes as an intraocular lens and interest in this area is growing rapidly.

The bone ends in movable joints are covered by articular cartilage, a natural hydrogel of mainly collagen and chondroitin sulfate. Its destruction in osteoarthritis, which results in debilitating pain and stiffness of the joints, has led to a search for suitable synthetic hydrogel replacements. Such materials must exhibit high tensile and compression strengths and low friction coefficients. Poly(vinyl alcohol), by being repeatedly subjected to a freeze–thaw process to increase its mechanical strength, has been successfully used as a replacement material for femur artroidal cartilage in rabbits. Poly(HEMA) sponges have also been used in similar studies. The mechanical properties of poly(HEMA) sponges are found to be critically influenced by pore size and pores of less than 50 μm are required to give the sponges the necessary weight-bearing characteristics. After some three months the implants become advantageously overgrown by cartilaginous or fibrocartilaginous tissue accompanied by ingrowth of chondroid tissue into the sponges.

The hydrated polymer network of hydrogels being swollen and heavily plasticized by its imbibed water is, in general, mechanically weak. Such a potentially limiting factor can be overcome by forming a composite of the hydrogel with a woven fabric or fiber. A composite of poly(ethylene terephthalate), fiber (Dacron) and poly(HEMA) has been used for synthetic tendons: the strength and stiffness of such tendons are, essentially, that of the fiber and can be adjusted to closely imitate the properties of natural tendons by altering the volume fraction of the fiber. It is important for the proper functioning of these tendons that they possess a very high creep resistance. Suitable poly(ethylene terephthalate) fibers have been made by the additional drawing of commercial textile fibers, texturized by false twist. A related composite in the form of a knitted Dacron material embedded in a cross-linked poly(HEMA) matrix has been studied for such applications as prosthetic leaflet heart valves and carotid artery implantations. It was found, in early studies, to show good fatigue strength and hemocompatibility.

Attempts have been made to create a hard tissue prosthesis material by incorporating inorganic fillers such as borates and silicones into poly(HEMA). These materials were found to be inferior to bone in their mechanical properties, but their use as dental root canal fillings for teeth has been examined. Guttapercha and other hydrophobic polymers have been studied for such dental applications, but have been found to be considerably less biocompatible than poly(HEMA)-based systems. In addition to their general biotolerance by the surrounding tissue, the ability of these materials to swell upon hydration ensures a tight seal within the tooth cavity.

Hydrophilic polymeric materials may exhibit the required mechanical performance characteristics for prosthetic implants, but their biotolerance is generally poor. This has led to attempts to render the surface of such materials more biocompatible by grafting a hydrogel onto their surface. Such composites tend to possess the mechanical properties of the underlying polymer substrate rather than that of their relatively thin hydrogel coating. Various techniques are used to graft the hydrophilic polymer onto its base. These usually involve generating radicals on the substrate surface via ionizing radiation, microwave-generated active hydrogen or electrical discharge, thus allowing the grafting of one or more polymerizable hydrophilic monomers to take place. Prostheses coated in this way have been examined for a variety of applications. Examples include polyacrylamide-grafted polyethylene for vascular prostheses, a cross-linked HEMA–methacrylic-acid copolymer grafted onto a polyurethane substrate for urological devices and cardiac pacing leads coated with poly(vinyl pyrrolidone) impregnated with iodine. In the case of hydrogel suture coatings, the incorporation of antibiotics into the gel can help prevent the development or spread of local infection. General reports suggest that the knotting ability, as well as the other mechanical properties, of sutures are unaffected by the presence of such coatings.

3. Wound Dressings

The development of synthetic occlusive wound dressings, used for the treatment of burns, granulation tissue, dermatitis, ulcerations, blisters, fissures, herpes and several other skin conditions, is currently a subject of great commercial interest. Before indicating the role of hydrogels in this area, it is appropriate to consider the properties that a successful wound dressing should possess. The material should be flexible, strong, nonantigenic and permeable to water vapor and metabolites, while securely covering the wound to prevent bacterial infection. Hydrogels possess many of the above properties to a degree and because of this they have been examined in various forms as wound dressing materials (see *Wound Dressings Materials*).

Several hydrogel polymers of natural origin have been investigated to determine their suitability for use as wound dressings. However, their poor mechanical strength usually necessitates the use of a fiber or polymer matrix as a support with these materials. Examples include dextran-based hydrogels reinforced with fine cotton gauze, hydrogels based on sugar and protein derivatives that are radiation cross-linked with acrylic acid and bonded to a supporting film, and copolymers of water soluble, linear anionic and cationic polyelectrolytes (for instance ammonium keratinate and chitosin or collagen). When hydrated, these materials are quite malleable and, therefore, can easily be shaped to the contours of the wound. The hydrogels can also be used for the release of appropriate topical agents such as gentamycin sulfate.

One of the most widely studied synthetic hydrogel-based wound dressings is based on poly-(HEMA). The dressing is formed directly on the burn wound from a two-component system comprising poly(HEMA) and, as the solvent, poly(ethylene glycol). Alternate layers of poly(ethylene glycol) and poly(HEMA) are applied to the wound. During this process the glycol dissolves the poly(HEMA), forming a saturated solution that converts into a solid film after some thirty minutes. There have been several clinical studies on this material with some conflicting reports on its suitability as a wound dressing. These studies have concluded that although the dressing is a valuable asset in the treatment of burns, it has several shortcomings. These include difficulty of application, poor adherence to the wound and cracking of the film. Additionally, as the dressing is translucent rather than transparent, it is not possible to visually monitor the wound healing and the dressing must be cut away to check the wound for infection.

Attempts to maintain the advantages of hydrophilic materials such as poly(HEMA) while overcoming its disadvantages have usually involved some form of composite material, such as blends or laminates. Examples include the use of laminates of poly(HEMA) with butadiene, colloidal poly(ethylene oxide) particles sandwiched between polyethylene film and hydrogels of various types coated onto poly(ethylene terephthalate) mesh. More complex systems include interpenetrating polymer networks based on acrylamide and agar together with the family of so-called "hydrocolloid" dressings. The latter typically consist of a blend of a swellable hydrocolloid (e.g., sodium carboxymethyl cellulose and gelatin) dispersed in an elastomeric gum (e.g., isobutylene or butadiene-based polymers) and coated onto a polyethylene or polyurethane support.

4. Controlled Release of Drugs

The ideal in drug administration is to achieve a level of the agent at the target site such that its effectiveness is optimized. The periodic application of drugs by conventional techniques, such as orally or by intravenous injection, means that plasma levels of the active agent will modulate about this optimal level. This runs the risk of achieving concentrations that are so high as to produce toxic side effects or so low as for the drug to be rendered ineffective. Poor patient compliance with the administration regime can exacerbate this problem. In addition, the large amounts of drugs that must be taken in order to obtain an effective level at the target site can be both wasteful and costly. It was for reasons such as these that so much interest in the development of devices for the targeted and controlled release of biologically active agents has been aroused (Langer and Peppas 1981). Such drug delivery vehicles are generally constructed from synthetic polymeric materials into which the drug is incorporated, either by being sequestered within the polymer matrix or chemically bound to the polymer backbone. Release of the agent can be achieved by having it diffuse out through the polymer matrix or having the device undergo some form of chemical erosion.

Hydrogels are often used in physically governed systems, because the water held within the swollen polymer network acts as a transport medium for hydrophilic species. Indeed, it is this permeability of the gel to ions and metabolites in body fluids that may be one of the contributory factors to the biotolerance of this family of materials. An understanding of transport, and thus an ability to influence permeability and selectivity, is important not only in drug delivery, but also in the design of membranes for applications such as reverse osmosis, kidney dialysis and biosensors. A universally satisfactory transport model has not yet been realized but most relate the permeability or diffusivity to some function of the overall amount of water in the gel matrix. The free-volume model (Yasuda et al. 1968) is, perhaps, the one that has been applied most successfully. This model applies to homogeneous water-swollen polymer matrices where it is assumed that there is neither macroscopic phase separation of the polymer and nonpolymer components nor any heterogeneity in these components. This means that there are no macroscopic pores or channels through which a permeant can be transported. On the molecular scale, however, nothing can be said to be purely homogeneous and in such hydrogels, pores are fixed neither in size nor location, but result from the random fluctuations of chain segments which may exhibit a high degree of mobility owing to the plasticizing effect of the water. The pores and channels in such systems are, therefore, in a constant state of flux. The free-volume model takes a partly thermodynamic, partly statistical approach. It assumes that the transported species is associated only with the water phase, its diffusion being dependent upon its probability of being located next to a suitable hole that is both unobstructed and large enough to accept the permeant. An equation can be derived from this model that predicts a linear relationship between ln P and $1/H$, where P is the permeability coefficient in the hydrogel and H is the degree of hydration.

By choice of monomer composition or polymerization technique, the amount of water, porosity and consequent permeability characteristics can be controlled. Indeed, heterogeneous hydrogels with a fixed macroporous structure can be generated as well as the homogeneous type mentioned previously. This allows hydrogels to be used not only for the release of low-molecular-weight drugs, but also of macromolecules such as the protein hormones. It has wider implications since materials have interesting sorption properties and are of potential value as artificial liver support hemoperfusion systems. This is typical of the more ambitious applications in which these unusual and versatile polymer systems are increasingly finding use.

See also: Acrylics for Implantation

Bibliography

Andrade J D (ed.) 1976 *Hydrogels for Medical and Related Applications*. American Chemical Society Symposium Series, Vol. 31. ACS, Washington, DC
Corkhill P H, Hamilton C J, Tighe B J 1988 Synthetic hydrogels (VI). Hydrogel composites as wound dressing materials and implants: A review. *Biomaterials* 10: 3–10
Langer R S, Peppas N A 1981 Present and future applications of biomaterials in controlled drug delivery systems. *Biomaterials* 2: 201–14
Pedley D G, Skelly P J, Tighe B J 1980 Hydrogels in biomedical applications. *Br. Polym. J.* 12: 99–110
Peppas N A 1982 Contact lenses as biomedical polymers. In: Hartstein J (ed.) 1982 *Extended Wear Contact Lenses for Aphakia and Myopia*. Mosby, St. Louis, MO, pp. 6–43
Peppas N A (ed.) 1987 *Hydrogels in Medicine and Pharmacy*. CRC Press, Boca Raton, FL
Tighe B J 1981 Contact lens materials. In: Stone J, Phillips A J (eds.) 1981 *Contact Lenses*, 2nd edn., Vol. 2. Butterworth, London, pp. 377–99
Tighe B J 1986 The role of permeability and related properties in the design of synthetic hydrogels for biomedical applications. *Br. Polym. J.* 18: 8–13
Wichterle O 1971 Hydrogels. In: Mark H F, Gaylord N G (eds.) 1971 *Encyclopaedia of Polymer Science and Technology*, Vol. 15. Interscience, New York, pp. 273–91
Yasuda H, Lamaze C E, Ikenberry L D 1968 *Makromol. Chem.* 118: 19–35

B. J. Tighe
[Aston University, Birmingham, UK]

I

Invasive Sensors

Physiologists, clinicians, scientists and engineers engaged in medical research, depend on the measurement of a wide range of quantities in order to advance their science and clinical practice. In the procedure of measurement, a "sensor" or "transducer" is needed to interface with the quantity of interest in order to convert it, or transduce it, into an electrical signal. The investigation of fundamental physiological phenomena, and the routine tasks of clinical diagnosis and therapeutic control, can often be greatly assisted through the use of sensors suitable for direct placement in an appropriate anatomical site. Measurements of physical or chemical variables made with such sensors can provide a dynamic indication of normal events or of pathological processes, thus allowing temporal relationships to be examined for phenomenological studies, or enabling rapid assessment of clinical condition and subsequent optimization of therapy.

Various anatomical sites may be used for sensor placement, the choice depending on both scientific and clinical factors. The options are:

(a) intracellular,

(b) tissue (e.g., intradermal and intratumor),

(c) intravascular,

(d) skin surface, and

(e) natural orifice (e.g., ear, airway, gastrointestinal tract and intraoral).

While some of these sites, such as the intravascular compartment, clearly require an invasive procedure to be used, others, such as the gastrointestinal tract, require at least an intrusive procedure. Furthermore, sensors designed to be used on the skin surface (e.g., for the monitoring of blood-gas levels in newborn babies) may be considered to be noninvasive. This distinction, particularly between invasive and noninvasive procedures, is especially important from the viewpoint of patient safety when considering sensors for clinical use since invasive techniques are clearly associated with finite risks. However, purely scientific considerations may often be as important because of the inevitable interference with the process being sensed by a device which "invades" the microenvironment of that process. This is the classical dilemma of measurement; that all too often a compromise must be found between, on the one hand, direct techniques which invade and disturb, and on the other hand, indirect methods which may not have sufficient accuracy or specificity for the variable of interest.

1. Sensors for Physical Variables

A great deal of ingenuity has been directed towards the problem of devising practical sensors for measurement of both physical and chemical variables. In spite of the diversity of practical devices which have emerged, a relatively small number of basic sensor principles is utilized, although in recent years novel chemical-sensing principles have been the subject of much important research.

Measurement of force, pressure, displacement and velocity (and its derivatives) are frequently of recurring interest. Perhaps the best known and most widely used principle on which practical devices have been based, is that of the strain gauge. The name of this device is derived from the definitions of the basic mechanical parameters which were first exploited in sensor design and construction. Strain is defined as the fractional change in length dL/L and can be measured with a device in which changes in length are transduced into an electrical signal. With resistive strain gauges, the change in length is arranged to produce a change in electrical resistance and this may be achieved, for example, by stretching a wire or film, the change in cross-sectional area then producing a change in resistance. Alternatively, a semiconductor material (e.g., n- and p-type silicon) can be used as the resistive element, these sensors having greater sensitivity than wire or film gauges, although basic linearity may not be as good. With both wire and film gauges and with semiconductor devices, the influence of temperature on resting resistance and on sensitivity must be taken into account, compensation techniques being used accordingly. Measurement of changes in resistance can be conveniently carried out with the Wheatstone bridge arrangement, and it is also possible to achieve temperature compensation with such a bridge.

Strain gauges, either using wire or film or using silicon, have been incorporated into transducers for implantation into animals for several months. These have been used for telemetering ventricular or arterial pressure. Catheter pressure transducers are used frequently for clinical diagnosis (Ko et al. 1979).

Electromagnetic coupling between two wire coils can be utilized as a means of sensing displacement, force or pressure. If one coil carries an alternating current, then this will induce a current in the second coil, the magnitude of the induced current being dependent on the magnetic coupling which in turn is influenced by coil spacing. The intercoil coupling can also be influenced by an armature passing along the axes of the coils, and such an armature can be affixed to, for example, a diaphragm, thus allowing the pressure applied to the diaphragm to be transduced into an electrical signal. These electromagnetic sensors can have good sensitivity and low temperature dependence, although they may be influenced by external magnetic fields, such as those used with magnetic resonance imaging and spectroscopy.

Very useful sensors can be constructed with piezoelectric crystals and ceramics. These materials have an asymmetrical charge distribution in their structure and become electrically polarized when mechanically strained in certain directions. The effect is reversible, so the application of a potential difference across the appropriate faces of a crystal will produce a dimensional change. One useful feature of some piezoelectric materials is that only a very small displacement is needed to produce an electrical signal.

Piezoelectric materials can be used to generate pressure waves, and these can be of appropriate frequencies, above say 17 kHz, to constitute ultrasound energy. Ultrasound may be used in a variety of ways to perform measurements of physical variables, especially displacement and velocity. Ultrasound can propagate through a material, such as biological tissue, and when it encounters a change in the density of the material it will be reflected and refracted. It is possible to transmit a short pulse of ultrasound, detect the occurrence of some of the transmitted energy as it is reflected from an interface between two materials and determine the distance x of the interface from the transmitter, knowing the ultrasound velocity v and the delay time t_d:

$$x = vt_d$$

Invasive ultrasound sensors based on this approach have been implanted for the measurement of cardiac and vascular displacement. Changes in distance can clearly also be monitored and the velocity of the interface calculated by taking the first derivative dx/dt. An alternative method for measuring velocity is to utilize the Doppler effect. With this approach, the frequency v of the ultrasound is changed when it strikes a moving interface; the frequency change is proportional to the velocity:

$$dv = 2vv_0 \cos \theta$$

where v_0 is the interface velocity and θ is the angle between the ultrasound beam and the direction of movement. Processing of the ultrasound signal can then be carried out to provide a directional indication of velocity. Once again, ultrasound crystals have been attached to the vascular wall for continuous invasive measurement of blood velocity.

1.1 Temperature Sensing

The measurement of temperature is important in its own right as a key factor relating to physiological processes, but it is also often necessary in order to correct the temperature dependence of certain sensors. As with other sensors, it is usually necessary to select a temperature-sensing principle that will allow an appropriately compact device to be constructed. There are five commonly used principles of temperature sensing relevant to invasive applications, although several other principles may be considered for specialist future applications.

The two most commonly used temperature-sensing techniques are the thermoelectric and the thermoresistive principles. When two different metals are connected together forming two junctions, and when the junctions are held at different temperatures, a current will flow in the circuit; this is the thermoelectric Seebeck effect. The potential difference producing the current flow is approximately proportional to the difference in the junction temperatures.

Devices constructed on this basis are thermocouples, and metal combinations such as copper–constantan and chromul–alumel can be used, producing 45 μV $^\circ$C^{-1} and 40 μV $^\circ$C^{-1}, respectively. Very fine wires, perhaps a few micrometers in diameter, can be used to produce small temperature sensors for implantable use.

Thermoresistive transducers employ the predictable changes of resistance with temperature exhibited by certain metals and semiconductors. Platinum has a temperature coefficient of resistance a of $+0.0039$, and the resistance R at temperature T is given by:

$$R_T = R_0[1 + a(T - T_0)]$$

where R_0 is the resistance at temperature T_0.

Certain ceramic-like semiconductors, or thermistors, also exhibit very significant changes of resistance with temperature, typically having values for a of -0.04. Note that this is a negative coefficient, but it is in fact also possible to obtain materials with positive coefficients. Thermistors exhibit a nonlinear temperature-resistance characteristic of an exponential form and linearization circuitry is needed. Thermistors can be constructed with small dimensions, in the order of 30 μm in diameter, and they have been used in hypodermic needles and a variety of canulae-type probes for invasive measurement.

Semiconductors are being used increasingly for sensor fabrication because they allow precise mass production of microminiature structures at low cost. The p–n junction diode can be operated at a constant current to produce a terminal voltage linearly related to temperature. Multiple temperature sensors can be integrated into a single device to provide back-up operation.

Electromagnetic energy is radiated by all bodies with temperatures above absolute zero, the total energy W_T being given by:

$$W_T = e\sigma T^4$$

where σ is the Stefan–Boltzmann constant and e is the emissivity of the surface of the body. At body temperature, the radiated energy is in the infrared part of the electromagnetic spectrum, and measurement of the energy allows temperature to be measured. Infrared radiation can be measured either by thermal detectors, such as the Golay cell, the pyroelectric type or the radiation thermopile, or by photon detectors such as photovoltaic or photoconductive cells.

2. Sensors for Chemical Variables

Sensors for the detection of chemical rather than physical variables can be used for an increasing range of physiologically and clinically important species. These include gases such as oxygen and carbon dioxide, ions such as H^+, K^+ and Na^+, catabolites such as glucose, urea and creatinine, and even drugs and hormones such as morphine and insulin. The techniques on which practical sensors are based, have emerged from collaborative ventures between engineers, physicists, chemists and biologists. The two most important chemical-sensing principles for invasive sensors are those using electrochemistry and optics.

2.1 Electrochemical Sensors

Electrochemical techniques are useful for the construction of invasive sensors, having the potential for microminiaturisation with a wide range of target analytes. There are two main classes of electrochemical sensors.

(a) Amperometric sensors are those in which a current is produced in proportion to the concentration (or partial pressure when considering a gas) of the chemical of interest.

(b) Potentiometric sensors produce a potential difference in proportion (usually logarithmic) to the analyte of interest.

The most widely used amperometric sensor is that for measuring oxygen. With this device, a noble metal (e.g., gold, silver or platinum) "working" electrode is held at a negative potential with respect to a reference electrode, usually silver–silver chloride (AgCl), the two being linked by an electrolyte solution, such as potassium chloride. Oxygen molecules in the vicinity of the cathode surface will be electrochemically reduced by the potential, and a flow of electrons produced. The consumption of oxygen by this process ideally produces zero oxygen concentration at the cathode surface; this establishes a concentration gradient from the cathode to the bulk of the sample solution and this ensures diffusion of oxygen molecules to the cathode. The magnitude of the cathode current can, by design, be made

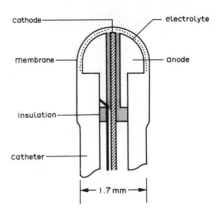

Figure 1
Construction of a catheter-tip P_{O_2} sensor; the cathode (silver, gold or platinum) and silver–AgCl reference anode are covered by electrolyte powder and a thin dip-coated polymer membrane

to bear a linear relationship to the oxygen partial pressure in the sample bulk. This is most easily achieved if the cathode, reference and electrolyte are surrounded by a thin gas-permeable polymer membrane which serves to control the gas diffusion in a reliable manner. In the ideal situation, the sensor current I will be related to the sample P_{O_2} by the equation:

$$I = nFADP_{O_2}/b$$

where F is Faraday's constant, A is the cathode area, b is the membrane thickness, D is the membrane oxygen diffusion coefficient and n is the number of electrons per molecule of oxygen. Membrane-covered oxygen sensors have been constructed as microneedles (Silver 1965) and as intravascular catheters (Rolfe 1988), as shown in Figs. 1 and 2.

Attempts have been made to develop amperometric sensors for other species and there has been particular interest in certain anesthetic agents. Nitrous oxide can be electrochemically reduced at a silver cathode with an applied potential of approximately $-1.5\,V$. It is theoretically feasible to use a single sensor for both oxygen and nitrous oxide measurement, by switching the cathode potential between the two appropriate voltages (Hahn et al. 1979). With more complex sensor designs, utilizing combined membrane structures, it may also be possible to measure other anesthetic agents, such as halothane and also perhaps carbon dioxide.

The family of potentiometric electrochemical sensors is typified by the glass-membraned pH electrode. This device uses a pH-sensitive glass bulb filled with electrolyte and a contact wire (Ag–AgCl). The bulb and a reference electrode, typically the standard calomel electrode, are placed in the fluid to be measured. The pH-sensitive glass then develops a potential difference between its two faces and this is measured between the internal contact wire and the reference electrode. The ideal theoretical relationship describing the membrane potential difference E is:

$$E = E_0 + \frac{RT}{F} \ln(a_i + K_{ij}a_j)$$

where a_i is the activity of the desired ion, a_j is the activity of an interfering ion and K_{ij} is the selectivity coefficient of the membrane.

In addition to being available for H^+ ions, glass membranes are available for a wide range of ions such as K^+, Na^+ and Ca^{2+}. There has also been significant activity in recent years to develop polymer ion-selective membranes, employing so-called ion carriers in a polymer matrix such as poly(vinyl chloride). The antibiotic valinomycin has a structure which allows potassium ions to be incorporated with high selectivity; this characteristic has been utilized in the design of potentiometric ion-selective sensors.

Electrochemical sensors for gas and ion measurement can also be used as the basis of sensors for more complex compounds, such as glucose, urea and creatinine. Such sensors often utilize biological principles for achieving specific detection of the target chemical, and this represents the family of biosensors. An important design approach is to incorporate an enzyme into the sensor that will promote a specific reaction of the target chemical such that reaction products can be measured directly with gas or ion sensors. To measure glucose, the enzyme glucose oxidase may be immobilized in cellulose acetate and this "membrane" coated over

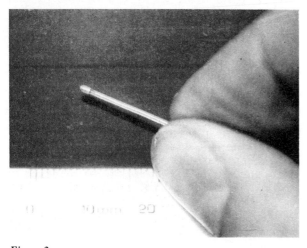

Figure 2
The catheter P_{O_2} sensor described in Fig. 1

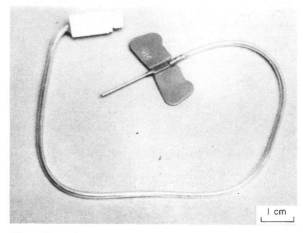

Figure 3
A needle glucose sensor attached to an infusion catheter; the sensor contains glucose oxidase and is covered by a polyetherurethane membrane

the surface of an electrochemical sensor (Rea et al. 1985). Glucose reacts with oxygen producing hydrogen peroxide which can be measured at a platinum anode. There has been considerable interest in developing glucose sensors for long- and short-term invasive use in patients with diabetes, and devices in the form of needles (see Fig. 3), intravascular catheters and tissue "disk" sensors have been produced, but with only limited success.

2.2 Optical Sensors

Optical techniques for sensor design have attracted considerable interest in recent years and certain configurations are potentially applicable to invasive use. The availability of small optical fibers has been an important factor in attracting this interest. Perhaps the simplest optical sensing approach is merely to employ an optical fiber to convey light to and from the measurement site, with the optical measurement being made remotely at the proximal end of the fiber. This method effectively enables invasive spectrophotometry to be performed.

Hemoglobin has an optical absorption spectrum which is oxygen dependent and this feature allows the degree of oxygenation of blood to be determined invasively using optical fibers. Two plastic optical fibers may be positioned within an arterial or venous catheter and light at three wavelengths transmitted down one fiber into the blood. Light backscattered and reflected from the blood is collected by the second fiber and measured externally. Hemoglobin oxygenation is then calculated from light-absorbance measurements at the particular wavelengths. This method is referred to as oximetry.

Light-absorbance measurement can also be employed to measure pH with an invasive sensor.

Once again optical fibers are used, but in this case a chemical reagent is attached to the distal end of the fiber. Typically, an H^+-permeable tube containing a pH-sensitive dye is affixed to the fiber tip, and either the absorbance or fluorescence are measured, depending on the dye type, in order to monitor pH. Oxygen partial pressure can also be measured with this arrangement, using a dye which exhibits an oxygen-dependent fluorescence (Peterson et al. 1984).

3. Problems of Practical Devices

Invasive sensors must of course be reliable in operation, producing measurements of the desired variable with acceptable precision and accuracy. However, since the device will be inserted directly into a patient, it must also be safe.

Invasive sensors can be constructed in a variety of forms in order to match the particular application; the precise geometrical design and the materials used will have a profound effect on the degree to which successful and safe operation is achieved. The attainment of the desired performance should be based on a realistic appreciation of the operational time needed; this may range from a few minutes to several weeks or months. The reliable operational lifetime of invasive sensors may depend on the volume of reagents contained within chemical devices which are consumed during normal operation. However, it is more usually determined by chemical interactions that occur between the sensor surface and the surrounding biological fluid, be it blood or interstitial fluid. Proteins are adsorbed to the sensor surface, and when the device is used in blood this may initiate a sequence of events leading in the worst case to the formation of a clot. The identification of materials and surface characteristics which will ensure appropriate "biocompatibility" is a major aspect of invasive-sensor design and development.

Catheter-based sensors are popular for relatively short-term monitoring of a wide range of variables including pressure, flow and temperature, and many chemical species. Polymers such as poly(vinyl chloride), polytetrafluoroethylene, polyethylene and polyetherurethane have been used as catheter material with varying degrees of success. Of these materials, polyetherurethane is currently the most successful and, for example, this material can be used for both the diffusion membrane and the supporting catheter of sensors for the measurement of oxygen and glucose. Biocompatibility is improved when a smooth surface finish is produced.

Although the potential hazards to patients of invasive sensors are of major concern, reliable quantitative performance of the sensor remains an important goal. Undoubtedly, one of the most critical aspects of this is the matter of sensor encapsulation, the purpose of which is to ensure that body fluids do not pass into the sensor so as to compromise electrical operation. Ingress of water vapor is the most likely mechanism leading to sensor failure, and the likelihood of this may be reduced by using appropriate materials for the encapsulation. Epoxy resins and various polymers have been employed for this purpose, as too has hermetic sealing with metal cans. Extreme care is also needed in the fabrication process to ensure cleanliness and sterility.

Many invasive sensors may be sterilized with γ irradiation before implantation; this approach has been used for catheter oxygen sensors. Some devices such as semiconductors, however, will be adversely affected by radiation, and chemical sterilization must be used instead. Low-cost fabrication methods make it feasible to consider some sensors as single-use disposables.

4. Conclusions

Invasive sensors can be used for the measurement of a wide range of both physical and chemical variables, to enable basic research in physiology to be carried out and to assist in clinical diagnosis and treatment. Sensors based on electrical, electromechanical, electrochemical and optical principles are being used and advanced fabrication technologies are being applied to the task of producing reliable reproducible devices at low cost. The *in situ* performance of many invasive sensors is not yet adequate to ensure long-term use; this is largely due to the interactions which occur between sensor surfaces and biological fluids, especially blood. New materials and fabrication techniques will be needed to overcome these problems in order that invasive sensors achieve their full potential.

Bibliography

Hahn C E W, Brooks W N, Albery W J, Rolfe P 1979 O_2 and N_2O analysis with a single intravascular catheter electrode. *Anaesthesia* 34: 263–6

Ko W H, Hynecek J, Boettcher S F 1979 Development of a miniature pressure transducer for biomedical applications. *IEEE Trans. Electron Devices* 26: 1879

Peterson J I, Fitzgerald R V, Buckhold D K 1984 Fibre-optic probe for *in vivo* measurement of oxygen partial pressure. *Anal. Chem.* 56: 62–7

Rea P, Rolfe P, Goddard P J 1985 Assessment and optimisation of dip-coating procedure for the preparation of electro-enzymic glucose transducers. *Med. Biol. Eng. Comput.* 23: 108–15

Rolfe P 1988 Review of chemical sensors for physiological measurement. *J. Biomed. Eng.* 10: 138–45

Silver I A 1965 Some observations on the cerebral cortex with an ultra-micro, membrane covered oxygen electrode. *Med. Biol. Eng.* 3: 377–87

P. Rolfe

[University of Keele, Stoke-on-Trent, UK]

Iron-Based Alloys

As early as 1666, Fabricius described the use of iron wire loops for wound closure, and in 1886, Hansmann reported the application of nickel-plated sheet steel to fracture fixation. However, it was not until the development of the corrosion-resistant stainless steels, between 1900 and 1915, that iron-based alloys had the potential to be successfully employed as surgical implant materials.

Steels are iron-based alloys numbering in the thousands. Plain carbon steel is converted into alloy or special steel when sufficient concentrations of other elements have been introduced to alter the properties significantly. Stainless steel is one of the best known of the special steels and is made by the addition of at least about 12% chromium. This minimum confers tarnish and corrosion resistance in gaseous and liquid media which would attack plain carbon steels. In 1947, the American College of Surgeons first recommended two types of stainless steel for implantation and today this series of alloys is one of the primary materials for biomedical applications.

1. Composition, Atomic Structure and Microstructure

The American Iron and Steel Institute (AISI) uses a three-digit system to separate standard grades of wrought stainless steels into four general classes, based on composition: series 200 (chromium, nickel and manganese), series 300 (chromium and nickel), series 400 (chromium) and series 500 (low chromium). The last two digits in each series indicate type, while a letter suffix is used to indicate a modification within a type (e.g., L: extra-low carbon). Precipitation-hardenable and duplex stainless steels are frequently designated by proprietary names or trademarks. Specifications generally used for cast alloys are those adopted by the Alloy Casting Institute (ACI). The compositions and structures of representative alloys are given in Table 1.

Addition to iron of body-centered-cubic (bcc) chromium, a ferrite former, produces an alloy system which is predominantly ferritic, restricting the γ (austenite) phase to a limited field or "loop." The resulting ferritic grade stainless steels range in chromium content from approximately 14.5% to 27%. While no desirable hardening heat treatment is possible, these alloys can be work hardened. Their corrosion resistance is generally superior to that of the martensitic grade.

The martensitic grade ranges from approximately 11% to 18% chromium, with sufficient carbon present to enlarge the γ field. Chromium concentrations are minimized to avoid ferrite and to promote the γ–α' martensite (bcc) transformation, while nickel is restricted to avoid austenite retention. Strength and hardness are therefore maximized, but at the expense of corrosion resistance.

The high chromium (~16–19%) and nickel (~6–12%) concentrations of the austenitic type 300 series produce some of the most corrosion-resistant stainless steels, which also possess high ductility. Although these steels cannot be heat treated, they can be significantly strengthened by cold working. For stable austenitic alloys containing more than about 11% nickel (e.g., type 316), however, the transformation to an acicular martensite structure is suppressed, thereby diminishing the strengthening effect of cold working.

Table 1
Representative stainless steel compositions and structure

| Alloy designation | Composition[a] (%) | | | | | | | | | Structure |
	Carbon	Man- ganese	Phos- phorus	Sulfur	Silicon	Chromium	Nickel	Molyb- denum	Others	
Wrought										
302	0.15	2.00	0.045	0.03	1.00	17.0–19.0	8.0–10.0			Austenitic
303	0.15	2.00	0.20	≤0.15	1.00	17.0–19.0	8.0–10.0	0.6[b]		Austenitic
304	0.08	2.00	0.045	0.03	1.00	18.0–20.0	8.0–10.5			Austenitic
305	0.12	2.00	0.45	0.30	1.00	17.0–19.0	10.5–13.0			Austenitic
316	0.08	2.00	0.045	0.03	1.00	16.0–18.0	10.0–14.0	2.0–3.0		Austenitic
316L	0.03	2.00	0.045	0.03	1.00	16.0–18.0	10.0–14.0	2.0–3.0		Austenitic
317	0.08	2.00	0.045	0.03	1.00	18.0–20.0	11.0–15.0	3.0–4.0		Austenitic
431	0.20	1.00	0.04	0.03	1.00	15.0–17.0	1.25–2.50			Martensitic
Precipitation hardenable										
17-7 PH[c]	0.09	1.00	0.04	0.03	1.00	16.0–18.0	6.5–7.75		0.75–1.5 Al	Austenitic/ martensitic
Cast										
CF-8M	0.08	1.50	0.04	0.04	2.00	18.0–21.0	9.0–12.0	2.0–3.0		Austenitic/ ferritic

a single values are maximum values unless otherwise noted   b optional   c trademark of the Armco Steel Corporation

Carbon is the most potent solid-solution-strengthening element in austenite, but high levels are undesirable, due to the precipitation of chromium carbides ($M_{23}C_6$) in the temperature range 450–900 °C. Precipitation is favored along grain boundaries, thereby depleting the adjacent matrix in chromium. A marked depletion renders these areas susceptible to corrosion attack, as chromium is effective only when in solution. The alloy is then said to be sensitized. Precipitation is suppressed by quenching the alloy from the annealing temperature and is also further counteracted by use of extra-low-carbon alloys. Vacuum or electroslag remelting is now used to produce steels with extra-low carbon levels and exceptionally low nonmetallic inclusion contents. Vacuum-remelted stainless steels are frequently designated by manufacturers by a VM suffix, while the electroslag-remelted alloys are designated by the letters ESR.

Compositions of the precipitation-hardenable alloys are generally in the range 11–18% chromium and 4–10% nickel, with the addition of such elements as aluminum, copper, titanium and molybdenum. The strengthening mechanism is believed to involve the precipitation of very fine intermetallics such as $Ni_3(Al,Ti)$ along slip planes or grain boundaries. These alloys offer the advantages of ease of fabrication in the annealed state and strengthening by simple heat treatments, while minimizing the reduction in ductility and corrosion resistance that can occur in producing steels of comparable strength levels.

2. Fabrication and Finishing

The methods of fabricating and finishing stainless steels have wide-ranging effects on mechanical properties and corrosion resistance, owing to such factors as recrystallization, phase transformations, cold working, carbide precipitation and modification of the surface state. Hot working, involving rolling or forging operations, refines the original cast structure (e.g., slab or billet), thereby enhancing mechanical properties. Hot-worked products can be further processed by cold finishing or cold working, to raise mechanical properties or to adjust dimensions, or both. Stainless steels have excellent cold-working behavior and are among the most readily fabricated and machined of the major implant alloys. Applicable forming operations include forging, drawing and extrusion. Most stainless steel implants are fabricated from cold-worked mill products and at these high strength levels they nevertheless retain considerable ductility and toughness compared with other implant alloys. This fact is utilized during surgical application in such operations as wire bending and twisting or bone plate contouring.

The production of a smooth surface enhances the service behavior of stainless steels by eliminating surface irregularities which can act as occluded corrosion sites or stress concentrators. A smooth surface is produced by either mechanical polishing and buffing or electrolytic polishing. A roughened or textured surface may also be employed in some instances in an attempt to provide improved mechanical interlocking for cemented implants. Exceptionally smooth finishes are developed with fine abrasives for the articulating surface of joint prostheses (roughness 0.025–0.05 μm). Minimum generation of defects on fine diameter wire during drawing and handling operations is especially important, as irregularities can constitute a large portion of the surface area.

Acid cleaning, to remove processing debris, is frequently referred to as a passivation treatment, since a highly oxidizing bath is used. This treatment develops a passive film of greater stability than the air-formed film. Nitric acid is normally the solution of choice, and considerable latitude exists in the selection of concentration, bath temperature and immersion time. The avoidance of disturbing the surface film during subsequent handling operations, including surgical implantation, remains a controversial subject. Although *in vitro* tests demonstrate a temporary reduction in corrosion resistance as a result of scratching or abrasion of the passivated surface, the clinical significance of this reduction is not known.

3. Mechanical Properties

Representative mechanical properties of implant-quality stainless steel wire and bar are given in Table 2. Cold-worked austenitic stainless steels compare favorably with other implant alloys for high-stress applications, as they possess high yield strength and good fatigue resistance. However, for more extended implantation periods, maximum corrosion resistance becomes increasingly important and therefore influences metallurgical processing. For example, forged type 316L stainless steel hip-joint prostheses are usually annealed to remove the effects of plastic deformation and variations in grain size. Larger cross-sectional areas are therefore required, to compensate for the reduction in yield strength and fatigue resistance caused by the annealing process. In small cross sections, such as intramuscular electrode applications, stainless steel is often selected because of its breakage resistance during movement.

Fatigue is generally believed to be the most common cause of fracture in orthopedic implants, while pure overload fracture rarely occurs (Wright et al. 1985). Stress corrosion cracking has been reported (Hughes et al. 1987). For stainless steels, it is frequently observed in engineering applications that increasing tensile strength and pitting resistance, and decreasing grain size, improve fatigue resistance. In the laboratory, ion implantation has

Table 2
Representative mechanical requirements for wire and bar from ASTM designation: F 138-86 (American Society for Testing and Materials 1988)

Condition	Grade[a]	Diameter or Thickness (mm)	Ultimate tensile strength (MPa)	Yield strength (0.2% offset) (MPa)	Elongation[b] in 4D or 4W (%)	Brinell[c] Hardness (HB)
Hot worked[d]	1	all				≥275
	2	all				≥250
Annealed	1	all	≤515	≤205	≤40	
	2	all	≤480	≤170	≤40	
Cold worked	1	25.4–31.8	≤725	≤450	≤20	
	2	25.4–31.8	≤725	≤450	≤20	

a grades 1 and 2 are within the composition requirements for AISI types 316 and 316L stainless steels, respectively   b 4D = 4 × diameter; 4W = 4 × width   c 29 kN load   d typically supplied as hot-rolled bar for forging applications

been shown to improve fatigue resistance of type 316LVM stainless steel (Higham 1986). Retrieval analyses of implants reveal that fatigue failure of all alloys is primarily associated with material defects, implant design, unstable internal fixation, delayed or absence of union and premature weight bearing. These categories are not mutually exclusive. Considerable disagreement continues as to whether these types of failure may be environmentally promoted by corrosion.

For joint prostheses, metal-on-metal stainless steel counterfaces are not used because of high wear rates and galling. The total hip-joint prosthesis, for example, utilizes a type 316L stainless steel femoral component which articulates on an ultrahigh-molecular-weight polyethylene (UHMWPE) socket. *In vitro* and *in vivo* wear rates and coefficients of friction are low, and compare favorably with other alloy–UHMWPE pairs.

In terms of biomechanical compatibility, attention has been drawn to the high rigidity of metallic implants when used in stress-bearing applications involving bone plates, intramedullary rods or joint prostheses (Terjesen and Apalset 1988). This rigidity is exemplified by Young's modulus, which for stainless steel is approximately ten times the value for compact cortical bone from the femur. In the case of bone plate fixation, for example, it appears that at later stages a nonphysiological condition results, in which the healing site is overprotected from stress. Consequently, during long-term maintenance, the biomechanical feedback required for the remodelling phase may be suppressed, as a significant portion of functional loading is borne by the more rigid plate, to the detriment of the healing site. A recommended design change for stainless steel plates is a hollowed plate design which retains the high bending and torsional stiffness of the traditional solid plate, while possessing a low axial stiffness.

4. Chemical Properties

The corrosion resistance of stainless steel is conferred by a very thin passive film (1–5 nm) of low ionic conductivity. The film is generally considered to be composed of hydrous oxides, enriched primarily with chromium relative to the bulk composition. Its protective capacity is reduced at microstructural inhomogeneities such as inclusions, chromium carbides, second phases and grain-boundary precipitates, and by the effects of cold working. The nature of the film is altered by corrosion history and corrosive media (Sundgren 1985, Zabel et al. 1988).

General corrosion or uniform dissolution of the entire alloy surface is adequately prevented by the passive film. In contrast, localized film breakdown constitutes a problem. This is related to the close proximity of the corrosion or rest potential E_r to the critical breakdown potential of the film E_c in the presence of chloride ions. Small changes in solution composition, especially at film defects, cause E_c to be exceeded. The mechanism of chloride ion attack is, however, unknown. If sufficient oxygen does not reach the rupture site, the area remains anodic and localized corrosion of the underlying metal ensues. The surrounding unattacked area is cathodic and the resulting large cathodic–anodic area ratio contributes to acceleration of the process. Types of localized attack of stainless steel include crevice, pitting, intergranular and galvanic corrosion. In comparison with cobalt- and titanium-based implant alloys, stainless steel has lower resistance to pitting and crevice corrosion. It is not recommended for porous-surfaced implant design because of its susceptibility to crevice corrosion.

The incidence of crevice corrosion is very high on multicomponent devices, particularly orthopedic bone plates and screws, between the screw and countersunk hole site (Cook et al. 1987). It is frequently found accompanied by fretting corrosion (Fig. 1). Although crevice corrosion is most often cited as the predominant mechanism, Brown and Simpson (1981) proposed, on the basis of *in vitro* and *in vivo* studies, that crevice corrosion of bone plates and screws is initiated by fretting rather than by electrochemical film breakdown. Additionally,

Figure 1
Micrograph of a retrieved tibia plate after implantation for eight months, showing burnished perimeter and corroded area (lower left), suggestive of fretting and crevice corrosion, respectively (after Brown and Simpson 1981. © Wiley, New York. Reproduced with permission)

fretting corrosion is believed to be the most important cause of corrosion products associated with joint prostheses. The other localized corrosion types are much less frequently reported. Whatever their respective frequencies, however, corrosion reactions alone are seldom reported to cause implant failure, and this includes crevice corrosion. Nevertheless, the frequency of corrosion of stainless steel implants continues to raise concerns regarding its use for long-term implantation (Cook et al. 1987).

The combination of dissimilar metals in solution (i.e., mixing metals) has generally been avoided because of the concern for promoting galvanic corrosion. However, Mears (1979) has proposed that this practice is not always appropriate, considering the good corrosion resistance of present implant alloys. On the basis of electrochemical theory, Mears views galvanic coupling primarily as an increase in the less passive metal surface area which experiences passive dissolution and cathodic reduction. Consequently, an increased incidence or rate of corrosion should not be expected. The use of dissimilar metals is desirable as a means to optimize selection for mechanical properties, such as employing higher-strength stainless steel bone screws with less rigid Ti–6% Al–4% V alloy bone plates. Under conditions of fretting corrosion, the combination of type 316LVM stainless steel screws with MP35N (35.1% nickel, 20.5% chromium, 9.9% molybdenum and the balance, cobalt) or commercially pure titanium dynamic compression plates produces only minimal effects on *in vitro* corrosion rates (Brown et al. 1988). While supportive evidence has been reported for the safe mixed use of current implant alloys, some conflicting observations require that careful selection and scrutiny be continued.

The possible occurrence of environmentally assisted mechanical failures in the form of corrosion fatigue or stress-corrosion cracking remains controversial. While some retrieval analyses of load-bearing implants have purported to show that corrosion fatigue was the major mechanism operating to cause fracture, other reports have claimed that pure fatigue was the predominant fracture mode and that corrosion, even when present, did not contribute to the fracture process.

Stress-corrosion cracking of austenitic stainless steels is believed to be unlikely at 70 °C, even in chloride environments which are more aggressive than *in vivo* conditions. Nevertheless, observations of this failure mode are increasingly reported and it has been suggested that some failures attributed to corrosion fatigue may have originated as stress-corrosion cracking. Metallographic examination of what may be a mixed failure mode is difficult and the problem is compounded by the frequent fracture surface damage resulting from rubbing, pounding and additional corrosive attack after implant breakage. Resolution of this controversy therefore requires much additional research.

Conventional steam sterilization, electropolishing and passivation in nitric acid have each been shown to increase corrosion resistance *in vitro*. For passivation, enhancement of crevice corrosion resistance has also been demonstrated. Although the clinical significance of these treatments has not been established, their continued use is nevertheless warranted by the recognized susceptibility of stainless steel to localized attack. Laboratory results show promising evidence for increasing corrosion resistance by surface treatment methods such as ion implantation (Higham 1986) and nitriding (Watkins et al. 1986). For the present, retrieval analyses indicate for stainless steel internal fixation devices that stricter manufacturing standards and routine removal after fracture healing would improve clinical performance by reducing the frequency of corrosion (Cook et al. 1987).

5. Biocompatibility

All the alloying elements of austenitic stainless steels present in concentrations of at least 2% (iron, chromium, nickel, molybdenum and manganese) are essential in the human diet but are toxic in excess. These transition elements are strongly electropositive, exhibit variable oxidation states and can form strong complexes with inorganic and organic ligands. Of these five metals, the only specific detoxication mechanism known is for iron, involving the accumulation of iron compounds in lysosomes. For chromium, a possible mechanism is the increased formation of ribonucleoproteins in the liver. Otherwise, normal tissue levels must be maintained by homeostasis, whereby stability of the internal environment is achieved by control mechanisms which are activated by negative feedback.

Extensive clinical experience continues to show that dissolution products of austenitic stainless steels are generally well tolerated in the absence of a severe and prolonged corrosion process (Linden et al. 1985). Implantation frequently results in the short-term appearance of an encapsulating fibrotic membrane, approximately 1 mm thick. The capsule represents a nonspecific response to the presence of a solid nonporous foreign body, and increasing membrane thickness develops when corrosion rates are elevated above passive current densities. Histological response (e.g., severe inflammation, tissue necrosis) to gross corrosion, however, can necessitate premature implant removal. When applicable, wear particles must also be considered in terms of their size and shape, in addition to their enhanced corrosion rate due to a relatively high surface area. The following clinical and laboratory observations illustrate these points.

Winter (1974) analyzed tissue adjacent to orthopedic stainless steel devices which had been implanted in humans. The devices were plates, screws, nails, wires and joint prostheses, which are susceptible at times to severe crevice, fretting and pitting corrosion and the production of wear particles. The following observations are from cases which usually required a second operation and therefore represent a negative selection.

Three types of deposits were optically observed. Type I deposits were small opaque irregularly shaped particles, 0.3 mm–0.1 μm, and were believed to be alloy wear particles from joint prostheses. Smaller alloy particles were found primarily in macrophages, while the larger particles were surrounded by multinucleate giant cells. There was no granulomatous reaction and the impregnated tissue was remarkably healthy, considering the quantity of foreign material present. Type II deposits were large platelets (sometimes green), 5.0–0.5 mm, and highly variable in size and morphology. They were identified as alloy corrosion products containing iron, chromium, phosphorus and sulfur, and were observed in 11 out of 44 specimens. Type II deposits provoked a vigorous foreign-body giant-cell reaction, with fibrosis, and impregnated tissue frequently consisted of relatively acellular collagen or was necrotic. Winter proposed that type II deposits were cytotoxic, producing phagocytosis, cell death, fibrosis and lastly necrosis in the presence of large accumulations. Type III deposits were somewhat spherical, yellow-brown, hemosiderin-like granules, 3–0.1 μm. Tissue analysis revealed mixtures of two or more of the iron oxides α-Fe_2O_3 and γ-Fe_2O_3 and the hydrated oxides α-$FeOOH$ (goethite) and γ-$FeOOH$ (lepidocrocite). No recognizable disturbance was observed when few particles were present. However, with heavy contamination, abnormalities included cytotoxicity and fibrosis or necrosis. It was proposed that type III deposits

might have been the result of an iron detoxication mechanism, while in some cases, pathological changes indicated a local tissue iron overload or iron toxicity. Types II and III deposits were frequently found together, but the former were considered to be more cytotoxic.

Considering another aspect of biocompatibility, it it is recognized that nickel and chromium are capable of producing delayed metal-sensitivity reactions. However, the percentage of metal-sensitive individuals who actually respond is small. The proposed mechanism views the released metal ion as a hapten, which, when complexed with protein, stimulates the immune response. Because of the deep tissue situation of implant dissolution, a metal-sensitivity reaction is not detectable until it has progressed to cause pain, fluid accumulation or a dermatological disorder. Reports in the literature conclude that it is primarily reactions to nickel that are clinically significant. While it appears that several months' implantation is necessary to cause sufficient dissolution, visible corrosion of the implant may not necessarily occur. Removal of the implant relieves the symptoms. Recent studies indicate that *in vivo* corrosion may release a significant amount of chromium in the more biologically active form Cr^{VI} in addition to Cr^{III} (Merritt and Brown 1985).

Implantation of a foreign body is believed to increase the risk of late infection. For metallic implants, a facilitated disease process may result from a mechanical barrier effect or from inhibition of host defense systems and enhancement of bacterial virulence by corrosion products. Animal and bacteriological studies have provided evidence for the potential role of alloys and individual elements. Implantation of type 316L stainless steel microspheres in rabbits has been shown to produce a significantly higher incidence of clinical infection compared with unoperated or sham-operated controls. In another study, nickel was shown to inhibit the phagocytosis of *Staphylococcus aureus* and *Streptococcus faecalis* by macrophages *in vitro*. Whie Fe^{II} and Fe^{III} have been shown to alter the efficacy of some bactericidal mechanisms, it is recognized that decreasing iron content in the blood is a response to bacterial invaders, which deprives the organisms of this element essential to growth. Iron release from stainless steel implants may play a clinically significant role in counteracting this mechanism. For example, iron has been shown to increase the virulence of *Escherichia coli* and *Listeria monocytogenes* in guinea pigs and mice, respectively. Studies of patients with stainless and cobalt–chromium orthopedic implants indicated a migratory capacity defect for white blood cells in the peripheral blood (Merritt and Brown 1985). However, no indication of a higher infection rate was observed.

Chronic exposure to metal implant corrosion products introduces concern about chemical carcinogenesis, particularly since carcinogenic effects have been observed with nickel, chromium and iron. While the event is very rare, the literature records at least ten cases of malignant neoplasms associated with metallic surgical implants in humans (Tayton 1980, McDonald 1981, Dodian et al. 1983, Bago-Granell et al. 1984, Penman and Ring 1984, Swann 1984, Bauer et al. 1987) and at least nine cases in canines (Harrison et al. 1976, Sinibaldi et al. 1976). In humans, two of the cases involved stainless steels and six involved cobalt–chromium alloys. The compositions of the implants in the remaining two cases were unidentified in the reports. Six of the human cases are summarized in Table 3. It was reported in the first case involving stainless steel (1924) that dissimilar metals were used and that corrosion was significant. A measured potential difference of 80 mV between the retrieved plate and screws was considered to be a source of irritation for 30 years. It should be noted also that the screws contained only 12% chromium, the minimum required for corrosion resistance. In the second case involving stainless steel (1944), dissimilar metals were also used, but the observed corrosion was not considered significant. All nine cases in dogs reported in 1975 and 1976 involved stainless steel, and in most, significant corrosion was observed. The neoplasms comprised seven osteosarcomas and two undifferentiated sarcomas. The period between implantation and neoplasm detection was from one year to 11 years (average four years). These observations were of particular concern because the osteosarcomas were atypical for the canine breeds and sites, thus suggesting causality. This may be a species-specific effect.

For humans, the recorded incidence of malignant neoplasm development associated with metallic surgical implants is well below statistical expectations and no causal relationship has been identified. Nevertheless, neoplasm induction is a long term process, lasting perhaps seven or more years. Realization that stainless steels are highly susceptible to corrosion in multicomponent devices, clearly indicates that increased research is also particularly needed in this area of biocompatibility.

6. Applications

Iron-based alloys currently form one of the predominant groups of metallic materials for biomedical applications. Of this group, only the corrosion-resistant stainless steels, primarily the austenitic or semiaustenitic precipitation-hardenable types, are used in situations requiring fluid contact. For implantation, and particularly in those circumstances which include multicomponent weight-bearing or articulating devices, type 316L stainless steel is most frequently selected. This is because of its superior corrosion resistance in addition to good mechanical properties. Introduction of vacuum and electroslag remelting has considerably improved material performance. For less rigorous environments and functions, such as some dental applications, austenitic stainless steels of lower corrosion resistance are also used, thereby permitting a wider

Table 3

Malignant neoplasms reported associated with metallic orthopedic implants in humans

Year; patient's age at implantation (years)	Lesion type and site	Implantation period (years)	Implantation to detection of neoplasm (years)	Histology	Implant type and alloy
1924; 12	Fracture of humerus	30	30	Ewing's sarcoma[a]	Fixation plate: Fe–18% Cr–8% Ni; screws: Fe–12% Cr
1944; 58	Fracture of tibia and fibula	26	26	Hemangioendothelomia (tibia)	Fixation plate: type 316 stainless steel; screws (8): 6 type 316, 2 type 304 stainless steel
1953; 37	Fracture of tibia	3	3	Unclassified sarcoma	Fixation plate and screws (2): unidentified metal
1969; 4	Congenitally dislocated right hip and a subluxating left hip	1	7.5	Ewing's sarcoma (femur)	Fixation plate and screws (6): cast Co–Cr–Mo
1974; 66	Osteoarthritis, right hip	10	10	Telangiectatic osteosarcoma	Total hip prosthesis, Charnley–Mueller: cast cobalt–chromium
1979; 79	Osteoarthritis, left hip	2	2	Malignant fibrous histiocytoma	Total hip prosthesis, Charnley–Mueller; unidentified metal

a most probable diagnosis

Table 4
Representative biomedical applications of iron-based alloys

Application	Description[a]
Bone screws and pins	Internal fixation of diaphyseal fractures of cortical bone, and metaphyseal and epiphyseal fractures of cancellous bone: screw comprised of hexagonal or Phillips recess driving head, threaded shaft and self-tapping or non-self-tapping tip; type 316L stainless steel
Onlay bone plates	Internal fixation of shaft and mandibular fractures: thin narrow plate with slots or holes for retaining screws; type 316L stainless steel
Blade and nail bone plates	Internal fixation of fracture near the ends of weight-bearing bones: plate and nail, either single unit or multicomponent; type 316L stainless steel
Intramedullar bone nails	Internal fixation of long bones: tube or solid nail; type 316L stainless steel
Percutaneous pin bone fixation	External clamp fixation for fusion of joints and open fractures of infected nonunions: external frame supporting transfixing pins; stainless steel
Total joint prostheses	Replacement of total joints with metal and plastic components (shoulder, hip, knee, ankle and great toe): humeral, femoral (hip and knee), talus and metatarsal components; type 316L stainless steel
Wires	Internal tension band wiring of bone fragments or circumferential cerclage for comminuted or unstable shaft fractures; type 316L stainless steel
Harrington spine instrumentation	Treatment of scoliosis by application of correction forces and stabilization of treated segments: rod and hooks; type 316L stainless steel
Mandibular wire mesh prostheses	Primary reconstruction of partially resected mandible; types 316 and 316L stainless steel
Fixed orthodontic appliances	Correction of malocclusion by movement of teeth: components include bands, brackets, archwires and springs; types 302, 303, 304 and 305 stainless steel
Preformed dental crowns	Restoration for extensive loss of tooth structure in primary and young permanent teeth: preformed shell; type 304 stainless steel
Preformed endodontic post and core	Restoration of endodontically treated teeth: post fixed within root canal preparation, with exposed core providing a crown foundation; types 304 and 316 stainless steel
Retention pins for dental amalgam	Retention of large dental amalgam restorations: cemented, friction lock and self-threading pins, placed approximately 2 mm within dentin with approximately 2 mm exposed; types 304 and 316 stainless steel
Wire mesh	Inguinal hernia repair, cranioplasty (with acrylic), orthopedic bone cement restrictor; types 316 and 316L stainless steel
Sutures	Wound closure, repair of cleft lip and palate, securing of wire mesh in cranioplasty, mandibular and hernia repair and realignment, tendon and nerve repair; types 304, 316 and 316L stainless steel
Stapedial prostheses	Replacement of nonfunctioning stapes: various types comprised of wire and piston or wire and cup piston (Teflon–stainless steel piston, platinum and stainless steel cup piston, and all stainless steel prostheses); types 316 and 316L stainless steel
Neurosurgical aneurysm and microvascular clips	Temporary or permanent occlusion of intracranial blood vessels; tension clips of various configurations, approximately 2 cm or less in length and constructed of one piece or jaw, pivot and spring components (similar and dissimilar compositions); 17-7PH, 17-7PH(Cb), PH 15-7Mo and types 301, 304, 316, 316L, 420 and 431 stainless steel
Self-expanding stent	Treatment of tracheobronchial stenosis, tracheomalacia and air collapse following tracheal reconstruction: 0.457 mm stainless steel wire formed in a zig-zag configuration of 5–10 bends
Balloon-expandable stent	Dilation and post-dilation support of complicated vascular stenosis (experimental): stainless steel
Hydrocephalus drainage valve	Control of intercranial pressure: one-way valve; type 316 stainless steel
Trachea tube	Breathing tube following tracheotomy and laryngectomy: tube-within-a-tube construction; type 304 stainless steel
Electronic laryngeal prosthesis system	Electromagnetic voicing source following total laryngectomy: implanted unit comprised of subdermal transformer, rectifier pack and transducer encased in type 316 stainless steel, with spring steel diaphragm
Electrodes and lead wires	Anodic, cathodic and sensing electrodes and lead wires: intramuscular stimulation, bone growth stimulation, cardiac pacemaker (cathode), EMG, EEG and lead wires in a large number of devices; types 304, 316 and 316L stainless steel

Table 4—continued

Application	Description[a]
Arzbaecher pill electrode	Atrial electrocardiograms: swallowed sensing electrode of short metal tubing segments forced over plastic tubing; stainless steel
Cardiac pacemaker housing	Hermetic packaging of electronics and power source: welded capsule; type 316L stainless steel
Variable capacitance transducer	Measurement of pressure on sound: metal diaphragm, mounted in tension; stainless steel
Variable resistance transducer	Measurement of respiratory flow: metal arms supporting wire strain gauge; stainless steel
Intrauterine device (IUD)	Contraception: stainless steel (Majzlin spring, M-316, M device), stainless steel and silicone rubber (Comet, M-213, Ypsilon device), stainless steel and natural rubber (K S Wing IUD), stainless steel and polyether urethane (Web device)
Intrauterine pressure sensor case	Protective shroud for transducer: stainless steel
Osmotic minipump	Continuous delivery of biologically active agents: implanted unit comprising elastomeric reservoir, osmotic agent, rate-controlling membrane and stainless steel flow moderator and filling tube
Radiographic marker	Facilitation of postoperative angiography of bypass graft: open circle configuration of 25 gauge suture wires; stainless steel
Butterfly cannula	Intravenous infusion: stainless steel
Cannula	Coronary perfusion: silicone rubber reinforced with an internal wire spiral; stainless steel
Acupuncture needle	Acupuncture: 0.26 mm diameter × 5–10 cm length needles; stainless steel
Limb prostheses, orthoses and adaptive devices	Substitution, correction, support or aided function of movable parts of the body, and technical aids not worn by the patient: components such as braces, struts, joints and bearings of many items; steel and stainless steel

a stainless steel types other than those listed for each application may also be used

selection for other material properties. In the absence of fluid contact, especially for limb prosthetics and orthotics, both steel and stainless steel components are utilized. A representative list of material applications is presented in Table 4. All the surgical and nonsurgical iron-based alloys have good fabricability and are therefore generally employed in the wrought and machined forms.

In the vast majority of implant applications, stainless steels perform their intended function without adverse physiological effects and have consequently achieved a large degree of success. Although they are clearly not without shortcomings in their mechanical properties for some applications, their major problem rests with crevice and fretting corrosion. For example, a great number of total hip prostheses have been inserted to date, including an estimated 80 000 during 1976 in the USA alone. A large percentage of these implants utilized stainless steel for the femoral stem. Fracture of this component is reported to account for approximately 0.1–0.2% of failures for all applications and is primarily due to fatigue. In contrast, despite the very high clinical success rate of millions of stainless steel fracture fixation plates, retrieval analyses record a crevice and fretting corrosion incidence in excess of 90%, and in one report 100%. In another application area, stainless steel stimulating electrodes experience marked pitting corrosion when larger current densities are required (e.g., 0.5 A m^{-2}).

Recently, other stainless steels have been investigated in a search for increased corrosion resistance and improved mechanical properties in comparison with type 316L stainless steel (ASTM F-138). The higher molybdenum content of type 317 stainless steel may offer improved crevice corrosion resistance, but its pitting resistance may be no better. Improved mechanical properties are found with duplex (50% austenite plus 50% ferrite) and high-ductility transformation-induced plasticity (TRIP) stainless steels; currently, however, their corrosion resistance appears to be the same or inferior. Improvements in control of composition and metallurgical processing variables may ultimately raise corrosion resistance to acceptable levels. An Fe–Cr–Al alloy has been developed, with initial reports claiming superior *in vitro* corrosion resistance. An α-Al_2O_3 coating forms on the alloy when heated in air at high temperatures. The coating resists spalling during tensile testing and demonstrates adherence to connective tissue during *in vivo* testing. Testing of a 22Cr–13Ni–5Mn stainless steel with up to 0.40 wt% nitrogen and 0.20 wt% niobium has also been reported (Shetty et al. 1985). Results show significantly improved static and fatigue strengths, and increased pitting potential. The thermomechanical processing which is used to strengthen Co–Cr–Mo alloy has also been investigated for type 316L stainless steel (Bardos et al. 1984). The process consists of a rapid cold reduction of the hot-rolled material and cold dynamic processing at levels far exceeding the yield strength. Results show significant improvements in static and fatigue strengths. Improvements in fatigue properties of type 316L stainless steel by a new surface treatment, which adds abrasive and glass-bead blastings to the conventional treatment of mechanical polishing, electropolishing and passivation, have also been demonstrated experimentally for both annealed and cold-worked materials (Shetty et al. 1987).

While corrosion-mediated implant failure remains very infrequent, increasing attention is rightly being focused on the biological burden of dissolution products. Additionally, a more cautious approach is necessitated as younger patients become implant recipients and the average life span is extended. Increased quality control, informed material selection and implant usage and correct surgical application will all undoubtedly help to minimize the incidence and severity of corrosion and therefore enhance the continuing benefits derived from the present generation of stainless steels.

See also: Biocompatibility: An Overview

Bibliography

American Society for Testing and Materials 1988 Medical Devices, Vol. 13.01. *1988 Annual Book of ASTM Standards*. ASTM, Philadelphia, PA

Bago-Granell J, Aguirre-Canyadell M, Nardi J, Tallada N 1984 Malignant fibrous histiocytoma of bone at the site of a total hip arthroplasty. *J. Bone Jt. Surg.* 66B: 38–40

Bardos D I, Baswell I, Garner S, Wigginton R 1984 The development of a new high strength, cold forged 316LVM stainless steel. *Trans. Ortho. Res. Soc.* 9: 88

Bauer T W, Manley M T, Stern L S, Martin A, Marks K E 1987 Osteosarcoma at the site of total hip replacement. *Trans. Soc. Biom.* 10: 36

Brown S A, Hughes P J, Merritt K 1988 *In vitro* studies of fretting corrosion of orthopaedic materials. *J. Orthop. Res.* 6: 572–9

Brown S A, Simpson J P 1981 Crevice and fretting corrosion of stainless-steel plates and screws. *J. Biomed. Mater. Res.* 15: 867–78

Cook S D, Thomas K A, Harding A F, Collins C L, Haddad R J Jr, Milicic M, Fischer W L 1987 The *in vivo* performance of 250 internal fixation devices: A follow-up study. *Biomaterials* 8: 177–84

Dodian P, Putz P, Amiri-Lamraski M H, Efira A, De Martelaere E, Heimann R 1983 Immunoblastic lymphoma at the site of an infected vitallium bone plate. *Histopathology* 6: 807–13

Harrison J W, McLain D L, Hohn R B, Wilson G P, Chalman J A, MacGowan K N 1976 Osteosarcoma associated with metallic implants. *Clin. Orthop. Relat. Res.* 116: 253–7

Higham P A 1986 Ion implantation as a tool for improving the properties of orthopaedic alloys. In: Williams J M,

Nichols M F, Zingg W (eds.) 1986 *Biomedical Materials*, Materials Research Society, Pittsburgh, PA, pp. 253–61

Hughes P J, Brown S A, Ritter M A 1987 Failure analysis of trapazoidal-28 total hip prostheses. *Trans. Soc. Biom.* 10: 189

Linden J V, Hopfer S M, Gossling H R, Sunderman F W Jr 1985 Blood nickel concentrations in patients with stainless-steel hip prostheses. *Ann. Clin. Lab. Sci.* 15: 459–64

MacDonald I 1981 Malignant lymphoma associated with internal fixation of a fractured tibia. *Cancer* 48: 1009–11

Mears D C 1979 *Materials and Orthopaedic Surgery*. Williams and Wilkins, Baltimore, MD

Merritt K, Brown S A 1985 Biological effects of corrosion products from metals. In: Fraker A C, Griffin C D (eds.) 1985 *Corrosion and Degradation of Implant Materials: Second Symposium*. American Society for Testing and Materials STP 859, ASTM, Philadelphia, PA, pp. 195–207

Shetty R H, Gilbertson L N, Jacobs C H 1985 The 22-13-5 stainless steel—An alternative to hot forged 316L stainless steel in fracture fixation. *Trans. Ortho. Res. Soc.* 10: 246

Shetty R H, Gilbertson L N, Jacobs C H 1987 The new surface finish—A method of improving the properties of 316L stainless steel. *Trans. Soc. Biom.* 10: 233

Sinibaldi K, Rosen H, Liu S, De Angelis M 1976 Tumors associated with metallic implants in animals. *Clin. Orthop. Relat. Res.* 118: 257–66

Sundgren J-E, Bodo P, Lundstrom I, Berggren A, Hellem S 1985 Auger electron spectroscopic studies of stainless-steel implants. *J. Biomed. Mat. Res.* 19: 663–71

Sutow E J, Pollack S R 1981 The biocompatibility of certain stainless steels. In: Williams D F (ed.) 1981 *Biocompatibility of Clinical Implant Materials*. CRC Press, Boca Raton, FL, pp. 45–98

Swann M 1984 Malignant soft-tissue tumour at the site of a total hip replacement. *J. Bone Jt. Surg.* 66B: 629–31

Tayton K J J 1980 Ewing's sarcoma at the site of a metal plate. *Cancer* 45: 413–15

Terjesen T, Apalset K 1988 The influence of different degrees of stiffness of fixation plates on experimental bone healing. *J. Orthop. Res.* 6: 293–9

Watkins K G, Ben Younis S, Davies D E, Williams K 1986 A preliminary investigation of the electrochemical properties of a nitrided stainless steel for dental applications. *Biomaterials* 7: 147–51

Williams D F, Roaf R 1973 *Implants in Surgery*. Saunders, London

Winter G D 1974 Tissue reactions to metallic wear and corrosion products in human patients. *J. Biomed. Mater. Res. Symp.* 5(1): 11–26

Wright T M, Burstein A H, Bartel D L 1985 Retrieval analysis of total joint replacement components: A six year experience. In: Fraker A C, Griffin C D (eds.) 1985 *Corrosion and Degradation of Implant Materials: Second Symposium*. American Society for Testing and Materials STP 859, ASTM, Philadelphia, PA, pp. 415–28

Zabel D D, Brown S A, Merritt K, Payer J H 1988 AES analysis of stainless steel corroded in saline, in serum and *in vivo*. *J. Biomed. Mater. Res.* 22: 31–44

E. J. Sutow
[Dalhousie University, Halifax, Canada]

M

Maxillofacial Prostheses

Maxillofacial materials may be used to replace extraoral portions of the flexible tissue of the face such as the nose, ear, cheek and orbit (see Fig. 1). The color of maxillofacial materials should blend with the surrounding tissue and should have processing, physical and mechanical properties that are compatible with clinical use as well as chemical properties that result in biocompatibility with adjacent facial tissue.

1. Composition and Fabrication

The four main groups of commercial materials are silicones, polyurethanes, poly(vinyl chlorides) and chlorinated polyethylenes.

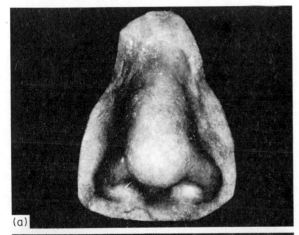

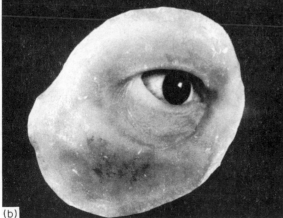

Figure 1
Examples of maxillofacial prostheses of (a) a nose and (b) an eye and orbit (after Craig 1978)

1.1 Silicones

The most commonly used silicone is supplied as two pastes, one containing a dimethylsiloxane with terminal vinyl groups and a chloroplatinic acid catalyst (50–100 ppm as platinum relative to the total weight) and the other containing a dimethylsiloxane with terminal silane groups. Microfine silica is included in the pastes as a filler to improve the physical and mechanical properties. Coloring of the pastes is accomplished by mixing earth pigments into the dispensed pastes. Ionic polymerization occurs after mixing the two pastes and the material sets to a rubber at room temperature in 24 h, but this time can be shortened to 15 min by setting at 378 K. This processing temperature permits the use of dental stone molds coated with a release agent.

Dimethylsiloxane oligomers with terminal hydroxy groups are polymerized by a condensation reaction at room temperature with stannous octoate and a cross-linking agent such as orthoethylsilicate with the formation of ethanol as a by-product. The material is supplied as two pastes, one containing the catalyst. Pigmentation can be accomplished by adding colorants to the base paste, then mixing with the catalyst followed by molding in dental stone molds.

A third type of silicone is a dimethyl and methyl–vinyl siloxane gum rubber and pigmentation is accomplished by milling. Microfine (fumed) silica is used as a reinforcing agent in amounts of 20–30%. A catalyst, bis(2,4-dichlorobenzoyl)peroxide is present in concentrations of about 0.5–1.5%. The milled material is packed into metal molds (linotype), molded under pressure, vulcanized for 10 min at 373 K by free radical attack, removed from the mold and postcured for 3 h at 533 K.

Silphenylene is a dimethylsiloxane oligomer with phenyl groups in the main chain and terminal hydroxy groups. It uses stannous octoate as a catalyst and orthopropylsilicate as a cross-linking agent. The material may or may not contain silica filler. The mixed material has a working time of 5 min and may be polymerized in dental stone molds in the temperature range 298–373 K.

1.2 Polyurethanes

These polymers are formed by the reaction of diisocyanates and polyols. The former may be aromatic or aliphatic diisocyanates and the latter may be a di- or trifunctional polyether, or a polyester macroglycol. The reaction may be accelerated by stannous octoate and must be conducted in the absence of moisture to avoid the formation of CO_2. Ultraviolet light stabilizers and antioxidants are mandatory for

Table 1
Mechanical properties of maxillofacial materials

	Silicones		Polyurethane (aromatic polyester)	Poly(vinyl chloride)
	RTV-addition type	Heat vulcanized		
Tensile strength (MN m^{-2})				
original	4.20	5.87	0.83	3.99
aged 900 h	3.82	6.20	0.44[a]	3.55
Permanent deformation after tensile rupture (%)				
original	0.24	0.26	0.36	13.8
aged 900 h	0.24	0.22	[b]	14.2
Elongation (%)				
original	445	440	420	215
aged 900 h	400	490	1330[a]	200
Tear strength (N cm^{-1})				
original	158		15.7	87.3
Tear energy, pants test[c] (N cm^{-1})				
original			66.7	42.8
aged 900 h			[b]	39.7
Hardness, Shore A				
original	32	45	6	53
aged 900 h	33	45	<1[a]	57
Flexural fatigue, cycles to failure when deformed from 0–100% at 1 Hz	12 × 10^6	11 × 10^6	9.5 × 10^6	9 × 10^3
Dynamic modulus (MN m^{-2})				
at 310 K	3.4	2.1	2.7	2.5
at 273 K	3.4	2.3	4.2	5.7
at 258 K	3.3	2.7	6.1	12.1

a tested at 600 h of aging since it disintegrated at 900 h   b not tested at 600 h since samples were too sticky and stretched as in tensile elongation or at 900 h because the samples disintegrated   c specimen is shaped like a pair of pants and is torn by pulling the legs in opposite directions; silicone samples did not tear but stretch as in tensile elongation

aromatic and advisable for aliphatic polyurethanes. Earth pigments are also used with these materials. Processing may be done in dry dental stone molds at 373 K or in metal molds. Since the reaction is stoichiometric, proportioning is critical as is the control of the processing temperature.

1.3 Poly(Vinyl Chlorides)
These are supplied as plastisols with poly(vinyl chloride) suspended in a plasticizer (an oily ester) as a finely divided solid. The plastisol is a viscous liquid to which pigments are added. When heated the polymer is partly dissolved and a gel is formed. Processing is carried out in a metal mold for 10 min once the mold reaches 350 K. The processing temperature and time schedule are critical if optimum properties are to be obtained in all portions of the prosthesis.

1.4 Chlorinated Polyethylene
The polyethylene polymer contains chlorine atoms

on the main chain as well as on the side chains and is compounded with low-density polyethylene, calcium stearate and soybean oil. Pigments are incorporated on a heated rubber mill, and the material is packed into metal molds and processed for 15 min at 463 K.

2. Physical and Mechanical Properties

Mechanical properties of four classes of maxillofacial materials are listed in Table 1. The tensile strengths, percent elongation, tear energy and hardness were tested after artificially aging the samples in a weathering chamber using a 2.5 kW xenon light with a spectral distribution similar to natural sunlight. The relative humidity was maintained at 90% and the temperature at 316 K. Samples were sprayed with water for 18 min out of every 102 min.

After aging, the tensile strength, elongation, tear energy or hardness did not change greatly except for

Table 2
Color parameters of maxillofacial materials[a]

	Silicones		Polyurethane (aromatic polyester)	Poly(vinyl chloride)
	RTV-addition type	Heat vulcanized		
Luminous reflectance				
original	6.7 (69.9)	8.9 (60.4)	6.2 (79.6)	7.1 (66.7)
aged 900 h	6.7 (78.3)	9.5 (64.5)	6.3 (69.6)	10.3 (62.9)
Dominant wavelength (nm)				
original	487.0 (586.0)	490.2 (585.8)	545.0 (583.2)	581.4 (584.0)
aged 900 h	487.8 (585.6)	489.0 (586.0)	573.2[b] (584.0)	501.4 (583.8)
Excitation purity				
original	0.096 (0.076)	0.102 (0.252)	0.008 (0.112)	0.252 (0.552)
aged 900 h	0.090 (0.106)	0.120 (0.178)	0.026[b] (0.164)	0.020 (0.198)
Contrast ratio[c]				
original	0.08	0.15	0.08[b]	0.11
aged 900 h	0.09	0.15	0.08[b]	0.16

a the first values are with a block background and the values in parentheses are with a white background
b tested at 300 h of aging because of degradation of samples at higher aging times of 600 h and 900 h   c ratio of the luminous reflectance with a black background to that with a white background

the polyurethane which disintegrated between 600 h and 900 h. The disintegration of the polyurethane is evident even at 600 h with dramatic decreases in mechanical properties. The tear strengths of silicones were superior to those of the polyurethane and poly(vinyl chloride) although the value for silphenylene was comparable to the polyurethane at about 16 N cm^{-1}. Chlorinated polyethylene had the highest tear strength at 184 N cm^{-1}. The tensile strengths and elongations of silphenylene and chlorinated polyethylene were comparable at about 9.5 MN m^{-2} and 980%, respectively. The Shore A hardness values of the silphenylene and the chlorinated polyethylene were 35 and 47, respectively.

The permanent deformation after tensile failure was least for the unaged silicones followed by the polyurethane. The poly(vinyl chloride) had a large value. No significant difference in permanent deformation resulted from 900 h of aging for the silicones. The polyurethane could not be tested even at 600 h because of severe degradation. There was a small but significant increase in permanent deformation of poly(vinyl chloride) after aging.

The flexural fatigue of the poly(vinyl chloride) material was substantially less than the other three classes of maxillofacial polymers with a low value of 9000 cycles between 0% and 100% elongation while the other materials were at least 1000 times more resistant to rupture by flexural fatigue.

The dynamic modulus values at 310 K were not greatly different for the four classes, but the poly(vinyl chloride) had major increases with decreasing temperatures and the polyurethane also had considerable increases. These increases are a result of glass transitions that are not present in the silicone elastomers in this temperature range.

The color of human skin was measured on whites, blacks, and orientals using reflection spectrophotometry. No major difference was found in dominant wavelength (hue), but major differences were detected in luminous reflectance (value) and excitation purity (chroma). The absorption by oxyhemoglobin was more prominent in whites, especially females, than in blacks or orientals.

A computer program for matching pigmented maxillofacial silicones to the color of a patient's skin has been developed. Color differences between pigmented maxillofacial formulation and the patient's skin varied from 1.5 to 3.5; this color difference would not be detected visually since the number would need to be at least 5 for a difference to be observed.

The color parameters in Table 2 indicate that only slight changes occur on the aging of either class of silicones and the changes that were observed by the spectrophotometer would not be apparent visually. Significant changes did occur with the polyurethane and the poly(vinyl chloride) and were more pronounced with the poly(vinyl chloride). The values determined with a black background are probably more important since that condition more nearly represents the clinical application with the facial side being more nearly black than white.

Similar color studies have been conducted on pigmented addition silicone materials using the entire range of earth pigments. Color changes after

900 h of aging occurred for the silicone containing white, yellow, yellow-orange and light-orange pigments. The changes in these pigmented samples were small, though visually observable. Thus, these measurements together with the color studies of the base polymers indicate that clinical changes in color probably result from staining rather than aging of the polymers and pigments. Staining and its removal will be discussed in Sect. 4.

3. Biocompatibility

Patch, cell culture and implant tests have been used to evaluate biocompatibility of maxillofacial materials. Patch testing on human skin (upper back or upper arm) showed that of eleven patients one showed irritation to bifaced (double-backed) tape, three to the rubber-type adhesive and one to the acrylic emulsion adhesive. Of this group of eleven patients none had a reaction to silicone or polyurethane and four showed irritation to poly(vinyl chloride).

Cell culture tests with human periodontal ligament fibroblasts and mouse fibroblasts showed highly positive tests to poly(vinyl chloride), polyurethane and silicones except for the type where polymerization was initiated with chloroplatinic acid. This latter silicone also evoked minimal foreign-body reaction when implanted in albino rats.

4. Clinical Results

Force-displacement measurements were made on various human facial tissues including the forehead, ear lobule and cheek bone, and on various ratios of uncatalyzed polydimethylsiloxane (U) and dimethylsiloxane triacetoxysilane (TA). Increasing the amount of U relative to TA increased the flexibility. For forces up to 0.4 N, combinations of U and TA could not reproduce the force-displacement properties of facial tissue; however, from 0.4 N to 1.2 N force combinations of U and TA could be formulated to match facial tissue.

Silicone materials are colored by intrinsic pigmentation using powders, pigmented silicone concentrates or dyed synthetic fibers. Their appearance can be further improved by extrinsic coloring with different colored silicone adhesives, but bonding of the various layers can present problems. Silicones can be characterized on the surface by tattooing with colored silicones, but in thin section the puncture holes produce stress raisers which reduce strength. Polyurethanes and poly(vinyl chloride) can be readily colored extrinsically, but the acceptance of dyes results in the prostheses staining more readily in service.

Cosmetics, dyes and body oils cause maxillofacial prostheses to become unsightly much more readily than exposure to ultraviolet light. Samples of the two classes of silicones and the poly(vinyl chloride) materials were coated with lipstick or soaked in tea or Bismark Brown disclosing solution for 24 h. Maxillofacial materials stained with tea or Bismark Brown were washed and dried and those stained with lipstick were cleansed with skin cream until no observable color was present on a cloth and then washed and dried. The room-temperature-vulcanized (RTV) and heat-cured silicones resisted staining with tea and disclosing solution better than poly(vinyl chloride), but did not have quite so good resistance to staining with lipstick.

Solvent extraction of RTV silicone severely stained with lipstick, cigarette smoke, Bismark Brown or methylene blue has been accomplished by extraction with either toluene, benzene, 1,1,1-trichloroethane or *n*-hexane. The swollen samples were deswollen by addition of methanol and dried under vacuum. All were equally effective in the removal of stains with no or insignificant changes between the three color parameters before staining and after cleaning. In addition, no significant changes occurred in dimension, hardness or tensile strength. Furthermore, all cell culture tests showed that the extracted samples resulted in a negative response as did the original samples. These results have been confirmed by the cleaning of several clinical prostheses, establishing that solvent extraction can extend the aesthetic service life of the RTV silicones. This technique is not appropriate with poly(vinyl chloride) products since the solvent extraction removes the plasticizer and the prosthesis becomes stiff. Unfortunately, the poly(vinyl chloride) prostheses stain readily, because stains readily dissolve in the plasticizer. Solvent extraction has not been conducted on polyurethanes, because of the observed degradation in the accelerated aging test.

Adhesives are supplied as a liquid or a bifaced tape. The liquid types contain natural rubber (with or without oxides of zinc or titanium) suspended in *n*-hexane, an emulsion of acrylic polymer in water or a polydimethylsiloxane. The *n*-hexane sets in 3–5 min by evaporation of the solvent and result in a porous film. The emulsion sets in about 15 min at 333–343 K. The adhesives can be readily removed from skin and prostheses made from poly(vinyl chloride) or silicone, but are difficult to remove from polyurethane. The tape types are rather stiff and difficult to manipulate and contamination by touching reduces the adhesiveness.

The adhesiveness has been evaluated by a peel test, where the force to peel 1 cm^2 of adhered prosthetic material off shaved rat skin was determined. For bifaced tape and rubber-type adhesives, poly(vinyl chloride) had the highest peel values followed by polyurethane and silicones. For the acrylic adhesive, polyurethane had the highest peel value followed by silicone and poly(vinyl chloride); these values were generally higher than for the other adhesives. A maximum peel value of $1.40\,\mathrm{N\,cm^{-2}}$ was obtained for polyurethane and the acrylic adhesive, and a minimum value of $0.04\,\mathrm{N\,cm^{-2}}$ was measured for silicone and bifaced tape. Peel values were usually in the range $0.30–1.00\,\mathrm{N\,cm^{-2}}$.

The peeling force per centimeter was determined for acrylic latex emulsions and a polydimethylsiloxane maxillofacial adhesive. The adhesives were used to bond strips or prosthetic grade silicone rubber to the human forearm and the strips were peeled off at $5\,\mathrm{cm\,min^{-1}}$. The peel force was $6.6\,\mathrm{N\,cm^{-1}}$ for the silicone adhesive and $0.6\,\mathrm{N\,cm^{-1}}$ for the acrylic latex emulsions. Failure occurred at the silicone–adhesive interface for the acrylic latex adhesives and at the skin–adhesive interface for the silicone adhesive.

See also: Biocompatibility: An Overview; Polysiloxanes; Polyurethanes

Bibliography

Chalian V A 1979 Treating the patient with facial defects. In: Laney W R (ed.) 1979 *Maxillofacial Prosthetics*. Postgraduate Dental Handbook Series, Vol. 4. PSG, Littleton, MA, pp. 279–307

Chalian V A, Drane J B, Standish S M 1971 *Maxillofacial Prosthetics: Multidisciplinary Practice*, Williams and Wilkins, Baltimore, MD

Craig R G (ed.) 1978 *Dental Materials: A Problem-Oriented Approach*. Mosby, St. Louis, MO, p. 195

Craig R G, Koran A, Yu R 1980 Elastomers for maxillofacial applications. *Biomaterials (Guildford)* 1: 112–17

Farah J W, Robinson J C, Hood J A A, Koran A, Craig R G 1988 Force-displacement properties of a modified cross-linked silicone compared with facial tissues. *J. Oral Rehabil.* 15: 277–83

Farah J W, Robinson J C, Koran A, Craig R G, Hood J A A 1987 Properties of a modified cross-linked silicone for maxillofacial prostheses. *J. Oral Rehabil.* 14: 599–605

Koran A, Powers J M, Lepeak P J, Craig R G 1979 Stain resistance of maxillofacial materials. *J. Dent. Res.* 58: 1455–60

Koran A, Powers J M, Raptis C N, Yu R 1981 Reflection spectrophotometry of facial skin. *J. Dent. Res.* 60: 979

Lewis D H, Castleberry D J 1980 An assessment of recent advances in external maxillofacial materials. *J. Prosthet. Dent.* 43: 426–32

Lontz J F, Mildan I, Nodijcka M D 1979 Polydimethylsiloxane for safe and effective orofacial prostheses, III. Assessment of pigmented formulation and fabrication variables for toxicity to human tissues. In: Cooke, F W, Johnson J K (eds.) 1979 *Trans. 11th Int. Biomaterials Symp.* Society for Biomaterials, San Antonio, TX, p. 146

Ma T, Johnston W M, Koran A 1987 The color accuracy of Kubelka–Munk theory for various colorants in maxillofacial prosthetic materials. *J. Dent. Res.* 66: 1438–44

Udagama A 1975 Biocompatibility and physical properties of adhesives used in maxillofacial prosthetics. Masters Thesis, Indiana University, IN

US Council on Dental Materials and Devices 1975 Maxillofacial prosthetic materials. *J. Am. Dent. Assoc.* 90: 844–8

Waranowicz M, Robinson J, Powers J M, Koran A 1983 Peeling energy of maxillofacial adhesives. *J. Dent. Res.* 62: 297

Yu R, Koran A, Craig R G 1980 Physical properties of maxillofacial elastomers under conditions of accelerated aging. *J. Dent. Res.* 59: 1041–7

Yu R, Koran A, Craig R G, Raptis C N 1982 Stain removal from a pigmented silicone maxillofacial elastomer. *J. Dent. Res.* 61: 993

Yu R, Koran A, Powers J M 1983 Effect of processing temperature on the properties of a polyvinyl chloride maxillofacial elastomer. *J. Dent. Res.* 62: 1098–1100

Yu R, Koran A, Raptis C N, Craig R G 1981 Stain removal from a silicone maxillofacial elastomer. *J. Dent. Res.* 60: 1754

Yu R, Koran A, Raptis C N, Craig R G 1983 Cigarette staining and cleaning of a maxillofacial silicone. *J. Dent. Res.* 62: 853–5

R. G. Craig
[University of Michigan, Ann Arbor, Michigan, USA]

Metals for Medical Electrodes

An important and challenging medical use of implanted electrodes is in prosthetic devices for neural control. These devices employ metal electrodes to transmit the current required for electrical stimulation of appropriate areas of the nervous system. Neural prostheses for direct control of peripheral organs include the cardiac pacemaker, the phrenic stimulator for respiratory control and spinal-cord stimulators for bladder control. More complex neural-control devices are auditory prostheses for the sensory deaf, experimental visual prostheses for the blind and neuromuscular prostheses for restoration of hand, arm or leg function in paralyzed individuals. The properties of a metal that are important to its performance as a stimulation electrode will depend to a great extent on the particular application.

1. Applications and Requirements for Stimulation Electrodes

All metals that are implanted in the body must be biocompatible; that is, they must not introduce toxic species into the body, nor should they be adversely affected by exposure to body fluids. Metals which pass current, as do electrodes for neural stimulation, have additional requirements. They must pass adequate coulombic charge without electrolyzing tissue components and they must have long-term mechanical and electrical stability. The relative importance of these requirements will depend on the particular application.

Stimulation of muscle fibers with intramuscular electrodes which must traverse multiple tissue planes, places a premium on mechanical strength and flexibility. Stainless steel electrodes have been used for short-term implantation, but these do not have adequate corrosion resistance for long-term use and are also limited in the quantity of charge that they can inject. In contrast, surface stimulation of the visual cortex demands little mechanical strength and pure platinum is commonly used. Similarly, stimulation of peripheral nerves uses large electrodes that are not subject to high mechanical or electrochemical stresses, hence platinum is also satisfactory for these applications. The stimulation charge densities used for cortical stimulation with surface electrodes or for peripheral nerve stimulation are moderate, so capacitor electrodes based on tantalum pentoxide may also be used successfully.

The most demanding application of all is for intracortical stimulation with microelectrodes which approximate the size of a neuron and have exposed surface areas of 10^{-4} mm^2, or less. These must have a high mechanical strength to penetrate cortical tissues without deformation and be able to pass currents that give rise to high current and charge densities at the electrode–tissue interface. For instance, the required stimulation currents of 2–20 μA delivered in 0.2 ms pulses, give rise to charge densities at a microelectrode surface in excess of 100 μC mm^{-2}. With conventional metal electrodes, most of this charge will flow via irreversible faradaic reactions which are unacceptable because of the chemical changes introduced into the tissue and the damage to the stimulation electrode.

2. Charge Injection

Electrical stimulation with metal electrodes requires the transfer of electric charge from the metal into the ionic medium, namely the biological tissue. This transfer of electric charge involves a change in charge carriers, from electrons in the metal to ions in the tissue electrolyte. The conversion of charge carriers occurs by two mechanisms. The first of these is a capacitive mechanism involving only the alignment of electrostatic and ionic charges at the electrode–tissue interface; that is, the charging and discharging of the electrode double layer. No electrons actually cross the electrode surface, so no chemical changes occur in the tissue. Capacitive charging is therefore the ideal mechanism of charge injection. However, the amount of charge that can be transferred by charging and discharging the double-layer capacitance is only of the order of 0.2 μC mm^{-2} of real area. Charge densities in excess of this will lead to the equivalent of dielectric breakdown and the onset of faradaic reactions.

Faradaic reactions constitute the second mechanism for conversion of charge carriers from electronic to ionic carriers at the electrode–tissue interface. By definition, faradaic processes involve exchange of electrons across the electrode–tissue interface and therefore necessitate that some chemical species be either oxidized or reduced. The chemical species may be part of the surrounding electrolyte, or it may be material from the electrode itself. Metal electrodes inject charge mostly by faradaic processes because the amount of charge required for electrical stimulation far exceeds that available from capacitive mechanisms. In the case of platinum and other noble metals, charge injection occurs by surface reactions of oxide formation and reduction, and hydrogen-atom plating and oxidation. These processes are called "reversible" because they are confined to the metal surface, do not produce new chemical species in the bulk of the solution and can be quantitatively reversed by passing a current in the opposite direction. Other reversible processes are the valence changes and the proton or hydroxyl-ion transfers that occur within multilayered oxide films, such as those produced under certain conditions on iridium and rhodium.

The electrolysis of water and oxidation of chloride ions are examples of irreversible faradaic reactions. These are undesirable because they alter the chemical composition of the tissue fluid, produce toxic products or generate extremes of acidity or alkalinity. Other irreversible faradaic reactions involve the dissolution of the metal electrode. Metal dissolution is almost always a charge-transfer process regardless of the stimulation charge density, but the severity of it depends on the particular metal and the shape of the stimulation waveform.

It is now well recognized that unidirectional currents will produce electrolytic reactions and be damaging to tissue. Thus, bidirectional current flow is the rule for safe electrical stimulation. This is achieved with charge-balanced biphasic rectangular waveforms, capacitively coupled pulses, or sinusoidal alternating currents. However, even avoiding dc currents does not guarantee electrochemically "safe" stimulation. An additional requirement for electrochemically safe stimulation is that the currents, and more precisely the pulse charge density, should not exceed the limit that can be delivered with reversible surface processes. This limit, which is called the reversible or electrochemically safe charge-injection limit, is expressed as a charge density referred to the real surface area of the electrode.

Different metals will inject charge by different processes and therefore will have different charge-injection limits. In addition, the charge-injection limits of any one metal will depend on the geometry of the stimulus waveform, the shape of the electrode, the surface morphology of the electrode and the chemical composition of the surrounding medium. The contribution of individual processes to charge injection during a stimulation pulse will depend strongly on the potential attained by the electrode during the pulse. Conversely, the potential attained by the electrode will depend on the quantity of charge injected and the capacity of the available processes to accommodate that charge. For instance, platinum, which has several reversible surface processes available for charge injection, can accommodate a higher charge density per real area with a smaller change in potential than is possible with stainless steel.

3. Noble Metals

The most frequently considered metals for electrical stimulation are the so-called noble metals: platinum, iridium, rhodium, gold and palladium. This is because of their resistance to chemical and electrochemical corrosion. However, all of these metals show corrosion effects during both *in vitro* and *in vivo* electrical stimulation. Corrosion effects include weight loss, formation of unstable surface films which tend to spall from the surface, and dissolution of metal (probably in the form of chloride complexes). With the exception of platinum, little work has been reported on use of noble metals for chronic long-term stimulation electrodes.

3.1 Platinum

Platinum is the most widely used noble metal for electrical stimulation and its charge injection properties have been reviewed extensively. The surface processes on platinum, plus a small contribution from double-layer charging, can accommodate about 3–3.5 μC mm^{-2} of real surface area when it is delivered with charge-balanced biphasic current pulses. This reversible limit has only been demonstrated for current densities up to 4.5 mA mm^{-2} of real area and pulse durations of 0.6 ms or more. Higher current densities, shorter pulses or a different waveform such as monophasic capacitively coupled current pulses may have different reversible charge-injection limits. Charge densities in excess of the reversible limits will lead to irreversible faradaic reactions involving the surrounding electrolyte.

Metal dissolution is a minor charge-transfer process on platinum electrodes at all stimulation charge densities. Platinum dissolution rates during biphasic pulsing in an inorganic saline solution lie in the range 100–1000 ng C^{-1} of the aggregate charge injected. (Aggregate charge equals the charge per phase multiplied by the number of biphasic pulses.) A dissolution "rate" of 100 ng C^{-1} corresponds to approximately 100 ppm of the injected charge. In the presence of protein, which inhibits dissolution, platinum dissolution rates are even lower, in the range 0.1–10 ng C^{-1}. Assuming a 10 ng C^{-1} dissolution rate, a platinum electrode which has a geometric surface area of 30 mm^2 and a thickness of 25 μm (a size approximating that used in peripheral nerve stimulation), injecting 0.4 μC per pulse would lose 10 ng of platinum in about 2.5×10^6 pulses. At a pulse repetition rate of 30 Hz this would require 23 h of continuous stimulation, and would represent a weight loss of only $6.2 \times 10^{-5}\%$ which would not appear to be a problem in terms of the lifetime of the electrode. However, dissolved platinum may form toxic complexes in the biological milieu, and if platinum were to be used as a microstimulation electrode, then a dissolution rate of 10 ng C^{-1} might lead to the loss of the electrode or significant changes in its surface properties.

Pure platinum is a soft metal, so it is not appropriate for all neural prosthetic applications. Thus it is frequently alloyed with iridium to obtain the required mechanical properties. The platinum–iridium alloys containing 10–30% iridium are indistinguishable from pure platinum with regard to reversible charge-injection limits. With regard to metal dissolution, both platinum and iridium dissolve during stimulation so that there will be two possibly toxic metals released into the tissue. Neither platinum nor its alloys are practical materials for the charge densities required of intracortical microstimulation electrodes.

3.2 Iridium

Pure iridium is a hard metal with mechanical properties which meet the physical requirements for intracortical electrodes of small size and great strength. Early studies indicated that the metal corroded badly when used as a stimulation electrode. However, when the metal is appropriately "activated" to form a multilayer surface oxide, it has highly advantageous electrochemical properties with respect to electrical stimulation.

When activated by repetitive potential cycling, iridium can store large amounts of coulombic charge in the form of reversible surface oxides. Charge injection with reversible surface oxides involves valence changes in the metal oxide coupled with proton or hydroxyl ion transfers. These reactions greatly enhance the limits for reversible charge injection, enabling up to a tenfold increase in anodic charge injection compared with bare iridium, platinum or platinum alloys, before encountering the formation of oxygen bubbles. A more conservative estimate of the safe charge-injection limits is based on observations of potential transients, or polarization curves, during stimulation and considers as safe only those charge densities at which the electrode potential does not surpass the potentials for water electrolysis. These conservative limits fall into the range 30–40 μC mm^{-2} geometric for anodic pulses and 10–20 μC mm^{-2} geometric for cathodic pulses, both of which are still significantly greater than the limits for platinum or platinum–iridium alloys. It is not possible to specify a definitive reversible charge-injection limit for an activated iridium electrode because that limit will depend primarily on the degree of activation; that is, the thickness of the oxide film and the number of iridium oxide sites that are accessible to the electrolyte. This is in marked contrast to the situation with platinum, for which the charge-injection limits are defined by surface processes on the bare metal.

Dissolution rates of activated iridium electrodes are about ten times less than those of platinum or bare iridium under the same stimulation conditions. They range from $0.1 \, \text{ng} \, \text{C}^{-1}$ to $1 \, \text{ng} \, \text{C}^{-1}$ depending on pulse polarity. This corrosion resistance of activated iridium might be a result of charge-transfer reactions involving only insoluble oxides of iridium with a minimum contribution from reactions on the bare metal.

3.3 Rhodium

Rhodium metal has not been studied extensively as an electrode material, but the results of the few *in vivo* studies and *in vitro* weight-loss studies indicate that it is comparable to platinum in its dissolution or corrosion properties.

Like iridium, rhodium forms a reversible surface oxide after appropriate treatment of its surface. Preliminary data on activated rhodium electrodes, indicate that it has charge-injection properties similar to those of activated iridium when stimulated with an appropriate stimulation waveform. The dissolution properties of rhodium or activated rhodium have not been examined in detail.

These activated noble metals which transfer charge via redox reactions within an oxide layer, are a very recent development in stimulation electrode materials research. Consequently, only a few studies are as yet published. Those that are available indicate the superiority of activated iridium over platinum–iridium alloy as an intercortical microelectrode, although the goal of stimulation charge densities of $100 \, \mu\text{C} \, \text{mm}^{-2}$ geometric without neural damage is yet to be realized (Agnew et al. 1986).

3.4 Metal Oxides

A promising new type of stimulation electrode under development might be considered as a kind of hybrid electrode because one metal comprises the body of the electrode, and this is coated with a metal oxide layer which constitutes the actual charge injection interface. An example of this is a titanium metal electrode which is coated with a film of iridium oxide. The titanium metal can be configured into the desired shape for a particular application prior to deposition of iridium oxide. The iridium oxide may be deposited either by a vacuum deposition method or by the thermal decomposition of an iridium salt. Such electrodes have the charge-transfer properties of the iridium oxide but the mechanical properties of the underlying metal. The approach enables the use of the high charge capacity of iridium oxide without the expense of pure iridium metal or the difficulty in forming the pure metal into the shape required for the stimulation electrode. As for activated iridium and activated rhodium, *in vivo* evaluations of these hybrid electrodes are just the beginning.

4. Other Metals

The other metals that have been studied as stimulation electrode materials include:

(a) the stainless steels 303, 316 and 316LVM;

(b) a nickel–cobalt alloy Elgiloy;

(c) a Co–Ni–Cr–Mo alloy MP35N; and

(d) zirconium, tungsten, tantalum and titanium.

The faradaic processes occurring on these metals and alloys are not as well characterized as those on platinum and other noble metals. Metals and alloys that rely on thin passive films (e.g., stainless steel, Elgiloy and MP35N) must primarily inject charge by faradaic processes involving reduction and oxidation of their passivating films. There are complications with these alloys that arise when their potentials are driven into the transpassive region in which breakdown of the passive layer occurs. At that point, charge injection must take place by irreversible processes involving metal dissolution; namely, corrosion and water electrolysis. The corrosion mechanisms of metals which are not noble are discussed in Sect. 5.

A common property of the stainless steels and cobalt alloys, is that they cannot withstand large anodic polarization without risk of rapid electrode failure owing to loss of passivity and dissolution. This is in marked contrast to platinum, iridium and rhodium where dissolution is only a minor charge-injection process with either polarity of stimulation, and is also in contrast to tantalum–tantalum pentoxide ($Ta–Ta_2O_5$) or titanium–titanium dioxide ($Ti–TiO_2$) capacitor electrodes, which operate best under anodic polarization and with virtually no faradaic reactions (Sect. 4.1). The stainless steels are the most susceptible to corrosion-related failure, even when used with cathodic-going pulses. Thus alternatives have been investigated. For cardiac pacemaker applications where monophasic capacitively coupled pulses are most commonly used, platinum or platinum–iridium alloys are used as the anode with Elgiloy serving as the stimulation cathode. However, special precautions are necessary in stimulator design to ensure that a small cathodic voltage is maintained across the Elgiloy between stimulus pulses to prevent corrosion.

Titanium, tantalum, zirconium, tungsten and tungsten bronzes give unacceptable performance under conditions of electrical stimulation with biphasic current pulses because of strong surface reactions and rapid changes in impedance. Rectangular pulses of anodal current on tantalum, tungsten and titanium result in increasingly large polarization voltages and large increases in impedance. These effects probably reflect the formation of insulating oxide films owing to the anodic polarization. This property of tantalum and titanium forms the basis of their development as capacitor electrodes (see Sect. 4.1).

To date, stainless steel 316LVM has given a moderate performance as an intramuscular electrode, primarily because of its mechanical strength and flexibility when fabricated into a coiled structure. However, the maximum charge density that can be accommodated with stainless steel is only 0.4–$0.5\,\mu C\,mm^{-2}$ geometric, and electrode breakage as a result of corrosion is a major deterrent to long-term implantation of intramuscular electrodes for motor prostheses.

Silver is not appropriate for use as an electrode under any conditions involving the passage of current. In a saline environment, silver dissolves readily, introducing potentially toxic ions into the tissue.

The Co–Ni–Cr–Mo alloy MP35N is presently being investigated as a replacement for stainless steel in intramuscular electrodes. While it has adequate mechanical strength and flexibility as an electrode lead material, its high electrical resistivity is a disadvantage. In addition, its corrosion properties under conditions imposed by electrical stimulation pulses are not fully detailed.

4.1 Capacitor Electrodes

Capacitor electrodes represent the ideal type of stimulation electrode. The introduction of a dielectric at the electrode–electrolyte interface, which will withstand substantial voltage without significant dc leakage, allows charge flow without the risk of faradaic reactions. Where moderate stimulation charge densities are required, Ta–Ta_2O_5 capacitor electrodes have been used successfully for neural stimulation.

Since the charge storage capacity of these electrodes depends on the effective surface area of the dielectric film, two approaches have been utilized to obtain the desired degree of surface-area enhancement. One is the sintered powder technology used by the electrolytic capacitor industry, in which a slurry of tantalum powder is pressed into a pellet, sintered and anodized. The other approach is to electrolytically etch tantalum wire to obtain a high surface roughness. The etched wire is then anodized to obtain the tantalum pentoxide dielectric. Charge injection with semimicro Ta–Ta_2O_5 capacitor electrodes is in the range 0.5–$1.5\,\mu C\,mm^{-2}$ geometric, so their use for neural stimulation is limited to applications requiring low-to-moderate charge densities.

An additional restriction on the use of Ta–Ta_2O_5 is its rectifying property: tantalum pentoxide loses its insulating ability when operated under a negative bias. To prevent electronic conduction across the oxide, tantalum capacitor electrodes must always operate at a positive potential with respect to the tissue. If cathodic stimulation is the physiological preference, then the tantalum capacitor electrodes must be biased to a positive voltage and pulsed cathodically.

Other capacitor electrode materials that have been investigated are anodized titanium to produce a Ti–TiO_2 capacitor electrode and the deposition of thin films of barium titanate on a noble metal. The dielectric constants k of these materials are substantially higher than that of Ta_2O_5: $k = 25$ for Ta_2O_5, $k = 100$ for rutile TiO_2, $k = 50$ for anatase TiO_2 and $k = 10^3$ for barium titanate. Thus, a significant increase in capacitance and charge storage should be possible. In fact, capacitor electrodes based on TiO_2 have higher charge storage than anodized tantalum but they also have substantially higher leakage current. Capacitor electrodes of barium titanate have lower charge storage than Ta–Ta_2O_5 electrodes, in spite of their high dielectric constant. This is because the barium titanate films are much thicker than those of tantalum pentoxide and are not pinhole-free.

In the final analysis, only capacitor electrodes based on Ta–Ta_2O_5 have adequate charge storage capacity and low leakage currents suitable for applications using electrodes of the order of $0.5\,mm^2$ geometric, or larger, and requiring moderate stimulation charge densities up to 1–$2\,\mu C\,mm^{-2}$ geometric. In view of the requirement for anodic bias, Ta–Ta_2O_5 would appear to be the ideal material to use as the large anode of a bipolar pair of electrodes in which the stimulation electrode receives monophasic capacitively coupled cathodic pulses.

5. Corrosion

Corrosion resistance in metals and alloys intended for chronic use in implanted medical devices, derives from intrinsic nobility or the formation of a passivating surface film. Metals used as neural stimulation electrodes are subjected to imposed electromotive forces. The imposition of an externally applied voltage and the demand for charge transfer at the electrode–tissue interface causes polarization of the electrode well beyond that encountered in passive corrosion. Faradaic charge transfer involving electrolyte species, often causes large diffusion-controlled changes in pH. The electrode must therefore survive in the *in vivo* environment with the additional demands of reversible charge transfer, a widely varying local pH and, in some applications, large mechanical loads. The severity of the stimulation environment limits the metals suitable for stimulation electrodes to platinum, iridium, rhodium, titanium, tantalum, 316LVM stainless steel and some alloys based on chromium, nickel, cobalt and molybdenum.

Perhaps the greatest difficulty in predicting corrosion under charge injection is the extremely transient nature of the imposed emf. Typical charge-injection schemes involve repetitive applications of very brief current pulses which produce rapid oscillations of the potential of the stimulation electrode. During the short interval between pulses, the electrode potential may be very different from the *in vivo* corrosion (or mixed) potential under non-charge-injection conditions. Under these conditions of rapidly changing electrode potential, the kinetics of the corrosion-related processes, such as anodic dissolution, loss of passivity, repassivation and pit initiation, becomes most significant in establishing the overall corrosion behavior of the electrode.

Corrosion as a degradation mechanism has been studied extensively in platinum stimulation electrodes. The review of Brummer and Turner (1977) presents a detailed description of the faradaic and nonfaradaic processes that occur during charge injection with platinum. The mechanism of platinum corrosion is one of dissolution activated by cyclical oxidation and reduction of the platinum surface. Other noble metals, particularly iridium and platinum–iridium alloys, exhibit the same general corrosion behavior as platinum. The few studies undertaken with gold electrodes indicate that they are more prone to dissolution than either platinum or iridium.

The corrosion of passivating metals and alloys is far more complex than that of the noble metals. In addition to the inherent questions of stability and formation of passivating films under an imposed emf, many of the alloys are used in electrode configurations that subject them to substantial mechanical stresses. The greatest *in vivo* corrosion resistance is obtained with tantalum and titanium and some alloys of titanium, most notably Ti–6Al–4V–ELI. These metals and alloys owe their excellent *in vivo* corrosion resistance to the formation of stable passivating oxide films. The critical pitting potentials are extremely noble (>12 V) in dilute chloride solutions at room temperature. The valve nature of the semiconducting oxide films allows these metals to withstand large anodic potential excursions without encountering deleterious faradaic charge-transfer reactions.

There have been no significant corrosion problems reported with tantalum and titanium capacitor electrodes. Cathodic faradaic processes, primarily hydride formation and hydrogen evolution, are encountered if the electrodes are polarized excessively in the cathodic direction. Tantalum is susceptible to embrittlement by hydrogen under extreme cathodic polarization. If titanium and tantalum are operated as stimulation electrodes without preanodization or anodic biasing, anodic polarization causes a significant increase in electrode impedance caused by thickening of the oxide layer. The extremely positive breakdown potentials of titanium and tantalum preclude pitting, crevice corrosion and anodic dissolution, in favor of anodization and oxygen evolution.

The most demanding electrode application with regard to mechanical loading is in intramuscular stimulation for restoration of muscle function in quadriplegic or paraplegic individuals. These electrodes are made from 316LVM stainless steel. They fail primarily by fatigue or corrosion fatigue exacerbated by the presence of crevice corrosion at the junction between the electrode and the insulation on the lead to the electrode. The role of an imposed emf in the failure of intramuscular electrodes has not been elucidated, but it is certain that the potential excursion of the electrode during stimulation exceeds the critical pitting potential of the stainless steel. Loss of passivity may also be induced by cathodic polarization, with ensuing questions concerning the rate of film breakdown and the rate of repassivation. The repassivation could result from the anodic half of a biphasic pulse or from self-passivation between the stimulation pulses.

Most of the questions raised about corrosion resistance of passive metals under an imposed emf have not received adequate theoretical or experimental consideration. It is apparent from clinical experience that stainless steel (exclusively the 316LVM) is barely adequate as an electrode for long-term functional electrical stimulation. Other biomedical alloys, such as MP35N and some Co–Cr–Mo alloys, are expected to perform better than 316LVM because their breakdown potentials are considerably higher.

An interesting approach to the problem of corrosion resistance arises from the use of valence-change oxides described in Sect. 3.4. These oxides can undergo considerable reversible faradaic charge transfer with only modest cathodic and anodic polarization. A metal electrode coated with iridium oxide, for example, may be protected from emf-induced corrosion if the polarization is within the stability range of the metal. Issues of crevice corrosion and galvanic coupling between the oxide layer and the metal have not been addressed.

The combination and probable synergistic effects of the *in vivo* environment, mechanical loading and an imposed emf on the corrosion of biomedical metals and alloys have only been investigated in a very superficial manner. An understanding of how these environmental conditions will influence the more subtle forms of corrosion-related failure such as crevice corrosion, corrosion fatigue, galvanic corrosion and stress-corrosion cracking have yet to be investigated. The neural prostheses currently being developed are becoming increasingly complex, involving the use of dissimilar metals, integrated circuit technology, new polymeric insulation materials and in-line electrical connectors requiring long-term *in vivo* survivability. Achieving long-term survival of neural prosthetic components in the chemically and mechanically severe *in vivo* environment, will require a more complete understanding of the corrosion phenomena than we now possess.

See also: Biocompatibility: An Overview; Cobalt-Based Alloys

Bibliography

Agnew W F, Yuen T G H, McCreery D B, Bullara L A 1986 Histopathologic evaluation of prolonged intracortical electrical stimulation. *Exp. Neurol.* 92: 162–85

Bernstein J J (ed.) 1977 Neural prostheses: Materials, physiology and histopathology of electrical stimulation. of the nervous system. *Brain, Behav. Evol.* 14: 1–160

Brummer S B, Robblee L S, Hambrecht F T 1983 Criteria for selecting electrodes for electrical stimulation: Theoretical and practical considerations. *Ann. N. Y. Acad. Sci.* 405: 159–71

Brummer S B, Turner M J 1977 Electrochemical considerations for safe electrical stimulation of the nervous system with platinum electrodes. *IEEE Trans. Biomed. Eng.* 24: 59–63

Dumbleton J H, Miller E H 1975 Failures of metallic orthopedic implants. In: American Society for Metals 1975 Failure Analysis and Prevention. *Metals Handbook*, 8th edn, Vol. 10. ASM, Metals Park, OH

Guyton D L, Hambrecht F T 1974 Theory and design of capacitor electrodes for chronic stimulation. *Med. Biol. Eng.* 7: 613–20

Hambrecht F T 1979 Neural prostheses, *Annu. Rev. Biophys. Bioeng.* 8: 239–67

Loeb G E, McHardy J, Kelliher E M, Brummer S B 1982 Neural prostheses. In: Williams D F (ed.) 1982 *Biocompatibility in Clinical Practice*, Vol. 2. CRC Press, Boca Raton, FL, pp. 123–49

Robblee L S, Mangaudis M J, Lasinsky E D, Kimball A G, Brummer S B 1986 Charge injection properties of thermally-prepared iridium oxide films. *Mater. Res. Soc. Symp. Proc.* 55: 303–10

Rose T L, Kelliher E M, Robblee L S 1985 Assessment of capacitor electrodes for intracortical neural stimulation. *J. Neurosci. Methods* 12: 181–93

Uhlig H H, Revie R W 1985 *Corrosion and Corrosion Control*, 3rd edn. Wiley, New York

Williams D F 1976 Corrosion of implant materials. *Annu. Rev. Mater. Sci.* 6: 237–66

L. S. Robblee and S. F. Cogan
[EIC Laboratories, Norwood,
Massachusetts, USA]

O

Ormosils: Organically Modified Silicates

A new group of materials, the organically modified silicates (or *ormosils*), is based on a silicate network into which organic radicals have been incorporated. These materials represent intrinsic combinations of inorganic silicate glasses with organic polymers.

Various fundamental ideas underlying the existence of these materials are quite old, but their practical realization came about only recently through progress in the sol–gel method, which allows the materials to be prepared at relatively low temperatures. This has permitted the "made-to-measure" synthesis of materials that combine the desirable properties of both inorganic and organic materials, bringing closer to reality a long-standing dream of materials designers.

1. Structural Fundamentals

Noncrystalline inorganic materials have many outstanding properties. It is well known that the silicate glasses, the principal group of such materials, have good transparency and high resistance to chemical attack and scratching. It is true, however, that the great brittleness of these glasses in normal use leads to mechanical fragility. In this respect, most organic polymers outperform inorganic glasses. It is thus a natural objective to incorporate organic components into inorganic materials with the aim of uniting the best properties of each. This has led to the creation of the organically modified silicates.

From a structural viewpoint, silica glass (vitreous silica) is the simplest glass. It consists of $[SiO_4]$ units; that is, of silicon atoms surrounded by tetrahedra of four oxygen atoms. These $[SiO_4]$ tetrahedra unite at their corners to form an irregular, but continuous, three-dimensional network. The high strength of the silicon–oxygen bond is the reason for the high glass transformation temperature of 1200 °C and thus for the great usefulness of silica glass.

The SiO_2 network can be modified in various ways. The schematic diagrams below show the unmodified network and two types of inorganic modification:

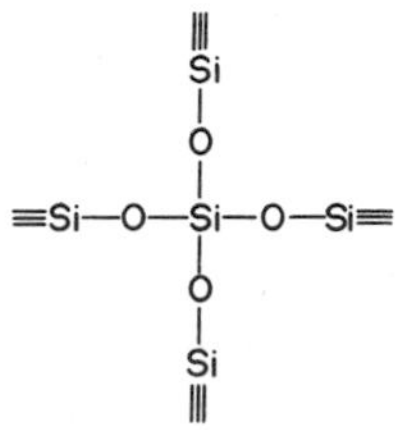

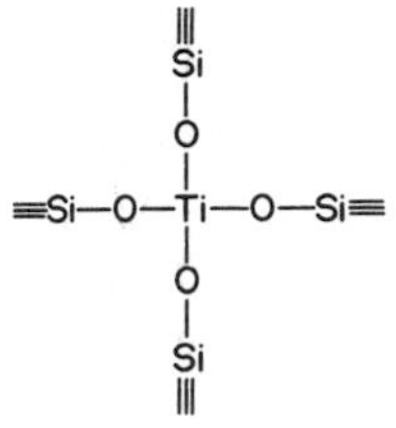

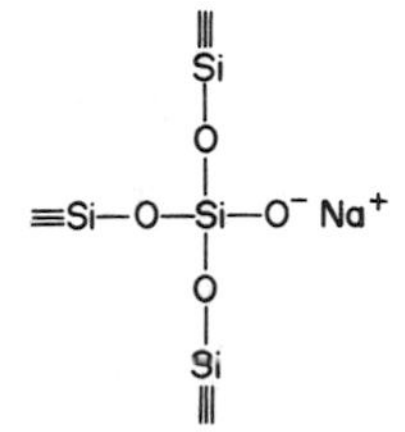

The middle example shows a substitution of a silicon atom by a four-valent titanium atom, which leaves the structure essentially unaltered. If, however, alkali metals are introduced, then the $\equiv$Si—O—Si$\equiv$ bridges are split and natural weak points are created at the singly bound oxygen atoms.

A similar introduction of weak points is to be expected if the structure is modified in such a way that direct silicon–carbon bonds are introduced, as shown in the following diagrams.

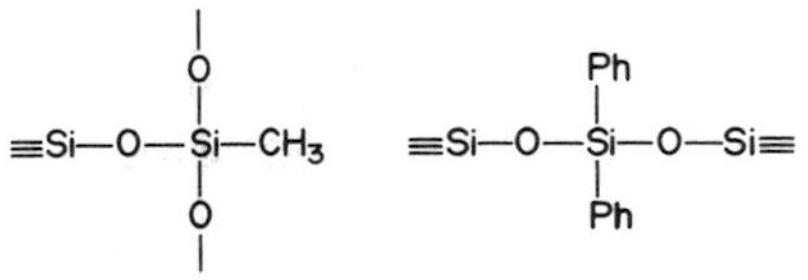

Great variability is possible in the nature of the incorporated organic radical. Further, it is possible to vary the number of organic radicals attached to a single silicon atom, as exemplified by the right-hand diagram above with two phenyl groups (Ph) per silicon atom. This further diminishes the degree of networking to the level of two connections per silicon atom. This means that only chain molecules can be formed. (This is also the basis of the family of the silicones.) This feature has a powerful effect on properties, as will be shown in Sect. 3. Evidently, the proportion of modified silicon atoms plays a major role in determining how far properties will be changed.

A further interesting possibility for modification arises from the incorporation of organic radicals that still carry functional groups. Such radicals include —OH, —NH$_2$, —COOH, —COH and —CH=CH$_2$. In this way, organic types of reactivity can be built into an inorganic network.

Some organic compounds are known to be polymerizable. This property can also be incorporated. If radicals attached to two neighboring silicon atoms can be linked, then the natural break point located at this site is bridged; that is, such radicals function as network formers. There are many ways of bringing about this state. In this example the zigzag line represents a C—H chain:

$$\equiv Si \sim CH=CH_2 + H_3C—Si\equiv$$
$$\rightarrow \equiv Si \sim CH_2—CH_2—CH_2—Si\equiv$$

The organic radicals are attached to silicon atoms. They can only react with each other if they happen to be near neighbors, which cannot always be achieved in practice. However, the possibility exists of bringing in a polymerizable purely organic compound to achieve internal bonding. In the following example, this is the ester of methacrylic acid:

A variant, similar in principle, of the synthesis of a siliceous material is represented by the internal combination of a purely siliceous network with a purely organic polymer, making use of the sol–gel process. A mixed solution in an organic solvent of two reactants, such as tetraethoxysilane and polyethylene glycol, undergoes a transesterification with separation of C$_2$H$_5$OH, with the consequence that the polyethylene molecules become linked with each other via [SiO$_4$] groups. In contrast to the types enumerated up to this point, the linkage is via Si—O—C groups.

There are also proposals for incorporating organic components still more loosely. The procedure follows that of the classical sol–gel process which is normally used to produce oxide networks, but involves the addition of organic compounds (e.g., dyes), which are then caged in the hollows of the gels.

The range of variability of ormosils becomes very wide when all these strategies are employed. Moreover, it is also possible to use the various strategies simultaneously in the same material, and in varying proportions. In this way, properties can be made to vary within wide limits (see Sect. 3) and can even be tailored to particular applications (see Sect 4); however, limits are imposed to what can be achieved by the conditions of production (see Sect. 2).

Structural determinations of short-range order have not yet been attempted. However, indications can be derived from the properties of the silicones, in which the silicon–oxygen bond length (~0.164 nm) is close to that in SiO$_2$ glass (0.161 nm). However, the silicon–carbon bond (~0.190 nm) is substantially greater; that is, an enlargement of the network is to be expected. The silicon atom will remain tetrahedrally coordinated, because the C—Si—C angle of approximately 110° has the same value as the O—Si—O angle in SiO$_2$ glass.

Among the property values listed in Sect. 3, the low density of the ormosils (1200–1500 kg m^{-3}) is particularly noteworthy. This indicates a low level of space filling. Approximate computations of volume requirements under the assumption of spherical atoms or radicals for the ormosil of composition Ph$_2$SiO.TiO$_2$ yield a theoretical molar volume of 51.5 cm^3 mol^{-1} oxygen, rising to approximately 68 cm^3 mol^{-1} oxygen when the free volume between the spherical atoms is taken into account. This agrees well with the experimentally determined volume of 67 cm^3 mol^{-1} oxygen. In this ormosil, the volume is determined to an extent of approximately 90% by the phenyl groups. In the structures of other ormosils, the organic radicals mostly have a smaller role in determining the volume, but taken together such structural considerations indicate that the organic radicals have large space requirements, which in turn increases their influence on properties.

2. Synthesis

The use of the sol–gel process in the synthesis of ormosils has previously been mentioned. The precursors for this process are compounds that are mutually soluble at room temperature. The precursor for SiO$_2$ is tetraethoxysilane Si(OEt)$_4$. The general formula of this type is Si(OR)$_4$; R is an organic radical, with a particular aliphatic or aromatic residue.

In going from the silicon ester to SiO$_2$ glass, several processes take place, beginning with a hydrolysis:

$$Si(OEt)_4 + H_2O \rightarrow (EtO)_3SiOH + EtOH$$

and followed by further analogous steps till the (hypothetical) end product Si(OH)$_4$ is attained. This last compound is not stable but condenses according to the reaction

$$\equiv Si{-}OH + HO{-}Si\equiv \longrightarrow \equiv Si{-}O{-}Si\equiv + H_2O$$

which involves the creation of Si—O—Si bridges, so that the molecules grow indefinitely. At first, these remain colloidally soluble in the sol, but later after further condensation a solid gel results.

The overall process can be written as

$$Si(OEt)_4 + 2H_2O \rightarrow SiO_2 + 4EtOH$$

It is influenced by many factors: water concentration, nature and quantity of the solvent, nature and quantity of the catalyst, temperature and nature of the organic residue R.

The rate of hydrolyzation is influenced particularly strongly by the catalyst and diminishes as the chain length of R increases. The first stage is always the most rapid; the subsequent stages become progressively slower, partly because of steric hindrance.

Condensation takes place in parallel with, and often more quickly than, hydrolysis. The conditions of synthesis, especially the catalyst, determine whether condensation is predominantly linear or three-dimensional; this has a considerable influence on the molecular structure and microstructure of the gel.

In synthesizing materials based on atoms other than silicon, the precursors used are again usually the corresponding esters. The majority of such esters hydrolyze faster and this carries the danger that the individual components hydrolyze at different rates so that an inhomogeneous product results. To overcome this problem, appropriate choice of the reaction conditions, especially of the catalyst, is required as a precondition for a reaction to place as uniformly as possible. The process can also operate in two steps by first condensing the slower components wholly or partially before adding the remaining components.

The gel thus obtained still contains solvents and water of reaction; after their removal a porous product remains. During this stage a shrinkage, which can be considerable, occurs and this generally leads to crumbling of monoliths. As a result, it is very difficult to manufacture large monoliths directly. Compaction is therefore usually achieved by tempering barely above the glass transformation temperature of the glass. External pressure accelerates compaction. If organic radicals are directly bonded to silicon, then tempering is no longer feasible.

Precursors with organic radicals have the general formula $R'_n SiX_{(4-n)}$, where R' denotes one or several organic radicals which may or may not be different. X is the hydrolyzable group (e.g., OR as in the example cited previously), but a halogen (generally chlorine) or OH may also be introduced. In contrast, the direct silicon–carbon bond is very stable and resistant to hydrolysis.

These precursors are subject to all the influences listed previously. Further, the nature and quantity of the radical R' also influence the sol–gel process. Investigations with the pure compounds $Me_n Si(OEt)_{(4-n)}$ (where Me is the methyl radical CH_3) have established that with an acid catalyst (0.002 mol HCl l^{-1}) hydrolysis of the pure ester (i.e., with $n = 0$) is relatively slow, while the presence of methyl groups greatly accelerates hydrolysis; it is of little account whether one, two or three methyl groups are present. However, with a basic catalyst with two moles of NH_3 per liter the pure ester hydrolyzes rapidly, while the presence of methyl groups markedly retards hydrolysis; retardation is greater for larger values of n.

Mixtures of $Si(OEt)_4$ and $Me_2 Si(OEt)_2$ hydrolyze almost completely with an HCl catalyst before any condensation is detectable and the product becomes transparent. When NH_3 catalysis is used, $Si(OEt)_4$ hydrolyzes and condenses substantially before the methylated precursor, so the product becomes inhomogeneous and imperfectly transparent.

Some of the cited reactions have a tendency to equilibrium states, so that the reverse reaction is also possible. Side reactions have also to be considered; for example, transesterification when alcohols are used as solvents. This process is also observed to involve R' when alcoholic groups are present, for example,

$$\equiv Si \sim CH_2OH + RO{-}Si\equiv$$
$$\rightarrow \equiv Si \sim CH_2{-}O{-}Si\equiv + ROH$$

This reaction leads to the loss of the (sometimes desirable) hydrophilic properties associated with the CH_2OH group.

In Sect. 1 it was shown that polymerization of the organic radicals can be used to achieve internal bonding. For this purpose, the use of polymerization catalysts that can be activated by heat or ultraviolet light is recommended.

In the production of ormosils by the sol–gel process there is apt to be shrinkage. However, this can be reduced or even prevented by using precursors that generate no reaction products during condensation as, for example, during internal bonding by organic polymerization, as just mentioned. It has also been found that shrinkage can be effectively reduced by adding titanic acid ester $Ti(OR)_4$ as the condensation catalyst. Water content must be carefully controlled when this is done, however, since $Ti(OR)_4$ hydrolyzes and condenses very quickly so TiO_2 separates out easily. Little water should be added at first, so that only a little $\equiv TiOH$ is formed, and this then has the opportunity, according to

Table 1
Some properties of vitreous-like ormosils (mainly at room temperature; Ph is the phenyl group, Epoxy is γ-glycidyloxypropyltrimethoxysilane)

Composition	Density $(kg\,m^{-3})$	Refractive index	Linear expansion coefficient $(10^{-6}\,K^{-1})$	Glass transformation temperature $(^\circ C)$
$0.2[PhSiO_{312}].0.8SiO_2$	1420	1.49	15	225
$0.35[PhSiO_{312}].0.65SiO_2$	1330	1.53	66	135
$0.5[Ph_2SiO].0.5TiO_2$	1380	1.64	62	40
$0.6[Ph_2SiO].0.4TiO_2$	1370	1.64	86	25
$0.65Epoxy.0.35TiO_2$	1450	1.55	106	180
$0.8Epoxy.0.2TiO_2$	1370	1.52	141	80
$0.8Epoxy.0.2ZrO_2$	1410	1.51	135	95

$$\equiv Ti—OH + RO—Si\equiv \;\rightarrow\; \equiv Ti—O—Si\equiv + ROH$$

to form Ti—O—Si bridges and thus incorporate titanium in the network.

Internal polymerization can also be effectively achieved by means of the epoxy group, according to

For this purpose, the commercially available γ-glycidyloxypropyltrimethoxy silane

is suitable. In an acid environment, water addition causes the epoxy ring to open and the corresponding glycol is formed. This can be avoided by proceeding with great care and by the use of a catalyst; $Ti(OR)_4$ and $Zr(OR)_4$ have been used. The epoxy silane is mixed with the ester in a one-to-one molecular ratio and 1/16 of the necessary water is added in a controlled manner by attaching it to silica gel. After stirring for 2 h a further 2/16 of the water is added. To this precondensate the remaining stoichiometric water quantity is added in the form of dilute HCl, the mixture is stirred for 2 h at room temperature and then heated for 30 min to 70 °C. After removal of the volatile components under vacuum the residual highly viscous fluid is cast into a mold and cured at 150 °C to form a hard monolith. The product is highly scratch resistant, even more so when $Ti(OR)_4$ is used rather than $Zr(OR)_4$.

Another strategy for combating highly variable reaction rates of several precursors is to generate the water needed for the reaction within the system itself; for example, by means of an esterification:

$$R^1OH + R^2COOH \rightarrow R^2COOR^1 + H_2O$$

This reaction allows precise control of the hydrolysis. The solvent can be used as the reacting alcohol and the catalyst can serve as the acid. A further advantage of this approach is that the water is formed in solution and is therefore distributed very uniformly. The use of this mode of synthesis has proved itself admirably in the production of several ormosils.

The curing step referred to previously is often the crucial end stage in the production of ormosils. During this step, residual solvent is removed and further condensation and polymerization take place; that is, the hardness increases.

As a further example, the principal steps in the production of a lacquer that has proved to adhere very firmly to antique glass surfaces will be cited. 65 mol% Ph_2SiCl_2 and 32.5 mol% $MeViSiCl_2$ (methylvinyldichlorosilane) are reacted under reflux in ethanol before the addition of 2.5 mol% $Si(OEt)_4$. After two hours, the solvent is removed under vacuum and a clear fluid of low viscosity remains. This fluid is washed with ethanol to neutral pH and then treated for 30 min with aqueous HCl at 70 °C. Evaporation is followed by curing for 2 h at 150 °C in a vacuum. The product is a very viscous resin that can be turned into a lacquer solution of low viscosity with acetic acid ethyl ester.

3. Properties

Since this family of materials has only recently been developed, few property measurements are available; in particular, hardly any measurements have been made with systematically varying composition. This state of affairs, together with the lack of knowledge about the structure of ormosils, prevents any confident assertions or predictions about properties and their dependence on composition. Nevertheless, some general relationships can be recognized. Table 1 lists some properties of glassy ormosils. These have been selected to include both heteroatoms (titanium or zirconium) and functional organic groups.

3.1 Aggregate Condition

The introduction of organic radicals signifies the creation of breaks in the silicate network. It is known for inorganic glasses that this causes them to soften at much lower temperatures. For example, the transformation temperature of silica glass falls from 1200 °C for the pure product to 440 °C for a glass containing one Na^+ ion per $[SiO_4]$ tetrahedron; that is, of composition $NaSiO_{2.5}$. The corresponding methylpolysiloxane $[CH_3SiO_{1.5}]_n$ is already resinous at room temperature, while dimethylpolysiloxane $[(CH_3)_2SiO]_n$ has a glass transformation temperature of -120 °C.

These figures show that organic radicals have a much more powerful softening action than inorganic network modifiers. Clearly, this action is greater the higher the concentration of such radicals. Further, the softening action is enhanced with increasing size of the organic radical. Thus the replacement of a methyl group in a glassy ormosil by an ethyl group often leads to an ormosil with resinous or highly viscous character. This influence can be interpreted in terms of the large spatial requirements of organic radicals, particularly notable in the case of the phenyl group which generates thermoplastic behavior and favors solubility in many organic solvents. Such solubility is greater the higher the content of difunctional silanes which have a reduced degree of network formation.

Table 1 exemplifies the great variability of the glass transformation temperatures of the ormosils. These temperatures show a certain degree of regularity when they are plotted as a function of the number of silicon–carbon bonds per network former. This plot yields, to a first approximation, a single curve. Deviations from this master curve indicate special influences.

3.2 Density

The low density of ormosils was emphasized in Sect. 1 and in this section the large space requirement of organic radicals has been stressed. This is confirmed by the values in Table 1. It should be particularly noted that increasing organic cross-linking through incorporation of epoxy silane leads to reduced density.

The open structure of the ormosils causes a noticeable permeability to gases. At room temperature for water vapor this is about $10^{-11}\,kg\,m^{-1}\,h^{-1}\,Pa^{-1}$ and for oxygen about $10^{-12}\,kg\,m^{-1}\,h^{-1}\,Pa^{-1}$.

3.3 Thermal Expansion Coefficient

An open structure generally implies a weak structure, marked by a high thermal expansion coefficient. The ormosils act in accordance with this generalization, as Table 1 shows; that is, ormosils have relatively large expansion coefficients. Whereas common calcium–sodium–silicate glasses have expansion coefficients of barely $10^{-5}\,K^{-1}$, this rises to about $10^{-4}\,K^{-1}$ for methylpolysiloxane and to as much as $3 \times 10^{-4}\,K^{-1}$ for dimethylpolysiloxane.

3.4 Optical Properties

As a result of their mode of synthesis, ormosils are noncrystalline and completely transparent in the visible region. With appropriate processing technique, good homogeneity is attainable.

However, it has already been mentioned that it is sometimes difficult to produce 100% dense monoliths, especially when the proportion of organic components is small. In such cases, the residual porosity renders the products opaque.

The refractive indices (see Table 1) are in the approximate range 1.5–1.6. It is advantageous that the refractive index is slightly variable so that products can be fitted to particular uses.

3.5 Mechanical Properties

In comparison with silicate glasses, the hardness of ormosils (between 3 and 5 on the Mohs scale) is low. Here also, it turns out that the type of organic radicals has a disproportionate effect on some mechanical properties. Thus, the tensile strength of an ormosil with 80 mol% of the previously mentioned epoxysiloxane and 20 mol% $Ti(OR)_4$ is $2\,MN\,m^{-2}$. Internal cross-linking by components containing vinyl groups leads to an increase in tensile strength to $5\,MN\,m^{-2}$.

Internal cross-linking also raises the scratch resistance to values that lie distinctly above those for the best polymers. Scratch resistance is also improved by curing for longer times or at higher temperatures.

3.6 Surfaces

As a result of their organic radical content, ormosils are, as a rule, hydrophobic. They can be made hydrophilic by incorporating organically bound OH groups, preferably glycol groups, via an epoxy compound. This will be clarified in Sect. 4.

Many ormosils show good adhesion to siliceous substrates. It is assumed that this is due to a direct binding to the always-available OH groups on the substrate, according to

$$\equiv Si\!-\!OR_{ormosil} + HO\!-\!Si\!\equiv_{glass}$$
$$\rightarrow \equiv Si\!-\!O\!-\!Si\!\equiv_{bound} + ROH$$

A bond of this kind is stable in the presence of moisture.

4. Applications

Individual opportunities for application have already been cited in preceding sections. When the advantages of the sol–gel process are considered together with the supplementary benefits derived from the introduction of organic components, a very broad field of possible applications can be detected.

The sol–gel process leads to the following benefits:

(a) preparation of products at relatively low temperatures;

(b) homogeneous mixing of several components;

(c) great purity of components; and

(d) exact dosage of trace elements.

Ormosil materials offer the further benefits of:

(e) consistency ranging from fluid to rigid, that is, from lacquer to a scratch-resistant layer;

(f) improved mechanical properties, for example, less brittleness;

(g) microstructure varying from porous to dense, for example, with a very high specific surface area; and

(h) feasibility of specific chemical reactions, useful when considered with a view to incorporating particular compounds.

The ormosils thus represent a class of materials the properties of which can be tailor-made within wide limits. It should, however, be noted that such tailoring requires much chemical finesse, because the individual reaction steps and their dependence on experimental variables have been little explored. Nevertheless, in several respects there is now a worthwhile degree of knowledge.

4.1 Sealings and Coatings

Sealing to glass surfaces presents several problems. The glass surface itself has properties dependent upon its history: it will be leached to a greater or lesser extent and will contain adsorbed water as well as Si—OH groups. Conventional sealing materials are of organic nature and are hydrophobic; that is, there are problems with wetting as well as adhesion problems.

These problems can be resolved with a sealing medium made from a precursor mix of 5 mol% $Si(OEt)_4$, 60 mol% $Ph_2Si(OH)_2$, 5 mol% $Ti(OEt)_4$ and 30 mol% $MeViSi(OEt)_2$. This material is soluble in acetone because, owing to the high proportion of difunctional silicon, the network is highly incomplete. It is suitable for coating aluminum foil and hot sealing this to glass containers; the phenyl group brings about the requisite thermoplastic properties. The material adheres very well as a result of the reactive bonding of the Si—OH groups of the glass surface to the residual Si—OH or Si—OR groups of the lacquer.

The hardness of the sealing materials is determined by (and can be adjusted by altering) the

proportion of $Si(OR)_4$, which makes for rigid network formation. Against that, the flexibility can be adjusted by controlling the proportion of $MeViSi(OR)_2$, because during the required thermal curing an internal cross-linking between the methyl and vinyl groups is established, according to

$$\equiv Si\text{—}CH_3 + CH_2\!\!=\!\!CH\text{—}Si\equiv$$
$$\rightarrow \equiv Si\text{—}CH_2\text{—}CH_2\text{—}CH_2\text{—}Si\equiv$$

The thermal curing also lowers the OH content; this effect must be optimized so that, according to

$$\equiv Si\text{—}OH + HO\text{—}Si \rightarrow \equiv Si\text{—}O\text{—}Si\equiv + H_2O$$

inorganic polymerization does not go too far. The OH content can also be regulated by inclusion of $Ti(OR)_4$, making use of

$$\equiv Si\text{—}OH + RO\text{—}Ti\equiv \rightarrow \equiv Si\text{—}O\text{—}Ti\equiv + ROH$$

to remove OH in a reactive manner.

The dependencies just described are equally valid in the development of coatings. Here, good scratch resistance is usually also required. Investigations have established that scratch-resistant coatings with a Mohs hardness around 4 can be obtained if, in addition to $Si(OR)_4$ and $Ti(OR)_4$, a further organic component in the form of epoxysilane is incorporated. It is necessary to precondense with carefully controlled addition of the stoichiometric amount of water to the point where the desired viscosity is attained. After coating, hardening is achieved in a few minutes at 110–120 °C. Such coatings are appropriate for the improvement of the scratch resistance of organic polymers.

For such products there are numerous applications. It is advantageous that appropriately developed lacquers are soluble in organic solvents even after use; for example, this is very important when old works of art are being restored.

The fact that these lacquers show some permeability of water vapor (see Sect. 3) must be taken into account and thus objects sensitive to moisture (e.g., ancient stained glass windows) cannot be completely protected by these lacquers. By admixture of lamellar inorganic barrier layers (e.g., mica or glass flakes) permeability to water vapor can be reduced to such a degree that the corrosion rate of the protected glass is reduced by a factor of 30. Ormosil lacquers have the further advantages for this application that they adhere very well (see Sect. 3) and that it is easy to match the refractive index (see Sect. 3).

4.2 Contact Lenses

Materials for contact lenses must satisfy many requirements. In addition to adequate strength and flexibility and appropriate optical properties, good wetting, high capacity to absorb water and good permeability to oxygen are all demanded. These requirements can be satisfied in large measure by ormosils. The starting point is the frequently mentioned epoxysilane, which after opening of the epoxy ring to produce glycol according to

$$-\overset{|}{C}H-CH_2 + H_2O \rightarrow -\overset{|}{C}H-\overset{|}{C}H$$
$$\underset{O}{\diagdown\diagup} \qquad\qquad \underset{OH\ \ OH}{|\ \ \ |}$$

gives expectations of good hydrophilic qualities. Epoxysilane alone, however, after hydrolysis and condensation, yields only a gel-like product that breaks apart into many pieces. Stronger network formation is therefore needed and this can be achieved by introducing $Si(OR)_4$. Hydrolysis must be performed in such a way the epoxy group does not open up prematurely. This can be ensured by treatment with HCl under reflux, which leads to an anhydrous condensation:

$$\equiv Si-OR + HCl + RO-Si\equiv$$
$$\rightarrow \equiv Si-O-Si\equiv + RCl + ROH$$

Condensation is then completed with aqueous HCl. Drying at 100 °C then leads to a clear solid product that, however, still tears into many pieces. This is mainly due to shrinkage, which can be prevented by replacing $Si(OR)_4$ by $Ti(OR)_4$. The flexibility can be increased by favoring an internal polymerization, for which purpose the use of methacryloxysilane in combination with monomeric methacrylate has proved effective.

Contact-lens materials based on ormosils thus consist of a precondensate of 90 mol% epoxysilane, 5 mol% methylacryloxysilane and 5 mol% $Ti(OEt)_4$, into which 20–30 mol% of monomeric methacrylate has been mixed together with catalytic quantities of a peroxide polymerization catalyst. Hardening follows at 150 °C. No appreciable shrinkage is observed.

The development of this contact-lens material is also a model for other possible uses of ormosils. It is to be expected that further developments will make large strides, not least in the direction of new materials that combine properties characteristic of inorganic and organic substances. It is particularly desirable that the development of such materials should go hand in hand with a more acute clarification of the reaction mechanisms involved in their synthesis.

See also: Glasses: Agricultural and Vetinary Applications

Bibliography

Andrianov K A, Zhdanov A A 1958 Synthesis of new polymers with inorganic chains of molecules. *J. Polym. Sci.* 30: 513–24

Avnir D, Kaufman V R, Reisfeld R 1985 Organic fluorescent dyes trapped in silica and silica–titania thin films by the sol–gel method. Photophysical film and cage properties. *J. Non-Cryst. Solids* 74: 395–406

Dislich H 1986 Sol–gel: Science, processes and products. *J. Non-Cryst. Solids* 80: 115–21

Gulledge H C 1950 Titanated organo–silicon–oxy compounds. US Patent No. 2,512,058

Mackenzie J D 1982 Glasses from melts and glasses from gels, a comparison. *J. Non-Cryst. Solids* 48: 1–10

Parkhurst C S, Doyle W F, Silverman L A, Singh S, Anderson M P, McClurg D, Wnek G E, Uhlmann D R 1986 Siloxane modified SiO_2–TiO_2 glasses via sol–gel. *Mater. Res. Soc. Symp. Proc.* 73: 769–73

Philipp G, Schmidt H 1984 New materials for contact lenses prepared from Si- and Ti-alkoxides by the sol–gel process. *J. Non-Cryst. Solids* 63: 283–92

Ravaine D, Seminel A, Charbouillot Y, Vincens M 1986 A new family of organically modified silicates prepared from gels. *J. Non-Cryst. Solids* 82: 210–19

Sakka S, Tanaka Y, Kokubo T 1986 Hydrolysis and polycondensation of dimethyldiethoxysilane and methyltriethoxysilane as materials for the sol–gel process. *J. Non-Cryst. Solids* 82: 24–30

Schmidt H 1985 New type of non-crystalline solids between inorganic and organic materials. *J. Non-Cryst. Solids* 73: 681–91

Schmidt H, Scholze H, Kaiser A 1984 Principles of hydrolysis and condensation reaction of alkoxysilanes. *J. Non-Cryst. Solids* 63: 1–11

Schmidt H, Scholze H, Tünker G 1986 Hot melt adhesives for glass containers by the sol–gel process. *J. Non-Cryst. Solids* 80: 557–63

Schmidt H, Seiferling B 1986 Chemistry and applications of inorganic–organic polymers (organically modified silicates). *Mater. Res. Soc. Symp. Proc.* 73: 739–50

Scholze H 1985 New possibilities for variation of glass structure. *J. Non-Cryst. Solids* 73: 669–80

Scholze H, Schmidt H 1980 Process for the production of silicic and heteropolycondensates useful as membranes and adsorbents. US Patent No. 4,238,590

Tünker G, Patzelt H, Schmidt H, Scholze H 1986 New approaches to the preservation of historic glass windows (in German). *Glastech. Ber.* 59: 272–8

Voronkov M G, Mileshkevich V P, Yuzhelevskii Y A 1978 *The Siloxane Bond.* Plenum, New York

Wilkes G L, Orler B, Huang H 1985 "Ceramers": Hybrid materials incorporating polymeric/oligomeric species into inorganic glasses utilizing a sol–gel approach. *Polym. Prepr.* 26: 300–2

Wills R R, Markle R A, Mukherjee S P 1983 Siloxanes, silanes, and silazanes in the preparation of ceramics and glasses. *Amer. Ceram. Soc. Bull.* 62: 904–15

Yoldas B E 1983 Modification of glass by non-compositional variations in the ultrastructure. *Glastech. Ber.* 56K: 1052–6

Zarzycki J 1985 The gel-glass process. In: Wright A F, Dupuy J (eds.) 1985 *Glass—Current Issues.* Nijhoff, Dordrecht, Netherlands, pp. 203–23

Zelinski B J J, Uhlmann D R 1984 Gel technology in ceramics. *J. Phys. Chem. Solids* 45: 1033–50

H. Scholze
[Würzburg, FRG]

P

Polyesters and Polyamides

The most widely used biomedical polyesters are the linear aromatic and aliphatic thermoplastic polymers, such as poly(ethylene terephthalate) (PET or Dacron), poly(glycolic acid) (PGA), poly(lactic acid) (PLA), polydioxanone, and poly-(glycolide–lactide) and poly(glycolide–trimethylene carbonate) copolymers. The Dacron polyesters manufactured by Du Pont are types 53, 68 and 73. The molecular weight of poly(ethylene terephthalate) that is capable of making Dacron fiber is about 20 000. Another commercially important thermoplastic polyester poly(butylene terephthalate) has a potential use in medical filters, medical disposables, absorbents, industrial dust and fluid filtration, and battery separators. Most linear aliphatic polyesters are biodegradable in physiological environments and their major uses as biomaterials are in wound closure, as components of surgical implants and in drug control/release devices. Several new aliphatic biodegradable polyesters are commercially available for biomedical uses. They are poly(β-hydroxy-butyrate) (Holmes 1985) and polyorthoesters (Sendelbeck and Girdin 1985).

Polyamides are equally successful as biomedical materials. Nylon is generally classified into two groups: aliphatic and aromatic. Numerous types of Nylon 66 such as type 702, 704, 705, 714 and 728 are available, with molecular weights of approximately 12 000–20 000. The two best-known aromatic polyamides, or aramids, are Nomex and Kevlar (Reimschuessel 1985). The types of Nomex currently offered are 430, 431, 433 and 434; the types of Kevlar are 29 and 49.

A polymer consisting of both ester and amide groups has been synthesized and tested both *in vitro* and *in vivo* (Barrows 1989). It appears to be a promising new class of degradable polymers.

1. Structure

In the aliphatic polyester, the polyglycolide chain conformation is a planar zigzag parallel to the *ac* plane. This is called "sheet structure." In the poly(glycolide–lactide) copolymer, the regularity of the chain is disturbed by $-CH_3$ groups. In the aromatic polyesters, particularly PET, the molecule has an extended rectilinear shape with the plane of the benzene rings parallel to the (100) plane. Since the chain is rigid and less flexible than in nylon and polyethylene, the melt-spun filaments are largely amorphous before drawing. The interplanar spacings, crystallinities and crystalline sizes of eight commercial PET fibers were recently determined by using a graphic multiple-peak resolution method of wide-angle x-ray scattering data. The physical structure of PET is also dependent on the winding speed during the melt-spinning process (Heuvel and Huisman 1978), physical aging and zone annealing. A zone-annealing method was able to make high-modulus high-strength PET film (Young's modulus of 14.5 GPa, tensile strength of 852.5 MPa). This was attributed to the dense packing of highly oriented amorphous chains.

Unlike PET, nylon fiber is fairly crystalline when it is spun. The crystal structures of aliphatic nylon fall, in general, into three categories: α, β and γ phases. The α form is the most stable physical structure for both Nylon 66 and Nylon 6 and consists of stacks of sheets of planar hydrogen-bonded chains in extended zigzag conformation (Reimschuessel 1985). Nylon 66 exists mainly in the α phase with the molecules in parallel alignment with the adjacent fully extended zigzag molecules (Reimschuessel 1985). Nylon 6 also exists in the γ phase and the molecules assume a puckered conformation, although the predominant structure of Nylon 6 is the α form with antiparallel alignment of the extended chain segments. The dimensions and shapes of Nylon 6 macrofibrils, microfibrils, amorphous and crystalline domains of microfibrils, and the spacing between the microfibrils were determined by $SnCl_2$-stained transmission electron microscopy. Among the aromatic polyamides, Kevlar 49 has a pseudo-orthorhombic crystal with unit cell dimensions of $a = 0.778$ nm, $b = 0.528$ nm and $c = 1.29$ nm. The aromatic rings orient in the direction of the fiber axis and give the fiber chemical stability and mechanical stiffness. Although hydrogen bonds link adjacent molecules within the same plane, they are considerably weaker than the covalent bonds along the direction of the fiber. However, there is no bonding between the two adjacent planes of molecules.

The temperature dependence of dye uptake has been used to probe the fine fiber structure of both Nylon 66 and Nylon 6. An anisotropy index defined as the ratio of the diffusion coefficient perpendicular to the fiber to the diffusion coefficient parallel to the fiber is able to distinguish any difference in fiber structure between Nylon 66 and Nylon 6 arising from different processing conditions.

2. Materials Fabrication Technology

Almost all aromatic polyesters are polymerized from a diol and a dicarboxylic acid. In the case of PET, the polymerization is conducted in two stages: the formation of low-molecular-weight oligomers and the polycondensation of these oligomers. PGA and PLA are synthesized by ring-opening polymerization of glycolide or lactide with organometallic catalysts. The resulting polyester chips are melt spun and hot stretched to about five times their original length.

Nylon 66 is polymerized from the polycondensation of adipic acid and hexamethylene diamine, while Nylon 6 is made from caprolactam by ring-opening polymerization. The resulting Nylon 66 or Nylon 6 chips are then melt spun and cold drawn to 3.5–4 times their original length. In the wet spinning of Nylon 66 and Nylon 6, fluted void structures were frequently observed and these were originated in a fingering phenomenon occurring at the interface. Aromatic polyamides are chiefly made from low-temperature polycondensation processes which have two principal types: interfacial and solution polycondensations.

3. Surface Characteristics and Treatment

The surface geometry of polyester and polyamide fibers is relatively simple and is characterized by a hydrophobic smooth surface with low reactivity. Polyester and nylon fibers have a circular cross-sectional shape. In biomedical uses, particularly vascular grafts, the polyester fibers are frequently texturized and the grafts crimped to make them soft, bulky and resilient.

In reference to surface wetting, nylon fiber has a higher work of adhesion ($0.106 \, J \, m^{-2}$). Both fibers have a similar hysteresis ratio.

4. Properties Relevant to Implantation

4.1 Mechanical Properties

In general, nylon is more extensible than polyester even though their tenacity is similar. Polyester fibers recover well from stretching, and similarly from bending, shearing and compressing. Polyester fibers have significantly higher moduli of elasticity than nylon, but nylon fibers exhibit better recovery. The mechanical properties of these fibers change after texturization and crimping.

4.2 Chemical Properties

Nylon is chemically stable. It is not seriously affected by diluted acids, but is hydrolyzed by concentrated acids like HCl. Nylon shows remarkable stability to alkalis. Aramids such as Nomex and Kevlar exhibit similar chemical properties.

Polyesters are more resistant to acids than polyamides. They are also more resistant to weak alkalis than to strong ones. Most alcohols, ketones, soaps and detergents have no chemical action on polyesters. In the body, aromatic polyesters are found to be more resistant to biodegradation than polyamides. Aliphatic polyesters, however, are easily degraded hydrolytically.

Contrary to the general belief that aromatic polyesters, such as poly(ethylene terephthalate), and aliphatic polyamides, such as Nylon 66 and Nylon 6, are stable in the presence of enzymes, an *in vitro* study of these polymers in several enzymes indicated that PET was affected by esterase and papain only, and Nylon 66 was affected by papain, trypsin and chymotrypsin.

4.3 Biocompatibility

In general, polyamides and polyesters are considered to be quite biocompatible with biological tissues. Numerous toxicity studies show that Nylon 66 and Nylon 6 inhibit the growth of L-929 mouse fibroblast cells only slightly or not at all. They are in cytotoxicity class I: noncytotoxic when compared with PGA control. Less lymphocytic foci have been found around nylon than around polyethylene and poly(methyl methacrylate). Nylon 66 shows deposits of thrombocytes and fibrin within 21 days of implantation. It has also been reported that Nylon 66 has an intermediate blood compatibility, while Nylon 6 has a poor one. Thus, nylon is considered to be biocompatible with tissue over short periods, but often degrades over extended periods.

Controversial data about biocompatibility of aromatic polyamides such as Kevlar have been reported and the issue has not been fully resolved. Kevlar fiber exhibits low tissue reactivity similar to that of Dacron. A study of Kevlar 49 fibers in rabbits' muscles indicated that the tissue response was similar to silicone control and hence was a biocompatible fiber.

The biochemical response of body tissues to a Dacron vascular prosthesis has been examined. The tissue compatibility of Dacron seems very good and the living body adapts to it well. A study on the healing characteristics within porous Dacron implants (Dacron velour fabric) in rabbits revealed that the newly generated connective tissue inside Dacron velour fabric was not identical to the normal connective tissue in terms of ground substance, collagen and reticular fibers after 10 days or 28 days of healing. This supports the hypothesis that healing within porous implants differs from that in normal connective tissues. The cell adhesion rate of polyester film has been found to depend on the total surface free energy of the film. The geometrical configuration of polyester fabric and collagen coating have been reported to play an important role in cell attachment. Scanning electron micrographs show that a tightly knit configuration produces maximum cell coverage, whereas loose knits and velour do not support cell growth. In a similar study, it was found that the geometric structure of fabrics influenced the uniformity of collagen coating, with woven and knitted Dacron fabric being least uniform and spun-bonded polyester being most uniform. The pattern of collagen deposition appeared to influence the adhesion and spreading of the cells on the substrate. Both the $CaCO_2$ and BHK fibroblasts were preferentially located on the regions of the greatest collagen deposition. BHK fibroblasts were capable of attaching to and growing on single filaments of the yarns. In terms of blood compatibility, woven Dacron used as angiographic catheters in dogs has been shown to be the poorest performer with respect to thrombotic and mean fibrinopeptide A deposits when compared with catheters made of polyethylene, polyurethane and polytetrafluoroethylene (PTFE). This parallels the findings in the *in vitro* study of vascular-graft-associated leukocyte adhesion. In that study, crimped Dacron Bionit showed higher complement-associated leukocyte adhesion than expanded PTFE or silicone rubber in the presence of heparinized whole blood. Since adherent leukocytes may be related to thrombosis and to the organization and healing of the pseudo-intima of vascular grafts, initial interaction between leukocytes and vascular grafts may be important in total blood biocompatibility.

Protein and platelet absorption on polymer surfaces have been frequently associated with the blood biocompatibility of polymers. The major conclusion from the existing studies, based on the surface hydrophobicity of the substrate and absorbed components in the blood, is that the degree of protein adsorption can be predicted on thermodynamical grounds. Polymers with higher surface tensions (i.e., more hydrophilic) such as Nylon 66 and PET have a lower amount of blood protein adsorption than those with lower surface tensions (i.e., more hydrophobic) such as Teflon.

Endothelial cells seeded grafts have been used to improve the patency rates and to reduce the thrombogenic character of synthetic vascular grafts like Dacron. The extent of surface hydrophobicity of a polymer substrate has also been found to be related to its degree of endothelialization.

All of the commercially available linear aliphatic polyesters and their copolymers, such as PGA, poly(glycolide–lactide) and poly(glycolide–trimethylene carbonate) copolymers, and polydioxanone, show very good to excellent biocompatibility in various types of tissues despite their biodegradability (Chu 1991). Their tissue reactions are comparable to some of the well-known biocompatible polymers such as polyethylene, polypropylene and Nylon 66. Aliphatic polyesters, such as polylactide, are able to enhance wound healing and soft tissue regeneration. Similar effects of this class of linear aliphatic polyesters on osteogenic potential have also been found.

4.4 Water Absorption

The presence of large quantities of water in biological tissues makes the moisture property of biomedical polymers important. Under 65% relative humidity at 21 °C, the moisture regains of both nylon and Dacron are low: 4.0–4.5% and 0.4–0.8%, respectively. This suggests that the absorbed water molecules are mainly on the surface of the polymers. Linear aliphatic polyesters and their copolymers, however, are expected to have better water absorption than their aromatic counterparts. For example, PGA sutures exhibited an average of 44% water uptake after one hour of immersion in distilled H_2O.

5. Applications

Polyester and polyamide are mainly used where firm connective tissue fixation is desirable. Their main applications are in wound closure, orthopedics, cardiovascular areas, body-wall covering, and drug control/release devices.

5.1 Wound Closure Applications

Polyester and polyamide play a very important role in wound closure management. The four synthetic absorbable sutures (Dexon, Vicryl, PDS and Maxon) are made from linear aliphatic polyesters and their copolymers. PET, Nylon 66 and Nylon 6 are used to make nonabsorbable sutures.

PGA and its lactide copolymer sutures are stronger and retain their strength better *in vivo* than catgut sutures. They also elicit much less tissue reaction. Unlike catgut sutures, they are degraded by hydrolysis and hence are pH dependent.

Polydioxanone (PDS) sutures retain strength much longer than PGA and its lactide copolymer sutures: more than half of its original tensile strength is retained after four weeks of implantation, whereas no strength is retained in either PGA or Vicryl sutures. The degradation of PDS sutures, however, was found not to be influenced by enzymes as profoundly as PGA sutures and only a slight enzymatic effect was found when the suture was γ irradiated at a low dosage level. The formation of surface cracks and chips on PDS sutures and their subsequent peeling off from the suture were the major morphological changes upon hydrolytic degradation.

Poly(glycolide–trimethylene carbonate) copolymer sutures (Maxon) have a degradation rate similar to PDS sutures. Experiments have shown retentions of 81%, 59%, and 30% of original tensile breaking strength at 14, 28 and 42 days of implantation, respectively (Katz et al. 1985). It would take seven months for a 2/0 size suture to be completely and grossly absorbed.

Both coated and uncoated PET sutures are available, but the former shows less tendency to drag in tissue and a better knot property. Carbon-coated Dacron sutures have also been evaluated, but performance was not significantly better than the uncoated type. Carbon-coated polyester sutures showed a greater degree of tissue ingrowth than both uncoated and polybutylate-coated polyester sutures.

After an initial drop of tensile strength, nylon sutures retain a relatively constant strength.

Monofilament nylon sutures are slightly better than multifilament types. Their use in ophthalmology has been questioned, because of the observed transverse surface cracks due to degradation. A silver-compound-coated multifilament nylon suture has been reported to have good antibacterial property when tested *in vitro* and to exhibit biocompatibility in rats similar to the commercial nylon suture, Nurolon.

Synthetic absorbable linear aliphatic polyesters (mainly PGA, its lactide copolymer and PDS) have also been used as vascular ligating clips or surgical staples (Wheeless 1986). These are Lactomer (70 wt% lactide–30 wt% glycolide copolymer), Absolok (PDS) and Polysorb. They have the advantages of being radiotransparent so as to not interfere with CT diagnosis, and are more degradable than conventional metal clips, thus eliminating chronic foreign body reactions. In addition, the fast closure of wounds and the ability to reach inaccessible locations make absorbable staples and clips attractive.

The degradation and tissue reactions of biodegradable ligating clips and staples are similar to their corresponding absorbable sutures, except for their rate of degradation. Contrary to suture degradation, Absolok clips degraded significantly faster than Lactomer clips. The two mating surfaces of a clip appeared to be the site most prone to degradation. All absorbable ligating clips showed parallel surface fissures after being exposed to buffer solutions or tissues. The extent of tissue adhesion onto the surface of these clips had been suggested to influence their degradation, because of the possible blockage of the exposure of the clips to enzymes by fibrous capsules. The presence of bacteria also slowed down the degradation of absorbable clips in terms of their surface morphologic change. This is similar to the reported bacterial effect on PGA suture degradation.

The adequacy of using absorbable clips in the vicinity of the reproductive organs of patients who want to retain their reproductive potential, however, was questioned because of the reported frequent tissue adhesion in the uterine horns.

5.2 Orthopedic Applications

In the orthopedic area, Dacron alone or combined with other materials has been used for tendon reconstruction. A model synthetic tendon consisting of 20 vol.% textured Dacron fibers in a poly(2-hydroxyethyl methacrylate) matrix has been proposed. By varying the fiber-to-matrix ratio, it is possible to adjust the required mechanical properties of the tendon. Another composite tendon prosthesis consists of a Dacron tube or a tape impregnated with a silicone rubber to produce a smooth flexible nonporous central region. It is based on the concept of a passive gliding tendon system and the prosthesis acts as a template for the development of a mesothelium like sheath capable of producing a liquid similar to synovial fluid and it has been moderately successful. Dacron fabrics of four different weave configurations (felt, velour, double velour and braid) have also been used. At the musculotendinous junction, the strength of the ingrowth (and hence of adhesion) into these fabrics was found to be related to the surface area and internal porosity of the implant, but not to the surface morphology. A 10 mm Dacron aortic graft has been used to reconstruct the patellar tendon, and an excellent return of function has been reported due to the strength of the Dacron graft and the formation of a new tendinous sheath. Dacron mesh in tube form covered by a silicone sheet has been used to reconstruct the achilles and patellar tendons, but the results have been unsatisfactory. Segments of knitted Dacron arterial graft have been employed to construct digital pulleys after laceration of a flexor tendon.

Dacron fabrics in either woven, knitted or velour form have been used in the prosthetic replacement of ligaments, particularly anterior cruciate ligament. Unfortunately, the long-term results have been unsatisfactory due to the insufficient strength of the materials and variability in the placement of the prosthesis. Bicomponent braids of Dacron–Kevlar and Dacron–Lycra have been proposed as a new design concept for anterior cruciate ligament replacement. The strength of the braid can be adjusted by the incorporation of chemically different bicomponent materials and by the changing of the composition of the biocomponents, the helix angle of the sheath and the core material. A loosely knitted Dacron–Hydron sponge composite has been used for cruciate ligament substitution, the results being mixed. Other orthopedic uses of Dacron include the repair of an acromioclavicular separation, the formation of a bony union between the ulnar and radius, and the filling of a bone gap.

Most aliphatic nylons are unsuitable for orthopedic use, because they deteriorate over the long term. Despite this mechanical deficiency, molded nylon non-weight-bearing prostheses have been used, but these were unsuccessful due to the formation of wear particles. Nomex fiber has been used as the core material of a cruciate stent for providing temporary knee stability; the results were satisfactory in only 50% of anterior cruciate stents and in 80% of posterior cruciate stents. Aromatic polyamides like Kevlar, however, have been tested as the core material of an active gliding tendon prosthesis. It appears to be a promising material, because of its strength, strength retention and biocompatibility. Other uses of nylon include the replacement of the femoral head, intramedullary nails and membrane for arthroplasty of the knee in chronic arthritis.

Biodegradable polyesters, such as PGA, PLA and

their copolymers, are becoming increasingly important in orthopedic use. They have been used either alone or with other nonabsorbable materials to repair injured tendons or ligaments (Rodkey et al. 1985). The biodegradable materials can be used either as a coating on nonabsorbable materials such as carbon fiber to improve the handling of the device or as a tissue scaffold for promoting the regeneration of injured tissues.

Degradable polyesters have also been molded as bone plates and screws for the internal fixation of injured bone or as the fillers for defected bone. The main obstacle in using biodegradable polymers in this area is in achieving an adequate degradation rate such that the device can maintain suitable mechanical properties for at least four months, thus allowing adequate mineralization of bone callus. The main advantage of using biodegradable polymers for bone fixation is that additional surgery is not needed to remove the device once the fracture is healed.

Biodegradable polyesters and copolymers have also been found in many other orthopedic uses; for example, as a PLA intramedullary plug for holding the bone cement in the femur during total hip replacement, as a tendon gliding device for minimizing tissue adhesion and as a spacer for inhibiting scar formation after laminectomy.

5.3 Cardiovascular Applications

The most successful medical application of Dacron is in the area of cardiovascular surgery, including the prosthetic heart valve and vascular graft, where it is generally more acceptable than polyamides. In the prosthetic heart valve, Dacron cloth (velour) covers the orifice and cage of the heart valve in order to encourage tissue ingrowth for better anchorage and thrombembolic resistance. A great deal of success has been reported. Dacron felt is used as the surface of the artificial atria in the Jarvik-5 heart. The fabric encapsulation of a prosthetic heart valve is, however, restricted to those valves with rigid retaining cages and movable occluders as opposed to those with flexible leaflets. In addition, complications such as restenosis due to excessive tissue ingrowth, severe hemolysis and deterioration of the cloth have been observed.

At present, Dacron fabric is still the major constituent of vascular prostheses (Guidoin et al. 1983). Dacron vascular grafts are classified according to their porosity as either microporous or macroporous. They can be constructed in woven and conventional knit, and velour configurations. Dacron vascular grafts are also available in crimped and noncrimped form. Tightly woven Dacron crimped grafts require minimal preclotting and, hence, can be handled easily and directly. They are mainly used in the thoracic aorta, where bleeding is a primary concern, with high rates of success.

However, their relative rigidity and lack of conformity of the woven graft make them unsuitable for small arteries or for a bypass across the knee joint. It is also difficult to suture. A bicomponent woven graft of Dacron and Spandex fibers has been reported to reduce the problems of rigidity and lack of conformity of conventional woven grafts. The graft was able to change its diameter in response to changes in blood flow and pressure after implantation in dogs.

Knitted Dacron grafts are easier to suture, more comfortable and heal better. The first knitted vascular graft was reported by DeBakey in 1957. A smooth-walled heavy knitted Dacron, it served very well as a replacement for large-caliber arteries, but relatively poorly for medium-caliber arteries. In 1962, a lighter-weight very-smooth-walled warp-knitted Dacron prosthesis called Weavenit was introduced, which had better suturability and more conformability. A still lighter weft-knitted Dacron prosthesis called Microknit was introduced in 1966. It was even more suturable and conformable, but preclotting was occasionally difficult. These prostheses are preferred both for bypass procedures and for replacement in aneurysmal and occlusive diseases of the abdominal aorta, visceral arteries and proximal peripheral arteries. In 1969, the DeBakey Vascular D Graft was introduced. This graft is a weft-knit Dacron with internal velour (loops extend into the lumen). It has been used successfully in the aortoiliac area and has been successful in about 70% of five-year-old cases of femoropopliteal bypass. In the 1970s, Sauvage introduced the Sauvage Filamentous Vascular Prosthesis, which is a weft knit with full-wall filamentousness of differential distribution. The total wall thickness is 500 μm and the effective internal filament height is in the range 25–50 μm. This prosthesis has an efficient fibrin trap and a good tissue trellis, which form a blood-tight but thin wall at preclot and provide maximum potential for tissue ingrowth. It has been used for aortic bifurcation grafts (98.6% patency), aortofemoral grafts (85.6% patency), femorofemoral grafts (86.7% patency), axillaryfemoral grafts (71.1% patency) and femoropopliteal grafts (65.5% patency, above the knee joint level). Their use in below-knee femoropopliteal and femorotibial bypasses, however, was not satisfactory with five year patency ranging from 20% to 40%, and 5% to 10%, respectively. The low patency rate was partially attributed to the kinking and compression of the graft as a result of knee, hip or torso flexion. A knitted noncrimped Dacron filamentous graft with an externally supported coil was developed in 1978 to remedy the problem. It resulted in a significant improvement in patency rate for the below-knee femoropopliteal bypass (five year patency of 63%) and the below-knee femoropopliteal bypass (four year patency of 25%).

265

The concept of using bicomponent fabrics for achieving low bleeding porosity at implantation and high healing porosity during the healing stage was first reported in 1963. However, because of the uneven and ultimate absorption of the collagen component, the loss of structural integrity after collagen absorption or the strong tissue reaction to collagen absorption, this first generation of bicomponent fabrics was not successful, even though some of the specimens had a 500-fold increase in water porosity. Another type of Dacron-tanned collagen prosthesis was constructed so that complete collagen absorption would occur after six weeks.

To advance the composite vascular graft concept further, the incorporation of more biocompatible linear synthetic biodegradable aliphatic polyester fibers (e.g., polylactide, polyglycolide and polydioxanone) as the component yarns in vascular graft fabrics was reported. Ruderman et al. (1972) reported that after approximately 100 days implantation in a canine abdominal aorta, all of the PLA composite grafts were patent with no microscopic indication of calcification. The histological data revealed excellent ingrowth and adherence of external tissue with a very thin well-adhered pseudoendothelium.

Bowald et al. (1980) used poly(glycolide–lactide) copolymer mesh fabrics or a combination of the copolymer and Dacron mesh fabrics as a patch graft to repair cardiac defects and as a tube to replace aortic defects. In all of their cases (pigs), the biodegradable and the composite mesh fabrics allowed the arterial tissue regeneration with the formation of a smooth nonthrombogenic luminal surface following absorption of the copolymer. This regeneration capability existed even with a graft as long as 15 cm. The newly formed tissue wall appeared to have adequate strength throughout the experimental period and was more resistant to lipid and calcium deposition than the fabrics made from Dacron or PTFE.

This arterial tissue regenerative capability over biodegradable scaffolds was further demonstrated by studies in rabbit aortas. Woven fabrics made from either polyglycolide or polydioxanone were implanted in rabbit aortas for 7.5 months. The results indicated that the biodegradable grafts were replaced by regenerating endothelialized blood vessels without thrombosis. The regenerated tissue consisted of an endothelial-lined neointima, circumferentially oriented smooth muscle like myofibroblasts and dense fibroconnective tissue underneath the neointima, and an outer adventitial vascularized connective tissue layer. All regenerated graft specimens were able to withstand three to five times systolic pressure after between one week and three months, irrespective of the continuous degradation of PGA fibers during this period.

A woven fabric made of PGA and Dacron, and coated with poly(lactide–glycolide) copolymer has been implanted in sheep. The PGA component yarns were replaced by well-vascularized host tissue with a mild inflammatory response. Organization of the inner capsule with a pseudoneointima on the luminal side was found at 12 week postimplantation in dogs' abdominal aortas. Smooth muscle cells in the layers closely surrounding the original vascular graft also appeared at this time.

Gogolewski et al. (1983) advanced the composite vascular graft concept to a totally different aspect. Instead of varying the yarns of a fabric in order to achieve biodegradability, they produced a biodegradable microporous prosthetic vascular graft. Formation of the neoartery composed of intima, media and adventitia was found to be associated with the gradual biodegradation of the prosthesis. The neoartery has a cellular structure similar to that of the original natural rat artery, particularly with regard to the presence of elastin fibers in the media.

In an *in vitro* study, Chu and Lecaroz (1987) designed, fabricated and evaluated a series of knitted biocomponent fabrics made of Dacron and PGA for vascular surgery. It was found that the overall properties of these materials depended on the amount of PGA incorporated into the material. The most significant finding was the achievement of an increasing water porosity with time of immersion in buffer solutions. Some of the bicomponent fabrics exhibited close to or greater than 100% increases in water porosity. An *in vivo* study of these materials, however, is needed to evaluate the significance of these findings.

Despite the success of Dacron grafts, they present numerous problems. So far, only one case has been reported in which complete healing at points far removed from the suture line have been observed. An incomplete healing imposes the danger of thrombotic occlusion, particularly in medium- and small-caliber blood vessels and in certain pathological conditions, such as the drop of cardiac output. Guidoin et al. (1983) have examined commercial Dacron vascular grafts of woven, knitted and velour types using canine models. They found that initial water porosity of the graft did not significantly influence the healing process; that is, it is far more host related than material dependent. This finding is contrary to most other studies. A more recent study linked the four types of failure (thrombosis, pseudoaneurysm, dilation and infection) of 22 cases of Dacron vascular devices retrieved from humans to the changes of the chemical and physical properties of the devices. All retrieved prostheses, particularly those exhibiting false aneurysm, infection and thrombosis, showed lower mechanical properties ranging from 2% to 75% of their original values. Similarly, the level of crystallinity of explanted prostheses decreased as a result of implantation and the values ranged from 56% to 100% of their original glutaraldehyde fixed prostheses. Morphologically, breakage of constituent fibers within the yarns and the subsequent migration of the fiber debris were observed. This Dacron fiber deterioration after implantation was also found in lightweight arterial prosthesis after between 3 and 15 years postimplantation.

A composite of host vascular tissue and Dacron velour fabric acting as a reinforcing agent has been suggested, and a successful human use has been recorded. A similar design using a tissue tube containing Dacron fabric in its wall has been reported. Since Dacron is thrombogenic, various coatings such as polyurethane, PTFE, graphitebenzalkonium and heparin have been used. The recent employment of a poly(hydroxyethyl methacrylate) coating on Dacron indicates that an endothelial-like cell layer can be established within 21 days. The prosthesis may lead to an accelerated endothelial development without the need of preclotting.

Dacron velour or flock has been used as the sewing ring and to make an antithrombogenic lining for the artificial heart chambers and the circulating assist pump, in order to promote the pseudoneointima or fibrin layer formation. This can reduce the danger of clotting, but raises a problem for long-term use. For the left ventricular assist device, an inflow tube is made of a woven Dacron arterial graft. The use of a stretch Dacron pouch covering the pacemaker tends to reduce postoperative oozing and to facilitate better fixation of the pacemaker. Firm fixation is important for accurate measurement of the amplitude of the electrical impulse of the pacemaker. A Dacron patch graft is frequently used to avoid compromising the arterial lumen following an endarterectomy. A Dacron patch coated with silicone rubber has been found to serve as a partial substitute for the pericardium in reducing postoperative adhesion over the heart and great vessels. Dacron mesh has also been used as the reinforcing fabric for the outside housing of a Jarvik-7 total artificial heart. Dacron has been laminated onto the surface of a cardic pumping diaphragm in order to maintain a totally biolized surface.

Although nylon velour laminated inside a silicone tube is used for vascular grafts, the use of nylon alone in cardiovascular systems is not as promising as Dacron. The major drawback is the loss of up to 85% of tensile strength during the first six months after implantation.

5.4 Body-Wall Covering Applications

As a result of radical resection of a cancer tumor, congenital lesions, infection and subsequent necrosis and debridement, violent crimes, accidents or large hernia body tissue defects are frequently produced. Surgical mesh fabrics have been used to reinforce tenuous aponeurotic closures and/or to bridge large defects in the abdominal or chest walls. Dacron

mesh has shown quite satisfactory results in repairing inguinal and ventral hernias, and tissue defects of the abdominal wall. The most common complication of using Dacron mesh in this type of repair is seromas. Although using nylon for repairing inguinal hernias has resulted in certain improvements, its loss of strength with time still makes Dacron mesh the better choice.

Synthetic absorbable polyesters like PGA and its lactide copolymer have been made in mesh form for the repair of contaminated abdominal wall defects and traumatized spleens (Dayton et al. 1986). Rapid simple splenic injuries can be healed more quickly by using PGA mesh.

The morphological and mechanical properties of Dacron, Teflon and polypropylene surgical meshes have been reported. It was found that Mersilene mesh (Dacron) had the highest relative porosity (51%) and the lowest bursting strength (1.3×10^4 kg m^{-2}), and tensile properties at both wale and course directions among the three types of mesh fabrics tested. Mersilene was also the softest mesh with flexural rigidity of 30.40 mg-cm.

5.5 Drug Control/Release Applications

The use of biodegradable microcapsules as a sustained drug control/release device has many advantages, such as the elimination of gastrointestinal absorption and the prolonged administration at therapeutic levels with minimal side effects. Although other biodegradable polyesters such as polyorthoesters have been used as drug carriers, poly(glycolide–lactide) copolymer and polylactide are the two most frequently used polyesters for this application. Their uses include narcotic and alcohol antagonist, antibiotics, antitumor, contraceptive, the incorporation of anesthetics for local anesthesia, antimalarial drugs, and vaccines, hormones and enzymes.

Since the release of embedded drugs in the bioerodible delivery system depends on the degradation of the bioerodible polymer matrix, a thorough understanding of the degradation mechanism of the polymer matrix is essential to the constant-rate release of drugs. Most of the reported biodegradation studies of glycolide–lactide microcapsules have been based on the measurement of the loss of radioactivity of ^{14}C-labelled microcapsules. It was found that the rate of degradation of the biodegradable glycolide–lactide copolymer capsules increased as the glycolide composition increased. For example, a 26% glycolide–74% lactide microcapsule would complete its degradation within 30 weeks, while a 50% glycolide–50% lactide microcapsule would be completed within 9 weeks (Visscher et al. 1985). This compositional effect of glycolide–lactide copolymer on its degradation rate must be attributed to its structure characteristic and chemical property. The level of crystallinity, the melting point and the percentage of water uptake of the glycolide–lactide copolymer depend on the percentage of the glycolide component. Although glycolide–lactide copolymers with the composition ratios of both 26:74 and 50:50 are amorphous, the one with the higher glycolide composition (up to 70 mole ratio) has a higher percentage of water uptake than the one with the lower glycolide composition. Since hydrolytic degradation depends on the amount of water available in a polymer, a higher water uptake should lead to a faster rate of hydrolysis, as is observed. The hydrolysis of this class of copolymer proceeds with bulk erosion, as is evident in their morphologic change. As a result, undesirable non-zero-order drug release was frequently observed. Another disadvantage of bulk hydrolysis is the massive disintegration of the device within a relatively short time period. This could cause a sudden release of the remaining drugs to boost the drug concentration in a living system to an undesirable level. To design polymers that undergo a heterogeneous hydrolysis with surface erosion has been suggested to be a better approach to achieve zero-order drug release kinetics (Heller 1980). The morphological change of the glycolide–lactide microcapsules was well correlated to the loss of radioactivity of ^{14}C-labelled copolymer (Visscher et al. 1985).

In addition, Visscher et al. reported that the size and drug concentration of the microcapsule were found to affect its degradation rate. A smaller size microcapsule with a higher concentration of drug accelerated the rate of degradation of the microcapsule.

Nonbiodegradable polymers, such as polyamides, have also been used as drug control/release devices. Nylon-made microcapsules are generally limited to the encapsulation of large biologically active compounds such as proteins and enzymes to treat enzyme deficiencies. These compounds are immobilized on nylon to prevent their escape to the outside of the microcapsule, causing immunological reactions. In the meantime, the immobilized enzymes can react with small molecules which diffuse across the semipermeable nylon membrane. The drugs that have been encapsulated in nylon include sulfathiazole sodium, pentobarbital sodium and phenothiazines. For the purpose of increasing the flow property of these capsules during nylon coating, different matrices (formalized gelatin, calcium alginate and calcium sulfate) have been incorporated into the nylon membrane, and drugs have been embedded in the matrix surrounded by a nylon membrane. Microcapsules of very small diameters (about 50 µm) could be made in this way by controlling the stirring speed during nylon formation. The effect of different matrices and pH on the retention of drugs and the surface morphologic change of the microcapsules has also been examined.

Although most drug control/release devices are in spherical microcapsule form, biodegradable and nonbiodegradable fibers in suture form have also been used to deliver radiation treatment to cancers. In general, an array of I-125 seeds were evenly embedded in a suture so that the suture could be sewn into the tumor. The key advantages were the even distribution of radioactivity and the easy, rapid, precise and nonmigrating placement of the radioactive source to maximize radiation therapeutic efficiency with minimal side effects. The use of synthetic absorbable sutures such as Vicryl would also eliminate the need for resurgery to remove the device. This approach appears to improve radiation therapy of cancer.

5.6 Other Applications

Among respiratory prostheses, a Dacron aortic graft has been used for tracheal replacement with unsatisfactory results. Knitted Dacron was found to work satisfactorily for the repair of the anterior wall of the cervical trachea after stenosis or malignant invasion. Dacron fabric has frequently been combined with stainless steel mesh or silicone rubber for reconstructing trachea. Crimped nylon reinforced with stainless steel has been used to treat circumferential defects of the trachea.

In the digestive system, a woven Dacron patch has been satisfactory for small-bowel replacement, while nylon mesh containing 42×32 threads cm^{-2} was found to be a great success in replacing esophageal musculature and esophageal wall defects. Other uses include Dacron mesh as the reinforcement cuff in the prosthetic sphincter system used in the gastrointestinal tract. Absorbable polyesters were molded into biodegradable rings for sutureless bowel anastomosis. These rings were found to have very low overall tissue necrosis and consequently high initial bursting strength.

In the genitourinary system, Dacron velour has been used to cover and make liquidproof the suture rim of an all-plastic spermatocele (a device that makes artificial insemination possible using an all-plastic reservoir either on caput or cauda epididymis). A Dacron loop is used in a gel-filled silicone testicular prostheses for fixation to the scrotal tissues to prevent its migration. Dacron-velour-lined silicone Dacron-reinforced silicone tubes have been used as urethral prostheses. The former was successful while the formation of calculus in the latter caused certain problems. Double-knitted siliconized Dacron used as a urethral prosthesis could facilitate the development of urethral lining. However, it has been reported that Dacron prostheses with a porous meshwork are unsuitable owing to the penetration of connective tissue and the obliteration of lumen. Silicone rubber covered with Dacron velour could assist bladder regeneration of the urinary bladder after cystectomy. The concept behind this approach is to actively induce the deposition of a fibrin-based layer which would be gradually replaced by a pseudoneointima. PET fibers of 250 µm in length and 25 µm in diameter were electrostatically adhered onto a silicone rubber bladder. Unsatisfactory results were reported, because of a nonuniform coverage and the detachment of PET fibers from the underlying substrate. Bladders fabricated from a random polyester fiber-flocking method had a maximal safe period of six months. More recently, an improved design has been used with nylon fibers of 300 µm in length and 3 denier as the die to cast a negative impression on the inner surface of a bladder so that an integral flock is formed on the substrate without the disadvantage of fiber detachment. Dacron straps attached to silicone rubber capsules to confer initial fixation of the device and Dacron-reinforced silicone rubbers have been used as prosthetic urinary sphincters for treating urinary incontinence. Dacron-mesh-reinforced silicone elastomer was used as an inflate–deflate mechanism for a silicone elastomer penile prosthesis. Dacron felt cuffs are also found in a subcutaneous catheter. Nylon 66 velour used as inner urinary and outer peritoneal interface cloth in urinary bladder reconstruction seems to be compatible with urine.

More recently, synthetic absorbable polyesters and their copolymers have found applications in the genitourinary system. Instead of the traditional silicone rubber stents, PGA stents have been used successfully in dogs' uretero-ureterostomy to assist in accurate approximation, to minimize urinary extravasation and to facilitate adequate lumenal space for epithelium regeneration. PLA tubes have also been used as a ureteral graft in order to overcome the problems of migration, chronic infection and incrustation associated with the stents made of nonabsorbable polymers. Ideally, the graft should function for at least six months until the postoperative periureteral fibrosis can be stabilized. The particles from the disintegration of PLA ureteral grafts, however, obstructed the ureter and caused the total loss of the kidney function in 70% of the testing animals. A collagen–polyglyactin 910 mesh was used to repair full-thickness defects in the urinary bladders of rabbits (Monsour et al. 1987). Regenerations of smooth muscle and urothelium lining were observed. This material was suggested to augment contracted bladders or to repair bladder fistulae.

In the repair of nerve systems, it is recommended to minimize fibroblast tissue reaction since it impedes the regeneration of damaged nerves. Consequently, various types of devices such as tubes and cuffs have been used to minimize the scar formation during neural healing. PGA, PLA and their copolymers were the first types of synthetic biodegradable polymers used for facilitating neural regeneration. They had the advantage of being a rapid reconstruction of injured nerves without the use of sutures and the results were comparable to those repaired by standard microneurorrpaphy. PGA has been found to evoke the least amount of scar formation in nerve repair, and its degradation and absorption do not interfere or damage the neighboring nerve (DeLee et al. 1977). These biodegradable conduit devices, however, have not been able to facilitate nerve regeneration beyond a gap greater than 1 cm.

In the epidermal system, there has been a continuous search for satisfactory artificial skin to treat burns for minimizing wound infection, fluid loss and the trauma of early wound excision. Nylon seems to be more successful as a temporary skin graft than Dacron. Both polypeptide- and silastic-laminated nylon velour can function as a temporary skin substitute for the purpose of reducing fluids, electrolyte loss, sepsis and local discomfort, and for enhancing the development of a wound bed acceptable for delayed autografting. A similar synthetic skin consisting of an impermeable layer of poly(vinyl fluoride) sandwiched between a layer of nylon velour fabric was tested in rats. The mortality rate was significantly lower than that for untreated or excision treatment only groups, regardless of the postburn waiting period before treatment. Wound sepsis was the sole cause of all deaths. Dacron flocking has been used as the inner layer of artificial skin, but this material adheres largely by incorporating coagulum, which may lead to infection. Polycaprolactone has been found to possess the physical properties of a temporary skin wound covering.

In ophthalmology, Nylon 6 plates were used to treat blowout fracture of orbit. Dacron mesh bonding to Nylon or silicone plates was used to promote tissue ingrowth for preventing orbit implant from migration.

In the other areas of tissue reconstruction, neither Dacron nor nylon has been used to construct a major component of a prosthesis. Dacron or nylon fabrics have been used either as a lining in the case of reconstructing an ear, as a backing in a breast prosthesis for better fastening of the prosthesis in the surrounding tissues, as a reinforced fabric for dura material repair, or as a covering for a percutaneous energy transmission system to provide a tenacious bonding of the system to the underlining tissues to relieve the strain experienced by the system.

A further unique application of Dacron as a surgical implant is in the replacement of intervertebral disks. The basic design is a silicone central core sandwiched between two reinforced layers made of siliconized woven Dacron mesh. However, complications such as resorption of adjacent bone, reactive bone formation and infection were observed.

Among extracorporeal devices, nylon–wool and Dacron–wool filters have been used for removing microaggregates of amorphous materials of stored blood during blood transfusions. The Dacron–wool filter seems more efficient than the nylon–wool filter. However, nylon fiber filtration is more effective in procuring human granulocytes during a transfusion. Both Nylon 6 and Nylon 11 have been used as a plasma thromboplastin component (PTC) drainage tube and intravenous catheter during organ transplantations and vascular surgery. UroKinase, a plasminogen activator, can be immobilized on a nylon tube surface in the hope of maintaining local fibrinolytic activity for antithrombogenic purposes. Enzyme-immobilized nylon tubes show better patency than uncoated ones for intravenous hyperalimentation during digestive tract surgery or for intra-arterial catheters. A nylon tube can also be conjugated to a variety of proteins, heptens and heptan–protein conjugates. Such a nylon catheter system may be applicable to the removal of circulating antigens and antibodies. Two uses are in immunotherapy of malignancies and in the treatment of autoimmune disease. More recent development of this enzyme-immobilized device has resulted in a Dacron–wool packed-bed extracorporeal reactor system which has effectively and safely lowered plasma L-asparagine levels by circulating the blood through this unit. Nylon-based hemodialysis membrane has shown greater dialyzing capacity than cuporphan (Luttinger and Cooper 1970). Nylon mesh rather than Mylar (PET film) should be used in the treatment of couching during isocentric megavoltage therapy. Nylon fibers have been chemically modified by bioactive compounds such as glutaraldehyde, nitrovin, nitrofurazone and nitrofurglacrolein for antimicrobic purposes. The work is still in progress. Nylon-6-coated charcoal granules were found useful for the direct detoxification of blood in cases of uremia and drug overdose. Kinetic studies have shown that the selectivity of charcoal granules of phenolic compounds was increased by Nylon 6 coating.

6. Sterilization Considerations

Polyesters and polyamides are preferably sterilized by ^{60}Co irradiation, because of the convenience of the method and the general improvement in mechanical properties at the normal sterilization dosage level (25 kGy), with the exception of impact strength and elongation (Bruck and Mueller 1988). Other methods such as autoclaving and ethylene oxide sterilization can be used. Heating in the presence of moisture, however, should be avoided because of possible adverse changes in some physical and mechanical properties.

See also: Biocompatibility: An Overview; Suture Materials

Bibliography

Barrows T H 1989 Bioabsorbable poly(ester–amides). In: Huang S J (ed.) 1989 *Biodegradable Polymers*. Hanser, Vienna

Bowald S, Busch C, Eriksson I 1980 Absorbable material in vascular prostheses: A new device. *Acta Chir. Scand.* 146: 391–5

Bruck S D, Mueller E P 1988 Radiation sterilization of polymeric implant materials. *J. Biomed. Mater. Res. A* 22: 133–44

Chu C C 1991 Degradation and biocompatibility of synthetic wound closure polymeric materials. In: Williams D F (ed.) 1991 *Degradation and Biocompatibility of Synthetic Degradable Polymers*. CRC Press, Boca Raton, FL

Chu C C, Lecaroz L E 1987 Design and *in vitro* testing of newly made biocomponent knitted fabrics for vascular surgery. In: Gebelein C G (ed.) 1987 *Advances in Biomedical Polymers*, Polymer Science and Technology, Vol. 35. Plenum Press, New York, pp. 185–214

Chu C C, Louie M 1985 A chemical means to examine the degradation phenomena of polyglycolic acid fibers. *J. Appl. Polym. Sci.* 30: 3133–41

Dayton M T, Buchele B A, Shirazi S S, Hunt L B 1986 Use of an absorbable mesh to repair contaminated abdominal wall defects. *Arch. Surg.* 121: 954–60

DeLee J C, Smith M T, Green D P 1977 The reaction of nerve tissue to various suture materials: A study in rabbits. *J. Hand Surg.* 2: 38–43

Gogolewski S, Pennings A J, Lommen E, Wildevuur C R H, Nieuwenhuis P 1983 Growth of a neo-artery induced by a biodegradable polymeric vascular prosthesis. *Makromol. Chem. Rapid Commun.* 4: 213–19

Guidoin R, Gosselin C, Martin L, Marois M, Laroche F, King M, Gunasekera K, Domurado D, Sigot-Luizard M F, Blais P 1983 Polyester prosthese as substitutes in the thoracic aorta of dogs I: Evaluation of commercial prostheses. *J. Biomed. Mater. Res.* 17: 1049–77

Heller J 1980 Controlled release of biologically active compounds from bioerodible polymers. *Biomaterials* 1: 51–7

Heuvel H M, Huisman R 1978 Effect of winding speed on the physical structure of as-spun poly(ethylene terephthalate) fibers, including orientation-induced crystallization. *J. Appl. Poly. Sci.* 22: 2229–43

Holmes P A 1985 Applications of PHB—A microbially produced biodegradable thermoplastic. *Phys. Technol.* 16: 32–6

Katz A R, Mukherjee D P, Kaganov A L, Gordon S 1985 A new synthetic monofilament absorbable suture made from polytrimethylene carbonate. *Surg. Gynecol. Obstet.* 161: 213–22

King M, Blais P, Guidoin R, Prowse E, Marcois M, Gosselin C L, Noel H P 1981 Polyethylene terephthalate (Dacron) vascular prostheses—Material and fabric construction aspects. In: Williams D F (ed.) 1981 Biocompatibility of Clinical Implant Materials, Vol. 2. CRC Press, Boca Raton, FL, pp. 178–207

Luttinger M, Cooper C W 1970 Parameters affecting preparation of nylon-based hemodialysis membranes. *J. Biomed. Mater. Res.* 4: 281–93

Monsour M J, Mohammed R, Gorham S D, French D A, Scott R 1987 An assessment of a collagen/Vicryl composite membrane to repair defects of the urinary bladder in rabbits. *Urol. Res.* 15: 235–8

Reimschuessel H 1985 Polyamide fibers. In: Lewin M, Pearce E M (eds.) 1985 *Fiber Chemistry*, Handbook of Fiber Science and Technology, Vol. 4. Dekker, New York, pp. 74–169

Rodkey W G, Cabaud H E, Feagin J A, Perlik P C 1985 A

partially biodegradable material device for repair and reconstruction of injured tendons. *Am. J. Sports Med.* 13: 242–7

Ruderman R J, Hegyeli A F, Hattler B G, Leonard F 1972 A partially biodegradable vascular prostheses. *Trans. Amer. Soc. Artif. Int. Org.* 28: 30–3

Sendelbeck S L, Girdin C L 1985 Disposition of a carbon-14 labelled bioerodible polyorthoester and its hydrolysis products, 4-hydroxybutyrate and cis, trans-1,4-bis-(hydroxymethyl) cyclohexane in rats. *Drug Metab. Dispos.* 13: 291–5

Sigot-Luizard M F, Domurado D, Sigot M, Guidoin R, Gosselin C, Marois M, Girard J F, King M, Badour B 1984 Cytocompatibility of albuminated polyester fabrics. *J. Biomed. Mater. Res.* 18: 895–909

Visscher G E, Robison R L, Maulding H V 1985 Biodegradation of and tissue reaction to 50:50 poly(DL-lactide-co-glycolide) microcapsules. *J. Biomed. Mater. Res.* 19: 349–65

Wheeless C R 1986 Use of absorbable staples for closure of proximal end of ileal loops. *Obstet. Gynecol.* 67: 280–3

C. C. Chu
[Cornell University, Ithaca,
New York, USA]

Polyethylene

Polyethylene (PE) is produced in different types which are known as low-density (LDPE), linear low-density (LLDPE) and high-density (HDPE), according to their crystallinity. These PE types are among the most widely used thermoplastics. Typical applications of LDPE and LLDPE include films, containers, tubing and insulating material for a whole range of medical products. Where higher stiffness, strength and chemical stability are required, the more crystalline HDPE is used. Most of the PE implants are manufactured from ultrahigh-molecular-weight PE (UHMWPE). This material is an HDPE with molecular weight in the range 2–10 million. UHMWPE exhibits very interesting properties. Advantageous properties are:

(a) very good sliding properties;

(b) exceptional impact strength, even at low temperatures;

(c) good cyclical fatigue resistance; and

(d) good human body compatibility.

Disadvantageous properties are:

(a) high creep compliance;

(b) low stiffness (only valid for certain applications);

(c) poor wear performance (which is very good among other polymers, but not sufficient for orthopedic implants); and

(d) poor oxidation stability.

Table 1
Applications for PE as an implant material

Joint replacement[a]	Other
knee	esophagus segments
elbow	coatings and insulations for
finger	pacemakers
wrist	dentures
hip (acetabulum)	catheters

a only UHMWPE

Typical applications in the field of medical implants are shown in Table 1. One of the most demanding applications is the use for joint replacements. In this field, UHMWPE has become one of the most important implant materials. In spite of several advantageous properties, the service life of joint replacements made from UHMWPE is limited. Complications occur with about 30% of hip-cup implants after implantation for 11 years (Gierse and Schramm 1984). Comprehensive investigations carried out to analyze the damage to more than 250 explanted hip cups and tibial plateaux, showed a considerably changed property profile compared with new material in terms of morphology and mechanical properties. The combined action of chemical demand and mechanical stress causes local changes in the density of the UHMWPE (Eyerer and Ke 1984). This can be explained by postcrystallization, which is caused by an increased level of molecule mobility brought about by oxidative chain scission. This results in brittle material behavior coupled with lower fracture toughness and reduced damping capacity. Both of these are to be regarded as negative property changes.

In the search for means of countering negative changes of this type, four different approaches have emerged (Eyerer et al. 1987):

(a) improvement of initial properties by variation of synthesis conditions;

(b) optimization of processing parameters in respect of properties by the use of very high pressure;

(c) investigation of the influence of different sterilization methods on the properties; and

(d) development of a test method for *in vitro* simulation of the *in vivo* degradation processes.

1. Variation of Synthesis Conditions

A further development of the synthesis of UHMWPE should provide more favorable starting properties for applications in endoprostheses. This should be done by increasing its mean molecular weight. Particular value is placed here on:

(a) increasing the resistance to wear;

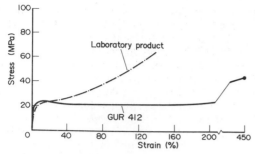

Figure 1
Stress–strain diagram of a UHMWPE laboratory product compared with the commercial product GUR 412

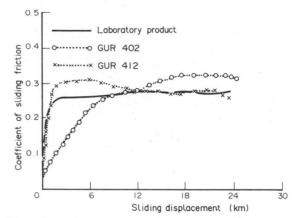

Figure 2
Sliding-friction behavior of a UHMWPE laboratory product compared with standard commercial GUR materials

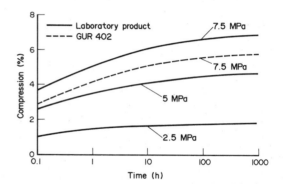

Figure 3
Creep in compression behavior of UHMWPE under different loads (37 °C in Ringer's solution)

(b) enhancing stiffness and strength;

(c) reducing the tendency to creep; and

(d) improving aging stability facing the body medium.

In the case of HDPE, it is known that resistance and stability increase with molecular weight.

Standard commercial UHMWPE has a mean molecular weight of some 4000 kg mol^{-1}. If synthesis is carried out at a reaction temperature of 70 °C using a heterogeneous Ziegler catalyst, this results in PE with a mean molecular weight in excess of 8000 kg mol^{-1} (viscosity number ~3 m^3 kg^{-1}). The degree of polymerization can be controlled by varying the temperature and modifying the cocatalyst (Sinn and Kaminsky 1980, Kaminsky 1986).

Figure 1 shows a typical curve for the stress–strain diagram of a laboratory product compared with a commercially available UHMWPE (Hostalen GUR, made by Hoechst, Frankfurt). For a virtually identical mean molecular weight (4000 kg mol^{-1}) the ultimate tensile strength of the laboratory product is one-third higher, while its elongation at break is three-times lower than that of the standard commercial UHMWPE (Table 2). There is only slight

Table 2
Properties of a laboratory product in comparison with Hostalen GUR 402/412

Property	Laboratory product	Hostalen GUR 402/412
Weight-average molecular weight (kg mol^{-1})	~4000	~4000
Density (kg m^{-3})	~930	~940
Crystallinity (%)	~46	~56
Yield point (MPa)	22	22
Elongation at yield (%)	11.5	9.5
Ultimate tensile strength (MPa)	66	44
Elongation at break (%)	140	450
Elastic modulus (MPa)	1070	930

evidence of a yield point in the modified material. The yield point value of 22 MPa is identical in both cases and hence, by way of a first approximation, independent of the synthesis conditions.

Figure 2 shows the sliding behavior that is of importance for prosthetics applications. A laboratory product, even with a slightly lower molecular weight (2600 kg mol^{-1}) compared with GUR 412 or GUR 402, displays good tribological properties. A further increase in the molecular weight and, under some circumstances, copolymerization with butene-1 or hexene-1, ought to produce even better tribological properties.

If UHMWPE is to be used as a material for replacement joints, favorable wear behavior as well as good creep resistance are required. Figure 3 shows the creep in compression behavior of a laboratory product under different loads in comparison with GUR 402. Direct comparison of these

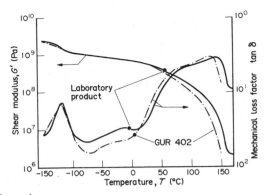

Figure 4
Shear modulus and mechanical loss factor (tan δ) of
UHMWPE vs temperature

two high-molecular-weight PEs with almost identical
mean molecular weight (4000 kg mol^{-1}) reveals a
slightly lower tendency to creep on the part of
GUR 402, on account of its higher crystallinity (see
Table 2).

Dynamic mechanical spectroscopy tests operating
with forced oscillation (1 Hz) provide significant
information on the temperature dependence of
mechanical behavior (see Fig. 4). When the glass
transition temperature of the two materials is
around −120 °C, the laboratory product displays
higher damping at low temperatures (−100 °C to
+10 °C). At temperatures of above 30 °C, it displays
stiffer behavior. The crystalline melting point is
some 10 °C higher, which is important for the heat
distortion temperature, particularly with eventual
steam sterilization instead of irradiation
sterilization.

From these examples, it becomes clear that the
properties for use of UHMWPE—especially in
medical applications—can be further refined
through the optimization of its synthesis.

2. Optimization of Processing Parameters

The very high mean molecular weight of UHMWPE
causes its unusually high melt viscosity. Thus it is
difficult to process (e.g., by injection molding) and
semifinished products are chiefly processed by pressure sintering—and at times by ram extrusion—
of the polymerized powder with pressures of
10–20 MPa and temperatures of about 200 °C.

These processes, however, can have negative
influences on the properties of the semifinished
product. On the one hand, calcium stearate is added
as an anticorrosion agent and a processing aid (lubricant and release agent); this leads to surface defects
in the form of notches, caused by the partial separation of particle boundaries and the formation of

small accumulations of calcium stearate. On the
other hand, it is still possible to detect particle
boundaries in conventionally sintered or ram-
extruded material using an electron microscope. In
addition, if PE macromolecules are processed without the exclusion of atmospheric oxygen, they will
suffer some oxidative degradation during sintering.

The drawbacks listed above can be avoided by
using material free from calcium stearate, and by
processing in an inert atmosphere. In addition,
by application of much higher sintering pressures
(and higher temperatures) it is possible to improve
the properties of the semifinished product by changing its morphology (e.g., by increasing the degree of
crystallinity).

Many studies (e.g., Basset and Turner 1974,
Takemura 1979, Leute and Dollhopf 1980, Wunder
1981) dealing with PE of low and medium molecular
weight (HDPE), show that hot sintering under high
pressure leads to an increase in crystallinity as well
as to a more ordered morphology. This mechanism
is set out schematically in Fig. 5. At normal (low)
crystallization pressure, macromolecules fold at regular distances and form crystals with lamellar structures and amorphous material in between the lamellae. For linear polyethylene, the thickness of the
lamellae normally varies between 40 nm and
100 nm. If the crystallization pressure is increased
above a molecular-weight-dependent pressure limit
of some 200–300 MPa (the transition pressure),
molecules crystallize in a more-or-less extended
form depending on their molecular weight and build
so-called extended-chain crystals with a banded
structure. The thickness of extended-chain crystals
should, by definition, be greater than 200 nm. For
linear UHMWPE this thickness can reach to several
hundred nanometers, depending on molecular
weight and crystallization conditions. The extended-
chain morphology improves stiffness and creep
behavior, whereas for normal, low or medium
molecular-weight PE, toughness is considerably
diminished. The reason is probably a lack of inter-
crystalline linking created by tie molecules. On the
other hand, UHMWPE, given suitable crystallization conditions, is expected to preserve first a sufficient quantity of tie molecules and second a large
number of molecular-chain entanglements and
hence good toughness. These assumptions were supported by the first study carried out into mechanical
properties of high-pressure crystallized UHMWPE
(Lupton and Regester 1974).

For closer investigation of the influence of the
high-pressure crystallization conditions on mechanical and application properties of UHMWPE a
very-high-pressure mold was designed and constructed at the Institute for Polymer Testing and
Polymer Science at the University of Stuttgart (Ellwanger et al. 1987). It is able to produce samples
35 mm in diameter and up to 65 mm in length, and to

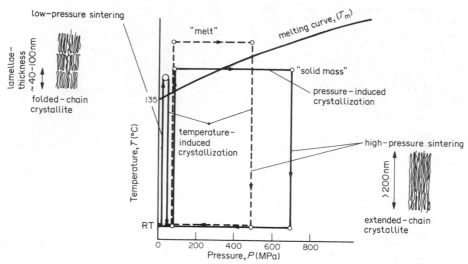

Figure 5
Crystallization procedures and structures of UHMWPE

operate with vacuum or an inert gas atmosphere at temperatures of up to 300 °C and pressures of up to 1 GPa. Specimens can be produced by two types of crystallization (Fig. 5): on the one hand, normal, or temperature-induced, crystallization (i.e., through virtually isobaric cooling, as in the conventional sintering process) and on the other hand, through pressure-induced crystallization (i.e., through a virtually isothermal pressure increase leading to crystallization). In the case of pressure-induced crystallization, the pressure-dependent crystallite melting point of PE is used to advantage. In contrast to the standard procedure, the specimen can crystallize at all points simultaneously, since there is nearly no temperature gradient. This leads to materials with greater homogeneity.

A number of differences in the properties of low-pressure and very-high-pressure sintered materials are listed in Table 3. The increase in density, crystallinity and crystalline melting point, and particularly in the elastic modulus, is clearly pronounced.

The banded structure with lamellae thicknesses in excess of 200 nm that forms at crystallization

pressures above about 300 MPa and indicates extended-chain crystallites, can be detected in electron micrographs (see Fig. 6).

Above the transition pressure of about 250 MPa, the elastic material characteristics increase distinctly with increasing crystallization pressure. The yield point (see Fig. 7), for example, is enhanced by about 40% as compared with low-pressure sintering. The curve of the stiffness vs sintering pressure shows the same strong increase in the range of the transition pressure as the yield point. Simultaneously, the mechanical loss factor, strain and strength at break, and the notched impact strength are clearly diminished, whereby strength at break and notched impact strength show a peak in the range of the transition pressure (see Fig. 8). The mechanical long-term performance is also improved. In comparison with low-pressure treatment (20 MPa), high-pressure crystallized UHMWPE exhibits much better resistance to creep in compression with rising crystallization pressure (see Fig. 9).

An additional example for the influence of high pressure modification is shown in Fig. 10. If the tests

Table 3
Properties of low-pressure (10 MPa) and very-high-pressure (300–1000 MPa) sintered UHMWPE

Property	Low-pressure sintering	Very-high-pressure sintering
Density (kg m^{-3})	~935	~965–990
Degree of crystallinity (%)	~50–60	~70–90
Crystalline melting point (°C)	~135–138	~145–155
Elastic modulus (MPa)	~900–1200	~1800–2700

are performed without temperature control, rising sintering pressure decreases the coefficient of sliding friction but increases the wear. In tests with temperature control (37 °C, body temperature), the slight increase in wear, coupled with the fall in the coefficient of sliding friction, can certainly be regarded as an improvement. Only a slight deterioration in friction behavior but a clear increase in wear

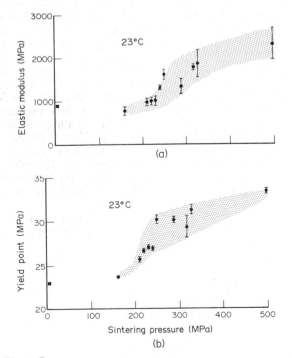

Figure 6
Transmission electron micrograph of UHMWPE molded at 450 MPa

Figure 7
(a) Elastic modulus vs sintering pressure for UHMWPE, and (b) yield point vs sintering pressure for UHMWPE

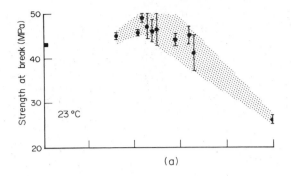

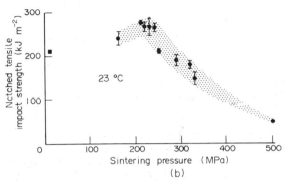

Figure 8
(a) Strength at break vs sintering pressure for UHMWPE, and (b) notched tensile impact strength vs sintering pressure for UHMWPE

is caused by γ radiation, such as is used to sterilize artificial hip joints. This shows that optimized processing under high pressure improves those properties of UHMWPE that have been unsatisfactory to date: especially creep resistance and stiffness, but also sliding performance and crystalline melting point (i.e., heat-distortion temperature).

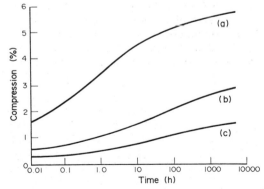

Figure 9
Curves of creep in compression of UHMWPE at 23 °C with 7.5 MPa load for the sintering pressures: (a) 10 MPa, (b) 320 MPa and (c) 500 MPa

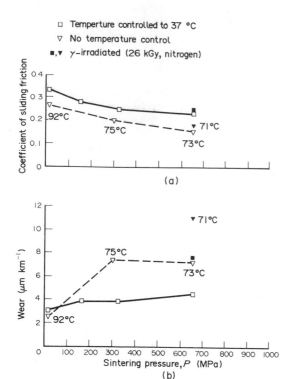

Figure 10
(a) Coefficient of sliding friction of UHMWPE vs sintering pressure and (b) wear of UHMWPE vs sintering pressure (pin-on-disk tests, nonlubricated, contact pressure 5 MPa, test velocity 0.05 m s⁻¹)

3. Sterilization of UHMWPE

Implant components made from UHMWPE are usually radiation sterilized in multilayer packaging on an industrial scale. This is done using ⁶⁰Co sources with different dosage rates, and it can take as long as 20 h for the minimum absorbed dosage of 25 kGy to be attained. During this time the ionizing radiation sets off a large number of primary and secondary reactions which can lead to predamage as a function of thickness (Rose et al. 1977). For the further development of products made of UHMWPE, it is thus essential to know the influence of high-energy radiation on properties of the material.

PE is one of the polymers that primarily cross-links by radiation under the exclusion of oxygen. In the presence of atmospheric oxygen, by contrast, this cross-linking is accompanied by oxidative degradation of the polymer. With the exclusion of oxygen, the formation of free radicals during irradiation (De Vries and Smith 1980) causes a radiation-induced cross-linking reaction, particularly in the amorphous regions (Böhm 1967). Hot extraction in *o*-xylene shows an initially clear increase in the cross-linked

gel fraction with increasing radiation dosage (see Fig. 11), although increasing the dosage still further does not lead to complete cross-linking. Since the cross-linking reaction always has chain scission of polymer molecules superimposed on it, optimum cross-linking is achieved with a relatively low dosage of some 30–50 kGy.

The presence of oxygen has an inhibiting effect on cross-linking and, apart from the peroxygenation of the material that sets off the oxidation, it also causes a considerable change in the molecular-weight distribution (MWD). Investigations using high-temperature gel permeation chromatography (GPC) reveal a widening of the MWD and a reduction in the mean molecular weight. Mean molecular weights established on irradiated UHMWPE powder by means of a viscosimeter, show the extraordinarily high sensitivity to oxidation for irradiation in the presence of oxygen (see Fig. 12). Significant property changes must be expected. UHMWPE that has been radiation-cross-linked in nitrogen reveals a wear resistance up to 60% better than that of nontreated material (see Fig. 13). The creep behavior in compression becomes more favorable if UHMWPE has been irradiated in nitrogen as well as in air (see Fig. 14). Impact strength and ultimate tensile strength are also improved to a certain degree, although the elongation at break decreases.

Since high-energy electrons essentially set off the same material reactions as γ rays, an optimum ratio between the degree of cross-linking and the reduction in molecular weight can be achieved by precisely dosed electron-beam sterilization. By varying

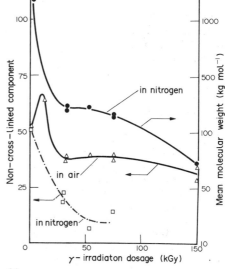

Figure 11
Non-cross-linked component and mean molecular weight of GUR 402 vs γ-irradiation dosage in air or in nitrogen

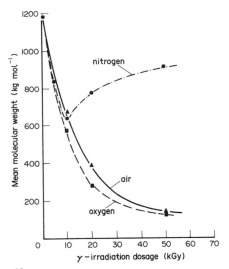

Figure 12
Mean molecular weight of GUR 402 powder vs γ-irradiation dosage in oxygen, air or nitrogen

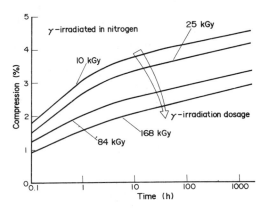

Figure 14
Compressive creep behavior of GUR 402 (37 °C in Ringer's solution, load 7.5 MPa after irradiation in nitrogen)

the accelerating potential, the changes in properties can be limited to regions close to the surface (see Fig. 15). By changing the irradiation time and the dose rate, as well as using a protective atmosphere, it is thereby possible to achieve reliable sterilization and to minimize material damage.

4. *In Vitro Simulation of In Vivo Degradation Processes*

As long ago as 1955, Oppenheimer et al. (1955) observed in tracer experiments on rats that PE undergoes degradation on account of the body medium of the test animals. The ultimate tensile stress of PE decreases by about 30% after implantation for 17 months and the elongation at break

falls to 54% of the initial value (Leininger et al. 1964). The modulus of elasticity close to the surface increases by some 90%, and the material becomes brittle.

Chemical testing methods show that this change in material properties correlates with changes in chemical structure. Using infrared spectroscopy, it is possible to detect the formation of hydroxyl and carbonyl groups in the polymer. GPC measurements on polypropylene clearly showed that the oxidation processes are coupled with a reduction in molecular weight (i.e., with degradation of the polymer) (Liebert et al. 1976). Kopecek and Ulbrich (1983) compare the *in vivo* degradation of PE with thermo-oxidation. They give a degradation rate which is higher than that of thermo-oxidation at 50–60 °C.

The causes of biodegradation are to be found in the composition of the body medium. Comprehensive

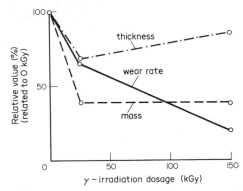

Figure 13
Wear behavior of GUR 402 (ring-on-disk test at 37 °C in Ringer's solution)

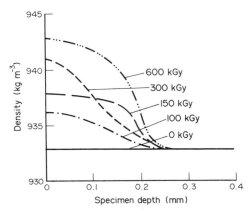

Figure 15
Density vs specimen depth of GUR 402 UHMWPE following electron irradiation in air (250 keV)

biochemical investigations on granulocytes—a cell type that is found in increased numbers in the area surrounding the implant—were able to show that highly oxidizing products of metabolism are formed in this area. Granulocytes react to implants—which are seen as foreign bodies—with a "metabolic burst" (i.e., an acceleration of the metabolism with a sharp increase in the amount of oxygen consumed). More than half of the oxygen used is converted into H_2O_2, with superoxide (O_2^-) as an intermediate product. Hydroxyl radicals, singlet oxygen and other highly reactive oxidizing agents are formed, in parallel and consequent reactions (Tschesche and McCartney 1981). The reasons behind the development of these aggressive products of metabolism are still largely unknown. It is, however, probable that they are closely connected with the microbiocidal function of the granulocytes.

Use has generally been made of Ringer's solution or pseudoextracellular fluid for *in vitro* simulation of the degradation processes to date. These test media are isotonic saline solutions and are suitable for the simulation of hydrolyzing processes. Oxidative degradation processes, such as those observed with polyolefins, cannot be reconstructed using these fluids. The newly developed method, by contrast, possesses both hydrolyzing and oxidizing properties. The test medium is based on H_2O_2 solutions.

The effect of the oxidizing fluid was investigated on unstabilized UHMWPE films, on which physico-chemical and mechanical property changes have been tested together. Figures 16 and 17 show changes in mechanical properties together with crystallinity, which was determined from the density. Obviously, strong property changes appear after an induction period of about 20 days. Strength, elongation at break and toughness decrease while the

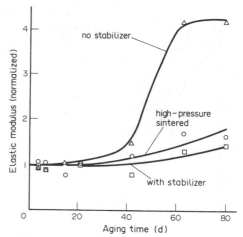

Figure 17
UHMWPE aging test with 3% hydrogen peroxide: elastic modulus (normalized)

modulus increases. The property changes which are the same as seen in explants can be explained by the increase of crystallinity. Infrared data (see Fig. 18) show the formation of an oxidation product (ester group). From these data it is very likely that the postcrystallization process is caused by a oxidative degradation.

Figures 17, 18 and 19 also show the stability of stabilized low-pressure sintered UHMWPE (0.2% IRGANAOX 1010, Ciba Geigy Co.) and the unstabilized high-pressure material (270 MPa). The unstabilized high-pressure sintered material proves to be significantly more stable against oxidation than the UHMWPE which was processed conventionally. As expected, the stabilized low-pressure material

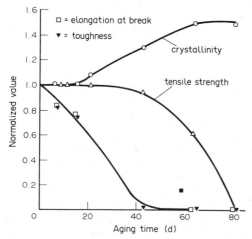

Figure 16
UHMWPE aging test with 3% hydrogen peroxide: mechanical properties and crystallinity (normalized)

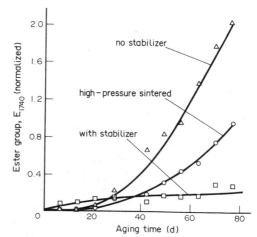

Figure 18
UHMWPE aging test with 3% hydrogen peroxide: ester group (normalized)

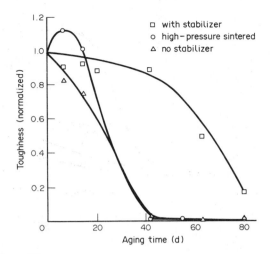

Figure 19
UHMWPE aging test with 3% hydrogen peroxide: toughness (normalized)

shows even better results. Again, the mechanical property changes can be explained by the increase of crystallinity (see Fig. 20). From these results, compared with the data on explants, it can be seen that a realistic method for oxidation-stability testing has been developed. Thus, oxidation stability of new materials can be examined, as well as the influence of processing (e.g., sintering and irradiation) parameters.

On the other hand, it is obvious that *in vivo* stability of UHMWPE can be improved by variation of processing parameters as well as by adding stabilizers. The question of biocompatibility of stabilizers, however, remains unanswered.

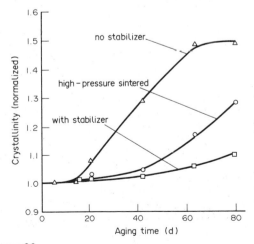

Figure 20
UHMWPE aging test with 3% hydrogen peroxide: crystallinity (normalized)

5. Summary

PE remains one of the most important implant materials, predominantly for orthopedic implants. The results presented show different means of modifying the quality of UHMWPE for use in artificial joints. Several advantageous properties (e.g., biocompatibility) are weighed against poor properties (e.g., creep behavior). The poor properties, however, can be improved by better synthesis as well as by a modified processing of the material.

Bibliography

Basset D C, Turner B 1974 On the phenomenology of chain-extended crystallization in polyethylene. *Phil. Mag.* 29: 925–55

Böhm G A 1967 Diffusion theory of the post-irradiation oxidation of polyethylene. *J. Polym. Sci.* 5: 639–52

De Vries K L, Smith R H 1980 Free radicals and new end groups resulting from chain scission. I: γ irradiation of polyethylene. *Polymer* 21: 949–55

Ellwanger R, Eyerer P, Siegmann A 1987 Very high pressure molding of ultra high molecular weight polyethylene (UHMWPE). *SPE Tech. Pap.* 33: 572–4

Eyerer P 1986 Kunststoffe in der Gelenkendoprothetik. *Z. Werkstofftech.* 17: 384–91; 422–8; 444–8

Eyerer P, Ke Y C 1984 Property changes of UHMWPE polyethylene hip cup endoprostheses during implantation. *Biomed. Mater. Res.* 18: 1137–51

Eyerer P, Kurth M, Ellwanger R, Mädler H, Federolf H -A, Siegmann A 1987 Ultrahochmolekulares Polyethylen für Gelenkendoprothesen. *Kunststoffe* 77(6): 617–22

Gierse H, Schramm W 1984 Nachuntersuchung von 997 Hüftendoprothesen unter besonderer Berücksichtigung der Spätergebnisse 9–11 Jahre post operationem. *Z. Orthop.* 122: 784–9

Kaminsky W 1986 Preparation of special polyolefins from soluble zirconium compounds with aluminoxane as cocatalyst. In: Keii T, Soga K (eds.) 1986 *Catalytic Polymerization of Polyolefins.* Elsevier-Kodansha, Tokyo, p. 293

Kopecek J, Ulbrich K 1983 *Biodegradation of biomedical polymers.* In: Jenkins A D, Stannet V T (eds.) 1983 *Progress in Polymer Science*, Vol. 9. Pergamon, New York, pp. 2–58

Leininger R I, Mirkovitch V, Peters A, Hawks W A 1964 Change in properties of plastics during implantation. *Trans. Am. Soc. Artif. Intern. Organs* 10: 320–1

Leute U, Dollhopf W 1980 A review of the experimental data from the high pressure phase in polyethylene. *Colloid Polym. Sci.* 258(4): 353–9

Liebert T C, Chartoff R P, Cosgrove S L 1976 Subcutaneous implants of polypropylene filaments. *J. Biomed. Mater. Res.* 10: 939–51

Lupton J M, Regester J W 1974 Physical properties of extended-chain high density polyethylene. *J. Appl. Polym. Sci.* 18: 2407–25

McKenna G B, Bradley G W, Dunn H K, Statton W O 1979 Degradation resistance of some candidate composite biomaterials. *J. Biomed. Mater. Res.* 13: 782–89

Oppenheimer B S, Oppenheimer B T, Donishefsky I, Stout A P, Eirich F R 1955 Further studies of polymers as carcinogenic agents in animals. *Cancer Res.* 15: 333–40

Rose R M, Goldfarb E V, Ellis E, Crugnola A 1977 The use of UHMW-polyethylene in articular prostheses. I: Effects of fabrication and γ sterilization on polymer characteristics. *Org. Coat. Plast. Chem.* 37(2): 280–4

Sinn H, Kaminsky W 1980 Ziegler-Natta-Catalysis. *Adv. Organomet. Chem.* 18: 99

Takemura T 1979 Structure and physical properties of high polymers under high pressure. *Polym. Prep., Am Chem. Soc., Div. Polym. Chem.* 20: 270–3

Tschesche H, McCartney H W 1981 A new principle of regulation of enzymic activity. *Eur. J. Biochem.* 120: 183–90

Wunder S L 1981 Raman spectroscopic study of the high-pressure phase of polyethylene. *Macromolecules* 14: 1024–30

P. Eyerer, R. Ellwanger, H. -A. Federolf, M. Kurth and H. Mädler
[Universität Stuttgart, Stuttgart, FRG]

Polymers for Controlled Drug Delivery

Since the late 1960s, there has been a rapid development of innovative techniques for the controlled delivery of drugs. This has been a result of the growing awareness that efficient therapy of an ailment with minimal adverse effects requires optimal delivery of drugs through proper design of the dosage form. The increased concern of pharmaceutical scientists over the daily administration of high doses of conventional drugs, as well as the increased cost of developing newer drug moieties, has been the impetus towards the development of rate-controlled drug delivery systems. A rate-controlled drug delivery system is a system that can regulate or control the release of a therapeutic agent. Such systems are advantageous over their conventional counterparts in that they deliver the drug at a predetermined rate, through which they maintain the steady-state drug concentration within a narrow therapeutic range, for a prolonged period of time.

The delivery of drugs from a rate-controlled drug delivery system can be regulated by use of a properly selected polymer (or combination of polymers), which permit the drug to be delivered at an accurate, reproducible and predictable rate. Polymeric materials were first introduced for controlling the release of drugs in the early 1960s in their application as film coating or matrix bases in sustained release tablets. By the early 1970s, the Alza Corporation, Palo Alto, California, had developed two commercial rate-controlled drug delivery systems, Occusert and Progestasert, both of which used the ethylene–vinyl acetate copolymer as the rate-limiting membrane to control the release of drug from a reservoir. Since then, there has been a surge of polymeric materials with potential

Table 1
Some polymers used in rate-controlled drug delivery systems[a]

Natural polymers

cellulose acetate phthalate	zein
hydroxypropyl cellulose	gelatin
carboxymethyl cellulose	natural rubber
ethyl cellulose	guar gum
methyl cellulose	gum agar
collagen	albumin

Synthetic polymers

Elastomers

silicone rubber	polybutadiene
polysiloxane	polyisoprene

Hydrogels

poly(hydroxyalkyl methacrylates)	alginates
poly(vinyl alcohol)	polyacrylamide
poly(vinyl pyrrolidone)	

Biodegradable

poly(lactic acid)	polyurethanes
poly(glycolic acid)	polyanhydrides
poly(alkyl 2-cyanoacrylates)	polyorthoesters

Adhesives

polyisobutylenes	silicones
polyacrylates	

Others

poly(vinyl chloride)	polyethylene
poly(vinyl acetate)	polyurethanes
ethylene–vinyl acetate	

a from Kydonius A 1980, Fundamental concepts of controlled release. In: Kydonieus A F (ed.) 1980 *Controlled Release Technologies: Methods, Theory and Applications*, Vol. 1. CRC Press, Boca Raton, FL

applications in rate-controlled drug delivery devices. Table 1 illustrates some of the polymers that have found use in the construction of rate-controlled drug delivery devices.

Physicochemical properties of the polymer such as its molecular weight, glass transition temperature and solubility parameter may be important considerations in the design of drug delivery systems. The structure of the polymer itself dictates its properties. For example, vinyl polymers can possess a wide variety of properties depending on the nature of the substituent on the polyethylene backbone. Thus, polymers such as polyethylene (semicrystalline), poly(methyl methacrylate) (glassy), poly(vinyl chloride (lipophilic), poly(vinyl alcohol) (hydrophilic) and poly(methyl 2-cyanoacrylate) (biodegradable) can result from the same polyethylene backbone. In addition to the physicochemical properties mentioned, polymers must have the appropriate diffusion and solubility characteristics with the active agents, must be easy to fabricate with high mechanical strength and must be stable to the drug as well as the environment. The toxicity, biocompatibility and immunogenicity of these materials are critical, because these devices interface directly with the microenvironment in which they are inserted, implanted or injected.

1. Commonly Used Polymers

1.1 Ethylene–Vinyl Acetate Copolymer

Ethylene–vinyl acetate copolymer (see Fig. 1) is perhaps the single most commonly used polymer in controlled drug delivery. This copolymer has been shown to possess good biocompatibility, has excellent flexibility and is capable of producing membranes with selective permeability for application in membrane permeation-controlled drug delivery systems. It consists of a polyethylene backbone copolymerized with vinyl acetate. Depending upon the vinyl acetate content (normally varying in a range of 0–40 wt%), this polymer can be tailored to deliver drugs at the required rate (Kim et al. 1980). The copolymer is available commercially as Elvax and Nipoflax, and is used in several commercial products such as Occusert, Transderm Nitro and Progestasert®.

1.2 Polydimethylsiloxane

Polydimethylsiloxane (see Fig. 2) is a very important component of several implants, artificial organs and other biomedical products (Arkles 1983). These polymers are stable, biocompatible and nonimmunogenic, and have been applied in the fabrication of several implantable and transdermal drug delivery systems. Silastic medical grade products (Dow Corning Corporation, Midland, Michigan) are commonly fabricated by copolymerizing dimethylsiloxane with varying amounts of methylvinylsiloxy units using vulcanizing agents such as dichlorobenzoyl peroxide (Arkles 1983). The resulting rubber has been used for the controlled delivery of drugs via subcutaneous, intrauterine and vaginal routes (Chien 1982). The Nitrodisc system uses one of the silicone elastomers to form a microsealed drug delivery system (Chien 1985a). More recently, several resins have been copolymerized on the polydimethylsiloxane backbone to yield adhesives that are useful for transdermal drug delivery systems (Pfister 1989).

1.3 Hydrogels

Hydrogels are hydrophilic polymers that are glassy in the dehydrated state and swell in the presence of

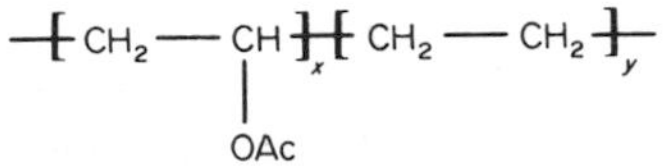

Figure 1
Chemical structure of ethylene–vinyl acetate copolymer

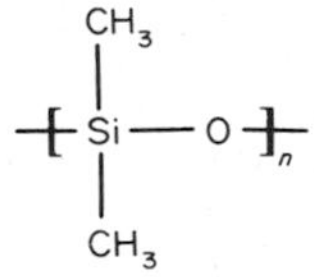

Figure 2
Chemical structure of polydimethylsiloxane

water to form an elastic gel. Their relatively high water content and their soft rubbery nature in the swollen state enables these polymers to entrap drugs in an aqueous medium and regulate their release (Lee 1988). Many polymers including poly(vinyl alcohol), poly(N-vinyl-2-pyrrolidone), polyelectrolyte compounds and poly(hydroxyalkyl methacrylates) may fall into the definition of hydrogels (see Fig. 3). Hydrogels may also arise from natural biopolymers such as dextrans and collagens. Hydrogels are hydrolytically stable, tissue compatible and can be used for controlling the release of both lipophilic and hydrophilic drugs. These polymers have been used experimentally for the delivery of steroids (Chien and Lau 1976, Zentner et al. 1978), fluorides (Cowsar et al. 1976) and some proteins such as heparin (Mori et al. 1978). More recently, Horbett and Ratner (1985) designed a self-regulating hydrogel-based system to regulate the delivery of insulin in response to glucose levels.

1.4 Biodegradable and Bioerodible Polymers

As the name suggests bioerodible polymers slowly breakdown in the biological microenvironment, thus releasing the drug by a combination of diffusion and degradation. The synthesis of biodegradable polymers for achieving controlled drug release is based on two broad approaches (Heller 1987). In one method, the drug is dispersed in the polymer and the release of the drug is achieved by the degradation of the polymer. In the other, the drug is attached to the polymer through a hydrolytically labile linkage, and the drug release is controlled by both the hydrolysis of the drug–polymer linkage and the diffusion through the polymer matrix.

Three main classes of polymers can be used for fabrication of bioerodible drug delivery devices (Heller 1984):

(a) water-soluble polymers that are rendered insolubilized via hydrolytically unstable cross-links,

(b) water-insoluble polymers that become soluble following a hydrolysis reaction, but retain their integrity, and

(c) water-insoluble polymers that become soluble via backbone cleavage reactions.

Some commonly used bioerodible polymers include poly(lactic acid), poly(glycolic acid), polylactide

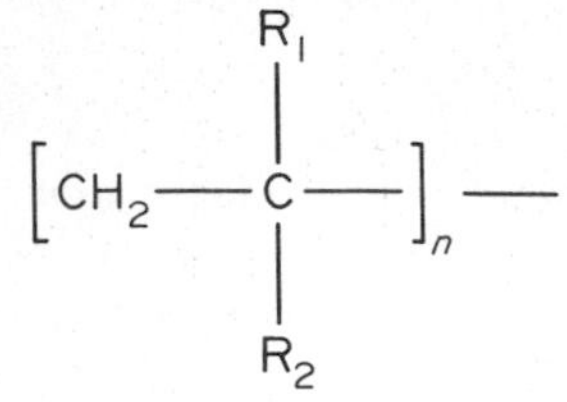

$$\left[CH_2 - \underset{\underset{R_2}{|}}{\overset{\overset{R_1}{|}}{C}} - \right]_n -$$

Polymer	R_1	R_2
Poly(vinyl alcohol)	—H	—OH
Poly(N—vinyl—2—pyrrolidone)	—H	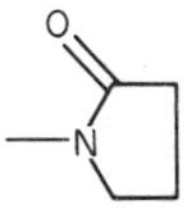 (N-pyrrolidone ring)
Poly(vinyl acetate)	—H	—O—C(=O)—CH₃
Poly(hydroxyethyl methacrylate)	—CH₃	—C(=O)—OCH₂CH₂OH
Polyacrylamide	—H	—CONH₂

Figure 3
Chemical structure of some hydrogels commonly used in controlled drug delivery

co-glycolide, poly(ε-caprolactone), polyorthoesters and polyurethanes. Another interesting application of bioerodible polymers would be to attach the drug covalently to the polymer backbone which would then cleave *in vivo* and release the drug in a controlled fashion (Heller 1987).

1.5 Cellulose Derivatives

Although many polymers have been used for film coating in oral controlled release products, the most widely used are the cellulose derivatives, such as methyl cellulose (MC), hydroxypropyl cellulose (HPC), hydroxypropyl methyl cellulose (HPMC), ethyl cellulose (EC) and cellulose acetate phthalate (CAP) (see Fig. 4, where R can be any functional group or combination on any of the three positions of the cellulose ring. The ratio of one functional group to another will depend upon the degrees of polymerization and will give rise to various grades of

the polymer). They are all derived from the cellulose backbone which consists of a basic repeating structure of anhydroglucose units, each unit having three replaceable hydroxyl groups. These polymers are generally odorless, tasteless, stable, biologically compatible and are useful for the film coating of pellets as well as for tablets (McGinity 1989).

HPMC is the most widely used water-soluble polymer and is available commercially as Opadry and Pharmacoat. Ethyl cellulose is a water-insoluble polymer and is available under the trade name Sulrelease.

1.6 Other Polymers

(*a*) *Acrylic resins.* Marketed under the trade name Eudragit, these polymers (see Fig. 5) are widely used as film coatings and as matrix bases for oral controlled release systems. They are usually copolymers of acrylic and methacrylic acid esters. Several

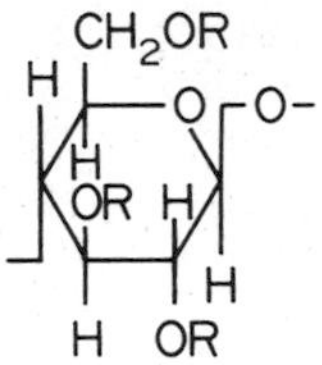

Polymer	R		
Methyl cellulose	—H	$—CH_3$	
Ethyl cellulose	—H	$—CH_2CH_3$	
Hydroxypropyl cellulose	—H	$—CH_2—CH(OH)—CH_3$	
Hydroxypropyl methyl cellulose	—H	$—CH_2—CH(OH)—CH_3$	$—CH_3$
Cellulose acetate phthalate	—H	$—C(=O)—CH_3$	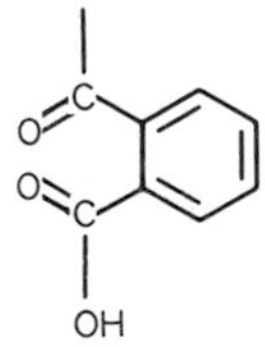

Figure 4
Cellulose derivatives for film coating or matrix bases in oral controlled drug delivery systems

grades such as RL, RS, RL PM, L30D and E30D are available with a wide spectrum of applications (Lehmann and Dreher 1969). These polymers are generally water insoluble, pH independent and are useful coating materials for lipophilic and hydrophilic drugs.

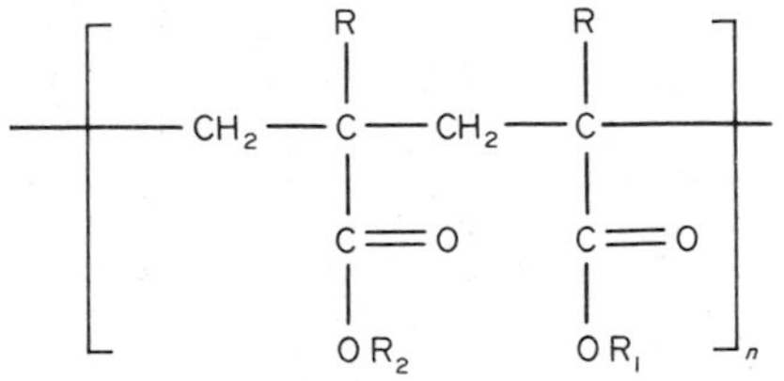

$R = H, CH_3$

$R_1 = CH_3, C_2H_5$

$R_2 = H, CH_2—CH_2—\overset{+}{N}(CH_3)_3$

Figure 5
Chemical structure of acrylic resins

(*b*) *pH-sensitive polymers*. Most of the polymers that have been described previously operate independent of the pH of the environment. Polymers such as CAP, hydroxymethyl cellulose and certain half esters of maleic anhydride when coated onto tablets provide an "enteric coating." Such coatings are used to protect acid-labile drugs from the acidic environment of the stomach ($pH < 3$), but contain acidic groups that ionize in the higher-pH environment in the intestine allowing swelling and dissolution of the polymer (Wood 1984).

(*c*) *Ion-exchange polymers*. Drugs that are either acidic or basic can form ion pairs with anionic or cationic ion-exchange resins. The release of drugs from such a drug–resin complex is governed by the ionic environment of the gastrointestinal tract.

2. *Development*

During the 1980s there was a surge in the development of rate-controlled drug delivery systems, and this new area of pharmaceutics research has increased in popularity and grown in demand in the medical, veterinary and dental fields. Controlled release products are now manufactured for every purpose, from pesticides (e.g., roach control strips) to the long-term delivery of insulin, by implantable devices, in humans. Table 2 illustrates some of the polymer-based controlled release products that have been successfully marketed for medical therapy.

2.1 Ocular Drug Delivery

Drugs can be applied for local effect of medication onto either the surface or the interior of the eye. A great part of any medication applied to the eye is normally diluted immediately by tears and removed by the nasolacrimal drainage systems (Chien 1982). Consequently, systems that are designed for ocular delivery must be capable of remaining in prolonged contact with the ocular tissue. A promising development in this field is that of the Ocusert system, an ocular insert unit for the controlled delivery of pilocarpine for the long-term continuous management of glaucoma. It consists of a pilocarpine–alginate core sandwiched between two semipermeable ethylene–vinyl acetate copolymer membranes and is designed to deliver, at a zero-order rate, 20 µg h^{-1} or 40 µg h^{-1} for a period of 5–7 days.

Other developments in this field include the development of hydrophilic contact lenses made of 2-hydroxyethyl methacrylate (a hydrogel-type material) which are soaked in the drug solution before application into the eye. Bioerodible systems that need not be removed after use have also been fabricated as ocular inserts. Such ocular devices are made of polysaccharides or cross-linked polypeptides in which the drug is homogeneously impregnated.

2.2 Nasal Drug Delivery

Although the nasal route is commonly used for the delivery of drugs to alleviate histamine-triggered allergic and related symptoms resulting from local infections and inflammation, there is a growing interest in using this route for the delivery of systematically effective drugs. A wide range of β blockers (e.g., propranolol), steroids (e.g., progesterone, estradiol) and proteins (e.g., insulin, interferon, vasopressin, oxytocin, LHRH analogs, thyroxine releasing hormone (TRH)) have been investigated and have been found to be absorbed from the nasal mucosa. Nasal drug delivery thus forms a potential route for the administration of a large variety of drugs using devices such as metered-dose nasal sprays and aerosols (Chien and Chang 1987). Hydrogels, such as Carbopol 940 and cellulose derivatives, have been reported to be useful for

Table 2
Some marketed and investigational polymer-controlled drug delivery systems

Rate-controlling polymer(s)	Site of application	Trade name	Drug	Manufacturer/ developer
EVAc[a]	ocular	Ocusert	pilocarpine	Alza Corp.
EVAc	IUD	Progestasert	progesterone	Alza Corp.
EVAc	transdermal	Transderm Nitro	nitroglycerin	Alza Corp.
EVAc	transdermal	Transderm Scop	scopolamine	Alza Corp.
EVAc	transdermal	Estraderm	estradiol	Alza Corp.
Polypropylene	transdermal	Catapress TTS	clonidine	Alza Corp.
PVA/PVP[b]	transdermal	Nitrodur	nitroglycerin	Key Pharm.
Acrylic adhesive	transdermal	Nitrodur II	nitroglycerin	Key Pharm.
Polyisobutylene	transdermal	Deponit	nitroglycerin	Lohmann
Olefinic copolymer	transdermal	NTG	nitroglycerin	Hercon Labs
Silicone elastomer	transdermal	Nitrodisc	nitroglycerin	G. D. Searle
Silicone elastomer	transdermal	investigational	levonorgestrel/ estradiol	Rutgers Univ.
Silicone elastomer	subdermal	Norplant	levonorgestrel	Leiras
Polyurethane	topical	Spandra	gentamycin	Thermedics
MC[c]	ocular	Artificial Tears	—	Ciba
HPC[d]	ocular	Lacrisert	soluble HPC	Alza Corp.
HPMC[e]	oral	Valrelease	diazepam	Roche
Cellulose triacetate	oral	Acutrim	phenylpropanol-amine	Alza Corp.
Cellulose triacetate	topical	investigational	gentamycin/ clindamycin	Moleculon
Lactide–Glycolide copolymer	subdermal	investigational	zoladex	ICI

a ethylene–vinyl acetate b poly(vinyl alcohol)/poly(vinyl pyrrolidone) c methyl cellulose d hydroxypropyl cellulose e hydroxypropyl methyl cellulose the preparation of sustained release nasal drugs, which maintain a long-term intimate contact of therapeutic agents with the nasal mucosa (Hussain et al. 1980).

2.3 Intrauterine Drug Delivery

Contraceptive pills and intrauterine devices (IUDs) are the two most effective methods currently available for reversible contraception. IUDs are commonly medicated with antifertility agents such as copper (e.g., Cu-7 and Multiload Cu 250) or progesterone (e.g., Progestasert) (Chien 1982). A Cu-7 IUD is a 7-shaped polypropylene frame with 89 mg of copper wire winding around the vertical limb, giving an effective copper surface of $200 \, mm^2$. A mean daily dose of about 10 µg of copper is released *in utero* for a period of about three years.

The Progestasert device consists of a medicated core with progesterone suspended as microcrystals in a silicone medical fluid, which is then encapsulated in a rate-limiting nonporous ethylene–vinyl acetate copolymer membrane. The device continuously delivers approximately 65 µg day^{-1} of progesterone into the uterine cavity for one year.

2.4 Subcutaneous Drug Delivery

Subcutaneous implantation is useful for drugs that need to be administered continuously for long-term therapy. Certain contraceptive steroids, antineoplastic agents and proteins are suited for this purpose. The polymers used for preparation of subdermal implants should be biocompatible, bioerodible (preferably but not necessarily) and easily sterilizable. Poly(lactic acid), poly(glycolic acid) and their copolymers have been used as bioerodible polymers for the controlled delivery of narcotic drugs such as naltrexone and contraceptive peptides such as synthetic LHRH analogs. Vapor pressure-activated drug delivery devices such as Infusaid and osmotic pressure-activated drug delivery devices such as the Alzet osmotic minipump can be used as implants for the controlled delivery of proteins such as heparin and insulin as well as for drugs such as propranolol. These implants may be designed in such a way as to release the drug in a circadian manner, so as to deliver drugs in compliance with the biological rhythms.

2.5 Rectal Drug Delivery

The use of the rectum for the local delivery of drugs using cocoa butter suppositories has been used for several centuries; however, its use for systemic drug delivery has only recently been investigated. Rectal drug administration offers some obvious advantages in cases of nausea and vomiting or unconsciousness of the patient. In addition, the "first pass" elimination of high-clearance drugs is partially avoided, because the lower part of the rectum is directly connected to the systemic circulation and bypasses the liver. Nevertheless, two important characteristics of rectal administration are obvious: the lack of patient compliance and the removal of the system by defecation. The Osmet osmotic pump system has been used for the rectal delivery of drugs, such as antipyrine and theophylline, to achieve constant plasma concentrations that are comparable to intravenous infusions, for a period of 24 h to 40 h (DeLeede et al. 1982). The same group of researchers also studied the rectal delivery of theophylline using a cylindrical hydrogel preparation.

2.6 Oral Drug Delivery

This is the oldest and perhaps the most well-accepted route of drug delivery currently available and the system design for this mode of administration has undergone a major metamorphosis from enteric coated tablets to zero-order drug delivery systems. Conventional sustained release oral delivery systems in clinical use are usually one of two types: matrix-embedded or polymer-coated. Matrix systems can be made of inert or noninert bioerodible materials.

Reproducible gastric emptying rates can be achieved by using multiunit delivery systems, a good example of which is Theo-Dur. A Theo-Dur tablet consists of many small cores (consisting of sugar beads surrounded by theophylline, wax and cellulose acetate phthalate), embedded in a theophylline-containing granulated matrix that serves as an initial dose. Another example is that of the Oros osmotic system designed to release drugs at a constant rate. This system has been used for drugs such as phenylpropanolamine, nifedipine, indomethacin and potassium chloride.

More recently, hydrodynamically balanced systems have been designed to float in the stomach and slowly release the drug (e.g., diazepam) into the gastrointestinal tract. The intragastric floating tablet contains hydroxypropyl cellulose and hydroxypropyl methyl cellulose as the polymeric materials which form a colloidal gel barrier when in contact with the gastric fluid.

2.7 Buccal and Sublingual Drug Delivery

Buccal and sublingual tablets as well as lozenges have been used for several decades for the local or systemic delivery of drugs. These drugs are absorbed quickly into the reticulated vein that lies under the oral mucosa and enter the systemic circulation directly, bypassing the liver. They have been used for the delivery of drugs such as testosterone, nitroglycerin and buprenorphine as well as for macromolecules such as insulin, LHRH analogs and TRH. Nagai and Machida (1985) delivered insulin buccally using a dome-shaped mucosal adhesive device consisting of a core of insulin, cocoa butter and sodium glycocholate, as the enhancer, and a bioadhesive peripheral layer made of hydroxypropyl cellulose and Carbopol 934. The dosage form adhered tightly to the oral mucosa of dogs in a gel-like swollen state without disruption and its shape was kept intact for more than 6 h.

2.8 Transdermal Drug Delivery

Delivery of drugs through the skin has advanced more than any other field of controlled drug delivery. This is apparent from the number of drug delivery systems that are marketed (see Table 2). There are three major technologies that are available for the preparation of these systems, all of them involving drug release from a polymer and subsequent permeation through the skin (Chien 1987).

(*a*) *Membrane permeation-controlled*. This type of system is exemplified by the Transderm Nitro, Transderm Scop, Catapress TTS and the Estraderm systems. It consists of a drug reservoir, a rate-controlling membrane (typically made of ethylene–vinyl acetate) and an adhesive (for intimate contact with the skin) sandwiched between a release liner and a backing membrane. The drug diffuses through the rate-controlling membrane, passes through the adhesive layer and then permeates through the skin to reach the blood circulation.

(*b*) *Matrix diffusion-controlled*. This type of system consists of a drug reservoir fabricated by homogeneously dispersing the drug in a rate-controlling polymer matrix that is made from either a lipophilic or a hydrophilic polymer. The Nitro-Dur system consists of an aqueous combination of polymers such as poly(vinyl alcohol), glycerol and poly(vinyl pyrrolidone) with nitroglycerin dispersed in the polymer gel. The device is fabricated to release a daily dose of 0.5 mg cm^{-2} for the treatment of angina pectoris. A newer type of drug delivery system is the Nitro-Dur II which is fabricated by dispersing the drug directly in an acrylic-adhesive matrix. This system has the advantage that it is self-adhesive and does not require a separate adhesive layer for intimate contact with the skin. Similarly, adhesives have also been used to prepare the Deponit system which is a multilaminated system consisting of a gradient of drug loading dispersed in a polyisobutylene adhesive.

(*c*) *Microsealed*. This polymeric system is exemplified by the development of Nitrodisc, in which the drug reservoir is a suspension of solid drug particles in an aqueous solution of a water-miscible polymer, which in turn forms a dispersion of microscopic reservoirs in a silicone-polymer matrix (Chien 1985a). The release of drugs from such a device can follow either partition control or a matrix-diffusion control (Chien 1984).

3. Kinetics

3.1 Membrane Permeation-Controlled Systems

Control of drug release by its permeation through a polymeric membrane from a drug-containing reservoir is a useful means by which to achieve drug release at a constant rate (zero-order kinetics). The underlying theory for the diffusion of a drug across a polymeric membrane is based on Fick's first law of diffusion (Crank 1975) which states that the flux J or rate of solute transfer across a plane of unit area is proportional to the concentration gradient across a diffusional path x. This can be represented mathematically as

$$J = -D\frac{\delta c}{\delta x} \tag{1}$$

where D is the diffusion coefficient or diffusivity, and $\delta c/\delta x$ is the concentration gradient. The negative sign denotes that the material flow is from a region of higher concentration to a region of lower concentration. For an isotropic (or homogeneous) film with a thickness l and a cross-sectional area A, at steady state, Fick's law can be simplified to

$$J = \frac{AD\,\Delta C_\mathrm{m}}{l} \tag{2}$$

Under sink conditions, this equation can be further simplified to

$$J = \frac{ADKC_\mathrm{s}}{l} \tag{3}$$

where K is the partition coefficient for the interfacial partitioning of drug molecules from the reservoir to the membrane, and C_s is the saturated solubility of the drug in the reservoir, whereas ΔC_m is the concentration difference across the polymeric membrane. This equation is applicable to a system of planar geometry. In the case where drug molecules permeate through a porous polymeric membrane via diffusion through the pore channels only, Eqn. (3) becomes

$$J = \frac{\varepsilon A \varepsilon/\tau DKC_\mathrm{s}}{\tau l} = \frac{A_\mathrm{e}D_\mathrm{e}KC_\mathrm{s}}{l_\mathrm{e}} \tag{4}$$

where ε is the porosity, τ is the tortuosity, and A_e, l_e and D_e are the effective surface area, effective thickness of the porous membrane and effective diffusivity of the drug molecules, respectively. Equation (4) is applicable, for example, to Occusert, Transderm-Nitro, Transderm Scop, Catapress TTS and Estraderm systems.

Table 3
Equations for monolithic systems in which the drug is dissolved in the polymer[a]

Device	Release rate	
	Early time approximation	Late time approximation
Slab[b]	$2\left(\dfrac{D}{\pi r^2 t}\right)^{1/2}$	$\dfrac{8D}{h^2}\exp\left(\dfrac{-\pi^2 Dt}{h^2}\right)$
Cylinder[c]	$2\left(\dfrac{D}{\pi r^2 t}\right)^{1/2} - \dfrac{D}{r^2}$	$\dfrac{4D}{r^2}\exp\left(\dfrac{-(2.405)^2 Dt}{r^2}\right)$
Sphere[c]	$3\left(\dfrac{D}{\pi r^2 t}\right)^{1/2} - \dfrac{3D}{r^2}$	$\dfrac{6D}{r^2}\exp\left(\dfrac{-\pi^2 Dt}{r^2}\right)$

a modified from Baker R W, Lonsdale H K 1974 Controlled release: mechanism and rates. In: Tanquery A C, Lacey D R (eds.) 1974 *Controlled Release of Biologically Active Agents*. Plenum, London, p. 15   b early time approximation is valid for up to 40% release, late time approximation is valid for greater than 40% release   c early time approximation is valid for up to 40% release, late time approximation is valid for greater than 60% release

For a cylindrical device coated with a polymeric membrane, such as the Progestasert IUD, the flux can be expressed as

$$J = \frac{2\pi h D K C_s}{\ln(r_0/r_1)} \tag{5}$$

where h is the length of the cylinder and r_1 and r_0 are the inner and outer radii.

Similarly, for a spherical device coated with a polymeric membrane, such as a microcapsule, the release rate is given by

$$J = \frac{4\pi D K C_s r_0 r_1}{(r_0 - r_1)} \tag{6}$$

3.2 Matrix Diffusion-Controlled Systems (Monolithic Devices)

(a) *Drug dissolved in the polymer*. In this type of polymeric matrix system, the drug is present in the polymer in a totally dissolved state. This system is generally prepared by soaking the polymer film in the drug solution, for a certain period of time, to impregnate the polymer with the drug. The release rates of drug from such a system of varying geometry are summarized in Table 3. Examples of this type of polymeric matrix system include norgestomet-releasing Syncro-Mate B subdermal implants for estrus synchronizaton in livestock (Chien and Lau 1976).

(b) *Drug dispersed in the polymer*. In this type of polymeric matrix system, a loading dose of drug

A, which is greater than the polymer solubility C_p of the drug, is dispersed in the polymer matrix. Thus, the amount of drug released Q is a function of the square root of time t and can be described by the following equation (Higuchi 1961):

$$Q = [(2A - C_p)C_p Dt]^{1/2} \tag{7}$$

Examples of this type of polymeric matrix system include Nitro-Dur and Nitro-Dur II transdermal systems, Valrelease tablets and Compudose implants.

If the release occurs from a porous device, such as a tablet, Eqn. 7 can be modified to (Higuchi 1963, Flynn et al. 1974)

$$Q = \left[\frac{\varepsilon}{\tau}(2A - \varepsilon C_p)C_p Dt\right]^{1/2} \tag{8}$$

where ε is the porosity and τ is the tortuosity of the matrix.

3.3 Bioerodible Systems

For a bioerodible/biodegradable delivery system, the drug is released only when the matrix hetero-geneously erodes. Hopfenberg (1976) modelled the release from cylindrical, spherical and flat matrices and found that a single relationship could be used to describe all three shapes:

$$\frac{Q_t}{Q_\infty} = 1 - \left(1 - \frac{Kt}{Aa}\right)^n \tag{9}$$

where Q_t is the amount of drug released from the device at time t, Q_∞ is the total amount of drug released when the device is exhausted, K is the erosion rate constant, A is the drug loading dose in the matrix, a is the radius of a sphere/cylinder or the half thickness of a slab, and n is a shape factor with a value of 3 for a sphere, 2 for a cylinder and 1 for a slab. Thus, a slab gives a zero-order drug release profile while a sphere or a cylinder have release rates that decrease with time. Equation (9) ignores the contribution of diffusion to the release. Lee (1980) developed a general model that takes into account erosion as well as diffusion. The analytical solution is complicated, but it predicts that the release profile would approach zero-order kinetics if the drug loading is much greater than the drug solubility in the matrix.

3.4 Hydrogel-Type Systems

These polymers can be stored in their fully hydrated state, such as in the case of ophthalmic drug-containing contact lenses, or in the dry glassy state, as in the case of oral dosage forms (Andrade 1976). The release kinetics of drugs from a fully hydrated state of a hydrogel will follow the equations given in Table 3 if the drug is dissolved in the hydrogel, whereas if the drug is dispersed in the hydrogel the kinetics will be similar to those given by Eqns. (7,8). The drug release from a dry hydrogel in contact with aqueous fluid is more complicated as it involves a combination of diffusion and swelling of the hydrogel boundary at the same time. The solution of the moving boundary has been solved by Good (1980) and can be expressed by

$$Q_t = Q_\infty \left\{ 1 - \frac{8}{\pi^2} \sum_0 \frac{\exp[-(2n+1)^2 \pi^2 f(T)]}{(2n+1)^2} \right\} \quad (10)$$

where $f(T)$ is a time function which can be further simplified at short times to

$$f(T) \sim \frac{Dt}{l^2} - \frac{D}{12D^*} \quad (11)$$

where D^* is the solvent diffusivity and l is the thickness of the hydrogel membrane.

3.5 Osmotic Pressure-Controlled Devices

Such devices contain an osmotically active agent that coats the external surface of the drug reservoir compartment and is enclosed in a rigid housing. The drug reservoir is contained inside a collapsible compartment. The wall of the rigid housing is constructed from a semipermeable membrane consisting of a polymer, such as cellulose triacetate, so that when the device is exposed to an aqueous environment, water will be imbibed into the device to dissolve the osmotically active agent and the osmotic pressure generated in turn will force the active agent through a delivery orifice (Chien 1985b). There are two major types of osmotic pressure-controlled drug delivery systems. The first is an osmotic pump (e.g., Alzet osmotic pump) in which the drug is in the form of a solution (Fara and Mitchell 1987). In such a system the release rate of the drug can be given by

$$\frac{dQ}{dt} = \frac{P_w A_m}{h_m} (\pi_s - \pi_e) \quad (12)$$

where P_w, A_m and h_m are the water permeability, the effective surface area and the thickness of the semipermeable membrane, respectively, and $(\pi_s - \pi_e)$ is the differential osmotic pressure between the delivery system and the environment.

In the second type of osmotic system (e.g., Acutrim, an appetite suppressant tablet), the drug is present as a solid mixed with the osmotically active agent. The rate of drug release from such a system can be defined by

$$\frac{dQ}{dt} = \frac{P_w A_m}{h_m} (\pi_s - \pi_e) S_d \quad (13)$$

where S_d is the aqueous solubility of the drug in the solid reservoir and the other terms are as in Eqn. 12.

4. Summary

The rate-controlled polymeric drug delivery systems outlined in this article have been steadily introduced into the biomedical community since the mid-1970s. It is expected that many more conventional drug delivery systems will be gradually replaced by these high-technology-based drug delivery systems.

Bibliography

Andrade J D (ed.) 1976 *Hydrogels for Medical and Related Applications*, American Chemical Society Symposium Series No. 31. ACS, Washington, DC
Arkles B 1983 Look what you can make out of silicones. *Chemtech.* 13(9): 542–55
Chien Y W 1982 *Novel Drug Delivery Systems: Fundamentals, Developmental Concepts and Biomedical Assessments.* Dekker, New York
Chien Y W 1984 Microsealed drug delivery systems. In: Anderson J M, Kim S W (eds.) 1984 *Recent Advances in Drug Delivery Systems.* Plenum Press, New York, pp. 367–88
Chien Y W 1985a Microsealed drug delivery systems: Methods of fabrication. In: Widder K J, Green R (eds.) 1985 *Drug and Enzyme Targeting, Methods of Enzymology*, Vol. 112. Academic Press, Orlando, FL, Chap. 34
Chien Y W 1985b Polymer-controlled drug delivery systems: Science and engineering. In: Gebelein C G, Carraher C E (eds.) 1985 *Polymeric Materials in Medication.* Plenum Press, New York, pp. 27–46
Chien Y W 1987 *Transdermal Controlled Systemic Medication.* Dekker, New York
Chien Y W, Chang S-F 1987 Intranasal drug delivery for systemic medications. *CRC Crit. Rev. Therapeutic Drug Carrier Systems* 4(2): 67–194
Chien Y W, Lau E P K 1976 Controlled drug release from polymeric delivery devices (IV): *In vitro–in vivo* correlation on the subcutaneous release of Norgestomet from hydrophilic implants. *J. Pharm. Sci.* 65: 488–92
Cowsar D R, Tarwater O R, Tanquary A C 1976 Controlled release of fluoride from hydrogels for dental applications. In: Andrade 1976, pp. 180–97
Crank J 1975 *The Mathematics of Diffusion*, 2nd edn. Oxford University Press, New York
DeLeede L G, De Boer A G, Van Velzen S L, Breimer D D 1982 Zero order rectal delivery of theophylline in man with an osmotic system. *J. Pharmacokinet. Biopharm.* 10: 525–37
Fara J, Mitchell C 1987 Osmotic systems for rate controlled drug delivery in preclinical and clinical research. In: Struyker-Boudier H A J (ed.) 1987 *Rate Controlled Drug Administration and Action.* CRC Press, Boca Raton, FL

Flynn G L, Yalkowsky S H, Roseman T J 1974 Mass transport phenomena and models: Theoretical concepts *J. Pharm. Sci.* 63: 479–510

Good W R 1980 Diffusion of water soluble drugs from initially dry hydrogels: In: Baker R (ed.) 1980 *Controlled Release of Bioactive Materials.* Academic Press, New York, pp. 139–45

Heller J 1984 Zero order drug release from bioerodible polymers. In: Anderson J M, Kim S W (eds.) 1984 *Recent Advances in Drug Delivery Systems.* Plenum Press, New York, pp. 101–22

Heller J 1987 Use of polymers in controlled release of active agents. In: Robinson J R, Lee V H L (eds.) 1987 *Controlled Drug Delivery: Fundamentals and Applications,* 2nd edn. Dekker, New York, pp. 139–212

Higuchi T 1961 Rate of release of medicaments from ointment bases containing drugs in suspension. *J. Pharm. Sci.* 50: 874–5

Higuchi T 1963 Mechanism of sustained action medication. *J. Pharm. Sci.* 52: 1145–9

Hopfenberg H B 1976 Controlled release from erodible slabs, cylinders and spheres. In: Paul D R, Harris F W (eds.) 1976 *Controlled Release Polymeric Formulations,* American Chemical Society Symposium Series No. 33. ACS, Washington, DC, pp. 26–42

Horbett T A, Ratner B D 1985 Enzymatically controlled drug release systems. In: Widder K, Green R (eds.) 1985 *Drug and Enzyme Targeting, Methods of Enzymology* Vol. 112. Academic Press, Orlando, FL, pp. 484–95

Hussain A, Hirai S, Bawarshi R 1980 Absorption of propranolol from different dosage forms in rats and dogs. *J. Pharm. Sci.* 69: 1411–13

Kim S W, Peterson R V, Feijen J 1980 Polymeric drug delivery systems. In: Ariens E J (ed.) 1980 *Drug Design,* Vol. 10. Academic Press, New York, pp. 217–19

Lee P I 1980 Diffusional release of a solute from a polymeric matrix—approximate analytical solution. *J. Membrane Sci.* 7: 255–62

Lee P I 1988 Synthetic hydrogels for drug delivery: Preparation characterization and release kinetics. In: Hsieh D (ed.) 1988 *Controlled Release Systems,* Fabrication Technology, Vol. 2. CRC Press, Boca Raton, FL

Lehmann K, Dreher D 1969 Permeable acrylic resin coatings for the manufacture of depot preparations of drugs. *Drugs Made Ger.* 12: 59–71

McGinity J W (ed.) 1989 *Aqueous Polymeric Coatings for Pharmaceutical Dosage forms.* Dekker, New York

Mori Y, Nagaoka S, Masubuchi J, Tanzawa H, Itoga M, Yamada Y, Yonaka T, Watanabe H, Idezuki Y 1978 The effect of released heparin from the heparinized hydrophilic polymer (HRSD) on the process of thrombus formation. *Trans. Am. Soc. Artif. Intern. Organs* 24: 736–45

Nagai T, Machida Y 1985 Mucosal adhesive dosage forms. *Pharm. Int.* 6: 196–200

Pfister W R 1989 Customizing silicone adhesives for transdermal drug delivery systems. *Pharm. Tech.* 13(3): 126–38

Wood D A 1984 Polymers in drug delivery systems. In: Florence A T (ed.) 1984 *Polymeric Materials used in Pharmaceutical Formulations.* Blackwell, New York, pp. 99–119

Zentner G M, Cardinal J R, Kim S W 1978 Progestin permeation through polymer membranes II: Diffusion studies on hydrogel membranes. *J. Pharm. Sci.* 67: 1352–7

R. Toddywala and Y. W. Chien
[Rutgers University, Piscataway, New Jersey, USA]

Polysiloxanes

Polysiloxanes are widely used as biomedical materials. Available in a variety of material types, they have unique chemical and physical properties, excellent biocompatibility and biodurability, and are versatile as a biomedical material for construction of medical devices or other applications. Material types include elastomers, gel-consistency materials, fluids, lubricants, antifoams, adhesives and others. Polysiloxanes are typically chemically stable and unreactive with other materials, including most drugs and biological substances. They are unchanged by moisture, they are excellent electrical insulators, they are resistant to oxidation, ultraviolet light and high temperatures and their properties change little at low temperatures. They are unchanged by clinical ultrasound and radiation, and can be sterilized with either steam autoclave, ethylene oxide or radiation. Ethylene oxide is highly soluble in most polysiloxanes and, when used for sterilization, adequate time must be allowed before use to ensure that there is no residual ethylene oxide present. When they are sterilized by radiation, the properties of polysiloxanes may change, because of dose-related intermolecular cross-linking. Thus, any time a polysiloxane material or device is to be sterilized by radiation, the critical properties important to its performance should be carefully retested in specimens subjected to the maximum exposure inherent in the process.

Of the many synthetic polymers developed by modern technology, polysiloxanes are among the few that are adequately biocompatible and biodurable for use in implant or other applications involving intimate contact with various tissues and fluids of the body. They are the only elastomeric materials generally accepted for use in implants or other uses where the biocompatibility and biodurability requirements are highly demanding. Implants of polysiloxane elastomers are used in the reconstructive procedures of nearly all surgical specialties. Soft implants consisting of a polysiloxane elastomer envelope filled with a soft polysiloxane gel are used when extra softness is desired, such as in breast reconstruction. Polysiloxane elastomer envelopes are also used in tissue expanders. Polysiloxane elastomers highly resistant to flaw propagation and flexion fatigue have been developed for use in the flexible hinge and spacer implants used by orthopedic, hand and foot surgeons for reconstruction of the smaller joints of the extremities. Hydrocephalus shunts fabricated from polysiloxane elastomer have been used for more than 30 years to treat this brain-destroying life-threatening condition.

Polysiloxane fluid serves as a safe effective lubricant for disposable hypodermic needles and syringes. Polysiloxane antifoams, known generically as simethicone, are used as an antiflatuent additive in most antacid preparations. The antifoams also suppress foam and improve the manufacturing efficiencies of a variety of pharmaceutical manufacturing processes. A variety of polysiloxanes, including elastomers, fluids and adhesives, are used in drug delivery systems. A nonirritating nonsensitizing polysiloxane pressure-sensitive adhesive may be used to affix drug repositories to the skin for transdermal drug delivery, such as nitroglycerin for the control of angina or scopolamine for the control of motion sickness. With some drugs, polysiloxane elastomer may be used as the material of construction for the drug repository, either for transdermal or implant delivery systems. Silicone fluid may also be useful as a nonreactive stable diluent for drugs delivered transdermally from a repository.

In these and many other applications, polysiloxanes have become widely used as a biomedical material and almost everyone in the developed countries of the world will have had polysiloxanes used in one manner or another to improve the quality of their personal health care.

1. History

Precursors to modern polysiloxanes were prepared by Kipping (1907) in the early 1900s. He reacted Grignard reagents with tetrachlorosilane to yield organochlorosilanes:

$$2\ CH_3CH_2MgBr + SiCl_4 \longrightarrow Cl - \overset{\overset{\displaystyle CH_3}{|}\ \overset{\displaystyle CH_2}{|}}{\underset{\underset{\displaystyle CH_3}{|}\ \underset{\displaystyle CH_2}{|}}{Si}} - Cl + 2\ MgBrCl$$

Diorganodichlorosilane, isolated from the Grignard reaction, was reacted with water to yield a mixture of oily polydiorganosiloxanes:

$$x\ Cl - \overset{\overset{\displaystyle CH_3}{|}\ \overset{\displaystyle CH_2}{|}}{\underset{\underset{\displaystyle CH_3}{|}\ \underset{\displaystyle CH_2}{|}}{Si}} - Cl + x\ H_2O \longrightarrow \left[\overset{\overset{\displaystyle CH_3}{|}\ \overset{\displaystyle CH_2}{|}}{\underset{\underset{\displaystyle CH_3}{|}\ \underset{\displaystyle CH_2}{|}}{Si}} - O\right]_x + 2x\ HCl$$

The polymers were called "silicones," because elemental analysis matched that for silicon-containing analogues of organic ketones.

$$\underset{\text{acetone}}{\overset{\overset{\displaystyle CH_3}{|}}{\underset{\underset{\displaystyle CH_3}{|}}{C}} = O} \qquad \underset{\text{polydimethylsiloxane}}{\left[\overset{\overset{\displaystyle CH_3}{|}}{\underset{\underset{\displaystyle CH_3}{|}}{Si}} - O\right]_x}$$

At the time it was not known that silicon does not form covalent double bonds with oxygen. If it did, silicon dioxide (molecular weight 50) would be a gas! Kipping found the chemistry and materials sciences of these new materials fascinating, but he found no practical applications.

The early research by Hyde at Corning Glass Works and the later participation by co-workers at Mellon Institute during the 1930s developed the technology that made modern commercial polysiloxanes possible. The processes developed for synthesis included techniques to control composition and molecular weight of the resulting polysiloxanes.

One of the earliest proposed uses of polysiloxanes was as a high-temperature electrically insulating resin to impregnate the newly developed glass fibers. However, during World War II the allies had a great need for an ignition-sealing compound to improve the performance of aircraft engines at high altitudes. Polysiloxane grease was found to be effective and the construction of the Dow Corning Plant as a joint venture between Corning Glass Works (polysiloxane research) and Dow Chemical Company (chemical processing know-how) received high priority from the government. Research continued and in 1945, as World War II ended, uses of polysiloxanes for nonmilitary applications could again be considered. The applications that soon developed included polysiloxane resins to impregnate fiberglass for high-temperature electrical insulation, polysiloxane fluid-impregnated lens tissue for cleaning eyeglass lenses, and the use of resins, fluids, antifoams and elastomers in a wide variety of applications where the emerging high technologies placed ever-increasing demands on materials. Animal studies, conducted in the 1940s to assess the biological properties of these new materials, showed that nonvolatile methyl- and mixed phenyl-methyl-polysiloxanes were essentially nontoxic. When data were published by Rowe et al. (1948, 1950), medical scientists recognized the biomedical material potential of these new materials and the new opportunities offered to researchers. Holter's construction of a hydrocephalus shunt in 1955 from dimethylpolysiloxane elastomer, for use in his own hydrocephalic son (LaFay 1957), demonstrated that a polysiloxane elastomer implant could facilitate treatment of a health condition that otherwise had no effective treatment. The hydrocephalus shunt became the treatment of choice for this health defect by 1957 and initiated the era of implant development that continues today. During the 1960s, Cronin and Gerow (1964) developed the polysiloxane elastomer/gel breast implant, and Swanson (1973) developed the finger joint and other implants for use in reconstruction of upper and lower extremities. Throughout the 1960s, medical researchers considered the potential of using implants from polysiloxane elastomer in reconstructive surgery to treat a wide variety of health conditions where various components of the body were diseased or defective and not otherwise repairable. Polysiloxane elastomer implants are now used in nearly every surgical specialty.

During the 1960s, polysiloxane fluid was recognized as a safe effective lubricant for hypodermic needles and syringes, facilitating the development of disposable devices. The value of polysiloxane antifoams as an antiflatuent additive to antacid preparations became known and widely used. Many biomedical uses for polysiloxanes were identified by the end of the 1960s.

During the 1970s and the 1980s, new biomedical materials applications for polysiloxanes included tissue expanders, drug delivery systems, inflatable urinary sphincters and inflatable penile implants. The properties of biomedical polysiloxanes continued to be improved by ongoing research. New elastomers, highly resistant to tearing, tear propagation and flexion fatigue, were developed for use in the Swanson flexible bone and joint implants (Swanson 1973, Swanson et al. 1982). New polysiloxane elastomer envelopes were developed for polysiloxane gel-containing implants that inhibit osmosis of fluid polymers from polysiloxane gel to the outside of the implant envelope. Implant designs were improved and refined, and new and improved elastomers were developed for use in tubing for blood pump and other applications. The need for continuing research to develop new and ever improved biomedical polysiloxane materials and devices is compelling. There is no ideal biomedical material and no ideal medical device. The demand for new and better health care techniques for patients with defective or destroyed parts of the body is always present.

2. Chemistry

The solid crust of the earth averages approximately 28 wt% silicon, second only to oxygen in the quantity present. Silicon has not been found as a free element on earth, but it is abundantly available in a variety of compounds (predominantly silicon dioxide and inorganic silicates). The processes for making polysiloxanes start with naturally occurring silicon dioxide. Silicon dioxide is chemically stable in earth's ambient environment, because of the high energy of the silicon–oxygen bond. The heat of formation of silicon dioxide is approximately $840 \, \text{J mol}^{-1}$. The heat of formation of carbon dioxide, by comparison, is less than $420 \, \text{J mol}^{-1}$. Silicon dioxide, typically as quartzite rock, is heated at high temperature with carbon

$$SiO_2 + C \overset{\Delta}{\to} Si + CO_2 \uparrow$$

At high temperature an equilibrium develops between the oxides of carbon and the oxides of silicon. Those of carbon are gaseous and leave the reaction mix and so the reaction goes to completion, yielding molten elemental silicon.

At ambient temperature elemental silicon is a hard, crystalline, brittle substance with a blue-gray metallic luster. This elemental silicon is pulverized and reacted directly with methyl chloride at elevated temperature:

$$Si + CH_3Cl \overset{\Delta}{\to} \begin{array}{l} SiCl_4 \\ CH_3SiCl_3 \\ (CH_3)_2SiCl_2 \\ (CH_3)_3SiCl \\ (CH_3)_4Si \end{array}$$

A mix of methylchlorosilanes is obtained, ranging from tetrachlorosilane to tetramethylsilane. Reaction conditions are usually adjusted to optimize the yield of dimethyldichlorosilane, the basic building block for polydimethylsiloxane. The various compounds are separated by fractional distillation. Dimethyldichlorosilane is then reacted with water to obtain polydimethylsiloxane prepolymer:

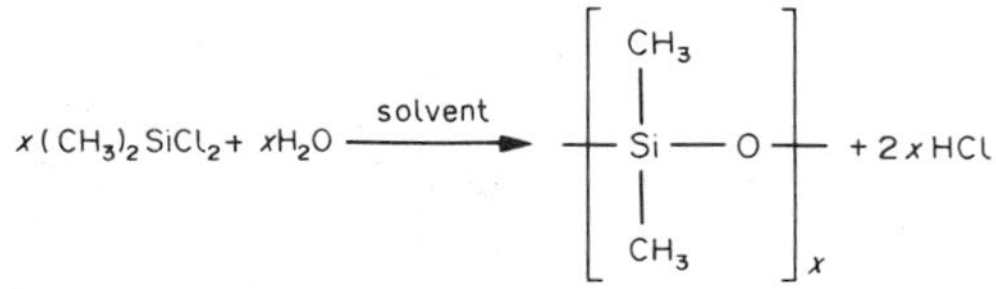

The dimethyldichlorosilane is normally diluted with solvent to help control the reaction rate and to dissipate heat. Excess water is added and the prepolymer solvent solution is separated from the hydrochloric acid solution. After washing to remove residual acid, solvent is removed by distillation. The prepolymer thus prepared contains a mix of dimethylpolysiloxanes with octamethylcyclotetra-

Table 1
Typical properties of polydimethylsiloxane fluid (Dow Corning 360 Medical Fluid)

Property	Value	Test method
Specific gravity at 298 K		
$0.2 \times 10^{-4}\,m^2\,s^{-1}$	0.949	ASTM D1298
$1.0 \times 10^{-4}\,m^2\,s^{-1}$	0.965	
$\geqslant 3.5 \times 10^{-4}\,m^2\,s^{-1}$	0.970	
Weight loss to circulating air over 24 h at 423 K $1.0 \times 10^{-4}\,m^2\,s^{-1}$	0.5%	
Refractive index	1.400–1.405	ASTM D1298
Pour point	186–227 K	ASTM D97
Viscosity/temperature coefficient	0.60	ASTM D445
Coefficient of expansion		
$0.20 \times 10^{-4}\,m^2\,s^{-1}$	$0.00107\,ml\,ml^{-1}\,K^{-1}$	
$1.00 \times 10^{-4}\,m^2\,s^{-1}$	$0.00096\,ml\,ml^{-1}\,K^{-1}$	
Surface tension at 298 K	$2.1 \times 10^{-2}\,N\,m^{-1}$	
Thermal conductivity at 323 K	$0.184\,J\,m^{-1}\,s^{-1}\,K^{-1}$	
Specific heat at 373 K	$1695\,J\,kg^{-1}\,K^{-1}$	

siloxane, a thermodynamically stable compound, predominating.

The prepolymer is repolymerized or copolymerized with other polysiloxanes, using various techniques depending upon the end product desired. Polysiloxanes may vary in molecular weight (average and distribution), filler, presence or absence of other additives and the chemistry of ligands attached to silicon, including chemically reactive ligands. Stable silicon ligands of commercial importance in polysiloxane technology include methyl (CH_3—), phenyl (C_6H_5—) and 3,3,3-trifluoropropyl ($CF_3CH_2CH_2$—). Reactive silicon ligands include vinyl ($CH_2{=}CH$—), hydroxyl (OH—), methoxy [$CH_3O({=}O)$—], acetoxy (CH_3CO—), hydride (H—), amine (NH_2—) and chloride (Cl—).

Uses for reactive silicon ligands include intermolecular cross-linking of elastomers, resins and gels, chemical reaction to yield new ligands and adhesion promotion. Polysiloxanes used as biomedical materials have been essentially limited to polydimethylsiloxanes. Polysiloxanes with other silicon ligands may have equal safety and offer advantages in specific applications, but essentially only polymethylsiloxanes have had a substantial history of biocompatibility testing and clinical use when manufactured under controlled conditions.

3. Fluids

Polydimethylsiloxane fluids have a variety of biomedical materials uses. They are clear colorless odorless oily liquids. They are highly chemically inert, hydrophobic, resistant to heat and oxidation, and good lubricants for glass, plastics and most elastomers except polydimethylsiloxane. Their properties change minimally when they are heated or chilled, they have low surface tension, low orders of toxicity and skin sensitization, and can be prepared in various viscosities. A biomedical-materials grade Dow Corning 360 Medical Fluid is manufactured and tested under controlled conditions to assure that its purity and biocompatibility characteristics have been duplicated with each batch produced. The typical physical properties of this polydimethylsiloxane fluid are shown in Table 1.

3.1 Chemistry

Polydimethylsiloxane fluids contain repeating dimethylsiloxy groups in polymer chains, end-blocked with trimethylsiloxy groups. They are prepared by equilibration copolymerization of polydimethylsiloxane prepolymer (the source of dimethylsiloxy) and a polysiloxane rich in trimethylsiloxy groups such as hexamethyldisiloxane.

The materials are mixed with a fugitive catalyst and reacted under conditions that result in random rearrangement of the siloxane bonds. The trimethylsiloxy terminal groups become randomly distributed to block the ends of polydimethylsiloxane chains. The average molecular weight of the resulting fluid is determined by the ratio of the two materials. Fugitive catalysts are removed when polymerization is finished:

hexamethyldisiloxane   polydimethylsiloxane prepolymer

polydimethylsiloxane

The resulting polymers have a nearly perfect statistically predictable mix of molecular weights. Distillable polysiloxanes, those with molecular weight less than 800, are removed by high-temperature low-pressure distillation.

3.2 Biocompatibility

Pure polydimethylsiloxane fluid, such as Dow Corning 360 Medical Fluid, elicits no cytotoxicity reaction when a sterile filter disk saturated with the fluid is evaluated by direct contact cell culture. It is nonpyrogenic. It elicits minimal reaction when tested subdermally, intraperitoneally or intramuscularly. Ingestion typically elicits no adverse effects other than a mild cathartic effect from larger doses. In the eye, transitory mild irritation is produced that normally subsides in a few hours. The irritation is believed to be biophysical rather than biochemical.

Hine injected ^{14}C-tagged Dow Corning 360 Medical Fluid into the spine of rats, rabbits and monkeys. Rats received 0.05 ml and 0.10 ml, rabbits 0.30 ml and monkeys 0.5 ml. Rats and rabbits were tested for 30 or 90 days and monkeys were tested for 90 days. Most of the fluid remained in the brain, spinal cord and tissues adjacent to the spinal canal. The presence of the fluid did not cause adverse reactions and there was no evidence of neurological changes as evidenced by behavioral or micropathologic examinations.

3.3 Applications

Polydimethylsiloxane fluid is widely used as a lubricant for glass, plastic or elastomer (except for silicone elastomer) components of medical devices. Hypodermic needles lubricated with this fluid insert into the skin and subcutaneous tissues with less force and pain than unlubricated needles. The availability of polydimethylsiloxane fluid for use as a needle lubricant contributed significantly to the development of disposable hypodermic needles. Disposable needles must be inexpensive and thus are usually not as sharp nor as highly polished as the older, more expensive, reusable needles. When lubricated, needle sharpness and polish are not as important to the ease of penetration and reduction of pain. With each injection given by a lubricated needle, a small quantity of fluid remains in tissue. In trace quantities, even when injections are given repeatedly, polydimethylsiloxane fluid does not appear to cause adverse reactions.

Disposable hypodermic syringes are also lubricated with polydimethylsiloxane fluid. Reusable ground glass syringes and plungers of the past were expensive. Disposable syringes and plungers are produced inexpensively by plastics molding. The plungers have an elastomeric seal which must fit tightly in the syringe barrel to permit filling the syringe by aspiration without admitting air and to permit forceful injection, when necessary, without leakage. This seal must also slide freely in the syringe barrel to permit accurate control of the material injected. The coefficient of friction between an unlubricated elastomer plunger and the plastic syringe barrel is too high for practical use. Polydimethylsiloxane fluid is the only lubricant that satisfies the requirements for excellent biocompatibility, biochemical compatibility with a wide variety of drugs, sterilizability, durability and stability such that syringes may be stored essentially indefinitely with no loss of lubricating characteristics.

Polydimethylsiloxane fluids can be used as lubricants for many instruments that must be inserted into a natural orifice of the body such as the rectoscope or urethral catheter. It has been used as a lubricant for the sockets of artificial limbs and eyes. Films of polydimethylsiloxane fluid are permeable to water vapor, but hydrophobic to liquid water. Thus, barrier creams and sprays can be formulated to protect the skin from water and water-borne irritants.

Polydimethylsiloxane fluid is not an effective lubricant for metal-to-metal applications. It provides only short-term benefit when used as a lubricant with polydimethylsiloxane elastomers since the fluid dissolves into the elastomer and lubricity is lost with time.

Subdermal injection of polydimethylsiloxane fluid for any purpose, other than in approved research studies, is not recommended. When injected into the bloodstream in experimental animals, the material appeared to behave as an embolus. In an artery the material appears to cause capillary occlusion and even in relatively small quantities may cause serious adverse reactions or death. If injected into a vein, the fluid embolus is apparently carried to the lung where capillary occlusion occurs. Lung capacity generally exceeds basic needs, thus the venous quantities needed to produce symptoms are larger than for arterial injection. Historically, polysiloxane fluids were sometimes misused for injection into humans to augment soft tissue for cosmetic purposes without appropriate research protocol or controls. The identity and purity of the fluids were often unknown and some were not polysiloxanes. A wide variety of local and systemic complications and adverse reactions have been reported in these patients. Thus, until the safety and effectiveness of polysiloxane fluid injections for soft-tissue augmentation (or any other injection use) have been established by careful controlled studies, and a specific product approved for distribution by a pertinent governmental regulatory agency, polysiloxane injections for other than legally approved research should not be carried out.

3.4 Antifoams

Polydimethylsiloxane antifoams are chemically similar to fluids, except that they contain 4–7 wt% fumed silica of high surface area, intermixed and processed in the polymer to improve defoaming properties. Polydimethylsiloxane fluids such as Dow Corning 360 Medical Fluid have inherent antifoam properties that are greatly enhanced by appropriate incorporation of silica. Antifoams that satisfy the standard *United States Pharmacopoeia* (1985) requirements may be labelled "Simethecone".

The low surface tension of polydimethylsiloxane fluid is important to the mechanism of defoaming action. Fluid-coated silica particles seem to become distributed in the films of foam bubbles and, because of the inherent low surface tension and low cohesive force of the fluid, it easily separates from the silica particles, causing foam bubbles to rupture.

Water emulsions of polydimethylsiloxane antifoam can be prepared. In these the emulsifiers must be carefully selected to yield emulsions that have stability adequate for storage and shipment, but that are also unstable so that the antifoam is released from the emulsion when diluted in use without loss of defoaming properties. Polydimethylsiloxane antifoams are generally effective at low concentrations; only 1–50 ppm are usually required.

Biomedical materials applications of polydimethylsiloxane antifoams include use as a deflatuent, often in combination with antacids that are available in either tablet or liquid form. They can also be used to reduce foam and eliminate gas in the gastrointestinal tract during gastroscopic examinations and x-ray studies. Pharmaceutical applications include suppression of foam in processes such as fermentation, maceration, percolation and mixing. In many applications, antifoams can also be used to suppress foam and improve the efficiency of filling drug-containing ampules and bottles where foaming is a problem.

4. Elastomers

Polysiloxane elastomers consist of cross-linkable polysiloxanes, reinforcing filler and an appropriate catalyst to initiate the formation of intermolecular cross-links. Fumed silica with 400 m² per gram surface area is generally preferred as the filler. High-molecular-weight polysiloxanes (> 350 M) are used in high-consistency elastomer compounds. These viscous materials flow only under pressure and do not soften much when hot. Thus, they are fabricated by pressure techniques such as compression molding, transfer molding, calendering and extrusion. They may also be used from solvent dispersion. Lower-molecular-weight polysiloxanes are used to make low-consistency elastomer compounds that are flowable and so can be fabricated by casting, reactive injection molding and similar processes. High-consistency polysiloxane elastomer compounds have low green strength and thus blow-molding and vacuum-forming have only limited application. Vulcanization systems include peroxide cure, rare metal (such as platinum) catalysis and room-temperature moisture-curing systems (the so-called RTVs).

4.1 Chemistry

The elastomers used as biomedical materials are primarily polydimethylsiloxanes that have been copolymerized with trace quantities of polysiloxanes containing reactive silicon ligands to facilitate cross-linking. Materials catalyzed with trace quantities of rare metal, such as platinum, are typically formulated as two-component materials. With high-consistency materials the two components must be mixed on a two-roll elastomer mill prior to use, while the low-consistency materials may be simply mixed with a spatula. The polysiloxane in one component contains a small amount of vinylmethylsiloxy units in an otherwise dimethylpolysiloxane chain. The second component contains a small amount of methylhydrogensiloxy units, also in an otherwise polydimethylsiloxane chain. When the two materials are blended with a trace of platinum catalyst and then heated, hydrogen from methylhydrogensiloxy contained in one polymer adds across vinyl contained as methylvinylsiloxy in the second polymer. This results in a dimethylene cross-link, covalently bonded between silicon atoms in separate polymer chains:

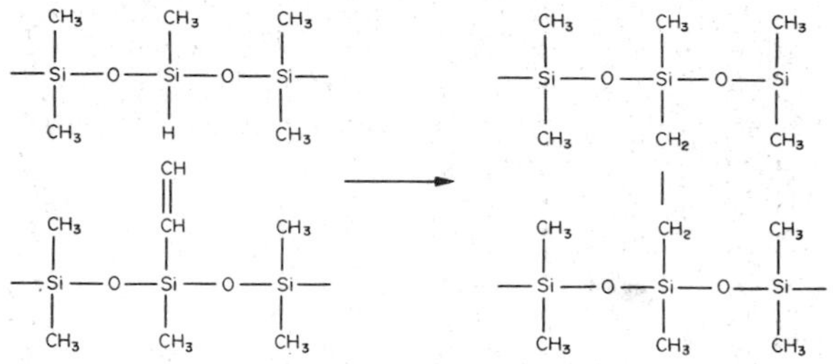

Low-consistency elastomers usually require a higher cross-link density than high-consistency elastomers, since polymer entanglement in the former contributes minimally to physical properties. The silica content typically varies from 10–40 wt%. Low-consistency materials typically contain less silica than high-consistency materials. Elastomers with a high durometer (hardness) value typically contain more silica than elastomers with a lower durometer value.

The peroxide vulcanizing systems used in biomedical materials applications are typically formulated with 2,4-dichlorobenzoyl peroxide. The free radicals produced at higher temperature remove hydrogen from silicon–methyl ligands and the free radicals interact with silicon–vinyl ligands, resulting in a trimethylene cross-link. Some dimethylene cross-links also form by interaction between two silicon-methyl free radical ligands.

SILASTIC Medical Adhesive Type A is a one-component biomedical material silicone elastomer that self-cures when exposed to moisture. It contains polydimethylsiloxane chains end-blocked with methyldiacetoxysilyl groups. It has a grease-consistency, because of the presence of a high-surface-area silica. When it is exposed to air, moisture slowly reacts with acetoxy ligands to form free acetic acid and hydroxy–silicon ligands. Hydroxy–silicon ligands then react with acetoxy ligands on adjoining polymer chains to form silicon–oxygen–silicon cross-links and additional acetic acid. These reactions continue and the resulting elastomer becomes a cross-linked network, with all acetoxy ligands eventually reacting to form acetic acid that then evaporates from the elastomer if in air, or is leached out if in an aqueous medium:

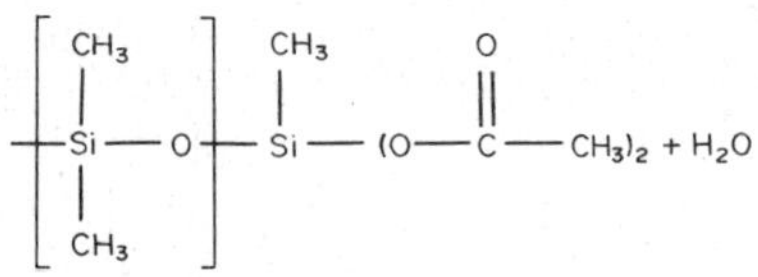

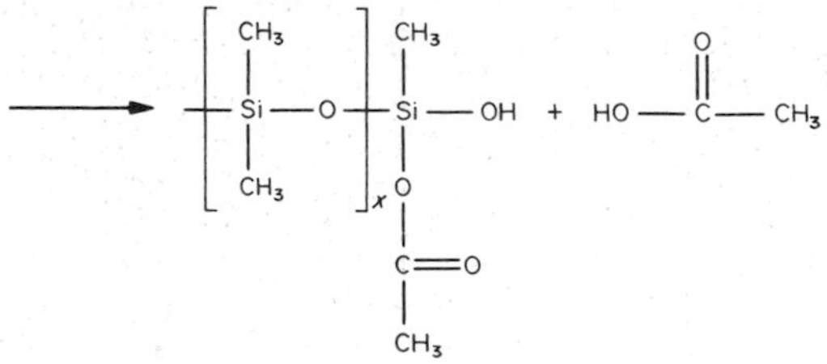

and

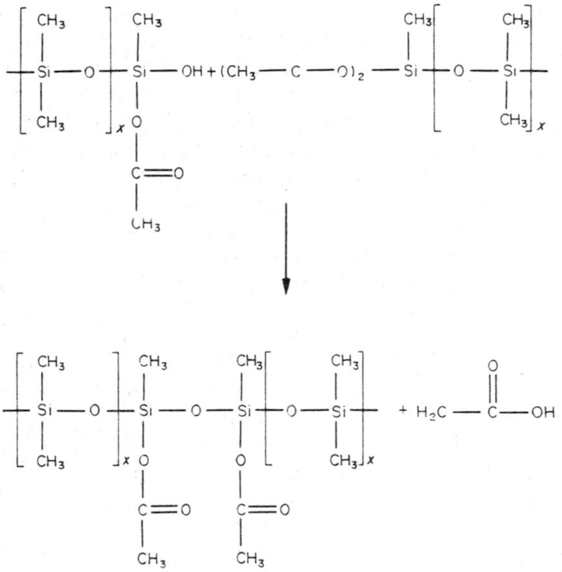

4.2 Physical Properties

Shore A durometer values may vary from 20 to 80. The tensile and tear strengths of high-consistency elastomers are generally higher than those available in low-consistency elastomers. The tensile strength of thermoset elastomers varies from a minimum of approximately $3.8\,MN\,m^{-2}$ for a low-consistency material to a maximum of approximately $9.5\,MN\,m^{-2}$ for a high-consistency elastomer. Elongation varies from 350% to a maximum of 1200%. Tear propagation strengths vary from approximately $0.3\,MN\,m^{-2}$ to around $2.1\,MN\,m^{-2}$. Specific gravities vary between 1.1 and 1.2. Proprietary formulations have been developed to satisfy specific implant requirements. SILASTIC brand high-performance elastomers with a combination of high resistance to both flaw propagation and fatigue flexion have been developed for use in implants where resistance to fatigue flexion is important, such as in the flexible hinge implants developed by Swanson. The current material used in flexible hinges has a typical crack growth of only

0.25 mm per 10^6 flexion cycles when tested in accordance with the American Standard ASTM D813 "Standard method of test for crack growth of rubber."

4.3 Biocompatibility

Biocompatibility testing and assurance of polysiloxane elastomers, as with other biomedical materials, varies according to application. The requirements for a material used to construct permanent implants are obviously more stringent than for a material used for contact against intact skin. The general requirements for biocompatibility are specified in the American Standard ASTM F748 "Standard practice for selecting generic biologic test methods for materials and devices." However, ASTM F748 does not address materials characterization and duplication. Findings from historic biocompatibility testing are valid to current material only when current material precisely duplicates test material. Polysiloxane elastomers are uncharacterizable once they are formulated or fabricated into finished biomedical devices. Thus, in applications where chronic biocompatibility data are important to performance, such as with implants, to ensure that the materials have been duplicated and remain unadulterated, all formulating must start with characterizable materials and all compounding, processing and fabricating must be done under carefully controlled conditions in plants or laboratories. These should adhere closely to the "Good manufacturing practice" requirements mandated by US government regulation.

Implant grades of SILASTIC brand polysiloxane elastomers, manufactured under appropriately controlled conditions, have been evaluated in two-year implant studies to assess host reaction and biodurability. The host reaction is typically limited to a mild foreign-body reaction resulting in fibrous tissue encapsulation of the implant specimen. Evaluations of implanted and control specimens found no evidence of biodegradation or loss of physical properties. In shorter-term biocompatibility testing, the materials meet or exceed the standard *United States Pharmacopoeia* 21, Class VI plastics testing. They are also nonpyrogenic, do not sensitize the skin, are noncytotoxic in direct contact cell culture testing and the results of 90-day implant studies are similar to the results of chronic studies with no evidence of systemic reaction in any of the major organ systems. These materials are manufactured with care and controls that satisfy the "Good manufacturing practice" requirements of the US Food and Drug Administration to ensure lot-to-lot duplication and freedom from adulteration.

SILASTIC Medical Adhesive Type A has similar biocompatibility characteristics when fully cured. Prior to cure, while acetic acid is evolving, the material is a strong transitory irritant. Implantation prior to the complete absence of acetic acid is not recommended.

In addition to SILASTIC brand implant grades of polysiloxane elastomer, medical grades are also available for use in nonimplant applications. Medical grade elastomers have not been tested in two-year animal implant studies, but otherwise have quality assurance similar to implant grades. Clean-grade and commercial-grade SILASTIC brand polysiloxane elastomers are available for use as biomedical materials in applications where the requirements for assured biocompatibility are minimal or where batch-to-batch acute biocompatibility testing can be carried out by the end-user. Clean-grade elastomers, but not commercial grades, have lot-to-lot quality control, lot traceability, product certification, physical properties audited on a regular basis and straining through a 200 mesh screen. Polysiloxane elastomers for biomedical materials applications are thus available with various degrees of quality assurance documentation, and biocompatibility and biodurability testing and assurance to satisfy the various needs as determined by the nature of the end application.

4.4 Applications

Polysiloxane elastomers are used as an alloplast in a variety of implants to facilitate improved health care in treatment of health conditions where no effective treatment is otherwise available. Holter's development of the polysiloxane hydrocephalus shunt in 1955 represented the first functional polysiloxane implant of major importance and demonstrated the potential value of these polysiloxanes as biomedical materials. In the early 1960s, Cronin and Gerow (1964) developed an implant for breast reconstruction containing a polysiloxane elastomer outer envelope filled with a soft polysiloxane gel. Cronin's work also led to the development of a number of other implants used in plastic surgery including those used for reconstruction of the ear, nose and chin (Cronin 1966). Lincoff et al. (1965) developed a polysiloxane implant to buckle the sclera for correction of detached retina. Swanson and co-workers developed flexible polysiloxane implants for use in reconstruction of various joints of the extremities including the finger, wrist, elbow and toe (Swanson 1973, Swanson and Le Beau 1974, Swanson et al. 1982). Their work enabled many individuals with destroyed joints, particularly those with rheumatoid arthritis, to have relief from pain, correction of deformity and continued joint function for use in activities of daily living, and thus improved quality of life. High-performance polysiloxane elastomers have now been developed to provide extended flexural durability in flexible hinge implants.

More recently, tissue expanders fabricated of polysiloxane elastomer have been developed. These implants consist of an empty elastomer envelope (fabricated from dispersion by dipping a mandrel) with an attached reinjectable port. They are implanted next to a defect and then slowly inflated with a saline solution from a hypodermic syringe using a percutaneous hypodermic needle puncture into the filling port. As the device is inflated over several weeks, the skin and subcutaneous tissue are slowly stretched. When stretched adequately, the defect (scar, tattoo, etc.) is excised and the extra skin generated by the stretching process is used at surgery to cover the defect. This technique is particularly useful for reconstruction of facial or scalp defects, and for facilitating postmastectomy reconstruction.

Medical devices other than implants made of polysiloxane elastomer include catheters, tubes, drains and other devices where the excellent stability, biocompatibility and biodurability of these materials are important to the quality or function of the device. Medical devices fabricated from polysiloxane elastomer are now used in nearly every medical specialty. No other elastomeric material is available with proven and documented biocompatibility and biodurability characteristics and a long history of safe medical use.

SILASTIC Medical Adhesive Type A may be used as an adhesive in biomedical materials applications. It is a particularly good adhesive for polydimethyl–polysiloxane elastomer. It also bonds well to glass and to some metals and plastics. In all applications the durability and strength of the bond should be evaluated, not only in the laboratory, but also when exposed to conditions of the intended use. In biomedical materials applications the adhesive should be completely cured prior to contact with tissue or implantation to avoid irritation from the evolving acetic acid.

4.5 Polysiloxane Gel

Polysiloxane gels may be considered with elastomers, because of similarities in polymer composition and the cross-linking chemistry. Polysiloxane gels are clear: they contain no fillers. The vulcanization chemistry is identical to that for elastomers; however, the cross-linked density of gels is much lower in order to maintain the soft consistency of gel materials. With polysiloxane elastomers, nearly all molecules become covalently bonded in a matrix. Those not bonded in the matrix can be extracted by a solvent. With elastomers, the extractable silicone is approximately 2 wt%; with gels the extractable content is much higher, in some instances 90 wt%. The high quantity of polysiloxane extractable from gel accounts for the "bleed", a layer of polysiloxane fluid that develops on the outside of a polysiloxane envelope in breast implants as the result of osmosis of unbound polymer from the gel through the envelope. SILASTIC brand implants (SILASTIC II) are now available containing a special silicone barrier in

the envelope to prevent or reduce bleed substantively. The unbound polymer in the gel is not soluble in the barrier layer, thus osmosis is prevented.

Polysiloxane gels are not recommended for implant use except when contained in a silicone elastomer envelope. Owing to the softness and the low strength of gels, migration and phagocytosis have been reported in the absence of an intact envelope. The desirability of having a polysiloxane gel with no extractable polysiloxane is well known. However, all attempts to increase cross-link density to decrease the amount of extractable polysiloxane have led to hard or firm materials unsuitable for use as soft implants.

5. Adhesives

A pressure-sensitive polysiloxane adhesive (Dow Corning 355 Medical Adhesive) has been specifically designed for biomedical applications. The material is contained in fluorocarbon solvent 18.5 wt%. It is designed to temporarily affix materials to skin. It retains its adhesive qualities in the presence of moisture or perspiration, during normal temperature variation and with time. It is an effective adhesive for use on many surfaces including skin, metals, glass, paper, fabric, plastics, polysiloxane and organic elastomers. The adhesiveness may be lost after a period of time when used with polysiloxane elastomers since some of the components of the adhesive may dissolve into the elastomer. The adhesive is nonsensitizing. When tested in keeping with the US Federal Hazardous Substances Act, the material was classified nontoxic and not highly toxic when administered by inhalation to albino rats. It had an LD_{50} (lethal dose for 50% of rats tested) greater than $2 \, g \, kg^{-1}$ when administered dermally to albino rats and was neither an eye nor a skin irritant in albino rabbits. Repeated human patch tests indicated that the material is not a primary skin irritant and does not have skin-fatiguing or skin-sensitizing properties. Applications include use to attach various devices to the skin, including transdermal drug delivery repositories.

Dow Corning 355 Medical Adhesives requires no vulcanization or cure and its function depends upon the stickiness and adhesive qualities of the polysiloxane layer left on a surface when the solvent has evaporated. The adhesive may be removed by a suitable solvent. For skin, a solvent that is nonirritating, such as trichlorotrifluoroethane, must be used. Other commercially available tape and adhesive removers (some containing ether or acetone) may also be effective for removing the adhesive providing the skin is intact and not sensitive. The adhesive may be removed from materials other than skin by a variety of solvents including isopropyl alcohol, ethyl alcohol, naphtha, methylchloroform or hexane. Some of these solvents are highly flammable. Additionally, a small amount of solvent should be tested on the substrate prior to use, to make certain the solvent does not cause deterioration.

6. Biomedical Material Considerations

Biomedical materials, including polysiloxanes, are far from ideal. Ideal materials or devices would restore health and function to equal or exceed that expected in normal health. Rarely, if ever, has this occurred. Biomedical materials and devices have all of the drawbacks inherent in mechanical devices, including potential for breakage, wear, fatigue, malfunction and failure to perform intended function. Even materials that are highly biocompatible, such as implant grades of polysiloxane elastomers, are not entirely accepted when implanted.

Biologically, the surgical creation of a cavity, even if filled with an implant, initiates a healing reaction. In a healthy individual the biological healing process will continue for as long as the cavity exists and the implant is in place. The development of fibrous tissue to fill the cavity is a normal healing response with or without an implant. The development of a fibrous tissue layer around an implant is physiological, but has been termed a foreign-body reaction, even though it represents a minimal reaction to any implant. Fibrous tissue contraction is also physiological and part of the wound-healing processes that aim to eliminate the cavity if possible. With many implants, fibrous tissue contraction causes no problem. However, soft polysiloxane gel-containing breast implants may be reshaped to a sphere by this physiological contracture. When spherical, the implant feels firm even at low biological intraimplant pressures, which are generally less than 20 cm water pressure, because it can no longer be distorted. A sphere has minimal surface area for the volume contained, fibrous tissue is essentially inelastic, and polysiloxane gel is noncompressible.

Implants cause an associated risk of late infection at the site of the implant. The interface around the implant with tissue provides ideal culture conditions including unlimited nutrients, ideal culture temperature and darkness. The body's defenses are poorly suited to dealing with bacteria growing in the serous fluid at the implant–tissue interface. Bacteria circulating systemically from a distant infection may become lodged next to an implant, multiply and require implant removal for treatment. Thus, in addition to all of the hazards and risks inherent in any surgical procedure, the use of a biomedical material implant adds additional risk even if constructed from the most appropriate material available.

See also: Adhesives in Medicine; Biocompatibility: An Overview

Bibliography

Cronin T D 1966 Use of a Silastic frame for total and subtotal reconstruction of the external ear: Preliminary report. *Plast. Reconstr. Surg.* 37(5): 399–405

Cronin T D, Gerow F J 1964 Augmentation mammaplasty: A new "natural feel" prosthesis. *Int. Congr. Ser. Excerpta Mec.* 66: 41–9

de Nicola R R 1950 Permanent artificial (silicone) urethra. *J. Urol.* 63: 168–72

Frisch E E 1983 Technology of silicones in biomedical applications. In: Rubin L R (ed.) 1983 *Biomaterials in Reconstructive Surgery.* Mosby, St Louis, MO, pp. 73–90

Frisch E E 1984 Silicones in artificial organs. In: Gebelein C G (ed.) 1984 *Polymeric Materials and Artificial Organs,* American Chemical Society Symposium Series No. 256. ACS, Washington, DC, pp. 63–97

Frisch E E, Langley N R 1985 Biodurability evaluation of medical-grade high-performance silicone elastomer. In: Fraker A C, Griffin C D (eds.) 1985 *Corrosion and Degradation of Implant Materials: Second Symposium,* ASTM STP 859, American Society for Testing and Materials, Philadelphia, PA, pp. 278–93

Hine C H, Elliott H W, Wright R R, Cavalli R D, Porter C D 1969 Evaluation of a silicone lubricant injected spinally. *Toxicol. Appl. Pharmacol.* 15: 566–73

Hyde J F 1944 Improved process for the manufacture of organo-silicon polymers and products produced thereby. British Patent No. GB 0,561,226

Kipping F S 1907 Organic derivatives of silicon, part II: The synthesis of benzylethylpropylsilicol, its sulphonation, and the resolution of the dl-sulphonic derivative into its optically active components. *J. Chem. Soc.* 91: 209–40

Kossovsky N, Heggers J P 1987 The bioreactivity of silicone. *CRC Crit. Rev. Biocompat.* 3: 53–85

LaFay H 1957 A father's last-chance invention saves his son. *The Reader's Digest (UK)* September: 127–30

Lincoff H A, Baras I, McLean J 1965 Modifications to the Custodis procedure for retinal detachment. *Arch. Opthalmol. (Chicago)* 73: 160–3

Piechotta F U 1979 Silicone fluid, attractive and dangerous. *Aesth. Plast. Surg.* 3: 347–55

Rand R W, Mosso J A 1973 Treatment of cerebral aneurysms by stereotaxic ferromagnetic silicone thrombosis. Case report. *Bull. Los Angeles Neurol. Soc.* 38: 21–3

Rowe V K, Spencer H C, Bass S L 1948 Toxicological studies on certain commercial silicones and hydrolyzable silane intermediates. *J. Ind. Hyg. Toxicol.* 30(6): 332–52

Rowe V K, Spencer H C, Bass S L 1950 Toxicologic studies on certain commercial silicones. II: Two year dietary feeding of "DC Antifoam A" to rats. *Arch. Ind. Hyg. Occup. Med.* 1: 539–44

Safian J 1966 Progress in nasal and chin augmentation. *Plast. Reconstr. Surg.* 37: 446–52

Snyder G B, Courtiss E H, Kaye B M, Gradinger G P 1978 A new chin implant for microgenia. *Plast. Reconstr. Surg.* 61: 854–61

Swanson A B 1973 *Flexible Implant Resection Arthroplasty in the Hand and Extremities.* Mosby, St Louis, MO

Swanson A B, de Groot Swanson G, Frisch E E 1982 Flexible (silicone) implant anthroplasty in the small joints of extremities: Concepts, physical and biological considerations, experimental and clinical results. In:

Rubin L R (ed.) 1983 *Biomaterials in Reconstructive Surgery*. Mosby, St Louis, MO, pp. 595–623

Swanson J W, LeBeau J E 1974 The effect of implantation on the physical properties of silicone rubber. *J. Biomed. Mater. Res.* 8: 357–67

United States Pharmacopoeia 21st revision. 16th edn, 1985. United States Pharmacopoeial Convention, Rockville, MD

Wilsnack R E 1976 Quantitative cell culture biocompatibility testing of medical devices and correlation to animal tests. *Biomater. Med. Dev. Artif. Organs* 4: 235–61

Wilsnack R E, Meyer F S, Smith J G 1973 Human cell culture toxicity testing of medical devices and correlation to animal tests. *Biomater. Med. Dev. Artif. Organs* 1: 543–62

E. E. Frisch
[Dow Corning Corporation, Hermlock, Michigan, USA]

Polytetrafluoroethylene

During the early development of implantable polymers, the requirement for inertness within the body was obviously an important issue. Along with polyethylene, polypropylene, silicones and acrylics, certain perfluorocarbon polymers were seen to be of considerable potential in this respect because of their chemical stability and apparently good biocompatibility.

Fluorocarbon polymers comprise a small group of polymers based on a carbon backbone and fluorine, or fluorine plus other halogen side groups. Polytetrafluoroethylene (PTFE), that is, $(CF_2—CF_2)_n$, represents the extreme of this series, being fully fluorinated. Other examples are polytrifluorochloroethylene, poly(vinyl fluoride), poly(vinylidene fluoride) and fluorinated ethylene–propylene copolymers. Since the carbon–fluorine bond is exceptionally strong, the greatest stability is achieved with PTFE, so it is this polymer which has received most attention as an inert biomaterial. In practice, commercial formulations approximate to PTFE and are often described as such, but they may be copolymers of TFE and other fluorocarbons, designed to optimize the properties.

PTFE has not been the easiest polymer to prepare or fabricate because of the nature of the carbon–fluorine bond. It is a thermoplastic but is difficult to fabricate by normal thermoplastic routes. The polymer itself is prepared by an addition reaction using free-radical polymerization in aqueous dispersion under pressure. The polymer is highly crystalline and the molecular weight ranges from 5×10^5 to 5×10^6. The crystalline melting point is 600 K, and even at this temperature the material is extremely viscous, preventing molding and extrusion. The material is, therefore, normally fabricated by sintering techniques, where the granulated powder is pressed into shape at around 50 MPa pressure. The resulting sintered product may be readily machined.

The extreme inertness of PTFE is responsible for most of the applications, since it is able to withstand working temperatures from 570 K down to near absolute zero. It is unaffected by most chemical environments, being attacked only by alkali metals and a few fluorinated solvents. In addition, it has an extremely low coefficient of friction and it is very difficult for any material to be bonded to it. The major deficiency is found with the mechanical properties. The tensile strength is in the region 15–35 MPa while the compressive strength is as low as 10 MPa. Ductility is typically 300%, and impact strength 150 J m^{-1}. The addition of fillers to the polymer may improve some of these properties.

The biomedical uses of PTFE have largely arisen for the reasons described above; that is, the extreme inertness has led to the belief that the material should display stability in the body and, therefore, good biocompatibility. The poor mechanical properties have provided problems however, and because of this the material has acquired an undeserved poor reputation under some circumstances.

1. Joint Prostheses

One of the major failures of biomaterials used in implant surgery concerns the fate of PTFE as a bearing surface in total hip replacement. Although this is a matter of history, it is important to recognize the causes of failure and to distinguish these from the features of the performance of the material in other circumstances. Because, intuitively, any artificial joint should have a low coefficient of friction, one of the first attempts to replace the hip joint with a total prosthesis utilized PTFE as the acetabular component, bearing against a metal, stainless steel. While it is true that the coefficient of friction is extremely low, the material does not perform well under high, or even moderately high, stresses, and the PTFE suffered extensive wear during sliding. The femoral component of the hip joint rapidly wore its way through the PTFE, generating considerable amounts of wear debris. Although PTFE is known to be extremely well tolerated as a monolithic solid implant, the release of particulate matter into the tissue provoked an extensive tissue response (see *Biocompatibility: An Overview*). The reaction was in fact so severe, constituting a granuloma, that the prostheses had to be removed from many of the patients. PTFE is no longer used in this situation, and it is deemed to be unsuitable for this type of load-bearing application within the body. It is emphasized, however, that the response was due to particulate material, where the major factors were the particle size and the rate of release of the particles. The fact that these particles were of PTFE was irrelevant as far as the response of the tissue was concerned (Charnley 1963).

2. Ligament Prostheses

One situation in which PTFE is used under stress is in the treatment of ligament injuries. Although there have been attempts over many years to use a variety of biological tissues for ligament replacement, few were successful and, with an increasing number of sports injuries affecting the knee ligaments, considerable attention has been paid to this subject in recent years. At the present time, four classes of device may be distinguished: the biological replacement derived, for example, from xenogenic tendon (i.e., tendon tissue derived from a different species), carbon fibers, inert polymers and resorbable polymers (Claes et al. 1987). None of these is ideal or truly successful, and particular difficulties arise from the failure to match the deformation characteristics of natural ligament, and from the attrition, abrasion and fatigue that occur at the points of entry into and attachment to the bone.

The inert polymers currently in use are the aromatic polyester Dacron and the expanded PTFE Gore-Tex. The structure of the expanded PTFE in Gore-Tex ligaments, is that of a microporous composite arrangement of very fine oriented fibrils of the polymer, held together by solid nodes of the same material. The fibrils are between 5 μm and 10 μm in diameter and are about 60–100 μm long between the nodes. The material may be used as a microporous three-dimensional solid or as fibers wherein the fibrils are oriented coaxially with the fiber length. Single-filament ligament prostheses have been made from this fibrous material, but it is more usual to provide a braided configuration of a single continuous fiber. The structure of the braid is arranged so as to optimize the longitudinal mechanical characteristics. At the ends of the prosthetic ligament, fiber bundles are formed into eyelets to provide initial fixation. A typical ligament will be 0.16 m long, with 160 strands. A density of 520 kg m^{-3} indicates a porosity in excess of 65%. The ultimate tensile breaking load of such a prosthesis will be around 5000 N (Bolton and Bruchman 1985).

These prostheses may be secured to the bone by means of screw fixation, the strength of attachment increasing with time due to tissue ingrowth into the porous PTFE at the attachment sites. The intra-articular segments of the prosthesis are invaded by synovial and fibrous tissue. It has been claimed that this is healthy fibrous tissue with no signs of inflammation.

Clinical results with the Gore-Tex prosthetic ligament are now emerging (Glousman et al. 1988, Ferkel et al. 1989, Indelicato et al. 1989). They are variable and, indeed, at the present time, no

material has been shown to be ideal in this situation. Problems include the abrasion of the prostheses at the point of entry, leading both to malfunction and excessive tissue response to the debris, and the creep of the polymer that radically alters the ability to replicate the natural mechanical response to loads, thereby leading to instability. PTFE will no doubt continue to be used until significantly better materials and constructions emerge.

3. Vascular Prostheses

One of the most difficult areas of reconstructive surgery involving the use of biomaterials is that of vascular replacement (see *Arteries, Synthetic*). For many years, the woven Dacron vascular graft dominated the surgical reconstruction of large-diameter blood vessels, although it was recognized at an early stage that the extension of this technology to the small vessels of the lower limbs and, most importantly, to the coronary arteries, would not be possible with that type of material. Partly to improve the performance with the large-diameter replacements and partly to attempt the reconstruction of smaller vessels, a type of PTFE vascular graft was introduced in the 1970s (Florian et al. 1976).

It is a requirement of vascular prostheses, that the material from which the tubular structure is made allows for the generation of a new inner tissue lining which is resistant to the formation of thrombi or emboli and which can function mechanically and hemodynamically in a manner analogous to the natural vessel. Dacron prostheses are able to satisfy this requirement, to a certain degree, through the use of the microporous structure provided by weaving or knitting the polyester fibers; this structure allows blood to permeate the porosity in the very early stages where it clots. This is subsequently reorganized and a type of lining, the so-called neointima, is generated on the inner surface. A combination of the microstructure provided by the weaving or knitting process and the overall architecture provides the type of mechanical behavior appropriate to the passage of blood, in a pulsatile manner, through the vessel that could not be obtained in a rigid-walled tube. The problems with the Dacron that prevent its use in smaller vessel replacements, are related to the difficulty of restricting the growth of the neointima and the accurate matching of the mechanical compliance, both of which become increasingly important as the vessel diameter gets smaller.

The PTFE used in vascular prostheses is essentially the same expanded microstructure used on the ligament prosthesis described in Sect. 2. The objectives behind the introduction of the expanded PTFE into vascular replacement, were that the combination of the greater inertness of the material itself and the different type of nonwoven microstructure should provide a better basis for the generation of a well-structured neointima that would not progressively thicken, and that, furthermore, the microporous material should be more compliant.

This approach has met with good clinical success in large-diameter vessel replacement, and apparently better performance than Dacron in medium-sized vessels. However, the extent to which the expanded PTFE really does meet the requirements is a matter of some controversy and it is certainly clear that it is not successful in those situations involving a diameter of less than 6 mm. Perhaps of greatest significance here is the exact nature of the neointima. The ideal structure of this tissue layer is one where the covering of the polymer is complete and uninterrupted, and consists of endothelium, the normal cellular lining of blood vessels. Moreover, this layer should be of self-limiting thickness. Some investigators have claimed that the PTFE grafts are associated with such a lining (Soyer et al. 1972, Baker et al. 1980), while others have reported incomplete tissue covering (Campbell et al. 1979), extensive fibrous tissue rather than endothelium (Selman et al. 1980), progressively thickening pseudointima (Fujiwara et al. 1980) and, indeed, intimal hyperplasia that leads to clinically significant narrowing of the lumen (Echave et al. 1979). In an extensive experimental study, Chignier et al. (1983) demonstrated that expanded PTFE prostheses were characterized by limited or even absent neointimal tissue proliferation, the surface remaining devoid of true endothelial-like structures. An incomplete intimal layer was observed in places consisting of fibroblastic cells in a collagenlike ground substance, while in many areas, the uncovered PTFE could still be observed in contact with the blood.

PTFE, therefore, remains as a clinically acceptable material for large-diameter blood-vessel replacement, but is unlikely to be successful in the more demanding small-vessel situation. This view is expanded in the report of Whittemore et al. (1989).

4. *General Surgical Applications*

PTFE has been selected for a wide variety of miscellaneous clinical applications, again for its inertness, where simple space filling without significant mechanical loading is required. In this situation, it is often seen as a competitor to silicone elastomers. In particular, PTFE is often used for small-volume implants in facial reconstruction. It has also been widely used in middle-ear surgery for stapedectomy prostheses. It is, on occasion, used for tubular structures, but is too rigid in comparison with silicones, polyurethanes and plasticized poly(vinyl chloride) for the majority of catheters and tubes and, indeed, its difficulty of manufacture and greater cost tend to

preclude use in these commodity applications. There is the possibility that the tissue ingrowth capability with expanded PTFE may be utilized with certain tubular structures that need integration into the tissue (Goodson 1987).

5. *PTFE Composites*

Very few materials have been specifically developed for implantation within the body, the vast majority being adapted from formulations and structures intended to meet other, nonmedical specifications. One of the materials coming into this category, however, is a PTFE-based composite, known commercially as Proplast. The history of the development and clinical applications of this material is complex and controversial, and again may serve to place PTFE in a poor light unless considered in context. The composite in question is a porous material comprising PTFE and either graphite fibers or alumina particles. The original rationale for this development, was that a low-modulus material was required as a coating on orthopedic prostheses to act as a resilient interface between the rigid metals and bone. The original PTFE–graphite combination was chosen on the basis of the inertness and putative biocompatibility of the components, and the porosity was introduced in order to control the elastic modulus and to allow for tissue ingrowth. It was postulated that the composite structure, developed as tissue ingrowth took place, would optimize the stress distribution at the interface (Homsy and Anderson 1976). It soon became clear that the mechanical properties of this material would not be appropriate for this type of stressed application, and attention moved away from the orthopedic field to other applications in less mechanically demanding areas, such as those in middle-ear surgery and maxillofacial surgery. The porosity was still maintained in order to provide the ingrowth capability, but for most applications the graphite filler was replaced with alumina to give better aesthetics. With respect to the latter point, some of the tissue reconstruction or augmentation procedures in the oral and facial areas were rather superficial, and the black color of the PTFE–graphite material was unacceptable.

Although this material has been used extensively by some surgeons, it has never received widespread support and usage. The reservations held by many people concerning its suitability for these applications, have unfortunately appeared to be well founded, and there are now many reports of poor clinical performance and, indeed, overt clinical failure with a requirement for implant removal. This has especially involved the use of the material in total or partial replacement of the temporomandibular joint. Although the symptoms indicating that removal was necessary have included the signs of a foreign-body host response, it has again been the mechanical deficiencies which have been the real cause of failure. Neither the graphite nor the alumina have been able to alter substantially the fact that PTFE is a weak material and any application where even moderately high stresses—especially involving sliding stresses—are involved, is unlikely to succeed (Timmis et al. 1986, Florine et al. 1988).

6. Perfluorocarbons as Blood Substitutes

One final use of substances of the PTFE family which requires a brief mention is that of blood substitutes involving perfluorocarbons. These are cyclic or straight-chain hydrocarbons, in which some or all of the hydrogen atoms have been replaced by fluorine atoms and which, in contrast to PTFE itself, exist as liquids. The main examples of these substances are perfluorodecalin, perfluoro octyl bromide, perfluoro tripropylamine and perfluoro tributylamine. The molecular weights are 462, 499, 521 and 671 respectively. The most important property, as far as blood substitutes are concerned, is the gas solubility, and with these substances, oxygen solubility is in the range 40–50%, and carbon dioxide 140–230%.

Perfluorocarbon liquids are generally immiscible in aqueous systems and, consequently, are poor solvents for physiological solutes. They must, therefore, be emulsified in an electrolyte solution with surfactants. A considerable amount of development work has been undertaken in this respect, particularly to increase the gas-carrying properties of the emulsions and also their *in vivo* stability and biocompatibility. Perfluorodecalin is generally regarded as having the superior biocompatibility, although it does not emulsify as well. The most commonly used surfactant for perfluorocarbon emulsification is the nonionic polyoxyethylene–polyoxypropylene copolymer Pluronic F-68. Concern has been expressed over the biocompatibility of the surfactants, especially in view of some impurities and other emulsifiers, such as lecithins, have been introduced.

The biocompatibility issues are obviously a little different to those involved with solid perfluorocarbon polymers. The substances are retained in the tissue for a reasonable length of time and emulsion particles may be seen in tissue macrophages. Clearance is mainly via the lungs. Areas of concern and under investigation include the effects on liver biochemistry and on the immune system. Despite these concerns, the perfluorocarbons are seen as very attractive substances for medical use and applications, such as tissue imaging and perfusion, organ preservation and cancer therapy, are under consideration (Lowe 1988, Chang and Geyer 1989).

Bibliography

Baker W H, Hadock M M, Littody F N 1980 Management of polytetrafluoroethylene graft occlusions. *Arch. Surg.* 115: 508–13

Bolton C W, Bruchman W C 1985 The Gore-Tex expanded polytetrafluoroethylene prosthetic ligament. *Clin. Orthop. Relat. Res.* 196: 202–13

Campbell C D, Brooks D H, Webster M W, Diamond D L, Peel R L, Bahnson H J 1979 Expanded microporous polytetrafluoroethylene as a vascular substitute. *Surgery* 85: 177–83 ·

Chang T M S, Geyer R P 1989 *Blood Substitutes*. Dekker, New York

Charnley J 1963 Tissue reactions to polytetrafluoroethylene. *Lancet* 1379–80

Chignier E, Guidollet J Heymen Y, Serres M, Clendinnen G, Louisot P, Eloy R 1983 Macromolecular, histologic, ultrastructural and immunocytochemical characteristics of the neointima developed within PTFE vascular grafts. *J. Biomed. Mater. Res.* 17: 623–36

Claes L, Durselen L, Kiefer H, Mohr W 1987 The combined anterior cruciate and medial collateral ligament replacement by various materials. *J. Biomed. Mater. Res. Appl. Biomaterials* 21(A3): 319–43

Echave V, Koornick A R, Haimov M, Jacobsen J J 1979 Intimal hyperplasia as a complication of the use of PTFE grafts in femoropopliteal bypass. *Surgery* 86: 791–8

Ferkel R D, Fox J M, Wood D, Del Pizzo W, Friedman M J, Snyder S J 1989 Arthroscopic "second look" at the Gore-Tex ligament. *Am. J. Sports Med.* 17(2): 147–53

Florian A, Cohn L H, Dammin G J, Collins J J 1976 Small vessel replacement with Gore-Tex, *Arch. Surg.* 111: 267–70

Florine B L, Gatto D J, Wade M L, Waite D E 1988 Tomographic evaluation of the temporomandibular joint following placement of PTFE implants. *J. Oral Maxillofac. Surg.* 46: 183–8

Fujiwara Y, Cohn L H, Adams D, Collins J J 1980 Use of Gore-Tex graft for the replacement of the superior and inferior venae cavae. *Surgery* 87: 774–9

Glousman R, Shields C, Kerlan R, Jobe F, Lombardo S, Yocum L, Tibone J, Gambardella R 1988 Gore-Tex prosthetic ligament in anterior cruciate deficient knees. *Am. J. Sports Med.* 16(4): 321–6

Goodson W H 1987 Application of expanded polytetrafluoroethylene tubing to the study of human wound healing. *J. Biomater. Appl.* 2: 101–17

Homsy C A, Anderson M S 1976 Functional stabilisation of soft tissue and bone prostheses with a porous low-modulus materials system. In: Williams D F (eds.) 1976 *Biocompatibility of Implant Materials*. Sectoe, London, pp. 85–93

Indelicato P A, Pascale M S, Huegel M D 1989 Early experience with the Gore-Tex anterior cruciate ligament prosthesis. *Am. J. Sports Med.* 17: 55–62

Lowe K C (ed.) 1988 *Blood substitutes; Preparation, Physiology and Medical Applications*. Ellis-Horwood, Chichester

Selman S H, Rhodes R S, Anderson J M, De Palma R G, Clowes A W 1980 Atheromatous changes in expanded polytetrafluoroethylene grafts. *Surgery* 87: 630–7

Soyer T, Lempinnen M P, Cooper P, Norton L, Eiseman B 1972 A new venous prosthesis. *Surgery* 72: 864–72

Timmis D P, Aragon S B, Van Sickels J E, Aufdemorte

T B 1986 Comparative study of alloplastic materials for temporomandibular replacement in rabbits. *J. Oral Maxillofac. Surg.* 44: 541–4

Whittemore A D, Kent K C, Donaldson M C, Couch N P, Mannick J A 1989 What is the proper role of polytetrafluoroethylene grafts in infrainguinal reconstruction? *J. Vasc. Surg.* 10: 299–305

D. F. Williams
[University of Liverpool, Liverpool, UK]

Polyurethanes

Polyurethanes are polymers that contain the urethane group,

Behind this very simple statement, however, is a very complex array of chemistry which reflects the fact that this group can be contained within a wide variety of polymeric structures which display interesting, but quite diverse properties. Depending on the precise structure, the materials may exist as rigid plastics, thermoplastic elastomers, fibers, sealants, adhesives, foams and so on.

The urethane group is best considered as resulting from the reaction between an isocyanate and an alcohol:

The polyurethane is then represented by the structure

The versatility of this family of polymers lies with the fact that R and R′ may be substituted by a variety of groups. R is usually oligomeric typically consisting of hydroxyl-terminated polyethers or polyesters of molecular weight in the range 200–4500. R′ may be either aliphatic or aromatic.

The most useful type of polyurethane in general, and certainly the most widely used in medical applications, is the elastomer. However, there are many variations within this group, ranging from the low-modulus sealant type of elastomer to the thermoplastic elastomer. Although there are several methods of polyurethane synthesis, it is most common with these elastomers to use a two-stage procedure involving the preparation of a low-molecular-weight prepolymer followed by chain extension and/or cross-linking.

The prepolymer is typically formed by the reaction between an excess of a diisocyanate with a polyol. The main high-performance prepolymers are based on TDI (2,4-toluene diisocyanate) or MDI (4,4′-diphenylmethane diisocyanate). Poly(ether urethanes) are usually based on poly(tetramethylene oxide) (PTMO), poly(propylene oxide) (PPO) or poly(ethylene oxide) (PEO), while polyester polyols such as polycaprolactones provide the basis for the poly(ester urethanes). Chain extension may be achieved with compounds such as glycols or diamines, 1,4-butanediol and ethylene diamine providing the best examples. If cross-linking is also required, the functionality of the chain extender may be increased. The nature of the chain extender is extremely important in determining the properties of the polymer, since it determines molecular chain flexibility, intermolecular attraction and microstructure. It is particularly important to note that as well as the urethane group, —NH—CO—O—, it is possible to have urea groups, —NH—CO—NH— (obtained when diamines are used as chain extenders), so that with the ether and ester groups it is possible to have combinations such as polyurethane–urea, polyurea–ether, polyurea–ester, polyester–urethane and so on. Hydrogen bonding between groups on adjacent chains is extensive and this has an important bearing on the properties.

One other important feature of the polyurethane elastomers is that they are generally two-phase block copolymers. Due to this, many of the polymers have become known in the medical field as segmented polyurethanes. The continuous phase, which is usually the softest and therefore provides the "soft segments", is the polyester or polyether polyol. The discontinuous dispersed phase is more rigid (giving the "hard segments"), and consists of the isocyanate and chain extender groups. The hardness and modulus of the polymer will both increase with greater levels of isocyanate and of chain extender.

1. Biomaterial Applications

Certain types of polyurethane were studied experimentally and used clinically in the 1960s. Although some good results were reported, the performance was generally considered to be very poor, with a severe tissue reaction being seen in association with the materials on many occasions. It is now clear that the causes of the problems lay with the very imprecise characterization of the polymers and with the fact that most formulations employed at that time were of the polyester type. It became fairly obvious, and is now well recognized, that the ester group is quite susceptible to hydrolysis and these early poly(ester urethanes) suffered degradation *in vivo*. This was sufficient to cause the aggressive tissue response, especially with those forms used as open-cell porous materials in an attempt to encourage tissue ingrowth (Mandarino and Salvatore 1959, Redler 1962, Thompson and Sezgin 1962, Walter and Chiaramonte 1965).

Due to these problems, polyurethanes in general suffered from a poor reputation in the biomaterials field for some time (Williams and Roaf 1973). However, the realization that poly(ether urethanes) should be more stable led to the development of segmented ether-based polymers for the medical field in the 1970s and 1980s (Boretos and Pierce 1967, Boretos 1981). Although many polyurethanes have been made commercially available to the medical device industry, the most commonly used and discussed over this period have been Biomer, Pellethane and Tecoflex.

Biomer is most widely used in the solution cast grade although it is also available in an extrusion form. It is synthesized from MDI and poly(tetramethylene ether glycol) (PTMEG) the solution grade being extended with a diamine such as ethylene diamine and the extrusion grade being extended with water (Lelah et al. 1983). The solution grade is soluble in a variety of polar solvents although *N,N'*-dimethylacetamide is the most common.

Pellethane is a very widely used form of poly(ether urethane) and it is available in injection molding and extrusion forms. Catheters are frequently made from Pellethane. It is synthesized from the same MDI and PTMEG as Biomer, although differences in proportions and the nature of the chain extenders yield a variety of products, usually characterized by their hardness. The Pellethane 2300 series is particularly attractive to the medical catheter and tubing industry with a range of hardnesses and, therefore, flexibilities being available (e.g., Shore hardness 55D, 80A, 90A).

Tecoflex is a further formulation of poly(ether urethane) that has been employed medically, being produced from a hydrogenated MDI, PTMEG and butanediol. It is available either as a solution cast grade or in extrudable form.

It is not easy to summarize the properties of this group of materials, because of their considerable variability and versatility. Their attractiveness in respect of biomaterials applications lies with their combination of elastomeric qualities and biocompatibility, especially blood compatibility. There are many applications where polymeric materials are required to be inserted into the vascular system, either temporarily or permanently, where flexibility coupled with reasonable strength and a resistance to thrombus formation are necessary, or at least desirable. Catheters, pacemaker leads, vascular prostheses, flexible leaflet heart valves, and the housings of artificial hearts and ventricular assist devices all come into this category. While some polyurethanes are the only biomedical polymers that possess this combination of properties (with the possible exception of silicone elastomers under some circumstances) there are some drawbacks which limit their applications. The biggest difficulty is their susceptibility to degradation, for it is abundantly clear that this problem is not confined to the poly-(ester urethanes). This susceptibility has become manifest in failures of pacemaker leads, of vascular prostheses and of other devices, and several mechanisms for the degradation have been proposed (see *Biodegradation of Medical Polymers*).

2. Blood Compatibility

According to most of the criteria for blood compatibility, certain polyurethanes would appear to be among the best materials available. As might be expected, however, there is a range of behavior to be seen, the most important determinants being the nature of the soft segments and the soft segment to hard segment ratios. Cooper et al. (1988) have demonstrated that increasing the hard segment to soft segment ratio increases the degree of phase separation and the blood compatibility of PPO- and PTMO-based materials. In addition, the PPO-based material is less thrombogenic than the PTMO-based material, probably because of the lower surface–water interfacial tension of the PPO segments resulting in less protein adsorption and platelet attachment.

The role of protein adsorption is known to be of considerable significance, although there are by no means unequivocal relationships between the various parameters of protein adsorption and blood compatibility. It is generally agreed that polymers that adsorb fibrinogen preferentially when exposed to protein solutions such as plasma are more thrombogenic than those with a preference for albumin and several studies have demonstrated that segmented polyurethanes have a greater affinity for albumin than do other polymers in this respect (Ito and Imanishi 1989). It is also possible for there to be variations in the tendency for the adsorbed polymers to be denatured when attached to the different types of segment in the polyurethane.

As good as the blood compatibility properties appear to be with some of these polymers, there have been many attempts to produce improvements through the use of surface modifications. Indeed, it is likely that one of the most important characteristics of these materials is the ease with which they may be modified, either chemically or pharmacologically, in comparison to most other polymers.

Among the most extensively investigated surface modifications have been those involving heparin (see *Heparinized Materials*) and poly(ethylene glycol). The principle for the former is very straightforward since heparin is a natural anticoagulant and it is a reasonably logical argument to suppose that some of this activity might be transferred to a polymer surface if the heparin could be immobilized there. Many methods for this heparin immobilization have been reported. A typical example consists of taking a sample of Pellethane, treating the surface with hexamethylene diisocyanate to introduce free isocyanate groups into the surface and then immobilizing the heparin to these groups by a formamide treatment (Han et al. 1989). Heparinized surfaces are quite variable in their effect on blood compatibility parameters, but significant reductions in platelet adhesion and in the rates of activation of the intrinsic clotting factors may be seen.

The situation with PEO is also unclear. Attachment to polyurethane surfaces via the same free isocyanate groups, mediated by a stannous octoate catalyst can result in a dramatic reduction in platelet adhesion to the polymer surface. This becomes increasingly obvious as the molecular weight of the PEO is increased. This type of surface does not necessarily have any effect on the clotting proteins, but it is possible for heparinized PEO surfaces to be particularly effective in both reducing platelet adhesion and activation, and in reducing activation of the clotting proteins, possibly because of a synergistic effect. It is interesting that the modification of poly(ether urethanes) by the incorporation of PEO macroglycols in the polymerization reaction itself does not necessarily result in improved blood compatibility; indeed, high-PEO-containing polymers tend to be more thrombogenic than pure PTMO polymers.

It has been noted previously that surfaces that preferentially adsorb albumin tend to be less thrombogenic. Attempts have been made with some polymers, including polyurethanes, to modify their surfaces to increase the affinity for albumin. Possibly the most widely investigated method has been the alkylation of the surfaces, first described by Munro (Munro et al. 1981). It has been shown that the coupling of alkyl chains of 16 or 18 carbon residues to polyurethane significantly enhances the albumin adsorption and simultaneously inhibits the fibrinogen adsorption (Kim 1985). This may be achieved by several methods including a two-stage substitution reaction involving a sodium ethoxide treatment of the surface to form an imide ion intermediate, followed by reaction with an appropriate alkyl halide.

Albuminated surfaces appear to exert their thromboresistant effect via a reduction in susceptibility to platelet adhesion, whereas heparin is effective in resisting activation of the clotting system. Several groups have tried to combine antiplatelet activity with inhibition of the clotting process by using surfaces onto which heparin–albumin conjugates have been covalently bonded (Hennick et al. 1983).

It is also possible to attach a variety of drugs to the surface of polyurethanes in attempts to improve blood compatibility (see *Drugs: Attachment to Polymers*). Further discussions may be found in Kim (1985) and Ito and Imanishi (1989).

3. Uses in the Cardiovascular System

Due to the good blood compatibility outlined in Sect. 2, polyurethanes have found extensive use within the heart and blood vessels (Planck et al. 1984).

The total artificial heart represents one of the most difficult of all organ or tissue replacements and some of the attempts to do so have received much attention and publicity. The Jarvik heart developed in the USA was largely constructed of a solution poly(ether urethane) and the performance of the material has been extensively documented (Coleman et al. 1986). The Penn State total artificial heart also utilizes a segmented polyurethane (Geselowitz 1988). Artificial heart programs in Japan also make use of polyurethanes in the construction of the housing. For example, Iwaya (1989) has described the use of a Toyobo segmented polyurethane in an artificial heart. Most designs of artificial heart have employed traditional mechanical valves, such as the Bjork–Shiley or the Medtronic–Hall valves, but some groups have been attempting to incorporate flexible leaflet valves within the structure, with polyurethanes again being the preferred choice (Fukumasu et al. 1988).

Vascular prostheses are also of considerable importance. The woven polyester Dacron is the most popular material for the construction of the larger-diameter vessel replacements, with increasing competition from expanded polytetrafluoroethylene (PTFE), especially in the small to medium size range. There is, as yet, no satisfactory material for the construction of artificial arteries of diameter less than 6 mm, but the material that has figured most prominently in the experimental work in this area is poly(ether urethane). The reason for this preference of the polyurethane over polyesters or PTFE lies with the elastomeric properties described previously, making it possible to prepare polyurethane vascular prostheses with good mechanical characteristics, especially compliance and kink resistance equivalent to those of natural vessels (How and Clarke 1984, How 1989). Among the vascular prostheses constructed of polyurethanes which have been described are those of Hayashi (Hayashi et al.

1989), Guidoin (Martz et al. 1987), Lyman (Lyman et al. 1977) and Annis (Annis et al. 1984). These different prostheses utilize different types of

polyurethane and different methods of construction. In the Annis prosthesis, for example, a technique of electrostatic spinning has been developed and employed for the preparation of the microporous prosthesis. Several different morphologies have been produced, an example being shown in Fig. 1.

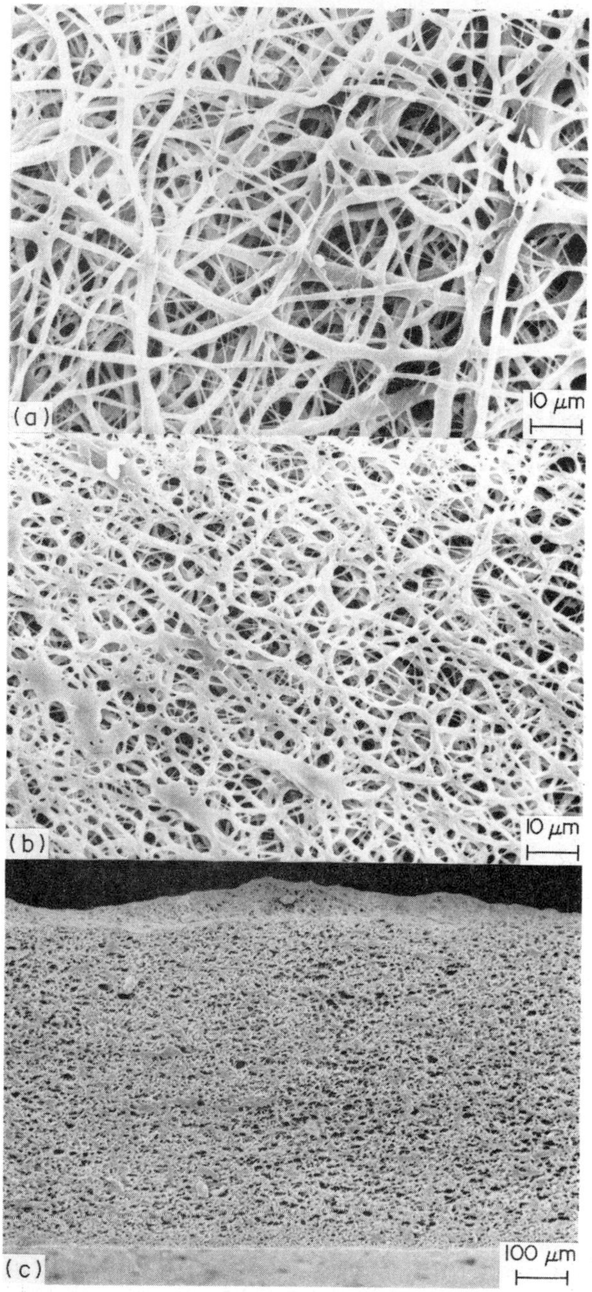

Figure 1
Scanning electron micrograph of electrostatically spun polyurethane vascular prosthesis: (a) outer surface, (b) inner surface, and (c) longitudinal section through wall

Bibliography

Akutsu T (ed.) 1988 *Artificial Heart 2*. Springer, Tokyo

Annis D, How T V, Fisher A C 1984 Recent advances in the development of artificial devices to replace diseased arteries in man. In: Planck, Egbers and Syre 1984, pp. 287–300

Boretos J W 1981 The chemistry and biocompatibility of specific polyurethanes for medical use. In: Williams D F (ed.) 1981 *Biocompatibility of Clinical Implant Materials*, Vol. 2. CRC Press, Boca Raton, FL, pp. 127–44

Boretos J W, Pierce W S 1967 Segmented polyurethane: A new elastomer for biomedical applications. *Science* 158: 1481–5

Coleman D L, Meuzelaar H L C, Kessler T R, McLennen W H, Richards J M, Gregonis D E 1986 Retrieval and analysis of a clinical total artificial heart. *J. Biomed. Mater. Res.* 20: 417–24

Cooper S L, Giroux T A, Grasel T G 1988 Blood material interactions: *Ex vivo* investigations of polyurethanes. In: Akutsu 1988, pp. 3–12

Fukumasu H, Yuasa S, Iwaya F, Tatemichi K 1988 New valve-containing systems for the total artificial heart. In: Akutsu 1988, pp. 333–42

Geselowitz D B 1988 Engineering studies of the Penn State artificial heart. In: Akutsu 1988, pp. 283–94

Han D K, Jeong S Y, Kim Y H 1989 Evaluation of blood compatibility of PEO grafted and heparin immobilised polyurethanes. *J. Biomed. Mater. Res. Appl. Biomater.* 23: (A2) 211–28

Hayashi K, Takamizawa K, Saito T, Kira K, Hiramatsu K, Kondo K 1989 Elastic properties and strength of a novel small-diameter compliant polyurethane vascular graft. *J. Biomed. Mater. Res. Appl. Biomater.* 23: (A2) 229–44

Hennick W E, Feijen J, Ebert C D, Kim S W 1983 Covalently bound conjugates of albumin and heparin. *Thromb. Res.* 29: 1–9

How T V 1989 Haemodynamic performance of arterial prostheses. In: Williams D F (ed.) 1989 *Current Perspectives on Implantable Devices*, Vol. 1. JAI Press, Greenwich, pp. 267–314

How T V, Clarke R M 1984 The elastic properties of a polyurethane arterial prosthesis. *J. Biomech.* 17: 597–608

Ito Y, Imanishi Y 1989 Blood compatibility of polyurethanes. *CRC Crit. Rev. Biocompat.* 5: 45–104

Iwaya F 1989 Development of a total artificial heart in Japan. In: Williams D F (ed.) 1989 *Current Perspectives on Implantable Devices*, Vol. 2. JAI Press, Greenwich, pp. 253–66

Kim S W 1985 Surface modification of polymers for improved blood compatibility. *CRC Crit. Rev. Biocompat.* 1: 229–60

Lelah M D, Lambrecht L K, Young B R, Cooper S L 1983 Physicochemical characterisation and *in vivo* blood tolerability of cast and extruded Biomer. *J. Biomed. Mater. Res.* 17: 1–22

Lyman D, Albo D, Jackson R, Knutson K 1977 Development of small diameter vascular prosthesis. *Trans. Am. Soc. Artif. Intern. Org.* 23: 253–61

Mandarino M P, Salvatore J G 1959 Ostamer: A polyurethane polymer. *J. Bone Jt. Surg.* 41A: 1542–7

Martz H, Paynter R, Forest J-C, Downs A, Guidoin R 1987 Microporous hydrophilic polyurethane vascular grafts as substitutes in the abdominal aorta of dogs. *Biomaterials* 8: 3–11

Munro M S, Quattrone A J, Ellsworth S R, Kulkarni P, Eberhardt R C 1981 Alkyl substituted polymers with enhanced albumin affinity. *Trans. Am. Soc. Artif. Intern. Org.* 27: 499–503

Planck H, Egbers G, Syre I (eds.) 1984 *Polyurethanes in Biomedical Engineering.* Elsevier, New York

Redler I 1962 Polymer osteosynthesis: A clinical trial of Ostamer. *J. Bone Jt. Surg.* 44A: 1621–4

Thompson F R, Sezgin M Z 1962 Polyurethane polymer: An experimental investigation of its adjunct value in the treatment of fractures. *J. Bone Jt. Surg.* 44A: 1605–9

Walter J B, Chiaramonte C G 1965 Tissue responses of the rat to implanted Ivalon, Etheron and Polyfoam plastic sponges. *Brit. J. Surg.* 52: 49–58

Williams D F, Roaf R 1973 *Implants in Surgery.* Saunders, London

D. F. Williams
[University of Liverpool, Liverpool, UK]

Porcelain–Metal Bonding in Dentistry

Porcelain is used in dentistry to meet the exacting aesthetic requirements of restoring anterior teeth. The complete porcelain crown, or porcelain jacket crown, however, has a tendency to fracture. Fusing the porcelain on a metal coping, combines the strength and accuracy of a cast dental alloy with the aesthetic properties of the porcelain.

1. Clinical Application

The restoration of morphology, function and contour, of the damaged coronal portion of a tooth is accomplished by cementing a veneer crown, a type of restoration designed to protect the remaining hard tissues from further damage. When teeth are missing, they have to be replaced to restore function, to maintain the teeth adjacent to the space in their respective positions, and to prevent the opposing teeth from erupting into the space. This may be done by cementing onto the teeth adjacent to the space (the abutment teeth) a particular type of restoration called a fixed bridge, which includes the retainers (to fit on the abutments), the pontic (to replace the missing tooth) and the connectors (to unite the pontic with the retainers).

The restoration of anterior teeth by means of crowns or bridgeworks imposes exacting aesthetic requirements. The mechanical stresses created during mastication or resulting from incidental impacts, and the possible abrasion from toothbrushing, define further requirements. Dental porcelain has constantly been improved to reach a level of quality which meets some of these requirements: it is possible to build full veneer crowns such as the porcelain jacket crowns (see *Dental Porcelain*), which simulate the shades of the adjacent teeth. These jacket crowns, however, are subject to very stringent rules in relation to tooth preparation and the size and design of the restoration, to prevent them from fracturing under mechanical stresses. The porcelain may be strengthened by the addition of alumina microcrystals, but this improvement does not eliminate the difficulty of achieving a perfect seal at the margin of the restoration: since porcelain contracts heavily under firing, the marginal fit is rarely as good as with metallic restorations. Furthermore, the making of long-span bridgework is extremely difficult, and the use of an inner core of high-density alumina may bring additional strength. Such restorations are very arduous to prepare and require a highly skilled dental technician to master the problems of color matching and proper fitting. Since about 1960, however, a new technique has been introduced to dentistry: the porcelain-fused-to-metal (PFM) restoration. By baking the porcelain directly on a metallic structure, it is possible to greatly increase the overall strength of the restoration, and to obtain a very tight fit of the margins, as with metallic crowns.

The technique had to overcome many difficulties prior to enjoying its present popularity among practitioners and patients. The major problem was matching two materials with properties which were apparently not compatible, and to achieve an effective bond between the two of them. Since the melting range of the usual dental gold alloys was much lower than the firing temperature of the porcelain, it was necessary to create new alloys and new dental ceramics, so that the metal casting could withstand the high temperature of the porcelain firing. The thermal coefficients of expansion also had to be finely adjusted to avoid chipping of the porcelain by the generation of internal stresses during cooling. A special porcelain, the opaquer, has been developed to mask the dark surface of the metal coping. A body porcelain, fired on the thin layer of opaquer, gives the proper shade and level of translucency necessary to simulate the adjacent natural teeth. Because of the opacity of the metal coping, it is impossible to recreate the transmission of the light as through the natural teeth, and only the incisal edge of an anterior ceramic-metal composite restoration transmits the incident light. The other cervical two-thirds of the crown can only reflect the incident light. This increases the difficulty of achieving a satisfactory simulation of the natural teeth.

The PFM technique has been steadily improved since 1960, and its use has been so extended that the porcelain jacket crown has been restricted to cases where aesthetics dominates other factors. PFM restorations are now indicated for crowns and bridgeworks on the anterior teeth and the premolars. In some instances, they may also be used to restore first molars. In all cases, the metal coping is so designed that it comes in contact with the cervical limits of the prepared tooth to ensure the tightest fit. The extent of the porcelain veneer on the vestibular (external) surface of the restoration is still disputed among restorative dentists: some of them ask for a complete covering of the metal coping down to the cervical limit, whereas others recommend maintaining a metal collar in the cervical region, where the restoration is in contact with the gingiva. This may be somewhat less aesthetic, since the metal becomes visible, but it is claimed to be much healthier for the soft tissues.

Despite the difficulty in obtaining wholly satisfactory aesthetic results, PFM restoration has gained its unparalleled popularity because of its greater strength and resistance to fractures: the combination of the two different materials, bonded together, is stronger than the porcelain alone. The porcelain–metal bond, therefore, plays a key role in PFM restorations.

2. Gold-Based Alloys

PFM restorations were developed at a time when the price of gold was low enough to make it the unchallenged metal of excellence for dental metallurgy. Since then, less expensive metals have been tried as alternatives to the dental gold alloys. The conventional gold alloys are designed around the gold–copper system, using the ordering reaction to produce hardening. Since copper creates unwanted colorations with the dental porcelains, it could not be used for PFM alloys. The increase in the melting range was obtained by adding platinum and palladium to gold, and the hardening by adding small amounts of iron, to form $FePt_3$ microprecipitates. Base metals such as indium and tin were also added to promote the porcelain bonding. A typical composition of such high-gold PFM alloys is: 87%Au–6%Pt–5%Pd–1%Ag–0.5%In–0.4%Sn–0.1% Fe. Alloys with a reduced gold content were later developed in an attempt to reduce cost. They belong to two different systems: Au–Ag–Pd and gold–palladium. In both cases, the gold content is reduced to around 50 wt%, with an increased content in base metals (up to 8–10 wt% in total).

The specific requirements for PFM alloys are:

(a) a melting range starting at no less than 1100 °C;

(b) a coefficient of thermal expansion closely matched to that of the low-fusing-point dental porcelains (960–980 °C) developed for the PFM technique—the values for these coefficients should stay in the ranges $12.7–14.8 \times 10^{-6}$ °C^{-1} for the alloys, and $10.8–14.6 \times 10^{-6}$ °C^{-1} for the porcelains;

(c) minimal creep or sag during firing of the porcelain; and

(d) good wetting of the alloy by the porcelain.

The wettability of the porcelain is achieved by the formation of a layer of oxides at the surface of the alloy through a preoxidation treatment. These superficial oxides are essential to the porcelain–metal bond.

2.1 Nature of the Interface

The usual procedure to prepare the metal coping for the first porcelain application includes a careful cleaning to eliminate any trace of organic materials, and a degassing/oxidation at high temperature (960–980 °C) under reduced pressure for about 10 min. During this thermal treatment, base metals, such as indium, iron and tin, migrate to the surface of the alloy and oxidize, as seen in Fig. 1. The gases trapped in the metal structure during the casting, partly because of the palladium, are then eliminated from the outer layers of the alloy.

These oxides markedly enhance the wettability of the porcelain on the alloy, as indicated by the acute contact angle observed with sessile-drop experiments. A true interface between porcelain and metal is thus achieved. The nature of the bond created between the opaque porcelain and the superficially oxidized alloy was much disputed in the early days of the PFM technique. Pure adhesion, chemical bonding or mechanical interlocking had been proposed as the main mechanism, but very little evidence was available. The use of the electron microprobe (Nally and Meyer 1969) has produced some experimental facts. When analyzing the distribution of the base metals in the vicinity of the porcelain–metal interface, it was clearly visible (see Fig. 2) that they accumulated in a bonding zone with a width of about 2 μm, thus confirming the observation of the oxide layer formed at the surface of the alloy during the oxidation treatment.

Since such images were not sufficient to prove that the superficial oxides did penetrate the contacting porcelain, PFM composites were fired five and ten times longer than usual, and then observed with the microprobe. The concentration profiles of gold, indium and iron across the interface (see Fig. 3) demonstrate their increasing penetration into the porcelain with increasing baking times.

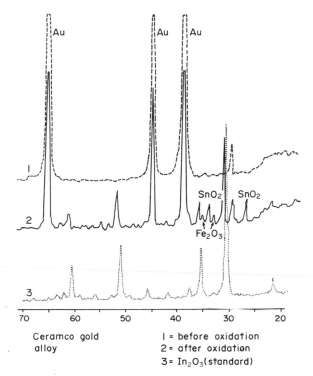

Figure 1
X-ray diffraction spectra of the surface of a high-gold
PFM alloy, before and after the oxidation treatment
(note the formation of indium, tin and iron oxides)

A mechanism leading to the formation of a chemical bond has been proposed (see Fig. 4). Owing to the preoxidation treatment, oxides of base metals are formed on the surface of the cast alloy. When the opaque porcelain is then fired on the oxidized metal coping, the indium, iron and tin oxides diffuse into the vitreous phase of the porcelain. A continuous solid is formed, with chemical bonds between metals, metal oxides and aluminosilicates of the feldspathic porcelain.

Pask (1977) and Cascone (1977) have described the fundamentals of bonding between dental alloys and dental ceramics. The superficial oxides are dissolved by the porcelain, leading to interfacial reactions between the metal, a minimum molecular oxide layer and the ceramic, which has the characteristics of a glass. This glass becomes saturated with the dissolved oxides. To form a true chemical bond, stable chemical equilibrium should be reached at the interface, relative to the lowest-valent oxide of the substrate metal with which the ceramic and the metal at the interface are saturated. This equilibrium is obtained by saturation of the metal and the glass at the interface with an oxide of the metal substrate, and with redox reactions such as

$$In^+ + Sn^{4+} \rightarrow In^{3+} + Sn^{2+}$$

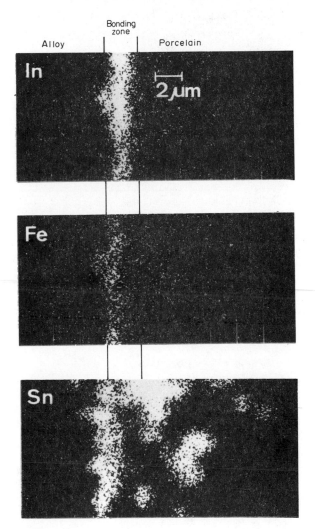

Figure 2
Electron microprobe x-ray images showing the
distribution of indium, iron and tin in the vicinity of
the porcelain–metal interface of a high-gold dental
PFM composite

It is postulated that the minimum molecular oxide layer, or transition layer, in equilibrium with both the metal and porcelain is SnO in the case of high-gold PFM composites.

2.2 Strength of the Bond

When a true chemical bond is formed at the interface, failure of the PFM composite is cohesive (i.e., through the porcelain), and the strength of the bond is usually greater than the strength of the porcelain itself. Six types of failure are recognized in theory:

(a) metal–porcelain,

(b) metal oxide–porcelain,

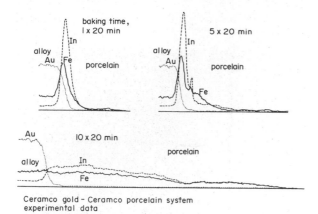

Ceramco gold – Ceramco porcelain system
experimental data

Figure 3
Uncorrected electron microprobe experimental data
showing concentration profiles of gold, indium and
iron across the interface of dental PFM composites, at
various porcelain baking times (note the diffusion of
indium and iron into the porcelain with increasing
baking times; the maximum penetration depth is
about 25 μm)

(c) cohesive within porcelain,

(d) metal–metal oxide,

(e) metal oxide–metal oxide, and

(f) cohesive within metal.

Types (a)–(c) can be observed with high-gold PFM
composites, while types (d) and (e) are more frequent with porcelains fused to nickel–chromium
alloys. The design of a test to evaluate the strength
of the porcelain–metal bond is not easy. Three types

have been proposed: shear or pull-through tests,
tensile tests and flexure tests. Each type of test gives
specific information about the nature of the failure
and the resistance of the PFM composite to the
particular applied stress. The shear test has been
refined over the years and proposed as a standard. A
small collar of porcelain is baked around a cast alloy
rod, in the usual way used in the dental laboratory.
The height of the collar is measured to determine
the surface of bonding. The remaining parts of the
rod are waxed to prevent adhesion with the large
cylinder of dental stone cast around the whole assembly. After a proper setting time, the end of the
rod protruding above or below the large cylinder is
either pushed or pulled, until the porcelain breaks
near or at the bonding zone with the cast alloy rod.
The shear stress at failure is used to measure the
strength of the bond. A proposed specification,
using this test, sets the minimum value at 69 MPa.
Up to 80 MPa can be obtained with some high-gold
PFM composites, and a value of 92 MPa has been
reported with a nickel-based PFM composite. The
strength of the porcelain veneer of actual PFM
restorations is influenced by the possible concentration of internal stresses, as well as by minute flaws in
the porcelain, which may propagate suddenly under
fatigue or impact. The bonding of the veneer to the
metal structure may be interrupted locally by surface
contaminations of the metal coping, or by gas
bubbles formed at the interface, thus decreasing the
overall strength of the bond. Improper oxidation or
building of too thick an oxide layer, as frequently
observed with some nickel–chromium alloys, may
fundamentally alter the formation of the chemical
bond and therefore change the type of failure, which
may become catastrophic.

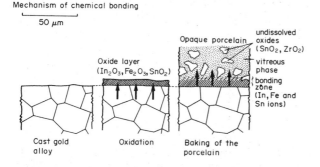

Figure 4
Proposed mechanism of chemical bonding: the oxides
of the base metals accumulated on the surface of the
cast alloy during the oxidation treatment diffuse into
the opaque porcelain during baking; a continuous
solid is thus formed, with successive chemical bonds
between metals, metal oxides and aluminosilicates
(feldspathic opaque porcelain)

3. Palladium–Silver Alloys

The surge in the price of gold has brought new types
of alloys, in an effort to keep the price of a PFM
restoration to a reasonable level. In the palladium–
silver alloys, gold has been completely eliminated. A
typical composition is 60% Pd–30% Ag–10%
(In + Sn). These alloys have key physical properties,
such as hardness and yield strength, equal to or
greater than high-gold PFM alloys, and the laboratory technique is identical to that for gold–palladium
alloys. The base metals (indium and tin) play a
similar role in the porcelain–metal bond, which also
results from the saturation of both metal and ceramic at the interface with an oxide of the metal
substrate. However, the higher silver content
increases the propensity for a sudden and unwel-
come appearance of a greenish-yellow coloration of
the porcelain. Its cause is not clearly known at this
time, but it is speculated that colloidal silver is
formed in the ceramic. The coloration seems to be

more intense in porcelains with a higher sodium content (lighter shades), probably because the silver diffuses faster in sodium-containing ceramics than in potassium-containing ones. This behavior has seriously limited the use of such alloys for PFM restorations. The same problem arises with low-gold Au–Ag–Pd alloys, but to a lesser extent. Gold–palladium alloys have been proposed to overcome this difficulty, but they usually are too brittle and can fracture when taking the casting out of the investment. These three types of PFM alloys (Pd–Ag, Au–Ag–Pd and Au–Pd) behave similarly to high-gold alloys as far as porcelain bonding is concerned.

4. Nickel–Chromium Alloys

Nickel–chromium alloys (see *Base-Metal Casting Alloys for Dental Use*) had been used in the PFM technique, even before the great increase in the gold price, although to a very limited extent. Since then, economic pressure has stimulated the development of low-cost nickel–chromium alloys; they account today for about 75% of all crowns and bridge units in the USA. Dozens of nickel–chromium alloys are being advertised for dental use, and there is a great deal of confusion about their relative merits. They usually have a much higher modulus of elasticity than high-gold PFM alloys (120–240 GPa), and the same elastic limit (400 MPa); therefore, the same level of rigidity in a given bridgework can be obtained with a thinner metal structure. More space is then available for building the porcelain veneer. Some problems, however, have been encountered with them, such as poor castability, ill-fitting castings, excessive hardness, and porcelain bond failure through an oxide layer (type (d) or (e) failure, Sect. 2.2). An additional problem is the wide variation in the composition of nickel–chromium alloys, which results in unique multiphase microstructures for each alloy. The adherence of porcelain to these alloys therefore becomes much more complex, depending on the type of metal oxide formed and the rate of interdiffusion at each zone where phase transitions occur. Figure 5 illustrates the considerable depth of penetration of manganese, a minor element (4 wt%) in an alloy containing 70%Ni–16%Cr–5%Mo–2.6%Al–0.8%Si–1.6% others. The depletion of the manganese concentration near the bonding zone demonstrates that this oxide does not saturate the porcelain at the interface, and cannot produce the required chemical equilibrium. Aluminum and chromium oxides seem to be responsible for porcelain bonding, but no unified theory suitable for all nonprecious PFM composites can be proposed because of the multiphase nature of nearly all nonprecious PFM alloys. Several nickel–chromium alloys utilize an aluminum-containing

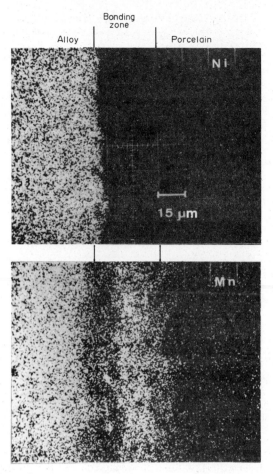

Figure 5
Electron microprobe x-ray images showing the distribution of nickel and manganese near the porcelain–metal interface of a nonprecious PFM composite (note the deep penetration of about 40 μm of manganese into the ceramics, and the low concentration directly at the interface)

bond coat as a preliminary step, prior to application of porcelain. This is a slurry of aluminum which, on firing, produces aluminum oxides which are evidently important in forming the bond.

4.1 Biological Considerations

Occasional allergic responses of individuals to nickel, and the need to be cautious in processing because of the beryllium content of the early nickel–chromium PFM alloys, have focused attention on the potential hazards related to these alloys. Beryllium has been eliminated in most of the nickel–chromium alloys available today. Adequate ventilation where beryllium-containing alloys are melted, ground and finished, and the wearing of a dust-mask, offer satisfactory protection to the dental technician. The beryllium in cast alloys is harmless for the patient.

Nickel ranks third among the five most common causes of allergic contact dermatitis, and women are more sensitive to it than men. Some individuals are far more sensitive to nickel than others, and the possible causes of sensitization in everday life are numerous. Since the ability of an alloy to induce a dermatitis seems to be related to its mode of corrosion, and since the corrosion behavior of nickel–chromium dental alloys differs widely as a function of their multiphase microstructures, no general pattern can be predicted (see *Corrosion of Dental Materials*). Nickel-containing dental alloys can therefore pose a threat to sensitized patients, particularly when a corrodible cast restoration is placed in an area where gingival tissues have been incidentally damaged during the tooth preparation. A careful selection of the most corrosion-resistant alloys, and a short inquiry about a possible previous nickel sensitization, may help the practitioner to minimize the risk of contact dermatitis for his patient.

See also: Biocompatibility of Dental Materials

Bibliography

Barreiro M M, Reijgo O, Vicente E O 1989 Phase identification in dental porcelains for ceramo-metallic restorations. *Dent. Mater.* 5: 51–7

Cascone P J 1977 The theory of bonding for porcelain-to-metal systems. In: Yamada H N 1977, pp. 109–17

Dehoff D H, Anusavice K J 1989 Effect of viscoelastic behaviour on stress development in a metal ceramic system. *J. Dent. Res.* 68: 1223–30

Fairhurst C W, Hashinger D T, Twiggs S W 1989 The effect of thermal history on porcelain expansion behaviour. *J. Dent. Res.* 68: 1313–15

Mezger P F, Stols A C H, Vrijhoef M M A, Greener D H 1989 Metallurgical aspects of palladium–silver porcelain bonding alloy. *J. Dent.* 17: 90–3

Nally J N, Meyer J M 1969 Electron microprobe investigations on the ceramic-to-metal interface to explain the nature of the chemical bonding. *Program and Abstracts of Papers of the 47th General Meeting of the IADR, 1969*, International Association for Dental Research Abstract No. 592. IADR, Washington, DC

O'Brien W J 1977 Cohesive plateau theory of porcelain-alloy bonding. In: Yamada H N 1977, pp. 137–41

Pask J A 1977 Fundamentals of wetting and bonding between ceramics and metals. In: Valega T M 1977, pp. 235–54

Rosenstiel S F, Porter S S 1988 Apparent fracture toughness of dental porcelain with a metal substructure. *Dent. Mater.* 4: 187–90

Shell J S, Nielsen J P 1962 Study of the bond between gold alloys and porcelain. *J. Dent. Res.* 41: 1424–37

Valega T M (ed.) 1977 *Alternatives to Gold Alloys in Dentistry.* US Department of Health, Education and Welfare Publication No. (NIH) 77-1227. US Government Printing Office, Washington, DC

Yamada H N (ed.) 1977 *Dental Porcelain: The State of the Art—1977.* University of Southern California, Los Angeles, CA

J.-M. Meyer
[Université de Genève, Genève, Switzerland]

Porous Biomaterials

Secure tissue-prosthesis attachment is a necessary requirement for the successful performance of most surgical implants. Orthopedic implants for bone and joint replacement are effective only if they can be firmly fixed within the host bone. Similarly, endosseous dental implants that are placed in the jaw bone must become and remain securely fixed under the repeated action of masticatory and occlusal loads. Even for implants that are not intended primarily for load bearing, such as cardiac pacer electrodes, prosthetic heart valves or vascular grafts, implant stabilization is critical for their reliable use. Porous-structured implants offer a means of achieving the necessary long-term implant fixation through the ingrowth of tissue into the pores of the implant.

The use of porous implants in orthopedics, dentistry and cardiovascular surgery is stressed in this article. Because implants fixed by tissue ingrowth are intended for permanent placement, they must be of the highest quality. Should mechanical failure of a tissue-ingrown porous-structured implant occur in service, then removal and replacement would pose a serious surgical problem, especially for implants fixed by bone ingrowth. For load-bearing orthopedic and dental implants, corrosion-fatigue strengths must be sufficiently high to ensure long-term implant survival, often under more severe loading conditions than would normally be applied to conventional, non-porous-structured implants. Even for porous cardiovascular implants (vascular grafts, artificial heart valves and pacemaker electrodes) that are not primarily intended for load bearing, mechanical breakdown of the porous structure represents a serious problem since this could introduce small thromboembolic fragments to the bloodstream. Additionally, mechanical breakup of porous-structured implants (or parts of implants) could lead to chronic inflammatory responses with the loose particulate matter acting as a cause of chronic inflammation.

1. Totally Porous vs Porous-Coated Implants

Early studies with porous implants showed that tissue ingrowth was an effective means for achieving reliable implant fixation through mechanical interlock of the implant and host tissue. For a number of applications, an additional important requirement for success of the implant is good mechanical strength, and fatigue strength in particular. Porous structures are inherently weak and display poor fracture resistance compared with fully dense materials. For this reason, dual-structured implants were developed consisting of porous surface coatings bonded to solid high-strength substrates. Currently, these porous-coated implants represent the most commonly used porous-structured implants in orthopedics, where highly loaded implants are used most often for skeletal joint replacements. Porous-coated prostheses are also being tested as dental implants (again, mechanical strength is an important requirement for this application).

2. Orthopedic Applications

2.1 Attachment by Bone Ingrowth

By the late 1960s, loosening of conventional (cemented) joint-replacement implants was recognized as the major problem in their use. This resulted in studies of alternative approaches for implant fixation to bone including the use of porous-coated implants. Metal, polymer and polymer-based composite porous surface coatings were investigated and this has led to the current widespread use of porous-coated joint-replacement implants, principally porous metal-coated hip and knee joint implants (Pilliar 1987).

Although a fundamental knowledge of the factors influencing bone ingrowth was lacking at the time, animal studies in the late 1960s and early 1970s suggested the necessary conditions for implant fixation by bone ingrowth. It was shown that for bone ingrowth to occur, gross movement of the implant relative to the host bone had to be avoided during the initial healing period. The occurrence of movement resulted in fibrous tissue rather than bone ingrowth or, in the extreme, fibrous tissue encapsulation of the implant without any tissue–implant attachment. Although a quantitative description has not been made of how early implant movement affects cell differentiation (i.e., the expression of phenotype and tissue formation in and around porous-coated implants), the accepted approach in using these devices in humans is to place the implants as snugly as possible and to avoid excessive early loading of the implant in order to achieve bone ingrowth and rigid implant fixation.

A number of studies have reported on the effect of pore size on bone ingrowth. It has been shown that bone ingrowth will be delayed if pore size is smaller than about 50 μm, even if no movement occurs between implant and bone. Studies using cylindrical implants placed transversely through cortical long bones, suggested an optimum pore size of 50–300 μm to achieve the fastest rate of bone

ingrowth. A maximum shear strength of attachment was reached by about 12 weeks. The time sequence for this to occur, as well as the effect of movement on bone ingrowth, suggests a similar mechanism for bone ingrowth and normal fracture healing of bone. It has been suggested that strong fixation requires the formation of Haversian bone (osteons) within the pores of the coating. This requires pores greater than about 100 μm. Most porous-coated joint-replacement implants currently in clinical use, are made with pores greater than 100 μm.

Porous-coated orthopedic implants have been made using a number of different materials by a variety of processes. As a result, porous coatings of different pore shapes and sizes have been studied *in vivo*, including coatings made by sintering powders, wires (or fibers), and wire mesh structures made of metals, polymers, ceramics and composites (carbon-fiber-filled polytetrafluoroethylene (PTFE)). Bone ingrowth has been demonstrated with most of these systems.

High-temperature sintering of metallic powders, wires or fibers has been used to form porous surface coatings (35–50 vol.% porosity) firmly bonded to solid substrates. The porous coatings so formed are typically 500–1000 μm thick (two to three particle layers) and consist of a regular three-dimensional interconnected porous structure. Tissue ingrowth into this three-dimensional porous network results in resistance to shear, compressive and tensile distractive forces at the tissue–implant interface. In addition to the high-temperature sintering processes, titanium wire-made coatings have been prepared using lower-temperature pressure sintering.

Processes other than sintering have been used to form so-called porous surface coatings. These include plasma spraying and procedures in which sacrificial constituents are melted or evaporated from the structure, thereby leaving a porous layer (void metal composites). There have been no reports of the use of void metal composite orthopedic implants in humans. Light and scanning electron microscopic examination of plasma-sprayed coatings indicate differences in structure compared with sintered coatings. Plasma-sprayed coatings, like sintered coatings, are 500–1000 μm thick, but they do not form a regular three-dimensional interconnected array of pores. The plasma-sprayed coatings essentially form irregular surfaces with very little interconnected porosity throughout the thickness of the coating.

A major concern with the clinical use of porous-coated orthopedic implants is the effect on tissues, both local and remote to the implant, of higher rates of metal ion release with these implants. Laboratory corrosion studies have shown higher corrosion rates for porous-coated implants, directly proportional to their higher surface areas. This has led to studies of the use of thin corrosion-inhibiting layers deposited onto the surface of porous-coated implants. Thin pyrolitic carbon coatings on porous titanium, Ti–6 wt%Al–4 wt%V and cobalt-based alloy coatings can reduce corrosion rates to levels similar to those for non-porous-coated implants. The effectiveness of such coatings has yet to be demonstrated *in vivo*.

Another concern with the use of porous-coated implants in highly loaded applications is the effect that the porous surface layer might have on the mechanical properties of the implants, particularly fatigue strength. Studies have shown that titanium alloy implants, in particular, experience drastic reductions in fatigue strengths following porous coatings (fatigue endurance strength <200 MPa, cf. >600 MPa for conventional non-porous-coated Ti–6 wt%Al–4 wt%V). The cause of this loss in properties has been attributed to easier fatigue crack initiation at the regions of particle–substrate bonding. These contact zones present regions of high stress concentration and, possibly, significant altered composition (higher interstitial impurity levels) and microstructure. Titanium is known to be notch fatigue sensitive, so that the topology alone could cause the much lower fatigue strengths. The current practice in designing porous-coated titanium alloy implants, is to avoid porous coatings on surfaces that might be subjected to substantial tensile stresses in service. This limitation is unfortunate since titanium alloys offer significant advantages over other metals in terms of corrosion resistance, metal ion release and biocompatibility.

The earliest porous-coated joint-replacement implants were made of cast cobalt-based alloys and most porous-coated orthopedic implants are still of this composition. The effect of the higher-than-normal heat treatment, used to sinter alloy powder particles to the cast cobalt alloy substrate, is a modified substrate microstructure due to local incipient melting of solute-enriched interdendritic zones. This embrittling structure can be avoided through appropriate heat treatments following the sintering operation (Pilliar 1987). Of prime concern is the effect of processing and structure on fatigue properties (as with the titanium alloys). With proper heat treatment, fatigue properties of porous-coated cobalt alloys are not very different from conventionally treated cast cobalt alloys, although it should be noted that the fatigue strengths of the cast and solution-annealed cobalt alloys are low compared with alternate surgical implant alloys. A potential benefit of good implant fixation (through bone ingrowth) is that lower stresses should result due to the integration of the implant and surrounding bone. Nevertheless, low fatigue strengths of porous-coated implants are a concern since fracture *in vivo* would require the removal of at least part of a well-fixed bone-ingrown implant. This can be achieved, but with considerable difficulty.

Porous-coated stainless steel implants, because of their poorer crevice-corrosion properties, are not suitable for use in humans.

A number of polymer-based porous coatings on metal substrates have been studied for orthopedic implants. These include porous polysulfone, polyethylene and carbon-fiber-filled PTFE (Spector 1982a). Generally, these coatings are similar in structure to the porous metal-coated systems. An acclaimed advantage of porous polymer-coated implants is that the lower-stiffness coating results in stresses acting on the tissues around and within the porous coating closer to normal physiological stresses. As a result, abnormally high and low stresses that can be detrimental to wound healing and cause undesirable bone remodelling are avoided. This has been demonstrated in animal studies using totally porous nonfunctional polyethylene implants. However, the advantage has not been demonstrated with functional implants. A problem with some of the polymers used for porous coatings is that they distort too easily and can, during placement, deform resulting in closing of pores, thereby preventing osseous tissue ingrowth. Additionally, poor creep properties of polymers could result in implant deformation and fracture in service, although it has been suggested that once bone ingrowth occurs, the polymer–bone composite should provide adequate strength. Polysulfone has been proposed as the preferred choice for porous polymer coatings because of its superior mechanical properties compared with other polymers. Porous polysulfone, polyethylene and carbon-reinforced PTFE-coated metal substrate implants have been used in human clinical studies, although to a far lesser extent than porous metal-coated implants. One obvious advantage of these polymer-coated implants is their easier removal after bone ingrowth, should that be necessary. An additional concern is the nature of the bond at the polymer coating–metal substrate interface. For the carbon-filled PTFE coatings this has proven to be a major problem, with reports of coating detachment from the metal substrate leading to loosening of femoral hip implant components in humans.

Particle debonding and shedding, either at the time of implant placement or at some later time, have also been reported for porous metal-coated implants. Structural examination of porous-coated cobalt alloy implants has shown that brittle phases can form at the bond junctions and, further, improper processing can lead to inordinately small particle-to-substrate contact zones for both the cobalt alloy powder and titanium fiber-made coatings. A concern with particle or fiber debonding is the possibility of fragments migrating into the joint space resulting in a highly undesirable three-body wear situation. This has been reported to occur with porous-coated knee implants.

Human clinical studies of porous-coated implants have identified other concerns with these designs. Femoral bone resorption caused by enhanced stress shielding has been reported to occur in the first year or so after implantation of porous-coated hip implants, and some cases of apparent fibrous tissue attachment with associated thigh pain have also been reported (Engh et al. 1987). The question of long-term effects of metal ion release and accumulation in tissues, as noted previously, remains a major concern. This very important issue will require continuing long-term clinical follow-up studies involving close monitoring of patients for signs of increased metal ion uptake, possibly leading to unacceptable tissue changes.

Porous ceramic and carbon coatings, although also permitting bone ingrowth, are not used as coatings on orthopedic implants because of their limited mechanical strength and fracture resistance. However, thin surface layers (less than a few micrometers thick) of carbon or calcium phosphate deposited onto the pore surfaces of porous metal coatings have been studied (e.g., pyrolitic carbon to reduce corrosion rates). Calcium phosphate has been proposed as an osteoconductive layer that would speed the rate of bone formation within the pores. This would result in faster implant fixation, thus allowing earlier implant loading. Studies of corrosion inhibitory or osteoconductive layers on porous-coated implants have been limited to animal studies only.

In addition to forming coatings for bone ingrowth, totally porous structures have been studied as possible bone substitute implants (Holmes et al. 1984, 1986). These wholly porous structures behave identically to the porous coatings discussed above in terms of tissue ingrowth. They are, however, weaker and, therefore, cannot be relied on to support large stresses. Such implants have been proposed for use in the treatment of patients who have had regions of bone ressected because of bone tumors. The ideal bone-substitute material for these cases is one that would act as a scaffold for bone formation, perhaps even encouraging it to form (i.e., be osteogenic), and then completely resorb leaving viable intact bone. Studies using corals, specifically the corals *Porites* and *Goniopora*, in their natural form as $CaCO_3$ (aragonite), have demonstrated their suitability for bone ingrowth as well as their biodegradability. The coral formations are characterized by a uniform interconnected distribution of pores with pore size and interconnections being unique to the individual corals. For the genus *Porites*, this corresponds to a pore size of 140–160 μm, while the coral *Goniopora* consists of a coarser-pored structure (pore size equal to 600 μm with pore interconnections of 240 μm). Strength increases of these structures can be achieved by coating the structure with a very thin polymer layer of dilactic–polylactic acid.

Presumably, this thin surface coating inhibits crack initiation, a factor that limits the fracture properties of brittle ceramics.

These corals have also been used as preforms for formation of calcium hydroxyapatite structures (through the hydrothermal conversion of the $CaCO_3$ to $Ca_{10}(PO_4)(OH)_2$) and inert ceramic or metal alloy forms, through the replication of the coral or a CaO derivative form (White et al. 1975). The totally porous calcium hydroxyapatite implants have been used in humans for replacement of diseased or severely traumatized bone segments. The porous structures act as a scaffold for the ingrowth of new bone over quite extensive regions. Studies have explored the use of protein coatings (noncollagenous proteins, most notably bone morphogenic protein (BMP)) in combination with porous $CaCO_3$ for encouraging bone formation in desired locations. The addition of the BMP appears to result in faster bone formation in the pores.

2.2 Attachment by Nonosseous Tissue Ingrowth

There have been limited studies on the use of porous implants for replacement of regions of degenerate hyaline cartilage. Porous TiO_2, Al_2O_3 and hydroxyapatite implants formed by the Replamineform method were used in a model study by Chiroff et al. (1977) for replacing 4 mm diameter full-thickness surgically created defects in rabbit hyaline cartilage. The implants were made with pores in the range 100–300 μm and after placement they became fixed by subchondral bone ingrowth. For the porous hydroxyapatite implants, hyaline cartilage formed over their articulating surfaces. For the Al_2O_3 and TiO_2 implants, fibrocartilage only formed on this surface. Later studies by Mears (1979) used porous metal implants with the bone-interfacing surfaces designed for bone ingrowth (pore size > 100 μm), while the opposite articulating surface was made with finer pores (< 50 μm). This finer surface was designed to discourage bone ingrowth while allowing hyaline cartilage ingrowth. The formation of hyaline cartilage was promoted through the seeding of the finer porous surface with autogeneous chondrocytes grown *in vitro*. Although initial reports of both these studies were encouraging, further long-term evaluation of the viability of the ingrown tissues has not been reported. The approach depends on careful control of conditions acting at the articulating surface shortly after implant placement but, unfortunately, these conditions have yet to be defined.

Another use of porous-structured implants in orthopedics is for artificial ligament and tendon replacements (Goodship et al. 1985). Studies have been reported on the use of filamentous structures formed from carbon or polymeric fibers for this application. Collagen formation within the interfilament space occurs *in vivo*, with the structure of the ingrown tissue being a function of the density of the filaments and their compliance; a looser more compliant filament structure results in a more natural crimped appearance of the ingrown collagen fibers. Studies have shown that although the elastic properties of the currently available ligament and tendon replacement implants in the as-fabricated condition are unlike those of natural ligaments and tendons, some of the fiber implants approach the natural tissue characteristics after about a year. This appears to be due to filament breakup and remodelling of the ingrown fibrous tissue, so that with time the properties of the implant are determined to a greater extent by the ingrown tissue. In the region of ligament-to-bone attachment, bone ingrowth occurs into the porous artificial ligament structures. One concern with the carbon-filament implants, is the fragmentation and migration of carbon particles into the joint space and surrounding soft tissues. Polymer-coated carbon-fiber systems have been proposed to overcome this problem.

3. Dental Applications

Totally porous biomaterials have been used in dentistry for alveolar ridge augmentation and filling periodontal defects (Piecuch et al. 1984), while porous-coated implants have been studied as endosseous dental implants (Pilliar 1986).

3.1 Ridge Augmentation

Porous hydroxyapatite, in either bulk or particulate form, has been used for alveolar ridge augmentation. The hydroxyapatite blocks or granules (320–600 μm) that are used are made by thermal decomposition of coral, genus *Porites* for this particular application (White et al. 1975). Animal studies have shown that bone ingrowth occurs equally into both forms, so that the choice of bulk or particulate is related primarily to surgical convenience. The particulate form, for example, has been reported to be more suitable for placement in interproximal periodontal defects. The major concern in using the particulate form for alveolar ridge augmentation is migration of particles from the desired site. The internal structure of the granular and bulk forms are the same (pore interconnections of 190–230 μm).

3.2 Endosseous Dental Implants

The necessary requirements for a successful endosseous dental implant are secure fixation in the jaw bone and formation and maintenance of a suitable permucossal seal. Porous-coated dental implants have been shown to satisfy these requirements in animal tests as well as in some human applications. As with the orthopedic implants, bone ingrowth will occur if relative movement of implants and host alveolar bone is limited during the post-implantation healing period, and pore size is sufficiently large to allow bone ingrowth. The condition of initial implant stability is conveniently achieved with porous-coated dental implants by using a multiple-component design and a two-stage placement technique, similar to that used with threaded implants that become "osseointegrated." The percutaneous nature of dental implants increases the probability of infection, since the porous surface offers a nidus for entrapment of bacteria residing in the mouth. For this reason, these implants are designed with smooth surfaces at the superior region of the root component so that the porous structure need not directly contact the oral environment.

Titanium implants with titanium wire-made porous coatings, and titanium alloy implants with titanium alloy powder-made porous coatings have been used in humans. There have also been reports of the use of porous-coated Al_2O_3 implants in humans. However, the most extensively used porous-coated dental implants are plasma-sprayed titanium implants. Although, as noted in Sect. 2.1, these implants are not porous in the sense of exhibiting regular three-dimensional pore interconnection throughout the coating thickness, bone does grow into the surface irregularities resulting in rigid implant fixation.

4. Cardiovascular Applications

4.1 Vascular Grafts

The usefulness of a porous fabric structure as a vascular graft was first demonstrated in the early 1950s. Today, most vascular grafts are made of polyester (Dacron) or PTFE (Teflon) (Sauvage et al. 1987). The grafts are formed by weaving, knitting or a complex weave–knit procedure and are designed to provide an open lattice structure through which fibrous connective tissue can grow, thereby securing them to the surrounding perimplant tissues. This also reduces the probability of infection at the implant site. The transmural growth of fibrous connective tissue and small blood vessels, enables fluids and solutes to be transported away from the graft site. The size and extent of the interstitial spaces in the fabric used for graft fabrication, determine the nature of the ingrown connective tissue. A looser structure, as provided by a knitted graft, allows the formation of a more mature collagen fiber structure, while the tighter structure formed in woven grafts results in less mature fibrous tissue. This tighter weave structure prevents excessive bleeding and eliminates the need for preclotting the graft prior to placement. While the formation of an endothelial-like tissue lining would be highly desirable and, in fact, has been observed to form on such grafts placed in animals, it does not appear to form regularly on grafts placed in humans. For large-diameter high-flow-rate artery replacements, the woven and knitted grafts made of Dacron and Teflon have been found suitable for use in humans with a compacted cross-linked fibrin layer forming an acceptable blood-interfacing surface. The interconnected pore structure provided by the textile graft, enables secure attachment of this fibrin layer, both at pre-clotting and after implantation. A true endothelia-lized surface is observed to form on grafts placed in humans within a few millimeters of the graft anasto-moses, but this is due to pannus ingrowth from the adjacent ends of the host artery. The use of these "larger-pored" knitted or woven grafts is limited to high-flow-rate vessels. Unfortunately, for flow rates below about $60-80\,l\,m^{-2}\,s^{-1}$ (vessel diameters $\sim <6\,mm$), excessive buildup of the fibrin layer leads to unacceptable graft occlusion. Because of this, other microporous-structured grafts have been investigated for use in low-flow situations. These include expanded PTFE, microporous polyurethane grafts formed by different variants of textile spinning processes and grafts of controlled porosity and size range formed by the Replamineform process.

Expanded PTFE is formed by subjecting PTFE to a tensile stress during formation of the implant, resulting in a structure consisting of an array of fine PTFE fibrils, interconnected at regular nodal points. The spacing of the nodes and size of the fibrils can be controlled through the choice of processing para-meters. For vascular graft applications, the fibrils are $15-30\,\mu m$ long with interfiber spacing of $1-3\,\mu m$. To provide sufficient mechanical strength (creep resistance in particular), expanded PTFE is re-inforced, either with an outer thin ply of expanded PTFE oriented normal to the inner expanded PTFE layer, or by an outer fabric sleeve. This does not appreciably alter the effective pore structure of the graft material. The fine porosity of the material compared with the coarser pores of the knitted or woven grafts, allows expanded PTFE to be used without preclotting. After implantation, connective tissue infiltration from the surrounding tissues is observed, but appears limited to a depth of about $150\,\mu m$. Unlike the conventional fabric grafts, only a sparse penetration of fibroblasts occurs through the wall with the bulk of the porosity being filled with fibrin, and some erythrocytes and leukocytes. Little transmural capillary penetration is observed and a thin firmly attached proteinaceous blood-interfacing layer is observed to form. Endothelialization of the luminal surface of such implants placed in humans does not occur except for the short distance near the anastomoses. The restricted transmural connective tissue proliferation, presumably because of the very fine porosity of the expanded PTFE, is thought to be responsible for the success of small-diameter grafts made from this material.

Microporous polyurethane represents another material that has been proposed for the fabrication of small-diameter grafts. The grafts are made from polyurethane fibers $1-2\,\mu m$ in diameter that are spun onto a rotating mandril to form a graft with pores in the $5-10\,\mu m$ size range. An electrostatic spinning process, or a process that relies on surface polymeri-zation and fiber bonding during the spinning pro-cess, have been reported for forming such grafts. The control of spinning parameters allows the for-mation of grafts of specific compliance as deter-mined by the orientation, diameter and spacing of the fibers. These types of grafts have been tested in animals and, to a limited extent, in humans.

The formation of vascular grafts of controlled porosity using the Replamineform process, has allowed a controlled study of the effect of pore size on tissue ingrowth into vascular grafts (Wright et al. 1986). These implants are formed using the calcite spines of sea urchins as templates for injection molding of a polymer. The calcite is dissolved using hydrochloric acid leaving the replicated graft struc-ture with its interconnected porosity of a specific size range. Animal studies showed that the nature of the ingrown tissues differed, depending on whether the resulting pores were greater or less than $45\,\mu m$. With the finer-pored implants, a more immature fibrous tissue network developed within the pores with many fibrohistiocytes being present, while a well organized collageneous structure formed with few histiocytes within the coarser-pored implants. Because compliance is considered important for maintenance of graft patency, the argument has been made that the finer-pored systems are pre-ferred since a decrease in graft compliance, due to the development of mature tissues transmurally, is less likely with the finer pore structures.

It is recognized that the establishment and main-tenance of an endothelial surface at the lumen wall is the most likely feature to guarantee long-term patency of any of these vascular grafts. With this as the objective, studies have been initiated on endo-thelial cell seeding of grafts (Stanley et al. 1987). One method of achieving the seeding is by the addition of endothelium to the blood used for preclotting the grafts. Early animal studies suggest higher patency rates for grafts prepared in this way.

4.2 Cardiac Pacing Applications

Porous-structured electrodes have been shown to provide significant advantages for cardiac pacing by virtue of enhanced attachment to the endocardium or epicardium (MacGregor et al. 1979). Fully porous electrodes are made of fine platinum–iridium wire ($25\,\mu m$ and $40\,\mu m$ in diameter) sintered to form a 90% totally porous electrode with pores approxi-mately $100-150\,\mu m$ in size, while porous-coated electrodes using either cobalt-based alloy or plati-num–iridium alloy powders are formed using pro-cedures similar to those for porous coating orthopedic implants except that finer powders ($<25\,\mu m$) are used. These coatings are about

100 μm thick, 30–35% porous, with a three-dimensional interconnected network of pores less than 25 μm in size. After implantation, fibrous tissue ingrowth occurs thereby inhibiting movement of the electrode relative to the heart muscle, resulting in the development of a significantly thinner fibrotic tissue layer at the implant–heart muscle interface. This results in significant improvement in stimulation and sensing characteristics for these implants. Comparison of porous-coated and totally porous electrodes, suggests some benefits of the porous-coated type, although both designs exhibit significant advantages over nonporous electrodes. Further, direct comparison of porous-coated cobalt alloy and platinum–iridium alloy electrodes in animal studies, has suggested advantages of the platinum–iridium system. The fine porosity associated with the porous-coated electrodes limits the strength of attachment of the electrode to the heart muscle (by virtue of the ingrowth of a loosely structured fibrous connective tissue). This is an advantage in this application since electrodes can be easily removed if necessary.

4.3 Heart-Valve Applications

The long-term fixation of prosthetic heart valves relies on fibrous tissue ingrowth into annular fabric sewing rings commonly made from Dacron. The use of porous fabric structures for this application has proved effective through long-term use.

Porous structures have also been studied as coatings or covers on blood-interfacing metallic components of prosthetic heart valves. The porous structure provides a scaffold for the attachment and growth of a pseudoendothelium, thereby creating a more thromboresistant synthetic heart valve. In the late 1960s, porous textile sleeves placed over metallic struts of ball-in-cage type heart valves, were tested in animals. The tests showed that with time the fabric sleeves became ingrown with fibrous tissue and an endothelial-like tissue formed on the blood-interfacing surface of the fabric. Unfortunately, the mechanical wear and tear caused by the occluder impacting the fabric, led to breakup of the cloth sleeves and subsequent implant failure. The concept of a more durable metal-made porous coating was studied in the late 1970s. The results of animal studies indicated that with a very fine metal powder-made porous coating on selected regions of metallic heart valve prostheses, an endothelial-like tissue layer could develop over these regions resulting in a more thromboresistant implant (MacGregor et al. 1979). The porous coatings were made using fine spherical metal powders (<20 μm in diameter) that were sintered to the solid substrate of the valve, thereby providing a well-bonded porous surface layer. Subsequent studies of functioning porous-coated valves in goats, indicated the potential advantage of this approach for tilting disc valves

(Bjork et al. 1988). A major concern exists as to whether endothelialization of porous-coated valves would occur in humans in view of the unpredictable endothelialization of porous textile and expanded PTFE vascular grafts in humans.

5. Percutaneous Components

The possibility of fibrous connective tissue ingrowth into porous-structured percutaneous implants limiting epithelial migration along the implant surface, has been explored by a number of investigators. Early studies suggested that mature connective tissue would grow into coatings with pores greater than 40 μm in size. However, porous-structured percutaneous implants have yet to prove successful over the long term, with eventual extrusion of porous-structured implants invariably occurring. The mechanism of extrusion is believed to involve migration of epithelial cells into the pores, followed by their migration to the tissue surface causing upward forces to act on the implants. This leads to eventual extrusion of the implants. The use of porous-structured implants as percutaneous implants warrants further investigation.

6. Summary

Porous-structured implants have found important applications in orthopedics, dentistry and cardiovascular surgery. The tissue response to such implants appears to be strongly influenced by the structure and properties of the porous surface layer (pore size and compliance, especially). This class of implants appears to offer a new dimension in implant use that should result in extended applications in the future.

Bibliography

Bjork V O, Wilson G J, Sternlieb J J, Kaminsky D B 1988 The porous metal-surfaced heart valve. Long-term study without long-term anticoagulation in mitral position in goats *J. Thorac. Cardiovasc. Surg.* 95: 1067–82

Chiroff R T, White R A, White E W, Weber J N, Roy D 1977 The restoration of articular surfaces overlying Replamineform porous biomaterials. *J. Biomed. Mater. Res.* 11: 165–78

Ducheyne P 1984 Biological fixation of implants. In: G W Hastings, P Ducheyne (eds.) 1984 *Functional Behavior of Orthopedic Biomaterials: Applications*, Vol. 2. CRC Press, Boca Raton, FL, pp. 163–99

Engh C A, Bobyn J D, Glassman A H 1987 Porous-coated hip replacement—The factors governing bone ingrowth, stress shielding, and clinical results. *J. Bone J Surg., Br. Vol.* 69: 45–55

Goodship A E, Wilcock S A, Shah J S 1985 The development of tissue around various prosthetic implants used as replacements for ligaments and tendons. *Clin. Orthop. Relat. Res.* 196: 61–8

Guillemin G, Patat J-L, Fournie J, Chetail M 1987 The use of coral as a bone graft substitute *J. Biomed. Mater. Res.* 21: 557–67

Homes R, Bucholz R, Mooney V 1986 Porous hydroxy-apatite as a bone-graft substitute in metaphyseal defects. A histomeric study. *J. Bone J. Surg., Am. Vol.* 68: 904–11

Holmes R, Mooney V, Bucholz R, Tencer A 1984 A coralline hydroxyapatite bone graft substitute. Preliminary report. *Clin. Orthop. Relat. Res.* 188: 252–62

MacGregor D C, Wilson G J, Lixfield W, Pilliar R M, Bobyn J D, Silver M D, Smardon S, Miller S L 1979 The porous-surfaced electrode. A new concept in pacemaker lead design *J. Thorac. Cardiovasc. Surg.* 78: 281–91

Mears D C 1979 *Materials in Orthopaedic Surgery.* Williams and Wilkins, Baltimore, MD, pp. 722–33

Piecuch J F, Goldberg A J, Chakrakody V S, Chrzanowski R B 1984 Compressive strength of implanted porous Replamineform hydroxyapatite. *J. Biomed. Mater. Res.* 18: 39–45

Pilliar R M 1986 Implant stabilization by tissue ingrowth. In: D van Steenberghe (ed.) 1986 *Tissue Integration in Oral and Maxillo-facial Reconstruction.* Excerpta Medica, Amsterdam, pp. 60–73

Pilliar R M 1987 Porous-surfaced metallic implants for orthopaedic applications. *J. Biomed. Mater. Res.* 21 (A1): 1–33

Sauvage L R, Davis C C, Smith J C, Rittenhouse E A, Hall D G, Mansfield P B, Schultz G A, Wu H-E, Ray L I, Mathisen S R, Usui Y 1987 Development and clinical use of porous Dacron arterial prostheses. In: P N Sawyer (ed.) 1987 *Modern Vascular Grafts.* McGraw-Hill, New York, pp. 225–55

Spector M 1982a Bone ingrowth into porous polymers. In: D F Williams (ed.) 1982 *Biocompatibility of Orthopaedic Implants*, Vol. 2. CRC Press, Boca Raton, FL, pp. 55–88

Spector M 1982b Bone ingrowth into porous metals. In: D F Williams (ed.) 1982 *Biocompatibility of Orthopaedic Implants*, Vol. 2. CRC Press, Boca Raton, FL, pp. 89–128

Stanley J C, Graham L M, Burkel W E 1987 Endothelial cell seeding of synthetic vascular prostheses. In: P N Sawyer (ed.) 1987 *Modern Vascular Grafts.* McGraw-Hill, New York, pp. 191–214

White E W, Weber J N, Roy D M, Owen E L, Chiroff R T, White R A 1975 Replamineform porous biomaterials for hard tissue implant applications. *J. Biomed. Mater. Res. Symp.* 6: 23–7

Williams D F 1982 Prosthesis stabilization by tissue ingrowth into porous ceramics. In: D F Williams (ed.) *Biocompatibility of Orthopaedic Implants*, Vol. 2. CRC Press, Boca Raton, FL, pp. 49–54

Wright C B, White R A, Hiratzka L F, Morin R P, Mitts D L 1986 Small caliber vascular grafts-Alternatives. In: H E Kambic, A Kantrowitz, P Sung (eds.) 1986 *Vascular Graft Update: Safety and Performance*, American Society for Testing and Materials STP 898, ASTM, Philadelphia, PA, pp. 60–7

R. M. Pilliar
[University of Toronto, Toronto, Canada]

S

Silver in Medical Applications

Silver is not merely an ornamental material—it is probably the most powerful antimicrobial metal ion and has a remarkably low human toxicity. Silver salts, complexes and the metal itself have played an important part in the development of medicine. Interest in the use of silver is being renewed in a wide variety of biomedical applications where it can prevent the growth of microorganisms responsible for disease. Silver has a broader spectrum of activity and is less likely to engender high levels of resistance than most modern antibiotics. A major problem in medicine is the large number of infections associated with implanted and indwelling devices. The increased use of silver in sophisticated delivery systems offers the possibility of a large reduction in the incidence of these infections.

1. Historical Uses

The history of the use of silver in medicine is summarized in Table 1. Silver has been known to humanity for thousands of years and was revered by early civilizations for its purity and luster. There are anecdotal reports of its use in early civilizations to maintain the quality of stored water, and doctors in the Middle Ages recommended silver salts for a variety of ailments including epilepsy and other nervous complaints.

The development of bacteriology in the second half of the nineteenth century led to a better understanding of the role of bacteria in disease and to the introduction of a number of antimicrobial strategies, among them Crédé's prophylaxis. This involved the application of 1% silver nitrate to the eyes of newborn infants to prevent ophthalmia neonatorum, an

Table 1
History of silver in medicine

4000 BC Egyptians,	First silver artefacts
Alexander the Great	Pure water
980 AD	Avicenna—silver filings to purify blood
1884	Crédé's prophylaxis
1893	Von Nägeli's "oligodynamic" metals
1893 onwards	Profusion of silver preparations used for many purposes
1950s	Introduction of antibiotics
1965	AgNO$_3$ compresses in burns units
1968	Silver sulfadiazine
1980s	Vitacuff Catheter treatments

infection caused by *Neisseria gonorrhoeae*. Von Nägeli in 1873 coined the phrase "oligodynamic" to describe the remarkable antimicrobial properties of highly diluted silver in water and work over the following 20 years confirmed the effectiveness and low toxicity of silver preparations as topical antimicrobial agents. Prior to the widespread introduction of antibiotics in the 1950s, a large number of silver-containing treatments were available. Some of these are still in use; for example, lunar caustic (toughened silver nitrate pencils), silver citrate and silver–protein combinations. Other compounds (e.g., silver aresphenamine, which was used in the treatment of syphilis) have been replaced by more effective and less toxic antimicrobials.

There was a revival of interest in silver in the 1960s when silver nitrate was shown to be highly effective in preventing burn infections, closely followed by the development of silver sulfadiazine. This compound avoided the problems associated with silver nitrate. More recently, interest has centered upon the use of silver for the prevention of infection associated with indwelling devices such as catheters, tubes and prostheses.

Silver is used in medical applications both as a metal and in a number of organic and inorganic salts and complexes.

2. Silver-Based Materials

Metallic silver is used in a number of surgical applications, both for structural devices (e.g., cranial support plates, suture wire, aneurysm clips and tracheostomy tubes) and for prostheses (e.g., the Jonas silicone–silver penile prosthesis).

It provides the advantage of good tissue biocompatibility, with corrosion resistance and physical strength. The antimicrobial properties of the metal are generally limited by its poor solubility. A preparation that surmounts this problem is a combination of micronized silver metal and benzoyl peroxide (Katoxyn). This product has been used by Italian investigators to treat burns and gynecological disorders.

Silver metal also has a high conductivity and is used for numerous external electrical devices. Its use for internal electrical devices is limited by the high rate of dissolution of silver electrodes in physiological fluids.

3. Biological Activity

The antimicrobial properties of silver are utilized in the form of silver salts and complexes that break down to release Ag$^+$. The inorganic salts include silver nitrate, silver carbonate and the silver halides. Their main use is as topical antimicrobials for the treatment of minor skin complaints. The astringent nature of silver nitrate in the form of lunar caustic pencils makes it ideal for removing blemishes such as warts while cauterizing the wound and minimizing the risk of infection. The use of silver nitrate (1% solution) against ophthalmia neonatorum is still common, especially in the developing world and in regions where penicillin-resistant strains of *Neisseria gonorrhoeae* are encountered. Silver halides, with their reduced solubility, are much less caustic.

Organic salts of silver that have been used as topical antimicrobials include silver lactate, silver acetate, silver citrate, silver picrate, and, more recently, silver–antibiotic salts or complexes such as silver sulfadiazine. The latter was developed by Charles Fox in 1968 and is used in the prophylactic treatment of burns, leg ulcers and pressure sores. Other silver–antibiotic combinations have been reported as potentially useful antimicrobials including silver mafenide, silver norfloxacin and silver oxacillin. Other organic salts mentioned in the recent literature include silver phosphanilides, silver uracil, silver allantoinate and the disilver salt of dinaphthylamine disulfonate (Methargen).

A strategy for reducing the caustic effects and prolonging the useful antimicrobial activity of silver salts has been to combine them with macromolecules. Examples of this approach are silver–protein, silver–kaolin and silver–pectin, which have all been used as treatments for infected lesions of the skin and of mucous membranes. The Merck index gives two different formulations for silver–protein. Mild silver protein is prepared using silver oxide and either gelatine, serum albumin, casein or peptone. Strong silver protein, in contrast, is prepared using silver nitrate and peptone, and despite containing only 7.5% to 8.5% silver is more caustic than mild silver protein.

Colloidal preparations are well known. These include silver iodide in gelatine, colloidal silver chloride containing 10% silver nitrate stabilized with sucrose, and colloids containing metallic silver.

Antimicrobial silver ions can also be generated by passing a low-intensity direct current through a silver anode. This approach has been used to treat infected bone fractures, where reversing the polarity of the current also acts to stimulate bone growth.

Silver can also be incorporated into polymeric materials, in particular nylon (which has potential as a suture or wound dressing material), either alone or as part of an electrically driven Ag^+-releasing system. Silver has also been incorporated into poly-(methyl methacrylate) bone cement and pluronic F127 "artificial skin."

More recent attempts to prevent the development of nosocomial infections related to the implantation of medical devices (including urinary catheters,

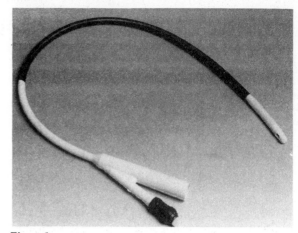

Figure 1
"AgX" Foley catheter

venous cannulae, etc.) have also utilized silver-releasing materials. These include the "AgX" Foley catheter (see Fig. 1), which employs a coating based on silver oxide, and the Vitacuff (see Fig. 2), which is made of silver–protein-impregnated collagen and is intended to form a barrier to infection at the entry site of vascular access devices. In controlled trials, this device has been shown to reduce rates of catheter-associated bacteraemia by nearly a factor of 4 (Maki et al. 1988). Improvements in medical devices of this nature can be expected in the near future with the introduction of new technology aimed at providing increased longer-term protection coupled with improved cosmetic appearance.

3.1 Antibacterial Activity

Silver ions (Ag^+), whether released from inorganic or organic salts, from macromolecular complexes or

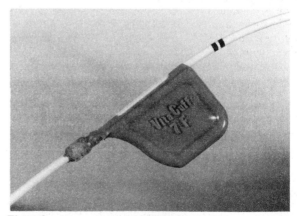

Figure 2
Vitacuff device—an infection barrier at the entry site of vascular access devices

by electrical means, have potent bactericidal activity against Gram negative and Gram positive bacteria.

The introduction in 1965 of 0.5% silver nitrate compresses for prophylaxis of infection in severely burned patients led to a large reduction in *Pseudomonas aeruginosa* positive swabs and *Ps. aeruginosa* septicaemia in a hospital burns unit. Some disadvantages are associated with this treatment, notably disturbance of electrolyte balance and a poor activity against some species of Gram negative bacteria (especially *Klebsiella sp.*). Silver sulfadiazine, developed by Fox in 1968 and applied as an ointment, overcame these problems, but has also been associated with the development of a transferable plasmid-borne sulfonamide resistance in at least one burns unit. Suspension of all sulfonamide use in the unit, including silver sulfadiazine, and the introduction of a 0.5% silver nitrate–0.2% chlorhexidine gluconate cream resulted in an overall fall in the number of antibiotic-resistant strains isolated. No silver-resistant bacteria were observed.

Silver sulfadiazine is a highly potent antimicrobial with a broad spectrum of activity. Its low solubility acts to reduce its systemic toxicity, although the possibility of increased sulfonamide resistance must be considered.

3.2 Fungi, Viruses and Protozoa

Silver ions are capable of inhibiting the growth of a number of pathogenic fungi. They are fungicidal against *Candida albicans* and inhibit the growth of the spores of *Neurospora, Rhizopus, Fusarium* and *Aspergillus* species.

Antiviral effects of silver have also been reported, with silver nitrate showing *in vivo* and *in vitro* activity against herpes simplex virus type I (although it has reduced activity against type II). Recent work on water-purifying agents containing silver show mild antiviral activity against vaccinia, influenza A and pseudorabies viruses in water. Silver sulfadiazine is claimed to be cidal against human immunodeficiency virus, the virus responsible for acquired immune deficiency syndrome, with a 1.0% suspension achieving a 2.25 log cycle kill in 10 min.

Silver also shows activity against the cysts of *Entamoeba histolytica* in water while a $6 \mu g \, ml^{-1}$ solution of silver nitrate rapidly kills amoebae, paramecium and euglena in water. Silver sulfadiazine also shows activity against protozoa, including trypanosomes and *Trichomonas sp.*

3.3 Mechanism of Action

The silver cation Ag^+ is a highly reactive chemical species and binds strongly to electron donor groups containing sulfur, oxygen or nitrogen. Bacterial cell surfaces, and biological molecules generally, contain all these components in the form of thiol, amino, imidazole, carboxylate and phosphate groups. In addition, silver can also act by displacing other essential metal ions such as Cu^{2+} or Zn^+.

In *Escherichia coli*, silver nitrate has been shown to inhibit a number of important transport processes, including phosphate uptake and succinate uptake into membrane vesicles, as well as the oxidation of glucose, fumarate, glycerol and lactate. There are two possible sites of interaction with the respiratory chain: between cytochromes b and a_2 or between NADH–succinate and flavoprotein (NADH is the reduced form of nicotinamide adenine dinucleotide), although more recent work has suggested that the nitrate counterion may be at least partly responsible for these results. Electron microscopy of *Ps. aeruginosa* cells grown in 3 ppm silver nitrate reveal abnormalities of cytoplasm, membrane and envelope which have resulted in a failure of daughter cell separation. Studies on the binding of Ag^+ to nongrowing cells of *E. coli* suggest that two distinct Ag^+ binding sites exist, binding to one of which is not reversed by washing with 0.1 M nitric acid and is thought to be intracellular. Silver ions can bind to microbial DNA *in vitro*. *In vivo*, when subinhibitory concentrations of $^{110}AgNO_3$ are added to a growing culture of *Ps. aeruginosa*, peak binding of the label to DNA occurs within 2 h and then declines rapidly. Below a value of 80 base pairs per silver atom growth resumes, suggesting that the interaction is chemically and biologically reversible. It has been known for some time that Ag^+ can inhibit a wide range of important enzymes. These include ATPase, urease, β-galactosidase and a number of dehydrogenases.

It is not clear which of these interactions is the most important for the antimicrobial action of silver ions. In studies of the disinfection kinetics of $AgNO_3$, a concentration exponent of approximately 1 is found; that is, the rate of kill is directly proportional to the silver concentration. This is typical of antibacterials acting at multiple targets (e.g., chlorine and bromine, and other metal ions such as Hg^{2+}) and it is likely that silver exerts its effects by interfering with a number of vital cell surface and cytoplasmic processes.

The mode of action of silver sulfadiazine has been studied in detail to determine what part, if any, the sulfonamide plays. It appears that the active species present is the free Ag^+. Unlike sulfadiazine alone, *p*-aminobenzoic acid does not act as an antagonist. The role of the sulfadiazine moiety appears to be to create a complex polymeric system with a low aqueous solubility which delivers a sustained release of Ag^+ and possibly enables transport of the Ag^+ into the microorganism. Silver sulfadiazine binds to DNA *in vitro*, but it is not clear whether this is significant *in vivo*.

Despite the many uncertainties surrounding the mechanism of action at the molecular level, it is apparent that the Ag^+ ion is the most effective antimicrobial metal ion. This is despite the fact that it has a slightly lower affinity for anions such as S^- and acetate than mercury, and points to the importance of steric and kinetic factors.

3.4 Microbial Resistance to Ag^+

Silver is widely distributed in the environment and is frequently found in high concentrations; for example, in silver mines, photographic processing plants and hospital burns units. In these situations, selective pressures force bacteria to develop resistance mechanisms that protect them from high Ag^+ concentrations, just as bacteria can become resistant to other metal ions and antibiotics.

It is difficult to assign a silver concentration that distinguishes resistant from sensitive strains. This can be compounded by variations between media which affect the amount of free Ag^+ available and so appear to alter the measured minimum inhibitory concentration.

Silver resistance has been identified in bacteria from the soil, in *Pseudomonas stutzeri* isolated from a silver mine and in a variety of Gram negative bacteria, including *Salmonella typhimurium*, *Enterobacter cloacae*, *Escherichia coli*, *Klebsiella pneumoniae* and *Pseudomonas aeruginosa*.

The mechanism of silver resistance in bacteria is less well understood than those for other heavy metal ions. A number of workers have identified differences between the cell walls of sensitive and resistant strains. The binding of Ag^+ by resistant strains is lower than that of sensitive strains, but no difference in Ag^+ uptake can be demonstrated in spheroplasts. It has been suggested that silver resistance is dependent on halide ion concentration, but other workers have not reproduced this.

Plasmids appear to play an important role in bacterial resistance to silver. One group isolated 2 plasmids from a strain of *E. coli* resistant to Ag^+. Curing of one of these plasmids (of size 83 kb) resulted in an Ag^+-sensitive strain. It did not prove possible to transfer the plasmid into a previously sensitive organism. The resistant organism possessed a slightly different outer membrane protein profile, lacking 2 protein bands present in the sensitive strain, which may be the basis for the differences in binding and/or transport of Ag^+ into the cell. The resistant strain was also capable of reducing methylene blue and produced H_2S; this could also be a defence mechanism against Ag^+. Deposits containing reduced silver are frequently observed in silver-treated bacteria.

Microorganisms deal with other heavy metals in a variety of ways, for example, by enzymatic reduction (for mercury) and by specific efflux pumps for cadmium and arsenate. A common observation about silver-resistant organisms is their decreased affinity for the silver ion. The precise mechanism for this has not been elucidated and other processes may also be important.

The main resistance problem with silver sulfadiazine appears to be resistance to the sulfadiazine moiety. Silver-sulfadiazine-resistant organisms isolated from burns units also show resistance against $AgNO_3$, however, and this is further evidence that the Ag^+ ion is the active agent in silver sulfadiazine.

4. Toxicology and Biocompatibility

Despite the fact that silver ions possess antimicrobial effects equal to or greater than other heavy metal ions, it is clear that their toxic effects on mammals are considerably lower. Table 2 lists the available acute toxicity data in animals. The human toxicity of silver and its compounds was authoritatively reviewed by Hill and Pillsbury (1939), when the use of silver preparations in medicine was near its peak. They concluded that the major adverse effect associated with exposure to silver and its salts or complexes was argyria. This is the buildup of silver in subcutaneous tissue which causes a grey or black discoloration. It can be localized in specific areas (e.g., hands or eyes) or generalized over the whole body. While this is an unpleasant condition which is irreversible, it is thought to be a purely cosmetic problem and no significant disturbances in function have been observed in any affected organs.

Between 1921 and 1930, 42 cases of argyria related to silver were reported in the literature. This was against a background of 3.7 million prescriptions for silver preparations in the USA in 1927 alone. Argyria is also associated with long-term use of silver-containing preparations. 93% of all the cases identified by Hill and Pillsbury (1939) were in persons continuously exposed to silver for more than one year. It is difficult to assign a dose of silver that is capable of causing argyria. The physical form of the silver, the route of administration and the metabolism of the subject are three important variables. Hill and Pillsbury identified 42 cases in which the total dose of silver compound ingested was known. The mean dose of silver compound received by these individuals was 156 g, by a variety of routes and in a number of formulations. The lowest dose recorded as causing argyria was 1.4 g (calculated as metallic silver). In a series of 18 patients treated for syphilis using intravenous silver arsephenamine, all of whom developed generalized argyria, the mean dose of silver (calculated as metallic silver) received was 2.46 g. The minimum dose of silver causing argyria was 0.91 g.

In industries where silver and silver salts are widely used, occupational argyria is a relatively rare event, usually associated with silver nitrate manufacture or photographic processes. Calculations at the

Table 2
Acute toxicity of silver salts[a]

Compound	Route of administration	Animal	Dose criterion	Toxic dose $(mg\,kg^{-1})$
Silver nitrate	oral	mouse	LD_{50}[b]	50
		dog	LD_{50}	20
		rabbit	LD_{Lo}[c]	800
	intraperitoneal	guinea pig	LD_{Lo}	216
		mouse	LD_{50}	34.5
	intravenous	rabbit	LD_{Lo}	8.8
	subcutaneous	guinea pig	LD_{Lo}	62
Colloidal silver	oral	mouse	LD_{50}	100
	intravenous	rabbit	LD_{Lo}	49
Silver sulfadiazine	oral	mouse	LD_{Lo}	5000
	intraperitoneal	mouse	LD_{50}	100
		rat	LD_{50}	95
Silver acetate	intraperitoneal	mouse	LD_{50}	343
Silver fluoride	subcutaneous	frog	LD_{Lo}	224
Silver cyanide	oral	rat	LD_{50}	123

a from Sweet 1989   b dose causing 50% of animals to die   c lowest dose causing death

Health and Environment Laboratories of the Kodak company suggest that 24 years' continuous exposure to the TLV (threshold limit value, currently $0.1\,mg\,m^{-3}$ for metallic silver and $0.01\,mg\,m^{-3}$ for soluble silver compounds) would be necessary before a worker is at risk of developing argyria.

Recommendations made by Hill and Pillsbury (1939) aimed at preventing argyria concerned increased education of medical staff, controls on self-medication using silver salts, adequate labelling and extra care when administering silver preparations to broken skin or inflamed mucous membranes.

Silver is used in smoking deterrent products in the form of silver-acetate-containing chewing gum. It causes an unpleasant taste in the mouth when a cigarette is smoked. The safety of silver acetate gum has recently been assessed by an advisory review panel appointed by the United States Food and Drug Administration who found it to be safe for over the counter use provided a dosage of $36\,mg\,day^{-1}$ for no longer than three weeks is not exceeded. Clinical trials of silver acetate chewing gum have shown that the product can safely be used for 12 weeks (mean dose $191\,mg\,week^{-1}$). During this period, high plasma levels of silver ($40\,\mu g\,l^{-1}$ mean) were achieved with no ill effects or argyria.

Experiments in a rat model have shown that orally ingested silver can be demonstrated in the cornea, peripheral nervous tissue and connective tissue. When silver disks were implanted intramuscularly into rats, significant silver levels were found in the blood within a few days. Subsequently, levels decreased and were back to normal in three months. Virtually all the silver was excreted in the faeces.

The toxicity of silver ions towards eukaryotic and prokaryotic cells has been determined, and a selective toxicity against prokaryotes was found. Mouse bone marrow cells are not affected by 4 ppm Ag^+. In experiments involving direct contact between a growing cell line and the metal, silver disks produced a zone of lysis, but no such zones were observed from a matrix of silicone rubber–titanium dioxide containing 2.5% silver.

At concentrations below $20\,\mu M$, silver lactate produced no ultrastructural abnormalities in mouse peritoneal macrophages. Silver was found in dense bodies within the cell, but no impairment of phagocytosis, migratory or interferon production was observed at these levels.

When silver metal is implanted into experimental animals, the local host response is generally a mild inflammation, similar to that produced by stainless steel. When silver and silver–titanium-coated silicone were implanted intramuscularly in rats, mild tissue reactions were initially observed. At 84 days postimplantation, enzyme histochemistry demonstrated a minimal response, while after five months the most prominent feature was a thick avascular collagenous capsule surrounding the implant, and there was little evidence of inflammation or necrosis (Williams et al. 1989). This data contradicts earlier studies suggesting that metallic silver may have carcinogenic potential when implanted in rats.

Bibliography

Fox C L 1968 Silver sulfadiazine—A new topical therapy for pseudomonas in burns. *Arch. Surg.* 96: 184–8

Grier N 1983 Silver and its compounds. In: Block S (ed.) 1983 *Disinfection, Sterilisation and Preservation*, 3rd edn. Lea and Febiger, Philadelphia, PA, pp. 375–89

Hill R W, Pillsbury D M 1939 *Argyria. The Pharmacology of Silver*. Bailière Tindall and Cox, London

Maki D G, Cobb L, Garman J K, Shapiro J M, Ringer M, Helgerson R B 1988 An attachable silver impregnated cuff for prevention of infection with central venous catheters: A prospective randomized multicenter trial. *Am. J. Med.* 85: 307–14

Starodub M E, Trevors J T 1989 Silver resistance in Escherichia coli R1. *J. Med. Microbiol.* 29: 101–10

Sweet D V (ed.) 1989 *Registry of Toxic Effects of Chemical Substances*. National Institute for Occupational Safety and Health, Cincinatti, OH

Williams R L, Doherty P J, Vince D G, Grashoff G J, Williams D F 1989 The biological properties of silver. *CRC Crit. Rev. Biocompat.* 5(3): 221–43

Wysor M S 1983 Silver sulfadiazine. *Antibiotics* 6: 199–232

G. J. Grashoff and R. O. King
[Johnson Matthey, Reading, UK]

Soluble Polymers in Drug Delivery Systems

Although many new pharmaceutical agents have been discovered during the twentieth century, and a multitude of common and often life-threatening diseases have been either brought under control, or in some cases even eradicated, there is still a pressing need to improve the available chemotherapy. Two groups of diseases are responsible for the largest percentage mortality in the West: coronary diseases, and the group of diseases which, for simplicity, may be called "cancers." New strategies for therapy and/or new drugs are urgently needed.

Most drugs are supplied as over-the-counter medicines, and are taken orally. Although acceptable oral formulations now exist for many drugs, oral availability of many compounds is still very poor, and gastrointestinal tract-related side effects are still commonplace if drugs are prescribed over long periods. The last decade has seen the emergence of the biotechnology industry, and with it the production of large quantities of peptide and protein drugs. To facilitate routine patient use of such compounds, it is essential that oral delivery formulations are devised that can protect proteins and peptides from the harsh degradative environment present in the gastrointestinal tract, and promote absorption of a pharmacologically active dose of drug.

Apart from actively seeking new agents for use in chemotherapy, many new strategies are currently under development to improve the delivery of drugs awaiting clinical use, and those already available in the clinic. Such strategic developments fall into two broad categories: controlled release formulations which are designed to optimize the rate of drug appearance in the target tissue and optimize per dose the duration of drug activity; and drug-targeted formulations which are designed to deliver drug uniquely to the target organ or disease (e.g., bacterial infection) such that active drug is only liberated at that site within the body. Soluble polymers have found use in both types of application.

Soluble polymers have several roles in drug delivery systems (Duncan and Kopeček 1984):

(a) as polymeric drugs which are pharmacologically active;

(b) as carriers to which drug is covalently or noncovalently bound, for purposes of drug targeting and/or controlled drug delivery;

(c) as polymers which can be bound to the surface of particulate drug delivery systems to control biodistribution following parenteral administration, or alternatively, ones that can be used to coat the surface of oral drug delivery formulations to permit controlled delivery of drugs, both temporally and spatially, in the gastrointestinal tract; and

(d) as polymers which can be used as simple excipients in drug delivery systems, to provide both bulk for the system and also improve drug formulation properties.

In general, the polymers used in drug delivery systems fall within three categories. Synthetic polymers which are truly artificially synthesized, natural polymers extracted (and purified) from a variety of biological systems including plant, fungal and animal products (e.g., proteins, dextrans and cellulose) and pseudosynthetic materials which are either derivatized natural products (e.g., oxidized dextrans), or materials synthesized to mimic the properties of natural products (e.g., synthetic poly(amino acids)).

1. Selection of Polymers for Use in Drug Delivery Systems

Any new material destined for use in humans or animals is subject to strict regulatory consideration before permission can be granted to allow widespread use. This is true not only for new pharmaceutical agents, but also for biomaterials designed for use as prostheses or as drug delivery systems. The regulatory rules governing multicomponent drug delivery systems are still being formulated.

When considering polymers for use in drug

Table 1
Structures of polymers used in drug delivery systems

Type of polymer	Structure
Homopolymer	A–A–A–A–A–A–A–A–A
Copolymer (statistical or random distribution of monomer units)	A–A–B–A–A–B–B–B–A–B–A
Alternating copolymer	A–B–A–B–A–B–A–B–A–B–A
Terpolymer (statistical distribution)	A–B–C–C–A–B–C–A–C–B–C
Block copolymer	AAAAAAAAAA–BBBBBBBBB
Alternating block copolymer	AAAAAAAAAA–BBBBBBBBB–AAAAAAAAAA

delivery systems, two important factors must be considered:

(a) general physicochemical properties (reproducibility from one batch to another, and methods available for characterization), and

(b) biocompatibility.

Any material designed for human use must be carefully defined, and products must be manufactured to specific degrees of purity and sterility. To understand the complexity of such requirements in relation to polymers developed for use in drug delivery systems, it is necessary to consider polymer structure in more detail. Polymers used in drug delivery systems have a variety of structures, shown in Table 1.

Synthetic polymers differ from natural polymers, such as proteins, in two fundamental respects. The primary amino acid sequence of a particular protein will always be the same. Unless a synthetic polymer polymerizes to form an organized alternating structure, this is unlikely to be the case, and the sequence of monomer units along the polymer backbone will display a random (sometimes called a statistical) distribution. If a reproducably defined product is required for use in drug delivery systems, this will raise problems unless a homopolymer or alternating system is used. A natural polymer, such as a protein, is also rigidly defined in terms of its molecular weight. Other natural polymers, such as dextrans

and certainly synthetic polymers, are not monodisperse in terms of their molecular weight, and this can have an important bearing on the biological properties of the polymer (see Table 2).

It is essential that all polymers developed for human use are characterized in terms of their mean molecular weight and polydispersity.

The weight-average molecular weight M_w is given by

$$M_w = \frac{\sum w_i M_i}{\sum W_i}$$

where a single polymer chain having i repeating units in the chain is said to have a degree of polymerization equal to i, and the molecular weight of such a chain is M_i. The total mass of molecules of size i in the sample is denoted W_i and the number of molecules of weight M_i in the polymer sample is N_i. The number-average molecular weight M_n is given by

$$M_n = \frac{\sum N_i M_i}{\sum N_i}$$

and the polydispersity is M_w/M_n.

Table 2
Biological processes affected by polymer molecular weight average

Process	Polymers studied
Distribution in the body	Poly(vinyl pyrrolidone), N-(2-hydroxypropyl)methacrylamide copolymers
Membrane binding	Dextrans, poly(amino acids)
Pinocytic capture by cells	Dextrans, poly(vinyl pyrrolidone), N-(2-hydroxypropyl)methacrylamide copolymers
Rate and mechanism of degradation	Dextrans, proteins
Pharmacological activity	Divinyl ether–maleic anhydride

Table 3
Polymers used in human applications

Polymer	Applications
Poly(vinyl pyrrolidone)	Plasma expander
Poly(ethylene glycol)	Oral formulations
	Protein conjugates with reduced immunogenicity and increased half life, lotions for topical use
Dextran 70 or 75, 40	Plasma expander, in blood substitute formulations
Hydroxyethyl starch	Plasma expander
Cellulose esters	Oral formulations including cellulose acetate, cellulose triacetate, éthylcellulose
Poly(acrylic acids)	Laxatives (e.g., polycarbophil)
Poloxamers	Laxatives

Materials destined for human or animal use must also display good biocompatibility. The term biocompatibility is defined by Williams (1987) as "Ability of a material to perform within an appropriate host response in a specific application."

There are relatively few soluble polymers that have been tested in humans, and have proven biocompatibility in respect of specific applications.

Some examples are shown in Table 3. Many other soluble polymers are currently under development for use in a wide range of drug delivery systems. These materials are listed in Table 4, and specific applications are also indicated. Although many polymers show considerable promise in *in vitro* and *in vivo* experimental trials, it remains to be seen whether many of these polymeric systems can meet

Table 4
Soluble polymers currently being developed for use in drug delivery systems

Polymer	Application/experimental approach
Synthetic	
Poly(ethylene glycol)	Reduces immunogenicity of conjugated proteins
N-(2-hydroxypropyl)methacrylamide copolymers	Targeted and controlled release of anticancer agents, proteins, antibiotics and immunomodulators
Divinyl ether–maleic anhydride	Member of a family of biologically active polyanions, immunopotentiator used as a carrier of anticancer agents
Poly(styrene-co-maleic acid)	Carrier of neocarzinostatin
Pseudosynthetic	Carriers of anticancer agents
Poly(L-Lysine)	
Poly(L-aspartic acid)	
Poly(L-α-glutamic acid)	
Poly(glutamylhydrazide)	
Natural molecules	Developed as carriers of anticancer and antiviral agents, immunomodulators
DNA	
Dextran	
Proteins	
albumin	
antibodies	
asialoglycoproteins	
low-density lipoprotein	

the stringent regulatory requirements in terms of their reproducible synthesis, characterization and lack of toxicological activity in the body.

2. Polymeric Drugs

Specific polymers have been proposed as pharmaceutical agents in their own right. These include polymers whose monomeric components are themselves active agents (polymerized 3,4-dihydroxyphenylalanine (DOPA), copolymers of alprenol and acrylamide, polymer analogues of *cis*-dichlorodiamineplatinum and arsenic-containing antibacterial polymers prepared by the reaction of triphenylarsenic dichloride and 2,4-diamino-6-mercaptopyrimidine). Certain polymers display inherent activity that is not shown by monomer alone, and many of these polymers have been implicated as potential antitumor agents. It has been suggested that cationic polymers are selectively antitumor (e.g., poly(ethylene imine), poly(L-lysine) and DEAE-dextran). However, a common difficulty is their profound toxicity which manifests itself against normal, as well as tumor cells.

Anionic polymers, including divinyl ether–maleic anhydride (DIVEMA, sometimes called pyran copolymer), display a wide spectrum of activities including antiviral, antitumor and antibacterial activity, all of which arise from the ability of such polymers to stimulate the immune system. In particular, they may induce interferon production and stimulate natural killer-cell activity. Early clinical trials against human cancer were discontinued due to prohibitively high DIVEMA-related toxicity. It has since become apparent that activity and/or toxicity is highly molecular-weight dependent, and preparations of DIVEMA of mean molecular weight less than 18 000 and of narrow polydispersity, have more recently shown acceptable toxicity in phase I clinical trial, although without significant therapeutic response against metastatic melanoma. Combination of the immune potentiating activity of poly(styrene-*co*-maleic acid) and the antitumor activity of neocarzinostatin in the form of a covalent conjugate called SMANCS, has produced a derivative with entirely new pharmacological properties with promising activity in human trials (Maeda et al. 1984). Of particular interest is the enhanced tumoritropism of this conjugate. Poly ribonucleotides (poly A–poly U and poly I–poly C) also induce interferon and have been evaluated in humans as an antitumor therapy. Poly I–poly C is a more effective interferon inducer than poly A–poly U and seems to show better antitumor activity in animal models. Studies in humans show that poly I–poly C cause a range of toxicities commonly associated with immunomanipulation (fever, hypotension, etc.) and a great variation in the maximum tolerated dose. Although

several trials have been undertaken against a range of tumor types (malignant melanoma, childhood leukemia, metastatic breast, metastatic ovarian, metastatic renal cancer, multiple myeloma and bladder cancer), any therapeutic effects have been marginal.

Clinical trials on cancer patients have also been undertaken using a copolymer of 1,3-bis(methylamine carboxy)-2-methylenepropane carbonate and *N*-vinylpyrrolidone (called coporithane). This polymer potentiates the immune system in animal models, but does not show therapeutic activity against metastatic melanoma (nor appreciable toxicity).

3. Polymers as Drug Carriers

Two methods have been used to produce drug-loaded polymeric carriers: noncovalent complexation (e.g., DNA-anthracycline antibiotic complexes and dextran-*cis*-platinum complexes), and covalent binding of drugs to the carrier. It is generally agreed that noncovalent binding produces a drug delivery system with limited stability. Several systems, including the DNA complexes, show rapid release of drug in animal model systems, and also in humans, with no marked therapeutic benefit. Lack of control, regarding both site and rate of drug delivery, has largely led to the abandonment of this approach, with the exception of polymeric carriers for those compounds which do not chemically easily lend themselves to covalent coupling.

A number of techniques have been used to covalently bind drugs to polymeric drug carriers, such as: derivatization of the drug with a polymerizable group so that a homopolymer (or copolymer) containing the drug can be prepared; or by a polymer analogous reaction (i.e., attachment of the drug to polymer by subsequent reaction). Using the latter approach, functional groups already present within the polymer chain can be used to facilitate drug binding. Alternatively, it is possible to introduce a comonomer into the polymerization mixture which will provide the necessary functional groups, or indeed, postpolymerization uses appropriate reagents to activate the polymer backbone. Both facilitate subsequent drug binding. A limited number of functional groups are routinely used to attach drugs to polymers and these are shown in Table 5.

3.1 Importance of Biodegradation for Drug Delivery

If the eventual target for the activity of a drug is a membrane receptor (externally disposed on the cell surface), it is possible to observe a pharmacological response, even if the drug–polymer linkage is not biodegradable. As long as the polymer backbone does not sterically hinder drug–receptor interaction,

Table 5
Functional groups used to bind drugs to polymeric carriers

Group	Structure
Ester	—CO—O—
Amide	—CO—N
Urethane	—O—CO—N
O-acylhydroxamic acid	—CO—NH—O—CO—
Hyrazone	—CO—NH—N=C
Thioether	—CH$_2$—S—CH$_2$—

activity will be seen (e.g., β-adrenergic drugs). However, most pharmacologically active agents must penetrate the cell membrane to exert activity via an intracellular receptor, and in their low-molecular-weight form this is readily achieved by mechanisms of active or passive transmembrane transport. Macromolecular drug conjugates cannot pass across cell membranes and their cellular internalization is limited to the mechanism of pinocytosis (Duncan 1987), thus polymeric drug carriers can only serve as prodrugs from which the agent must be liberated in order to exert intracellular activity. Many studies have served to illustrate this principle, but the observation that poly(L-lysine)–methotrexate conjugates show antitumor activity, while nondegradable poly(D-lysine) conjugates do not provide a useful example. Certain polymer–drug linkages have been designed with extreme care to optimize the site and rate of drug delivery. In particular protein and *N*-(2-hydroxypropyl)meth-acrylamide copolymer conjugates containing the antitumor agents adriamycin and daunomycin, have been synthesized to contain peptide polymer–drug conjugates (tri- and tetra-peptides). These peptide linkages are stable during transport of conjugates in extracellular fluids, but are degraded by specific lysosomal enzymes following pinocytic capture. Kopeček (1984) described the role of the lysosomal thiol-dependent proteinases in the degradation of such peptidyl linkages, and showed that their amino acid sequence can be chosen to optimize the rate of intracellular drug delivery. Peptidyl spacers have also been used to bind antitumor agents to other polymers including poly(L-lysine) and poly(L-glutamic acid).

Natural and pseudosynthetic polymeric carriers are often selected since their polymer backbone is biodegradable. Degradation of the main polymer chain will thus ultimately liberate drug. However, it has been shown that modification of such polymers (e.g., oxidization of dextran) can dramatically decrease the natural susceptibility to enzymatic degradation.

3.2 Cell-Specific Targeting

Polymers show a distribution in the body following parenteral administration which is highly molecular-weight dependent. Materials for which M_w is greater than 100 000, begin to promote phagocytic uptake by cells of the reticuloendothelial system and thus are recovered in the liver, bone marrow and spleen. Deposition in such cells increases as the molecular weight rises. Materials of molecular weight less than the renal threshold are rapidly excreted via the kidney. Nonionic polymers, such as poly(vinyl pyrrolidone) and *N*-(2-hydroxypropyl)methacrylamide, do not have affinity for cell plasma membranes, and they are only internalized by the mechanism of fluid-phase pinocytosis unless other moieties are included in their structure to promote membrane absorption. Thus, polymers of low molecular weight are not preferentially captured by any particular cell type.

Polymers that bear positively or negatively charged groups bind nonspecifically to cell membranes. Particularly polycations, DEAE dextran and modified *N*-(2-hydroxypropyl)methacrylamide copolymers, adsorb strongly to cell surfaces indiscriminantly. Polyanions have the ability to interact with macrophagelike cells. It has also been shown using modified proteins and polymers substituted with hydrophobic residues (e.g., tyrosinamide and tyramine) that increasing hydrophobicity of polymers increases nonspecific membrane interaction. The observation that increased substitution with charged or hydrophobic residues promotes cellular interaction has obvious implications relating to the optimum degree of loading with drugs displaying these characteristics.

For purposes of cell-specific drug targeting, both inherent properties of polymers, and conjugation to selected targeting residues have been used. Polymers composed of, or modified to contain, carbohydrates are recognized by particular cells. For example, cells of the reticuloendothelial system, including kupffer cells in the liver and macrophages in the peritoneum, lung and spleen, have cell surface receptors which recognize glucose or mannose presented in macromolecular form, and also galactose, when presented on the surface of a particle. Hepatocytes within the liver also have a surface receptor recognizing galactose and fucose, specifically when presented in macromolecular form. Natural polymers, pseudosynthetic polymers (neoglycoproteins) and *N*-(2-hydroxypropyl)methacrylamide copolymers containing pendant sugars, have all successfully targeted drugs to these cell types. In particular, the galactose-recognizing receptor of hepatocytes has been used to deliver therapeutically active doses of anticancer and antiviral agents to the liver. Such systems are beginning clinical evaluation against primary and secondary liver cancer.

To increase the possibilities for cell-specific targeting following parenteral administration, various ligands have been incorporated into polymer structure including hormones (e.g., melanocyte stimulating hormone) and proteins (e.g., transferrin). The success of this approach remains to be proven. Of far-reaching potential is the possibility of using monoclonal antibodies (or fragments of such) to address polymer conjugates for delivery to particular cell types. A number of antibody polymer–drug conjugates have already been described (Ghose and Blair 1986) and these show activity against tumor models in animals (e.g., dextran antibody conjugates and poly(glutamic acid) antibody conjugates). In addition, *N*-(2-hydroxypropyl)methacrylamide copolymers have been bound to antibodies and antibody fragments, and show potential as delivery systems for anticancer agents and immunomodulators. Protein–drug conjugates which include polymers display a number of advantages; in particular, that conjugation of drug to polymer before linking to antibody increases the drug payload that the antibody can carry. Direct coupling of drugs to antibodies quickly leads to loss of ability to interact with the target antigen. Synthetic polymers also have proven ability to reduce the immunogenicity of protein conjugates (see Sect. 4).

4. Polymer-Coated Drug Delivery Systems

Use of soluble polymers to modify the surfaces of, and hence recognition of, drug delivery systems has been reported. Two distinct applications are known: poly(ethylene glycol) has been bound to many model proteins and also pharmacologically important enzymes, including uricase and asparaginase (Davis et al. 1980); and antibody conjugates have also been described. It was immediately noted that the polymer adduct did not only display better formulation properties, but also showed reduced immunogenicity following repeat administration of the protein conjugate. Uricase and asparaginase conjugates have shown promise in clinical trials against hyperuricemia and gout, or leukemia and lymphoma, respectively. Recently, other polymers have displayed the ability to reduce the immunogenicity of antibodies and proteins. *N*-(2-hydroxypropyl)methacrylamide–antibody conjugates were up to 250 times less immunogenic than free antibody. The mechanisms of reduced immunogenicity are still unclear but, in part, involve the ability of the conjugated polymer to mask immunogenic determinants present on the protein surface.

Ability of polymers to mask biologically interactive surface determinants have also been put to good use in the surface modification of particulate drug carriers. Unlike soluble macromolecules, particles are readily opsonized in the circulation and, thus, avidly recognized by cells of the reticuloendothelial system. Particulates administered parenterally, usually target rapidly to the liver and spleen. Particles coated with a variety of macromolecules

(to reduce opsonization) have the ability to alter their distribution and rate of uptake by different organs (Tomlinson 1987). In particular, poloxamers have been studied. These are alternating block copolymers consisting of hydrophilic polyoxyethylene and hydrophobic polyoxypropylene blocks which interact with a hydrophobic surface, such that the hydrophilic chains radiate out from the surface and, therefore, sterically hinder approaching proteins (opsonins).

5. Polymers in Oral Drug Delivery Systems

Although many sophisticated multicomponent drug delivery systems are well under development for parenteral use, most soluble polymers used in pharmaceutical products are taken orally in very simple form. Such materials have been selected because of their relatively high molecular weight, their ability to form aqueous solutions which do not irritate mucous membranes and, moreover, because they often have the ability to mask the obnoxious tastes of drugs. As such, they prove suitable vehicles for drug delivery and include cellulose derivatives, sodium alginate, poly(ethylene glycols), polycarbophils and poloxamers. The poly(ethylene glycols), in particular, show increasing pharmaceutical use because of their wide ranging biocompatibility. These water-soluble materials have found uses in lotions (for topical use), suppositories, tablet coatings and as emulsifying and dispersive agents used to stabilize oily components.

Polymers will play an important role in the evolution of more effective oral drug delivery systems in the future, both for site-specific delivery within the gastrointestinal tract and also in regulating the transit of formulations so that they are retained for sufficient duration in the region most appropriate for drug absorption and/or activity. Nonspecific interaction of polymers with mucin (and the mucosal surface) has been termed bioadhesion. Numerous polymers have been studied in respect of such properties, and a range of polyanions, polycations and neutral polymers have displayed bioadhesive potential. Carboxymethyl cellulose, polycarbophil, sodium alginate and carbopol have shown the most promise. Once bioadhesive polymers become fully characterized in respect of their biocompatibility in defined delivery systems, they should find uses in oral, buccal and ocular drug delivery.

Design of polymers for specific receptor-mediated interaction (and possibly internalization) in the gastrointestinal tract is already underway, and the possibility of using regional differences in enzymatic activity to control the liberation of both protein (e.g., insulin) and low-molecular-weight drugs has been demonstrated in animal models.

6. Future Developments

Soluble polymers, both synthetic and natural, already play important roles in a variety of drug delivery systems for oral, topical and parenteral administration. At the present time, considerable resources are being invested in the development of such materials to provide the drug delivery systems of the future. Although tests in animal experiments show considerable promise, particularly the polymeric derivatives of anticancer agents, their potential in humans will not become obvious for some time. If soluble polymers are to achieve widespread use in parenteral formulations, there is an ongoing need for fundamental research to understand their biocompatibility. Multicomponent systems must be carefully defined in respect of their chemical synthesis and fate in the organism. Guidelines for preclinical toxicological evaluation of such materials are long overdue. The continuing use of polymers in oral formulations is guaranteed, and undoubtedly new polymers, and fabrication techniques, will continue to appear.

Bibliography

Davis F F, Abuchowski A, Palczuk N C, Savoca K, Chen R H-L, Pyatak P 1980 Soluble, nonantigenic polyethylene glycol-bound enzymes. In: Goldberg E P, Nakajima A (eds.) 1980 *Biomedical Polymers.* Academic Press, New York, pp. 441–52

Duncan R 1987 Selective endocytosis of macromolecular drug-carriers. In: Robinson J R, Lee V H (eds.) 1987 *Controlled Drug Delivery, Fundamentals and Applications*, 2nd edn. Dekker, New York, pp. 581–621

Duncan R, Kopeček J 1984 Soluble synthetic polymers as potential drug carriers. *Adv. Polym. Sci.* 57: 51–101

Ghose T, Blair A H 1986 The design of cytotoxic-agent-antibody conjugates. *CRC Crit. Rev. Ther. Drug Carrier Syst.* 3: 263–359

Kopeček J 1984 Controlled biodegradability of polymers— A key to drug delivery systems. *Biomaterials* 5: 19–25

Maeda H, Matsumoto T, Konno T, Iwai K, Ueda M 1984 Tailor-making of protein drugs by polymer conjugation for tumor targeting: A brief review on SMANCS. *J. Protein Chem.* 3: 181–93

Tomlinson E 1987 Theory and practice of site-specific drug delivery. *Adv. Drug Delivery Rev.* 1: 87–198

Williams D F (ed.) 1987 *Definitions in Biomaterials.* Elsevier, Amsterdam

R. Duncan
[University of Keele, Keele, UK]

Sterilization Using Ethylene Oxide

In hospitals, economic considerations have led to the repeated use of devices that are relatively expensive, thus necessitating repeated sterilization between use on different patients. Steam autoclaving

Table 1
The range of process parameters employed by the majority of commercial EO sterilizers

Prevacuum	up to 0.1 bar
Nominal temperature	20 °C to 60 °C
Relative humidity	greater than 50%
EO concentration	600 mg l^{-1} to 1200 mg l^{-1}
Sterilizing gas mixture and pressure	100% EO at 0.9 bar to, for example, 15% EO–85%CO_2 at 5.5 bar
Aftervacuum and airflushing	up to 0.1 bar

(at 121 °C or 134 °C) is the most frequently employed sterilization procedure; however, the introduction of new heat-labile materials and equipment has led to the development of an alternative low-temperature sterilization process, ethylene oxide gas (EO) sterilization. The essential parameters for achieving sterilization with EO (i.e., gas concentration, gas contact time, temperature, relative humidity) had been specified by 1950. In the 1960s, EO emerged as the most promising means of hospital sterilization at relatively low temperatures and by the 1970s it had become established as the principal method of sterilizing heat-sensitive supplies, particularly in the USA where in excess of 10 000 EO sterilizers are reported to be in use. EO is commonly used in industry to sterilize disposable materials, although some manufacturers are able to bear the large capital cost associated with facilities for sterilizing with ionizing radiation which also finds considerable application. In the 1970s a sterilization process that combined low-temperature steam and formaldehyde was investigated, but commercially produced machines have lacked reliability so the process finds little routine use.

1. The Process

Although advances have occurred in the instrumentation and control of EO sterilization cycles, the hardware has changed little since the mid-1970s (Caputo and Odlaug 1982). Generally, EO sterilizers may be divided into two categories: those that use EO in the pure form, operating at or near atmospheric pressure or under vacuum, and those that use a mixture with a diluent gas, such as freon or carbon dioxide, under pressure (up to 6 bar). A variety of sterilizers are on the market, each with a different combination of process parameters. However, the majority use parameter values within the range shown in Table 1. A typical sterilization cycle for both categories of machine is shown in Fig. 1 and includes the following steps.

(a) The articles are prewrapped in wrapping materials permeable to gas such as a paper/polythene film packet and then loaded into the sterilization chamber.

(b) The chamber door is closed.

(c) A vacuum is drawn, the load raised to temperature and the chamber is humidified.

(d) Gas is let into the chamber.

(e) The gas pressure inside the chamber is raised to the desired level.

(f) The gas is exhausted from the chamber.

(g) The chamber is purged, usually with repeated vacuum/air flushing.

(h) The door of the chamber is opened and the sterilized articles are removed.

A reliable method of effectively monitoring and controlling all the essential process parameters which would enable quality assurance has not yet been commercially developed. Therefore, quality control of the sterilization process is effected by biological indicators, as described in Sect. 3.

2. Potential Hazards

Evidence relating to the mutagenicity (Generoso et al. 1983, Yager et al. 1983) of EO as well as uncertainty regarding its carcinogenicity (Hogstedt et al. 1979) has resulted in a drastic reduction in occupational exposure limits and has fuelled concern over the potential exposure to patients. The number of exposed workers in medical and related facilities in the USA has been estimated to be in excess of 100 000 and the Occupational Safety and Health Administration (OSHA) standard for EO has been reduced from a time-weighted average (TWA) concentration over an 8 h work shift of 50 vpm (i.e., 50

volumes of EO per million volumes of contaminated air) in 1982 to a standard of 1 vpm in 1987. In order to take account of exposure to relatively high levels of EO over periods of a few minutes, which would still meet the standards for TWA exposure, OSHA has now established three different limits for EO exposure with corresponding monitoring requirements:

(a) PEL (permissible exposure limit) of 1 vpm over 8 h TWA,

(b) AL (action level) of 0.5 vpm over 8 h TWA, and

(c) EL (excursion limit) of 5 vpm over 15 min sampling period.

Occupational exposure can occur as a result of poor venting of sterilizers and also through degassing from items (see Sect. 2.2) that still contain EO while they are in storage.

Items from which the EO has not been completely eliminated can also present a hazard to patients as outlined in Sect. 2.1. Early reports detailed some deaths during surgery that have been attributed to EO residues in tubing used for extracorporeal circulation. Plastics sterilized by EO have caused serious tissue reactions, including hemolysis in pump oxygenators and blood administration sets, tracheal necrosis during intubation and burns from gloves, masks and gowns. More recently, case reports have concentrated on anaphylactoid reactions in dialysis patients after exposure to equipment sterilized with EO.

2.1 Chemical Residues in Sterilized Items

Residues of EO in polymeric materials have been measured after sterilization. For example, EO residues in polyvinylchloride have been found to be in excess of 10 000 ppm (i.e., micrograms of EO to grams of host material). Against the background of the carcinogenic and mutagenic properties of EO as reviewed by OSHA (Occupational Safety and Health Administration 1983), Italian and French requirements are a maximum EO residue limit of 2 ppm in medical devices. In the USA, the conclusion of the FDA in 1978 (Food and Drug Administration 1978) that for many medical devices there is no alternative to EO sterilization still remains. Nevertheless, it was recognized that EO residue, as well as its two major reaction products ethylene chlorohydrin and ethylene glycol, may produce toxic reactions in patients. The FDA proposed recommended maximum residue limits for EO, ethylene chlorohydrin and ethylene glycol in certain devices intended for human use: small implants (less than 10 g) which include sutures and contact lenses, medium implants (10–100 g), large implants (greater than 100 g), intrauterine devices, intraoccular lenses, devices contacting human mucosa (mouth, nose, trachea, urinary tract), devices contacting

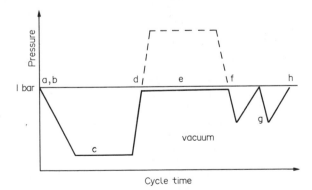

Figure 1
The dynamics of a typical EO sterilization cycle as exercised by the majority of commercial sterilizers

Table 2
FDA proposed residue limits (ppm)

Type of device	Ethylene oxide	Ethylene chlorohydrin	Ethylene glycol
Small implants (less than 10 g)	250	250	5000
Medium implants (10–100 g)	100	100	2000
Large implants (over 100 g)	25	25	500
Intrauterine devices	5	10	10
Intraocular lenses	25	25	500
Devices contacting blood (*ex vivo*)	25	25	250
Topical	250	250	5000
Surgical scrub sponges	25	250	500

blood but used outside the body (hemodialysis units, blood oxygenators, blood bags) and devices contacting normal skin (surgical drapes, bandages, surgical scrub sponges). These proposed residue limits are shown in Table 2 (Food and Drug Administration 1978). They are the maximum acceptable limits for medical devices in their market containers at the time of release for marketing. The limits were derived from values developed by a toxicity working group of the Association for the Advancement of Medical Instrumentation (AAMI) Ethylene Oxide (Z-79) Subcommittee from industrial data submitted to the FDA and from residue limits already established by current good manufacturing practice for similar products subject to approved new drug applications. For example, the proposed residue limits for intrauterine devices and surgical scrub sponges are the same as those being proposed for similar articles that are classified as drugs.

As in the case of drug products, residue limits established for certain medical devices would apply if EO was used as a sterilant during any part of the manufacturing process for any components of the device or the market container of the device. All limits are given as micrograms of EO per gram material (ppm) and are not related to the amount of EO to which the patient might be exposed depending on the use to which the device is put. The FDA considers that the maximum levels of exposure ($30 \mu g \, kg^{-1} day^{-1}$ for EO, $15 \mu g \, kg^{-1} day^{-1}$ for ethylene chlorohydrin and $2.5 \, mg \, kg^{-1} day^{-1}$ for ethylene glycol) proposed for drug products cannot reasonably be applied to medical devices for human use at this time. There is a lack of data on the rate of residue diffusion from various plastic materials and, given the nature, manner and frequency of use of many medical devices, it is considered impractical to

expect medical device manufacturers to work in concert with health professionals to restrict the total exposure from different devices used on the same day. None of the previously mentioned residue limits take material composition of the device into account. It is known that the rate of diffusion of EO in different materials is extremely variable. Therefore, dependent upon the use to which the device is put, differences in material composition might change the EO exposure to the patient by more than a factor of ten when measured as the amount of EO released for a given residue level in the device. However, an estimation of the exposure of the patient can be obtained either experimentally (Hoy and Handlos 1983) or by model calculations (Gibson et al. 1989).

It is well recognized that EO residuals in newly sterilized goods can lead to occupational exposure in the work environment as staff are repeatedly exposed to items during removal and transportation from the sterilizer. Therefore, to minimize the risk to patients and staff, it is necessary to reduce the level of residue in sterilized materials to as low a value as is practicably achievable.

2.2 Desorption of EO Residues

A number of publications describe the theory and mechanism of diffusion (i.e., absorption and desorption) of EO in polymeric materials (Crank and Park 1968). In practice, the factors that are important in desorption are the nature of the material, the concentration gradient of the diffusant, temperature and time. It has been claimed that negative pressure can also accelerate the rate of desorption (Thomas and Longmore 1971, Fischer-Bothof 1982), but this has proved to be untrue (Roberts and Rendell-Baker 1972).

To reduce the concentration of residual EO to a safe level, each of the sterilized articles needs to be adequately desorbed/aerated. The desorption time required for a particular article depends upon many variables; for example, the composition, geometry and mass of the article, and the type of EO sterilization process employed. Some plastics, such as polyvinylchloride, require a minimum of 7 days aeration at ambient temperature before they may be regarded as safe (Association for the Advancement of Medical Instrumentation 1981). The only commercially available method of shortening the aeration period is through the use of a heated aerator that circulates and continuously vents air at $50 \, °C$ to $60 \, °C$. This procedure reduces aeration time for most materials and geometries to between 8 h and 12 h. The airstream is discharged to the outside atmosphere to prevent exposure to staff in the sterilizer facility.

A novel combined sterilizer–desorber has been developed in which microwave irradiation of EO-sterilized materials is utilized to enhance the rate at which the gas is desorbed. This process is 400% faster than conventional aeration with warm air. The integration of the desorption process into the sterilizer chamber and the associated improvement in efficiency of desorption is of practical importance. The current situation in regard to EO sterilization is that with conventional aeration there may still be an unacceptable level of risk from occupational exposure as devices are transferred from the sterilizer to the aerator. Evidence suggests that short-term exposure at relatively high levels may be more harmful than long-term exposure for the same total dose (Generoso et al. 1983) and this may provide the impetus for implementing limits on short-term exposure. Therefore, a process that renders sterilized items residual free before the sterilizer door is opened guarantees that there will be no short-term exposure of operators. It therefore obviates attempts to regulate this situation by environmental control measures which incur substantial costs. Further, it will guarantee compliance with FDA standards on residue limits and consequently ensure that the risk to patients from this source is eliminated. Finally, the excessively large inventories required (to facilitate long-term aeration in order to comply with residue limits) could be drastically reduced thus resulting in considerable economic savings.

3. *Microbiological Testing of EO Systems*

The continuous measurement throughout the process of all the factors that are critical to achieving sterility (i.e., parametric monitoring) is an ideal that is well accepted for sterilization by high-pressure steam and ionizing radiation. However, continuous process monitoring of the factors necessary for EO sterilization is technically possible, but not economically feasible except on large industrial sterilizers. Before any sterilizer of whatever type is used, it must be validated (i.e., checked to ensure that the required lethality is obtained). Therefore, whereas steam sterilization is validated and monitored by physical measurements, EO sterilization is validated and monitored by biological measurements. This is achieved by demonstrating the killing of standardized preparations of microorganisms of known resistance to EO. Various organisms are employed that satisfy the criteria of resistance to EO, ease of sporulation and ease of storage (e.g., *B. subtilis var. niger*). Validation of the process is assessed through the generation of survivor curves for the chosen spore. This enables the sterilization parameters to be set at a suitable level for sterility. For routine process monitoring of production cycles after exposure to EO, the biological indicators are placed in culture media and incubated to determine growth and survival. The biological indicator is, in effect,

used as a biological instrument that integrates all the critical sterilization parameters into a single measurement. However, the absence of growth of all biological indicators from a load does not, in itself, guarantee that all items are sterile.

The British Pharmacopoeia refers to a Department of Health and Social Security specification that describes a method of producing and characterizing a biological monitor with reproducible characteristics, and in the USA the relevant specification is published by AAMI (Association for the Advancement of Medical Instrumentation 1985). The use of biological monitoring is, in principle, less satisfying than parametric monitoring, because of the inherent variability of biological monitors and the associated recovery systems.

However, in addition to process parameters, the efficacy of the process is also influenced by the nature of the product (i.e., by the "presterilization bioburden" or extent of microbial contamination), by other proteinaceous contaminants shielding the bioburden from the sterilant, by the type of packaging used and by the chamber loading configuration. Items that have been in contact with a patient may be extensively contaminated with body fluids and other proteinaceous material, as well as microbial contamination. Inadequate cleaning can give rise to the occlusion of microorganisms by a protective layer of material and, therefore, monitoring and appropriate control of cleaning procedures is essential for assurance of sterility. It is worth noting that biological indicators may pick up faulty procedures relating to packaging and loading, but they are incapable of indicating faulty cleaning procedures. Therefore, in order to verify that a particular process is adequate, it is necessary to know the following: the lethality of the process, the adequacy of wrapping and loading procedures, and the adequacy of cleaning procedures.

4. *Current Status*

The conclusion of the FDA in the late 1970s that for many medical devices there is no alternative to EO sterilization is still borne out by its continued widespread use. However, the issues of operator exposure and residue limits will continue to be a problem that will concern hospitals as well as the OSHA and the FDA until sterilizer manufacturers have implemented a solution that is economically viable. There is the additional considerable problem of EO emissions to the atmosphere from sterilizer facilities. In view of the environmental debate considering the deleterious effects of chlorofluorocarbons (CFCs), the Environmental Protection Agency (EPA) is committed to reducing 1986 levels by 50% by mid-1988. Consequently, the Du Pont Corporation, which is the largest producer of CFCs in the world, is phasing out production. Therefore, those EO sterilization processes that use CFCs in a mixture with EO will require allied equipment for their elimination or recycling. The EPA is also collecting data on EO emissions and plans to require EO emission reduction. Consequently, in the near future, sterilizer facilities will be required to employ emission cleaning technologies such as catalytic converters or reclamation units.

In view of these issues there is currently some attempt at defining alternative low-temperature sterilization processes, for example, using propylene oxide, chlorine dioxide, hydrogen peroxide or ozone. These all suffer some considerable disadvantages of toxicity and/or adverse effects upon biomaterials. In addition, in view of the turnaround/throughput time of the EO sterilization process including aeration, there is an opinion that because this is much greater than the period of reuse of certain devices between patients, disinfection should be accepted as an alternative to sterilization in these situations. In this connection it should be noted that problems of toxicity are not eliminated by disinfecting with low-temperature steam and formaldehyde or 2% glutaraldehyde. Indeed, in principle, any agent that is lethal to microorganisms is likely to be toxic to humans. Therefore, it would seem that it is preferable to consider whether there are possible improvements that can be made to the EO sterilization process that eliminate the problems mentioned previously at an acceptable economic cost.

5. Optimization

An economically viable improved process that will eliminate the problems of EO residues and operator exposure has been successfully tested (Samuel et al. 1988), but this is not yet commercially available. The cost of reducing discharges of EO to the atmosphere either by catalytic oxidation or by reclamation and subsequent disposal seem likely to be considerable for hospitals, even if industry can bear these costs. If the in-chamber concentration of EO could be measured for relatively low cost then this would, for the first time, make possible the recycling of the gas between loads to be sterilized. Low-cost methods of trapping and then recycling the EO gas as well as measuring the gas concentration in-chamber have been described, but await commercial implementation.

Once these solutions have been implemented commercially, there is still potential for further optimization of the process. Parametric monitoring of the physical conditions that ensure sterility can be implemented once an inexpensive sensor of in-chamber EO concentration is developed. In effect this will mean that EO sterilization will be placed on an equal footing with steam sterilization insofar as the assurance of the process factors will be determined instrumentally as they are for steam. Therefore, biological monitoring with its attendant variability and cost will only need to be used in the same role that it is employed for in steam sterilization, that is, in the monitoring of correct procedures (e.g., wrapping, load configuration) rather than in the monitoring of machine performance.

Since the concentration of EO in-chamber is not measured and since loads vary considerably in their ability to absorb EO, it is necessary to admit quantities of gas into the chamber that give rise to a concentration of EO that is sufficiently high to indirectly ensure the sterility of the load whatever its material composition. This, of course, has an adverse effect upon the levels of EO absorbed into materials and, consequently, the associated residues in materials poststerilization. Measurement of in-chamber EO concentration throughout the process should permit sterility assurance with a reduction of the in-chamber concentration by a factor of two with an attendant reduction in EO residues.

The prehumidification phase of the sterilization cycle is an essential part of the sterilization process as mentioned previously, insofar as it influences the equilibrium relative humidity of contaminating microorganisms. A potential problem for the EO sterilization process is that if any "cold spots" exist within a load then water will condense at these points and the relative in-chamber humidity will be reduced. There is the further problem that materials that are damp or wet may become contaminated with residues of ethylene glycol upon contact with EO. In addition, if EO is in solution with water but has not reacted to form ethylene glycol then it could subsequently evaporate upon opening the chamber which would give rise to occupational exposure.

Current process methods of heating the load rely upon heating elements in the chamber walls which can result in cold spots in the load. Therefore, the quantity of water admitted to the chamber is greatly in excess of the amount needed to generate the required relative humidity levels specified in Table 1, thus causing condensation to take place. In addition, it has been noted that in certain areas of the world at certain times of the year the atmospheric humidity is very low which causes materials and microorganisms to become desiccated which, in turn, is detrimental to the sterilization process. However, microwave radiation is known to interact with and mobilize water molecules very effectively. Application of a microwave field has been found to greatly enhance the prehumidification process as well as maintaining a greater uniformity of temperature throughout a load of materials compared to conventional heating processes. Thus, the application of microwave radiation during the prehumidification phase will increase the effectiveness of sterilization through enhancing the equilibrium relative humidity of microorganisms. Further, it will accomplish this using a much reduced quantity of water which will consequently reduce any potential problems of ethylene glycol residue production. Finally, preliminary indications are favorable that the sequential application of microwave radiation and EO during the sterilization phase may have a synergistic effect. If this is borne out by a full biological validation then it would have the important consequence of permitting even lower in-chamber concentrations of EO to achieve sterility. The levels absorbed and the desorption times required would then be much lower and, consequently, the efficiency of the whole process in terms of throughput of material would be further improved.

6. Conclusion

Despite the difficult issues (i.e., residue limits, environmental discharges, occupational exposure) associated with the EO sterilization process, there appears to be no alternative to its widespread use for many materials. Cost-effective technical solutions now exist for all of these problems and these need, therefore, to be commercially implemented by sterilizer manufacturers.

It also seems likely that the current focus, particularly in the USA, of the infection hazard and associated cost of disposal of hospital materials after patient use may result in resterilization gaining ground in comparison with single-use disposable items. Further, it is technically and economically feasible to achieve parametric monitoring of the process, and also to achieve considerable improvements in the throughput time of materials with attendant economies. When these technologies are commercially available, the sterilization of heat-sensitive medical supplies will be placed on a par with steam sterilization.

It would, therefore, seem ill conceived to place much effort into a search for alternative low-temperature sterilizing processes, particularly as these are likely to suffer from similar problems to those that currently beset EO sterilization.

The principle of "best practicable means" is well accepted in solving problems in public health. The improvements that can be achieved in the EO sterilization process will ensure that it will be the best practicable means of preventing infection associated with heat-labile materials and it will inevitably expand to replace much disinfection which is currently employed as the best practicable means in many hospital situations.

Bibliography

Association for the Advancement of Medical Instrumentation 1981 *Good Hospital Practice: Ethylene Oxide Gas-Ventilation Recommendations and Safe Use*, AAMI EO-VRSU-3/81. AAMI, Arlington, VA

Association for the Advancement of Medical Instrumentation 1985 *Good Hospital Practice: Performance Evaluation of Ethylene Oxide Sterilizers—Ethylene Oxide Test Packs*. AAMI, Arlington, VA

Caputo R A, Odlaug T E 1982 Sterilization with ethylene oxide and other gases. *Chemical and Physical Sterilization*. CRC Press, Boca Raton, FL

Crank J, Park G S 1968 *Diffusion in Polymers*. Academic Press, London

Food and Drug Administration 1978 Ethylene oxide, ethylene chlorohydrin and ethylene glycol. Proposed maximum residual limits and maximum levels of exposure. *Fed. Regist.* 43: 27474–87

Fischer-Bothof E 1982 Aeration methods. In: Ayliffe G A J, Cripps N F, Deverill C E A, George R H (eds.) 1982 *The Health Service Use of Ethylene Oxide Sterilisation*. United Birmingham Hospitals, Birmingham, UK

Generoso W M, Cumming R B, Bandy J A, Cain K T 1983 Increased dominant lethal effects due to prolonged exposure of mice to inhaled ethylene oxide. *Mutat. Res.* 119: 377–9

Gibson C, Matthews I P, Samuel A H 1989 A computerized model for accurate determination of ethylene oxide diffusion in sterilized medical supplies. *Biomaterials* 10: 343–8

Hogstedt C, Malmquist N, Wadman B 1979 Leukaemia in workers exposed to ethylene oxide. *J. Am. Med. Assoc.* 241: 1132–3

Hoy K, Handlos V 1983 Release of gas residues from an ethylene oxide sterilized haemodialyser. *Biomaterials* 4: 321–2

Occupational Safety and Health Administration 1983 Occupational exposure to ethylene oxide. Proposed rule. *Fed. Regist.* 48: 17284–319

Roberts R L B, Rendell-Baker L 1972 Aeration after ethylene oxide sterilization. Failure of repeated vacuum cycles to influence aeration time after ethylene oxide sterilisation. *Anaesthesia* 27: 278–82

Samuel A H, Matthews I P, Gibson C 1988 Microwave desorption: A combined sterilizer/aerator for the accelerated elimination of ethylene oxide residues from sterilized supplies. *Med. Instrum.* 22(1): 39–44

Thomas L C, Longmore D B 1971 Ethylene oxide sterilization of surgical stores. *Anaesthesia* 26: 304–7

Yager J W, Hines C J, Spear R C 1983 Exposure to ethylene oxide at work increases sister chromatid exchanges in human peripheral lympocytes. *Science* 219: 1221–3

I. P. Matthews, A. H. Samuel and C. Gibson
[University of Wales College of Medicine,
Cardiff, UK]

Surface Structure and Properties

Surface structure directs or strongly influences the biological reaction to materials. However, in order to observe unambiguously the influence of surfaces on biological reactions, materials must be free from leachable substances; where leachables are present, the biological response might be controlled by the

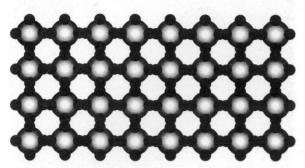

Figure 1
The creation of a surface leads to unfulfilled bonding. Note that in this two-dimensional represention of a crystal, the atoms at the surface can only participate in trivalent or divalent bonding, while those in the bulk are each bound to four other atoms

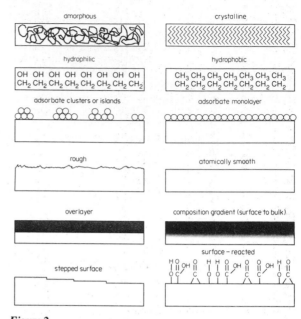

Figure 2
Polymer surfaces: some considerations

pharmacologic or toxicologic potential of those leachables. Biomaterials are expected not to leach substances except when specifically designed to do so (e.g., a drug release system). Other factors that influence the performance of biomaterials include the mechanical properties of the materials and the design of the implant fabricated from them. The focus of this review is on surfaces and their ability to induce biological reactions, and does not specifically address these other factors.

1. The Nature of a Surface

The surface zone of a solid has been described as a unique state of matter (Duke 1984). This is a reasonable assessment considering that the chemistry and structure of the outermost layers of a material are different from that in the bulk of the material, and that the particular combination of chemistry and structure found at the surface can exist nowhere else but at the surface.

The unique properties of surfaces can be attributed to three factors. First, there is a thermodynamic driving force to minimize the high energy situation created in simply producing a surface (Fig. 1). This minimization of surface energy occurs by translating and/or rotating lower energy molecular structures toward the outside (a number of examples of this are presented in this article). Second, the outside atoms are more exposed to the external world and, therefore, more subject to reaction (e.g., with oxygen) or to contamination (e.g., with hydrocarbons). Third, if the material contains diffusible components that are of lower energy than the bulk of the surface, these components (generally additives or contaminants) will migrate to the surface.

There are a number of factors that must be identified to adequately describe a surface. These factors

fall into two categories: chemical and morphological. Figure 2 outlines some of the considerations involved in a complete description of a surface.

2. Surfaces as the Trigger of Biological Reactions

Surfaces, partly because of their intrinsic energy, can effectuate chemical reactions. Synthetic catalysts (e.g., platinum and zeolites) are being exhaustively studied and are of industrial importance. The surfaces of these materials can lower reaction energy barriers, thereby raising reaction rates. Similarly, examples of biological materials acting as catalysts (i.e., exhibiting high specificity and speeding reactions) are plentiful (enzymes, antibodies, lectins, etc.).

Contemporary synthetic biomaterials do not generally exhibit the chemical specificity associated with synthetic or biological catalysts. However, processes do occur at their surfaces. These processes are often associated with adsorption of components. Upon placement of a material into a biological environment, rapid adsorption of water, ions and proteins to the surface takes place. The nature of the surface will dictate the materials that will adsorb and how they adsorb. Thus, different surfaces can fractionate the complex mixture of proteins that exist in the biological environment in unique ways, leading to differing combinations of proteins at the surface. The surface can also denature or alter the conformation of proteins adsorbed to it (Horbett and Brash

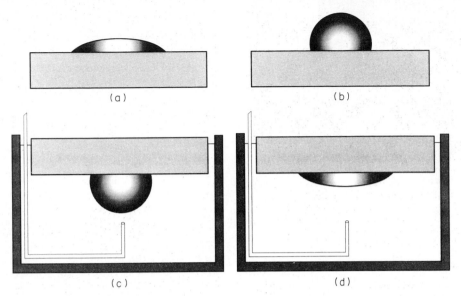

Figure 3
Possibilities observed during contact angle measurement: (a) hydrophilic surface, waterdrop in air, (b) hydrophobic surface, waterdrop in air, (c) hydrophilic surface, *n*-octane drop under water, and (d) hydrophobic surface, *n*-octane drop under water

1987). Since cells can respond with high specificity and reactivity (i.e., catalystlike behavior) to certain proteins, the nature of the proteins at the surface will control the biological reaction to the surface. Furthermore, since cellular reactions can be related to the tissue reactions observed by a physician, the biological activity of a surface can be described in terms of a hierarchical series of events: the surface influences the adsorbed protein layer, which influences the cellular reaction which can be related to the ultimate tissue reaction. Thus, the reactions observed by the physician can be attributed to the nature of the surface.

3. Surface Measurement

A wide variety of powerful tools are available to measure surfaces (Ratner et al. 1987, Briggs and Seah 1983). Great demands are placed upon these techniques because of the small total mass of material associated with surfaces (approximately 10^{14} atoms per cm^2) and because of the requirement that the surface atoms be uniquely resolved while in close proximity to an immensely larger population of atoms in the bulk phase. A few of the more important methods that are applicable to biomaterials are briefly described in this section.

3.1 Contact Angle Methods

Measurement of the angle of contact of a drop of liquid with a solid surface (Fig. 3) is one of the oldest, least expensive, and most informative methods available to probe the surface properties of a solid. The Young equation describes the equilibrium between the surface tension components interacting at the rim of any drop of liquid on a surface limiting its spread (Fig. 4). The Young equation is written as

$$\gamma_{SV} = \gamma_{SL} + \gamma_{LV} \cos \theta \qquad (1)$$

where γ_{SV} is the surface tension between the solid and the vapor, γ_{SL} is the surface tension between the solid and the liquid, γ_{LV} is the surface tension between the liquid and the vapor and θ is the contact angle in degrees. Ideally, one wants to know γ_{SV}, which is directly related to the wettability of the

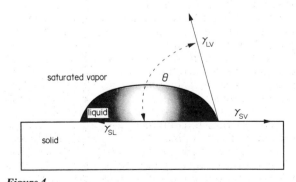

Figure 4
The set-up describing Young's equation by showing the equilibrium between surface tension components

material. The contact angle and the liquid–vapor surface tension are readily measured. This leaves two unknown quantities in this relationship. From this starting point, a number of schemes have been developed to calculate or approximate γ_{SV} from contact angle data. Certainly the most widely used is the scheme developed by Zisman, which allows the computation of a critical surface tension, γ_C; that is, an approximation to the solid surface energy and that correlates in most cases with observed wettability (Zisman 1964).

Other methods that have been used (or considered) for calculating an approximate surface energy value, or that have provided additional information on surface energetics, include underwater contact angle methods, computation of a dispersive and polar contribution to the surface energy, an equation of state approach and the calculation of a Lewis acid and base contribution to the surface energy. Much controversy exists over which method is most appropriate for a given situation, and also the measurements themselves and the meaning of the results are subject to artifact and interpretational difficulties. Therefore, the reader is directed to the many review articles and monographs published on this subject (e.g., Good 1979, Sacher 1988). Still, it is clear that relationships do exist between the biological reaction to surfaces and surface wettability.

3.2 Electron Spectroscopy for Chemical Analysis (ESCA)

ESCA (also called x-ray photoelectron spectroscopy or XPS) has rapidly become a key method for the surface characterization of materials intended for biomedical application. This is because of its information-rich character, its specificity in chemical identification, and its proven value in optimizing biomedical devices and in correlating surface properties to biological responses.

The ESCA method is based upon the photoelectric effect, the nature of which was explained by Einstein in 1905. When a surface is exposed to photons of sufficiently high energy (x rays, in this case), electrons are released. The energy of these electrons is directly related to the atomic or molecular environment from which they originated, while the number of emitted electrons is proportional to the surface concentration of the species from which the electrons originate. The electrons have a poor ability to penetrate through matter without losing energy to inelastic collisions. Therefore, only those electrons in the outermost surface of the material can escape from the surface without energy loss and be detected, hence the surface sensitivity of ESCA. ESCA usually samples a surface to a depth of approximately 100 Å. However, by using specialized x-ray anodes in the ESCA instrument, or by varying the angle of the sample with regard to the analyzer lens in the ESCA spectrometer, the effective depth

sampling for ESCA might be expanded over a 10 Å to 250 Å range.

The basic equation used for ESCA qualitative surface analysis is

$$BE = h\nu - KE \qquad (2)$$

where BE is the electron binding energy in the atom of origin, $h\nu$ is the energy of the x-ray source (a known value) and KE is the kinetic energy of the electron that is measured by the electron spectrometer. Thus, in principle, BE can be simply calculated, and from this value one can learn which elements are present and what kind of atomic or molecular environments they are in. In practice, there are many instrumental considerations that complicate the interpretation of these BE values. These, and other considerations important to qualitative and quantitative ESCA, are summarized in many books and review articles (Briggs and Seah 1983, Ratner 1983).

A wide range of information about the surface of a sample can be obtained from ESCA including

(a) which elements are present (except hydrogen and helium),

(b) approximate surface concentrations of elements ($\pm 10\%$),

(c) bonding state (molecular environment) and/or oxidation level of most atoms,

(d) information on aromatic or unsaturated structures from shakeup ($\pi^* \leftarrow \pi$) transitions,

(e) information on surface electrical properties from charging studies,

(f) nondestructive depth profile and surface heterogeneity assessment using
   (i) photoelectrons with differing escape depths, and
   (ii) angular dependent ESCA studies,

(g) destructive depth profile using argon etching (for inorganics),

(h) positive identification of functional groups using derivitization reactions, and

(i) "fingerprinting" materials using valence band spectra.

There are many applications areas in biomedical and biomaterials studies for which ESCA is useful, and these are discussed in the literature.

A diagram illustrating the components of an ESCA spectrometer is presented in Fig. 5.

3.3 Secondary Ion Mass Spectrometry (SIMS)

The SIMS method has only recently been applied to biomaterials problems, largely because of the advent of static SIMS, which permits highly surface localized (uppermost 10 Å) analysis of polymeric materials without massive surface damage.

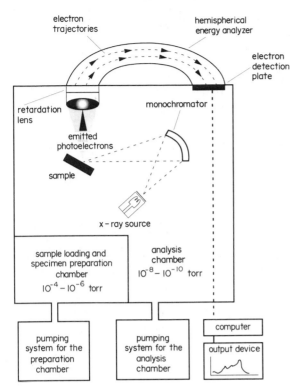

Figure 5
The components of an ESCA spectrometer system

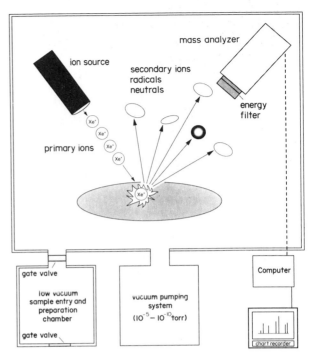

Figure 6
The SIMS technique measures the molecular weight of atoms sputtered from a surface by an energetic, focused ion beam. The experiment is conducted in a high vacuum chamber

The SIMS method, in essence, allows the acquisition of a mass spectrum of the outermost region of a material. Since mass spectra are rich in structural information, static SIMS has the potential to provide much previously unobtainable information about a surface. The principle of the SIMS method is illustrated in Fig. 6.

The number of publications describing the application of static SIMS to biomaterials problems has, to date, been limited. Static SIMS has shown particular utility for analyzing methacrylate and acrylate polymers, for detecting contaminants and surface localized additives and for analyzing the surface composition of polyurethanes. The potential to perform semiquantitative SIMS analysis has also been demonstrated. General review articles on static SIMS, particularly as applied to polymeric systems, have been published (e.g., Castner and Ratner 1988).

3.4 Surface Infrared Techniques

Infrared (ir) absorption spectroscopy provides much information on the molecular structure of materials. Achieving surface sensitivity with ir methods often presents experimental difficulties. Some of the modes by which surface-sensitive ir absorption measurements can be made are illustrated in Fig. 7.

The attenuated total reflection (ATR) method has received the most attention. However, the surface region sampled by the ATR method ranges from 1–10 μm, a deep penetration by surface analysis standards and, in addition, sample preparation can be troublesome. Recent developments in ATR that might have importance for biomaterials studies include the use of metal depositions on the ATR element to enhance sensitivity, the application of ATR to study surface crystallinity and the use of ATR to study the solid–aqueous interface.

The external reflection ir method (also called infrared reflection absorption spectroscopy) provides extreme surface sensitivity and also offers information on the orientation of molecules at the surface. However, this method requires a highly reflective metal substrate for the absorbing molecules and often suffers from a poor signal-to-noise ratio.

The diffuse reflectance infrared technique requires little or no sample preparation prior to analysis. However, with this method, the sampling depth is poorly defined.

3.5 New Developments in Surface Characterization

There are a number of new developments that might contribute strongly to the surface characterization of

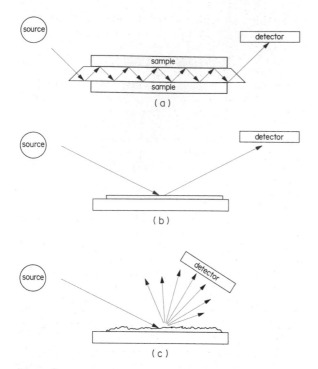

Figure 7
Modes for measuring surface-localized infrared absorption

biomaterials. Foremost among these are scanning tunnelling microscopy (STM) and atomic force microscopy (AFM). These methods have the potential to image surfaces and adsorbates on surfaces at the atomic level of resolution (Binnig and Rohrer 1985, Binnig et al. 1987). An STM image of a graphite surface is shown in Fig. 8. Atomic level imaging using STM has also been demonstrated under water.

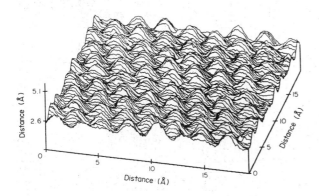

Figure 8
A scanning tunnelling microscopy (STM) image of a graphite surface

Surface-enhanced Raman spectroscopy (SERS) may permit high sensitivity Raman measurements of phenomena occurring at interfaces (Allara et al. 1981). However, a rough silver surface must be in close proximity to the interface being studied, a difficult requirement in many analysis situations relevant to biomaterials problems.

High-resolution electron energy loss spectroscopy (HREELS) has recently been applied to polymer surfaces (Pireaux et al. 1986). This method is highly surface sensitive and information-rich. However, even with new high-sensitivity detectors, there is concern over sample damage from the impinging electron source.

New developments in microscopy have also been applied to surface studies. In particular, the potential of transmission electron microscopy and low-voltage scanning electron microscopy for observing biomaterial surfaces is being explored (Goodman et al. 1988).

4. The Nature of Biomaterial Surfaces

A number of specific classes of biomaterials have been subjected to exhaustive surface characterization. This section will review some of the observations made in the study of these systems.

4.1 Polyurethanes

The interest in polyurethanes stems largely from their excellent mechanical properties, which has led to applications in artificial hearts, vascular prostheses, blood tubing, catheters, blood bags, and numerous soft tissue implants (Lelah and Cooper 1986). The strong interest in the surfaces of this class of polymers has its origins in two observations. First, the chemistry of the surfaces of polyurethanes has been shown to directly correlate with blood–polymer reactivity (Hanson et al. 1982). Second, the surface composition and structure of polyurethanes are generally different from those observed in the bulk of the material. A number of papers have been published that define the basic principles of polyurethane surface analysis (e.g., Ratner and McElroy 1986).

There are five important surface observations made on polyurethanes. These observations also define the primary areas of study of polyurethane surfaces. First, polyurethane surfaces are often compositionally enriched with low molecular weight, surface active additives that migrate from the bulk of the polymers. Thus, low molecular weight extrusion lubricants (*bis*-stearamides) and silicones are often found dominating or overcoating polyurethane surfaces. Second, in the absence of these additives, the surface of block copolyurethanes (i.e., the poly(ether urethanes) (PEUs) favored for medical applications) are dominated by the soft polyether

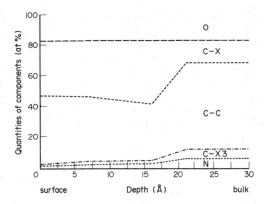

Figure 9
A depth profile diagram based upon angular
dependent ESCA data taken from a polyurethane
surface. C–C represents hydrocarbon-like
functionalities resolved from the carbon 1s spectrum.
C–X represents carbon species singly bound to
oxygen. C–X3 represents carbon species bound with
two or more bonds to oxygen or nitrogen. The high
concentration of C–X near the outermost surface
suggests polyether surface localization (depth profile
constructed by R. Paynter)

segment. This is illustrated in the ESCA depth
profile diagram in Fig. 9. Third, the interest in the
relationship between the domains (100 Å to 200 Å
phase separated regions) found in the bulk of PEUs
and such structures at the surfaces of PEUs has been
a fertile area for study. Some groups argue that
hard-segment and soft-segment domains exist at
PEU surfaces and can influence biological interactions, while others claim that the surface is totally
overcoated by only one component. In fact, PEUs
may have small amounts of hard-segment interspersed in a soft-segment-dominated surface. Fourth,
the chemical structures at the surfaces of PEUs seem
to possess ready mobility and can respond to the
external environment by moving to or from the
surface. The facile surface mobility of PEUs may
account for many of the contradictory observations
on phase separation at polyurethane surfaces.
Finally, there is a great interest in surface modification of polyurethanes to improve blood compatibility. Such modifications have been made by the
addition of specially engineered oligomeric components and by covalently incorporating specific side
chain or block components expected to be surface
active.

4.2 Hydrogel Surfaces
Hydrogel surfaces present a diffuse gradient boundary between the solid phase and the liquid phase
(Fig. 10). These surfaces are primarily characterized
by (a) their high polymer chain flexibility (rotational and translational), (b) gradients of diffusible,
lower molecular weight components from the bulk,
through the surface zone, and to the fluid phase and
(c) their relatively low adhesiveness to proteins and
cells (perhaps due to polymer chain flexibility).
Review articles on hydrogel surfaces have been
published (Ratner 1986, Silverberg 1976), and their
surface mobility has been studied. The extremely
low interaction of proteins and cells with poly(ethylene oxide) hydrogel surfaces has been the focus of
much study in recent years (Gombotz et al. 1988).

4.3 Plasma-Deposited Surfaces
The plasma-deposition method has many advantages for synthesizing surfaces useful for biomedical
materials and medical devices (Yasuda and Gazicki
1982, Ratner et al. 1989). In this method, a gas (the
monomer) is introduced into a rapidly oscillating
electrical field, leading to fragmentation and

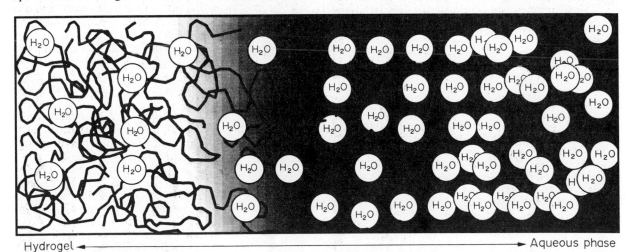

Figure 10
Hydrogels present a diffuse, gradient boundary between the solid phase and the liquid phase

ionization. The highly reactive species formed in this ionized gas phase environment can, under appropriate conditions, deposit on surfaces and recombine (or vice versa), leading to a surface film with unique characteristics. This is illustrated in the ESCA spectra in Fig. 11, which show the massive molecular rearrangement that occurs when tetrafluoroethylene gas is deposited as a film from a plasma environment. Surface characterization methods are important for understanding these films since (a) their chemistry is different from that of the starting gas and (b) they are so thin (often in the range 50–1000 Å) that the high sensitivity of surface methods is needed. The unique biological interactions observed with the surfaces of these films are of special relevance to this article. Thus, various plasma-deposited layers have been found to exhibit low thrombogenicity, a high affinity for albumin, a low cell adhesiveness and an ability to serve as effective substrates for cell growth. The nature of these unique surfaces has been under study using such methods as ESCA, chemical derivitization and SIMS. Some of the specific advantages of plasma-deposited thin films for biomedical applications are

(a) they are conformal,

(b) they are pinhole free,

(c) they can be coated on unique substrates (e.g., metals, glasses, polymers, ceramics and carbons),

(d) they show good adhesion to substrates,

(e) unique film chemistries can be achieved,

(f) they serve as excellent permeation barriers,

(g) they show low levels of leachables,

(h) they are relatively easy to prepare,

(i) a well-developed technology for performing these depositions already exists,

(j) they can be characterized, and

(k) plasma-deposited layers are sterile upon preparation.

4.4 Titanium Surfaces

Titanium surfaces, when prepared using specifically defined procedures, have been reported to bond directly to bone and tissue (Kasemo and Lausmaa 1988). Titanium is the only metal that does not become encapsulated by collagen and that exhibits the interfacially intimate bone-bonding property. Thus, by studying titanium surfaces, important insights might be gained into the nature of biomaterial interactions.

Titanium surfaces are dominated by a thin layer that is mostly titanium dioxide. After implantation, this oxide film thickens over time. In an ultrahigh vacuum chamber, titanium surfaces can be prepared that contain only titanium and oxygen (as measured by ESCA). However, the types of surfaces used successfully in a clinical setting are found to have measurable quantities of carbon, calcium, chlorine and nitrogen in addition to titanium and oxygen. These contaminants are associated with the recommended autoclave sterilization procedure. The carbon component is adventitious hydrocarbon, possibly similar to that found on all surfaces under atmospheric conditions. It is intriguing to note that optimum performance upon implantation may not be associated with the "cleanest" surface. Experiments have shown that titanium surfaces may pick up more fibronectin, a protein associated with cell adhesion, when hydrocarbon contamination levels are high (Lew 1986).

Because of the unique (and clinically efficacious) performance of titanium surfaces, these materials will remain an area for serious study for some time. Since titanium shows a clear correlation between surface preparation and clinical performance, it represents a prime example of the importance of surface properties for biomedical material performance and, possibly, a key to understanding the nature of these interactions.

5. New Developments

The surfaces used for biomaterials applications are generally amorphous or polycrystalline and exhibit little regularity or specificity. On the other hand,

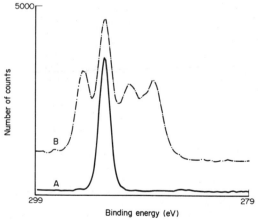

Figure 11
The plasma environment alters the nature of a monomer introduced into it, and yields a polymeric film with a structure different from that obtained by conventional polymerization methods. Lower curve: the carbon 1s ESCA spectrum of polytetrafluoroethylene (Teflon, PTFE), upper curve: the carbon 1s spectrum of an RF plasma-deposited film prepared using tetrafluoroethylene gas

surfaces found in living systems (cell surfaces, protein surfaces and antibody surfaces) have a high degree of order, leading to specificity and biological recognition. A new generation of biomaterials might attempt to mimic the natural processes and exploit this specificity so that interfacial reactions can be engineered to give the most appropriate results.

One method used to accomplish this involves immobilizing biomolecules that produce desired reactions to surfaces. Although the biospecificity of surfaces can be vastly enhanced by these methods, the long-term stability of the biomolecules at surfaces and the immunologic reactions to those molecules can impose limitations on their use. Another approach to engineering surfaces to give desired responses uses chemical surface modification methods. Surface modification has achieved significant success, but, in most cases, the surfaces still exhibit the nonspecific response associated with synthetic materials.

Many synthetic molecular systems have been shown to self-assemble or organize on surfaces and this may present new possibilities for developing biologically recognizable surfaces (Ringsdorf et al. 1988). Thus, molecular-level organized structures can be fabricated that have many of the desirable properties of nonbiological materials with the geometric organization of natural biological surfaces. Future biomaterials might be fabricated or surface-modified using Langmuir–Blodgett structures or surface-localized, self-assembled nanostructures (Lehn 1988). In addition, vesicles and microtubules might also assume increasing importance in biomaterial fabrication.

6. Conclusions

This article is centred upon the premise that the surface properties of materials govern biomedical performance. This thesis is still strongly held and ample evidence exists that surface properties can directly affect biological system–biomaterial interactions. However, a clear, predictive relationship that can be used in an engineering sense to design new materials has not been established. New developments in well-defined surfaces and in methods to characterize those surfaces (Ratner 1988) may lead to this understanding, which in turn should lead to rationally designed clinical prostheses that invoke specific, desired biological responses.

Acknowledgement

Support was received from NIH grants HL25951 and RR01296 during the preparation of this article and for some of the studies described herein.

Bibliography

Allara D L, Murray C A, Bodoff S 1981 Spectroscopy of polymer surfaces using the surface enhanced Raman effect. In: Mittal K L (ed.) *Physiochemical Aspects of Polymer Surfaces*. Plenum, New York

Binnig G, Gerber C, Stoll E, Albrecht T R, Quate C F 1987 Atomic resolution with atomic force microscope. *Euorphys. Lett.* 3: 1281–6

Binnig G, Rohrer H 1985 The scanning tunneling microscope. *Sci. Am.* 253: 50–6

Briggs D, Seah M P 1983 *Practical Surface Analysis*. Wiley, Chichester, UK

Castner D G, Ratner B D 1988 Static secondary ion mass spectroscopy: A new technique for the characterization of biomedical polymer surfaces. In: Ratner B D (ed.) *Surface Characterization of Biomaterials*. Elsevier, Amsterdam

Duke C B 1984 Atoms and electrons at surfaces: A modern scientific revolution. *J. Vac. Sci. Technol.* 2: 139–43

Gombotz W R, Guanghui W, Hoffman A S 1988 Immobilization of poly(ethylene oxide) on poly(ethylene terephthalate) using a plasma polymerization process. *J. Appl. Polym. Sci.* 35: 1–17

Good R J 1979 Contact angles and the surface free energy of solids. In: Good R J, Stromberg R R (eds.) 1979 *Surface and Colloid Science*. Plenum, New York

Goodman S L, Pawley J B, Cooper S L, Albrecht R M 1988 Surface and bulk morphology of polyurethanes by electron microscopies. In: Ratner B D (ed.) 1988 *Surface Characterization of Biomaterials*. Elsevier, Amsterdam

Hanson S R, Harker L A, Ratner B D, Hoffman A S 1982 Evaluation of artificial surfaces using baboon arteriovenous shunt model. In: Winter G D, Gibbons D F, Plenk H Jr (eds.) 1982 *Biomaterials 1980; Advances in Biomaterials*. Wiley, Chichester, UK

Horbett T A, Brash J L 1987 Proteins at interfaces: Current issues and future prospects. In: Horbett T A, Brash J L (eds.) *Proteins at Interfaces: Physicochemical and Biochemical Studies*, ACS Symposium Series. American Chemical Society, Washington, DC

Kasemo B, Lausmaa J 1988 Biomaterial and implant surfaces: On the role of cleanliness, contamination, and preparation procedures. *J. Biomed. Mater. Res: Appl. Biomat.* 22: 145–58

Lehn J M 1988 Supramolecular chemistry—scope and perspectives: Molecules, supermolecules, and molecular devices (Nobel lecture). *Angew. Chem. Int. Ed. Engl.* 27: 89–112

Lelah M D, Cooper S L 1986 *Polyurethanes in Medicine*. CRC Press, Boca Raton, FL

Lew P J 1986 Titanium Surfaces: ESCA analysis, protein adsorption, and cell interaction. M.S. Thesis, University of Washington, Seattle, WA

Pireaux J J, Thiry P A, Caudano R 1986 Surface analysis of polyethylene and hexatriacontane by high resolution electron energy loss spectroscopy. *J. Chem. Phys.* 84: 6452–7

Ratner B D 1983 Analysis of surface contaminants on intraocular lenses. *Arch Ophthal.* 101: 1434–8

Ratner B D 1986 Hydrogel surfaces. In: Peppas N A (ed.) *Hydrogels in Medicine and Pharmacy*. CRC Press, Boca Raton, FL

Ratner B D 1988 The surface characterization of biomedical materials: How finely can we resolve surface structure? In: Ratner B D (ed.) *Surface Characterization of Biomaterials*. Elsevier, Amsterdam

Ratner B D, Chilkoti A, Lopez G P 1989 Plasma deposition and treatment for biomaterial applications. In: D'Agostino R (ed.) *Plasma Deposition of Polymer Films*. Academic Press, New York

Ratner B D, Johnston A B, Lenk T J 1987 Biomaterial surfaces. *J. Biomed. Mater. Res: Appl. Biomat.* 21: 59–90

Ratner B D, McElroy B J 1986 Electron spectroscopy for chemical analysis: Applications in the biomedical sciences. In: Gendreau R M (ed.) *Spectroscopy in the Biomedical Sciences*. CRC Press, Boca Raton, FL

Ringsdorf H, Schlarb B, Venzmer J 1988 Molecular architecture and function of polymeric oriented systems: Models for the study of organization, surface recognition, and dynamics of biomembranes. *Angew. Chem. Int. Ed. Engl.* 27: 113–58

Sacher E 1988 The determination of the surface tensions of solid films. In: Ratner B D (ed.) *Surface Characterization of Biomaterials*. Elsevier, Amsterdam

Silverberg A 1976 The hydrogel–water interface. In: Andrade J D (ed.) *Hydrogels for Medical and Related Applications*. American Chemical Society, Washington, DC

Yasuda H, Gazicki M 1982 Biomedical applications of plasma polymerization and plasma treatment of polymer surfaces. *Biomaterials* 3: 68–77

Zisman W A 1964 Relation of the equilibrium contact angle to liquid and solid constitution. In: Fowkes F M (ed.) *Contact Angle, Wettability and Adhesion*, ACS Advances in Chemistry Series. American Chemical Society, Washington, DC

B. D. Ratner
[University of Washington, Seattle, Washington, USA]

Suture Materials

Virtually every surgical operation requires the use of suture materials to close the wound for subsequent successful healing. Sutures are also used to close wounds produced by trauma. The proper closing of the wound can influence the success of the operation. According to records, the ancient Egyptians used linen as a suture material as far back as 2000 BC. Today, the surgeon can choose from a large number of suture materials with various chemical, physical, mechanical and biological properties.

1. Composition

A suture consists of two components: the needle and the thread. The needle (normally made of stainless steel) must be able to carry suture materials through tissue with minimum trauma. The basic shapes of the needle point are tapered, spatula, cutting or blunt. A tapered point is used for less resistant tissues, such as intestine and dura mater. A spatula point is used in both soft and tough tissues. A cutting point is used in tough tissue, such as skin, tendon and ligament, and a blunt point is used where either cutting or piercing could damage the function of the tissue; for example, with parenchymatous and friable tissues.

Suture materials are generally classified into two broad categories: absorbable and nonabsorbable. The absorbable sutures are catgut (collagen sutures derived from sheep intestinal submucosa), reconstituted collagen, polyglycolic acid (PGA) and its lactide copolymers, polydioxanone (PDS), and poly(glycolide–trimethylene carbonate). The nonabsorbable sutures are divided into natural fibers (i.e., silk and cotton) and synthetic fibers (i.e., polyethylene, polypropylene, polyamide, polyester, polytetrafluorethylene, and stainless steel).

Suture materials are also classified according to their size. Currently, two standards are used to describe the size of suture materials: United States Pharmacopoeia (USP) and European Pharmacopoeia (EP). In USP standard, the size of sutures, except Gore-Tex suture, is represented by a series combination of two arabic numbers: a zero and any number other than zero; for example, 2/0 or 5/0. Because the porous nature of the Gore-Tex suture results in a suture diameter slightly larger than USP guidelines, the size of Gore-Tex suture is marked by a nontraditional manner, namely CV-2 (corresponding to 2/0), CV-3 (corresponding to 3/0), and so on. In EP standard, the code number ranges from 0.1 to 10. For example, a synthetic suture of 2/0 size (USP) has a range of diameter from 0.25 mm to 0.29 mm, whereas a suture of EP size 2 has a range of diameter from 0.20 mm to 0.24 mm (Chu 1983). Recently, it has been suggested that a more fundamental property, such as 60% of ultimate tensile strength, be used to classify suture materials.

Suture materials are classified in terms of their physical configuration as monofilament and twisted or braided multifilament. Suture materials made of nylon or stainless steel are available in both multifilament and monofilament forms. Catgut, reconstituted collagen and cotton, are available in twisted multifilament form, while PGA, poly(glycolide–lactide) and silk are available in braided multifilament configuration. Polypropylene, PDS, poly(glycolide–trimethylene carbonate), and polytetrafluoroethylene (PTFE) suture materials exist in monofilament form.

Multifilament suture materials are frequently coated with a polymer to facilitate their easy entrance through tissues during wound closure. Suture manufacturers prefer to use the same type of polymer for coating as the suture to be coated. Examples are polybutilate for Ethibond suture, and poly(glycolide–lactide) for Vicryl suture. Other types of coating materials that have been used are paraffin wax, silicone, Teflon, and ultralow temperature isotropic carbon.

2. Chemical Structure and Manufacturing Processes

2.1 Absorbable Suture Materials

The basic constituent of catgut is collagen. It is derived from the submucosa of sheep intestines, or serosa of bovine intestine. The resulting untreated catgut suture is called plain catgut. If the plain catgut is tanned in a bath of chromium trioxide, it is called chromic catgut. Depending on the concentration of the chromic bath, and the duration of chromicizing, mild- and extra-chromic catgut are available. Reconstituted collagen sutures are prepared from long flexor tendons of beef. The tendon is cleaned, frozen, sliced, treated with ficin, and then swollen in dilute cyanoacetic acid. The resulting viscous gel is extruded through a spinerette into an acetone bath for coagulation. The coagulated fibril is stretched, twisted, and dried or treated with chromic salts before twisting and drying.

PGA can be polymerized, either directly or indirectly, from glycolic acid. The direct polycondensation produces a polymer of molecular weight less than 10 000. PGA of molecular weight greater than 10 000 is prepared by ring-opening polymerization of the cyclic dimers of glycolic acid. For biomedical applications, preferred catalysts are stannous chloride dihydrate or trialkyl aluminum.

The resulting polymer having molecular weight from 20 000 to 140 000 is suitable for fiber extrusion and suture manufacturing (Frazza and Schmitt 1971). The fiber is stretched above its glass transition temperature (36 °C), heat set to improve dimensional stability and inhibit shrinkage, and subsequently braided into final suture forms of various sizes. The lactide–glycolide copolymer suture material, sometimes called Polyglactin 910, is randomly copolymerized in the same fashion.

Both PGA and Polyglactin 910 are semicrystalline polymers. The crystal structure of PGA exhibits an orthorhombic unit cell (Chatani et al. 1968). The planar zigzag chain molecules form a sheet structure parallel to the *ac* plane, and do not have the polyethylene-type arrangement. In the case of lactide–glycolide copolymer, the regularity of the chain stacking in the sheets is disturbed because of the presence of the lactide component.

PDS has been reported as a promising new synthetic absorbable suture material (Ray et al. 1981). It is polymerized in the presence of a diethylzinc catalyst. PDS is one of the two synthetic absorbable sutures in monofilament form.

Poly(glycolide–trimethylene carbonate) (Maxon) is the most recent commercially available synthetic monofilament absorbable suture. It is a linear block copolymer, and consists of 32.5 wt% trimethylene carbonate. Diethylene glycol is used as an initiator and stannous chloride dihydrate serves as the catalyst.

2.2 Nonabsorbable Suture Materials

Silk is one of the three major fibrous proteins; the others are wool and collagen. The actual fiber protein is called fibroin. Silk does not contain sulfur-containing amino acid residues like cystine, but is composed mainly of glycine, alanine and serine residues. The crystalline portion of the polypeptide chains in silk are in the extended zigzag β-conformation. Neighboring chains are antiparallel to one another, and a pleated sheet is formed. Different silkworm species produce fibrous proteins that contain different sequences and proportions of amino acids. These compositional differences, in turn, influence the mechanical properties of the fibers.

Cotton sutures are made from long staple fiber. The cross-sectional shape of cotton fiber varies from a U-shape to a nearly circular form. The fiber has a natural twist, called a convolution. Cotton is 10–20% stronger when wet than dry, and its strength can be increased by a process called mercerization. Linen is made from twisted long staple flax fibers.

The basic structural unit of cotton and flax is cellulose. The skeleton of the cellulose chain consists of successive β-linkages of glucose rings. Since it decomposes before it melts, cellulose cannot be melt spun, nor can it be dissolved in water or other common solvents.

Almost all polyester sutures are made of poly(ethylene terephthalate) (PET), which is produced by polymerization of ethylene glycol and terephthalic acid (see *Polyesters and Polyamides*). Polybutester, a block copolymer of poly(butylene terephthalate), served as the hard segment and poly(tetramethylene ether) glycol terephthalate served as the soft segment has also been used as a monofilament polyester suture (Novafil). The ratio of the hard and soft segments can be adjusted to achieve the desirable handling properties. Novafil suture has the ratio of hard to soft segments of 84–16% (Rodeheaver et al. 1987).

The molecular weight of PET that is required for making fibers is of the order of 20 000. Unlike the simple zigzag structure of polyethylene and nylon, the PET molecule has an extended rectilinear shape, the plane of the benzene rings being parallel to the 100 plane. The chain is rigid and less flexible than nylon and polyethylene. The melt-spun filaments are thus largely amorphous before drawing. The stereochemical structure of the PET chain changes during crystallization. The physical structure of PET is also winding-speed dependent during the melt spinning process (Heuvel and Huisman 1978).

Table 1
Mechanical properties of suture materials[a]

Suture (2/0)	Yield stress (GPD)	Breaking stress (GPD)	Yield strain (%)	Breaking strain (%)	Modulus of Elasticity (GPD)	Specific work of rupture (cN tex^{-1})
Dexon	0.80	6.30	1.9	22.6	55.0	6.63
Vicryl	0.97	6.55	1.8	18.4	67.5	5.46
Mersilene	1.20	4.20	2.7	8.0	53.0	1.32
Silk	1.33	3.43	1.9	11.5	79.0	2.36
Nurolon	0.34	3.80	1.6	18.2	21.0	2.80
Ethilon	0.41	6.25	2.2	33.0	20.0	8.96
Prolene	0.53	5.14	1.2	42.0	58.5	14.69

a from Chu (1981a)

Nylon is basically polymerized either from poly-condensation or through a ring-opening polymerization. Only nylon 66 and nylon 6 are used to make suture materials. Nylon 66 is made from adipic acid and hexamethylene diamine. It can also be made by interfacial polycondensation. Nylon 6 is made from caprolactam through ring-opening polymerization or by an anionic process. Whichever polymerization method is used, Nylon 6 contains about 10% extractable monomers and cyclic oligomers which act as plasticizers.

Unlike PET, both nylon fibers are fairly crystalline when they are spun. The crystal structures of nylon 66 and nylon 6 fall into three categories: α, β and γ phases. Nylon 66 exists mainly in the α phase, and is characterized by the progressive stacking of planar sheets (parallel to *ac* plane). The molecules in the crystalline region are in the fully extended zigzag conformation. The structure is triclinic, with one chemical repeat per unit cell.

Unlike nylon 66, hydrogen-bonded sheets of the γ phase of nylon 6 involve adjacent molecules arranged in opposite directions (antiparallel) so that all the hydrogen bonds are free of strain. The chains assume a puckered or pleated conformation.

Polypropylene suture materials are made from isotactic polypropylene, which is polymerized from propylene with a Ziegler–Natta catalyst. This catalyst consists of a transition metal halide (e.g., trialkyl aluminum). The stereoregularity of the resulting polypropylene is controlled by the propagation step, which consists of repeated insertion of the propylene monomer into the carbon–transition metal bond through the polarization mechanism. The resulting isotactic polypropylene has a molecular weight of about 80 000, and is melt spun and drawn into fiber form like polyamides and polyesters. In the isotactic form, the chain twists into the form of a helix with three chemical repeat units per turn and packs into a monoclinic unit cell. The helix form permits the best intermolecular and intramolecular packing of side groups.

Gore-Tex suture is manufactured from PTFE that has been expanded to produce a porous microstructure with about 50% porous volume. Due to the repulsion by fluorine atoms, the chain assumes a twisted zigzag conformation similar to a very loose helical structure with a repetition every 13 carbon atoms.

3. Mechanical Properties of Suture Materials

3.1 Tensile Properties

Detailed reports on the tensile properties of suture materials such as stress–strain curves, elastic moduli, work of rupture, and stress-relaxation are sparse (Chu 1981a, Holmlund 1976). These stress–strain data reveal a literal and overall discrimination of suture materials that is difficult to obtain by breaking strength and size specification alone. The results of Chu's study are summarized in Table 1.

It is evident that a wide range of stress–strain relationships were observed in the seven commonly used sutures. Suture materials can be grouped according to these mechanical properties. On the basis of yield stress, Nurolon, Ethilon and Prolene have the lowest, Dexon and Vicryl are in the middle, while Mersilene and silk have the highest. When grouped on the basis of breaking stress, Dexon, Vicryl and Ethilon are the strongest, Prolene is in the middle, while silk, Nurolon and Mersilene are the weakest. The order of increasing modulus of elasticity is Ethilon, Nurolon, Mersilene, Dexon, Prolene, Vicryl and silk; the values are in the range 20–79 grams force per denier (GPD) (where the denier is a weight unit of fibers in g per 9000 m). Dexon, Mersilene and Prolene have close moduli of elasticity. There is a wide difference in specific work of rupture among the seven sutures, and the values range from 1.32 cN tex^{-1} for Mersilene to 14.69 cN tex^{-1} for Prolene, more than a tenfold difference (tex defined as the weight in g per 1000 m fiber or yarn). Specific values of the work of rupture may exhibit a fourfold difference for sutures of the same chemical nature, but different geometrical configuration, such as nylon.

The data cited in Table 1 strongly suggest that sutures having a few similar or even identical mechanical properties are actually very different from each other in terms of overall performance. The ultimate solution for choosing suture materials of the right mechanical properties for closing specific wounds, might be to select a material whose stress–strain curve has the best match with the stress–strain curve of the tissue to be sewn. Of course, other properties such as tissue reactions, knot security and degree of absorption should also be considered, along with the mechanical properties of the suture.

Among the newly available sutures, Gore-Tex suture has a knot-pull tensile strength of 3.5 kg for a size CV-2 suture. Hence, monofilament Gore-Tex suture is stronger than monofilament polypropylene suture (Prolene), but weaker than monofilament nylon suture (Ethilon). Polybutester suture has breaking stress similar to polypropylene (Surgilene) suture and breaking elongation similar to nylon (Dermalon) suture. The new synthetic monofilament absorbable suture, poly(glycolide–trimethylene carbonate) has tensile strength quite similar to PGA and Polyglactin 910 sutures.

The retention of tensile strength and the rate of tensile strength loss in suture materials are vital factors in their usefulness in wound healing. Of the absorbable suture materials, Polyglactin 910 sutures retain better tensile strength than chromic catgut and PGA during the initial 14 days *in vivo*; PGA is slightly inferior to chromic catgut sutures during this period, and remains so until 21 days. Chromic catgut retains the best tensile strength after 15 days. It is generally recognized that the results of catgut are erratic. In contrast, Katz and Turner (1970) found that PGA sutures were two to three times stronger than the values obtained with medium chromic catgut at seven days and 11 days in all sizes. In studies *in vivo* in rat subcutis, the new synthetic absorbable suture material PDS retained an average of 58%, 41% and 14% of its original strength at four, six and eight weeks, respectively (Ray et al. 1981). Therefore, PDS is one of the slowest absorbable suture materials, by far. The newly commercially available poly(glycolide–trimethylene carbonate) synthetic absorbable suture showed a similar level of strength loss to PDS sutures. It retained 81%, 59% and 30% of its original values at implantation for two, four and six weeks, respectively, in the abdominal subcutaneous tissues of rats.

Of the nonabsorbable suture materials, braided polyester and monofilament polypropylene sutures have the best retention of tensile strength. There is virtually no loss of strength during a period of 390 days. Braided silk sutures, however, lost almost all of their strength in the same period, and this loss is more severe during the first 56 days (65% of original tensile strength lost). No strength could be measured after two years. Cotton sutures lose approximately half their tensile strength by the end of the first year, but the tensile strength remains more or less constant during the second year.

The degradation of nylon sutures has been reported by many investigators. After an initial drop of strength, they remain relatively constant thereafter. The percent strength remained ranged from as low as 20% to as high as 75% in nylon 66, and from 75% to 90% in nylon 6 sutures. Monofilament nylon sutures are slightly better than multifilament nylon. Furthermore, nylon 6 appears to retain strength slightly better than nylon 66 sutures. Morphological changes of nylon sutures, due to degradation when used in ophthalmology, have been reported, and transverse surface cracks have also been found.

Although polypropylene sutures show no sign of strength loss in most human tissues, they degrade in human eyes after a period of several years. Scanning electron microscopy observations reveal both the surface cracks perpendicular to the fiber axis of the suture and the fibrillar nature of the subsurface layer. Factors, such as enzymes, ultraviolet light, sterilization process and mechanical stress, have been suggested to be responsible for this morphological change of the suture. It is believed that synergestic effect might exist among those factors.

3.2 Knot Strength and Security

A knot is the weakest point in any suture material. Numerous wound disruptions are due to either the breakdown of the knot or the slippage and untying of the knot. Thus, knot strength and security play an important role in successful use of suture materials in wound healing.

A classification system for the knot of sutures has been proposed by Tera and Aberg (1976). It was found that both multifilament and monofilament stainless steel sutures have the best average knot efficiency for all 12 types of knots, while silk (all types) and polyethylene sutures have the poorest efficiency. Catgut (plain and chromic), Dexon, Mersilene and Prolene sutures are in the middle. Knot efficiency is independent of knot type, for steel and catgut sutures (plain and chromic) only. Irrespective of the nature of the suture materials, an increase in both turn and throw, increases the efficiency of the knot. In terms of knot type, parallel knots, on average, are stronger than crossed knots. A selection of knots for various suture materials is shown in Table 2.

With regard to knot security, both chromic and plain catgut sutures exhibit no slippage, regardless of knot type. Multifilament and monofilament stainless steel sutures rarely slip. Mersilene (uncoated Dacron) and Dexon sutures slip in slightly more than a quarter of all instances. Silk, monofilament

Table 2
Suitable knots for various suture materials[a]

Suture Materials	Type of knot
Catgut (plain and chromic)	1 = 1 = 1
	or 2 × 2
Dexon	1 = 1 = 1
Steel wire	1 = 1
Silk and braided polyester	1 = 1 = 1
	or 2 × 2
Coated polyester	2 × 2
Monofilament nylon, polyolefin	2 × 2

a from Holmlund et al. (1978)

nylon and polyethylene sutures slip in more than two-thirds of cases. It is generally true that coated suture materials are less knot secure than uncoated ones. Knot security is also improved with an increase in the number of turns and throws (Tera and Aberg 1976), or by welding the knot by heat or laser.

Since a knot is held in place by frictional forces, its security, in turn, depends on the coefficient of friction of the fibers. The importance of friction in suture knot security, has been subject to several studies. Gupta and co-workers measured the frictional characteristics of eight suture materials by the equation $\mu = \pi n \beta (\ln T - \ln T_0)$, where n, β, T_0, and T are the number of turns of the twist, twist angle, initial tension, and the tension activating slippage, respectively. It was found that the coefficient of friction μ is a function of suture materials, their construction and applied tension. An increase in applied tension would decrease μ. At low tension, monofilament polypropylene (Prolene) and nylon (Ethilon) sutures have the highest μ, while their μ became the smallest at high applied tension. The results of Gupta et al. further the general belief, held by surgeons many decades ago, that a braided uncoated suture has a better knot security than a monofilament coated suture. Welding the two ends of a suture knot by heat has been suggested to improve the security of the knot.

4. Tissue Response

A knowledge of tissue responses to suture materials is important because the wound healing process, its rate and complications are greatly influenced by the tissue reactions to suture materials. An excessive tissue reaction around suture materials could result in:

(a) cutting out sutures,

(b) delay in healing,

(c) granuloma and sinus formation, and

(d) predisposition to infection.

Cellular response to both absorbable and non-absorbable suture materials generally involves the formation of a band of fibrous connective tissue encapsulating the suture material. The width of the band and the amount of surrounding macrophage and giant cells are related to the severity of the tissue reaction.

4.1 Absorbable Suture Materials

Because of its foreign protein nature, catgut sutures elicit a far more intense tissue reaction than synthetic absorbable and synthetic nonabsorbable sutures. Plain catgut produces the worst reaction associated with an early, prolonged and marked exudate, with some tissue necrosis. Chromic catgut causes a less pronounced, but nonetheless persistent, tissue reaction. After complete absorption, it is replaced by a dense mass of macrophages. The absorption and degradation of catgut sutures involve mainly proteolytic enzymes of the phagocytes and other cells.

Both PGA suture materials and those of its lactide copolymer Polyglactin 910, elicit minimal tissue reactions (Katz and Turner 1970, Craig et al. 1975). The tissue reaction to PGA lasts substantially longer than that associated with Polyglactin 910. The latter is absorbed by 90 days, while considerable quantities of PGA remain at 120 days. Hydrolysis is apparently the chief mechanism of their degradation. Williams (1980) has demonstrated that certain enzymes and bacteria are able to influence the rate of hydrolysis of PGA under the appropriate conditions.

Like PGA and Polyglactin 910, PDS suture materials elicit minimal foreign-body reactions for periods of up to 168 days postimplantation in rat subcutis (Ray et al. 1981). After absorption, a few enlarged macrophages or fibroblasts were sometimes present between normal muscle cells.

The new poly(glycolide–trimethylene carbonate) monofilament absorbable suture elicits minimal tissue reaction and its absorption is achieved through the action of mononuclear and multinuclear macrophages. The suture site is replaced by a narrow band of fibrous tissue and collagen after the complete absorption of the suture, which takes about six to seven months.

4.2 Nonabsorbable Suture Materials

All nonabsorbable suture materials elicit less tissue reaction than catgut. In this group, nonsynthetic suture materials, such as silk, cotton and flax, induce somewhat stronger tissue reactions than synthetic ones. This intense interaction of silk sutures is due to the combined effects of physical configuration (braid) and chemical constituents (protein) of silk.

Synthetic nonabsorbable suture materials elicit far less tissue reaction than do the corresponding natural suture materials. Among the synthetic materials, multifilament configurations cause more tissue response than monofilament ones. By comparison, the tissue response to polypropylene monofilament suture is minimal. Monofilament nylon 66 and nylon 6 sutures behave very much like polypropylene sutures. Monofilament Gore-Tex suture shows a quite different degree of tissue reaction due to its porous structure. The internodal space of the suture permits the incorporation of fibroblasts and other inflammatory cells. As a result, firm tissue attachment between the suture and its surrounding tissue can be achieved. This attachment is beneficial for closing internal wounds and for fastening surgical implant. It, however, should not be used in skin or where a suture has to be removed, because the infiltration of tissue into the suture makes the removal of the suture difficult.

Although monofilament sutures, in general, elicit minimal tissue reaction in most tissues, they have been found to cause some complications when used in ophthalmic surgery due to their exposed suture ends in knots. A variety of symptoms, such as foreign-body sensation, pain, contact lens intolerance, giant papillary conjunctivitis, tarsal ulceration, conjunctival granuloma corneal infiltrate, and corneal vascularization, have been reported after cataract surgery, corneal transplantation, and pars plana vitrectome. Direct mechanical irritation of the surrounding tissue because of the stiff protruding ends of a monofilament suture, is thought to be responsible for the adverse tissue reactions. The reactions are immune nature. Multifilament nylon sutures induce a slightly greater tissue response than monofilament nylon. Uncoated braided polyester sutures exhibit a slightly wider connective tissue zone than monofilament nylon and polypropylene sutures. Teflon-coated braided polyester sutures differ from uncoated polyester sutures after only about one month. Finally, stainless steel sutures cause very little tissue response.

5. Wound Infection

It is important to know whether or not the presence of a suture material in a surgical wound increases the susceptibility of the wound to infection, and, if so, how it varies with different suture materials.

Alexander et al. (1967) found no significant difference in the severity of infection by *Staphyloccus aureus* among multifilament nonabsorbable sutures made of silk, braided nylon, twisted cotton and stainless steel. Sutures in multifilament form resulted in higher wound infection than the same suture in monofilament form. Twisted-silk sutures were associated with producing more infection than braided silk. This demonstrates that the physical structure of the nonabsorbable sutures is more important in inducing wound infection than the chemical constituents.

Edlich et al. (1973) observed both plain and chromic catgut sutures induce higher incidences of gross infection than PGA. Among nonabsorbable sutures,

braided nylon sutures have significantly lower infection rates in contaminated tissues than other multifilament nonabsorbable sutures, with polyester sutures the next best. Stainless steel sutures have a potentially greater incidence of gross infection than nylon and polyester sutures, but significantly less than silk and cotton sutures, which are the poorest in this group. Edlich et al. found that the physical structure of a suture is less important in enhancing infection than the chemical nature. This conclusion is different from the findings of Alexander et al.

The preferential adherence of bacteria to suture materials has been suggested to be associated with wound infection (Sugarman and Musher 1981, Bucknall and Ellis 1983, Chu and Williams 1984). Chu and Williams reported a wide range of bacterial adherence to suture materials, dependent on the type of materials and bacteria, and on the duration of contact; findings also confirmed in Sugarman and Musher's study. Chu and Williams also suggested that the process of bacterial attachment on a suture surface was a dynamic phenomenon and reversible. It has been reported that, when studied with knots, the different amounts of bacteria recovered from mouse wound were not significant among monofilament polypropylene, silicone and Teflon-coated braided polyester sutures.

Because of the close association between the bacterial adherence of sutures and wound infection, impregnation of sutures with a variety of antibacterial agents, such as neomycin palmitate, penicillin, sulfonamide, and disinfectants (e.g., chlorhexidine and iodine), have been investigated (Echeverria and Olivares 1959). They have achieved some degree of success in reducing wound infection in animals; however, their results are far from perfect for reasons such as the shortness of the effective domain and duration, and narrowness of spectrum of the antibacterial effect. A new type of approach for making antibacterial suture materials has been reported (Chu et al. 1987). A silver-compound coated nylon suture was tested *in vitro* against seven types of bacteria under the influence of weak direct current (0.4–400 μA). It was found that the new suture exhibited very good-to-moderate bactericidal property towards these seven bacterial species. The application of direct current through the suture positively enhanced their antibacterial property, and the degree of enhancement depended on the current level. The suture material elicited a lesser degree of inflammatory reaction than the commercial braided nylon suture (Nurolon) up to 60 days of postimplantation in rat gluteal muscle.

6. Carcinogenicity

Suture materials cannot only be vehicles for the mechanical transplantation of tumor cells, but can also interact biologically with the surrounding tissue and exert influence on cancer cells to determine the likelihood of their growth. The degree of potentiation of tumor growth by a suture material appears to parallel its effect in wound infection.

All tested sutures (chromic catgut, PGA, silk, nylon and steel) potentiate early tumor occurrences and growth when the inoculating dose is 10^5 cells per 0.2 ml of solution, or above. With the normal subclinical cell dose, namely 10^3 cells per 0.2 ml of solution, only silk and steel sutures consistently increase the rapidity of onset and the total number of tumors. In comparison, chromic catgut and PGA do not increase tumor occurrence at this cell dose level. Monofilament nylon suture is the only one which does cause early occurrence or a greater number of tumors.

A foreign body such as silk suture in the bladder wall of rats, significantly promotes but does not initiate the development of invasive tumors compared with the control bladder walls, which are originally exposed to a carcinogen only. The exact mechanism of this growth promotion by silk sutures remain unknown.

7. Thrombogenicity

A study of five suture materials (Prolene, Mersilene, Vicryl, silk and Ethilon) using scanning electron microscopy has shown that polypropylene is the most blood compatible material and silk is the most thrombogenic. Nylon 66 showed deposits of thrombocytes and fibrin during the first three postimplantation weeks. Uncoated polyester sutures exhibited a fairly satisfactory antithrombogenicity, Vicryl sutures showed a thrombogenicity similar to that of Mersilene at three days, and were covered extensively by thrombocytes and erythrocytes at seven days and thereafter.

8. Biodegradation of Synthetic Absorbable Sutures

The degradation mechanism of PGA *in vitro* has been proposed on the basis of a microfibrillar model of fiber structure (Chu 1981b). On the basis of this model, it is believed that degradation proceeds through two main stages, with the first stage in the amorphous region and the second in the crystalline region. Hydrolytic degradation starts in the amorphous regions, as the tie-chain segments, free chain ends, and chain folds in these regions degrade into fragments. As the degradation proceeds, the size of the fragments decreases to that which can be dissolved into the buffer medium, resulting in loss of material. As sufficient amounts of chain segments in these regions are removed, the spaces originally occupied by chain segments become vacant and large enough to be visible as cracks.

When all the amorphous regions have been removed by hydrolysis, the second stage of degradation starts. The degree of crystallinity reaches a maximum at the end of the first-stage degradation, and starts to decrease as hydrolysis destroys the crystalline lattice. During this first period, the loss of tensile strength is mostly due to the scission of tie-chain segments. This proposed degradation mechanism was recently confirmed by others in the study of glycolide and lactide copolymers (Fredericks et al. 1984). By using a chemical means, the existence and characteristics of the previously proposed two-stage degradation mechanism has also been demonstrated.

The hydrolytic degradation of synthetic absorbable sutures has been found to be influenced by the nature of the materials (i.e., morphology and chemical structure) and the environment (i.e., pH, temperature, enzymes, bacteria, etc.).

Browning and Chu (1986) examined the use of an annealing method to alter the hydrolytic degradation phenomena of PGA suture, because annealing could alter the ratio of crystalline-to-amorphous regions of the suture. It was found that annealing treatments, in general, did alter the mechanical properties of PGA sutures, as well as altering their degradation properties. Most of the annealing treatments resulted in lower tenacity and breaking elongation, when compared with the control samples. PGA sutures that have been exposed to any level of axial tension during annealing, however, exhibited more resistance towards hydrolysis than was shown by the freely hung specimens.

The role of enzymes in synthetic absorbable sutures degradation is controversial. Salthouse and Matlaga (1976) reported that Polyglactin 910 degradation was independent of enzyme activity, and the same conclusion was found by Ray et al. (1981) on the study of PDS sutures. However, in a series of studies it has been demonstrated that certain enzymes, under some conditions, can influence the degradation of PGA and PDS sutures (Chu and Williams 1983, Williams and Chu 1984). They also reported that enzymes such as trypsin, which did not affect PGA sutures initially, could begin to influence them after alterations to their physical and chemical structures by γ irradiation (Chu and Williams 1983).

The possible role of enzymes in the hydrolysis of synthetic absorbable sutures was further demonstrated in the study of the effect of enzymatic wound cleaning surgical practice on PGA sutures (Persson et al. 1986). By using urine as a model, Holbrook (1982) found that the addition of porcine esterase into a control urine did accelerate the breakdown of PGA suture; but amylase, hyaluronidase and urokinase failed to show enzymatic catalytic degradation.

Other factors, such as pH of the buffer medium, the external force that a suture experiences during the process of hydrolysis, the presence of bacteria, lipids have also been reported to influence the hydrolytic degradation of synthetic absorbable sutures, particularly PGA sutures (Chu 1982, Bucknall 1983, Millar and Williams 1984, Sebeseri et al. 1985).

9. Sterilization of Suture Materials

Basically, suture manufacturers use two sterilization methods: ionizing radiation and ethylene oxide treatment. The former is ^{60}Co γ irradiation, and a dose of 25 kGy is usually employed. The advantages of ionizing radiation are speed, convenience and reproducible control. The disadvantage is the change of properties after irradiation. The change of mechanical properties due to either chain scission or cross-linking is the most important concern in suture use. The only absorbable suture material that is sterilized by ionic radiation is catgut. All the synthetic nonabsorbable suture materials except polypropylene are also sterilized by γ irradiation.

For those polymeric materials that are severely damaged by radiation sterilization, ethylene oxide gas is used for sterilization. The major disadvantages are the length of time required and the residual. A minimum concentration of 450 mg l^{-1} is recommended. In addition, a period of "sitting" is required after exposure to the gas, in order to remove the gas trapped inside the material. PGA, Polyglactin 910, poly(glycolide–trimethylene carbonate), polypropylene and cotton are the suture materials which are sterilized by this gaseous method.

Bibliography

Alexander J W, Kaplan J Z, Altemeier W A 1967 Role of suture materials in the development of wound infection. *Ann. Surg.* 165: 192–9

Browning A, Chu C C 1986 The effect of annealing treatments on the tensile properties and hydrolytic degradative properties of polyglycolic acid sutures. *J. Biomed. Mater. Res.* 20: 613–32

Bucknall T E, Ellis T H 1983 The choice of a suture to close abdominal incisions. *Eur. Surg. Res.* 15: 59–66

Chatani Y, Suehiro K, Okita Y, Tadokoro H, Chujo K 1968 Structural studies of polyesters, I. Crystal structure of polyglycolide. *Makromol. Chem.* 113: 215–29

Chu C C 1981a Mechanical properties of suture materials: An important characterization. *Ann. Surg.* 193: 365–71

Chu C C 1981b Hydrolytic degradation of polyglycolic acid: Tensile strength and crystallinity study. *J. Appl. Polym. Sci.* 26: 1727–34

Chu C C 1982 A comparison of the effect of pH on the biodegradation of two synthetic absorbable sutures. *Ann. Surg.* 195: 55–9

Chu C C 1983 Survey of clinically important wound closure biomaterials. In: Szycher M (ed.) 1983 *Biocompatible Polymers, Metals, and Composites.* Technomic, Lancaster, PA

Chu C C, Tsai W C, Yao J Y, Chiu S S 1987 Newly-made antibacterial braided nylon sutures. I. *In vitro* qualitative and *in vivo* preliminary biocompatibility study. *J. Biomed. Mater. Res.* 21: 1281–1300

Chu C C, Williams D F 1983 The effect of gamma irradiation on the enzymatic degradation of polyglycolic acid absorbable sutures. *J. Biomed. Mater. Res.* 17: 1029–40

Chu C C, Williams D F 1984 Effects of physical configuration and chemical structure of suture materials on bacterial adhesion—A possible link to wound infection. *Am. J. Surg.* 147: 197–204

Craig P G, Williams J A, Davis K W, Magoun A D, Levy A J, Bogdansky S, Jones J P 1975 A biological comparison of polyglactin 910 and polyglycolic acid synthetic absorbable sutures. *Surg., Gynecol. Obstet.* 141: 1–10

Echeverria E, Olivares J 1959 Clinical and experimental evaluation of suture material treated with antibiotics. *Am. J. Surg.* 98: 695

Edlich R F, Panek P H, Rodeheaver G T, Turnbull V G, Kurtz L D, Edgerton M T 1973 Physical and chemical configuration of sutures in the development of surgical infection. *Ann. Surg.* 177: 679–88

Frazza E J, Schmitt E E 1971 A new absorbable suture. *Biomed. Mater. Symp.* 1: 43–58

Fredericks R J, Melveger A J, Dolegiewitz L J 1984 Morphological and structural changes in a copolymer of glycocide and lactide occurring as a result of hydrolysis. *J. Polym. Sci., Polym. Phys. Ed.* 22: 57–66

Holbrook M C 1982 The resistance of polyglycolic acid suture to attack by infected human urine. *Br. J. Urol.* 54: 313–315

Holmlund D, Tera H, Wiberg Y, Zederfeldt B, Aberg C 1978 *Sutures and Techniques for Wound Closure,* Naimark and Barba, New York

Katz A R, Turner R J 1970 Evaluation of tensile and absorption properties of polyglycolic acid sutures. *Surg., Gynecol. Obstet.* 131: 701–16

Miller N D, Williams D F 1984 The *in vivo* and *in vitro* degradation of poly(glycolic acid) suture material as a function of applied strain. *Biomaterials* 5: 365

Persson M, Bilgrav K, Jensen L, Gottrap F 1986 Enzymatic wound cleaning and absorbable sutures: An experimental study on Varidase and Dexon sutures. *Eur. Surg. Res.* 18: 122–7

Ray J A, Doddi N, Regula D, Williams J A, Melveger A 1981 Polydioxanone (PDS), a novel monofilament synthetic absorbable suture. *Surg., Gynecol. Obstet.* 153: 497–507

Reed A M, Gilding D K 1981 Biodegradable polymers for use in surgery—Poly(glycolic)/poly(lactic acid) homo and copolymers: 2. *In vitro* degradation. *Polymer* 22: 494–8

Salthouse T N, Matlaga B F 1976 Polyglactin 910 suture absorption and the role of cellular enzymes. *Surg., Gynecol. Obstet.* 142: 544–50

Sebeseri O, Keller U, Spreng P, Schull R T, Zingg F 1985 The physical properties of polyglycolic acid suture (Dexon) in sterile and infected urine. *Invest. Urol.* 12(6): 490–6

Sugarman B, Musher D 1981 Adherence of bacteria to suture materials. *Proc. Soc. Exp. Biol. Med.* 167: 156–60

Tera H, Aberg C 1976 Tensile strength of twelve types of knot employed in surgery, using different suture materials. *Acta Chir. Scand.* 142: 1–7

Williams D F 1980 The effect of bacteria on absorbable sutures. *J. Biomed. Mater. Res.* 14: 329–38

Williams D F, Chu C C 1984 The effects of enzymes and gamma irradiation on the tensile strength and morphology of poly(p-dioxanone) fibers. *J. Appl. Polym. Sci.* 29: 1865–77

C. C. Chu
[Cornell University, Ithaca, New York, USA]

T

Tantalum and Niobium

Tantalum was put forward as a suitable surgical implant material by Burke in 1940, mainly because the corrosion resistance of the metal led him to suppose that it would be extremely biocompatible. Until recently, however, the mechanical properties of the metal have restricted its application to diagnostic aids and low-stressed implants such as wires, staples and pliable sheets (Mears 1979). Niobium is a very similar metal.

1. Occurrence and Recovery

Tantalum (Ta) was discovered in 1802 by A K Ekeberg in Swedish ores: he named it after the mythological Greek king Tantalus, because of the "tantalizing" difficulties in dissolving the metal oxide in acids.

Niobium (Nb) was discovered in 1801 by W Hatchett and first named columbium (Cb) after the origin of the ore. It was subsequently rediscovered by H Rose in 1844, who called it niobium after Tantalus' daughter Niobe to indicate the close relation of the metal to tantalum. This name was internationally accepted in 1949.

Both metals are always found together, and the principal ores are called columbite and tantalite, depending on the prevalence of the respective metals. The natural abundance of tantalum in the earth's crust is 5 ppm, similar to that of uranium, whereas the abundance of niobium is 65 ppm, similar to that of copper and tungsten. This difference in availability is also reflected in the price of the metals.

The concentrates obtained from the ores are dissolved in hydrofluoric acid, and the mixed solutions of tantalum and niobium fluorides are either processed to master alloys or separated to pure metal powders by electrolytic or chemical purification.

2. Chemical Composition

Both tantalum and niobium exhibit excellent corrosion resistance and good mechanical properties only in a highly pure state (99.90% minimum). The requirements for surgical implant applications of tantalum, and the average impurity limits of commercially available tantalum and niobium are listed in Table 1.

3. Physical Properties

Tantalum and niobium belong to the fifth group in the periodic table of elements and are therefore refractory, highly reactive base metals. Their crystal structure is body-centered-cubic. The most important physical properties are listed in Table 2.

Table 1
American Society for Testing and Materials (1978) requirements for tantalum surgical implants and average chemical composition of commercially available high-purity tantalum and niobium[a]

Element	Tantalum surgical implants (ppm)	High-purity tantalum (ppm)	High-purity niobium (ppm)
Carbon	100	15	30
Oxygen	150	20	30
Nitrogen	100	5	20
Hydrogen	10	<5	<5
Niobium	500	150	balance
Tantalum	balance	balance	6000
Iron	100	120	200
Titanium	100	<5	20
Tungsten	300	200	150
Molybdenum	100	150	300
Silicon	50		
Nickel	100	30	150

a courtesy of Metallwerke Plansee, Reutte, Austria

Table 2
Physical properties of tantalum and niobium at 20 °C

Properties	Unit	Tantalum	Niobium
Mass			
atomic weight		180.95	92.91
density	$Mg\,m^{-3}$	16.60	8.57
Thermal properties			
melting point	°C	2996	2468
specific heat (at 100 °C)	$kJ\,kg^{-1}\,K^{-1}$	0.14	0.27
thermal conductivity	$W\,m^{-1}\,°C^{-1}$	54.4	52.7
coefficient of linear thermal expansion	$\mu m\,m^{-1}\,K^{-1}$	6.5	6.9
Electrical properties			
electrical conductivity	% IACS	13	13.2
electrical resistivity	$n\Omega\,m$	135.0	142.0
Nuclear properties			
isotopes		^{181}Ta ^{182}Ta ^{178}Ta[a]	^{95}Nb
Thermal neutron cross section (at 220 m s^{-1})	b	21.3	1.1

a half-life 9 min

4. Manufacturing Techniques and Workability

Tantalum, niobium and their alloys are not suitable for casting owing to their high melting points. They can be compacted either using powder metallurgical techniques (compression of the powder and sintering at 80–90% of the kelvin melting temperature under a high vacuum), or by arc melting or electron-beam melting under high-vacuum conditions.

A high-vacuum or an inert gas atmosphere are essential when manufacturing or working tantalum and niobium at higher temperatures, to avoid contamination with interstitial elements such as oxygen, hydrogen, nitrogen and carbon. Even small quantities of interstitial impurities result in some increase in strength, but produce a loss in ductility and workability. Alloying with vanadium, tungsten, molybdenum, zirconium, hafnium or titanium also results in an increase of ultimate strength, but without any great loss in ductility. The same effect can be achieved by dispersion hardening (e.g., by mixing the powder with TiO_2 hard-phase powder) or by heat treatment in a low-pressure gas atmosphere.

Pure tantalum and niobium remain ductile at very low temperatures, resulting in excellent workability. Sheet, foil, rod, wire and special parts can be formed using all conventional cold-working methods. The final product can easily be machined, even under operating-room conditions (Mears 1979).

Welding can be carried out using the tungsten inert gas method (TIG welding), electron-beam welding under vacuum (10^{-2} Pa min), or spot or laser welding.

It is essential that the metals are chemically cleaned before any heating operation such as annealing, heat treatment or welding, to avoid contamination by interstitial elements and metallic impurities. Thorough degreasing with detergents and pickling or electropolishing produce clean and highly polished surfaces. The self-passivating oxide film on the surface (see Sect. 6) can be reinforced by electrochemical anodic oxidation, the thickness corresponding to the voltage used and correlated with a certain interference color.

5. Mechanical Properties

The mechanical properties of tantalum and niobium are summarized in Table 3.

After annealing, pure tantalum and niobium are very soft and ductile, similar to pure iron and soft copper. In this condition they are suitable for low-stressed implants such as pliable sheets, nets, wires and staples.

The necessary increase in strength for load-bearing devices can be obtained over a wide range by cold working, but usually with a loss in ductility (see Fig. 1). Ultimate strengths of up to 1000 MPa and above can be achieved by dispersion hardening (see Fig. 1) or alloying without any great loss in ductility.

Satisfactory fatigue strength acquires even greater importance in heavy load-bearing applications. The *Metals Handbook* (1979) only contains sparse data on the fatigue strength of wrought tantalum. Preliminary investigations (Schider and Bildstein 1982) employing ultrasonic resonance loading in inert and corrosive media have shown that cold-worked tantalum attains the same range of fatigue strengths at a given elongation as the best cobalt-base alloys (see Fig. 2). However, a ratio of fatigue strength to ultimate strength of 0.6–0.7 can be obtained with tantalum and niobium, whereas this ratio is only about 0.5 for cobalt-base alloys. Thus, after appropriate manufacture, tantalum and niobium can be used for high-stressed implants despite their lower ultimate strengths: this is an alternative approach to the ultrahigh-strength alloys currently in use. What is more, the fatigue behavior of tantalum and niobium does not change significantly in a corrosive environment and both metals show a very low notch sensitivity and tendency to crack propagation. These are beneficial properties for prostheses that are subjected to conditions in the body that can lead to implant failure that originates at the implant surface.

The elastic moduli of tantalum and especially that of niobium are closer to that of bone than cobalt-based alloys (see Table 2), and this coupled with excellent biocompatibility will make it possible to

Table 3
Mechanical properties of differently worked tantalum and niobium

Properties	Tantalum			Niobium		
	annealed		cold-worked	annealed		cold-worked
Vickers hardness (HV 10)	80–110		120–300	60–110		110–180
Elastic modulus (10^3 MPa)		186–191			103–116	
Ultimate tensile strength (MPa)	200–300		400–1000	275–350		300–1000
Elongation (%)	20–50		1–25	25–40		1–25

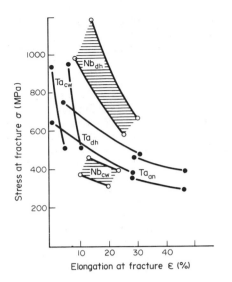

Figure 1
Ranges of ultimate tensile strength for differently
processed tantalum and niobium wires of diameter
0.8 mm to 1.5 mm: cw, cold-worked; an, annealed;
dh, dispersion-hardened. These values are
representative of what can be achieved with tantalum
and niobium and correspond to the ranges shown in
Table 3 (courtesy of Metallwerke Plansee, Reutte,
Austria)

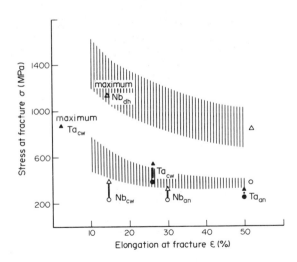

Figure 2
Ranges of ultimate strength △ and fatigue strength ○
of differently processed tantalum and niobium
specimens in ultrasonic resonance loading tests
(Schider and Bildstein 1982). The shaded areas are
the ranges for differently processed PROTASUL-10

design implants directly attached to the skeletal
system.

The friction behavior of tantalum and niobium
makes them unsuitable for use at articulating
surfaces.

6. Electrochemical Properties and Corrosion Behavior

Tantalum and niobium are highly reactive metals:
owing to their reactivity they are immediately
covered by a dense protective oxide layer in biological and most other environments. This passivating
film makes the metals chemically inert and imparts
excellent corrosion resistance to almost all chemical
attack except that of hydrofluoric acid, fuming sulfuric acid, and strong alkaline solutions and melts.
Corrosion resistance can even be improved by alloying with other refractory metals. In most cases,
tantalum exhibits better corrosion resistance than
niobium. Tantalum is able to repair the passivating
oxide film even in the absence of oxygen and this is
regarded as an essential condition for resistance to
crevice corrosion (Mears 1979). Even more important for certain implant situations, such as endosseous but also transmucosal dental implants, seems
the low current density, which tantalum and niobium share with titanium (Zitter and Plenk 1987)
and which should at least diminish negative influences of currents created by different redox
potentials or other galvanic elements in such a
situation.

This passivating oxide film can be reinforced by
anodic oxidation and has semiconductive properties,
which makes both metals suitable for stimulating
electrodes using both capacitive and inductive
means. Apart from titanium- and zirconium-base
materials, only tantalum and niobium can be combined with other metals without breaking down this
passive film (Mears 1979). Finally, tantalum may
become embrittled by hydrogen if it is used as a
cathode in a galvanic couple exposed to an acid
environment.

7. Biocompatibility

Since the biocompatibility of any metal depends
firstly on the rate of corrosion and the toxicity of the
metal ions (Williams 1981), the highly corrosion-
resistant inert metals tantalum and niobium are the
best candidate materials for biomedical applications.
When Burke (1940) introduced tantalum as a surgical implant material, he deduced from *in vitro* corrosion experiments and the undisturbed healing of
tissue around tantalum sutures in the body that the
metal would exhibit excellent biocompatibility. This
has been confirmed by numerous subsequent

reports, but some contradictory observations have also been published.

In the investigation by Laing et al. (1967) of the soft tissue response to different metallic implants, tantalum obtained only average ratings. Columbium and titanium got the best ratings, and it is interesting to note that columbium was rated much less favorably than niobium. This apparent contradiction is difficult to explain but material impurities may have accounted for these differences and also for the rapid *in vivo* corrosion and adverse reactions to tantalum. However, this problem has probably now been overcome by improved manufacturing techniques (Mears 1979). Surface preparation of tantalum also seems to play a role in the tissue response (Meenaghan et al. 1979).

Recent *in vitro* and *in vivo* studies have confirmed the excellent short- and long-term biocompatibility of pure tantalum and niobium (Zetner et al. 1980, Zitter and Plenk 1987). There is no significant growth inhibition of fibroblasts and tight bone contact develops around intraosseous implants, lasting up to a year under both "unloaded" and heavily loaded conditions. This direct bone contact has also been demonstrated for up to ten years in human dental implants made of tantalum (Grundschober et al. 1982, Schuh et al. 1988).

8. Toxicology

All the data relating to the toxic effects of tantalum and niobium (Luckey and Venugopal 1977, Oehme 1979) are derived from the investigations of Chochran et al. (1950) in rats.

The toxicity of tantalum depends on the solubility of the compound. Tantalum oxide is poorly absorbed and nontoxic perorally, and inhalation causes only transient inflammation. However, the cytotoxic effect of tantalum oxide on rabbit alveolar macrophages *in vitro* has been demonstrated (Matthay et al. 1978). Tantalum chloride, however, shows an LD_{50} (i.e., a median lethal dose fatal to 50% of test animals) of 985 mg kg^{-1} administered perorally, and inhalation causes alveolar thickening and histiocytosis. After parenteral administration (other than through the intestines) the LD_{50} is 38 mg kg^{-1} and the metal is deposited in the liver, kidneys and bones.

Niobium salts are well tolerated perorally, but parenterally showed a higher toxicity (LD_{50} of 13 mg kg^{-1}) than the salts of tantalum. Sublethal doses are severely hepatotoxic and nephrotoxic owing to the inhibition of succinic dehydrogenase. The metal is finally deposited in the bones and bone marrow.

Tantalum and niobium exhibit only an unspecific (surface) carcinogenic effect after implantation in animal tissue. No terratogenic effect is known.

There are no reports on the toxicity of tantalum and niobium in humans.

9. Biomedical Applications

Burke (1940) introduced tantalum for surgical implants such as sutures, bone screws and plates. Owing to the ductility and insufficient rigidity of the tantalum available until recently (Mears 1979), its application in the musculo-skeletal system has been confined to cerclage wire, nets to hold bone grafts in place, pliable sheets and plates for cranioplasty and reconstructive surgery, and implants for reconstruction of the ossicular chain in the ear. Most reports document an undisturbed healing reaction and an excellent long-term performance of the implants. The improvements in the manufacturing and cold-working techniques of tantalum and niobium now make it possible to produce even load-bearing implants for high-stress conditions (Schider and Bildstein 1982). Intramedullary nails with a special nonrotating profile have proved mechanically suitable and clinically very successful. Surface-grooved femoral shafts for total hip-joint replacements have been anchored in beagle dogs without the use of bone cement by direct bone contact for up to 13 months (Plenk et al. 1984).

Thus, the production of human joint endoprostheses that can be implanted without bone cement now seems possible. However, it is still necessary to prove the fatigue strength and consistent quality of the dispersion-hardened tantalum- and niobium-base materials which offer the highest ultimate strength while retaining some ductility. The emphasis should be placed on the niobium-base materials: these are new and more promising candidate materials for the manufacture of short- and long-term orthopedic implants, because they have half the specific weight and elastic modulus of tantalum and are more readily available.

It is believed that tantalum is still a suitable material for small highly stressed devices such as dental implants, where its corrosion resistance in the more aggressive oral environment and its excellent biocompatibility are of utmost importance. It has been used by Marziani for subperiosteal (in 1955) and screw-type endosseous implants (in 1965), by Scialom in 1962 for single and tripod needle implants, for Heinrich's "helicoidal screw" in 1971 and for the double blade vent implant after Herskovits in 1974 (Smith 1974, Schroeder 1974). While the clinical success of these implants has hitherto not been very convincing and tantalum is not recommended, or even mentioned, as a dental material in reviews, its suitability has now been demonstrated histologically by the long-term osseous anchorage of "helicoidal screws" (Grundschober et al. 1982) as well as of the double blade vent (Schuh et al. 1988). These favorable results have been attributed to the biomechanically adequate design of the components used to the biocompatibility of tantalum, making direct bone contact possible instead of the pseudoperiodontal ligament formation usually found around loaded metallic dental implants.

Tantalum implants have been used successfully in general surgery and neurosurgery as monofilament and braided suture wires for skin closure, tendon and nerve repair, as foils and sheets for nerve anastomoses, as clips for the ligation of vessels, bile duct, vas deferens and oviducts, as staples for anastomoses of the gut and as meshes for abdominal wall reconstruction after hernias. Good mechanical performance, undisturbed healing and resistance to infection are emphasized in all reports.

The excellent biocompatibility and electrochemical properties of tantalum are used for stimulating electrodes in cerebral and muscular tissue. Sintered tantalum capacitive electrodes with an anodized high surface area show no signs of corrosion *in vitro* and *in vivo*, and the physiological and histopathological results indicate that they are the safest electrodes yet tested (Johnson et al. 1977). While in previous studies tantalum has elicited the least reaction of all standard metallic implants in cerebral tissue, a more recent evaluation has shown a more pronounced mesodermal and foreign body reaction, which might be responsible for the rise in threshold which also occurs in tantalum pentoxide electrodes after prolonged capacitive stimulation.

The biocompatibility and inertness of tantalum coupled with its high density are the reasons for its widespread application as a diagnostic aid in radiology. Tantalum balls, pins and screws are implanted into the myocardium to monitor cardiac position, motion and function, or into bones to follow the growth of the skull by cephalometry and the possible migration of prostheses, employing specially developed roentgen stereophotogrammetric methods. Tantalum powder is mainly used for tracheography and bronchography of the respiratory tract. Owing to its high density, less material has to be inhaled than with other bronchographic materials. There is less interference with the long function and tantalum oxide powder was regarded as safe until its possible toxic effect on alveolar macrophages was observed (Matthay et al. 1978). Tantalum powder is also used for the retrograde filling of the biliary and urinary tracts, and in combination with cyanoacrylates and other polymers for transarterial embolization. Apart from radiodensity, this last application utilizes a possible thrombogenic effect, which has also been observed with tantalum rings in contact with the blood and used as markers in artificial heart valves.

Finally, tantalum and niobium isotopes are used as diagnostic aids and in radiotherapy. The short-lived [178]Ta and the [95]Nb microsphere method are used for cardiac imaging and blood flow monitoring. Inhalation of these isotopes in dogs served to investigate the fate of these bronchographic materials. [182]Ta interstitial implants have been used in the radiotherapy of head and neck tumors.

10. Sterilization Considerations

Tantalum and niobium implants can be sterilized either by steam sterilization or by γ radiation.

Acknowledgement

The considerable assistance rendered by the Technical Library of the 3M Company, St Paul, Minnesota, in the literature search for this article is gratefully acknowledged.

Bibliography

American Society for Testing and Materials 1978 *Standard Specification for Unalloyed Tantalum for Surgical Implant Applications*, ASTM F560-78. ASTM, Philadelphia, PA

Burke G L 1940 Corrosion of metals in tissues and an introduction to tantalum. *Can. Med. Assoc. J.* 43: 125–8

Chochran K W, Doull J, Mazur M, Dubois K B 1950 Acute toxicity of Zr, Cb, Sr, La, Ce, Ta and YHr. *Arch. Ind. Hyg.* 1: 637

Grundschober F, Kellner G, Eschberger J, Plenk H Jr 1982 Long-term osseous anchorage of endosseous dental implants made of tantalum and titanium. In: Winter G D, Gibbons D F, Plenk H Jr (eds.) 1982 *Biomaterials 1980, Advances in Biomaterials*, Vol. 3. Wiley, New York, pp. 365–70

Johnson P F, Bernstein J J, Hunter G, Dawson W W, Hench L L 1977 *In vitro* and *in vivo* analysis of anodized tantalum capacitive electrodes: Corrosion response, physiology and histology. *J. Biomed. Mater. Res.* 11: 637–56

Laing P G, Ferguson A B Jr, Hodge E S 1967 Tissue reaction in rabbit muscle exposed to metallic implants. *J. Biomed. Mater. Res.* 1: 135–49

Luckey T D, Venugopal B 1977 *Metal Toxicity in Mammals*. Plenum, New York, pp. 139, 227–31

Matthay R A, Balzer P A, Putman C E, Gee J B L, Beck G J, Greenspan R H 1978 Tantalum oxide, silica and latex: Effects on alveolar macrophage viability and lysozyme release. *Invest. Radiol.* 13: 514–18

Mears D C 1979 *Materials and Orthopaedic Surgery*. Williams and Wilkins, Baltimore, MA, pp. 53, 115, 123, 720

Meenaghan M A, Natiella J R, Moresi J L, Flynn H E, Wirth J E, Baier R E 1979 Tissue response to surface treated tantalum implants: Preliminary observations in primates. *J. Biomed. Mater. Res.* 13: 631–43

Metals Handbook, 9th edn., Vol. 2, 1979. American Society for Metals, Metals Park, OH, pp. 777–9, 799–804

Oehme F W 1979 (ed.) *Toxicity of Heavy Metals in the Environment*, Part 2. Dekker, New York, pp. 566–8

Plenk H Jr, Pflüger G, Schider S, Böhler N, Grundschober F 1984 The current status of uncemented tantalum and niobium femoral endoprostheses. In: Morscher E (ed.) 1984 *The Cementless Fixation of Hip Endoprostheses*. Springer, Berlin, pp. 174–7

Schider S, Bildstein H 1982 Tantalum and niobium as potential prosthetic materials. In: Winter G D, Gibbons D F, Plenk H Jr (eds.) 1982 *Biomaterials 1980, Advances in Biomaterials*, Vol. 3. Wiley, New York, pp. 13–20

Schroeder A 1974 Das Implantat nach Herskovits. Vorläufige Mitteilung über eine neue Implantatform. *Schweiz. Mschr. Zahnhlk.* 84: 742–7

Schuh E, Reichsthaler J, Plenk H Jr 1988 The tantalum double blade vent implant system "Implandent Austria" for the edentolous jaw. In: Watzek G, Matejka M (eds.) 1988 *Der zahnlose Unterkiefer*. Springer, Vienna, pp. 359–69

Sisco F T, Epremian E 1963 *Columbium and Tantalum*. Wiley, New York

Smith D C 1974 Materials used for construction and fixation of implants. *Oral Sci. Rev.* 5: 23–55

Williams D F (ed.) 1981 *Fundamental Aspects of Biocompatibility*, Vol. 1. CRC Press, Boca Raton, FL, p. 3

Zetner K, Plenk H Jr, Strassl H 1980 Tissue and cell reactions *in vivo* and *in vitro* to different metals for dental implants. In: Heimke G, De Groot K (eds.) 1980 *Dental Implants*. Hanser, Munich, pp. 15–20

Zitter H, Plenk H Jr 1987 The electrochemical behavior of metallic implant materials as an indicator of their biocompatibility. *J. Biomed. Mater. Res.* 21: 881–96

H. Plenk Jr
[Universität Wien, Vienna, Austria]

S. Schider
[Metallwerke Plansee, Reutte, Austria]

Titanium and Titanium Alloys

Rapid and significant progress has been made in the development of medical instrumentation and implants during the past two decades. As material requirements have become increasingly sophisticated, titanium and its alloys have been experimented with in numerous situations. Of the many titanium alloys that have been found to be suitable for medical applications, pure titanium and Ti–6 wt% Al–4 wt% V alloy have become most widely used. The clinical success of titanium alloys is due in no small measure to their outstanding mechanical properties, corrosion resistance and superior biocompatibility. Titanium has played a significant role in numerous surgical procedures in the field of orthopedic, cardiovascular and dental implantation.

1. Physical Metallurgy

Titanium was a laboratory curiosity until 1946, when Kroll developed a process for commercially producing titanium by reducing titanium tetrachloride.

Table 1

Chemical composition (wt%) and minimum mechanical properties of CP titanium

	Grade			
	1	2	3	4
Nitrogen, max.	0.03	0.03	0.05	0.05
Carbon, max.	0.10	0.10	0.10	0.10
Hydrogen, max.[a]	0.01	0.01	0.01	0.01
Iron, max.	0.20	0.30	0.30	0.50
Oxygen, max.	0.18	0.25	0.35	0.40
Titanium	bal.	bal.	bal.	bal.
Yield strength (MPa)	170	275	380	485
Ultimate strength (MPa)	240	345	450	550
Elongation (%)	24	20	18	15

a value for bar products 0.0125, for flat products 0.015

Since that time, the availability of titanium has prompted much work on the development of new and improved alloys as well as extensive evaluations of the properties of some 20 alloys.

Medical researchers evaluated commercially pure titanium and Ti–6 wt% Al–4 wt% V alloy and found them to be outstanding materials for surgical implant applications. Several other alloys were also found to be highly corrosion-resistant and suitable for medical applications, but widespread usage has been limited to commercially pure titanium and Ti–6 wt% Al–4 wt% V.

The term commercially pure (CP) titanium is applied to unalloyed titanium and designates several grades containing minor amounts of impurity elements, such as carbon, iron and oxygen. The amount of oxygen can be controlled at various levels to provide increased strength. The compositions and properties of the four grades of CP titanium listed in Table 1 illustrate this.

The microstructure of CP titanium is essentially all α titanium (hexagonal-close-packed crystal structure) with relatively low strength and high ductility. The material may be slightly cold worked for additional strength, but cannot be strengthened by heat treatment.

The addition of aluminum to titanium stabilizes the α phase, while vanadium addition stabilizes the β phase (body-centered-cubic structure). The combination of 6 wt% aluminum and 4 wt% vanadium enables both allotropic modifications to exist at room temperature; therefore, this alloy is classified as a two-phase α–β material. The alloying additions also contribute to increased strength by solid-solution strengthening mechanisms. The complex metallurgical transformations, as well as the relative amounts of α and β phases present in the material, can affect the mechanical properties. For applications where high strength and fatigue resistance are required, the material is annealed. The annealed

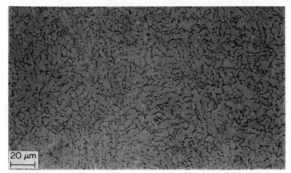

Figure 1
Typical microstructure of Ti–6 wt% Al–4 wt% V alloy in the mill-annealed condition, consisting of a uniform distribution of α (light) and β (dark) phases

Table 3
Typical mechanical properties of Ti–6 wt% Al–4 wt% V alloy for medical applications

Ultimate tensile strength (MPa)	965
Yield strength (MPa)	895
Young's modulus (GPa)	110
Elongation (%)	12
Fatigue endurance limit at 10^7 cycles (MPa)	515

microstructure corresponds to a uniform distribution of the α and β phases, as illustrated in Fig. 1. A coarse α network may embrittle the material and is carefully avoided. Similarly, heat-treated or cast and refined materials that exhibit higher tensile and fatigue strengths must be carefully controlled to avoid those microstructural features that are associated with low ductility or low fracture toughness.

The chemical composition of the material selected for surgical implant applications is given in Table 2. The extra-low interstitial (ELI) grade used in the USA and the standard grade used in the UK have demonstrated the same successful clinical performance characteristics.

The annealing of commercially available mill products of Ti–6 wt% Al–4 wt% V may be accomplished by heating the material to 700–820 °C for 0.5–2 h, depending on the size of the part. Air cooling from the annealing temperature may be used; however, variation in cooling rates has been found not to have a major effect on the mechanical properties. While the alloy may be cast or solution treated and aged for higher tensile strength, the mill-annealed condition described above has commonly been used for medical applications.

Table 2
Chemical composition (wt%) of Ti–6 wt% Al–4 wt% V alloy for surgical implant applications

	ELI grade	Standard grade
Nitrogen, max.	0.05	0.05
Carbon, max.	0.08	0.10
Hydrogen, max.	0.0125	0.015
Iron, max.	0.25	0.30
Oxygen, max.	0.13	0.20
Vanadium	3.5–4.5	3.50–4.50
Aluminum	5.5–6.5	5.50–6.75
Titanium	bal.	bal.

2. Mechanical Properties of Ti–6 wt% Al–4 wt% V Alloy

The mechanical properties of Ti–6 wt% Al–4 wt% V alloy (see Table 3) compare favorably with those of other implantable metal alloys. The yield strength is approximately the same as that of surgical quality 316L stainless steel and almost twice that of the familiar cast Co–Cr–Mo alloy used in orthopedic implants. The elastic modulus is approximately half that of the other common metal alloys used in surgery. The low modulus results in a material that is less rigid and deforms elastically under applied loads. These properties may play significant roles in the development of orthopedic products where a close match is desired between the elastic properties of long bone and the surgical implant.

The fatigue strength of the alloy is approximately twice that of stainless steel or cast Co–Cr–Mo alloy. The high fatigue properties are compromised by stress raisers, such as sharp covers, notches or mechanical damage. Therefore, it is important to avoid these features, not only in design and manufacturing, but also in the handling of surgical implants.

3. Surface Treatment

Both CP titanium and Ti–6 wt% Al–4 wt% V alloy have been used in various surface-treated conditions. One method of imparting a heavy oxide coating on the surface of a material is anodizing, an electrochemical process in which the material is immersed in an oxidizing medium and the surface oxide film is allowed to grow until the material appears to have various colors. Blue and yellow anodized materials can be obtained. Nitriding produces TiN and Ti$_2$N compounds on the surface, that are gold in color. The nitrided surface is extremely hard and ceramiclike.

The vast majority of surgical implants, however, are used without these surface treatments. The titanium material can be polished to a high luster, and it is customary to subject it to a passivation treatment similar to that used for austenitic stainless steels. The passivation procedure using nitric acid as the oxidizing medium helps to remove foreign contaminants which may be embedded in the surface, and also aids in the formation of a tenacious oxide film responsible for the high corrosion resistance of the material.

Titanium readily forms oxides; therefore, when the passive film is damaged in use, the material becomes rapidly repassivated. If titanium material is subjected to repeated abrasive action in the body, this process of removal of the oxide and rapid repassivation may produce sufficient quantities of oxide to darken the surrounding tissue. This phenomenon is believed to be harmless from a biological point of view. An anodized surface may also be abraded by tissue.

The surfaces of highly polished, as well as anodized or nitrided, implants are sensitive to certain contaminants. Discoloration of implants during steam autoclaving can occur as a result of surface reaction of titanium with impurities in the steam or the protective material contacting the implant. Such discoloration may range from slight tarnishing to vivid colors. The exact chemical composition of such stain spots is not easy to establish, but none of the suspected titanium compounds is known to be hazardous or cytotoxic. Careful attention to autoclaving conditions should eliminate this situation. Surgical implants sold in γ-ray sterilized condition are manufactured to provide optimum surface conditions.

4. Biocompatibility of Titanium and Ti–6 wt% Al–4 wt% V Alloy

The outstanding biocompatibility of these materials was recognized by early medical researchers. Titanium appears to have an extremely low toxicity and is well tolerated by both bone and soft tissue. Animal experiments have revealed that the material may be implanted for an extensive length of time; fibrous encapsulation of the implants is minimal to nonexistent. Histopathological examinations have failed to reveal any cellular changes adjacent to titanium implants. An increased concentration of metallic elements in adjacent tissue has been observed by spectrochemical analysis; however, no adverse clinical effects were detected. Some darkening of the soft tissue adjacent to CP titanium has been reported, which may be due to the low hardness and low abrasion resistance of the unalloyed material.

Titanium readily forms titanium oxide or complex oxide and hydride compounds. Therefore, it is possible that any material removed from the implant may immediately be stabilized by the formation of these inert compounds. The host tissue appears to exhibit no response to the chemically inert oxide of titanium. Titanium oxide is used in creams for dermatologic treatments, where it has been found to be both inert and nontoxic. Careful examination of tissues adjacent to Ti–6 wt% Al–4 wt% V alloy has

revealed neither giant cells nor macrophages, nor any other signs of inflammation. Aluminum and vanadium atoms are interspersed among titanium atoms in a random substitutional solid solution, and thus they are not free to leach out of the material at body temperature. Therefore, concerns about the biological effects of these elements are valid only to the extent that extremely minute quantities from the surface may enter the host tissue as a result of abrasive micromovement. The material has been found to be safe in intravascular applications, owing to its high electronegativity and passive surface.

5. Hypersensitivity

Some patients may be allergic to metals in contact with their skin. Similarly, allergic reactions of various severity are known to occur with implant materials, and cases of eczematous dermatitis related to internal exposure are known to occur. There is some evidence that even the corrosion-resistant materials used in orthopedic implants, such as stainless steel and cobalt–chromium alloy, produce some minute quantities of corrosion products which may be responsible for allergic reactions. The exact immunologic mechanism triggering allergic reaction in patients with implants is not clearly understood. Surgical removal of the implant frequently eliminates the symptoms, thus establishing a clear causative effect. All allergic cases reported involve stainless steel or cobalt–chromium alloys, which would indicate that the common elements in these alloys, such as cobalt, nickel or perhaps chromium, may be the sensitizing constituents. These metal ions may also form compounds which are known skin sensitizers.

Titanium does not cause hypersensitivity, nor do any of the alloying elements in Ti–6 wt% Al–4 wt% V alloy. Titanium alloy is the only alloy available which does not contain any of the known sensitizing elements. It is the metal of choice in patients suspected of being sensitive to metals. For several decades, special titanium implants have been used with outstanding success in patients with histories of severe allergic reactions.

6. Cardiovascular Implants

Titanium has been satisfactorily used in cardiovascular surgical devices for over two decades. Figure 2 illustrates the Starr–Edwards aortic valve. In patients with irreparable heart valve disease, the titanium cage holding the ball is sutured into the aorta by a Dacron sleeve. Annually, more than 60 000 patients receive artificial heart valves worldwide.

Heart attacks, and various other causes, create conditions of cardiac dysfunction where the patient

Figure 2
The Starr–Edwards aortic heart valve (courtesy
Baxter Healthcare Corporation, Edwards CVS
Division)

may require a temporary or permanent electronic
pacemaker system in order to lead an active life.
Figure 3 illustrates a pulse generator containing a
hermetically sealed lithium–iodine power source and
electronic circuitry protected by a hermetically
sealed titanium can. The pulse generator can be
implanted in the abdominal cavity and connected to
the heart muscles by pacing electrodes. Titanium
was selected for this application because of its superior corrosion resistance, low weight and manufactur-
ability. The thin-gauged titanium wall permits

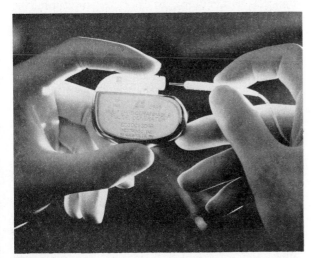

Figure 3
Pulse generator of an electronic heart pacemaker
system (courtesy Medtronic, Inc.)

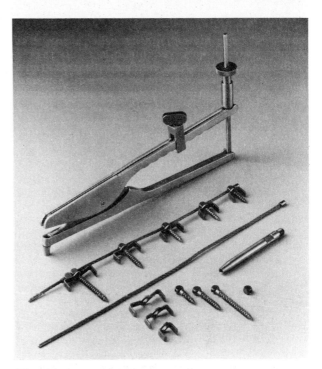

Figure 4
The Dwyer spinal correction cable system, showing
staples, screws, cable and (top) cable tightener

radioopaque identification markings to be placed
inside the can so that the surgeon may identify the
type of device through the use of common medical x
rays. This can be accomplished because titanium is
more radiolucent than other implant-quality metal
alloys.

7. Spinal Surgery

For the surgical correction of scoliosis, titanium
cables and screws have been used with significant
clinical success. The titanium devices known as the
Dwyer system and the instruments used in the surgical procedure are illustrated in Fig. 4. In this procedure, the screws and the staples are inserted into
the vertebrae of the area of curvature of the spine;
the cable is subsequently threaded through the eyes
in the heads of the screws. With the aid of the
tightener, the cable is then tightened to effect
"straightening" of the segment of the spine in question. Finally, the eyes of the screw heads are
crimped to secure the cable in the correct position.
The ductility of CP titanium enables this crimping to
take place without cracking of the screw heads.
Grade 2 CP titanium wire stranded to 7×19 cable
has the desired flexibility and strength for this application. Titanium and its alloys are not susceptible to

crevice corrosion at the various pH levels encountered *in vivo*. This provides an extra measure of safety for the multicomponent Dwyer system, where crevices are unavoidable design features.

8. Orthopedic Implants

The use of titanium orthopedic implants is most commonly indicated in disabling arthritis, usually of rheumatoid or degenerative origin. Titanium alloy prostheses have been used in total hip replacement, finger-joint replacement, total knee replacement, total elbow replacement and fracture fixation appliances.

The concept of total hip replacement is relatively simple. Various designs and materials of the ball-and-socket type have been tried, with varying degrees of clinical success. The high mechanical strength and fatigue endurance limit of titanium alloy, permit the design of stronger total hip prostheses for the active patient. Titanium alloy with its low modulus approximates the elastic behavior of the human bone more closely than other available implant material. Figure 5 illustrates a titanium alloy total hip stem with CP titanium beads firmly centered on the surface. As strong tissue grows between the beads, it provides a firm biological fixation eliminating the need for bone cement. The acetabular component is designed in a hemispherical shape with a pure titanium beaded alloy shell and polyethylene liner that can be fastened to the pelvic bone with specially designed screws. The low coefficient of friction and the rate of wear of the plastic acetabulum, permit easy articulation of the components and ensure a range of motion similar to that of the unimpaired leg.

Figure 5
The Opti-Fix total hip prosthesis made from high-strength Ti–6 wt% Al–4 wt% V alloy with CP titanium beads on its surface (courtesy Richards Medical Company)

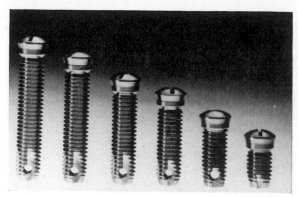

Figure 6
A dental implant utilizing the biocompatibility and strength of titanium (courtesy Nobelpharma USA, Inc.)

9. Intraoral Implants

CP titanium formed into perforated sheet has been used in reconstructive and plastic surgery in maxillofacial problems. The outstanding biocompatibility and the ease with which surgeons can form the material at the operating site, make it an ideal material for this application. The mesh can be secured to bone by small bone screws, and it can hold bone chips for bone grafting purposes.

Various materials and designs have been used in endosseous implants on which teeth or dentures can be constructed. Titanium alloy implants have been introduced recently to help overcome the problem of implant loosening and tissue resorption found with other alloys. Figure 6 illustrates a novel design that takes advantage of the high biocompatibility of titanium. Four to six weeks after insertion, the threaded portion of the implant is firmly stabilized in the bony structure (osseointegration) while the top protrudes through the gingiva and is ready for the construction of artificial teeth or dentures. Titanium permits soft tissue to grow around it, providing an effective shield against bacterial penetration from the oral cavity. The introduction of this device resulted in more favorable clinical results than any previously tried oral implant.

See also: Biocompatibility: An Overview

Bibliography

Albrektsson T 1987 Present clinical application of osseointegrated percutaneous implants. *Plast. Reconstr. Surg.* 79: 721–31
Brown S A, Mayor M B 1978 The biocompatibility of materials for internal fixation of fractures. *J. Biomed. Mater. Res.* 12: 67–82

Hill G H 1966 Titanium for surgical implants. *J. Mater.* 1: 373–83

Laing P G 1973 Compatibility of biomaterials. *Orthop. Clin. North Am.* 4: 249–73

Lemons J, Niemann K M W, Wiess A B 1976 Biocompatibility studies on surgical-grade titanium-, cobalt-, and iron-base alloys. *J. Biomed. Mater. Res.* 7: 549–53

Levanthal G S 1951 Titanium, a metal for surgery. *J. Bone Jt. Surg., Am. Vol.* 33: 473–4

Luckey H 1984 *Titanium Alloys in Surgical Implants, Symposium.* American Society for Testing and Materials, Philadelphia, PA

Meachim G, Williams D F 1973 Changes in nonosseous tissue adjacent to titanium implants. *J. Biomed. Mater. Res.* 7: 555–72

Mears D C 1979 *Materials in Orthopaedic Surgery.* Williams and Wilkins, Baltimore, MD, Chap. 7

Sarmiento A, Gnien T A 1985 Radiographic analysis of a low-modulus titatnium-alloy femoral total hip component. Two to six-year follow-up. *J. Bone Jt. Surg., Am. Vol.* 67: 48–56

Solar R J, Pollack S R, Korostoff E 1979 *In vitro* corrosion testing of titanium surgical implant alloys: An approach to understanding titanium release from implants. *J. Biomed. Mater. Res.* 13: 217–50

Williams D F 1976 Biomaterials and biocompatibility. *Med. Prog. Technol.* 4: 31–42

Williams D F 1977 Titanium as a metal for implantation. Part 2: Biological properties and clinical applications. *J. Med. Eng. Technol.* 1: 266–70

Williams D F, Roaf R 1973 *Implants in Surgery.* Saunders, London, Chap. 5, pp. 203–97

D. I. Bardos
[Smith & Nephew Richards, Memphis, Tennessee, USA]

Wound Dressings Materials

Wound healing in humans is essentially a repair process which has evolved, without the aid of dressings, to close the wound as rapidly and efficiently as possible.

Wounds that are left to heal naturally cannot always be said to heal optimally from either a functional or a cosmetic aspect. In addition, open wounds are susceptible to infection and can be unsightly. The contention by some authors and clinicians that the true functions of wound dressings are to prevent infection and to protect the patient from the "affront on his body's integrity" may have held some truth until recent times. Modern developments in our understanding of the biology of wound healing along with advances in polymer chemistry have, however, led to a revolution in wound dressings materials and a reassessment of the potential value of novel dressings. With these new insights has come a higher expectation for wound-healing performance and an understanding by health workers and industry that the function of a wound dressing is to assist and optimize the natural process of healing.

1. Historical Perspective and Traditional Dressings

A number of reviews outlining the early history of dressings from Ancient Egypt until the twentieth century have been written, perhaps the earliest documentation being the Ebers Papyrus (BC 1600–1500). In spite of this extended history of wound dressings and practices, the gauze dressings and cotton pads developed by surgeons of the nineteenth century are still familiar (and widely used). Listings of these and other traditional products along with specification standards can be found in the *British Pharmaceutical Codex* (1973), the *British· Pharmacopoeia* (1980) and the *United States Pharmacopoeia* (1985).

Traditional dressings fulfill a number of roles, including protection of the wound from physical damage, covering unsightly wounds, exclusion of infective microorganisms, absorption of excessive fluid from the wound (wound exudate), prevention of "strike-through" (the transfer of microorganisms from the wound to the external environment) and improvement of patient comfort. These functions might be regarded as mainly protective, having little to do with positive creation of optimized healing conditions. Among this family of dressings materials, only limited progress has been made towards further improvement in wound treatment.

Perhaps the most significant change has been the introduction of Tulle Gras (paraffin-impregnated gauze dressing). This material has been valuable in treatment of burns owing to its inherent nonadherent properties, allowing its removal with reduced pain and less rewounding of the delicate repair tissues than is possible with adherent gauze. Like other traditional materials, Tulle Gras can be effectively medicated.

2. New Materials

Synthetic polymer chemistry provided possibilities for new dressings biomaterials. It was not, however, until the classical observations of Winter (1962) that understanding of the healing phenomenon provided sufficient impetus for industry to seriously consider an assault upon the long-established treatment practices and materials. Winter's essential observation is that under a ventilated dressing or in an undressed wound, the formation of a scab forces the new epidermis (the outermost layer of skin which provides the primary biological "barrier") to migrate below the dried exudate of the scab, resulting in a loss of viable tissue, since all material above the new epidermis is now essentially extracorporeal.

Winter (1964) further demonstrated that in wounds covered by water-impermeable (occlusive) polyethylene, new epithelium covered the wound surface in only half the time taken for air-exposed wounds. These studies stimulated the current proliferation of new dressings materials, initially based upon these simple principles of occlusion, but later developing into dressings of more sophisticated design. These modern dressings often comprise both synthetic and naturally occurring materials and have been formulated to capitalize on the rapid progress that has been made in the understanding of the healing process.

3. Wound Dressings: Categories

Wound dressings might be categorized in several different ways based on composition, form and function. Any such categorization provides the compiler with a number of problems; most materials would appear in at least two lists. Rapid redefinition of functional terms by dressings manufacturers as the development of new materials accelerates adds to the difficulties. In this article, the general principles of novel dressing design will be reviewed under a classification based on "product type".

Table 1
Examples of film dressings

Dressing	Manufacturer	Description
Opsite	Smith and Nephew	adhesive-coated polyurethane film with "handles" to aid application
Bioclusive	Johnson and Johnson	adhesive-coated polyurethane film with perforated tabs to aid application
Tegaderm	3M	adhesive-coated polyurethane film with "frame" to aid application
Tegaderm Pouch Dressing	3M	bilayer polyurethane film perforated to allow drainage of exudate
Transign	Smith and Nephew	trilayer dressing— polyurethane outer layer, absorbent pad mid-layer and nonadherent wound contact layer with slits to allow passage of exudate
Ioban 2	3M	polyester incise drape containing iodine to control microorganisms during surgery
Omiderm	Omikron Scientific	self-adhesive sulfoethyl methacrylate-coated polyurethane film with properties of a hydrogel (see Sect. 3.4) when in contact with skin

3.1 Film Dressings

These are transparent adhesive-coated materials that are permeable to gases such as water vapor and oxygen, but impermeable to bacteria. Examples include Opsite, Bioclusive and Tegaderm, all ·of which are polyurethane based (see Table 1). An alternative generic description of these dressings is "semipermeable films." Modifications of the semipermeable film concept have appeared including Transigen, Tegaderm Pouch Dressing and Omiderm. The main differences between film dressings are in presentation and mode of application. Interestingly, the surface-modified polyurethane film dressing Omiderm may be further classified as a hydrogel dressing (see Sect. 3.4).

Film dressings tend to allow accumulation of large volumes of wound exudate beneath the dressing. The "modified film dressings" (see Table 1) have gone some way to alleviating this problem, but perhaps no longer belong within the original classification of film dressings.

Medicated film dressings have been used experimentally with some success. An incise drape containing iodine to control skin bacteria during surgery (Ioban 2) has recently been launched.

3.2 Foams

Foam products can include preformed foam sheets as well as dressings supplied in liquid form that "set" in the wound upon application. There are major performance differences between commercial foamdressings materials, reflecting differences in their composition and structure.

Release (Johnson and Johnson) is a sterile carboxylated styrene butadiene rubber latex foam sheet bonded to a nonwoven fabric coated with polyethylene film to render the surface less likely to adhere to a drying wound. Since the basic foam is hydrophobic, a surfactant is included to facilitate exudate absorption. This foam is primarily designed for low-exudate wounds.

Synthaderm (Armour Pharmaceutical) is a thin polyurethane foam sheet treated on one side to render it hydrophilic. The cellular structure is claimed to retain debris from the wound and the aqueous component of wound exudate is continuously lost by evaporation through the back of the dressing. Covaderm is essentially a secondgeneration dressing which is thinner, more comfortable and stronger than Synthaderm. Permeability of these two dressings is regulated by water uptake. The dressing shows an increase in surface area of about 20% as it hydrates. This increase in surface area is accompanied by an increase in water-vapor permeability, resulting in a relationship between permeability rate and exudate production in the wound site. However, owing to build up of exudate materials, these foams do not retain their high water-vapor permeability indefinitely.

Lyofoam (Ultra Laboratories) is a closed-cell polyether foam with a heat-modified contact layer. The contact layer is hydrophilic and about 0.5 mm thick. However, the outer layer of the dressing, which is approximately 5 mm thick, is hydrophobic and, as a result, absorption of wound exudate is regulated to an extent. Lyofoam C is a modified version of the material which contains charcoal to absorb odor.

Allevyn Burn Dressing (Smith and Nephew) is a hydrophilic polyurethane foam backed with a moisture/vapor-permeable polyurethane membrane. The foam is bonded to a polyurethane net to render it less adherent to the drying wound. It is permeable to water vapor and is claimed to possess characteristics of high absorbency while maintaining sufficient moisture to aid wound healing.

Silastic Foam Dressing (Dow Corning) is a medical grade silicone elastomer supplied in liquid form for premixing with a stannous octoate catalyst. The resultant foam is poured into the wound cavity to set and must be held in place with a secondary dressing. The foam is absorbent, nonadherent and permeable to air (see *Polysiloxanes*).

3.3 Polysaccharide Dressings

This chemical classification includes all dressings and wound dressings materials constructed or derived from naturally occurring polysaccharides. Examples include Sorbsan, Bard Absorption Dressing, Debrisan, Kaltostat, Iodosorb medicated dextranomer and Sorbact 10[5] (see Table 2). Some of the dressings materials have been functionally classified as "wound-cleansing agents" (Debrisan, Bard Absorption Dressing, Iodosorb and Sorbact 10[5]).

Perhaps the most important of the polysaccharide dressings are the alginates. Sorbsan and Kaltostat are produced mainly from calcium alginate (derived from brown seaweeds). In the wound environment, calcium–sodium ion exchange causes generation of relatively soluble sodium alginate and a dissolution of the dressing into the wound exudate. This results in the formation of a hydrophilic gel. Alginate dressings are said to be biodegradable. The fate of solubilized alginates from these dressings is the subject of some controversy. The manufacturers believe that dissipation of sodium alginate into body fluids provides distinct advantages over the risks of inclusion of fiber (e.g., from cotton or biological dressings) into the healing site with concomitant risk of hypertrophic scarring. Alginate dressings may also function as effective hemostats, arresting bleeding in wounds such as graft donor sites.

Table 2
Examples of polysaccharide dressings

Dressing	Manufacturer	Description
Bard absorption dressing	C. R. Bard	flakes of modified corn starch copolymer produced by graft copolymerization of carboxyl and carboxamide groups onto maize starch
Debrisan	Pharmacia	dextranomer beads or paste
Iodosorb	Pertsorp	dextranomer beads with bound cadexomer iodine as antiseptic agent
Sorbact 10[5]	LIC Hygiene	cellulose fibers impregnated with hydrophobic fatty acid to absorb bacteria
Sorbsan	NI Medical	calcium alginate dressing
Kaltostat	CAIR	calcium alginate dressing
Stop hemo	Windsor Pharmaceuticals	calcium alginate pack

Table 3
Examples of hydrogel dressings

Dressing	Manufacturer	Description
Geliperm	Geistlich	polyacrylamide/agar gel available as dry or preswollen sheets, or as granules
Vigilon	C. R. Bard	polyethylene mesh support with a colloidal suspension of radiation cross-linked polyethylene oxide which is used either as an occlusive or as a "totally breathable" dressing by removing one or both of the polyethylene film backings as necessary
Scherisorb	Schering AG	highly absorbent Graft T starch copolymer available in sachet form or as sheets
Omiderm	(see Table 1)	

3.4 Hydrogels

A hydrogel is a polymeric water-swollen network. Hydrogels swell in water to an "equilibrium water content" value, but are not soluble in water. They have a large number of biomedical applications, and their bulk and interfacial properties have been the subject of intensive research. The properties of hydrogels that make them attractive as dressings materials are biocompatibility, superficial resemblance to living tissue (attributable to equilibrium water contents up to over 99%), permeability control and design adaptability. By varying the nature of the polymer backbone composition, a range of water-binding behavior and subsequent mechanical, interfacial and permeability properties can be achieved.

Hydrogels as biomaterials were suggested by Wichterle and Lim (1960). Wichterle had developed DuPont's poly(2-hydroxyethyl methacrylate) (poly-(HEMA)) as a surgical material during the 1950s. Perhaps the most significant early hydrogel dressing was Opsite (see Table 1) originally launched as a poly(HEMA)-based film dressing (now classified as a film dressing having a polyurethane formulation). Examples of available hydrogel dressings include Vigilon, Scherisorb and Geliperm (see Table 3).

A useful three-way comparison of the physical properties of these dressings is given by Turner (1986). Absorption and water-vapor permeability properties of these dressings result in maintenance of a moist wound "with a sorption gradient which assists in the removal of toxic components from the wound area." Reduction of pain in wounds treated

Table 4
Examples of biological dressings

Dressing	Manufacturer	Description
Biobrane	Woodroof Labs	collagen-coated silicone—nylon film
E. Z. Derm	Genetic Labs	aldehyde cross-linked porcine collagen containing silver ions as antimicrobial agent. Stable at room temperature
Mediskin	Genetic Labs	forerunner of E. Z. Derm. unstable at room temperature
Corethium 1	Johnson and Johnson	porcine epidermis derived
Corethium 2	Johnson and Johnson	porcine dermal collagen derived
Meipac	Meija Seika	liquid collagen spun into a nonwoven dressing

with hydrogel dressings may be related to a cooling effect of the material, although this has not been fully demonstrated (see *Hydrogels*).

3.5 Biological Dressings

Biological dressings are at a much earlier stage of development than other dressings types. They comprise material derived from human or animal tissues, usually collagenous, or may be biosynthetic materials coated with treated collagen. The major application of these materials is expected to be for severe burns where they may be regarded as a temporary skin substitute. Indeed, in at least one instance, this principle has been taken to its logical conclusion by production of a "living skin equivalent" constructed to resemble living dermis and seeded with dermal fibroblast cells (Organogenesis Inc. LSE). Other examples are shown in Table 4.

A number of other biological materials are currently being investigated for commercial use, including collagen-based products, human amnion and chorion.

3.6 Hydrocolloid Dressings

As occlusive materials, perhaps the biggest disadvantage of film dressings is their nonabsorbency and, therefore, poor handling of exudate. Moreover, they have been criticized for being difficult to handle, adhering very easily to themselves when removed from the backing paper.

Hydrocolloid dressings take the principle of the occlusive dressing further by virtue of original design principles that incorporate the properties of adhesion (dry tack and wet tack), occlusion and absorbency. In addition, some of these dressings are easily removed from the healing wound without

damage to the newly formed granulation tissue and neoepidermis of the healing site.

The first hydrocolloid dressing (DuoDERM, also known as Granuflex) was introduced in 1983, DuoDERM comprises a layer of "hydrocolloid" material backed with a layer of cellular polyurethane foam.

It is claimed that the foam backing acts as a thermal insulator and as a barrier to the passage of gases, moisture and microorganisms in either direction, and also confers a degree of mechanical protection against pressure and shear forces. The inner layer consists of the hydrocolloid mass—essentially comprising gelatin, pectin and sodium carboxymethylcellulose dispersed within a hydrophobic matrix of polyisobutylene with plasticizer and tackifier. The effect of this is to produce controlled uptake of wound fluid in such a way as to sustain a moist wound surface without maceration of the surrounding skin. Adherence to surrounding intact skin ensures the maintenance of a physiological environment with isolation of the wound from the external atmosphere. Initial adhesion is by dry tack from the adhesive polyisobutylene contained

Table 5
Examples of hydrocolloid dressings

Dressing	Manufacturer	Description
Comfeel Ulcus	Coloplast	thin polyurethane film backing, sodium CMC particles embedded in styrene–isoprene copolymer adhesive elastic mass
Biofilm	Biotrol	nonwoven polyester backing, sodium CMC and Karaya gum in polyisobutylene adhesive mass
Intact	C. R. Bard	ethylene–vinyl acetate copolymer foam backing, polyvinyl alcohol and hydroxylated cellulose ether in synthetic adhesive elastic mass
Restore	Hollister	polyvinylchloride foam backing, pectin and sodium CMC embedded in ethylene–vinyl acetate elastomer
DuoDERM	ConvaTec	semiopen cell polyurethane foam backing with occlusive outer polyurethane film; pectin, gelatin and sodium CMC embedded in polyisobutylene elastomer

within the gel mass. As hydration occurs, this is gradually replaced by the more powerful adhesion force of wet tuck. Liquefaction of the gel over the moist wound surface, however, ensures no damage to developing epithelium or granulation tissue at the time of dressing removal.

A number of other hydrocolloid dressings of superficially similar design to DuoDERM are available, including Comfeel Ulcus, Dermiflex, Intact, Restore and Biofilm (see Table 5). Each of these dressings can be considered a hydrocolloid dressing yet there are a number of important fundamental differences in composition and properties.

The hydrocolloid classification of dressings is perhaps the first generation of dressings that not only provide the optimum healing environment (according to current knowledge of the repair process), but also interact with the wound site in such a way as to directly influence the healing cascade. This is particulary true of DuoDERM which hydrates in the aqueous environment of the wound to release hydrocolloids into the wound cavity itself so that wound exudate and dressing become a complex continuum of wound-derived and dressing-derived materials.

The observed clinical efficacy of hydrocolloid and other modern surgical dressings is now the subject of intensive scientific and clinical research. Early indications that such dressings might actively promote cellular and enzymatic events in healing chronic refractory wounds, such as leg ulcers, point the way forward to the next generation of wound dressings.

See also: Biocompatibility: An Overview; Hydrogels; Suture Materials

Bibliography

British Pharmaceutical Codex 1973 Pharmaceutical Press, London, pp. 609–36
British Pharmacopoeia 1980 Her Majesty's Stationery Office, London
Davies J W L 1983 Synthetic materials for covering burn wounds: Progress towards perfection. Part I. Short term dressings materials. *Burns* 10: 94–103
Developments in Wound Care: An Overview 1985 George Street Publications, London
Ebbell B 1937 *The Papyrus Ebers, The Greatest Egyptian Medical Document.* Levin and Munksgaard, Copenhagen
Gamgee S 1880 Absorbent and medicated surgical dressings. *Lancet* 24 January: 127–8
Lawrence J C 1982 What materials for dressings? *Injury* 13: 500–12
Ryan T J (ed.) 1985 *An Environment for Healing: The Role of Occlusion.* Royal Society of Medicine, London
Schmidt R J 1986 Xerogel dressings—An overview. In: Turner T D, Schmidt R J, Harding K (eds.) 1986 *Advances in Wound Management.* Wiley, Chichester, pp. 65–71
Thomas S 1986 The role of foam dressings in wound management. In: Turner T D, Schmidt R J, Harding K (eds.) 1986 *Advances in Wound Management*, Wiley, Chichester, pp. 23–9
Turner T D 1984 Semipermeable films for wounds. *Pharm. J.* 232: 452–4
Turner T D 1986 Hydrogels and hydrocolloids—An overview of the products and their properties. In: Turner T D, Schmidt R J, Harding K (eds.) 1986 *Advances in Wound Management*, Wiley, Chichester, 89–95
United States Pharmacopoeia, 21st revision, 16th edn., 1985. United States Pharmacopoeial Convention, Rockville, MD
Wichterle O, Lim D 1960 Hydrophilic gels for biological use. *Nature (London)* 185: 117–18
Wicks C J, Peterson H I 1972 Medicated wound dressings, a historical review. *Opusc. Med.* 17: 90–5
Winter G D 1962 Formation of the scab and the rate of epithelization of superficial wounds in the skin of the young domestic pig. *Nature (London)* 193: 293–4
Winter G D 1964 Movement of epidermal cells over the wound surface. In: Montagna W, Billingham R E (eds.) 1964 *Advances in Biology of Skin*, Vol. 5. Pergamon, Oxford, pp. 113–27

M. J. Lydon
[Newtech Clwyd, Clwyd, UK]

Wrought Dental Wires

Wrought wires, used primarily in the dental specialty of orthodontics, are cast alloys that have been cold worked and formed by mechanical processes such as rolling, extrusion and drawing. These processes establish the internal structure of the alloy as well as its mechanical properties. Depending on the alloy, the properties of the material may be altered by the clinician through heat treatment. Orthodontic wires function to deliver force to malaligned teeth in order to change their spacial configuration to approximate an ideal dental arch. This change is mediated by physiological remodelling of the supporting bones of the upper and lower jaws. How quickly teeth respond to applied forces depends on the age of the patient and the magnitude of the force. However, large forces may cause pathological changes as ischemic necrosis of the periodontal ligament (i.e., damage to the anatomical junction between the tooth and supporting bone). In addition, tooth root tip resorption may occur when large forces are used for extended periods of time.

In order to maximize the rate of tooth movement and minimize the dangers of pathological change, wires capable of delivering a force of approximately 0.9 N and maintaining this force during tooth movement, must be chosen. Initially, an alloy with a high modulus of elasticity can deliver a large force through a small wire deflection to move a tooth. However, the clinical effectiveness of the alloy diminishes rapidly. Typically, patient visits are scheduled at three-week intervals; wires capable of delivering a force in the range where remodelling can occur over this period of time should, therefore, be selected.

Table 1
Properties and nominal composition of wrought dental wires

Alloy	Yield strength (GPa)	Elastic modulus (GPa)	Springback (10^{-2})	Nominal composition (wt%)
Gold	0.71–0.96	0.97–1.20	0.73–0.8	65 Au–5 Pt–2 Pd–15 Cu–10 Ag
Stainless steel	1.9	160	1.2	Fe–18 Cr–8 Ni
Elgiloy	2.1	200	1.1	40 Co–20 Cr–15 Ni–15 Fe–7 Mo
Nitinol	1.7[a]	33		Ti–55 Ni
β-titanium	1.2	65	1.8	Ti–11 Mo–6 Zr–4 Sn
Ni–Co–Cr–Mo	1.8	230	0.78	Ni–35 Co–20 Cr–10 Mo

a tensile strength

The capability of maintaining remodelling activity is a function of the ratio of yield strength to elastic modulus for a material. The term "springback" is used to define this property. Materials with low elastic moduli (5×10^4 MPa) and high yield strength (10^3 MPa) are ideal. Formability is a measure of several mechanical properties, which include yield strength and ductility. Ductility must be sufficient to allow the fabrication of complex low-radius bends.

Since orthodontic appliances must function for long periods in the oral environment (i.e., at high relative humidity; a constant temperature of 37 °C; and moderately high acidity, pH 3–5) the material used must have good corrosion resistance.

Several alloy systems have been used, as described below and summarized in Table 1.

1. Gold Alloys

High cost and volatile market conditions have largely precluded the use of wrought gold alloy wires, although these materials have significant historical interest. They possess many near-ideal properties (*Metals Handbook* 1979) and until recently were the alloys of choice. Composition and physical property requirements for wrought gold wires are contained in the American National Standards Institute/American Dental Association ANSI/ADA Specification No. 7 for dental wrought gold wire alloy (Council on Dental Materials, Instruments and Equipment 1962). High (>75 wt% gold and platinum group metals) and low (<65 wt% gold and platinum group metals) gold alloys must exhibit minimum yield strengths of 862 MPa and 690 MPa, respectively; both alloy types must also exhibit a minimum of 15% elongation on a 50.8 mm gauge length for quenched specimens.

2. Stainless Steel

Austenitic stainless steel with a relatively low carbon content (<0.15 wt%), 18 wt% chromium and 8 wt% nickel is the most widely used alloy for orthodontic wires and may be the wire of choice in most treatment modalities. American Iron and Steel Institute (AISI) types 302 and 304 are widely used. In the annealed condition, these wires have a minimum yield strength of 205 MPa (*Metals Handbook* 1980). Cold-worked orthodontic wires have a yield strength of 1.9 GPa and an elastic modulus of 160 GPa (Goldberg et al. 1977), which produce a springback of 1.2×10^{-2}. Drawing to produce round wires and subsequent rolling to yield wires of rectangular cross section, leads to the necessary cold-worked condition. Stress relief at 450 °C for short periods (<10 min) is the only effective heat treatment for these alloys.

The high elastic modulus of stainless steel alloys means that small-diameter wires (0.45 mm) must be used where lower forces are to be generated to move teeth. The disadvantage of reducing wire size is that excessive rotation of the wire in the bracket may occur, owing to the difficulty in fabricating these wires to close tolerances. Larger wires may be used if the alloy chosen has a lower elastic modulus, and a concomitant improvement in control of wire rotation in the bracket results.

In addition to their suitable mechanical properties, austenitic stainless steel alloys are also favored for orthodontic applications because their corrosion resistance is well known, particularly in biological environments, and because in wire form they can easily be joined by soldering or electrical resistance welding. In fact, because of the wide acceptance and long history of successful orthodontic use of stainless steel, other dental alloys are generally evaluated against a stainless steel standard.

3. Cobalt–Chromium Alloys

Elgiloy, a cobalt–chromium alloy originally formulated for the watch industry for use as a main spring, is widely used as an orthodontic alloy. This alloy has a high elastic modulus (2×10^5 MPa) (Burstone and Goldberg 1980) and good corrosion resistance, and can be heat treated to increase the yield strength. Orthodontic appliances can readily be fabricated in the as-received condition and subsequently furnace heat treated at 480 °C to achieve desirable properties. Electrical-resistance and flash-paste heat treating are also effective. In the latter case, a pyrogenic paste is applied to the wire, which is then heated by passage of an electric current. The paste ignites, further heating the wire. Although this method is not as controllable as the furnace method, it is much faster.

Orthodontic appliances are easily fabricated from as-received Elgiloy wire, because of its ductility. The alloy can be joined by soldering, but this technique requires more skill than that needed for joining stainless steel. The ability to heat treat this alloy, to improve spring qualities, is dependent on the temper of the as-received alloy.

4. Nickel–Titanium Alloys

Nitinol, a stoichiometric alloy composed of 55 wt% nickel and 45 wt% titanium, is corrosion resistant and possesses a unique shape-memory property whereby plastically deformed wires can be returned to their original shape by an appropriate heat treatment. Although the shape-memory effect has not been employed by orthodontists, other desirable properties make this alloy attractive to clinicians.

Nitinol has a low elastic modulus (3.3×10^4 MPa) and a relatively high tensile strength (1.7×10^3 MPa) (Andreasen and Morrow 1978). Rather large deflections of standard-size wires, result in the generation of relatively low forces. Unfortunately, the alloy possesses very low plastic deformability and so does not lend itself readily to manipulation into dental appliances; it is available only in preformed arches.

5. β-Titanium Alloys

The first applications of β-titanium alloys to dental use as orthodontic wires were in the late 1970s. The body-centered-cubic structure of β-titanium is stabilized by alloying with 11 wt% molybdenum, 6 wt% zirconium and 4 wt% tin. The resulting alloy has a low modulus (6.5×10^4 MPa) and relatively high yield strength (1.2×10^3 MPa) (Goldberg and Burstone 1979), and therefore has high springback (1.8×10^{-2}). This latter value represents an approximate increase of 50% over that for stainless steel. A high springback permits relatively large deflections without the danger of placing excessively high force against the tooth. The range of control, or time-dependent effectiveness, of these wires is substantially improved in comparison with that of stainless steel. The plastic deformability of β-titanium alloys, although not as great as that of stainless steel, is

sufficient to allow fabrication of complex bends with small radii.

The combination of high springback, low modulus and formability renders this alloy near ideal for certain treatment modalities. Although the alloy cannot be soldered, joining by resistance welding may be easily accomplished.

6. Nickel–Cobalt Alloys

A recently developed quaternary alloy, composed essentially of nickel, cobalt, chromium and molybdenum, undergoes a phase transformation on cold working. The alloy in the annealed condition has a face-centered-cubic structure that can be partially transformed to a hexagonal-close-packed structure. The transformation cannot be induced by heat treatment; however, working can be performed at a temperature as high as 425 °C. Additional strength can be achieved by aging in the temperature range 425–650 °C, where the development of coherency strains between Co_3Mo precipitate particles and the matrix occurs. The presence of the aged structure cannot be detected by light microscopy.

The corrosion resistance of this alloy is similar to that of austenitic stainless steel. The ductility of this high-modulus (2.3×10^5 MPa) alloy is unusually high with 3–5% elongation, but this property is affected considerably by aging heat treatment (Smith and Yates 1968). Thus, orthodontic appliances can be fabricated readily and strengthened further by heat treatment.

Bibliography

Andreasen G F, Morrow R E 1978 Laboratory and clinical analyses of nitinol wire. *Am. J. Orthod.* 73: 142–51

Burstone C J, Goldberg A J 1980 Beta titanium: A new orthodontic alloy. *Am. J. Orthod.* 77: 121–32

Council on Dental Materials, Instruments and Equipment 1962 American Dental Association Specification No. 7 for dental wrought gold wire alloy. *J. Am. Dent. Assoc.* 64: 439–41

Goldberg J, Burstone C J 1979 An evaluation of beta titanium alloys for use in orthodontic appliances. *J. Dent. Res.* 58: 593–9

Goldberg A J, Vanderby R, Burstone C J 1977 Reduction in the modulus of elasticity in orthodontic wires. *J. Dent. Res.* 56: 1227–31

Kapila S, Sachdeva R 1989 Mechanical properties and clinical applications of orthodontic wires. *Am. J. Orthod.* 96: 100–9

Kusy R P, Stuch A M 1987 Geometric and material parameters of a NiTi and a beta-Ti orthodontic arch wire alloy. *Dent. Mater.* 3: 207–17

Metals Handbook, 9th edn., Vol. 2, 1979. Gold in dentistry. American Society for Metals, Metals Park, Ohio, pp. 684–7

Metals Handbook, 9th edn., Vol. 3, 1980. Wrought stainless steels. American Society for Metals, Metals Park, Ohio, pp. 18–19

Miura F, Mogi M, Ohura Y, Hamanaka H 1986 The superelastic properties of Ni–Ti alloy in orthodontics. *Am. J. Orthod.* 90: 1–10

Nelson K R, Burstone C J and Goldberg A J 1987 Optimal welding of beta titanium orthodontic wire. *Am. J. Orthod.* 92: 213–19

Smith G D, Yates D H 1968 High strength–ductility–corrosion resistance: Multi-phase alloys have all three. *Met. Prog.* 93: 100–2

Wilson D F, Goldberg A J 1987 Alternative beta-titanium alloys for orthodontic wires. *Dent. Mater.* 3: 337–41

J. L. Sandrik and L. Laub
[Loyola University, Maywood, Illinois, USA]

Z

Zirconia-Toughened Ceramics

In the 1960s total hip replacement (THR) became a widely used surgical procedure due, among other reasons, to the successful use of a materials combination leading to low-friction arthroplasty: metal ball against ultrahigh molecular weight polyethylene (UHMWPE) socket. However, after a ten-year period of use, drawbacks related to UHMWPE began to appear, such as creep, wear and tissue inflammation induced by the production of polyethylene wear debris. In this respect, there was a need for the use of materials exhibiting better performances for total joint replacement. Alumina ceramic was introduced in orthopedic surgery because of both its high wear resistance and biocompatibility. Since then, high medical-grade alumina has functioned in a satisfactory manner, either in alumina–UHMWPE or in alumina–alumina combinations. However, alumina exhibits a brittle behavior, with a low fracture toughness and a low tensile strength. It is sensitive to microstructural flaws and subsequently has a low resistance to stress concentration and mechanical impact in service. This is one of the major reasons limiting the diameter of most prosthetic alumina femoral balls to 32 mm in order to avoid the occurrence of brittle fractures. In this instance, toughened ceramics including zirconia-toughened ceramics appeared particularly efficient and attractive for the highly loaded environment found in joint replacement (Pascoe et al. 1979, Christel et al. 1988).

1. Properties of Toughened Zirconia and Material Processing

Zirconium oxide ceramics have three phases: monoclinic, tetragonal and cubic. The cubic phase is stable but brittle; the tetragonal phase is tough but unstable and may transform into the monoclinic phase. Zirconia exhibits phase transformations at high temperatures. At room temperature, the stable phase of zirconium oxide has a monoclinic symmetry. During heating, it first transforms into a tetragonal phase (in the 1000–1100 °C range), then into a cubic phase (above 2000 °C). Noticeable changes in volume are associated with these transformations. During the monoclinic-to-tetragonal transformation, which occurs when zirconium oxide is heated, there is a 5% volume decrease; conversely, a 3% increase in volume is observed during the cooling process. This last retransformation into the monoclinic phase is of the same nature as the martensitic transformation occurring in steel and can be compared with it.

These phenomena are detrimental to the mechanical behavior of the zirconium oxide because the stresses induced during the phase transformations results in crack formation. This undesirable phase transformation can be inhibited by the addition of stabilizing oxides (CaO, MgO, Y_2O_3) and this process has become common practice. Accordingly, in the presence of a small amount of stabilizing additives, tetragonal particles (provided they are small enough) can be maintained in a metastable state at temperatures below the tetragonal-to-monoclinic transformation temperature. The transformation of small tetragonal grains, which should result in a volume increase, is prevented by the compressive stresses applied on these grains by their neighbors. The corresponding class of ceramics is named partially stabilized zirconia (PSZ).

Two kinds of microstructures can be generated. In the ZrO_2–MgO or ZrO_2–CaO systems, materials are sintered in the cubic state and small tetragonal precipitates are formed during cooling as a result of partial transformation of the cubic phase. In the ZrO_2–Y_2O_3 system, the extent of the stability range of the tetragonal phase, in terms of temperature and amount of yttrium oxide as shown on the phase diagram in Fig. 1, allows sintering of fully tetragonal fine-grained materials. Thus, using Y_2O_3 as a stabilizing agent, it is possible to produce a zirconium oxide ceramic made of 100% small metastable grains, leading *de facto* to a fully stabilized zirconia.

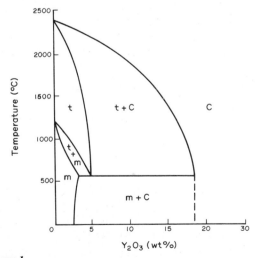

Figure 1
Y_2O_3–ZrO_2 phase diagram: the addition of less than 5% of Y_2O_3 to ZrO_2 allows the sintering of a fully tetragonal material (t = tetragonal phase; m = monoclinic phase; c = cubic phase)

The volume change related to the tetragonal-to-monoclinic phase transformation results in a pre-stressed material. In this respect, a propagating crack can release the stresses in the neighboring grains which then transform from the metastable state into the monoclinic phase. Because the volume of the monoclinic grains is larger than the tetragonal ones, the associated volume expansion results in compressive stresses at the edge of the crack front and extra energy is required for the crack to propagate further (Fig. 2). Thus it is believed that the main energy-absorbing mechanism is due to the martensitic-like transformation occurring at the crack tip.

Yttrium-oxide-stabilized zirconia is obtained by sintering of ultrafine zirconia powder (average particle size: $0.2\,\mu m$) with about 5 wt% Y_2O_3. This powder is highly adapted to sintering as its densification occurs in the 1400–$1500\,^{\circ}C$ range, corresponding to the tetragonal phase of the ZrO_2–Y_2O_3 system. The final microstructure consists of $0.5\,\mu m$ average diameter grains (Figs. 3 and 4). However, ultrafine powders are unsuitable for compacting; as a result, shaping of thick components such as hip prostheses balls can be difficult. Cold isostatic pressing is the most widely used process in the shaping of ZrO_2–Y_2O_3 ceramics. Densification can be achieved in two different manners: either pressureless sintering or sintering associated with hot isostatic pressing (HIP). This latter procedure consists of two stages: first, presintering the components without pressure up to about 95% of the theoretical density; second, removing the residual porosity in a complementary

Figure 3
Microstructure of yttrium oxide–partially-stabilized zirconia as shown by scanning electron microscopy (courtesy Céramiques Techniques Desmarquest)

HIP process. In this procedure, complete densification occurs with only limited grain growth, resulting in improved strength.

The materials properties of commercially available Y_2O_3–ZrO_2 ceramics, with reference to surgical-grade Al_2O_3 (International Organization for Standardization (ISO) standard ISO 130 or DIS standard DIS D13) are listed in Table 1. Accordingly, it is clear that the strength and toughness of transformation-toughened zirconia are much higher than those of alumina. In addition, the ZrO_2-based materials exhibit a lower Young's modulus, pointing to an interesting elastic deformation capability when compared with alumina (Christel et al. 1988).

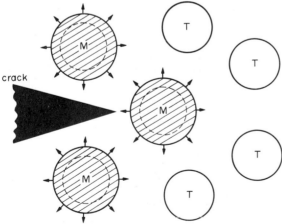

Figure 2
Diagram representing the mechanism of transformation toughening in partially-stabilized zirconia; the crack propagation induces the transformation of metastable tetragonal grain (T) into monoclinic phase (M); the monoclinic grains are larger than the tetragonal ones and stop the crack propagation

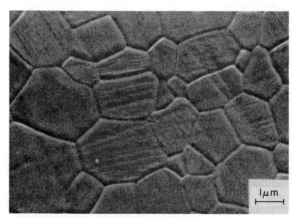

Figure 4
Microstructure of surgical-grade alumina shown for comparison with zirconia, with the same magnification as Fig. 3 (courtesy Céramiques Techniques Desmarquest)

Table 1

Comparison of several mechanical properties between surgical grade
alumina (ISO requirements) and two commercially available yttrium oxide–
partially-stabilized zirconia (Prozyr®, Céramiques Techniques Desmarquest,
France; Metoxit®, Metoxit AG, Switzerland)

Property	Alumina (ISO/DIS 130/D13 requirements)	Prozyr® YPSZ	Metoxit® YPSZ
Density (kg m^{-3})	3900	6100	*c.*6050
Average grain size (μm)	<7	<0.5	0.2
Vickers hardness (HV)	2000–3000	1000–1300	*c.*1200
Young's modulus (GPa)	380	200	150
Bending strength (MPa)	400	1200	*c.*800
Toughness, K_{lc} (mN m$^{-3/2}$)	5–6	9–10	*c.*7

2. Mechanical Performances of Zirconia Femoral Heads

Tateishi and Yunoki (1987) have studied the static
fracture strength of 22 mm zirconia balls fitted on
10 mm or 11 mm diameter tapers with a cross-head
speed of 0.5 mm min^{-1} with reference to 28 mm
alumina balls as the control. The testing was carried
out at room temperature, with the taper oriented
vertically. In order to avoid a direct contact between
ceramic and metal cross head, polypropylene sheets
were inserted at the interface.

The fracture of alumina heads occurred at loads in
the range 1.5–2.5 × 10^4 N, whereas the zirconia
heads broke at 4–5 × 10^4 N with the 10 mm taper and
3–4 × 10^4 N with the 11 mm taper.

Impact fracture tests were carried out by falling
weights, either on the 22 mm zirconia or the 28 mm
alumina heads at a constant height of 0.5 m. The
alumina balls, fitted on a 13 mm, taper fractured for
a 15 J impact energy; the zirconia heads, fitted with a
10 mm taper, exhibited an impact energy in the
range 50–100 J. Unfortunately, the impact at fracture cannot, at present, be compared between the
two ceramics because of the difference in taper size.
However, it is likely that the difference will not be
so great since for zirconia, increasing the taper
diameter reduced the impact strength.

3. Biodegradation of Toughened Zirconia

The degradation of toughened zirconia can be
secondary to aging, fatigue or wear occurring during
its *in vivo* service.

Aging has been studied through accelerated procedures in an autoclave, combining various degrees
of heat, humidity and pressure for different lengths
of time (Hanninck and Garvie 1982). It has been
shown that there is a strength loss of a few percent
with MgO–PSZ samples boiled in saline for 1000 h.
This has been related to the extension of the critical

flaw size by a stress corrosion mechanism. No significant drop has been found in the fracture toughness
of Y$_2$O$_3$–PSZ after γ sterilization or aging in Ringer's
solution (Christel et al. 1989). *In vivo* aging of
Y$_2$O$_3$–ZrO$_2$ samples implanted by Kumar et al.
(1989) in rabbit bone or subcutaneous tissue, up to
12 months, has shown no change in bending strength
of the implants. Simultaneously, a 2% transformation rate from tetragonal into monoclinic phase was
measured on the same implants by x-ray diffraction
analysis. It was identical for the specimens
implanted in bone or subcutaneous tissue, or stored
either in saline solution at room temperature or in
the atmosphere. Since theoretically up to 60% of
transformation occurs from tetragonal into monoclinic phase, there is no significant change in toughness of tetragonal zirconia (Sato and Shimada 1984):
this 2% per year transformation rate is without
consequence.

Fatigue tests have been carried out in Ringer's
solution for up to 10^7 cycles on 22 mm Y$_2$O$_3$–PSZ
femoral balls under an alternating load varying
between 0 kN and 10 kN at a frequency of 30 Hz.
The ceramic heads were placed vertically, fitted on
either 10 mm or 11 mm diameter Ti$_6$Al$_4$V taper
cones. Prior to the fatigue test, in order to fit the ball
with the cone, the ball was preloaded by 3 kN.
Under these test conditions the zirconia balls sustained 10^7 cycles without damages.

Wear tests of zirconia have been conducted using
either a zirconia–UHMWPE or a zirconia–zirconia
combination.

A bench test compared the deformation of
UHMWPE sockets bearing with 32 mm heads made
of either chromium–cobalt alloy or Y$_2$O$_3$–PSZ
ceramic in saline serum at 37 °C. The heads were
submitted to a bidimensional rotational movement
combining a ±45° tilt of the taper cone-head assembly with a 60 revolutions per minute axial rotation
under a constant load of 2 kN at a frequency of
0.5 Hz. After 10^6 cycles the UHMWPE socket
deformation, bearing against a chromium–cobalt

head, was 10.2×10^{-2}, whereas after 11×10^6 cycles with a Y_2O_3–PSZ ball, the plastic deformation was only 0.51×10^{-2}. Assuming that the creep of UHMWPE was identical in each test, the difference in socket deformation could be related to different wear rates in relation to the higher surface roughness of chromium–cobalt ball when compared with zirconia ceramic. In another study (Tateishi and Yunoki 1987) based on a joint simulator test to characterize the zirconia on polyethylene combination, a rotating movement at 0.125 Hz, ±10° in the frontal, transverse and sagital planes under an alternating load varying between 0 kN and 4 kN was used with pure water as a temperature-controlled lubricant at 25 °C. The socket was placed in the lower position. Measurements of frictional torques after 4×10^6 cycles showed a value of 2.6 N m for the Charnley prosthesis vs 2 N m for a 28 mm alumina ball bearing against a UHMWPE cup, and 1.5 N m for a 22 mm Y_2O_3–PSZ head and UHMWPE socket combination. However, no 28 mm Y_2O_3–PSZ head was tested.

Zirconia–zirconia tribological behavior has been compared by Sudanese et al. (1988) with alumina–alumina using a disk-on-ring test according to the ISO B474–1981 standard (angle of oscillation, ±25°; disk diameter, 30.4 mm; disk height, 4 mm; ring inside diameter, 14 mm; ring outside diameter, 20 mm; superficial pressure, 20–23 bar; frequency of oscillation, 1 Hz; and lubricant, Ringer's solution at 37 °C). The surface roughness of the Y_2O_3–PSZ disk was 0.01 µm Ra and 0.03 µm Ra for alumina. The wear rate of zirconia (16 mm³ h⁻¹) was 5 000 times greater than alumina (0.0033 mm³ h⁻¹). These results are difficult to extrapolate to THR because the lubrication conditions, loading and motion speed are quite different from the *in vivo* situation (Denape et al. 1984). In this test configuration, the heavy degradation of the zirconia–zirconia combination could be related to the low thermal conductivity of zirconia generating thermal shocks. Other ceramics combinations need further testing; for example zirconia/alumina–zirconia, zirconia–alumina and zirconia–silicon nitride.

4. Biocompatibility of Toughened Zirconia Ceramics

The biocompatibility of pure zirconia and PSZ is not widely documented. The tissue reaction to zirconia ceramics has been studied after implantation, either in soft tissues or in bony site. The tissue reaction to plasma- or flame-sprayed ZrO_2 on a 316L stainless-steel substrate has been evaluated by Bortz and Onesto (1973) by implantation of such tubes for tracheal or vein replacement in rabbits and dogs. Neither adverse tissue response nor blood clotting have been observed. MgO–zirconia samples have

been implanted by Garvie et al. (1984) in rabbits' muscles for six months, without control material. The observation of the tissue reaction, which was graded qualitatively, has led to the conclusion that the tissue response to the material was acceptable with no adverse tissue reaction. Also, no mechanical degradation of the MgO–zirconia has been observed after the six-month implantation period. Y_2O_3–ZrO_2 cylinders, differing in the methods of preparing the original powders, were implanted in the paraspinal muscles of rats for one, four and 12 weeks (Christel et al. 1989). The tissue reaction to these implants, for each implantation time, was studied by quantitative histomorphometry, in comparison with surgical-grade alumina (ISO requirements). For each cell type (macrophages, polymorphonuclear cells, giant cells, round cells or fibrocytes) the parameters of the cell distribution (amplitude, density and distances from the tissue–implant interface) and the thickness of the encapsulating membrane were computed. There was no statistical difference in any of the parameters of the distributions and membrane thickness of the cell types between Y_2O_3–ZrO_2 ceramics and alumina for each implantation time.

Implantation of Y_2O_3–zirconia cylinders in bone has been performed in rats and rabbits from two weeks up to one year, with reference to alumina. Most of the results have shown an equal amount of bone contacting the implants directly and an identical tissue reaction to both ceramics. The similarity of tissue reaction to zirconia and alumina may be explained by the fact that both materials are in the form of their most oxidated state, thus they lack the ability to leach any soluble component. In this instance, it has been shown by Soumiya (1984) when using x-ray diffraction analysis that after a four-week implantation of Y_2O_3–ZrO_2 powder and screws into the rabbit femur, no diffusion of yttrium or zirconium was detectable. However, at the time of writing there is no information on the long-term biocompatibility of bulk zirconia ceramics, nor on the tissue tolerance to submicrometer zirconia wear particles.

Bibliography

Bortz S A, Onesto E J 1973 Flame-sprayed bioceramics. *Bull. Amer. Ceram. Soc.* 52: 898

Christel P, Meunier A, Dorlot J M, Crolet J, Witvoet J, Sedel L, Boutin P 1988 Biomechanical compatibility and design of ceramic implants for orthopedic surgery. In: Ducheyne P, Lemons J (eds.) 1988 Bioceramics: Material characteristics versus *in vivo* behavior. *Ann. N.Y. Acad. Sci.* 523: 234–56

Christel P, Meunier A, Heller M, Torre J P, Peille C N 1989 Mechanical properties and short-term *in vivo* evaluation of yttrium oxide–partially-stabilized zirconia. *J. Biomed. Mater. Res.* 23: 45–61

Denape J, Lamon J, Broussaud D 1984 Friction wear of

ceramics: Theoretical and experimental studies. In: Vincezini P (ed.) 1984 *Sciences of ceramics*, Vol. 12, Elsevier, Amsterdam, pp. 529–35

Garvie R C, Urban C, Kennedy D R, McNeuer J C 1984 Biocompatibility of magnesia–partially stabilized zirconia (Mg–PSZ ceramics). *J. Mater. Sci.* 19: 3224–8

Hanninck R H J, Garvie R C 1982 Sub-eutectoid aged Mg–PSZ alloy with enhanced thermal up-shock resistance. *J. Mater. Sci.* 17: 2637–43

Kumar P, Shimizu K, Oka M, Kotoura Y, Nakayama Y, Yamamuru T, Yanagida T, Makinouchi K 1989 Biological reaction to zirconia ceramics. In: Oonishi H, Aoki H, Sawai L (eds.) 1989 *Bioceramics*. Ishiyaku Euro America Inc., Tokyo, pp. 341–6

Pascoe R T, Hughan R R, Garvie R C 1979 Strong and tough zirconia ceramics. *Sci. Sintering* 11: 185–92

Sato T, Shimada M 1984 Crystalline phase change in yttria–partially-stabilized zirconia by low temperature annealing. *J. Am. Ceram. Soc.* 67: C12–13

Soumiya M 1984 Study of zirconia ceramics. In: Oonishi H and Ooi Y (eds.) 1984 *Orthopaedic Ceramic Implants*, Vol. 4. Japanese Society of Orthopaedic Ceramic Implants, Osaka, pp. 45–9

Sudanese A, Toni A, Cattaneo G L, Ciaroni D, Greggi T, Dallari D, Galli G, Giunti A 1989 Alumina vs zirconium oxide: A comparable wear test. In: Oonishi H, Aoki H, Sawai L (eds.) 1989 *Bioceramics*. Ishiyaku Euro America Inc., Tokyo, pp. 45–9

Tateishi T, Yunoki H 1987 Research and development of advanced biocomposite materials and application to the artificial hip joint. *Bull. Mech. Eng. Lab. Jap.* 45: 1–9

P. S. Christel
[Université de Paris, Paris, France]

LIST OF CONTRIBUTORS

Contributors are listed in alphabetical order together with their addresses. Titles of articles that they have authored follow in alphabetical order. Where articles are coauthored, this has been indicated by an asterisk preceding the title.

Al-Lamee, K. G.
Institute of Medical and Dental Bioengineering
University of Liverpool
Duncan Building
Royal Liverpool Hospital
P O Box 147
Liverpool L69 3BX
UK
Drugs: Attachment to Polymers

Baier, R. E.
Health Care Instruments and Devices Institute
State University of New York
105 Parker Hall
Buffalo, NY 14214
USA
Adhesives in Medicine

Bamford, C. H.
Institute of Medical and Dental Bioengineering
University of Liverpool
Duncan Building
Royal Liverpool Hospital
P O Box 147
Liverpool L69 3BX
UK
Drugs: Attachment to Polymers

Bardos, D. I.
Smith & Nephew Richards Inc.
1450 Brooks Road
Memphis, TN 38116
USA
Titanium and Titanium Alloys

Belleville, J.
Institut National de la Santé et de la Recherche Médicale
INSERM U 37
18 Avenue Doyen Lepine
F-69500 Bron
FRANCE
Biomaterial–Blood Interactions

Bokros, J. C.
Carbomedics, Inc.
1300 East Anderson Lane
Austin, TX 78752
USA
Carbons

Braden, M.
Department of Material Science in Dentistry
London Hospital Medical College
University of London
Turner Street
London E1 2AD
UK
Elastomers for Dental Use

Causton, B. E.
King's College School of Medicine and Dentistry
King's College London
Bessemer Road
London SE5 9PJ
UK
Chemical Adhesion in Dental Restoratives

Charlesworth, D.
University Hospital of South Manchester
West Didsbury
Manchester M20 8LR
UK
Arteries, Synthetic

Chien, Y. W.
Controlled Drug Delivery Research Center
College of Pharmacy
Rutgers University
Busch Campus
P O Box 789
Piscataway, NJ 08855
USA
Polymers for Controlled Drug Delivery

Christel, P. S.
Laboratoire de Recherches Orthopediques
Faculté de Médecine Lariboisière-Saint-Louis
Université de Paris
10 Avenue de Verdun
F-75010 Paris
FRANCE
Zirconia-Toughened Ceramics

Chu, C. C.
Department of Textiles and Apparel
New York State College of Human Ecology
Martha Van Rensselaer Hall
Cornell University

Ithaca, NY 14853-4401
USA
Polyesters and Polyamides
Suture Materials

Clemow, A. J. T.
IatroMed, Inc.
4645 North 32nd Street
Suite 105
Phoenix, AZ 85018
USA
*Cobalt-Based Alloys

Cogan, S. F.
EIC Laboratories Inc.
111 Downey Street
Norwood, MA 02062
USA
*Metals for Medical Electrodes

Combe, E. C.
Unit of Biomaterials Science
Department of Restorative Dentistry
Turner Dental School
University of Manchester Dental Hospital
Higher Cambridge Street
Manchester M15 6FH
UK
Acrylic Dental Polymers

Courtney, J. M.
Bioengineering Unit
University of Strathclyde
Wolfson Centre
106 Rottenrow
Glasgow G4 0NW
UK
*Hemodialysis Membranes

Craig, R. G.
School of Dentistry
University of Michigan
Ann Arbor, MI 48109-1078
USA
Maxillofacial Prostheses

de Wijn, J. R.
Laboratorium voor Celbiologie en Histologie der
   Rijksuniversiteit
Rijnsburgerweg 10
Geb. 55
NL-2333 AA Leiden
THE NETHERLANDS
*Acrylics for Implantation

Duncan, R.
CRC Polymer-Controlled Drug Delivery Research Group
Department of Biological Sciences

University of Keele
Keele
Staffordshire ST5 5BG
UK
Soluble Polymers in Drug Delivery Systems

Earnshaw, R.
Department of Prosthetic Dentistry
University of Sydney
2 Chalmers Street
Sydney 2010
New South Wales
AUSTRALIA
Dental Plaster and Stone

Ellwanger, R.
Institut für Kunststoffprüfung und Kunststoffkunde
Universität Stuttgart
Postfach 80 11 40
D-7000 Stuttgart 80
FRG
*Polyethylene

Eloy, R.
Institut National de la Santé et de la Recherche Médicale
INSERM U 37
18 Avenue Doyen Lepine
F-69500 Bron
FRANCE
*Biomaterial–Blood Interactions

Erhan, S.
Center for Protein Research
Albert Einstein Medical Center
Korman Research Pavilion
York and Tabor Roads
Philadelphia, PA 19141
USA
Adhesives from Protein–Polymer Grafts

Eyerer, P.
Insitut für Kunststoffprüfung und Kunststoffkunde
Universität Stuttgart
Postfach 80 11 40
D-7000 Stuttgart 80
FRG
*Polyethylene

Federolf, H.-A.
Insitut für Kunststoffprüfung und Kunststoffkunde
Universität Stuttgart
Postfach 80 11 40
D-7000 Stuttgart 80
FRG
*Polyethylene

Frisch, E. E.
Health Care Products
Research & Development
Dow Corning Corporation
12334 Geddes Road
Hermlock, MI 48626
USA
Polysiloxanes

Gibson, C.
Department of Epidemiology and Community Medicine
University of Wales College of Medicine
Heath Park
Cardiff CF4 4XN
UK
Sterilization Using Ethylene Oxide

Grashoff, G. J.
Biomedical Technology
Johnson Matthey Technology Centre
Blount's Court
Sonning Common
Reading RG4 9NH
UK
Silver in Medical Applications

Harper, R. A.
Biomedical Engineering
Rensselaer Polytechnic Institute
Troy, NY
USA
Calcium Phosphates and Apatites

Haubold, A. D.
Carbomedics, Inc.
1300 East Anderson Lane
Austin, TX 78752
USA
Carbons

Heimke, G.
Department of Bioengineering
College of Engineering
Clemson University
301 Rhodes Engineering Research Center
Clemson, SC 29634-0905
USA
Aluminum Oxide

Hench, L. L.
Advanced Materials Research Center
University of Florida
1 Progress Blvd
#14
Alachua, FL 32615
USA
Glasses: Medical Applications

Irvine, K.
Bioengineering Unit
University of Strathclyde
Wolfson Centre
106 Rottenrow
Glasgow G4 0NW
UK
Hemodialysis Membranes

Jedynakiewicz, N. M.
School of Dentistry
University of Liverpool
P O Box 147
Liverpool L69 3BX
UK
Endodontic Materials

Katz, J. L.
Biomedical Engineering
Rensselaer Polytechnic Institute
Troy, NY
USA
*Acoustic Measurements of Bone and Bone–Implant
   Systems*
Calcium Phosphates and Apatites

King, R. O.
Biomedical Technology
Johnson Matthey Technology Centre
Blount's Court
Sonning Common
Reading RG4 9NH
UK
Silver in Medical Applications

Knott, P.
School of Materials
Division of Ceramics
University of Leeds
Leeds LS2 9JT
UK
Glasses: Agricultural and Vetinary Applications

Kurth, M.
Insitut für Kunststoffprüfung und Kunststoffkunde
Universität Stuttgart
Postfach 80 11 40
D-7000 Stuttgart 80
FRG
Polyethylene

Langeland, K.
School of Dental Medicine

University of Connecticut Health Center
Farmington, CT
USA
Biocompatiblity of Dental Materials

Laub, L.
Loyola University School of Dentistry
2160 South First Avenue
Maywood, IL 60153
USA
Wrought Dental Wires

Lloyd, C. H.
Department of Dental Prosthetics and Gerontology
University of Dundee
The Dental School
Park Place
Dundee DD1 4HN
UK
Fracture Toughness

Lydon, M. J.
Convatec Biological Research Laboratory
Newtech Clwyd Limited
Newtech Square
Deeside Industrial Park
Clwyd CH5 2NU
UK
Wound Dressings Materials

Mädler, H.
Insitut für Kunststoffprüfung und Kunststoffkunde
Universität Stuttgart
Postfach 80 11 40
D-7000 Stuttgart 80
FRG
Polyethylene

Marek, M.
Georgia Institute of Technology
School of Materials Engineering
Atlanta, GA 30332-0245
USA
Corrosion of Dental Materials

Matthews, I. P.
Department of Epidemiology and Community Medicine
University of Wales College of Medicine
Heath Park
Cardiff CF4 4XN
UK
Sterilization Using Ethylene Oxide

Meunier, A.
Laboratoire de Recherches Orthopédiques
Faculté de Médecine Lariboisière-Saint-Louis
Université de Paris

10 Avenue de Verdun
F-75010 Paris
FRANCE
Acoustic Measurements of Bone and Bone–Implant Systems

Meyer, J.-M.
Faculté de Médecine
Section de Médecine Dentaire
Université de Genève
19 rue Barthélemy-Menn
1211 Genève 4
SWITZERLAND
Porcelain–Metal Bonding in Dentistry

O'Brien, W. J.
Dental School
University of Michigan
Ann Arbor, MI
USA
Dental Porcelain

Okabe, T.
Baylor College of Dentistry
Department of Dental Materials
3302 Gaston Avenue
Dallas, TX 75246
USA
Dental Amalgams

Pilliar, R. M.
Faculty of Dentistry
University of Toronto
124 Edward Street
Toronto
Ontario M5G 1G6
CANADA
Porous Biomaterials

Plenk, H. Jr
Histologisch-Embryologisches Institut der Universität Wien
Schwarzspanierstrasse 17
A-1090 Wien
AUSTRIA
Tantalum and Niobium

Ratner, B. D.
Center for Bioengineering and Department of Chemical Engineering
BF-10
University of Washington
Seattle, WA 98195
USA
Surface Structure and Properties

Robblee, L. S.
EIC Laboratories Inc.
111 Downey Street
Norwood, MA 02062
USA
Metals for Medical Electrodes

Rolfe, P.
School of Postgraduate-Medicine and Biological Sciences
University of Keele
Department of Biomedical Engineering and Medical Physics
North Staffordshire Hospital Centre
Thornburrow Drive
Hartshill
Stoke-on-Trent ST4 7QB
UK
Invasive Sensors

Samuel, A. H.
Department of Epidemiology and Community Medicine
University of Wales College of Medicine
Heath Park
Cardiff CF4 4XN
UK
Sterilization Using Ethylene Oxide

Sandrik, J. L.
Loyola University School of Dentistry
2160 South First Avenue
Maywood, IL 60153
USA
Wrought Dental Wires

Schider, S.
Metallwerke Plansee AG
Reutte
AUSTRIA
Tantalum and Niobium

Schoen, F. J.
Brigham and Women's Hospital
75 Francis Street
Boston, MA 02115
USA
Heart-Valve Replacement Materials

Scholze, H.
Keesburgstraße 22
D-8700 Würzburg
FRG
Ormosils: Organically Modified Silicates

Sefton, M. V.
Department of Chemical Engineering and Applied Chemistry
University of Toronto
Toronto
Ontario
CANADA M5S 1A4
Heparinized Materials

Stafford, G. D.
Department of Prosthetic Dentistry
University of Wales College of Medicine
Heath Park
Cardiff CF4 4XY
UK
Denture Base Resins

Steinboch, A. F.
Whip Mix Corporation
Louisville, KY
USA
Dental Investment Materials

Sutow, E. J.
Faculty of Dentistry
Dalhousie University
Halifax
Nova Scotia B3H 3J5
CANADA
Iron-Based Alloys

Tesk, J. A.
Dental and Medical Materials Group
Polymers Division
United States Department of Commerce
National Institute of Standards and Technology
Gaithersburg, MD 20899
USA
Base-Metal Casting Alloys for Dental Use

Tighe, B. J.
Aston University
Aston Triangle
Birmingham B4 7ET
UK
Hydrogels

Toddywala, R.
Colgate Palmolive Co.
909 River Road
Piscataway, NJ 08855
USA
Polymers for Controlled Drug Delivery

Travers, M.
Akzo
Enka AG
Wuppertal
FRG
Hemodialysis Membranes

Turner, R. M.
ICI Materials Research Centre
Wilton
UK
Composite Materials

van Mullem, P. J.
Department of Oral Histology
University of Nijmegen
Nijmegen
THE NETHERLANDS
Acrylics for Implantation

Waterstrat, R. M.
Dental and Medical Materials Group
Polymers Division
United States Department of Commerce
National Institute of Standards and Technology
Gaithersburg, MD 20899
USA
Base-Metal Casting Alloys for Dental Use

Weinstein, A. M.
IatroMed, Inc.
4645 North 32nd Street
Suite 105
Phoenix, AZ 85018
USA
Cobalt-Based Alloys

Williams, D. F.
Institute of Medical and Dental Bioengineering
University of Liverpool
Duncan Building
Royal Liverpool Hospital
P O Box 147
Liverpool L69 3BX
UK
Biocompatibility: An Overview
Biodegradation of Medical Polymers
Collagen
Composite Materials
Dental Implants

Polytetrafluoroethylene
Polyurethanes

Williams, J. M.
Solid State Division
Oak Ridge National Laboratory
P O Box 2008
Oak Ridge, TN 37831-6057
USA
Beam Ion Implantations

Wilson, J.
Advanced Materials Research Center
University of Florida
1 Progress Blvd
#14
Alachua, FL 32615
USA
Glasses: Medical Applications

Wilson, N. H. F.
Unit of Conservative Dentistry
Department of Restorative Dentistry
Turner Dental School
University of Manchester Dental Hospital
Higher Cambridge Street
Manchester M15 6FH
UK
Dental Materials: Clinical Evaluation

Yapp, R. A.
Carbomedics, Inc.
1300 East Anderson Lane
Austin, TX 78752
USA
Carbons

Yasuda, K.
Department of Dental Materials Science
Nagasaki University School of Dentistry
Nagasaki 852
JAPAN
Gold Alloys for Dental Use

SUBJECT INDEX

The Subject Index has been compiled to assist the reader in locating all references to a particular topic in the Encyclopedia. Entries may have up to three levels of heading. Where there is a substantive discussion of the topic, the page numbers appear in *bold italic* type. As a further aid to the reader, cross-references have also been given to terms of related interest. These can be found at the bottom of the entry for the first-level term to which they apply. Every effort has been made to make the index as comprehensive as possible and to standardize the terms used.